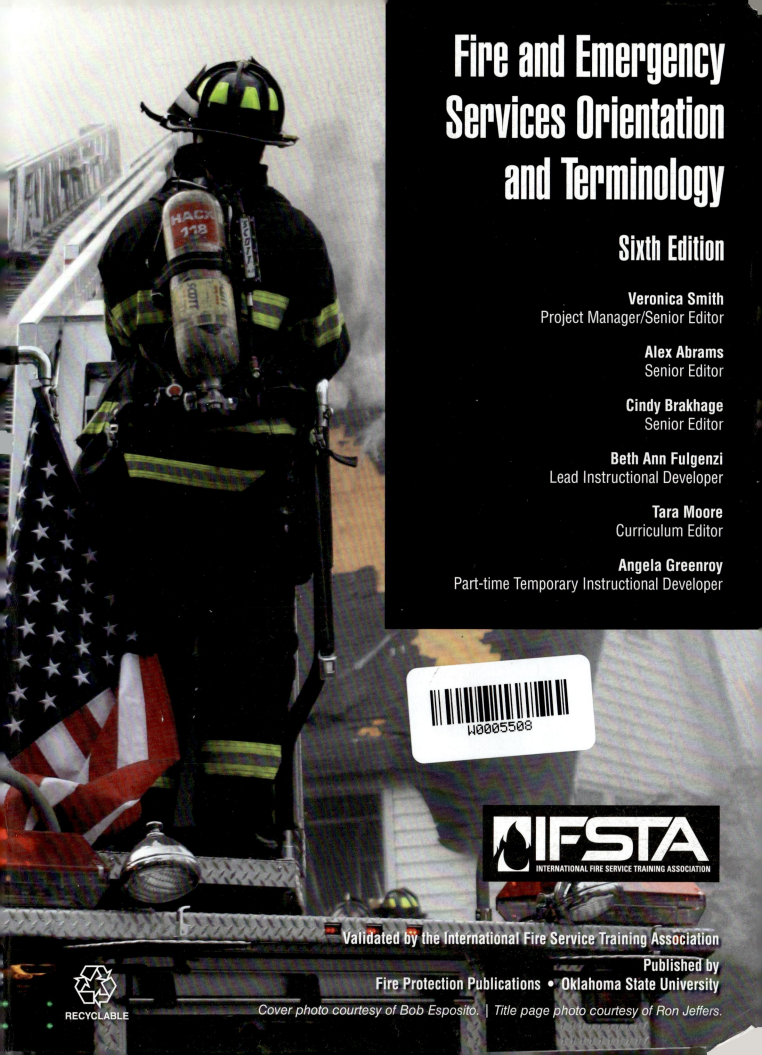

The International Fire Service Training Association (IFSTA) was established in 1934 as a *nonprofit educational association of fire fighting personnel who are dedicated to upgrading fire fighting techniques and safety through training.* To carry out the mission of IFSTA, Fire Protection Publications was established as an entity of Oklahoma State University. Fire Protection Publications' primary function is to publish and distribute training materials as proposed, developed, and validated by IFSTA. As a secondary function, Fire Protection Publications researches, acquires, produces, and markets high-quality learning and teaching aids consistent with IFSTA's mission.

IFSTA holds two meetings each year: the Winter Meeting in January and the Annual Validation Conference in July. During these meetings, committees of technical experts review draft materials and ensure that the professional qualifications of the National Fire Protection Association® standards are met. These conferences bring together individuals from several related and allied fields, such as:

- Key fire department executives, training officers, and personnel
- Educators from colleges and universities
- Representatives from governmental agencies
- Delegates of firefighter associations and industrial organizations

Committee members are not paid nor are they reimbursed for their expenses by IFSTA or Fire Protection Publications. They participate because of a commitment to the fire service and its future through training. Being on a committee is prestigious in the fire service community, and committee members are acknowledged leaders in their fields. This unique feature provides a close relationship between IFSTA and the fire service community.

IFSTA manuals have been adopted as the official teaching texts of many states and provinces of North America as well as numerous U.S. and Canadian government agencies. Besides the NFPA® requirements, IFSTA manuals are also written to meet the Fire and Emergency Services Higher Education (FESHE) course requirements. A number of the manuals have been translated into other languages to provide training for fire and emergency service personnel in Canada, Mexico, and outside of North America.

Copyright © 2015 by the Board of Regents, Oklahoma State University

All rights reserved. No part of this publication may be reproduced in any form without prior written permission from the publisher.

ISBN 978-0-87939-592-6   Library of Congress Control Number: 2015959098

Sixth Edition, Second Printing, September 2019   *Printed in the United States of America* 10   9   8   7   6   5   4   3   2

If you need additional information concerning the International Fire Service Training Association (IFSTA) or Fire Protection Publications, contact:

Customer Service, Fire Protection Publications, Oklahoma State University
930 North Willis, Stillwater, OK 74078-8045
800-654-4055   Fax: 405-744-8204

For assistance with training materials, to recommend material for inclusion in an IFSTA manual, or to ask questions or comment on manual content, contact:

Editorial Department, Fire Protection Publications, Oklahoma State University
930 North Willis, Stillwater, OK 74078-8045
405-744-4111   Fax: 405-744-4112   E-mail: editors@osufpp.org

Oklahoma State University in compliance with Title VI of the Civil Rights Act of 1964 and Title IX of the Educational Amendments of 1972 (Higher Education Act) does not discriminate on the basis of race, color, national origin or sex in any of its policies, practices or procedures. This provision includes but is not limited to admissions, employment, financial aid and educational services.

Chapter Summary

Chapters

1	Fire and Emergency Services as a Career	11
2	Roles of Fire and Emergency Service Personnel	47
3	Early Traditions and History	81
4	Roles of Public and Private Support Organizations	113
5	Fire Prevention, Life Safety Education, and Fire Investigation	145
6	Scientific Terminology	183
7	Building Construction	213
8	Fire Detection, Alarm, and Suppression-Systems	237
9	Fire and Emergency Services Apparatus, Equipment, and Facilities	259
10	Fire Department Organization and Management	321

Appendices

A	Chapter and Page Correlation to NFPA® 1001 Requirements	349
B	NFPA Professional Qualifications Standards	350
C	Firefighter Life Safety Initiatives	351
D	Incident Command System (ICS) Organizational Chart	352

Glossary ... 353
Index ... 584

Table of Contents

Preface ....
Introduction 1
Purpose and Scope 1
Terminology 2
Resources 2
Key Information 2
Key Terms 3
Signal Words 4
Metric Conversions 5

1 Fire and Emergency Services as a Career ... 8
Case History 11
Fire and Emergency Services Culture 12
Fire Protection Agencies and Other Emergency
    Response Organizations 13
Types of Fire Departments 13
    Jurisdiction 14
    Municipal 14
    County, Parish, or Borough 15
    Fire Districts and Fire Protection Districts 16
    Public Fire Departments 17
        Career Departments 18
        Volunteer Departments 18
        Combination Departments 19
        Paid On-Call Departments 19
        Public Safety Departments 20
Private and Other Government Departments 20
    Industrial Fire Departments 20
    Federal Fire Departments 21
    Commercial Fire Protection Services 21
    Airport Fire Departments 21
Firefighters as Public Figures 22
    The Firefighter Image 22
    Responsible Behaviors of Firefighters 22
    Conduct 23
    Teamwork 24
    Human Relations and Customer Service 24
    Uniform and Dress 25
    Physical Fitness 27
Training and Education 28
    Types of Training 28
    Departmental Training 28
    Regional/County and State/Provincial
        Training Programs 29
    Value of Education 30
    Colleges and Universities 31
    National Fire Academy 31
    Non-Traditional Training and Education 32

Firefighter Certification/Credentialing 33
    National Board on Fire Service Professional
    Qualifications (NBFSPQ or Pro-Board) 33
    International Fire Service Accreditation
    Congress (IFSAC) 34
    Center for Public Safety Excellence (CPSE) ... 34
Firefighter Selection Process 34
    Recruitment 35
    Preparation 35
    Application 35
    Written Examination 36
        Math 36
        Reading Comprehension 36
        Mechanical Aptitude 37
        Recognition/Observation 37
        Psychological Testing 38
    Physical Ability Test 38
    Interview Process 39
        Oral Interview 39
        Second Interview 39
    Probationary Period 39
    Volunteer Firefighter Selection 39
Fire Service Career Information 40
    Pay 40
    Hours of Duty 40
    Promotion 40
    Retirement 41
Chapter Summary 41
Review Questions 42
Chapter 1 Notes 42

2 Roles of Fire and Emergency
Services Personnel 44
Case History 47
Fire Companies 47
    Engine Company 48
    Ladder (Truck) Company 48
    Rescue/Squad Company 51
Fire and Emergency Services Personnel 51
    Firefighter 52
    Fire Apparatus Driver/Operator 53
    Company Officer 54
    Battalion/District Chief 55
    Training Division Personnel 57
        Instructor 57
        Training Officer (Chief of Training/
        Drillmaster) 58

- Administration 59
  - *Human Resources/Personnel Services*.......... 59
  - *Grants Administrator*.................................. 59
  - *Safety Officer*... 60
  - *Public Information Officer (PIO)* 61
  - *Assistant/Deputy Chief* 62
  - *Fire Chief/Chief of Department* 63
- Special Operations Personnel 64
  - *Airport Firefighter*....................................... 64
  - *Hazardous Materials Technician* 65
  - *Technical Rescuer (Special Rescue Technician or Technical Rescue Specialist)* 65
  - *Marine, Wildland, and Industrial Firefighters* 67
  - *Military Firefighter* 69
- Emergency Medical Services (EMS) Personnel... 69
  - *First Responder* 70
  - *Emergency Medical Technician* 70
  - *Paramedic*... 71
  - *EMS Chief/Officer*.. 71
- Other Fire Department Personnel 72
  - *Telecommunications/Dispatch Personnel* ... 72
  - *Information Technology Personnel* 73
  - *Facilities Maintenance Personnel*................ 74
  - *Apparatus/Equipment Maintenance Personnel* 74

Chapter Summary 75
Review Questions... 76

3 Early Traditions and History 78
Case History 81
Early Fire Services 81
- The First Fire Pump 82
- Extinguishing Techniques 82
- Fire Laws and Ordinances 83
- Causes of Fire 83
- Britain's First Fire Brigade 84
- Development of the Fire Engine 85

Fire Protection in Early America 86
Growth of the Volunteer Fire Service................... 87
- Volunteers, Patriots and Competitors 87
- First Hose Company 89
- Mutual Fire Society of Boston 90
- Philadelphia's Union Volunteer Fire Company . 90
- New York Volunteer Department........................ 90
- First Fire Insurance Companies 91
  - *The Philadelphia Contributorship of the Assurance of Houses from Loss by Fire* 91
  - *Mutual Assurance Company* 92

Key Developments in Fire Fighting 92
- Development of American Fire Engines 92
  - *New York Style Hand Engines*....................... 92
  - *Piano-Type Engines*...................................... 92
  - *Other Hand Engines* 93
  - *Improvements in Hand Engines* 94
- The Age of Steam 95
- Chemical Engines and Ladder Trucks 97
  - *Chemical Engine*... 97
  - *Ladder Trucks* 97
- Gasoline and Diesel Powered Equipment 98
- Improvements in Protective Clothing and Self-Contained Breathing Apparatus (SCBA).. 99
  - *Protective Clothing*....................................... 99
  - *SCBA Equipment*.. 100
- Fire Extinguishers... 102
- Alarm and Communications Systems 103
  - *Town Crier* 103
  - *Watchmen and Watchtowers*....................... 103
  - *Telegraph Alarm Systems*........................... 103
  - *Telephone Systems* 104
  - *Radio Communications*.............................. 104
  - *Pagers/Personal Alerting Systems* 105
  - *Mobile Data Communications Systems (MDCS)*... 105

Impact of Historic Fires on Firefighter and Public Safety in North America 105
- Significant Historic Fires in North America 105
  - *Iroquois Theatre Fire, Chicago, Illinois (1903)*.. 106
  - *Great Fire of 1904, Toronto, Canada (1904)*.. 106
  - *Triangle Shirtwaist Fire, New York City, New York (1911)*.. 106
  - *Cocoanut Grove Nightclub Fire, Boston, Massachusetts (1942)*................................. 106
  - *Ringling Brothers Barnum and Bailey Circus Fire, Hartford, Connecticut (1944)* 106
  - *Our Lady of the Angels School Fire, Chicago, Illinois (1958)* 107
  - *Hartford Hospital Fire, Hartford, Connecticut (1961)*.. 107
  - *Beverly Hills Supper Club Fire, Southgate, Kentucky (1977)*.. 107
  - *MGM Grand Hotel Fire, Las Vegas, Nevada (1980)*.. 107
  - *Station Nightclub Fire, West Warwick, Rhode Island (2003)* 107

Impact of Historic Fires on Firefighter Safety ... 108
Chapter Summary 108
Review Questions.. 109

4 Roles of Public and Private Support Organizations 110
Case History 113
Local Agencies and Organizations 114
    Local Government 114
    Local Law Enforcement 114
    Building Department (Code Enforcement) 115
    Water Department/Water Authority 115
    Zoning/Planning Commission 115
    Department of Public Works 116
    Utilities... 116
    Judicial System 117
    Office of Emergency Management 117
    Local Health Department................................ 118
    Nongovernmental Organizations 118
State/Provincial Agencies and Organizations 118
    State Fire Marshal or Commissioner 119
    State/Provincial Fire Training 119
    Fire Commission 119
    State Law Enforcement.................................... 119
    State Highway Department/ Department of Transportation/ Turnpike Commission........... 120
    State Environmental Protection Agency 120
    State Occupational Safety and Health Administration 120
    State Health Department................................. 121
    State Forestry/Department of Natural Resources.. 121
    Office of Emergency Management 121
    Special Task Forces 121
    State Firefighters Association........................... 121
    State Fire Chiefs Association 121
    State Fire Marshals Association 122
    Other State Agencies....................................... 122
Federal Organizations in the United States and Canada.. 122
    Army Corps of Engineers................................. 123
    Bureau of Alcohol, Tobacco, Firearms, and Explosives (BATFE)... 123
    Chemical Safety Hazard Investigation Board (CSB).. 123
    Consumer Product Safety Commission (CPSC).. 123
    Emergency Management Institute (EMI) 123
    Environmental Protection Agency (EPA)......... 124
    Federal Aviation Administration (FAA)........... 124
    Federal Emergency Management Agency (FEMA)... 124
    National Fire Academy (NFA) 125
    National Guard/Civil Support Team 125
    National Highway Traffic Safety Administration (NHTSA).. 126
    National Institute for Occupational Safety and Health (NIOSH) 126
    National Institute of Standards and Technology (NIST) 126
    National Transportation Safety Board (NTSB)... 127
    National Weather Service (NOAA) 127
    Nuclear Regulatory Commission (NRC).......... 127
    Occupational Safety and Health Administration (OSHA).. 127
    Public Safety Canada (PS) 127
    Transport Canada 128
    United States Coast Guard (USCG) 128
    United States Department of Energy (DOE) 128
    United States Department of Homeland Security (DHS).. 128
    United States Department of the Interior (DOI).. 129
    United States Department of Labor (DOL)...... 129
    United States Department of Transportation (US DOT) 129
    United States Fire Administration (USFA) 129
    United States Forest Service (USFS) 130
North American Trade, Professional, and Membership Fire and Emergency Services Organizations... 130
    American National Standards Institute (ANSI) 131
    Canadian Association of Fire Chiefs (CAFC) 131
    The Canadian Centre for Emergency Preparedness (CCEP) 131
    Canadian Fallen Firefighters Foundation (CFFF) 131
    Canadian Fire Alarm Association (CFAA)......... 131
    Canadian Fire Safety Association (CFSA) 131
    Canadian Standards Association (CSA)........... 131
    The Canadian Volunteer Fire Services Association (CVFSA) 132
    Council of Canadian Fire Marshals and Fire Commissioners 132
    FM Global Research (FM) 132
    Fire Prevention Canada 132
    Home Fire Sprinkler Coalition 132
    Insurance Services Office, Inc. (ISO)................ 132

International Association of Arson Investigators
(IAAI) 133
International Association of Black Professional
Firefighters (IABPFF) 133
International Association of Fire Chiefs
(IAFC) 133
International Association of Firefighters
(IAFF) 134
International Association of Women in Fire and
Emergency Services (iWOMEN) 134
International City/County Management
Association (ICMA) 134
International Code Council (ICC) 134
International Fire Marshals Association
(IFMA) 135
International Fire Service Training Association
(IFSTA) 135
International Municipal Signal Association
(IMSA) 135
International Society of Fire Service Instructors
(ISFSI) 136
National Association of Hispanic Firefighters
(NAHF) 136
National Association of State Fire Marshals 136
National Fallen Firefighters Foundation
(NFFF) 136
National Fire Protection Association®
(NFPA)® 137
National Interagency Fire Center (NIFC) 137
National Propane Gas Association (NPGA) 138
National Volunteer Fire Council (NVFC) 138
National Wildfire Coordinating Group
(NWCG) 138
North American Fire Training Directors
(NAFTD) 138
Society of Fire Protection Engineers (SFPE) 138
Underwriters Laboratories (UL) 138
Underwriters Laboratories of Canada (ULC) ... 139
Other Related Organizations 139
American Association of Railroads Bureau of
Explosives (BOE) 139
American Trucking Association (ATA) 139
National Tank Truck Carriers (NTCC) 139
Chapter Summary 139
Review Questions 140

5 Fire Prevention, Life Safety Education, and Fire Investigation 142
Case History 145
Principles of Fire Prevention 145

Need for Fire Prevention 146
Monetary Loss 146
Fire Hazards 147
Common Fire Hazards 148
Special Considerations 149
Target Hazard Properties 151
Fire Risk Analysis 152
Trends 152
Demographics 153
Components of Fire Prevention 153
*Surveys, Inspection, and Code
Enforcement* 153
Fire and Life Safety Education 153
Fire Investigation 154
Preincident Surveys 154
Facility Surveys 155
Personnel Requirements 155
Scheduling the Preincident Survey 155
Conducting the Preincident Survey 156
Maps and Sketch Making 157
Exit Interview 157
Residential Safety Surveys 158
Inspection and Code Enforcement 159
Codes and Standards 159
Consistent Codes 160
Effects On Construction 160
Building Design 160
Local Safety Amendments 160
Role of the Fire Inspector 161
Role in Fire Prevention 161
Fire Protection Engineer/Building Plans
Examiner 162
Role in Fire Prevention 162
Permits 162
Types 162
Process 164
Application 164
Review 164
Issuance 164
Expiration 165
Fire and Life Safety Education 165
Presenting Fire and Life Safety Information 165
*Role of the Fire and Life Safety
Educator* 166
Fire and Life Safety Presentation Topics 167
Stop, Drop, and Roll 167
*Smoke and Carbon Monoxide (CO)
Alarms* 168
Fire Station Tours 168
Fire Investigation 168

vii

Role of the Investigator 169
Observations Made by Emergency
    Responders 170
    Observations En Route 170
    Observations Upon Arrival 171
    Firefighter Observations 172
    Responsibilities After the Fire 174
Conduct and Statements at the Scene 174
Securing the Fire Scene 175
Legal Considerations 176
Protecting and Preserving Evidence 177
Chapter Summary **178**
Review Questions **178**

6   Scientific Terminology 180
Case History **183**
Properties of Matter **184**
Physical States of Matter 184
Specific Gravity and Vapor Density 184
Physical and Chemical Changes 185
Mass and Energy 185
Combustion **186**
Fire Tetrahedron 187
    Heat and Temperature 188
    Sources of Heat Energy 189
    Transmission of Heat 190
    Fuel 192
    Oxygen (Oxidizing Agent) 195
    Self-Sustained Chemical Chain Reaction ... 196
Stages of Fire Development 196
    Factors that Affect Fire Development 197
    Incipient 198
    Growth 199
    Flashover 199
    Fully Developed 200
    Decay 201
    Rapid Fire Development 201
Special Considerations **202**
Rollover 202
Thermal Layering of Gases 203
Backdraft 203
Smoke Explosion 205
Products of Combustion 205
Classification of Fires **205**
Class A Fires 206
Class B Fires 206
Class C Fires 206
Class D Fires 207
Class K Fires 207
Fire Extinguishment Theory **207**
Temperature Reduction 208
Fuel Removal 208
Oxygen Exclusion 209
Chemical Flame Inhibition 209
Chapter Summary **209**
Review Questions **209**

7   Building Construction 210
Case History **213**
Types of Building Construction **214**
Type I Construction 214
Type II Construction 217
Type III Construction 217
Type IV Construction 218
Type V Construction 219
Effects of Fire on Common Building Materials ... **220**
Wood 220
Masonry 222
Cast Iron 223
Steel 223
Reinforced Concrete 224
Gypsum 224
Glass/Fiberglass 224
Firefighter Hazards Related to Building
    Construction **225**
Dangerous Building Conditions 225
    Fire Loading 225
    Combustible Furnishings and Finishes 226
    Roof Coverings 226
    Wooden Floors and Ceilings 228
    Large, Open Spaces 228
    Building Collapse 228
Lightweight and Truss Construction Hazards ... **229**
New Building Construction Technologies 231
Construction, Renovation, and Demolition
    Hazards 231
Chapter Summary **232**
Review Questions **233**

8   Fire Detection, Alarm, & Suppression
    Systems 234
Case History **237**
Reasons for Installing Fire Detection, Alarm and
    Suppression Systems **237**
Types of Alarm Systems **238**
Heat Detectors 239
    Fixed Temperature Heat Detectors 239
    Rate-of-Rise Heat Detectors 239
Smoke Detectors and Smoke Alarms 240
    Photoelectric Smoke Detectors 240
    Ionization Smoke Alarms 241

Other Detectors	*242*
Automatic Alarm Systems	243
Auxiliary Services	244

Automatic Sprinkler Systems **244**
- Sprinkler Systems and Their Effect on Life Safety 246
- Sprinklers 246
- Sprinkler Position 246
  - *Control Valves* *246*
  - *Waterflow Alarms* *249*
- Applications of Sprinkler Systems 250
  - *Wet-Pipe System* *250*
  - *Dry-Pipe System* *250*
  - *Preaction System* *251*
  - *Deluge System* *251*
  - *Residential Systems* *251*

Standpipe Systems **251**
- Classes of Standpipe Systems 251
  - *Class I Standpipe Systems* *252*
    - *Class II Standpipe Systems* *252*
    - *Class III Standpipe Systems* *252*
- Types of Standpipe Systems 253

Chapter Summary **254**
Review Questions **254**

9 Fire & Emergency Services Apparatus, Equipment, and Facilities **256**

Case History **259**
Fire Department Apparatus **260**
- The Engine (Pumper) 260
- Smaller Fire Apparatus 263
- Mobile Water Supply Apparatus 264
- Wildland Fire Apparatus 264
- Aerial Apparatus 265
  - *Aerial Ladder Apparatus* *266*
  - *Aerial Ladder Platform Apparatus* *266*
  - *Telescoping Aerial Platform Apparatus* *268*
  - *Articulating Aerial Platform Apparatus* *268*
  - *Water Towers* *268*
- Quintuple Apparatus (Quint) 269
- Rescue Apparatus 269
  - *Light Rescue Vehicle* *270*
  - *Medium Rescue Vehicles* *270*
  - *Heavy Rescue Vehicle* *270*
- Fire Service Ambulances 271
- Aircraft Rescue and Fire Fighting Apparatus (ARFF) 273
- Hazardous Materials Response Unit 274
- Mobile Air Supply Unit 275
- Mobile Command Post 276
- Fire Boats and Search and Rescue Boats 276
- Power and Light Unit 277
- Mobile Fire Investigation Unit 277
- Fire Fighting Aircraft 278
- Other Special Units 278

Uniforms and Personal Protective Clothing **279**
- Uniforms 279
- Personal Protective Clothing 280

Breathing Apparatus **280**
- Air Purifying Respirators (APRs) 280
- Supplied Air Respirators (SARs) 280
- Self-Contained Breathing Apparatus (SCBA) 280

Tools and Equipment **281**
- Hand Tools 281
- Power Tools 281
- Equipment 281

Ropes, Webbing, Related Hardware and Harnesses **281**
Ground Ladders **282**
Fire Hose, Nozzles, and Hose Appliances and Tools **282**
Fire Department Facilities **287**
- Fire Stations 287
- Administrative Offices and Buildings 288
- Telecommunication Centers 289
- Training Centers 289
  - *Burn Building* *290*
  - *Drill Tower* *290*
  - *Smoke House* *290*
  - *Training Pads* *292*

Maintenance Facilities **292**
Chapter Summary **293**
Review Questions **293**

10 Fire Department Organization and Management **318**

Case History **321**
Principles of Organization **321**
- Unity of Command 323
- Span of Control 324
- Division of Labor 324
- Discipline 325

Local Government Structure **325**
- Commission 325
- Council (Board)/Manager 325
- Mayor/Council 325
- Fire Districts 326

Response Considerations **326**
- Automatic Aid 326
- Mutual Aid 326

ix

Fire Department Funding 328
    Tax Revenues 328
    Trust Funds.. 328
    Enterprise Funds 329
    Bond Sales 329
    Grants/ Gifts 329
    Fundraisers 329
    Subscriptions/Fees 330
Current Challenges Facing Fire Protection........ 330
    Funding/Budget Constraints 330
    Outdated Equipment 330
    Recruiting Personnel 331
    Retaining Personnel....................................... 331
    Adequate Water Supply 332
Policies and Procedures...................................... 333
Standard Operating Procedures (SOPs)/
    Guidelines (SOGs) 333
    Disciplinary Procedures 334
    Formal Communications 334
Incident Command System................................... 335
    Functional Areas of ICS 336
        Command.. 336
        Operations 337
        Planning 337
        Logistics.. 337
        Finance /Administration 337
    NIMS Terms 337
        Incident Commander (IC) 338
        Command Staff... 338
        Incident Action Plan................................ 339
        Section.. 339
        Section Chief 339
        Branch.. 339
        Branch Director 339
        Division... 339
        Division Supervisor................................. 339
        Group 340
        Group Supervisor 340
        Unit 340
        Unit Leader.. 340
        Strike Team /Task Force 340
        Strike Team / Task Force Leader.................. 340
        Resources... 340
    NIMS-ICS Training 341
    Emergency Operations 341
    Personnel Accountability Systems.................. 342

        Tag System 342
        SCBA Tag System.. 342
        Computer - Based Electronic
            Accountability... 343
    Rapid Intervention Crews (RIC) 344
    Critical Incident Stress Management /Exposure ..
        to Potentially Traumatic Events................. 344
Chapter Summary 345
Review Questions... 346

Appendix A
FESHE Course Requirements 349

Appendix B
NFPA Professional Qualifications Standards 350

Appendix C
Fire Fighters Life Safety Initiatives 351

Appendix D
Incident Command System (CIS)
    Organization Chart....................................... 352

Glossary 353

Index 584

List of Tables

Table 6.1	Table Common Products of Combustion and Their Toxic Effects	206
Table 6.2	Table of Flammable Gases and Liquids	207
Table 7.1	Fire Resistance Rating Requirements	215
Table 8.1	Sprinkler Ratings, Classifications, Color Coding	247
Table 9.1	Personal Protective Clothing (Structural, Wildland, and Proximity Protective Clothing, Hazardous Materials Protective Clothing, Hazardous Materials Protective Clothing)	294
Table 9.2	Breathing Apparatus (Air Purifying Respirators)	297
Table 9.3	Breathing Apparatus (Supplied Air Respirators (SARs)	298
Table 9.4	Breathing Apparatus (Self-Contained Breathing Apparatus-Open Circuit SCBAs)	299
Table 9.5	Breathing Apparatus (Self-Contained Breathing Apparatus-Closed Circuit SCBAs)	300
Table 9.6	Hand Tools (Striking Tools), Prying Tools, Cutting Tools, Lifting Tools, Stabilizing Tools, Specialized Hand Tools, Mechanic's Tools	301
Table 9.7	Power Tools (Pnuematic Powered, Hydraulic-Manually Operated, Hydraulic-Power Driven)	311
Table 9.8	Ropes, Webbing, Hardware and Harnesses (Natural and Synthetic Ropes, Hardware, Rescue Harnesses)	314

Acknowledgements

The sixth edition of the **Fire and Emergency Services Orientation and Terminology** manual is written to provide information for students about the history, organization, and work of the fire service. It will assist the student in meeting all of the learning outcomes in the FESHE *Principles of Emergency Services* course.

Acknowledgement and special thanks are extended to the members of the IFSTA validating committee who contributed their time, wisdom, and knowledge to the development of this manual.

Contract Writer

The sixth edition of the **Fire and Emergency Services Orientation and Terminology** could not have been completed without the skill and hard work of contracted technical writer Captain David DeStefano. He provided the early drafts of all manual chapters to the validation committee for their review, revision, and approval. He also assisted the committee and project manager with changes to chapter content and validation discussions.

IFSTA Orientation and Terminology Sixth Edition Validation Committee

Chair
Rick McIntyre
Assistant Fire Marshall
North Carolina Department of Insurance
Raleigh, NC

Vice Chair
Mark Pare
Director
Rhode Island Fire Academy
Plainville, MA

Secretary
Mark Butterfield
Principal Senior Instructor
Butterfield & Associates
Hutchinson, KS

Committee Members

Jimmie Badgett
Fire and Training Consultant
JLB Fire and Safety
Dallas, TX

Randy Novak
Director
Iowa Fire Service Training Bureau
Ames, IA

William Neville
Principal
Neville & Associates
Grass Valley, CA

Shawn Bayouth
Chief
Ames Fire Department
Ames, IA

Lauren Casady
Deputy Fire Marshall
Norwich Fire Department
Norwich, CT

IFSTA Orientation and Terminology Sixth Edition Validation Committee

Committee Members (cont.)

Tom Jenkins
Chief
Rogers Fire Department
Rogers, AR

Michelle Mallek
Firefighter/EMT
Vienna Volunteer Fire Department
Vienna, VA

Richard Merrell
Captain
Fairfax County Fire Rescue Department
Dumfries, VA

Geoffrey Herald
Chief (Retired)
Danbury Fire Department
Danbury, CT

Doug Bledsoe
Captain
Port Neches Fire Department
Groves, TX

Holly Nelson
Firefighter/Paramedic
San Bruno Fire Department
San Bruno, CA

Much appreciation is given to the following individuals and organizations for contributing photographs instrumental in the development of this manual:

- 10 Air Carrier, LLC
- American Emergency Vehicles (AEV)
- Eddie Avila
- Ted Boothroyd
- Frank Carter
- Wayne Chapdlaine
- Congressional Fire Services Institute
- Bob Esposito
- Dianna Gee, Federal Emergency Management Agency News Agency
- Fireman's Hall Museum
- Donny Howard
- Ron Jeffers
- Dayna Hilton, Johnson County RFD#1
- Las Vegas (NV) Fire and Rescue
- John Lewis
- Steve Lofton
- McKinney (TX) Fire Department
- Chris Mickal
- Ron Moore
- Donny Howard, Oklahoma City Fire Marshall
- National Fire Academy (NFA)
- National Firefighters Foundation (NFFF)
- National Institute of Standards and Technology (NIST)
- James Nilo, Richmond International Airport
- United States Fire Academy (USFA)
- Chief Mass Communications Specialist Bill Mesta, United States Navy
- Edwin A. Jones, Urban Search and Rescue (USAR)

Last, but certainly not least, gratitude is extended to the following members of the Fire Protection Publications staff whose contributions made the final publication of this manual possible.

Orientation and Terminology, Sixth Edition
Project Team

Project Manager
Veronica Smith, Senior Editor

Contract Writer
David DeStefano
Captain
North Providence Fire Department
North Providence, RI

Director of Fire Protection Publications
Craig Hannan

Curriculum Manager
Leslie Miller

Editorial Manager
Clint Clausing

Production Coordinator
Ann Moffat

Editors
Alex Abrams, Senior Editor
Cindy Brakhage, Senior Editor

Illustrator and Layout Designer
Clint Parker, Senior Graphic Designer

Curriculum Development
Beth Ann Fulgenzi, Lead Instructional Developer
Tara Moore, Curriculum Editor
Angela Greenroy, Part-time Temporary Instructional Developer

Photographers
Jeff Fortney, Senior Editor

Technical Reviewer
Wendy J. Gross
Captain
Orange County Fire Rescue
Orlando, Florida

Editorial Staff
Tara Gladden, Editorial Assistant

Indexer
Nancy Kopper

The IFSTA Executive Board at the time of validation of the **Fire and Emergency Services Orientation and Terminology, Sixth Edition** was as follows:

IFSTA Executive Board

Executive Board Chair
Steve Ashbrock
Fire Chief
Madeira & Indian Hill Fire Department
Cincinnati, OH

Vice Chair
Bradd Clark
Fire Chief
Ocala Fire Department
Ocala, Florida

Executive Director
Mike Wieder
Associate Director
Fire Protection Publications at OSU
Stillwater, Oklahoma

Board Members

Steve Austin
Project Manager
Cumberland Valley Volunteer Firemen's Association
Newark, DE

Mary Cameli
Assistant Chief
City of Mesa Fire Department
Mesa, AZ

Dr. Larry Collins
Associate Dean
Eastern Kentucky University
Safety, Security, & Emergency Department
Richmond, KY

Chief Dennis Compton
Mesa & Phoenix, Arizona
Chairman of the National Fallen Firefighters
Foundation Board of Directors

John Hoglund
Director Emeritus
Maryland Fire & Rescue Institute
New Carrollton, MD

Dr. Scott Kerwood
Fire Chief
Hutto Fire Rescue
Hutto, TX

Wes Kitchel
Assistant Chief
Sonoma County Fire & Emergency Services
Department
Santa Rosa, CA

Brett Lacey
Fire Marshal
Colorado Springs Fire Department
Colorado Springs, CO

Robert Moore
Division Director
Texas A&M Engineering Extension Services
College Station, TX

Dr. Lori Moore-Merrell
Assistant to the General President
International Association of Fire Fighters
Washington, DC

Jeff Morrissette
State Fire Administrator
State of Connecticut Commission on
Fire Prevention and Control
Windsor Locks, CT

Josh Stefancic
Division Chief
Largo Fire Rescue
Largo, FL

Board Members (cont.)

Don Turno
Program Manager
Savannah River Nuclear Solutions
Aiken, SC

Paul Valentine
Senior Engineer
Nexus Engineering
Oakbrook Terrace, IL

Steven Westermann
Fire Chief
Central Jackson County Fire Protection District
Blue Springs, MO

Dedication

This manual is dedicated to the men and women who hold devotion to duty above personal risk, who count on sincerity of service above personal comfort and convenience, who strive unceasingly to find better and safer ways of protecting lives, homes, and property of their fellow citizens from the ravages of fire, medical emergencies, and other disasters

...The Firefighters of All Nations.

Introduction

Introduction Contents

Introduction 1
Purpose and Scope 1
Book Organization 2
Terminlolgy 2
Resources 3
Key Information 3
   Safety Alert 3
   Information 3
   What This Means to You 3
   Key Term 3

Metric Conversions 4
   U.S. to Canadian Measurement
      Conversions 5
   Conversion and Approximation
      Examples 6

Introduction

Courtesy of Chris Mickal.

From a broad perspective, the purpose of the fire service is to protect life and property from the effects of fire and other hazards by such activities as fire prevention, fire suppression, fire and life safety education, fire investigation, hazardous materials mitigation, and emergency medical services. Clearly, firefighters today do much more than fight fires. Most of them are trained to provide emergency medical care, and many are additionally trained in special skills such as water rescue, heavy rescue, hazardous materials response, and disaster response. All of these areas require intensive training if firefighters are to respond efficiently and safely. For example, emergency medical incidents must be handled much like hazardous materials incidents in order to protect both victims and rescuers.

New firefighters must become acquainted with all of the responsibilities of their job. They must understand the organizational structure of their departments and learn how their departments interact with other local, state, or provincial agencies. It is also useful if the firefighter understands some of the history associated with the fire service.

Firefighter candidates can enter the fire service several ways. Candidates may be hired by one of the career departments that protect larger communities throughout North America. However, the majority of new firefighters are those who join the ranks of the volunteer service. Volunteer fire departments protect the major portion of North America. Others may join industrial fire brigades. Regardless of the mode of entry, all firefighters need proper orientation and training in basic skills.

Purpose and Scope

The purpose of this manual is to acquaint new firefighters with the history, traditions, terminology, organization, and operation of the fire and emergency services. In addition, the manual contains typical job and operation descriptions that should provide insight into the inner workings of the fire service. An extensive fire and emergency services glossary is included. The manual is also intended for college students who are beginning their Associate's level studies in fire-related fields.

The scope of this manual is to meet the learning outcomes of the National Fire Academy (NFA) Fire and Emergency Services Higher Education (FESHE) *Principles of Emergency Services* Associate's level course. In addition, information in this manual addresses numerous requisite knowledge items identified in Chapter 5, Fire Fighter I; and Chapter 6, Fire Fighter II; of NFPA® 1001, *Standard for Fire Fighter Professional Qualifications*. The manual is intended to be used in a college classroom, but it is also an excellent companion text to IFSTA's **Essentials of Fire Fighting and Fire Department Operations, 6th Edition.**

Terminology

This manual is written with a global, international audience in mind. For this reason, it often uses general descriptive language in place of regional- or agency-specific terminology (often referred to as *jargon*). Additionally, in order to keep sentences uncluttered and easy to read, the word *state* is used to represent both state and provincial-level governments (or their equivalent). This usage is applied to this manual for the purposes of brevity and is not intended to address or show preference for only one nation's method of identifying regional governments within its borders.

The glossary at the end of the manual will assist the reader in understanding words that may not have their roots in the fire and emergency services. The sources for the definitions of fire-and-emergency-services-related terms will be the *NFPA® Dictionary of Terms*.

Resources

To help you increase your knowledge of occupational safety and health issues, this manual contains references to additional materials. Books, articles, journals, and websites are included in the Suggested Readings section at the end of the manual. These materials were used in the development of this manual and are recommended by the members of the validation committee for your use.

Additional educational resources to supplement this manual are available from IFSTA and Fire Protection Publications (FPP).

Key Information

Various types of information in this book are given in shaded boxes marked by symbols or icons. See the following definitions:

Case History

A case history analyzes an event. It can describe its development, action taken, investigation results, and lessons learned.

Information

Information boxes give facts that are complete in themselves but belong with the text discussion. It is information that needs more emphasis or separation. (In the text, the title of information boxes will change to reflect the content.)

Key Term

A **key term** is designed to emphasize key concepts, technical terms, or ideas that the student in an associate's degree program in a fire-related field need to know. They are listed at the beginning of each chapter and the definition is placed in the margin for easy reference. An example of a key term is:

Fire Hazard — Any material, condition, or act that contributes to the start of a fire or that increases the extent or severity of fire.

> **Fire Hazard** — Any material, condition, or act that contributes to the start of a fire or that increases the extent or severity of fire.

Signal Words

Three key signal words are found in the book: **WARNING, CAUTION,** and **NOTE.** Definitions and examples of each are as follows:

- **WARNING** indicates information that could result in death or serious injury to the student in an associate's degree program in a fire-related field needs to be aware of. See the following example:

> **WARNING!**
> Truss constructed buildings have been known to fail in as little as 5 to 10 minutes of fire exposure. Firefighters should exercise extreme caution when operating in or on any building of lightweight construction in which the structural elements have been exposed to heat or fire.

- **CAUTION** indicates important information or data that the student in an associate's degree program in a fire-related field needs to be aware of in order to perform their duties safely. See the following example:

> **CAUTION**
> Cable and telephone lines are utility wires that generally are not thought of as posing a major hazard. However, these lines carry current and may pose a threat in certain hazardous environments. In addition, these wires are often run in close proximity to electrical lines, and under certain conditions they may become charged.

- **NOTE** indicates important operational information that helps explain why a particular recommendation is given or describes optional methods for certain procedures. See the following example:

NOTE: More information concerning fire and life safety education follows in a subsequent section.

Metric Conversions

Throughout this manual, U.S. units of measure are converted to metric units for the convenience of our international readers. Be advised that we use the Canadian metric system. It is very similar to the Standard International system, but may have some variation.

We adhere to the following guidelines for metric conversions in this manual:

- Metric conversions are approximated unless the number is used in mathematical equations.
- Centimeters are not used because they are not part of the Canadian metric standard.
- Exact conversions are used when an exact number is necessary such as in construction measurements or hydraulic calculations.
- Set values such as hose diameter, ladder length, and nozzle size use their Canadian counterpart naming conventions and are not mathematically calculated. For example, 1½ inch hose is referred to as 38 mm hose.
- Add metric notes particular to your manual.

The following two tables provide detailed information on IFSTA's conversion conventions. The first table includes examples of our conversion factors for a number of measurements used in the fire service. The second shows examples of exact conversions beside the approximated measurements you will see in this manual.

U.S. to Canadian Measurement Conversion

Measurements	Customary (U.S.)	Metric (Canada)	Conversion Factor
Length/Distance	Inch (in) Foot (ft) [3 or less feet] Foot (ft) [3 or more feet] Mile (mi)	Millimeter (mm) Millimeter (mm) Meter (m) Kilometer (km)	1 in = 25 mm 1 ft = 300 mm 1 ft = 0.3 m 1 mi = 1.6 km
Area	Square Foot (ft^2) Square Mile (mi^2)	Square Meter (m^2) Square Kilometer (km^2)	1 ft^2 = 0.09 m^2 1 mi^2 = 2.6 km^2
Mass/Weight	Dry Ounce (oz) Pound (lb) Ton (T)	gram Kilogram (kg) Ton (T)	1 oz = 28 g 1 lb = 0.5 kg 1 T = 0.9 T
Volume	Cubic Foot (ft^3) Fluid Ounce (fl oz) Quart (qt) Gallon (gal)	Cubic Meter (m^3) Milliliter (mL) Liter (L) Liter (L)	1 ft^3 = 0.03 m^3 1 fl oz = 30 mL 1 qt = 1 L 1 gal = 4 L
Flow	Gallons per Minute (gpm) Cubic Foot per Minute (ft^3/min)	Liters per Minute (L/min) Cubic Meter per Minute (m^3/min)	1 gpm = 4 L/min 1 ft^3/min = 0.03 m^3/min
Flow per Area	Gallons per Minute per Square Foot (gpm/ft^2)	Liters per Square Meters Minute (L/(m^2.min))	1 gpm/ft^2 = 40 L/(m^2.min)
Pressure	Pounds per Square Inch (psi) Pounds per Square Foot (psf) Inches of Mercury (in Hg)	Kilopascal (kPa) Kilopascal (kPa) Kilopascal (kPa)	1 psi = 7 kPa 1 psf = .05 kPa 1 in Hg = 3.4 kPa
Speed/Velocity	Miles per Hour (mph) Feet per Second (ft/sec)	Kilometers per Hour (km/h) Meter per Second (m/s)	1 mph = 1.6 km/h 1 ft/sec = 0.3 m/s
Heat	British Thermal Unit (Btu)	Kilojoule (kJ)	1 Btu = 1 kJ
Heat Flow	British Thermal Unit per Minute (BTU/min)	watt (W)	1 Btu/min = 18 W
Density	Pound per Cubic Foot (lb/ft^3)	Kilogram per Cubic Meter (kg/m^3)	1 lb/ft^3 = 16 kg/m^3
Force	Pound-Force (lbf)	Newton (N)	1 lbf = 0.5 N
Torque	Pound-Force Foot (lbf ft)	Newton Meter (N.m)	1 lbf ft = 1.4 N.m
Dynamic Viscosity	Pound per Foot-Second (lb/ft.s)	Pascal Second (Pa.s)	1 lb/ft.s = 1.5 Pa.s
Surface Tension	Pound per Foot (lb/ft)	Newton per Meter (N/m)	1 lb/ft = 15 N/m

Conversion and Approximation Examples

Measurement	U.S. Unit	Conversion Factor	Exact S.I. Unit	Rounded S.I. Unit
Length/Distance	10 in	1 in = 25 mm	250 mm	250 mm
	25 in	1 in = 25 mm	625 mm	625 mm
	2 ft	1 in = 25 mm	600 mm	600 mm
	17 ft	1 ft = 0.3 m	5.1 m	5 m
	3 mi	1 mi = 1.6 km	4.8 km	5 km
	10 mi	1 mi = 1.6 km	16 km	16 km
Area	36 ft^2	1 ft^2 = 0.09 m^2	3.24 m^2	3 m^2
	300 ft^2	1 ft^2 = 0.09 m^2	27 m^2	30 m^2
	5 mi^2	1 mi^2 = 2.6 km^2	13 km^2	13 km^2
	14 mi^2	1 mi^2 = 2.6 km^2	36.4 km^2	35 km^2
Mass/Weight	16 oz	1 oz = 28 g	448 g	450 g
	20 oz	1 oz = 28 g	560 g	560 g
	3.75 lb	1 lb = 0.5 kg	1.875 kg	2 kg
	2,000 lb	1 lb = 0.5 kg	1 000 kg	1 000 kg
	1 T	1 T = 0.9 T	900 kg	900 kg
	2.5 T	1 T = 0.9 T	2.25 T	2 T
Volume	55 ft^3	1 ft^3 = 0.03 m^3	1.65 m^3	1.5 m^3
	2,000 ft^3	1 ft^3 = 0.03 m^3	60 m^3	60 m^3
	8 fl oz	1 fl oz = 30 mL	240 mL	240 mL
	20 fl oz	1 fl oz = 30 mL	600 mL	600 mL
	10 qt	1 qt = 1 L	10 L	10 L
	22 gal	1 gal = 4 L	88 L	90 L
	500 gal	1 gal = 4 L	2 000 L	2 000 L
Flow	100 gpm	1 gpm = 4 L/min	400 L/min	400 L/min
	500 gpm	1 gpm = 4 L/min	2 000 L/min	2 000 L/min
	16 ft^3/min	1 ft^3/min = 0.03 m^3/min	0.48 m^3/min	0.5 m^3/min
	200 ft^3/min	1 ft^3/min = 0.03 m^3/min	6 m^3/min	6 m^3/min
Flow per Area	50 gpm/ft^2	1 gpm/ft^2 = 40 L/(m^2.min)	2 000 L/(m^2.min)	2 000 L/(m^2.min)
	326 gpm/ft^2	1 gpm/ft^2 = 40 L/(m^2.min)	13 040 L/(m^2.min)	13 000 L/(m^2.min)
Pressure	100 psi	1 psi = 7 kPa	700 kPa	700 kPa
	175 psi	1 psi = 7 kPa	1225 kPa	1 200 kPa
	526 psf	1 psf = 0.05 kPa	26.3 kPa	25 kPa
	12,000 psf	1 psf = 0.05 kPa	600 kPa	600 kPa
	5 psi in Hg	1 psi = 3.4 kPa	17 kPa	17 kPa
	20 psi in Hg	1 psi = 3.4 kPa	68 kPa	70 kPa
Speed/Velocity	20 mph	1 mph = 1.6 km/h	32 km/h	30 km/h
	35 mph	1 mph = 1.6 km/h	56 km/h	55 km/h
	10 ft/sec	1 ft/sec = 0.3 m/s	3 m/s	3 m/s
	50 ft/sec	1 ft/sec = 0.3 m/s	15 m/s	15 m/s
Heat	1200 Btu	1 Btu = 1 kJ	1 200 kJ	1 200 kJ
Heat Flow	5 BTU/min	1 Btu/min = 18 W	90 W	90 W
	400 BTU/min	1 Btu/min = 18 W	7 200 W	7 200 W
Density	5 lb/ft^3	1 lb/ft^3 = 16 kg/m^3	80 kg/m^3	80 kg/m^3
	48 lb/ft^3	1 lb/ft^3 = 16 kg/m^3	768 kg/m^3	770 kg/m^3
Force	10 lbf	1 lbf = 0.5 N	5 N	5 N
	1,500 lbf	1 lbf = 0.5 N	750 N	750 N
Torque	100	1 lbf ft = 1.4 N.m	140 N.m	140 N.m
	500	1 lbf ft = 1.4 N.m	700 N.m	700 N.m
Dynamic Viscosity	20 lb/ft.s	1 lb/ft.s = 1.5 Pa.s	30 Pa.s	30 Pa.s
	35 lb/ft.s	1 lb/ft.s = 1.5 Pa.s	52.5 Pa.s	50 Pa.s
Surface Tension	6.5 lb/ft	1 lb/ft = 15 N/m	97.5 N/m	100 N/m
	10 lb/ft	1 lb/ft = 15 N/m	150 N/m	150 N/m

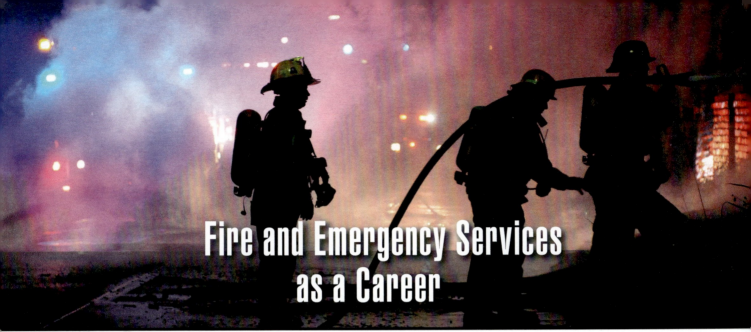

Courtesy of Chris Mickal.

Fire and Emergency Services as a Career

Chapter Contents

Fire and Emergency Services Culture 12	
Fire Protection Agencies And Other Emergency Response Organizations 13	
Types of Fire Departments 13	
Jurisdiction 14	
Municipal 14	
County, Parish, or Borough 15	
Fire Districts and Fire Protection Districts 16	
Public Fire Departments 17	
Career Fire Departments 18	
Volunteer Departments 18	
Combination Departments 19	
Paid On-Call Departments 19	
Public Safety Departments 20	

Private and Other Government Departments 20
    Industrial Fire Departments 20
    Federal Fire Departments 21
    Commercial Fire Protection Services 21
    Airport Fire Departments 21

Firefighters as Public Figures 22
    The Fire Fighter Image 22
    Responsible Behaviors of Firefighters 22
    Conduct 23
    Teamwork 24
    Human Relations and Customer Service ... 24
    Uniform and Dress 25
    Physical Fitness 27

Training and Education 28
    Types of Training 28
    Departmental Training 28
    Regional/County and State/Provincial Training Programs 29
    Value of Education 30
    Colleges and Universities 31
    National Fire Academy 31
    Non-Traditional Training and Education 32
    Firefighter Certification/Credentialing 33

Firefighter Selection Process 34
    Recruitment 35
    Preparation 35
    Application 35
    Written Examination 36
    Physical Ability Test 38
    Interview Process 39
    Probationary Period 39
    Volunteer Firefighter Selection 39

Fire Service Career Information 40
    Pay 40
    Hours of Duty 40
    Promotion 40
    Retirement 41

Chapter Summary 41
Review Questions 42
Chapter 1 Notes 42

chapter 1

Key Terms

Career Fire Department 18	Paid On-Call Firefighter 19
Career Firefighter 18	Public Safety Department......................... 19
Combination Fire Department................... 19	Volunteer Fire Department 18
Industrial Fire Brigades 20	Volunteer Firefighter 19

FESHE Outcomes

This chapter provides information that addresses the outcomes for the Fire and Emergency Services Higher Education (FESHE) *Principles of Emergency Services* course.

1. Illustrate and explain the history and culture of the fire service.

3. Differentiate between fire service training and education and explain the value of higher education to the professionalization of the fire service.

4. List and describe the major organizations that provide emergency response service and illustrate how they interrelate.

5. Identify fire protection and emergency service careers in both the public and private sector.

7. Discuss and describe the scope, purpose, and organizational structure of fire and emergency services.

11. Recognize the components of career preparation and goal setting.

12. Describe the importance of wellness and fitness as it relates to emergency services.

Fire and Emergency Services as a Career

Learning Objectives

After reading this chapter, students will be able to:

1. Describe the fire and emergency services culture. (Outcome 1)
2. Recognize why incidents may require response from fire protection agencies and other emergency response organizations. (Outcome 4)
3. Explain the different ways that fire departments may be categorized. (Outcome 7)
4. Recognize standards that firefighters as public figures are expected to maintain. (Outcomes 1, 12)
5. Differentiate between types of fire service training. (Outcomes 3, 11)
6. Explain the value of education in the fire service. (Outcomes 3, 11)
7. Identify education and training opportunities for the fire service. (Outcomes 3, 11)
8. Explain the importance of firefighter certification/credentialing. (Outcomes 3, 11)
9. Describe the firefighter selection process. (Outcomes 5, 11)
10. Recognize general fire service career information. (Outcomes 5, 7, 11)

Chapter 1
Fire and Emergency Services as a Career

Courtesy of Bob Esposito.

Case History

A high school senior asked a local firefighter about the fire service as a profession. The student was interested in the type of work a firefighter does, what it takes to become a firefighter, and what things he could do to improve his chances to get hired.

The firefighter provided the student with information about the professional fire service. They discussed the physical and mental aspects of being a firefighter. They reviewed educational opportunities at the local community college. They also talked about volunteering as a firefighter with the local volunteer fire department as a way of learning about and understanding the profession.

After the student's meeting with the local firefighter, the student was very enthusiastic about the fire service. The student was interested in learning more and was looking forward to meeting the fire chief of the local volunteer fire department.

Fire fighting is an occupation that, while dangerous, is a rewarding endeavor with a rich heritage of honor, sacrifice, and service. Members of the public hold firefighters in high regard, and they must always strive to meet the expectation of trust placed upon them (**Figure 1.1, p.12**).

People looking to join the fire and emergency services as a career will be joining an ever-evolving industry. The fire service profession includes a variety of emergency response agencies and types of fire departments. People can join the profession whether they live in a metropolitan or a rural area. Firefighters, as representatives of their departments, will take part in numerous training and educational opportunities. This chapter will further discuss the following topics:

- Fire and Emergency Services Culture
- Purposes of Fire Protection Agencies
- Other Emergency Response Organizations
- Types of Fire Departments
- Firefighters as Public Figures
- Training and Education
- Firefighter Selection Process

Figure 1.1 Firefighters fighting a structure fire. *Courtesy of Bob Esposito.*

Fire and Emergency Services Culture

The public expects members of the fire service to react to a chaotic situation with a calm sense of purpose, critical analysis, and decisive action. As a result of generations of heroic actions under adverse conditions, the fire service has built a reputation as one of the most highly regarded public service organizations. Fire fighting consistently tops lists of the most trusted and respected occupations.

The public, as well as other emergency response agencies, call the fire department to respond to almost every conceivable type of incident. In addition, fire departments are called for other emergencies that require the resources, expertise, and dedication to duty that the fire department will provide, such as:

- Medical responses
- Vehicle and industrial accidents
- Collapses
- Hazardous material incidents
- Water rescues
- Natural or human-caused disasters **(Figure 1.2a, b, and c)**

Figure 1.2 Firefighters respond to several types of emergencies, including vehicle and industrial accidents, hazardous materials incidents, and water rescues. *1.2a Courtesy of Bob Esposito. 1.2b Courtesy of Chris Mickal.*

Fire Protection Agencies and Other Emergency Response Organizations

Fire protection agencies, no matter what their affiliation, are tasked with protecting people and property from fire. The majority of these agencies are responsible for fire prevention and public education concerning fire dangers.

Fire service organizations often respond to incidents along with other public and private entities. Each organization has a specific skill set and area of responsibility to ensure the protection of lives and/or property. Common local response agencies may include police, emergency medical services (EMS), and public works and utility departments. During larger scale incidents, county, state, or federal response agencies may be called upon for assistance. Chapter 10 provides additional information concerning other emergency response organizations **(Figure 1.3)**.

Figure 1.3 USAR teams may assist local fire departments with emergency responses. *Courtesy of Ron Jeffers.*

Types of Fire Departments

Fire departments generally function under some form of government structure, although some departments are independent organizations. Some state governments may furnish fire protection, and in some cases, emergency medical

Figure 1.4 The federal government provides fire prevention and suppression for the U.S. Forestry Service and the National Park Service and often contracts with local fire departments for mutual aid. *Courtesy of 10 Tanker Air Carrier LLC.*

services for state turnpikes, forests and parks, or large state institutions. The federal government provides fire protection for national parks and forests and military installations. The National Park Service and U.S. Forest Service (USFA) provide fire prevention and suppression activities, often contracting for mutual aid with local fire departments **(Figure 1.4)**.

The term fire department is loosely applied to almost any agency providing fire protection. However, a more precise definition refers to the fact that the organization is a departmental division within a larger body. This body may be a municipal government, county, parish, or borough commission. There are more than 33,000 public fire departments in the United States.

Jurisdiction

The primary means of categorizing public fire and emergency services organizations is by the jurisdiction the organization serves. In the context of this manual, the term jurisdiction has two different connotations. First, it refers to the area that a fire or emergency services organization serves. Second, it refers to the authority that gives the organization the legal right to exist, provide fire protection, and ensure that protection is adequate.

Municipal

Municipalities are the most common jurisdiction for fire departments. Whether career, volunteer, or combination, this type of organization operates under the authority of the local government, and receives funding and oversight from that body.

Municipal fire department refers to a functional division of the local government, that the state or province authorizes to provide fire protection for functional divisions of the local government, and can include the following:

- City
- Town
- Township
- Village
- Other local governmental structure

As a department within the municipal government, most fire departments have an organizational structure that is similar to other jurisdictional agencies. The department head, generally called the chief or commissioner, serves as the primary interface between the fire department and the rest of the municipal government.

The citizens determine the level and type of services through their elected officials. When making this determination, certain factors must be considered, such as:

- Population
- Geographic size
- Predominant risk
- Any special fire protection requirements

A municipal fire department must maintain sufficient personnel, equipment, and fire stations to serve the jurisdiction properly. In smaller departments, all training and administrative functions may operate from a single building. Larger departments may feature separate facilities for separate functions such as:

- Training
- Administration
- Fire prevention
- Other support functions

Funding to support municipal fire departments is generally obtained through the collection of taxes levied on businesses and homeowners within the jurisdiction. However, some fire departments rely on fundraising efforts and grants, and may charge a separate subscription fee or at least a portion of the cost to provide an emergency response. This funding is most typical in emergency medical responses and non-emergency patient transfers. Most municipal fire department budgets are set on an annual basis and include all expenses, including the following:

- Personnel
- Equipment
- Maintenance
- Other operating expenses

County, Parish, or Borough

Fire departments operated by county, parish, or boroughs are becoming more common in the U.S. Often these departments begin with mutual aid agreements between communities in a county jurisdiction and progress to the sharing of resources. Communications, fire prevention, technical rescue and hazardous materials response capabilities may be shared to conserve and avoid duplication of resources.

The development of shared county facilities, such as airports, industrial complexes, or other infrastructure, may lead to consolidation of area fire departments. This arrangement may enable smaller towns and suburban or rural locations to benefit from a greater variety of resources without having to individually bear the total cost. The county fire department may simply augment municipal fire departments, or it may consolidate each department into one larger organization. These departments may also serve the unincorporated areas of a county.

Fire Districts and Fire Protection Districts

A fire district may serve the same purpose as a county fire department in some states, but it is not directly tied to the county, parish, or borough. These districts may overlap other jurisdictional boundaries in their fire protection services. The district is a state-authorized governing body established to provide fire protection within a specified area. Generally, the district operates under a board of commissioners or trustees who represent the residents of the district.

The fire district may feature a career, combination, or volunteer organization, and may also provide the following services:

- EMS
- Hazardous materials response
- Technical rescue
- Other specialty operations **(Figure 1.5)**

Figure 1.5 Fire districts may use specialized vehicles for each type of emergency. These can include EMS, haz-mat operations, rescue, and special operations. *Courtesy of Ron Jeffers.*

In most cases, the fire district oversees the duties generally associated with fire departments. However, depending on its organization, it may be responsible for other services, such as water supply or law enforcement.

Unlike other public fire departments, the fire district lacks other municipal departments to support its operation. It must provide or contract for services, such as apparatus and facility maintenance and repair.

The fire protection district, which is defined by state statute, varies slightly from the fire district concept. This form of organization shares some elements of a fire district. However, in some areas, it may not exist as a separate government agency. The fire protection district may be established when a group of people with a common need for fire protection, typically in a rural or seasonal community, petition an established fire department to extend protection to its property. For a fee, the local fire department provides this service under a contract. The fee collected will compensate the local fire department for any equipment or personnel required to provide the additional coverage.

A state or province may formally recognize and possibly provide some funding for the contract. In other cases, the only recognition is the contract itself, with the group of property owners serving as the legal entity.

Public Fire Departments

Jurisdictions receive fire protection in a variety of ways. The most prevalent systems include:

- Career Departments
- Volunteer Departments
- Combination Departments
- Paid On-Call Departments
- Public Safety Departments

Regardless of the jurisdiction or organization, a fire department must be properly staffed in order to function safely and effectively. This staffing may include career firefighters who earn a salary and are employed full-time in this capacity. Other firefighters are paid on-call or receive no monetary compensation. In this context, the term professional firefighter refers not to an amount of compensation, but a level of competence and expertise that may be applied equally to all firefighters.

U. S. Fire Department Profiles

Fire departments that are wholly or mostly volunteer comprise approximately 87 percent of all public fire departments. These departments protect approximately 35 percent of the entire U.S. population. Departments comprising career or mostly career members, totaling 13 percent of fire service organizations, protect approximately 66 percent of the U.S. population.

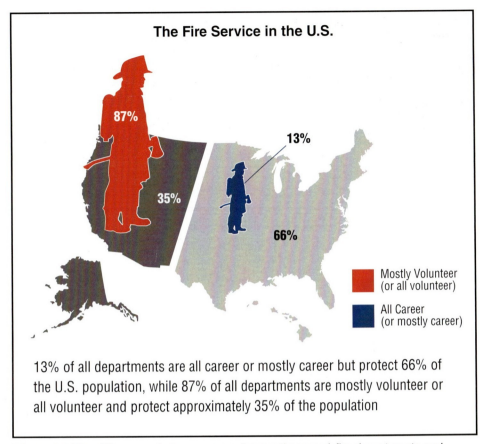

Figure 1.6 The differences between career (or mostly career) fire departments and volunteer departments in terms of percentage of the population protected.

Career Departments

Most large cities in the U.S. and Canada maintain **career fire departments**. Many counties and fire protection districts also operate career fire departments. These jurisdictions maintain full-time firefighters who are on duty in firehouses around the clock, working one of a variety of shift schedules (see the work schedule section in this chapter for examples). These **career firefighters** are paid a salary and may receive additional monetary incentives for fire or EMS certifications, college degree, rank, or longevity. Most career departments also employ firefighters and civilian personnel who work conventional business hours to staff administrative positions and support functions. In addition, some career fire departments may employ part-time or seasonal personnel to work during peak times of the year when call volume or a specific danger, such as wildfires, may be present **(Figure 1.6)**.

Volunteer Departments

Although **volunteer fire departments** protect some cities, they are typically found in smaller towns and rural communities. These departments may be organized in a variety of ways. Some departments operate as a department of local government, but others are totally independent organizations. In some jurisdictions, the government may provide a building and apparatus, while in other instances, the volunteer department owns the equipment, buildings, or both **(Figure 1.7)**. Without governmental support, the organization may operate on the following:

Career Firefighter — Person whose primary employment is as a firefighter within a fire department.

Career Fire Departments — Fire department composed of full-time, paid personnel.

Volunteer Firefighter – Active member of a fire department who many receive monetary compensation for on-call time and/or fire fighting duty time.

Figure 1.7 Volunteer fire departments are typically found in smaller towns and rural communities and may rely on donations to purchase apparatus and equipment. *Courtesy of Ron Jeffers.*

- Donations
- Subscription fees from residents
- Billing for response costs
- Fundraising efforts

Many volunteer departments combine several of the aforementioned sources to maintain services. Volunteer fire departments are overseen by the government entity of which they are a member, or officers of an independent association or board in the case of non-government organizations.

Some volunteer departments may have members on shift at the firehouse, while other personnel are alerted by pager, telephone, or community wide audible signal. **Volunteer firefighters** respond from home, leisure, or work in order to provide services. Depending on the jurisdiction, some members may respond to the firehouse to drive the apparatus, while other members report directly to the incident scene in their own vehicles.

Volunteer Fire Department – Organization of part-time firefighters who may receive monetary compensation for on-call time or fire fighting duty time.

Combination Departments

Combination fire departments exist in a wide variety of configurations. Some departments maintain a cadre of volunteers who supplement career firefighters with specific duties or major incidents. Other departments are primarily volunteer, with a handful of career firefighters working daytime hours when volunteers may be scarce. In some systems, control of the department is the responsibility of a volunteer chief and officers. Meanwhile, in other departments, career members direct operations and administration. Oversight and division of responsibilities as well as the percentages of volunteer and career members varies greatly among different jurisdictions.

Combination Fire Department – Organization in which some of the firefighters receive pay while other personnel serve on a voluntary basis. In Canada, a combination fire department is called a Composite Fire Department.

Paid On-Call Departments

Functionally, most paid on-call departments operate in a manner similar to volunteer departments. These organizations may have minimally staffed fire stations or rely totally on the response of paid on-call members. **Paid on-call firefighters** are generally paid with an hourly wage or set fee per response. If they are not in the firehouse, they may be summoned using the same methods as volunteers. This system may also be found in small career or combination

Paid On-Call Firefighter – Firefighter who receives reimbursement for each call that he or she attends.

departments as a method to maintain part-time personnel for use at large-scale incidents. Generally, the funding for a paid on-call department is derived from a local government or association.

Public Safety Departments

In most public safety departments, the fire command/management officers, company officers, and apparatus driver/operators are career fire personnel. On a given shift, they may function as police officers or as firefighters according to need.

Some jurisdictions combine the roles of law enforcement officers and firefighters by creating a position often referred to as a public safety officer. The public safety officer may operate in either capacity based on staffing or incident requirements. Depending on the organization, all personnel or only a portion of the staff may operate in this role.

> **Public Safety Department** — Organization that combines the administrative, financial, and technical service and support for functions such as fire and rescue services, police, ambulance and emergency medical services, and emergency communications.

Private and Other Government Departments

Fire departments may be organized outside the public sector as part of a private company. Some of these private fire departments may be employed by a single company or industrial complex, while others may be organized as private corporations that engage in contracting with other private companies or governments for the purpose of fire protection. The federal government also provides fire protection and emergency services as part of the Department of Defense as well as other civilian agencies that protect numerous federal facilities. Examples of these types of fire departments include:

- Industrial Fire Departments
- Federal Fire Departments
- Commercial Fire Protection Services
- Airport Fire Departments

Industrial Fire Departments

Some large industrial complexes, especially those whose facilities are especially hazardous (such as refineries or power plants), may operate their own industrial fire departments, fire brigades, or emergency response teams. Depending on the facility's size and requirements, these departments may be larger than the public fire department of the community in which the industry is located.

Members of industrial fire organizations must comply with the professional qualifications standards included in the National Fire Protection Association® (NFPA®) standard: NFPA® 1081, *Standard for Industrial Fire Brigade Member Professional Qualifications*. Additionally, in the U.S., these brigades must comply with Occupational Safety and Health Administration (OSHA) regulations contained in CFR 1910.156, which regulates the organization, training, and personal protective equipment of fire brigades established by employers.

Similar to all other fire departments, **industrial fire brigades** have specific structure, training, and policies for all aspects of their operation. For additional information concerning industrial fire fighting, refer to IFSTA's **Industrial Emergency Services Training: Incipient Level** and **Industrial Exterior and Structural Fire Brigades** manuals.

> **Industrial Fire Brigades** — Team of employees organized within a private company, industrial facility, or plant who are assigned to respond to fires and emergencies on that property.

Federal Fire Departments

The federal government operates many fire and emergency services departments on military and federal installations throughout the world. These departments may provide structural fire protections as well as aircraft rescue and fire fighting (ARFF) services and marine fire fighting for vessels and maintenance facilities. In addition, these departments may also provide mutual aid to local civilian fire departments in surrounding areas.

Commercial Fire Protection Services

Some jurisdictions may contract with private companies for fire protection, EMS, and other emergency services. Privatization of services once considered to be government functions has gained popularity in recent years. While privatization is not the norm in the fire service, jurisdictions may consider commercial fire protection as an alternative.

Airport Fire Departments

Some public, non-military airports rely partially or entirely on the local fire department for fire protection. Other airports may be required to operate their own fire departments or contract with commercial fire departments to provide this service. Regardless of funding or organization, the department must be certified to provide ARFF services. If the fire departments are responsible for airport buildings, they must also be trained and equipped to provide structural fire fighting. **(Figure 1.8)**.

Figure 1.8 Airport firefighters are specially trained and certified to fight aircraft fires.

Firefighters as Public Figures

As public figures, all firefighters are expected to maintain certain standards. These standards include the following:

- The firefighter image
- Responsible behaviors of firefighters
- Conduct
- Teamwork
- Human relations and customer service
- Uniform and dress
- Physical fitness

The Firefighter Image

The image of the fire department, to include its firefighters, in the eyes of the public affects the ability of firefighters to operate effectively in the community they serve. Rather than a notion of firefighters sitting in the firehouse waiting for an alarm, most fire departments seek to establish and communicate a proactive posture to the public.

In many neighborhoods, the fire department is known as the one government service that responds immediately to virtually any problem and will attempt to solve any issue in the best interest of those parties involved. The very nature of those individuals who become firefighters lends itself toward providing the highest level of service to the community no matter what the need. Maintaining, enhancing, and doing nothing to tarnish this well-deserved reputation is the responsibility of every firefighter.

Responsible Behaviors of Firefighters

The fire service has standards of behavior and a code of ethics to which compliance is expected. The following principles are important for transitioning into a career in the fire service:

- **Be sincere in your interest and dedication to the job** — Successful firefighters selflessly support the group effort. Cooperation and teamwork are an integral part of safe fire fighting and emergency operations. Members must be willing to work to the best of their ability in the position to which they are assigned, during emergency incidents, and routine duties.

- **Be loyal to the department and fellow firefighters** — Firefighters should be prepared to implement and support the policies and operations of their department whenever possible. Loyalty to other department members is a necessity as firefighters constantly depend on each other for safety and support on and off the fireground.

- **Aggressive pursuit of training and educational opportunities** — Firefighter training is a continual process based on changing equipment, evolving tactics, and expanded missions. Members must display a willingness to learn throughout their careers and be open to new ideas and procedures in order to operate safely at incidents that may pose new challenges.

- **Guard speech on and off duty** — While firefighters should be well-versed and able to readily discuss fire service history and fire safety with the public, there is confidential information that firefighters must not divulge in conversation. During the course of duty, firefighters may be privy to private health or personal information concerning citizens or firefighters. This information must be held in confidence and only relayed to appropriate authorities based on department policy and local and federal law.

- **Inspire confidence and respect** — Firefighters should strive to be honest, fair, and trustworthy in all dealings with fellow firefighters as well as members of the public. Showing other individuals proper respect and acting with personal integrity will help to enhance one's own image as well as the image of the fire service in general.

- **Accept praise modestly and criticism graciously** — Legitimate constructive commentary should be accepted with the thought towards improving future performance. Accolades, honors, and advancement often follow improvement. This acknowledgement should be accepted with humility and in a respectful manner.

- **Ask questions** — Communication only takes place when information is understood by the receiver. Firefighters must receive and process both routine and critical information from a variety of sources on a regular basis. If any portion of this information is incomplete or unclear, the firefighter must ask specific questions until an understanding is gained.

Health Insurance Portability and Accountability Act

The Health Insurance Portability and Accountability Act of 1996 (HIPAA) includes safeguards for patient privacy, including medical information. As caregivers, fire service members should be aware of the need to keep all patient information confidential and never discuss patient information.

Conduct

Firefighters must conduct themselves in accordance with the high standards to which public safety professionals are held both on and off duty, exercising good judgment and adhering to department policy. Members should report for duty in good mental and physical condition, ready to perform their assigned duties. Firefighters should never be in the firehouse, on or off duty, while intoxicated or under the influence of any illegal drug. Unprofessional conduct, such as the following, may shed an unfavorable light on the fire service in the eyes of the public and should not be tolerated in or around the firehouse:

- Horseplay
- Unruly behavior
- Loud swearing
- Rudeness

NOTE: No activity that may compromise firefighter or public safety or gives an unfavorable impression of the fire department can be tolerated.

Figure 1.9 Firefighters are expected to be respectful and courteous when interacting with citizens.

Figure 1.10 Team work is essential to safe fire fighting operatons. *Courtesy of Bob Esposito.*

Firefighters are expected to adhere to the following general rules of conduct:

- Be fully prepared to report for duty as required without specific permission for absence based on local policy.
- Arrive on time for duty shifts, training, meetings, and other activities.
- Follow orders to the best of your ability per organizational policy.
- Be respectful and courteous to the public and all department members **(Figure 1.9)**.
- Establish and maintain the best possible working relationship with fellow firefighters, other public employees, and the general public.

Teamwork

Fighting fires and meeting the challenges of other emergencies is a team effort. Individuals learn this basic concept during recruit training, where they are taught that the safety of each member is dependent on one another, and the operation will succeed only if all members are working in a coordinated effort toward a common goal.

Effective teams are built through training and working together. These teams must constantly train together in order to ensure safe and effective operations. A team develops and maintains enthusiasm and confidence as it works successfully at numerous incidents over a period of time. Each member must contribute effort in order to ensure the strength and success of the team. Basic factors for team success include the following:

- All members should work hard to earn the trust of other firefighters.
- All members must consistently demonstrate a high level of integrity.
- Each member must contribute to the team.
- Each firefighter should know his/her job as well as the jobs of other group members in order to function most efficiently.
- Each member must communicate well with the team **(Figure 1.10)**.

Human Relations and Customer Service

Firefighters must function in a diverse world, both inside and outside the firehouse. Members of the modern fire service have a more diverse background than ever before. Many firefighters work in neighborhoods that reflect a wide variety of cultural backgrounds. In public and in the firehouse, firefighters may encounter diversity, such as:

- Gender
- Race
- Religion
- Ethnicity
- Socioeconomic status
- Sexual preference

In order to maintain a cohesive workplace, firefighters must be prepared to work in a diverse environment. Respecting individual differences and cooperating with others is essential to providing effective services to the public.

Uniform and Dress

Department policy usually defines appropriate dress. Most organizations have several uniforms that are worn based on functionality, appearance, and safety requirements. A dress uniform (sometimes called class A) is often worn for special events, public appearances, inspections, parades, or certain social functions. This uniform generally consists of uniform pants, dress shirt (white or blue per policy), dress jacket, tie, black shoes, appropriate uniform insignia, and uniform hat **(Figure 1.11)**.

Station work uniforms should comply with NFPA®1975, *Standard on Station/Work Uniforms for Emergency Services*. The style of the uniform may vary based on local preferences and the type of duty the wearer is expected to perform. Common styles include collared shirts, polo shirts, or tee shirts with uniform trousers. In some instances, coveralls may be worn while particularly dirty jobs are undertaken. Clothing that does not meet the fire resistance standards of the fire department should not be worn during fire response.

Volunteer firefighters generally do not wear a specific duty uniform when responding to an incident. Jurisdictions may recommend clothing made of natural fibers as opposed to synthetic material because natural material will not melt under heat conditions like synthetic fibers. In addition, this clothing often has a patch or logo to identify responders as members of the fire department **(Figure 1.12)**.

Figure 1.11 A dress uniform is often worn for public appearances and events.

Figure 1.12 All PPE must be stored properly.

Figure 1.13a. Reflective trim catches light and reflects it to enhance the visibility of responders at night or in low-light conditions.

Figure 1.13b Wristlets are built into the ends of coat sleeves to provide a protective interface between the sleeves and gloves.

Figure 1.13c Coat collars prevent embers, water, and other debris from getting under the coat.

Figure 1.13d A coat closure system may include more than one mechanism to ensure a complete seal.

Figure 1.13e A loop on the back of protective clothing serves as a handle to aid in the rescue of a distressed firefighter.

Figure 1.13f Due to their weight, protective trousers are supported in place with suspenders.

Figure 1.13g Protective gloves cover the hands and coat wristlet to provide protection against common hazards.

Figure 1.13h Fire fighting boots protect the foot and ankle from hazards routinely found at an incident scene.

Figure 1.13i Styles of fire fighting boots may vary, but their common purpose is to protect a firefighter's feet and lower legs.

All firefighters must wear personal protective clothing while working at an incident. Full personal protective equipment (PPE) includes the following:

- Turnout coat and pants
- Boots
- Helmet
- Protective hood
- Eye protection
- Gloves
- Hearing protection **(Figure 1.13 a-i)**

When working in atmospheres that are or may become hazardous, firefighters must also wear self-contained breathing apparatus (SCBA) equipped with a personal alert safety system (PASS) device. Additional information concerning station uniforms, PPE, and SCBA may be found in Chapter 3.

Physical Fitness

In order to safely perform many physically demanding tasks efficiently during an incident, firefighters must have agility, skill, and coordination. Members must maintain proper physical conditioning in order to perform these tasks without endangering themselves or others.

In order to gain entrance to the fire service, most candidates must meet certain physical standards. This criterion is often based on NFPA® 1582, *Standard on Comprehensive Occupational Medical Program for Fire Departments*. Applicants who are selected generally participate in a strenuous training academy where physical conditioning is combined with fire service training. In recent years, a greater emphasis has been placed on continued physical conditioning throughout a firefighter's career. Because of the physically demanding nature of fighting fires, a lack of fitness endangers the individual and the entire team to which they are assigned **(Figure 1.14)**.

Figure 1.14 Physical fitness must be maintained to safely perform fire fighting tasks.

Physical Fitness

Maintaining physical fitness is an important part of ensuring the safety and success of a fire department. Recent health and wellness initiatives have begun to aid firefighters in this goal by assisting with dietary and fitness programs for members of all ages.

Training and Education

Training and education is just as important to veteran firefighters as it is to a new recruit. The depth and complexity of training will vary based on experience. However, learning new skills and maintaining current in a changing profession requires constant attention to education and training. This section will discuss the following topics:

- Types of training
- Departmental training
- Regional/county and state/provincial training programs
- Value of education
- Colleges and universities
- National fire academy
- Non-traditional training and education
- Firefighter certification/ credentialing

The Difference Between Training and Education

Training, also called drilling, is a supervised activity or process designed to achieve and maintain proficiency through instruction and hands-on practice in the operation of equipment and systems that are expected to be used in the performance of assigned duties. Education is the process of teaching new skills or additional knowledge in preparation for some kind of activity. Education is the process by which teachers bring about learning in their students. This process may involve formal classroom instruction or other processes.

Types of Training

Fire service training primarily consists of classroom study and training drills. Classroom work is the fundamental method used to acquire the knowledge that can be further developed by practical application during a training drill. Practical skill training drills enable firefighters to rehearse and master fire-ground evolutions in a controlled environment. Training and practice is necessary to ensure firefighters are able to perform safely and efficiently during actual incidents and build necessary teamwork and proficiency in their skills.

Departmental Training

Most of the training that firefighters receive throughout their careers comes from within the fire department. Beginning with recruit training in the career fire service, a trainee is placed in a departmental training academy. Upon successful completion of the academy, a new firefighter may then be assigned to a fire company to serve a probationary period as a firefighter. During this time, and for the rest of their careers, firefighters will take part in in-service training at the company level. This activity may be in the form of individual

Figure 1.15 Live fire training drills allow firefighters to practice what they have learned.

company or multi-company drills, or it may involve returning to the training division for updates on new equipment or procedures.

Some departments have a staff of firefighters, fire officers, and/or civilian personnel dedicated to training. Often a chief or other fire officer is in command of this staff. Personnel assigned these responsibilities coordinate and instruct recruit training as well as other department training and drills. Smaller career or volunteer departments may have a single person assigned the role of training officer. This person may fulfill all the duties associated with training, and in some small departments may also maintain additional responsibilities of a chief or company level officer **(Figure 1.15).**

Volunteer fire departments must attempt to coordinate their training schedule at the convenience of their members. Drills are often held during evening hours or on weekends. Training academies established for new members of these organizations must usually meet during these hours as well. Members must not respond to an emergency until they have received proper training in the duties that they will be expected to perform.

Occupational Health and Safety
NFPA® 1500, *Standard on Fire Department Occupational Safety and Health Program*, states that no member (career or volunteer) shall be allowed to engage in structural fire fighting until they have met the recommendations contained in NFPA® 1001, *Standard for Fire Fighter Professional Qualifications*.

Regional/County and State/Provincial Training Programs
Virtually every level of government sponsors fire training programs. The primary agencies responsible for providing this instruction vary based on location and may range from occasional seminars or classes to full-time academies with fixed facilities, full-time staff, and a yearly course calendar.

In jurisdictions that feature numerous small departments, the county, parish, or borough may administer programs that include a full slate of courses from recruit training to advanced level instruction. These courses may be

Figure 1.16 Fire training programs include fixed structures and various types of apparatus.

comparable to the programs held at state sponsored academies. A growing trend toward regional training centers features several jurisdictions combining resources to operate a shared facility for mutual training programs that benefit fire departments that closely border each other.

Some states or provinces provide fire training that operates under the jurisdiction of the state or provincial fire marshal's office or other governmental agency. Training programs are also offered at community colleges or vocational training centers. These programs may have full- or part-time instructional staff that develop and deliver courses at a fixed campus or at field locations. Some organizations with large training centers offer dormitory accommodations that allow students to attend weekend classes or week-long programs **(Figure 1.16)**.

Value of Education

The value of obtaining education and continuing training in the fire service is immeasurable in terms of its potential to impact the safety and effectiveness of a firefighter as well as the members and civilians with whom he or she interacts. Acquiring a quality education and maintaining a current skill set is vital to a firefighter's career development.

Beginning as a recruit, the firefighter must read, train, and practice skills to become proficient and learn the techniques associated with equipment and procedures. This training is an ongoing process as the field of fire fighting and emergency response is vast and continually evolving with emerging hazards and specialized equipment. Firefighters should not presume they have learned, studied, or trained enough to meet the challenges of every scenario.

Colleges and Universities

Firefighters are finding the pursuit of higher education increasingly valuable for a career in the fire service. Many colleges and universities offer programs at the Associate's, Bachelor's, or Master's degree level that will increase a student's knowledge base relative to fire science, emergency management, or administration of a public organization, such as a fire department. An increasing number of departments require some level of a college degree or credits in order to obtain employment or gain promotion.

Community colleges throughout North America offer many traditional fire science degree programs in the classroom as well as an increasing number of online opportunities to advance a fire service education. Both platforms for learning usually receive the recognition of an Associate's degree upon successful completion.

Likewise, colleges and universities in the U.S. and Canada also provide offerings with Bachelor's or graduate degree programs in fire service management, fire protection engineering, or similar fields of study. Schools such as Oklahoma State University, University of Maryland, and University of New Haven are well known for their commitment to the fire service **(Figure 1.17)**.

Figure 1.17 Oklahoma State University provides fire service training for fire departments as well as several degree programs in fire service management and fire protection engineering.

College level programs are usually designed to supplement local department, county, and state fire academy offerings. Most college curriculum contains arts and science components as well as fire protection courses that are beyond the scope of basic fire training programs. These courses often emphasize supervisory and management skills and are designed to prepare students for entry into or promotion within a fire department. Some institutions may offer programs that emphasize fire protection engineering, building design, or code enforcement.

FESHE

The National Fire Service Programs Committee (NFSPC) developed the Fire and Emergency Services Higher Education Model (FESHE) curriculum to create a national standard for the fire service in higher education. Degree programs based on FESHE curriculum allow students to more easily transfer between schools and create a national continuity in fire service education.

National Fire Academy

The United States Fire Administration (USFA) operates the National Fire Academy (NFA) as a part of the Department of Homeland Security (DHS). This facility offers a wide variety of courses in the areas of incident management, fire technology, and fire prevention. With the goal of improving professional development in the fire service, the academy delivers courses through on-campus programs as well as an extensive outreach effort.

The on-campus resident programs are conducted at the National Emergency Training Center (NETC) in Emmetsburg, Maryland. Students are primarily fire service members from around the nation who attend courses at minimal expense due to a stipend program that pays the cost of tuition, travel, and accommodations in campus dormitories. Most on-campus courses are one or two weeks in duration and are linked for student advancement through increasing levels of complexity in a particular subject or general knowledge in a variety of subject areas.

The NFA outreach program provides for short duration (generally two day) classes in locations throughout the U.S. State training agencies usually coordinate these programs, which are often scheduled for weekends to allow maximum participation from both the career and volunteer sector.

Non-Traditional Training and Education

Distance learning through online courses is a method of training and education that is growing in popularity. Many colleges and universities offer a wide variety of degree programs that are partially or entirely based on online courses. Based on the institution's accreditation, these programs may range from continuing education credits to degree-awarding programs. Degrees earned online from accredited schools carry the same validity as those degrees that students earn in classroom programs.

These programs offer the advantages of allowing students to work at a time of day convenient to their schedules. They also reduce or eliminate travel to a physical classroom and enable a student to receive an education from an institution in a distant location without leaving their homes. This type of education is particularly appealing to students already employed as firefighters who may need to balance a shift schedule and family responsibilities with class requirements **(Figure 1.18)**.

The NFA also offers a program of online courses. Most of these offerings feature continuing education units or college credit. In addition, the Degrees at a Distance Program (DDP) is an effort coordinated by the NFA with a number of colleges and universities across the country that offer either a program of online or correspondence study that, upon completion, may lead to a Bachelor's degree as well as recognition from the NFA.

Figure 1.18 Some institutions offer online training, which is well suited to students already employed as firefighters.

A wide variety of classes, conferences, and seminars are sponsored by all levels of government as well as numerous fire service organizations and private concerns. These programs may vary in the following ways:

- Topics
- Venues
- Cost
- Quality

An interested student may want to speak to someone who has previously attended the event and conduct his/her own research to determine if the offering is a good value. Some important questions may include:

- Did the advertised instructors actually teach/speak?
- What was the quality/availability of instructional materials?
- Did the program content live up the advertising?
- Were continuing education units (CEU) or a certificate of completion provided?

Firefighter Certification/Credentialing

When discussing credentials for fire fighting skills, the term certification attests that an individual has met the criteria specified in a particular standard. The certification of firefighters is important for a variety of reasons, including the aspect of liability. A firefighter who is certified in a particular set of skills should be ready and able to perform those skills safely and effectively during an incident.

The fire service typically uses consensus standards that committees of fire service and industry professionals develop and the National Fire Protection Association® (NFPA®) adopts. These documents are commonly called Professional Qualifications Standards (Pro-Qual), which set minimum competency standards for a wide variety of fire fighting and related activities.

Because they establish minimum requirements, some jurisdictions may choose to adapt and increase the criteria to safely and effectively meet local protection requirements. In these instances, the training of local agencies may be more comprehensive or rigorous than the guidelines set forth in the NFPA® standard.

In order to become nationally certified in any Pro-Qual standard, a firefighter must pass a certification examination subsequent to completing the specified training program. The agency that provided the training must be accredited by one of the two entities that accredit training organizations. The National Board on Fire Service Professional Qualifications (NBFSPQ or Pro-Board) and the International Fire Service Accreditation Congress (IFSAC) provide national recognition of a firefighter's certification.

National Board on Fire Service Professional Qualifications (NBFSPQ or Pro-Board)

The National Board on Fire Service Professional Qualifications (Pro-Board) was an outgrowth of the Joint Council of Fire Service Organizations that was conceived in 1972 to provide minimum performance standards with which firefighter's skills could be evaluated. In addition, the Pro-Board was tasked with accrediting the certifying agencies. In the mid-1980s, the responsibility for creating and maintaining professional qualifications standards was given to the NFPA®. The Pro-Board began to concentrate solely on the accreditation of fire service programs. Many agencies throughout the U.S. and Canada, as well as overseas hold Pro-Board Accreditation for their programs.

International Fire Service Accreditation Congress (IFSAC)

The International Fire Service Accreditation Congress (IFSAC), based at Oklahoma State University (OSU), was created as a result of action by the National Association of State Directors of Fire Training (currently the North American Fire Training Directors). Organized in 1991, the purpose of IFSAC is to provide a self-governed and peer-driven system that accredits fire service certification programs as well as fire-related degree programs. IFSAC is organized into four sections: Administration, Certificate Assembly, Degree Assembly, and the Congress. The Administration is provided by staff at OSU, and the Congress is composed of assemblies drawn from agencies representing certification and degree programs.

Center for Public Safety Excellence (CPSE)

The Center for Public Safety Excellence (CPSE) provides training and career resource information to agencies and individuals who are striving to meet international performance standards. Fire departments can implement a strategic self-assessment model to aid in achieving improved service and accreditation through the Commission on Fire Service Accreditation International (CFAI).

The Commission on Professional Credentialing (CPC) provides guidance to individuals for career development and recognizes through credentialing fire service professionals who have achieved excellence.

Firefighter Selection Process

Potential firefighters must possess a variety of skills and attributes as well as good physical conditioning in order to be successful candidates for selection to a fire department. Excellent interpersonal skills vital for interaction with members of the public and other firefighters are sought, along with technical aptitude and the ability to work safely in a physically demanding profession. Firefighters must also be trustworthy and adhere to a code of ethics expected by the public as well as other members. The ability to present a positive image of the fire service on and off duty is an important consideration during the firefighter selection process.

The specific details of selecting firefighter candidates vary by region and jurisdiction. Typically, an advertisement for the position of firefighter will begin the application period of the hiring process. A written examination, physical agility test, and oral interview are often the primary steps to thin the field of applicants to those candidates most likely to succeed as firefighters. The remaining group of applicants may be subjected to a physical and psychological examination as well as background screening and drug testing. A list of potential trainees may be developed based on these criteria, and the desired number of applicants selected to proceed to the training academy.

Recruitment

Many career and volunteer fire departments use various forms of advertising to elicit a sufficient number of applicants for firefighter positions. Jurisdictions have used traditional media outlets, such as television, radio, and newspapers, to advertise for applicants. Some departments place advertisements in trade publications to reach those individuals who may be members of departments in different geographical areas. An increasing number of departments use their official website or other popular fire service websites to convey their recruiting message.

Fire Cadet/Explorer Program

Many career and volunteer fire departments sponsor cadet or Explorer Scout programs for the youth in their jurisdiction. These programs educate young people (generally ages 14-21) about careers in the fire service and may act as a catalyst for the recruitment efforts of the sponsoring agency. Firefighters often serve as mentors in these programs and conduct introductory training courses on fire service topics. In some programs, older cadets or Explorers may assist with routine maintenance duties in the firehouse.

Preparation

Heavily populated jurisdictions often have a large pool of applicants for relatively few firefighter positions. Applicants should consider a regimented plan of preparation in order to maximize their potential for acceptance into a training academy. Potential candidates often participate in a physical conditioning program in order to excel during the physical agility testing that is often required. Some candidates work toward a college degree or certification as a firefighter or Emergency Medical Technician (EMT) as these are sometimes prerequisites for applying for a position. Participating in practice interviews or pre-tests are also worthwhile methods of preparation.

Application

The initial step for potential candidates usually involves filing an application with the organization with which they seek employment. Prerequisites vary widely by jurisdiction but usually require an applicant be at least 18 years of age, hold a valid driver's license, and have a high school diploma or General Equivalency Diploma (GED).

Some departments may have more stringent requirements for application. Specifying that applicants hold a certificate in compliance with the NFPA® 1001, *Standard for Firefighter Professional Qualifications,* an EMT or paramedic license, or other fire service credentials help narrow the number of applicants in many jurisdictions. Fire departments may also require applicants to possess a commercial driver's license or obtain a certain number of college credits or Associate's degree prior to applying.

Written Examination

Subsequent to the application process, a written examination is usually administered. An independent testing agency contracted by the fire department often creates this test. These examinations are usually multiple choice and are designed so that a candidate with no prior fire fighting experience will not be at a disadvantage. Written tests generally feature sections on the following:

- Math
- Reading comprehension
- Mechanical aptitude
- Recognition/observation
- Psychological testing

Math

Entrance examinations for fire fighting positions often contain word problems that must be broken down into a mathematical equation and solved. These problems simulate the type of work fire apparatus driver/operators need to be able to perform quickly in their heads on the fireground. These calculations may be needed to determine pump discharge pressures or friction loss for different hose layouts **(Figure 1.19)**.

Figure 1.19 Good math skills are required to calculate proper pump discharge pressures.

Reading Comprehension

Throughout their careers, firefighters must read the following:

- Training bulletins
- Policies
- Manuals
- Study materials for promotional examinations

In order to gauge reading comprehension, many firefighter examinations require the candidate to read a short selection of material. The candidate is then required to answer questions relating to the content of that material. This type of question helps determine the ability of the candidate to retain written material and recognize important content.

Figure 1.20 During the size-up process, firefighters must quickly observe and interpret incident factors. *Courtesy of Bob Esposito.*

Mechanical Aptitude

In the course of their duties, firefighters are required to operate a wide variety of tools under both emergency and non-emergency conditions. A basic understanding of tools and mechanics allows the firefighter to operate more efficiently and may enable him/her to adapt or improvise a solution where the desired tool is not readily available. This aptitude is often tested by presenting questions that consist of a diagram in which the candidate must determine the effect of an event given a series of gears, pulleys, or cylinders.

Recognition/Observation

Many written tests evaluate a candidate's ability to recognize numerous details in a short period of time by presenting a picture or illustration. After a prescribed period of time, the picture is no longer available for reference and the candidates are required to answer questions about specific details in the picture. Minor points, such as the time on a clock or the number of books on a table, may be used to test observation skills. These observations must then be used to formulate a fire attack plan without delay.

In the process known as "size-up," a firefighter must quickly observe and interpret the following **(Figure 1.20)**:

- Construction features of a building
- Location and extent of the fire
- Occupants
- Hazards
- Means of access and egress
- Numerous other factors

Psychological Testing

Due to the nature of their job, firefighters may be subject to stressful events. In an attempt to ensure candidates are psychologically compatible with fire fighting duties, a psychological screening process is often used during the pre-employment evaluation.

Physical Ability Test

The physical ability or agility test is commonly used to measure a candidate's strength and aerobic condition. The particular evolutions may vary by jurisdiction but are usually designed to approximate the types of activities most common in the field of fire fighting. Based on the jurisdiction, the tests may be pass/fail or scored based on time. Generally, the sponsoring agency provides material that explains the various parts of the test long before it is administered. Some agencies even provide a day when candidates may practice on the actual test course. These tests are physically demanding. Candidates should train well in advance to ensure their best possible performance and lessen a chance of injury.

Candidate Physical Ability Test (CPAT) Program

The International Association of Fire Chiefs and the International Association of Fire Fighters jointly developed the widely used Candidate Physical Ability Test (CPAT). The CPAT examination is comprised of eight stations that must be completed in one continuous sequence. It is a pass/fail evaluation. The maximum time allowed for passing is 10 minutes and 20 seconds. CPAT has been validated as an acceptable method for testing the physical requirements for being a firefighter.

During the test, the candidate must wear long pants, appropriate footwear, a safety helmet with chin strap, gloves and a 50-pound (25 kg) weighted vest to simulate the weight PPE with SCBA. An additional 25 pounds (12.5 kg) (12.5 lbs. per shoulder (6.25 kg) will be added during the first event, but will be removed at the completion of it. Watches and loose or restrictive jewelry may not be worn during the test.

The events are placed in a sequence that best simulates fire scene events while allowing an 85-foot (25 m) walk between events. This walk allows for approximately 20 seconds for recovering and regrouping before the next event. Running is not allowed between events. The events are as follows:

- Stair climb
- Hose drag
- Equipment carry
- Ladder raise and extension
- Forcible entry
- Search
- Rescue
- Ceiling breach and pull

Interview Process

Subsequent to successful completion of a written and physical test, a number of candidates may be contacted to participate in an oral interview process. This process varies by jurisdiction, but it generally consists of two interviews.

Oral Interview

During an oral interview, often referred to as an *oral board*, candidates are required to answer a variety of questions asked by multiple interviewers. Interviewers are evaluating not only the answers, but the candidate's appearance, demeanor, and attitude.

In order to maintain a fair process, all candidates are usually asked the same list of questions and are rated by each member of the interview panel. The interviewers will often discuss their impressions in order to arrive at a final score and eliminate subjectivity.

Second Interview

In some cases, a candidate may be asked to return for a second interview. This next step is sometimes referred to as a Chief's Interview as it is often conducted by the department chief or senior staff members. Usually not scored, this meeting may be used to gain a better perspective of the candidate's suitability for the position and compatibility with the department. Upon successful completion of this interview, the candidate is likely to proceed to the final stages of the process, which generally include physical and psychological evaluations and a background check.

Probationary Period

New firefighters in both career and volunteer fire departments may expect to serve a probationary period of one to three years. The requirements of this period vary widely based on jurisdiction. Some periods extend from successful completion of a training academy through a specified period of service. Other systems may include the time spent in the training academy as part of a probationary period. Whatever its length, the probationary period is intended to allow the new firefighter or candidate an opportunity to demonstrate a desire to be a member of the fire service and successfully complete the training and other tasks to which he or she is assigned. Failure to do so may result in a request for resignation at any time during the probationary period.

Volunteer Firefighter Selection

Due to the many obligations to careers and families, volunteer firefighters may be difficult to recruit and retain in some communities. The selection process for volunteer firefighters differs greatly from that of career firefighters. Some volunteer fire departments may require a physical examination or written test. Meanwhile, in other departments, an interested person may complete an application and interview with the fire chief or a committee of department members in order to be accepted. Prior certification or fire service knowledge is usually not required as most volunteer departments sponsor their own training programs or send members to a county or state fire academy.

Like career firefighters, volunteers represent their department to the community. They are likewise required to maintain a standard of ethics and professionalism. A background check prior to acceptance as a volunteer is often required in order to conform to the standards that the local jurisdiction requires.

Fire Service Career Information

Those considering a career in the fire service often have questions regarding compensation, work schedules, promotions, and retirement. While these details are unique to each jurisdiction, some general information is outlined in the following sections.

In addition to salary and retirement benefits, career firefighters generally receive additional forms of compensation. This compensation may include life and health insurance as well as paid vacation and sick leave. Many departments offer uniform allowance and incentives for firefighters to further their education. These benefits may be provided under a contract that a labor union negotiates on behalf of the workers.

Pay

Fire service salaries vary widely by location, organization, and rank. Certifications, education, and length of service typically factor into rates of pay as well. Generally, firefighters in more affluent or larger metropolitan areas earn more than those firefighters in small towns and less affluent communities.

Hours of Duty

Career firefighters work a variety of schedules. Very few firefighters work a typical eight-hour day, five-days-per week schedule found in other professions. Generally, only members of a fire department's administrative staff work traditional business hours. Most firefighters work a longer rotating shift schedule. Generally, a rotating shift schedule is made up of three or four shifts in the public fire service. Military firefighters may work a two shift rotation with straight 24-hour periods on and off duty.

Much of the North American fire service operates a three shift system. Several work schedules are possible using three shifts. However, the most common comprises a straight 24-hour duty shift with 48 hours off duty.

Other jurisdictions may use a four-shift system that revolves around a 10 hour day shift and a 14-hour night shift. Commonly, firefighters will work two 10 hour days, then two 14-hour nights, followed by a cycle of four days off.

Promotion

A promotion in the fire service entails increased pay and responsibility. Each fire department may employ different criteria to gain a promotion; however, most departments require an exam or other certification of knowledge and skills. Specialized training and advanced education may also factor into the promotional process as well as credit for length of service. Additional information regarding the roles of various fire service positions follows in a subsequent section of this chapter.

Retirement

Numerous retirement programs are in place for career and even some volunteer firefighters that vary widely between states and provinces. Some combine years of service and a firefighter's age to determine retirement compensation. Additionally, retired firefighters may receive health care and some form of life insurance coverage for a spouse or other beneficiary.

Some jurisdictions offer volunteer firefighters a retirement package based on years of service to the department. These benefits are commonly called length of service awards programs (LOSAP), which are often partially funded by a state, province, or county government.

Chapter Summary

Fire departments across the nation are organized according to local needs. There are numerous systems in use to provide fire protection, including career, volunteer, paid on-call, and combination fire departments. In addition, many large industrial plants operate their own fire departments to protect large facilities or those involved in hazardous operations.

Firefighters must train and learn new skills throughout their careers. This action may include training at local or state training academies or attending seminars and workshops. In order to advance in the fire service, many firefighters seek a degree from a college or university. Although the application and hiring process for firefighters varies by jurisdiction, this process usually consists of an application period before a written examination and physical agility test. Applicants successful in these stages of the process may undergo a physical examination, psychological screening, and a background check. Members of a panel may also interview the candidates for their compatibility with the goals of the fire service. Furthermore, candidates for firefighter positions generally serve a probationary period in which their skills and general aptitude and attitude for fire fighting may be evaluated.

In the volunteer sector, an entrance examination is not usually required. However, a potential member is often subject to a background check and an interview with the fire chief or committee of members. This process is undertaken to ensure the volunteer is likely to uphold the image and trust the department has built in the community it serves.

Review Questions

1. What are examples of types of incidents that the fire service may respond to?
2. What are some ways that fire departments are categorized by jurisdictions?
3. What are the different types of public fire departments?
4. What are examples of private and federal government fire departments?
5. What are some examples of responsible behaviors and appropriate conduct for firefighters?
6. Why is physical fitness an important aspect of being a firefighter?
7. What are different ways that fire service personnel receive training and education?
8. Why is firefighter certification/credentialing important?
9. What steps might a potential firefighter go through in the selection process?
10. What are some general facts about compensation, work schedules, promotions, and retirement for the fire service?

Chapter 1 Notes

1. Michael Karter and Gary P Stein, "U.S. Fire Dept. Profile," NFPA®, Qunicy, MA., Oct. 2013.

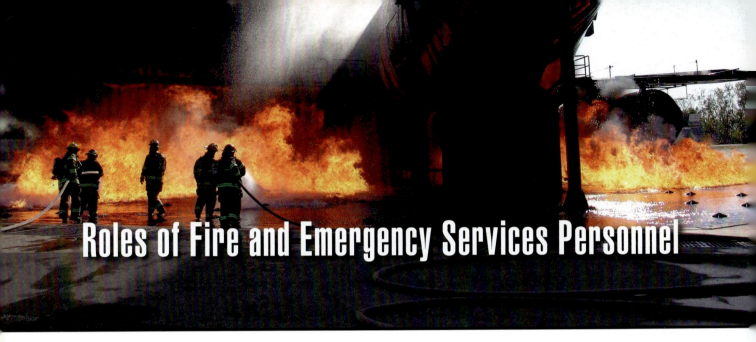

Roles of Fire and Emergency Services Personnel

Chapter Contents

Case History **47**	Battalion/District Chief 55
Fire Companies **47**	Training Division Personnel 57
Engine Company 48	Administration 59
Ladder (Truck) Company 48	Special Operations Personnel 64
Rescue/Squad Company 51	Emergency Medical Services (EMS) Personnel 69
Fire and Emergency Services Personnel **51**	Other Fire Department Personnel 72
Firefighter 52	**Chapter Summary** **75**
Fire Apparatus Driver/Operator 53	
Company Officer 54	

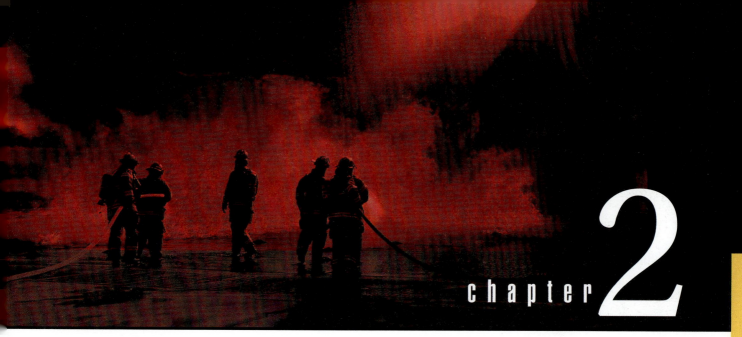

Chapter 2

Key Terms

Airport Firefighters..................................64	Marine Firefighter.....................................67
Company..47	Paramedic..71
Company Officer.....................................54	Public Information Officer.......................61
EMT..70	Rescue Companies.................................51
Engine Company....................................48	Safety Officer..60
Fire Apparatus Driver/Operator.............53	Technical Rescuer..................................66
First Responder.....................................70	Training Officer.......................................58
Hazardous Materials Technician...........65	Ladder (Truck) Company........................48
Industrial Firefighter...............................68	Wildland Firefighter................................67

FESHE Outcomes

This chapter provides information that addresses the outcomes for the Fire and Emergency Services Higher Education (FESHE) *Principles of Emergency Services* course.

5. Identify fire protection and emergency-service careers in both the public and private sector.

7. Discuss and describe the scope, purpose, and organizational structure of fire and emergency services.

10. Identify the primary responsibilities of fire prevention personnel including code enforcement, public information, and public and private protection systems.

Roles of Fire and Emergency Service Personnel

Learning Objectives

After reading this chapter, students will be able to:

1. Distinguish between the different types of fire companies. (Outcome 7)
2. Describe the different duties of personnel in the fire and emergency services. (Outcomes 5, 10)

Chapter 2
Roles of Fire and Emergency Services Personnel

Case History

Captain Lynch explained that when she joined the department, only first aid was taught. She went on to say that not long after she began her career, the department began to preplan and perform fire inspections in the buildings in its district. Other significant changes that she introduced:

- A few years later, the department began training their firefighters to become paramedics and run the ambulance service.
- The next big change was learning how to safely fight chemical fires and releases. SARA Title III and the Occupational and Safety Act required various levels of training in safely sizing up a scene, entry into chemical incidents, protecting life and property, and often basic cleanup of chemical spills.
- The fire service then trained to assist law enforcement in responding to law enforcement situations, natural and manmade disasters, and viruses affecting public health.

Firefighters must be able to change quickly to provide protection to themselves and the people they serve. Because of this challenge, it is an interesting and rewarding profession.

In order to fully serve its numerous roles in the community, the fire department is organized based on function and specialty to enhance efficiency, supervision, and administration. Firefighters often participate in more than one activity depending on the size and requirements of the jurisdiction in which they serve. This chapter describes the basic organization and duties of fire companies and discusses other roles common to the fire service.

Fire Companies

The basic operating unit of a fire department is a fire **company**. This unit consists of a number of firefighters and officers assigned to a particular piece or pieces of apparatus. The firefighters and officers are divided into shifts based on the local work schedule. It is common to have a fire officer responsible for supervising the company. The firefighters are usually distributed evenly on each shift to operate the company. Company staffing is an issue that is determined at the local level based on a variety of factors, including the type of company and number of personnel available.

Company — (1) Basic fire fighting organizational unit consisting of firefighters and apparatus; headed by a company officer. (2) Term that encompasses the whole crew of a vessel.

Fire companies are organized, equipped, and trained for specific functions on the fireground and other emergency scenes. The specific details of a company's assignment may vary based on staffing, local policy, and the type of equipment available. The type of apparatus and equipment corresponds to the purpose and assigned duties of that company. The following descriptions of basic types of fire companies illustrate the purpose and role of each type at fires and other emergencies **(Figure 2.1)**.

Task Force
Task Force: (1) A group of individuals convened to analyze, investigate, or solve a particular problem. (2) Any combination of single resources within a reasonable span of control assembled for a particular tactical need with common communications and a leader.

Strike Team
Strike Team: Specified combinations of the same kind and type of resources with common communications and a leader; usually composed of either engines, hand crews, or bulldozers, but may be composed of any resource of the same kind and type.

Engine Company

> **Engine Company** — Group of firefighters assigned to a fire department pumper who are primarily responsible for providing water supply and attack lines for fire extinguishment.

The primary duties of an **engine company** during structural fire fighting revolve around applying water to the fire and/or establishing and maintaining a continuous water supply to the fire scene as required. These duties may involve stretching and operating hoselines or master stream appliances, or supplying water to elevated master streams. In addition, engines may supply standpipe or sprinkler systems through a fire department connection. Engine companies may also lay supply lines from a pressurized water supply or draft from a static water source. Engine company members, in the absence of sufficient truck companies, may also be assigned to perform the following:

- Search and rescue
- Forcible entry
- Ventilation
- Other truck company functions

Engine companies also respond to non-structural fires, such as automobiles, trash, or natural cover fires. Based on jurisdictional policy, engine companies may also handle the following:

- Medical emergencies
- Auto and industrial accidents
- Various other citizen assistance calls

Ladder (Truck) Company

> **Truck (Ladder) Companies** — Group of firefighters assigned to a fire department aerial apparatus equipped with a compliment of ladders; primarily responsible for search and rescue, ventilation, salvage and overhaul, forcible entry, and other fireground support functions.

Ladder (truck) companies perform numerous duties on the fireground, such as:

- Ventilation
- Forcible entry **(Figure 2.2, p. 50)**
- Search and rescue

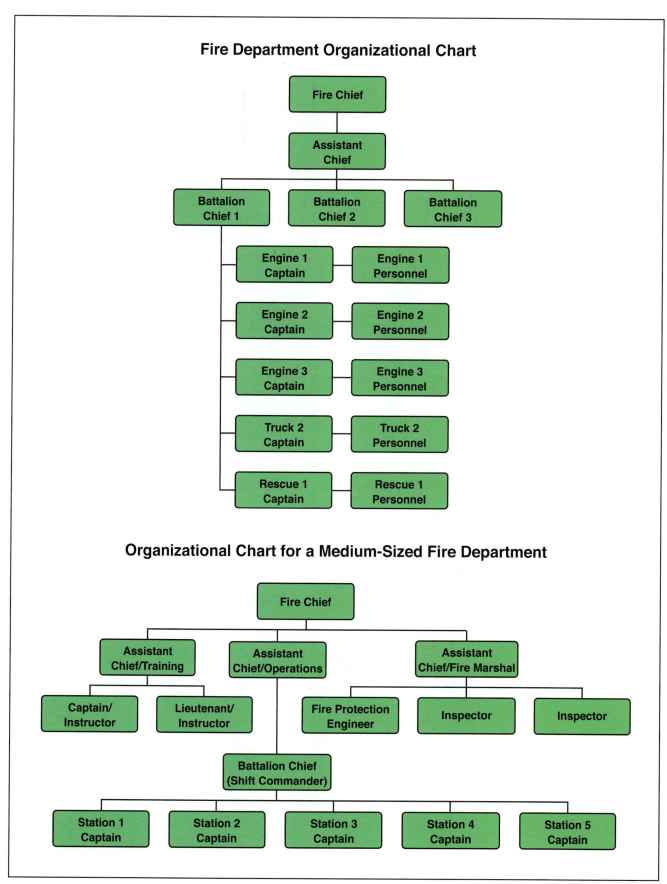

Figure 2.1 Example of fire department organizational charts

- Salvage
- Overhaul
- Control of utilities

In some fire departments, truck companies respond to medical emergencies and other citizen assistance calls in addition to fire incidents. Truck company members may also be called upon to assist engine companies in advancing hose lines under certain circumstances.

Ladder (truck) companies use apparatus that feature a large number of storage compartments, which many companies often use to store rescue equipment for vehicle extrication or other specialty rescues.

Figure 2.2 A forcible entry may be necessary to enter a structure. *Courtesy of Bob Esposito.*

These apparatus are also equipped with a variety of ground ladders and an aerial device for access to roofs and upper floors. The aerial device may also be used to operate elevated master stream appliances **(Figure 2.3)**.

Figure 2.3 A mechanical spray hose can be used to access upper floors. *Courtesy of Ron Jeffers.*

Rescue/Squad Company

The role and apparatus style of a rescue or squad company varies widely between departments. The primary purpose of this company is generally to rescue victims from areas of danger or entrapment **(Figure 2.4)**. Companies may perform rescue operations in a variety of ways, such as:

- Vehicle extrication, including industrial or agricultural vehicles and machinery
- Trench rescue
- Structural collapse rescue
- Rope rescue

Some **rescue companies** operate apparatus that carry only their specialized equipment and basic fire fighting tools. Other rescue/squad companies operate apparatus that are equipped with a fire pump, and may be assigned engine company functions at fires and other emergencies.

Fire and Emergency Services Personnel

The primary mission of a fire department should include community risk reduction and fire prevention. However, when fire suppression is necessary, the safest and most efficient methods to mitigate an incident include coordination of the following functions:

- Rescue
- Exposure protection
- Fire confinement
- Fire extinguishment
- Ventilation
- Salvage and overhaul
- Securing utilities
- Fire cause determination

Figure 2.4 Fire fighters use special equipment for vehicle extrication.

Rescue Companies — Specialized unit of people and equipment dedicated to performing rescue and extrication operations at the scene of an emergency. *Also known as* Rescue Squad or Rescue Truck.

Suppression personnel primarily meet these objectives. The common positions held by these personnel include:

- Firefighter
- Fire Apparatus Driver/Operator
- Company Officer
- Battalion/District Chief
- Training division personnel
- Administration
- Special Operations personnel
- Emergency Medical Services (EMS) personnel
- Other fire department personnel

Figure 2.5 Properly worn SCBA protects firefighters from injuries. *Courtesy of Bob Esposito.*

Firefighter

Fire fighting requires extensive skill in preventing and combating fires as well as responding to a variety of emergency calls and requests for service. Firefighters operate in hazardous environments, exerting themselves wearing personal protective equipment (PPE) and self-contained breathing apparatus (SCBA) **(Figure 2.5)**.

Although emergency response activities are the most hazardous part of a firefighter's work, a large portion of any shift is typically spent performing routine duties, such as inspections, training, and station and equipment maintenance. Typical duties of a firefighter include the following:

- Attend training courses.
- Read and become familiar with policies, procedures, or general orders issued through the chain of command.
- Don PPE/SCBA.
- Respond to various types of fire or other incidents as assigned by the local jurisdiction.
- Operate equipment as required, including hose, nozzles, ladders, power saws, and fire extinguishers.
- Ventilate buildings and force entry as required.
- Remove people and animals from all manner of dangerous situations.
- Administer emergency medical care to patients.

- Perform salvage and overhaul operations.
- Respond to, and relay as required, instructions and information concerning alarms received from the dispatcher.
- Care for assigned apparatus, equipment, and facilities.
- Perform fire inspections as assigned.
- Interact with the public concerning fire and life safety questions.
- Know department organization, operational policies, and procedures.
- Know the district or city street layout.
- Have the ability to climb ladders and operate at considerable heights.
- Have the ability to understand and follow oral and written instructions.
- Have the aptitude to perform routine assigned maintenance duties.

In order to complete the aforementioned duties safely and efficiently, firefighters should meet the recommendations of NFPA® 1001, *Standard for Fire Fighter Professional Qualifications*. Firefighters should also meet the minimum health and fitness requirements of the authority having jurisdiction (AHJ). Information on health and fitness requirements can be found in the following NFPA® standards:

- NFPA® 1500, *Standard on Fire Department Occupational Safety and Health Program*
- NFPA® 1582, *Standard on Comprehensive Occupational Medical Program for Fire Departments*
- NFPA® 1583, *Standard on Health-Related Fitness Programs for Fire Department Members*

Fire Apparatus Driver/Operator

A **fire apparatus driver/operator**, under the direction of a company officer, is responsible for safely driving the apparatus and operating its pump, aerial device, or other mechanical equipment as required. Depending on jurisdictional policy, the driver/operator may also be responsible for apparatus and equipment maintenance **(Figure 2.6a and b)**.

NOTE: In some fire departments, fire apparatus driver/operator may be the first level of promotion in the chain of command.

> **Fire Apparatus Driver/Operator** — Firefighter who is charged with the responsibility of operating fire apparatus to, during, and from the scene of a fire operation, or at any other time the apparatus is in use. The driver/operator is also responsible for routine maintenance of the apparatus and any equipment carried on the apparatus. This is typically the first step in the fire department promotional chain.

Figure 2.6a and b In some jurisdictions, the driver/operator is responsible for apparatus maintenance.

Personnel take on these driver/operator responsibilities in addition to the general responsibilities of a firefighter. Driver/operators may also be expected to perform the following duties:

- Keep any inventory or other records required by local jurisdiction.
- Keep the apparatus in a state of readiness and report any mechanical deficiencies using the appropriate method of communication.
- Maintain a state or provincial license to operate fire and emergency services vehicles.
- Know the location of streets, hydrants, and fire department connections.
- Know the types of building construction.
- Know the target hazards in the district.
- Know and understand the mechanical principles involved in the operation of fire apparatus and assigned equipment.
- Know the driver/operator-specific policies and procedures.
- Have the ability to mentally calculate solutions to water flow problems by applying standard formulas.

The driver/operator must have technical knowledge beyond that of a regular firefighter. In order to fulfill this important role, a firefighter should meet the recommendations of NFPA® 1002, *Standard for Fire Apparatus Driver/Operator Professional Qualifications.*

For more detailed information concerning the role of the driver/operator, consult the IFSTA **Pumping Apparatus Driver/Operator Handbook** or **Pumping and Aerial Apparatus Driver/Operator Handbook**.

Company Officer

The **company officer** supervises the activities of a fire company in the firehouse and at incident scenes. These duties include oversight of the maintenance and operation of the firehouse, apparatus, and equipment.

Company officers routinely perform a wide variety of duties that encompass fire fighting and routine responsibilities. The officer supervises company members and operates within the department's guidelines and under the direction of a chief officer. Lieutenant and Captain are some of the most common titles for company officers throughout North America.

Typically, company officers perform the following duties:

- Respond to all alarms assigned to the company.
- Advise the driver/operator regarding response to the scene.
- Perform scene size-up.
- Direct initial company operations.
- Direct subordinates at the scene until command is transferred to a senior officer or chief operating within the Incident Command System (ICS).
- Request additional resources as required.
- Inspect and monitor conditions at a fire scene to prevent re-ignition.
- Assist in determining origin and cause of the fire.
- Provide safe company operations and citizen safety.

> **Company Officer** — Individual responsible for command of a company. This designation is not specific to any particular fire department rank (may be a firefighter, lieutenant, captain, or chief officer if responsible for command of a single company).

Figure 2.7 A company officer inspecting a new construction.

- Inspect and oversee maintenance of equipment and facilities to ensure proper order and condition.

- Inspect public buildings as ordered to identify hazards or conditions that may be dangerous to life and property, and make written and oral reports of such conditions with appropriate recommendations for corrective action **(Figure 2.7)**.

- Present fire prevention and community risk reduction information to the public.

- Prepare and present training within the company and at the fire training division as required.

- Attend to administrative and personnel matters as required, including filing appropriate departmental reports.

- Assign and supervise non-emergency duties at the fire station.

- Possess knowledge of streets, target hazards, hydrants, and other city infrastructure.

- Possess knowledge and ability in fire fighting practices, fire prevention, and emergency medical care.

- Quickly and accurately recognize dangerous conditions, and take appropriate immediate action to protect life and property.

The job requirements and expectations for company officers are extremely high. As the first level of supervision in the chain of command, the company officer plays a pivotal leadership role in the safe and efficient operation of the company. Many experienced fire chiefs view the company officer's position as vital to the fire department's success.

Company officers should meet the recommendations found in NFPA® 1021, *Standard for Fire Officer Professional Qualifications*. IFSTA's **Fire and Emergency Services Company Officer** manual contains in depth information regarding the role of a company officer.

Battalion/District Chief

The first level of chief may be referred to as a Battalion, Deputy, or District Chief depending on the terminology of a fire department's organization. The number of companies and personnel a Battalion Chief supervises depends

Figure 2.8 Battalion Chiefs manage emergency operations. *Courtesy of Ron Jeffers.*

on the size of the department. In smaller departments, one Battalion Chief may supervise all on-duty personnel and act as the shift commander. Larger departments may be divided into numerous battalions, each having a chief who may be supervised by one or more levels of chief officer above the battalion **(Figure 2.8).**

A Battalion Chief supervises the administrative and emergency operations of a number of fire companies within a geographic region of a jurisdiction. These battalion chiefs normally respond to incidents that require the operation of three or more companies, as well as other situations where they are requested. Some Battalion Chiefs may supervise administrative functions within a fire department, such as fire prevention, training, or fleet maintenance.

Battalion Chiefs are required to perform the following duties:

- Manage incidents within the battalion per department policy.
- Review records and reports pertaining to battalion operations and administration.
- Inspect fire stations, equipment, and apparatus to ensure compliance with department standards.
- Direct company officers as required during emergency operations and administrative matters.

- Assist other battalion chiefs engaged at major incidents in their battalion's per department policy.
- Provide assistance with special projects or administrative functions to chiefs of superior rank as requested.
- Evaluate fires and other emergencies recognizing hazards and developing strategies to safely mitigate incidents.
- Possess thorough knowledge of fire fighting tactics combined with the ability to employ the National Incident Management System (NIMS), or incident command system to organize emergencies.
- Possess extensive knowledge of the fire department's policies and operations, as well as the location of streets, water supply, and major hazards of the jurisdiction.

The position of Battalion Chief requires considerable attention to administrative detail as well as fire fighting expertise. As such, the battalion chief must possess considerable technical and managerial skill. A Battalion Chief's qualifications include meeting the recommendations of the NFPA® 1021, *Standard for Fire Officer Professional Qualifications*.

Training Division Personnel

In order to properly prepare firefighter candidates, maintain the skills of current firefighters and remain current with emerging technology, fire departments generally staff a training division. Smaller departments may have a single dedicated training officer who oversees administrative functions and conducts much of the actual training. This person may be supplemented by firefighters or officers who are qualified instructors and are assigned training as a collateral duty aside from serving in a fire company. Larger fire departments may have more officers and firefighters on staff to perform training. The following section separates the role of instructor and training officer in the training division.

Instructor

The fire service instructor is responsible for delivery of training programs ranging from recruit academies to in-service training with experienced members of the department. The instructor may deliver these programs at a classroom in the division of training, a training facility, in a firehouse, or other suitable location **(Figure 2.9 a and b, p. 58)**. Fire service instructors take direction from the training officer, or chief of training, and are expected to perform the following duties:

- Deliver prepared lesson plans using a variety of classroom and practical skills delivery methods.
- Prepare curriculum for new courses at the direction of the training officer.
- Monitor practical skills training for safety hazards and immediately remedy any hazardous conditions.
- Complete the appropriate reports to document the progress and completion of training per department policy.
- Possess knowledge of the jurisdiction's operating procedures and the ability to convey them in conjunction with training programs.

Figure 2.9 a and b Firefighters receive both classroom and on-site training. *Courtesy of Ron Jeffers.*

- Demonstrate excellent verbal and written communication skills.
- Possess thorough knowledge of all the skills and theories required to be conveyed in specific training programs.

The fire service instructor should meet the recommendations of NFPA® 1041, *Standard for Fire Service Instructor Professional Qualifications*. For additional information, refer to the IFSTA **Fire and Emergency Services Instructor** manual.

Training Officer (Chief of Training/Drillmaster)

The **Training Officer** (Chief of Training/Drillmaster) generally oversees the administration of all intra-departmental training activities. The Training Officer may report directly to the Fire Chief or to another higher ranking chief in a larger organization. The Training Officer is expected to perform the following duties:

- Plan, coordinate, and supervise the work of subordinate fire service instructors.
- Evaluate new techniques, procedures, and methods in order to maintain currency in course curriculum.
- Conduct risk assessments of all training programs to ensure the safety of participants and instructors.
- Research and secure the necessary resources to achieve training goals.
- Oversee personnel training records.
- Supervise the maintenance of training facilities and equipment.
- Pursue training in educational theory and methodology.
- Possess knowledge of record-keeping requirements.
- Maintain an accurate record-keeping system.
- Interact with administrators within the department as well as outside agencies to determine training requirements and coordinate courses. The training officer should meet the recommendations of NFPA® 1041, *Standard for Fire Service Instructor Professional Qualifications*.

> **Training Officer** — Individual responsible for running a fire department's training program, to include the administration of all training activities; typically reports directly to the fire chief.

Administration

Management and leadership of all department functions are provided through an administrative division. In smaller departments, this leadership may consist of a small number of individuals who work full-time conducting administrative duties. This staff may be supplemented by firefighters or officers from fire companies who are assigned certain administrative functions as collateral duties in addition to fire fighting responsibilities.

In larger departments, administration may be divided internally into various divisions with numerous personnel assigned to fulfill specific responsibilities. Many of the positions described in the following section are filled based on the fire department's size and complexity of the administrative function.

Human Resources/Personnel Services

Often a liaison with a municipal human resources department, fire department human resources personnel may be responsible for portions of the following functions:

- Recruitment and hiring of uniformed and non-uniformed personnel.
- Salary and promotion practices.
- Management of benefits, such as retirement and health, life, and disability insurance.
- Disciplinary action.
- Maintenance of personnel records.
- Supervision of non-uniform support personnel.

Grants Administrator

Fire departments of all sizes routinely seek monetary awards called grants to offset the cost of programs, equipment, and personnel. Larger fire departments may employ a civilian or uniformed member of the department in the capacity of grant administrator. Other jurisdictions may have a municipal employee available to any city agency to assist with grant proposals and submission. There are also private companies that contract their services as grant writers for specific projects to municipal or state agencies who require assistance.

The responsibilities of a grant administrator include:

- Identify sources of funding that meet department needs.
- Prepare and submit grant proposals to potential funding sources in accordance with source and department requirements.
- Once a grant is awarded, monitor implementation to ensure it is in accordance with all legal, ethical, and funding source guidelines.
- Prepare and submit any subsequent documentation that the grant source or fire department requires.

Grants may be provided through local, state, provincial, national, or international sources. Private organizations or non-profits may provide some grants, while various levels of government fund other grants. This funding includes the Assistance to Firefighters Grants (AFG) awarded by the Federal Emergency Management Agency (FEMA).

Figure 2.10 The Safety Officer oversees incident safety functions. *Courtesy of Ron Jeffers.*

Safety Officer

The **Safety Officer** has oversight responsibilities for safety and health initiatives within the fire department. This individual may hold the rank of Company Officer or Battalion Chief. In some jurisdictions, this individual responds to each emergency requiring an Incident Safety Officer.

In other, larger departments, the Safety Officer oversees a division with other members qualified to perform incident safety functions. All members who fulfill safety roles must have, and maintain, the respect of all ranks within the organization in order to be most effective **(Figure 2.10)**.

The Safety Officer is required to perform the following duties:

- Formulate and administer occupational health and safety programs.
- Maintain an accident investigation and record-keeping system.
- Coordinate safety inspections of department facilities, apparatus, and equipment.
- Respond (based on organizational structure) to incidents with the goal of maintaining the safety of personnel.
- Coordinate with the Incident Commander
- Modify operations as necessary when safety concerns arise.
- Possess extensive knowledge of fire fighting operations, building construction, hazardous materials, and incident management.

> **Safety Officer** — (1) Fire officer whose primary function is to administer safety within the entire scope of fire department operations. *Also known as* Health and Safety Officer. (2) Member of the IMS command staff responsible to the incident commander for monitoring and assessing hazardous and unsafe conditions and developing measures for assessing personnel safety on an incident.

- Communicate with members at all levels of the department as well as various outside agencies.
- Compile and review research and maintain an effective record-keeping system for firefighter health and safety.

In order to fulfill a wide scope of assignments, the Safety Officer should meet the recommendations of NFPA® 1521, *Standard for Fire Department Safety Officer Professional Qualifications*. Additional information on the role of the safety officer may be found in IFSTA's **Fire Department Safety Officer** manual.

Public Information Officer (PIO)

The **Public Information Officer (PIO)** is responsible for maintaining a positive relationship between the fire department, the media, and the general public **(Figure 2.11)**. Based on jurisdiction, the PIO may hold the rank of Company Officer, Battalion Chief, or may be a civilian employee. The PIO is required to perform the following duties:

- Act as the department's primary contact with media representatives.
- Coordinate all news releases and public service announcements issued by the department.
- Respond to major emergencies to act as liaison between the Incident Commander and the news media.
- Serve as primary department spokesperson.
- Coordinate any additional interaction between the fire department and community and civic organizations.

Public Information Officer (PIO) — (1) Member of the command staff responsible for interfacing with the media, public, or other agencies requiring information direct from the incident scene. *Also known as* Information Officer (IO). (2) Those who have demonstrated the ability to conduct media interviews, prepare news releases, and advisories.

Figure 2.11 The public information officer is responsible for relaying information about emergency incidents to the media.

- Effectively convey information via media sources to the general public in easily understood terms without extensive scripting or advance preparation.
- Separate information that is in the public interest from confidential information that should be withheld from public knowledge.
- Establish and maintain a good rapport with members of the media as well as the general public and fire department personnel.
- Convey information clearly and concisely in verbal and written communication.

The PIO is often the only direct contact members of the public have with the fire department. It is important that this individual meets the recommendations of NFPA® 1035, *Standard on Fire and Life Safety Educator, Public Information Officer, Youth Firesetter Intervention Specialist and Youth Firesetter Program Manager Professional Qualifications*. Additional information concerning the role of the public information officer may be found in the IFSTA **Public Information Officer** manual.

Figure 2.12 An Assistant Chief is responsible for some administrative duties at the fire station.

Assistant/Deputy Chief

The title Assistant or Deputy Chief is often used interchangeably in the fire service. Larger departments often have several Assistant Chief positions that perform senior management functions. An Assistant Chief may be assigned to operations, administration, fire prevention, planning, or another major division **(Figure 2.12)**. Assistant chiefs are generally expected to perform the following duties:

- Manage the division or function to which they are assigned. This task may include budgeting, policy development, research, and personnel matters.
- Respond to major incidents as required by department policy and participate in the command structure as required.
- Assume the duties of the Fire Chief as required.
- Have a college degree appropriate for management level fire service responsibilities (some jurisdictions require advanced degrees).
- Possess extensive knowledge of the fire department's administrative structure and process.
- Demonstrate excellent written and oral communication skills.

- Possess thorough knowledge of fire and rescue strategy and tactics.
- Organize an incident using the NIMS and assume the duties of the chief of department as required.

In order to fulfill the duties of Assistant Chief, an individual should meet the recommendations of NFPA® 1021, *Standard for Fire Officer Professional Qualifications*.

Fire Chief/Chief of Department

The Fire Chief, known as the Chief of Department, is the chief executive officer ultimately responsible for the operation of the department. Depending on jurisdictional organization, the Fire Chief may answer to a mayor, manager, city council, or district board members.

Figure 2.13 A Fire Chief meets with a local government official.

Although the overall responsibilities remain the same, a Fire Chief, in a small department may have daily interaction with line personnel and handle daily functions that would be the responsibility of an Assistant Chief or other staff officer in large organizations **(Figure 2.13)**. In large fire departments, the Fire Chief is primarily an administrator who has little daily interaction at the company level. The fire chief is expected to perform the following duties:

- Interact with other department heads and government officials.
- Respond to a major incident to assist in the command structure per department policy.
- Provide leadership and direction, developing policies and plans for the betterment of the organization.
- Participate in the budget process for the department as required by jurisdictional authority.
- Have a college degree appropriate for management level fire service responsibilities.
- Possess extensive knowledge of the fire department's administrative process and structure.
- Demonstrate excellent oral and written communication skills.
- Possess thorough knowledge of fire and rescue strategy and tactics.
- Organize an incident using the Incident Command System (ICS).
- Integrate with jurisdictional government officials.
- Manage personnel matters, including labor negotiations with labor unions in many career fire departments.

Figure 2.14 Airport firefighters must possess knowledge of aircraft types and traits.

The position of Fire Chief is a challenge that requires great preparation, extensive qualifications, and meeting the recommendations of NFPA® 1021, *Standard for Fire Officer Professional Qualifications*. The IFSTA **Chief Officer** manual contains more detailed information regarding the role of the Fire Chief.

Special Operations Personnel

Many fire departments provide a wide variety of services and protect jurisdictions containing numerous special hazards. In order to safely mitigate many emergencies, personnel with special training and equipment are required. In many, but not all cases, members with this specialized training may serve as both regular firefighters and specialists in a particular discipline. The following section addresses several types of special operations personnel.

Airport Firefighter

Airport firefighters are responsible for fire and life safety involving airport and aircraft operations. Airport firefighters operate fire apparatus to standby during aircraft emergency landings and airport fuel spills, and perform fire and rescue operations on downed aircraft **(Figure 2.14)**.

Some airport firefighters also provide structural fire fighting and emergency medical services at the airport facility. Airport firefighters are generally responsible for the following duties:

- Respond to all aircraft incidents.
- Position apparatus according to policy, including considering the size and type of aircraft as well as environmental factors, such as wind.
- Perform fire fighting, rescue, and emergency medical services for aircraft emergencies.
- Operate aircraft rescue fire fighting (ARFF) apparatus and equipment.
- Use appropriate extinguishing agents as required.
- Perform structural fire fighting, emergency medical services, fire prevention duties, and other functions per organizational policy.
- Attend regular training and drill as required.
- Possess knowledge of the proper use and care of ARFF equipment.

> **Airport Firefighters**— Firefighter trained to prevent, control, or extinguish fires that are in or adjacent to aircraft.

- Possess knowledge of tactics used in ARFF operations as well as structural fire fighting (if required).
- Possess knowledge of aircraft types and characteristics.
- Possess knowledge of airport policies and regulations.

Airport firefighters must have all the skills of structural firefighters if they are to engage in conventional fire fighting. In addition, to perform their role as aircraft rescue firefighters, they should meet the recommendations of NFPA® 1001, *Standard for Fire Fighter Professional Qualifications* and NFPA® 1003, *Standard for Airport Fire Fighter Professional Qualifications*.

The IFSTA **Aircraft Rescue and Fire Fighting** manual provides additional information on the role of an airport firefighter.

Hazardous Materials Technician

Responding to incidents involving hazardous materials (haz mat) is a part of the responsibility of virtually all fire service organizations. Some departments train firefighters in engine, truck, and rescue companies to the level required to mitigate haz mat incidents, while other departments use dedicated haz mat response teams **(Figure 2.15)**. These teams may respond with the support of other fire companies to handle such incidents. **Hazardous materials technicians**, called haz mat techs, typically receive extensive specialized training and can be expected to perform the following duties:

- Conduct hazardous materials inspections for compliance with applicable codes and standards.
- Respond to and mitigate haz mat incidents.
- Assist with the development of department programs related to haz mat.
- Participate in specialized haz mat training.
- Possess extensive knowledge of chemistry and other physical sciences.
- Possess knowledge of departmental procedures for haz mat and CBRNE (chemical, biological, radiological, nuclear, explosive) incidents.

In order to obtain and maintain the competence required to mitigate incidents of a highly technical nature, the haz mat tech should meet the recommendations of NFPA® 1072, *Standard for Hazardous Materials/Weapons of Mass Destruction Emergency Response Personnel Professional Qualifications*. Additional information may be found in the IFSTA **Hazardous Materials for First Responders and Hazardous Materials: Managing the Incident** manuals.

Figure 2.15 Chemical protective clothing is designed to protect against chemical, physical, and biological hazards.

Hazardous Materials Technician — Individual trained to use specialized protective clothing and control equipment to control the release of a hazardous material. Hazardous materials technicians can specialize in four areas: Cargo Tank Specialty, Intermodal Tank Specialty, Marine Tank Vessel Specialty, and Tank Car Specialty.

Technical Rescuer (Special Rescue Technician or Technical Rescue Specialist)

Fire departments are called to respond to a wide variety of emergencies that require technical rescue. These emergencies may require special skills and equipment beyond the capabilities of the traditional fire company **(Figure 2.16, p. 66)**. Examples of these emergencies include the following:

- High or low angle rescue
- Trench collapse
- Confined space entry

Figure 2.16 Some fire departments have technical rescue teams.

- Industrial and agricultural accidents
- Transportation incidents
- Building collapses
- Swift water rescue

Some fire departments establish specialized teams for each type of emergency, while other departments may employ a rescue or squad company to respond to all specialized incidents. No matter how a response is organized, the **technical rescuer** is generally expected to perform the following duties:

- Respond to technical rescue incidents in accordance with fire department policy.
- Maintain rescue equipment in a state of readiness.
- Participate in technical rescue training.
- Provide input to department administrators regarding equipment and training needs.
- Have specialized training on the specific equipment and rescue incidents that the rescuer may encounter.
- Have mechanical skills to adapt to dynamic incident conditions.

Technical Rescuer — Individual who has been trained to perform or direct a variety of unique and/or complex rescue situations, such as rope rescues (low- and high-angle), confined space, trench and excavation, structural collapse, mine and tunnel, and other rescue types.

Figure 2.17 Marine firefighters entering a boat cabin on fire. *Courtesy of John Lewis.*

Most technical rescuers are firefighters who have received additional training to meet the NFPA® 1006, *Standard for Technical Rescuer Professional Qualifications*. The IFSTA **Principles of Vehicle Extrication**, **Fire Service Search and Rescue**, and **Technical Rescue for Structural Collapse** manuals contain additional information regarding technical rescue.

Marine, Wildland, and Industrial Firefighters

Other areas of specialization within the fire service include marine, wildland, and industrial firefighters. **Marine firefighters** are land-based firefighters who specialize in operating in the marine environment **(Figure 2.17)**. A marine environment may include:

- Various sizes and types of vessels
- Marinas
- Piers
- Docks
- Wharves

Marine firefighters may also have traditional structural fire fighting responsibilities and be quartered in firehouses where watercraft and allied facilities are in their primary response area. Qualifications for these firefighters may be found in the NFPA® 1005, *Standard for Professional Qualifications for Marine Fire Fighting for Land-Based Fire Fighters*. Additional information on this specialty may be found in the IFSTA manual: **Marine Fire Fighting for Land Based Firefighters**.

Some **wildland firefighters** respond only to incidents involving this specialty, while other personnel may be cross-trained as structural firefighters and employ the required equipment and tactics for either domain. The professional qualifications for wildland firefighters may be found in NFPA® 1051, *Standard for Wildland Fire Fighter Professional Qualifications*, which details the levels of advanced training and responsibility **(Figure 2.18, p. 68)**.

Marine Firefighter — Firefighter personnel assigned to work on a fireboat for the purpose of extinguishment of fires in the marine environment.

Wildland Firefighters — Person trained to function safely as a member of a wildland fire-suppression crew. NFPA® identifies four levels of progression, Wildland Firefighter I through Wildland Firefighter IV; each level denotes a higher level of capability, supervision, or management.

Figure 2.18 Wildland firefighters working a fire line.

> **Industrial Firefighters** — Full-time emergency response fire fighters that provide fire suppression, rescue, and related activities at a commercial, institutional, or industrial facility or facilities under the same ownership and management.

Industrial firefighters serve the fire fighting requirements of large manufacturing, storage, or refinery facilities. These departments are necessary in some cases to supplement the local public fire department because of the hazards present at a specific facility **(Figure 2.19)**. IFSTA's **Industrial Emergency Services Training: Incipient Level and Industrial Exterior and Structural Fire Brigades** manuals provide additional information on this subject.

Figure 2.19 Industrial fire brigade teams practicing foam application techniques.

Military Firefighter

The United States military operates fire protection services on military installations throughout the world. Some firefighters are members of a branch of the armed forces. These military firefighters may provide ARFF protection, marine, or structural fire protection, or a combination of services depending on the requirements of the installation at which they serve. Often these firefighters provide technical rescue and hazardous materials response and may have mutual aid agreements with surrounding communities.

Emergency Medical Services (EMS) Personnel

The fire service has a long history of providing care and transportation to sick or injured people, from the basic first aid measures offered at the beginning of the 20th century to the advanced life support that today's fire department members offer in many jurisdictions. Local requirements and practices have driven the delivery of emergency medical care and the level of fire department involvement. Some fire departments do not provide EMS responsibilities, while in other jurisdictions the fire department is solely responsible for all patient care and transport. There are also numerous variations of service delivery that involve fire department first response to high priority calls, or in some cases, all calls requesting medical assistance (**Figure 2.20**).

Fire departments may provide emergency medical services in a variety of ways. In some jurisdictions, firefighters are cross trained as first responders, EMTs, or paramedics.

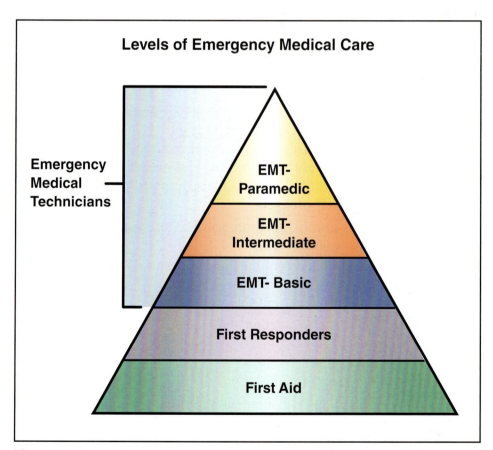

Figure 2.20 Levels of emergency medical care.

The following sections highlight some of the capabilities of first responders, emergency medical technicians, and paramedics. The authority having jurisdiction determines the specific skills assigned to each qualification over emergency medical care where the member is licensed.

First Responder

The primary function of the **first responder** is to maintain life until the more highly trained personnel arrive on scene. First responders have basic training and equipment and are usually not responsible for patient transport to the hospital. Some of the conditions or injuries that the first responder may stabilize include the following:

- Respiratory distress or arrest
- Cardiac arrest
- Extensive bleeding
- Poisoning
- Shock
- Emergency childbirth
- Stroke
- Diabetic emergencies
- Fractures
- Seizures
- Severe exposure and burns

The Brady/IFSTA **Fire Service First** manual contains additional information and the National Highway Traffic Safety Administration provides curriculum for first responders.

Emergency Medical Technician

In most jurisdictions, the **Emergency Medical Technician (EMT)** provides basic life support (BLS) to sick or injured patients. As compared to the first responder, the EMT is responsible for providing more advanced patient care and will require significantly more training and specialized equipment to treat the patient. In many cases, the EMT is also responsible for transporting the patient to a hospital **(Figure 2.21)**.

In addition to the duties of a first responder, the EMT is generally expected to provide the following level of care:

- Conduct examinations to determine the nature and extent of an illness or injury.
- Stabilize the patient prior to transport.
- Employ equipment, such as suction and oxygen, to support respiration or ventilation.
- Obtain vital signs.
- Immobilize fractures.
- Use appropriate methods in moving patients from the incident scene to an ambulance, and provide safe transport to hospital.
- Maintain awareness of incident scene safety.
- Use of proper communication and record keeping procedures.

> **First Responder** — (1) First person arriving at the scene of an accident or medical emergency who is trained to administer first aid and basic life support. (2) Level of emergency medical training, between first aider and emergency medical technician levels, that is recognized by the authority having jurisdiction (AHJ).

> **Emergency Medical Technician (EMT)** — Professional-level provider of basic life support emergency medical care; requires certification by some authority.

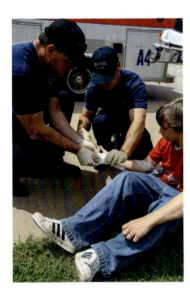

Figure 2.21 EMTs perform medical aid on an injured person.

These duties outline only those expected of basic level EMTs. Many jurisdictions have multiple levels of certification within the specialty of emergency medical technician. These levels allow the EMT to perform additional duties that are more advanced and require a higher degree of knowledge, training, and more sophisticated equipment.

NOTE: The Brady/IFSTA **Fire Service Emergency Care** manual contains additional details on this subject.

Paramedic

Paramedics provide the highest level of patient care in the emergency medical service. Training for paramedic positions requires classroom and practical instruction as well as a field internship. Paramedics are often dispatched to high priority calls where their skills may best be utilized. They respond to medical emergencies with more in-depth knowledge and qualifications to perform a higher level of patient care than first responders or EMTs.

In addition to the duties of first responders and EMTs, paramedics must be able to perform the following advanced skills:

- Recognize and assess medical emergencies.
- Decide priorities of medical treatment and communicate pertinent information to a supervising physician.
- Follow the direction of the supervising physician and provide reports on patient status.
- Exercise proper judgment using standard of care protocols when out of contact with a physician in order to properly treat patients.
- Administer intravenous fluids and narcotics per physician orders and standard protocols.
- Manage a patient's airway, including intubation.
- Direct and supervise the appropriate transport for a patient.
- Maintain required records and inventory of equipment and supplies.

Paramedic — (1) Certified emergency medical professional who provides advanced life support. (2) Professional level of certification for emergency medical personnel who are trained in advanced life support procedures.

EMS Chief/Officer

The EMS Chief or other ranking officer is responsible for supervision of EMS functions in quarters and at an incident scene. Depending on the organizational structure, this position may include supervision of maintenance and operation of the fire station as well as EMS unit and equipment.

The EMS officer generally performs many duties that include inspection of equipment and personnel in accordance with department policy and under the direction of a ranking officer as provided for by organizational guidelines. Depending on rank and organization, some EMS officers may respond to supervise routine medical calls while those of higher rank may respond only to major incidents and primarily coordinate administrative functions on a daily basis. EMS officers may hold rank titles similar to fire officers (Lieutenant, Captain, and/or Chief) and are generally responsible for the following duties based on their rank and department structure:

- Reviews reports and statistical data.
- Maintains training files to ensure compliance with jurisdictional standards.
- Coordinates field care audits and continuing education.

- Acts as a liaison with hospitals, licensing authorities, and other EMS agencies.
- Develops and implements EMS policies as necessary.
- Assists in the preparation of an EMS program budget.
- Purchases, evaluates, and maintains EMS supplies and equipment.
- Responds to major incidents to manage EMS function.
- Acts as Infection Control Officer for the department.

The EMS officer must have extensive qualifications that escalate based on rank and organizational responsibility. General qualifications may include:

- Knowledge of laws and regulations pertaining to the delivery of emergency medical services.
- Understanding of specific practices employed with EMS in regard to fire departments.
- Knowledge of and the ability to teach EMS training procedures.
- Ability to conduct data collection and supervise the implementation of quality assurance/improvement programs.
- Knowledge of administrative functions with the ability to coordinate work of subordinates in order to achieve organizational goals.
- Ability to develop and maintain working relationships with hospitals, regulatory boards, mutual aid partners, and others with whom cooperation is necessary.
- Ability to possess and maintain appropriate licensure to provide supervision to members of the EMS staff.

Other Fire Department Personnel

In order to achieve organizational goals, fire departments rely on personnel who have special skills other than fire fighting or emergency medical services. These employees are often considered support personnel as their work enables firefighters to conduct operations safely and efficiently. In some organizations, firefighters are assigned these non-fire fighting roles. Meanwhile, in other departments, civilian employees perform these functions.

Telecommunications/Dispatch Personnel

The communications system is the nerve center of any fire department. Telecommunicators who operate this system may be sworn fire department members or civilian personnel. In some areas, a regional dispatch center processes calls for several communities and serves numerous emergency services, including police, fire and EMS **(Figure 2.22)**.

The duties of the telecommunicator generally include the following:

- Answer and process emergency and non-emergency phone calls.
- Dispatch units to incidents per department policy.
- Maintain communication with units operating in the field, providing support as requested.
- Follow department policy regarding requests for information by members of the media.

Figure 2.22 Dispatchers receive emergency calls from the public and dispatch emergency responders.

- Record information regarding calls and dispatch information per department policy.
- Achieve Emergency Medical Dispatch (EMD) qualifications where necessary.
- Remain calm and communicate clearly and concisely.
- Possess knowledge of fire department procedures and common terminology.
- Operate computers, radios, phone systems, and other dispatching equipment.

A telecommunicator should meet the recommendations of NFPA® 1061, *Professional Qualifications for Public Safety Telecommunications Personnel* and APCO International Minimum Training Standards for Public Safety Telecommunicators (APCO ANS 3.103.1- 2010). Additional information regarding the role of a telecommunicator is available in the IFSTA **Telecommunicator** manual.

Information Technology Personnel

Fire departments have increasingly complex information technology requirements. The computer software and hardware required to administer daily operations, monitor systems, and maintain records require the services of information technology (IT) professionals. These personnel may be fire department employees or work for a municipal IT department that serves the needs of numerous city or county agencies. Some jurisdictions have taken to outsourcing this function to private technology companies.

Information technology may be used in the fire service to perform the following functions:

- Computer-aided dispatch
- Apparatus location systems
- Mapping
- Real-time building data collection
- Telemetry of patient medical information to hospitals
- Shift scheduling/personnel data management

Facilities Maintenance Personnel

Although firefighters generally provide routine maintenance in the firehouse where they are assigned, administrative offices and other fire department facilities require routine maintenance and all buildings require repair services at some point. As with other support functions, facilities maintenance may be a division within the fire department, a separate city/county department, or contracted to a private company.

Some of the maintenance and repair functions of facilities maintenance personnel include:

- Repair of all systems in firehouses, administration, and training facilities.
    - HVAC
    - Electrical
    - Plumbing
    - Carpentry
- Maintenance of vehicle fueling systems.
- Maintenance of outside areas and parking lots on fire department property.

Apparatus/Equipment Maintenance Personnel

Special skills and equipment are required in order to perform certain maintenance and repair service to fire apparatus and other fire fighting equipment. In some jurisdictions, the fire department operates a repair facility, while other areas utilize a city or county wide maintenance or "shop" facility for all types of vehicles **(Figure 2.23)**. However they are organized, maintenance personnel must be specially trained in order to maintain and repair the complex systems that comprise modern fire apparatus and equipment. Generally, maintenance personnel must maintain not only the automotive systems, but the pumps, aerials, and portable equipment assigned to fire apparatus.

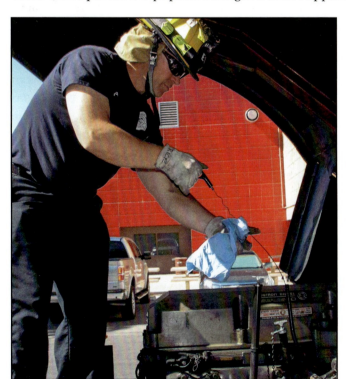

Figure 2.23 Some departments have maintenance personnel that do all vehicle maintenance and repair.

The typical duties of maintenance personnel include the following:

- Perform maintenance and repair services to automotive systems, fire fighting systems, and portable equipment as required.
- Perform service testing on aerials and pumps.
- Inspect vehicles per state or provincial regulations.
- Record and maintain documentation of all repairs.
- Possess certification by the National Association of Emergency Vehicle Technicians.
- Possess certification to perform state or provincial safety inspections.
- Have training in manufacturer's recommendations for the specific apparatus in the fire department inventory.
- Operate towing equipment.

NOTE: Apparatus maintenance personnel should meet the recommendations of NFPA® 1071, *Standard for Emergency Vehicle Technician Professional Qualifications*.

The maintenance and repair of portable equipment may be completed at the same facility as apparatus by some of the same technicians, or personnel specifically assigned equipment repair tasks. These duties often include:

- Performing maintenance and repair of hose, ladders, generators, and power tools according to manufacturer's guidelines.
- Conducting service testing portable equipment per manufacturer's guidelines and department policy.
- Maintaining inventory and a record-keeping system for equipment repair.

Personnel responsible for equipment maintenance should possess the following qualifications:

- Manufacturer certification for repair of specialized fire department equipment.
- Knowledge of recommended testing procedures for equipment in fire department inventory.
- Ability to analyze equipment problems and take appropriate corrective action.

Chapter Summary

Many roles exist within the fire and emergency services. Many firefighters will be assigned to a fire company where they will be responsible for performing a variety of fire fighting duties, such as fire extinguishment, forcible entry, and ventilation. Many other roles are needed within the fire department. Firefighters can promote to driver/operators, company officers, and even to the role of District Chief or Fire Chief. Firefighters could also specialize in a given field, earning titles such as hazardous materials technician and paramedic. There are also a number of administrative and support roles, such as Training Officer and Public Information Officer. All of these roles must work cohesively within the fire and emergency services organization in order to provide the best possible service to the surrounding communities.

Review Questions

1. What are the primary duties of an engine company, a ladder company, and a rescue/squad company?
2. What are some typical duties of a firefighter?
3. What are some duties of a fire apparatus driver/operator?
4. What is the difference between a company officer and a battalion/district chief?
5. What are two types of training division personnel?
6. What are some positions that assist in the administrative duties of a fire department?
7. What are some types of special operations personnel?
8. How do the duties of a first responder, an emergency medical technician, and a paramedic differ?
9. What are the duties of telecommunications/dispatch personnel?
10. What are some typical duties of apparatus/equipment maintenance personnel?

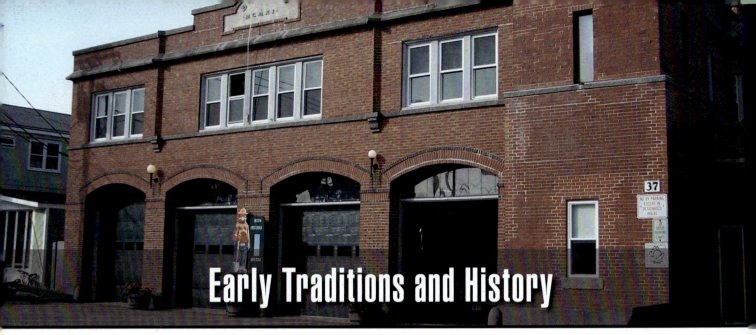

Courtesy of Ron Jeffers.

Early Traditions and History

Chapter Contents

Case History 81	Key Developments in Fire Fighting 92
Early Fire Services 81	Development of American Fire Engines 92
The First Fire Pump 82	The Age of Steam 95
Extinguishing Techniques.............................. 82	Chemical Engines and Ladder Trucks......................... 97
Fire Laws and Ordinances 83	Gasoline and Diesel Powered Equipment 98
Causes of Fire 83	Improvements in Protective Clothing and Self-Contained Breathing Apparatus (SCBA) 99
Britain's First Fire Brigade 84	Fire Extinguishers.......................................102
Development of the Fire Engine.................... 85	Alarm and Communications Systems 103
Fire Protection in Early America 86	**Impact of Historic Fires on Firefighter and Public Safety in North America**................105
Growth of the Volunteer Fire Service....... 87	Significant Historic Fires in North America...............105
Volunteers, Patriots, and Competitors........................87	Impact of Historic Fires on Firefighter Safety 108
First Hose Company 89	**Chapter Summary** 108
Mutual Fire Society of Boston 90	
Philadelphia: Union Volunteer Fire Company 90	
New York Volunteer Department 90	
First Fire Insurance Companies................................. 91	

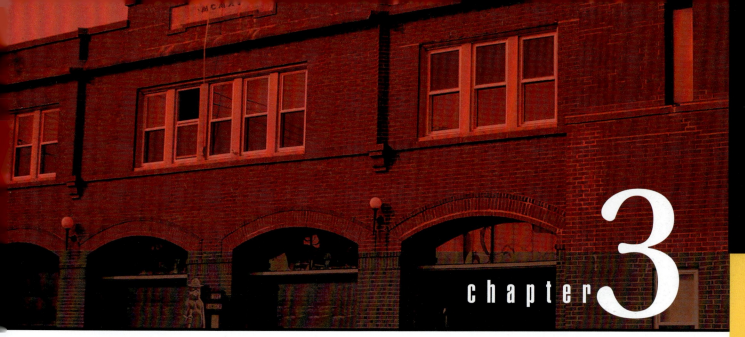

chapter 3

Key Terms

Brands 97	Fire Pump 82
Conflagrations 83	Fire Wards 86
Fire Brigades 85	Fire Wardens 86
Fire Mark 85	Party Wall 83
Fire Hook 83	Turnout 88

FESHE Outcomes

This chapter provides information that addresses the outcomes for the Fire and Emergency Services Higher Education (FESHE) *Principles of Emergency Services* course.

1. Illustrate and explain the history and culture of the fire service.

Early Traditions and History

Learning Objectives

After reading this chapter, students will be able to:

1. Describe the early history of fire services. (Outcome 1)
2. Summarize facts about fire protection in early America. (Outcome 1)
3. Explain how the growth of the volunteer fire service affects modern-day fire and emergency services. (Outcome 1)
4. Identify key developments that have affected modern-day fire and emergency services practices. (Outcome 1)
5. Explain how the modern-day philosophy of public safety and firefighter safety has been impacted by historic fires in North America. (Outcome 1)

Chapter 3
Early Traditions and History

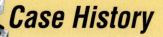

Case History

Amy's uncle is a career firefighter. Since she is considering a similar career path, he invited her to his fire station for a visit. During her visit, she saw numerous historic photographs on the station's walls. Many of the photos were black-and-white, and the people and equipment in them looked different than what Amy had seen in books and her uncle's station. Amy also noticed an old engine that crews called "'The Antique," and some old tools hanging on the wall in the apparatus bay. Until then, Amy had focused her studies on the skills and science of fire fighting and had not realized that the fire service has changed over the years.

At lunch, Amy asked her uncle and his crew about the photos, "The Antique" engine, and the old equipment on the walls. These questions spurred a lively and inspired conversation about the long-standing traditions that have been with the department since before any of the crew members could remember. Amy was excited to gain a better understanding of the roots of some fire service traditions and changes made over the years. As she said goodbye to her uncle at the end of the day, she felt committed to furthering her insight about the fire service and learning more about its history.

The fire service enjoys a rich and colorful heritage that is interwoven with some of the most notable figures in history. The origin and development of modern fire department equipment and organization may be traced back through the ages as people learned how to better protect themselves from the ravages of fire. The following chapter discusses the path of the fire service from the first organized firefighters battling ancient conflagrations to the lessons learned from recent fire losses.

Early Fire Services

After experiencing fires that destroyed large sections of Rome around 24 BCE, Emperor Caesar Augustus organized a band of servants, known as the Familia Publicia, to be stationed at the city's gates for the purpose of fighting fires. Following another fire in 6 BCE, the Emperor formed the Corps of Vigiles, a nightly patrol of slaves who protected Rome from fire for 500 years. This Corps watched for fires, alerted citizens, and enforced fire safety. They were equipped with buckets and axes for fire fighting and empowered to lash those individuals who disregarded fire safety. The force of 7,000 men was organized using a chain of command similar to a modern fire department with officers and firefighters.

Despite the efforts of the Corps of Vigiles, a devastating fire occurred on the night of July 18, 64 AD. Windswept flames spread through Rome, consuming structures for six days and seven nights. Although commonly criticized for inaction during the fire, there is evidence that Emperor Nero actually traveled to Rome from his palace in Antium to personally direct fire fighting efforts.

The First Fire Pump

Fire Pump — (1) Water pump used in private fire protection to provide water supply to installed fire protection systems. (2) Water pump on a piece of fire apparatus. (3) Centrifugal or reciprocating pump that supplies seawater to all fire hose connections aboard a ship.

The exact date of the introduction for the first **fire pump** is not definitively known. However, there is a detailed description from the 3rd Century BCE of a machine consisting of two brass cylinders with pistons that drew water through valves at their base and pushed it through outlets into a chamber. Rising water in the chamber compressed the air inside, causing it to force the water to eject in the form of a stream through a pipe or nozzle. Long handles operated the action of the pistons. This pump, the remains of which have been found in Roman excavation sites, was likely invented by an Alexandrian Greek named Ctesibius. Known as a siphona, the principle of this simple double-force pumping device remained the basic mechanical method for hand pumps until modern times **(Figure 3.1)**.

With the fall of Rome, the vigiles and the siphona were disbanded. With their passing, Europe would not see another well-organized and equipped fire department for a thousand years.

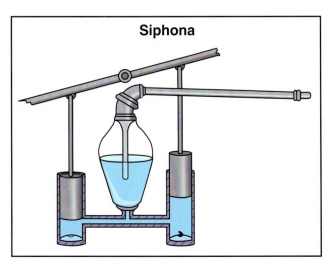

Figure 3.1 The first fire pump, known as the Siphona Pump, was hand operated and worked as a double-force pump.

Extinguishing Techniques

People learned quickly that water was the most plentiful, cheapest, and effective agent for fighting fires. However, other unusual methods were employed throughout the ages in an effort to combat fires. In Japan, if fire threatened the emperor's palace, hundreds of people with large fans would line the walls of the palace grounds in an effort to fan flying embers away from the structures within. In Europe during the 7th Century, some people held the belief that ringing church bells in reverse would control large fires. This method, while ineffective for fire control, at least warned citizens of a fire and caused them to gather at the church where they could be put to use in fire fighting efforts. Ringing church bells has been used to summon volunteer firefighters up until the modern era of electronic communications.

Figure 3.2 Fire hooks could pull thatch off of a roof or be attached to a ring set in the gable of a house and used to pull a building down.

Fire Laws and Ordinances

Laws and ordinances enacted to prevent or reduce the devastation of fires may be traced back to the Norman conquest of England in 1066. A law enacted by William the Conqueror required that all fires in the household had to be extinguished at nightfall. The simplest way to accomplish this task was to cover the open fireplace hearth with a metal cover to exclude the air. In Norman French, this cover was known as a couvre feu. This word later became the English word curfew. The bell that tolled at nightfall as a signal to extinguish the fires became known as the curfew bell.

In 1189, the first lord mayor of London issued a series of orders that addressed several issues of fire safety. All houses were required to be built of stone, with the roof made of slate or burnt tile, as opposed to the highly combustible thatched roofs. **Party walls** between attached buildings were required to be at least sixteen feet (5 m) high and three feet (1 m) in breadth. The cost of construction was to be borne equally by parties on either side of the wall. In addition, citizens were required to be prepared should a fire occur. Each home was to have a ladder and barrel of water set by the front door. Fire hooks were to be provided to hook onto gables or other structural members in order to pull down a structure to create a fire break. Some structures were built with an iron ring in the gable end into which a **fire hook** could be inserted (**Figure 3.2**). The fire service still uses the fire hook, a tool dating back to Roman times, in various forms. Some early fire apparatus carried fire hooks and ladders, coining the term "hook and ladder truck."

Causes of Fire

During the 17th Century, many fires in England were the result of unsafe chimneys or careless methods of transporting burning sticks of peat from one dwelling to another to light a fireplace. The congestion of cities, as well as building construction methods and narrow streets that allowed rapid fire spread, compounded the problems associated with an unorganized method of fire fighting to allow many **conflagrations** to occur during this time.

Party Wall — Dividing wall that stands between two adjoining buildings or units, often on the property line, and is common to both buildings. A party wall is almost always a load-bearing wall and usually serves as a fire wall.

Fire Hook — Hook, on the end of a rope or pole, used to pull combustible roofing materials such as thatch from burning buildings. These devices could also be used to hook into the gables of a building, or a ring installed on the gable, to pull that end of the building down. Because these hooks were carried on early ladder trucks, these trucks became known as "hook and ladder companies". The pike pole is a modern variation of the fire hook.

Conflagration — Large, uncontrollable fire covering a considerable area and crossing natural fire barriers such as streets; usually involves buildings in more than one block and causes a large fire loss. Forest fires can also be considered conflagrations.

Britain's First Fire Brigade

In September 1666, London suffered a conflagration that raged out of control for four days, destroying more than eighty percent of the city. The fire fighting efforts included the use of buckets, fire hooks, fire squirts, and some primitive fire engines. Some areas of the city were served with lead or wooden pipes that supplied water.

Fire Fighting Syringes

In the 16th Century, a barrel and plunger combination known as a fire squirt was used as a fire fighting appliance. This device could be hand-held or mounted on wheels. In order to achieve a flow of water, the barrel was pulled toward the holder, causing the plunger rod to expel the water. **(Figure 3.3)**.

Figure 3.3 In the 16th Century, syringes were used for fire fighting.

Subsequent to the fire, city authorities recommended numerous improvements to fire fighting methods. However, many citizens believed that the "hand of God" was at work in large scale fire disasters, while other citizens held that the cost of organizing a fire fighting force was too high.

Although many ordinary citizens were indecisive about fire protection, the owners of the fire insurance companies that were created after the London fire were concerned about protecting their interests. In order to prevent the destruction of their client's property from haphazard fire fighting efforts, they developed Britain's first organized **fire brigades**. The insurance company provided its policyholders with a plaque representing the company's name or design mark. These "fire marks" were posted conspicuously at the insured's property as the fire brigade would only fight fires at properties displaying their company's **fire mark**.

Fire Brigades in the United States and the United Kingdom

In the United States, the term fire brigade typically refers to an industrial fire brigade; a team of employees organized with a private company or industrial plant that is assigned to respond to fires or emergencies on that property. In the United Kingdom and elsewhere, a fire brigade refers to a municipal fire department.

Fire Brigade — (1) Organization of industrial plant personnel trained to use fire fighting equipment within the plant, and who are assigned to carry out fire prevention activities and basic fire fighting duties and responsibilities. The full-time occupation of brigade members may or may not involve fire suppression and related activities. (2) Term used in some countries, outside the U.S., in place of fire department.

Fire Mark — Distinctive metal or wooden marker once produced by insurance companies for identifying their policyholders' buildings.

Development of the Fire Engine

The London fire and the competition between the insurance companies' fire brigades hastened the development of the fire engine. In Holland, Amsterdam's municipal fire brigade boasted approximately sixty fire engines. Captain of the brigade, Jan Van der Heijden was an accomplished engineer who was instrumental in the design and invention of a fire engine that used the advantage of a continuous water supply to make a close attack on burning buildings.

Van der Heijden also designed and invented a fire hose made from pieces of leather. In 1690, he wrote a manual called *A Description of the Newly Invented and Patented Fire Engines with Flexible Hose, and Their Manner of Extinguishing Fires*. This book contained pictures and explanations of the advantages of the close attack method he helped pioneer.

By the end of the 17th Century, several English manufacturers were engaged in a heated competition producing fire engines. In 1721, Richard Newsham developed a beautifully crafted yet durable fire engine that consisted of a rectangular box or tub mounted on wheels. Members of a bucket brigade filled the tub with water, while men operating large pumping handles supplied the power to produce water pressure for a nozzle **(Figure 3.4, p. 86)**.

Newsham's design was the first to feature long wooden bars for pumping handles. Known as brakes, these smooth oak handles extended up to fifteen feet and allowed as many as ten men to operate the pumping action on each side of the engine. Working the brakes up and down like a seesaw moved a metal beam that operated a piston pump. Newsham's fire engine was capable of providing a continuous stream of water through a gooseneck nozzle with enough pressure to reach the height of many roof tops. In his advertising, Newsham claimed his engine could pump a stream of water to a height of 165 feet (50 m).

Figure 3.4 A Newsham pumper circa 1764. *Courtesy of Fireman's Hall Museum.*

Over time, many improvements were made to the Newsham engine. Some models were built with bent axles, placing the body closer to the ground and making it easier to operate the brakes. Other engines featured treadles that allowed firefighters to use their legs as well as their arms to assist in pumping. Other developments included the introduction of a suction hose and three-way fitting that allowed the tub to be filled with water directly from a nearby well or reservoir.

Fire Protection in Early America

In 1608, Jamestown, the first permanent English settlement in America, suffered a disastrous fire that burned down most of the dwellings and destroyed many of the provisions required to sustain the colony. Without proper supplies, many colonists died during the winter months following the fire.

In the years that followed, Europeans began to arrive in New World settlements at an ever-increasing pace. The main concern in these burgeoning towns was meeting the new settlers' immediate need for shelter. Little thought was given to long-term planning or fire protection needs. Towns expanded aimlessly with the only structured fire fighting efforts consisting of town leaders organizing bucket brigades and issuing orders to pull down burning buildings that may threaten other structures.

In the bucket brigade, villagers would form two lines between the water source and the fire. Men would pass full buckets in one line, with women and children forming a return line of empty buckets. **(Figure 3.5)**

In North American cities of the early 1600s, **fire wards**, **fire wardens**, and building surveyors provided some measure of fire protection. Boston, America's fastest growing city of the period, began to experience devastating fires just after it was settled in 1631. After enduring numerous fires, the fire city leaders took steps to help reduce the likelihood of fire by ordering a ban on wooden chimneys and thatched roofs. They also appointed fire wards for each section of the city.

Fire Wards — Individuals appointed by the city of Boston after 1631 in an attempt to prevent fires. They were provided badges and staffs of office and assigned to different parts of the city. Similar to the fire wardens of New Amsterdam and other early American cities.

Fire Wardens — Individuals appointed by the city of New Amsterdam and other early American cities in an attempt to prevent fires by inspecting chimneys and hearths. They were empowered to cite residents for failing to meet the city fire codes. Similar to the fire wards of early Boston.

Figure 3.5 Bucket brigades were formed with one line of men to advance the full pails of water and a second line of women and children to return the empty pails.

In 1647, Peter Stuyvesant arrived from Amsterdam to become the governor of New Amsterdam (later called New York). He reorganized many aspects of government, including surveying, mapping, and establishing a building code.

A new position was created to serve the need for fire protection. Fire wardens patrolled the streets of New Amsterdam from dusk until dawn carrying wooden noisemakers called rattlers to rouse citizens if a fire was discovered. Although these men held no authority to command fire fighting efforts, they enforced fire prevention laws and imposed fines on violators. This organization, Surveyors of Buildings, was the first of its kind in North America.

Before the first fire departments were established, other cities appointed fire wardens to inspect chimneys and hearths. Fines could be levied upon those who failed to keep their chimneys clean or who improperly stored straw or other combustible material. In addition, homeowners were often required to keep a leather bucket close at hand for immediate fire fighting efforts. Public chimney sweeps, who charged a standard fee based on the height of a structure, were appointed to maintain safety standards and keep chimneys free of a dangerous buildup of soot.

Growth of the Volunteer Fire Service

As the 21st Century began, there were more than 30,000 fire departments in the United States and approximately 3,500 fire departments in Canada. These departments had begun modestly enough just after the turn of the 18th Century as mutual fire societies and then volunteer fire companies, as described in the sections that follow.

Volunteers, Patriots, and Competitors

After the Revolutionary War, volunteer fire fighting spread across the country as many veterans returned home and joined a fire company, making the firehouse their clubroom. Prideful members cared for beautifully crafted and ornate pieces of apparatus as rivalries increased between local companies for the best looking equipment **(Figure 3.6, p. 88)**.

Figure 3.6 Volunteer fire companies in the late 1700s showed their pride by elaborately decorating and painting their engines.

Many famous Americans served as volunteer firefighters, including:

- George Washington
- John Hancock
- Alexander Hamilton
- Samuel Adams
- Paul Revere

Turn Out — Alerting of a fire company for a response.

Changes in the volunteer fire service were happening rapidly as the United States continued to grow. Better building regulations, improved water supply systems, and new methods for fighting fires were introduced. The competitive spirit to outdo other companies helped fuel innovation among apparatus manufacturers as volunteers practiced with their new engines to achieve faster times turning out of the firehouse and setting up at fires. Some companies established a system of "bunking out" or sleeping at the firehouse to **turn out** more efficiently.

During these years of volunteer service, firefighters came to be looked upon as heroes for making daring rescues and gallant efforts to contain huge blazes. The heroism of these firefighters was captured in song and verse, as well as the popular drawings of Currier and Ives.

In the early years, ranks of volunteers were filled with men of means who maintained socially exclusive firehouses. The status of belonging to a fire company was desirable, even though these men were expected to endure the hazards and physical challenges of fighting fires. Members of these exclusive organizations were responsible to the company, the fire department, and the city. A chain of command from the department chief down to company foreman maintained order and supervised fire fighting efforts. Fines were levied for various infractions, including:

- Nonattendance at a fire
- Being out of uniform
- Being intoxicated
- Using profanity in the firehouse
- Using tobacco products in the firehouse

Figure 3.7 Hose wagons were developed to make maximum use of the early water systems.

The legendary competitions of the volunteers grew more heated as time passed. Rival companies attempted to pass each other on the way to a fire, knocking over anyone in their way. Firefighters not only wanted to be first on the scene, but first to have water on the fire as well. This rivalry led to companies in some areas employing a child to dash to the scene of a fire and cover the nearest fire plug with a barrel until his/her companies arrived. The use of these "plug uglies" sometimes resulted in fist fights between companies, as the fire continued to burn.

First Hose Company

Along with the rivalries and spirit of competition, several major advances helped shape the way fire fighting operations are organized to this day. Water systems improved through necessity in order to serve ever-increasing populations. With these improvements came a water supply system of hollowed out logs that featured a hole every half block or so that was fitted with a wooden plug that was removed to obtain water (a fire plug). This system was first installed and operated in Philadelphia. The city's volunteer firefighters wasted no time in developing a major water supply innovation, the hose wagon **(Figure 3.7)**.

Although flexible hose of crude design could be traced back to the ancient Greeks, the Philadelphia volunteers developed a riveted hose that was a major improvement over the old, leaky sewn hose. They designed a hose reel, paid to have it built, and outfitted it with 600 feet (180 meters) of the new hose and other fire fighting equipment. Hose No. 1, with a limited roster of 20 men, was formed. This company had an agreement with Engine No. 1 to supply it with water from a fire plug by laying its hose as a supply line. During one fire, Engine 1 was late arriving and the members of Hose 1 used a hose supplied directly from the fire plug to operate an effective stream on the fire without the need for a pumping engine. The hose company concept became popular, and many other hose companies were established in Philadelphia and elsewhere.

Figure 3.8 The bed key was a very important early fire fighting tool. If firefighters could quickly dismantle a bed and carry it outside, they could save one of the family's most valuable possessions.

Mutual Fire Society of Boston

As a result of the devastating fires it experienced, Boston was the first city in North America to establish a fire department. Years ahead of others, Boston recognized that fire engines, buckets, and other fire fighting equipment were vital to the city's existence. Despite this protection, some Bostonians, fearful of another conflagration, banded together to form the first fire society on September 30, 1718. This fire society was the beginning of the age of the volunteer firefighter.

Mutual fire societies were independent organizations that offered assistance to regular Boston firefighters. Members carried a list containing the names, home addresses, and places of business of the entire membership. Upon hearing an alarm of fire, all nearby members responded with personal equipment consisting of a bucket and equipment bag emblazoned with the society's emblems: a bed key and screw driver. In case of fires in or near a member's property, fellow fire society members would ensure all occupants were evacuated, and then because the bed was generally the most valuable item in a home, the bed key was used to dismantle it and remove the pieces from harm's way. In addition, whatever household valuables could be saved were placed in the society members' bag and taken to safety. After assisting the fire department in extinguishing the blaze, a member would guard the property from looters. **(Figure 3.8)**.

Philadelphia's Union Volunteer Fire Company

After a devastating fire in 1730, Philadelphia resident Benjamin Franklin researched fire prevention and protection methods **(Figure 3.9)**. After visiting Boston's Mutual Fire Society, Franklin wrote an article in his newspaper detailing that organization's efforts. The following year, he and other residents established the first fire organization in Philadelphia, the Union Volunteer Fire Company. Unlike the Boston society, which responded only to member's properties, the Philadelphia volunteers responded to any fires in their vicinity. This organization received such an enthusiastic response, Franklin soon added additional companies to form an actual fire department **(Figure 3.10)**.

New York Volunteer Department

In 1737, the Volunteer Fire Department of the City of New York was formed with 35 members. By 1786, the department had been reorganized with increased membership and numerous districts under separate leadership. Every volunteer company vied for the most ornate fire engine adorned with gleaming metal and hand-painted murals by well-known artists. The firefighters competed to be the quickest to the fire and wore colorful uniforms at parades and public contests between companies.

The fire department continued to grow with the city and boasted 25 engines with a crew of ten to twenty men for each by the early 1800s. The crew size was often determined by the size and weight of the machine and sometimes its geographic location within the city. The department was organized around a chief engineer (fire chief) and six assistant engineers. During this time, the chief engineers and their assistants were mainly figureheads occasionally inspecting apparatus while each individual company continued to function on its own.

Figure 3.10 The Philadelphia-style engine developed in 1794 was an excellent double-piston pump.

Figure 3.9 A portrait of Benjamin Franklin, Father of the Fire Service. © by Chris Fagan. *Courtesy of the Congressional Fire Services Institute.*

The autonomy of the individual fire company came to an end in 1811, when Thomas Franklin became the chief engineer and assumed an active role as chief of the fire department. With Franklin's sincere personality and strong leadership, his authority was accepted and his orders followed by each company foreman (an elected position in the company similar to a company officer) in the city. For the next 25 years, Franklin and his successors operated the only fire department in North America with a unified command structure.

First Fire Insurance Companies

In its earliest North American form, the first fire insurance company not only insured against fires but responded to them. Benjamin Franklin and the city of Philadelphia were also at the root of this colonial innovation.

The Philadelphia Contributorship of the Assurance of Houses from Loss by Fire

Benjamin Franklin and several other businessmen founded the Philadelphia Contributorship of the Assurance of Houses from Loss by Fire in 1752. This company's fire mark consisting of four men gripping each other's wrists was called the "hand-in-hand" as it symbolized unity and strength. The hand-in-hand symbol is still in use today as the logo of The Contributorship Companies **(Figure 3.11)**.

Figure 3.11 The first successful American fire insurance company, known as the Philadelphia Contributorship of the Assurance of Houses from Loss by Fire, used the "hand-in-hand" symbol. This symbol is still in use today.

Mutual Assurance Company

Benjamin Franklin believed that trees attracted lightning, so the Philadelphia Contributorship refused to insure buildings that were close to trees. Eventually, a portion of the Philadelphia Contributorship split to form the Mutual Assurance Company. In 1784, its fire mark, the green tree, symbolized the fact that this company would insure buildings in close proximity to trees.

Insurance companies, and sometimes town or city treasuries, would authorize payment to the first-arriving volunteer fire companies who put water on the fire. This system set into play fierce rivalries between volunteer companies who raced to be first at a blaze not just for pride but payment as well.

Key Developments in Fire Fighting

With the increasing population, fire departments have had to expand significantly to properly serve the public. In order to provide this service, fire departments have relied upon technological advancements in the fire service. These key developments in fire fighting include:

- Development of American fire engines
- The age of steam
- Chemical engines and ladder trucks
- Gasoline and diesel powered equipment
- Improvements in protective clothing and self-contained breathing apparatus
- Fire extinguishers
- Alarm and communications systems

Development of American Fire Engines

Little detailed information exists about the first engine used in North America. It is known that city leaders in Boston signed a contract with Joseph Jynks to construct a fire engine in 1653. However, no description of the engine, record of its construction, or stories of its use in fighting fires exists today.

Slightly more information may be found concerning a subsequent effort to obtain a fire engine for Boston from an apparatus builder in England. American merchants had high praise for an engine they had seen while visiting London, so in 1676, Boston placed an order with the English manufacturer. Upon the arrival of what is believed to have been a hand tub apparatus in 1678, thirteen men were appointed under chief engineer Thomas Atkins. The city paid this force to maintain the engine, tow it to fires, and operate the pump.

New York barrel maker and boat builder Thomas Lote is credited with manufacturing the first successful fire engine in North America with his delivery to the city of New York in 1743 **(Figure 3.12)**. Lote's engine was nicknamed "Old Brass Backs" due to the lavish use of brass on the box of the engine. Prior to purchasing this engine, New York had imported the first Newsham fire engines from England in 1731.

New York Style Hand Engines

The competition between New York's volunteer fire companies led to the development of ever more powerful hand tubs. One such design, the "New York style" hand engine featured improvements on Newsham's use of the side stroke

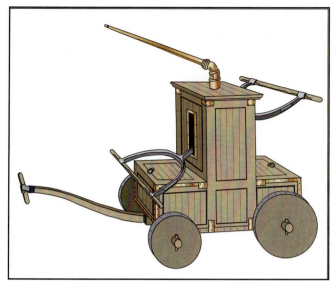

Figure 3.12 The first engine built in the United States was delivered to New York City in 1743 by Thomas Lote.

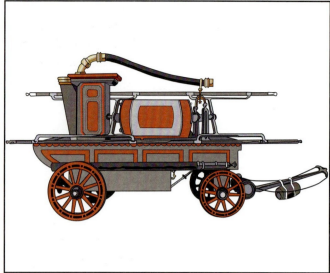

Figure 3.13 The New York style engine was famous for its gooseneck discharge on top of the machine.

pumping method. The city of New York replaced antiquated engines with this newer, more powerful design that featured a pipe with a prominent bend protruding from the air chamber. Nicknamed the "gooseneck," this engine became the most popular new style of apparatus **(Figure 3.13)**.

One drawback to the gooseneck engine was its size and weight. It was too heavy for quick maneuvering and too large for narrow streets and alleys. As a result, firefighters experimented with attaching a threaded hose line instead of a nozzle to the gooseneck. This major advancement, called leading hose, allowed the hose line to be taken close to (or even in) the fire building with the engine parked some distance away.

Piano-Type Engines

From the gooseneck design, the even more powerful piano-type engines were developed. The box-like body of these engines resembled a piano, resulting in the name. A major improvement on these engines was the permanent attachment of a suction hose placed in such a way that it was referred to as a squirrel tail **(Figure 3.14, p. 94)**.

Other Hand Engines

Inventors and manufacturers throughout North America were soon developing their own fire engine designs. One engine, known as the "coffee mill," featured a rotary type engine crank on each side of the machine. Another unique rotary design, the "cider mill" was operated like the capstan of a ship with men pushing a bar around the body of the engine **(Figure 3.15, p. 94)**.

In 1794, Pat Lyon produced the Philadelphia style engine. This machine was a powerful and reliable double-piston, double-deck, end stroke engine that many considered to be the best and most successful hand engine developed in North America. Meanwhile, in Boston, William Hunneman produced a small, compact, end stroke engine that was well suited for the narrow streets of Boston and smaller towns and villages **(Figure 3.16, p. 94)**.

Figure 3.14 The piano-style pumper was distinct because of the preconnected "squirrel-tail" suction hose.

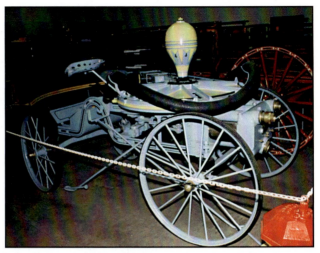

Figure 3.15 The "cider mill" was powered by firefighters walking around the engine.

Figure 3.16 The Hunneman hand tub was used extensively in New England because its compact size made it easy to maneuver.

Improvements in Hand Engines

Hand engines continued to become larger and more powerful as time went on. Designs improved and pumping capacity increased. The suction hose replaced the bucket brigade as the preferred form of water supply to the hand tubs. Newer model fire engines featured lengthened handles or brakes that ran the length of the machine, allowing for fifteen or more men to pump on each side.

The engines were normally operated at approximately sixty strokes per minute. Although, on occasion, speeds of 120 and even 170 strokes per minute were recorded. A stroke consisted of a full up-and-down motion of the brake. At normal pace, an efficient crew might work for about ten minutes before being relieved by a fresh team. This work was not only exhausting but dangerous as well. Many firefighters got torn fingers or broken arms as the crews attempted to change places while the engine was being operated at a fast pace.

Engines were sometimes classified by the size of their pump cylinder. The size of a cylinder might vary from 5 to 10 inches (125 to 250 mm) in diameter, with the stroke of the piston rods varying from 8 to 18 inches (200 to 450 mm). The length of the brakes might be anywhere from 16 to 25 feet (5 to 7.5 m). A first-class engine with a 9- or 10-inch (225 to 250 mm) cylinder would often require forty to sixty men to haul and operate.

94 Chapter 3 • Early Traditions and History

The Age of Steam

Early steam engines can be traced back to 1829 in London. The construction of workable steam engines in North America in 1840 marked the beginning of the end for the hand-operated fire engine. A group of insurance companies commissioned Paul Hodge of New York to build a steam fire engine in 1840. After a year of work, he introduced a self-propelled machine with a horizontal boiler that successfully pumped a stream of water over the cupola of city hall. Although the engine operated successfully, the volunteer firefighters wanted no part of it. They made every effort to discredit the machine and its performance. The sponsoring insurance companies were not fooled by the volunteers, but realized they relied upon these men and their companies for fire protection. Because they needed the services of the volunteer firefighters, the insurance companies relented and eventually sold the Hodge steam engine to power a box factory **(Figure 3.17)**.

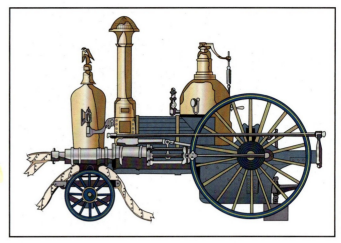

Figure 3.17 Paul Hodge built the first steamer in 1841. However, it failed its initial test.

Captain John Ericsson, a Swedish engineer and designer of the USS Monitor (the first ironclad steamship in the U.S. Navy), also designed a steam fire engine. Like the Hodge engine, Ericsson's steamer was never used due to the disapproval of the volunteers.

Officials at the time did not look at the rejection of these inventions as a serious problem, as the hand-operated engines with their large crews of firefighters were capable of handling the fires that occurred. City leaders began to re-think this strategy after an increasing number of wild brawls between rival volunteer companies became so intense that fires burned uncontrollably while men fought in the streets.

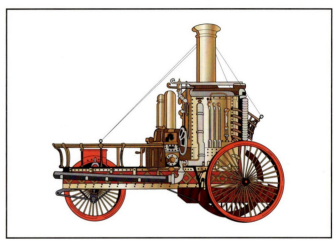

Figure 3.18 The rivalry between volunteer organizations led to the purchase of steam engines, such as the popular Latta.

Latta Steam Engine

Jacob Wykoff Piatt, a city councilman in Cincinnati, was appointed as head of a committee to implement the Latta steam engine for the fire department. Already a long-time opponent of the volunteer system, Piatt had been harassed continually by the volunteers for his outspoken views during council meetings. Police escorted him to and from council sessions, and he was once burned in effigy at his home by a mob of volunteer firefighters. His efforts in the council and appointment of a well-respected volunteer to operate the Latta engine helped usher in the era of paid firefighters in Cincinnati **(Figure 3.18)**.

After one such fight in Cincinnati, city leaders purchased a Latta steam engine. The volunteer firefighter's response was a physical attack on the machine. Ordinary citizens became outraged and rescued the steamer. The Latta engine remained on the job, and Cincinnati became the first all-steam fire department in North America. In 1853, Cincinnati also became the first fully paid fire department. After the acceptance of steam engines in Cincinnati, other cities rapidly followed suit.

Figure 3.19 The Dalmation has been associated with the fire service for many years and can still be found in some fire stations as mascots.

Some of these first steam pumpers were self-propelled, but they were slow and difficult to maneuver. Large groups of men were required to pull the heavy steamers that lacked self-propulsion. This practice was found to be inefficient, and teams of horses were settled on as the best method to pull the first steamers.

Along with the horses, dogs were introduced to the fire service. Canines served multiple purposes that complemented the use of horses. The dogs acted as a warning to firefighters in the event anyone attempted to steal the team of horses from the firehouse. Perhaps the best known role of dogs in the fire service was to clear the way in front of a responding steamer. The dogs would keep people and other dogs away from the street and rally the team of horses to run faster. Although all breeds of dogs were used for this type of service, none is more associated with fire fighting than the Dalmatian. This breed formed a special bond with the teams of horses and has an inherent love of running alongside carriages. Dalmations continue to serve as a mascot in some firehouses in the present day **(Figure 3.19)**.

In order to have the steam engine ready to pump water immediately upon arriving at a fire, the steamer's boiler was connected to a stationary boiler in the firehouse with hot water continuously circulating through both systems. Upon receipt of an alarm, the stoker would disconnect the stationary boiler and light the steamer's boiler by igniting a pack of bound matches on a piece of sandpaper mounted to its exterior. These matches touched off a bed of excelsior and small kindling wood in the firebox. When the engine arrived at the scene, a determination would be made if the pump was to be used. If required, the stoker could throw a shovel of cannel coal (a hot, quick burning fuel) on the fire to get a quick head of steam for pumping. Firefighters would then make the appropriate hook-ups and the nozzleman would call for water. To supply the hoselines, the engineer would open the throttle, allowing the pistons to move and the flywheel to turn. As the machinery gained speed, the firefighters at the nozzle prepared to receive water.

As this action was taking place, the horses were led away from the engine to a safe area where the shower of sparks from the steamers smokestack would not endanger them and the vigorous action of the machinery would not spook them.

While in operation, the engineer monitored the action of the steam engine and the stoker monitored the coal supply in the firebox. When the coal supply was low, the stoker gave a blast on the whistle to summon a coal wagon. The coal wagon driver would be standing by to respond to the whistle of each steamer as the fire fighters began to exhaust their fuel supply at major fires.

Chemical Engines and Ladder Trucks

As fire fighting technology improved, new chemical agents were used to supplement water. These agents were particularly useful when fighting smaller fires. As fire suppression capabilities improved, fire departments could undertake additional operations.

Ladder trucks performed many of these additional operations, such as:

- Raising ladders
- Search and rescue
- Ventilation
- Overhaul
- Operating specialized equipment

Chemical Engine

The first chemical apparatus came soon after the development of the steamer. Generally, a wagon pulled by two horses often carried up to four 50-gallon tanks of soda (sodium bicarbonate) and a carbonic acid preparation.

When the operator mixed the ingredients with a front-mounted crank, the carbonic acid caused a chemical reaction that produced carbon dioxide gas with sufficient pressure to expel the mixture. The streams were delivered through ¾-inch or 1-inch (20 or 25 mm) rubber hose and mainly used to extinguish incipient fires.

Sometimes, a chemical engine would respond behind a steamer in case its boiler sparks ignited a small fire. However, the chemical engine was designed to respond quickly in an attempt to extinguish a fire in its early stage before it grew too large. During large fires, the primary role of the chemical engine was to patrol for flying **brands** that may ignite secondary fires.

Brands — Large, burning embers that are lifted by a fire's thermal column and carried away with the wind.

Ladder Trucks

Ladder work and the demand for ladder companies did not begin to develop until the increase in multi-story buildings. These early ladder companies consisted of groups of men who responded to specified locations throughout the city to retrieve ladders, ropes, hooks, and other equipment stored for fire fighting purposes. Initially, men carried the ladders by hand to the fire. Later, a hand drawn hook and ladder truck that carried several different length ladders was developed. Its hooks were used to pull down sections of a building by "hooking" a ring attached to the gable end of the structure **(Figure 3.20)**. As ladder lengths increased, the maneuverability of these first ladder trucks became more difficult. To solve this problem, a tillering device on the rear wheels was added.

All types of ladders, including 65- or 75-foot (20 or 23 m) extension ladders, were typically raised by hand. These ladders were not only heavy and time-consuming to raise, but they were also hazardous because their bases were not secure. Daniel Hayes, a former New York City firefighter and master mechanic for the San Francisco Fire Department, devised a method for attaching a ladder to a truck bed and lifting it mechanically. By 1870, with the assistance of San Francisco firefighters, Hayes perfected the first aerial ladder truck. Known as the "Hayes Aerial," this apparatus initially featured

Figure 3.20 The pompier ladder was once used by firefighters to scale the outside of tall buildings. *Courtesy of Frank Carter.*

hand cranks to raise the ladder. Later, a set of springs was employed to raise the aerial from the stowed position. As technology progressed, compressed air and then hydraulic power was used to operate aerial devices. The ladder truck also evolved to carry several different lengths of ground ladders and a wide variety of other lifesaving equipment.

Pompier Ladder and Life Net

The pompier ladder and life net now rarely seen in the fire service were common pieces of equipment on ladder trucks into the 1970s. The pompier or scaling ladder is constructed of a single beam with rungs on either side and a large metal hook used to secure the ladder over window sills. Firefighters could advance from floor to floor on the face of a taller building using this ladder. The canvas life net consisted of a folding circular metal frame with springs connecting the canvas. This net was used to catch people jumping from buildings at the fourth floor or below.

Gasoline and Diesel Powered Equipment

With the automobile gaining in popularity at the turn of the 20th Century, it was a natural progression to apply this technology to fire apparatus. Firefighters had grown attached to their horse-drawn engines and initially were against a transition to gasoline-powered apparatus. In the firefighter's eyes, sputtering and foul-smelling gasoline engines were far less effective than a traditional, speedy team of horses **(Figure 3.21)**.

In time, the advantages of motor-driven apparatus were recognized, especially for long-distance runs that would tire a team of horses. However, early equipment failed to have a reliable shifting mechanism that would transfer

Figure 3.21 Horse drawn steamers were developed due to the heavy weight of the apparatus and a lack of manpower.

the engine's power from road gear to pump gear. Failures and frequent breakdowns led many fire chiefs to specify that steamers be pulled by gas-powered tractors. This change enabled the engine to travel to the fire efficiently and operate the pump using the proven technology of steam power.

Eventually, engineers worked out the problems associated with power transfer and reliable gasoline-driven apparatus replaced the aging steamers. By the 1920s, automotive and fire fighting technology had given rise to vehicles with more power and greater reliability.

Today, most modern apparatus feature the technology of diesel engines. Technology has allowed the size of a fire pump to be decreased while increasing its capacity. Most current engines carry their own hose, pump, ladders, and a wide variety of fire fighting equipment based on the needs of the jurisdiction they serve.

Improvements in Protective Clothing and Self-Contained Breathing Apparatus (SCBA)

Like apparatus design, technological advancements also improved firefighter's equipment. These improvements were especially dramatic in protecting clothing and **self-contained breathing apparatus (SCBA)**, as the improvements offered better protection from a variety of hazards.

PPE and SCBA Improvements
Many improvements in firefighter personal protective equipment (PPE) and SCBA have been the result of technology first developed for the space program. Other innovations in fire fighting equipment and tactics have been gained by the experiences of various branches of the military.

Protective Clothing
During the early 1800s, larger fire departments adopted standard uniforms to identify their members. The protective clothing available during this time period was made of wool or other natural fibers, making them heavy, especially when wet. This material also placed tremendous heat stress on the wearer during warm weather. Eventually, rubber coats became the norm because of their water repellent nature. These coats subjected firefighters to rapid fatigue and offered little protection from heat. Trends later reverted back to woven fibers, with coats being constructed of a heavy cotton duck canvas material. This material also offered little resistance to flames, leading to the development of manufactured fibers, such as Nomex and PBI Gold (polybenzimidazole), which provide increased protection.

Helmets made of leather were used for protection as early as 1740, with their design evolving to the now traditional shape over the next 60 years. By 1820, the helmet featured a front shield and often with the company number and rank designation as well as the familiar extended scoop on the back to carry water off and protect the wearer's neck from hot brands **(Figure 3.22)**.

Figure 3.22 An old style leather fire helmet.

SCBA Equipment

Recognition of the need for respiratory protection in dangerous environments can trace its roots back two thousand years when Pliny, a Roman writer, made reference to loose fitting animal bladders used to protect Roman miners against the inhalation of red oxides of lead. During the 1700s, the ancestors of modern respiratory protection devices were developed. Although these designs have changed dramatically, the performance of these devices is still based on two basic principles:

1. Purifying air by removing contaminants
2. Providing clean breathing air from an uncontaminated source

In 1814, the first air purifying filter (in a rigid container) was developed. In 1854, activated charcoals began to be used as a filtration medium. Several designs were developed and used in the United States and Europe in the latter part of the 1800s. These designs included an air purifying mask patented by Peter Ackerman of Bangor, Maine in 1872 and a facepiece designed by Captain E.M. Shaw of the London Fire Brigade coupled with a filter that Professor John Tyndall created. The Shaw design was the forerunner of the Type N Universal gas mask that many fire departments used into the 1960s. Firefighters also used masks manufactured by the Draeger company for miners in the late 1800s and early 1900s.

Many improvements in respiratory protection have been made since the end of World War II, mostly due to workplace safety regulations resulting from the Occupational Safety and Health Act of 1970. In order to comply with these requirements, manufacturers have developed systems to meet specific hazards. In the process, the equipment has become enhanced with better fitting face pieces featuring improved fields of vision and masks that interface with other protective equipment.

The first known self-contained breathing apparatus (air-supplied system) was developed in Germany in 1795. In England, mining engineer Henri Fayol developed what may have been the first pressure demand device. The unit had a bellows type air bag filled with clean air capped with a lead lid. The weight of the lid forced air into the user's mouthpiece for breathing and supplied air to keep a lamp burning in oxygen deficient atmospheres.

Self-Contained Breathing Apparatus (SCBA)

The contemporary SCBA used by the fire service is a respirator that supplies the user with a breathable atmosphere that is either carried in or generated by the apparatus independent of the ambient atmosphere. Respiratory protection, commonly known as an air pack, is worn in all atmospheres that are considered to be immediately dangerous to life and health (IDLH).

The United States had few dependable manufacturers of breathing devices before World War I. As late as 1910, many cities required firefighters to maintain beards at least six inches long. The firefighters would dip their beards in water and fold them into their mouths for use as a smoke filter for breathing.

The earliest known American-made SCBA was a device created by A. LeCour in 1863. This apparatus consisted of a leather bag with two tubes leading from the bag to the mouth. One tube connected to the top of the bag for inhaling air, and the other tube went into the bottom of the bag for exhaling carbon dioxide. During the same time period, smoke hoods containing compressed air or oxygen were also in use **(Figure 3.23)**.

Early Attitudes Regarding SCBA Use

Past generations of firefighters were reluctant to use SCBA due to peer pressure and a culture of bravado in which firefighters referred to themselves as "smoke-eaters" or "leather lungs." There has since been a major cultural change toward safety and health awareness regarding SCBA use and other aspects of fire service life. Increased education and knowledge that exposure to smoke can lead to cancer, heart disease, and other serious health conditions have helped to drive this change of attitude.

Supplied air equipment saw limited service in the United States beginning in the late 1800s. The Merriman's Smoke mask from 1892 featured an air hose located inside a water hose. The Nealy Smoke Mask operated using sponges that kept moist in a water-filled box on the user's chest.

The first SCBA to enjoy approval in the United States was the Gibbs closed circuit oxygen breathing apparatus produced by Mine Safety Appliance company in the early 1920s. This apparatus was the first to use a completely lung-governed principle of operation.

Figure 3.23 An early SCBA design.

The use of air supplied respiratory protection equipment in North America dates to the 1950s. During this time period, devices were expensive and generally seen as unnecessary. Departments using this equipment often assigned it to ladder trucks where the SCBAs were used for search and rescue operations.

The need for expanded use of SCBA became apparent with the recognition of smoke toxicity and the increase in hazardous materials in the environment. Widespread use of plastic and other synthetic materials in homes and businesses as well as the use of increasingly toxic substances in manufacturing led to the creation of more stringent government regulations. Occupational Safety and Health Administration (OSHA) regulations and National Fire Protection Association® (NFPA®) standards provide guidance on the use of SCBA.

With increased hazards and new government regulations, the fire service adopted more stringent procedures for the use of SCBA. Fire apparatus were outfitted with an SCBA for each firefighter riding the vehicle, and spare air cylinders were carried to augment the supply of safe breathing air. Most fire departments developed and implemented policies that required the use of SCBA in any environment that was immediately dangerous to life and health (IDLH). By the 1990s, firefighters in many jurisdictions were issued their own SCBA facepieces that were fit tested and worn only by the firefighter to which they were issued. In addition, annual medical evaluations and physical fitness programs were instituted to ensure firefighters were physiologically capable of operating in hazardous environments with SCBA.

Figure 3.24 A modern fire extinguisher. The pictures on the instruction plate identify the types of fire that it will extinguish.

The following technological advances have been incorporated into modern air supplied respiratory equipment:

- Use of heat and fire resistive material for harnesses and facepieces
- Increased air cylinder capacity
- Use of lighter weight materials
- Integrated personal alert safety system (PASS)
- Positive pressure regulators
- Rapid fill
- Development of buddy breathing techniques, called the Emergency Breathing Support System (EBSS)
- Integrated communications systems with breathing air apparatus

Fire Extinguishers

The first portable fire extinguishers may be considered the buckets of water homeowners were required to keep at the ready beside their doors. A true fire extinguisher was finally patented sometime in the late 1800s. Some historians credit Alanson Crane with creating the first fixed location extinguisher in 1863, while other historians state that it was not until 1872 that Thomas J. Martin patented the fire extinguisher. These early devices contained only water and were effective on burning water, paper, and cloth, materials now known as Class A fuels.

Later versions of the fire extinguisher featured the chemical carbon tetrachloride as an active ingredient **(Figure 3.24)**. These models have been declared obsolete for the following reasons:

- These extinguishers cannot be turned off once they are activated
- The agent they contain is more corrosive than water
- The extinguisher is potentially dangerous to the operator during use
- The tank may have corroded over time, resulting in a violent failure upon pressurization that could result in injury or death

Modern Fire Extinguishers

Depending on their intended use, modern fire extinguishers may contain the following:

- Water
- Carbon dioxide
- Dry chemical
- Dry powder
- Clean extinguishing agents

Fire extinguishers are rated according to their intended use and fire fighting capability on the five classes of fire (A, B, C, D, and K). Some modern fire extinguishers may be used effectively on multiple classes of fire. All fire extinguishers should be clearly labeled for their intended uses.

Alarm and Communications Systems

The earliest requirements of fire service communications were to warn people of a fire and summon them to assist with its extinguishment. As fire fighting practices and technology progressed, signaling devices as diverse as church bells and cannons were used to notify the public and fire fighting forces that a fire had broken out in the vicinity.

Town Crier

The town crier may have been the first organized method of notifying the public of emergencies. Originally tasked with walking the streets at dusk to light street lamps and then again at dawn to extinguish them, the town crier would also run up and down the streets to warn people if a fire broke out or if a crime was being committed.

The role of the town crier later expanded in many communities to include a number of individuals who were charged with monitoring conditions around a town on a 24-hour basis. Eventually, advances in alarm notification and watchmen or police patrols supplanted the need for a town crier.

Watchmen and Watchtowers

Evolving from the town crier, watchmen were often located in a "watchtower" in the center of town. Charged with the responsibility to warn townspeople of a fire or other emergency, the watchmen employed various signaling devices that produced a response from the public to begin fire fighting efforts or restore law and order. Many wildland fire protection agencies continue to employ the watchtower as a means to report the location of fires.

Telegraph Alarm Systems

Major innovations in emergency notification came during the mid-1800s with the introduction of the wired telegraph system. This system allowed the fire department to be notified immediately when an alarm occurred. Initially, fire departments operated a separate network of telegraph lines that connected several locations throughout a city where personnel were on duty to sound bells or whistles to announce an alarm.

In June 1845, Dr. William Francis Channing adopted Morse Code for use with his patented fire alarm system. Realizing the extraordinary potential for profit in this system, John Nelson Gamewell purchased the patent rights for the United States. By 1904, the Gamewell Alarm System enjoyed 95 percent of the country's fire alarm market.

Boston was the first city to adopt the fire alarm telegraph in 1852. Soon street boxes, accessible to the public for reporting fires, became a common fixture. Circuits were extended to firehouses, and fire departments established fire alarm offices for the receipt and transmittal of alarms. Initially, fire alarm boxes transmitted only the location of the district where the box was transmitted. Later, alarm boxes became capable of sending a specific number that corresponded to their exact location **(Figure 3.25)**.

Figure 3.25 A street type pull station box used to send telegraph alarms to fire department receiving panels

Many cities eliminated their telegraph box systems as telephones became prevalent in communities across the country. However, some cities, particularly those along the East Coast, continue to operate fire alarm telegraph systems for street and/or building fire alarm systems.

Telephone Systems

With the widespread availability of public telephones, citizens could speak to a telephone company operator who would transfer the call or relay information to the local fire department. When telephones spread to many private residences, automated dialing exchanges were created to allow calls to be placed directly between two parties without the assistance of an operator. Speaking directly to the fire department improved efficiency and allowed fire department personnel to question callers and obtain more detailed information. This technology aided in dispatching the appropriate resources to mitigate the incident.

An estimated seventy percent of 9-1-1 calls are placed by wireless phones. That percentage is growing each year, making these devices a vital tool with which to notify public safety agencies of emergencies.

The Federal Communications Commission (FCC) is in the process of implementing increasingly stringent requirements regarding 9-1-1 notification capabilities for wireless devices. In addition to requiring the transmission of all 9-1-1 calls regardless of service subscription, enhanced 9-1-1 rules require the wireless carrier to provide public safety telecommunicators with the phone number and location of the cell site transmitting the call. In order to provide more accurate location data, 9-1-1 also requires the wireless carrier to provide location information based on latitude and longitude that is accurate to within 165-1,000 ft. (50-300 meters) outdoors.

Universal Emergency Number

The first emergency system began operation on February 16, 1968, in Haleyville, Alabama. Since that time, the concept of a universal emergency number has spread worldwide. Technological advances have made an enhanced version possible, giving telecommunicators information about the caller. This feature allows dispatchers to send assistance even if the caller does not know the address or is unable to speak.

Radio Communications

Radio performs a dispatch and communication role in the modern fire service, allowing efficient transmittal of information to companies operating over a large area. In this way, resources may be deployed more effectively and monitored from a central location.

Law enforcement agencies were the first to use wireless radio. Initially, these radios were only receivers. After a transmission, the police officer needed to stop at a telephone to report back to the station. Later, the fire service followed by installing radio receivers in chief's cars and fireboats. Two-way radios were eventually installed in all fire apparatus, and over time, radio systems have changed and more frequencies have been added to serve the needs of the modern fire service.

Radio systems have also come into widespread use for dispatching fire apparatus, gradually phasing out telephone or telegraph systems. Tone activated receivers can alert fire stations, volunteers in their homes, or other staff personnel. Pagers tied to the radio system can be used to notify members away from quarters, their offices, or residences.

Pagers/Personal Alerting Systems

The earliest pagers used for public safety were tone-only pagers, which set off an alert upon receipt of a radio signal. The user then needed to call in by phone or radio to obtain information regarding the emergency. Later, tone and voice pagers were developed that provided an alert as well as a voice message containing response instructions. In addition, display pagers feature a visual message containing words and numbers as well as an audible tone and voice message. Some of these devices allow the user to acknowledge the message with a pre-programmed response.

Smartphones, in widespread use by civilians as well as firefighters, allow communication with any member not in the firehouse. Many fire departments subscribe to services that allow their members to receive information about alarms and other data via text message. Some systems also allow firefighters to indicate if they are responding to the incident or are en route to the firehouse.

Figure 3.26 A mobile data and communications system used by modern firefighters.

Mobile Data Communications Systems (MDCS)

Mobile data devices have the ability to send and receive data between a telecommunicator and remote user. The original purpose of these devices was to reduce radio traffic between the dispatcher and units operating in the field or local firehouses. These units may be fixed in a building, mounted in a piece of fire apparatus, or carried as a portable unit **(Figure 3.26)**. Modern data systems can relay dispatch information, confirm the status of fire companies, and provide GPS mapping data and full Internet access. Many fire departments link their pre-plan and building survey information to these networks for quick reference in the field during emergencies.

Impact of Historic Fires on Firefighter and Public Safety in North America

Over time, numerous fires have occurred that have had a major impact on building practices, fire safety, and code enforcement. Due to the magnitude of a fire, the number of people killed, or large monetary loss, many safeguards have been put in place to avoid similar circumstances in the future.

Significant Historic Fires in North America

The following fires have resulted in changes to fire and life safety regulations:

- Iroquois Theatre Fire, Chicago, Illinois (1903)
- The Great Fire of 1904, Toronto, Ontario, Canada (1904)
- Triangle Shirtwaist Fire, New York City, New York (1911)
- Cocoanut Grove Nightclub Fire, Boston, Massachusetts (1942)
- Ringling Brothers Barnum and Bailey Circus Fire, Hartford, Connecticut (1944)

- Our Lady of the Angels School Fire, Chicago, Illinois (1958)
- Hartford Hospital Fire, Hartford, Connecticut (1961)
- Beverly Hills Supper Club Fire, Southgate, Kentucky (1977)
- MGM Grand Hotel Fire, Las Vegas, Nevada (1980)
- Station Nightclub Fire, West Warwick, Rhode Island (2003)

Iroquois Theatre Fire, Chicago, Illinois (1903)

A fire in a theatre that was designated to be "fireproof" claimed the lives of 602 people and injured 250 more. An electric spotlight ignited combustible scenery, curtains, and interior finishes. Most of the people killed were suffocated or trampled in the panic to escape. In response to the fire, laws were enacted requiring panic hardware on exit doors, which were also required to swing outward. Additional laws restricted the number of theatre seats in an aisle and the amount of combustible material allowed in places of assembly.

The Great Fire of 1904, Toronto, Ontario, Canada (1904)

A fire of undetermined origin destroyed 104 buildings in downtown Toronto. The fire spread rapidly, outstripping the ability of the fire department to contain it. Unprotected stairways and elevator shafts contributed to the rapid spread within buildings. The fire led to major changes in Toronto's building code, such as:

- Requirements for fire doors, walls, and sprinkler systems
- Larger water mains
- Additional fire alarm boxes
- Removal of overhead obstructions along streets

   NOTE: The size of the fire department was also increased.

Triangle Shirtwaist Fire, New York City, New York (1911)

A fire in a 10-story building used for manufacturing clothing claimed the lives of 146 employees. The employees, mostly young women, leapt to their deaths as the fire spread through the combustible contents of a building that was thought to be fireproof. The fire resulted in the NFPA® 101 Life Safety Code® (originally titled Building Exits Code), which established requirements for means of egress.

Cocoanut Grove Nightclub Fire, Boston, Massachusetts (1942)

A fire in a single-story nightclub killed 492 people. The high death toll was caused by overcrowding, the use of combustible interior finishes and decorations, and the lack of emergency lighting. This tragedy led to stricter fire and life safety requirements for assembly-type occupancies.

Ringling Brothers Barnum and Bailey Circus Fire, Hartford, Connecticut (1944)

Fire in a circus tent that claimed the lives of 168 people, many of them children. Paraffin (wax) and gasoline had been used to make the tent waterproof, and exits were inadequate or blocked. This incident resulted in the development of life safety standards to regulate the manufacture of tents for public occupancy.

Our Lady of the Angels School Fire, Chicago, Illinois (1958)
This fire killed 92 children and three teachers. Unprotected stairwells, a combustible interior finish, and the lack of a fire alarm system were blamed for the loss of life. The fire brought attention to the need for improvements in the design of school buildings, the requirement for fire detection and alarm systems, and the need for enclosed stairwells. New laws also required schools to conduct fire evacuation drills throughout the academic year.

Hartford Hospital Fire, Hartford, Connecticut (1961)
A smoldering cigarette discarded in a trash chute caused a fire that spread up the chute from the basement to the ninth floor, killing 16 people. This fire led to major changes in hospital fire codes around the United States. The changes included additional requirements for automatic sprinklers and non-combustible interior finishes.

Beverly Hills Supper Club Fire, Southgate, Kentucky (1977)
This fire killed 162 people and injured over 200, including several firefighters. Investigators concluded that the fire was started by an electrical short and may have burned unnoticed inside a heavily plastered wall in one of the cabaret rooms for more than an hour. The fire spread due to the following:

- Overcrowding
- Inadequate fire exits
- Faulty wiring
- Lack of firewalls
- Poor construction
- Numerous safety code violations
- Lack of supervision by regulatory agencies

As a result, many fire safety building requirements were added, including the following:

- Improved exiting systems
- Safer interior finishes
- Emergency planning
- Installation of fire alarms and automatic fire suppression systems

MGM Grand Hotel Fire, Las Vegas, Nevada (1980)
This fire in a 26-story, unsprinklered hotel/casino killed 85 people and injured 679. Smoke spread into the tower via stairwells, elevator shafts, and ventilation and pipe shafts. Smoke inhalation killed the majority of victims, mainly on upper floors. Local laws were enacted that required all medium- and high-rise buildings to install sprinkler systems.

Station Nightclub Fire, West Warwick, Rhode Island (2003)
One hundred patrons died and 256 were injured when stage pyrotechnics ignited a fire in the crowded nightclub. The lack of a sprinkler system allowed fire to spread rapidly through combustible interior finishes. The primary cause of fatalities was because occupants attempted to exit through the main

entrance rather than using nearby emergency exits. A bottleneck at the main exit caused many people to be trampled to death and many more individuals to die of smoke inhalation. If they had used other exits, patrons would likely have been able to escape more quickly. Fewer deaths from smoke inhalation and none from trampling would likely have occurred.

One result from the fire was a greater awareness of the need for occupants of any type of assembly occupancy to be aware of and use the closest exit. The NFPA® also made changes to sprinkler and crowd management recommendations for nightclubs and other assembly occupancies.

Impact of Historic Fires on Firefighter Safety

The fire service has also learned many difficult lessons through scores of line of duty deaths (LODD) and serious injuries over the years **(Figure 3.27)**. Many changes in equipment and procedures have resulted from fires where members have been killed or injured in the line of duty. Some fires that have had a particular impact on firefighter safety include the following:

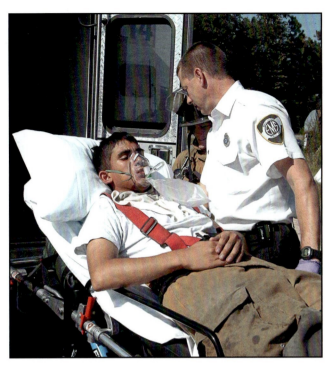

Figure 3.27 Firefighters are susceptible to injuries that range from minor to life-threatening over the course of their careers.

- Mann Gulch Wildfire, Montana (1949) - thirteen firefighter fatalities
- Boiling Liquid Expanding Vapor Explosion (BLEVE) during propane transfer operation, Kingman, AZ (1973) - eleven firefighter fatalities
- One Meridian Plaza, high rise fire, Philadelphia, PA (1991) – three firefighter fatalities
- Food Plant fire, Seattle, WA (1995) - four firefighter fatalities
- Cold Storage warehouse fire, Worcester, MA (1999) - six firefighter fatalities
- Southwest Supermarket fire, Phoenix, AZ (2001) – one firefighter fatality
- World Trade Center Attack, New York City (2001) - 343 firefighter and paramedic fatalities
- Super Sofa Store fire, Charleston, SC (2007) - nine firefighter fatalities
- West Fertilizer Company, West, TX (2013) – ten firefighter fatalities

Chapter Summary

Throughout history, mankind has been plagued by fires that have destroyed property and killed or injured countless numbers of people. Typically, after experiencing a devastating loss, members of the community will enact changes to prevent such a disaster from reoccurring. From ancient Rome with the Corps of Vigiles to modern day requirements for sprinklers and fire alarm systems, people have sought protection from the ravages of fire.

Modern firefighters receive much more extensive training and operate equipment far superior to that available at the dawn of the American fire service. However, certain aspects of fire fighting remain unchanged since colonial times. The dedication and teamwork necessary to fight fires and the pride for the traditions and accomplishments of the fire service are common traits firefighters have displayed since the first fire company was organized.

Review Questions

1. What were some key developments in the early history of fire services?
2. How did settlements in early America address fire protection efforts?
3. Why was there competition among early fire service volunteers?
4. What was the first city in North America to establish a fire department?
5. What were the first fire insurance companies in North America?
6. What are some key developments in fire fighting?
7. What are three significant fires in North America and how did they change fire and life safety regulations?
8. What are two fires that have had a significant impact on firefighter safety?

Roles of Public and Private Support Organizations

Courtesy of Ron Jeffers.

Chapter Contents

Case History 113
Local Agencies and Organizations 114
    Local Government 114
    Local Law Enforcement 114
    Building Department (Code Enforcement) 115
    Water Department/Water Authority 115
    Zoning/Planning Commission 115
    Department of Public Works 116
    Utilities 116
    Judicial System 117
    Office of Emergency Management 117
    Local Health Department 118
    Nongovernmental Organizations 118

State/Provincial Agencies and Organizations 118
    State Fire Marshal or Commissioner 119
    State/Provincial Fire Training 119
    Fire Commission 119
    State Law Enforcement 119
    State Highway Department/Department of Transportation/Turnpike Commission 120
    State Environmental Protection Agency 120
    State Occupational Safety and Health Administration 120
    State Health Department 121
    State Forestry/Department of Natural Resources 121
    Office of Emergency Management 121
    Special Task Forces 121
    State Firefighter Association 121
    State Fire Chiefs Association 121
    State Fire Marshals Association 122
    Other State Agencies 122

Federal Organizations in the United States and Canada 122
    Army Corps of Engineers 123
    Bureau of Alcohol, Tobacco, and Firearms (ATF) 123
    Chemical Safety Hazard Investigation Board (CSB) 123
    Consumer Product Safety Commission (CPSC) 123
    Emergency Management Institute (EMI) 123
    Environmental Protection Agency (EPA) 124
    Federal Aviation Administration (FAA) 124
    Federal Emergency Management Agency (FEMA) 124
    National Fire Academy (NFA) 125
    National Guard/Civil Support Team 125
    National Highway Traffic Safety Administration (NHTSA) 126
    National Institute for Occupational Safety and Health (NIOSH) 126
    National Institute of Science and Technology (NIST) 126
    National Transportation Safety Board (NTSB) 127
    National Weather Service (NOAA) 127
    Nuclear Regulatory Commission (NRC) 127
    Occupational Safety and Health Administration (OSHA) 127
    Public Safety Canada (PS) 127
    Transport Canada 128
    United States Coast Guard (USCG) 128
    United States Department of Energy (DOE) 128
    United States Department of Homeland Security (DHS) 128
    United States Department of Interior (DOI) 129
    United States Department of Labor (DOL) 129

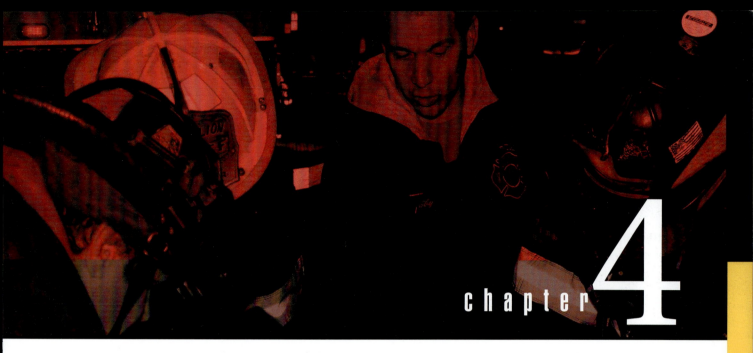

chapter 4

United States Department of Transportation
(US DOT) 129
United States Fire Administration (USFA) 129
United States Forest Service (USFS) 130

North American Trade, Professional, and Membership Fire and Emergency Services Organizations 130
American National Standards Institute (ANSI) 131
Canadian Association of Fire Chiefs (CAFC) 131
The Canadian Centre for Emergency
Preparedness (CCEP) 131
Canadian Fallen Firefighters Foundation (CFFF) 131
Canadian Fire Alarm Association (CFAA) 131
Canadian Fire Safety Association (CFSA) 131
Canadian Standards Association (CSA) 131
The Canadian Volunteer Fire Services
Association (CVFSA) 132
Council of Canadian Fire Marshals and
Fire Commissioners 132
FM Global Research (FM) 132
Fire Prevention Canada 132
Insurance Services Office Inc. (ISO) 132
International Association of Arson
Investigators (IAAI) 133
International Association of Black Professional
Fire Fighters (IABPFF) 133
International Association of Fire Chiefs (IAFC) 133
International Association of Fire Fighters (IAFF) 134
International Association of Women in Fire and Emergency
Services (iWOMEN) 134
International City/County Management Association
(ICMA) 134
International Code Council (ICC) 134

International Fire Marshal's Association (IFMA) 135
International Fire Service Training Association
(IFSTA) 135
International Municipal Signal Association
(IMSA) 135
International Society of Fire Service Instructors
(ISFSI) 136
National Association of Hispanic Firefighters
(NAHF) 136
National Association of State Fire Marshals 136
National Fallen Firefighters Foundation (NFFF) 136
National Fire Protection Association® (NFPA®) 137
National Interagency Fire Center (NIFC) 137
National Propane Gas Association (NPGA) 138
National Volunteer Fire Council (NVFC) 138
National Wildfire Coordinating Group (NWCG) 138
North American Fire Training Directors (NAFTD) 138
Society of Fire Protection Engineers (SFPE) 138
Underwriters Laboratories (UL) 138
Underwriters Laboratories of Canada (ULC) 139

Other Related Organizations 139
American Association of Railroads Bureau of Explosives
(BOE) 139
American Trucking Association (ATA) 139
National Tank Truck Carriers (NTCC) 139

Chapter Summary 139
Review Questions 140

Roles of Public and Private Support Organizations

Key Terms

Building Department 115
Fire Protection Engineers 138
Health Department 118
Judicial System 117
Utilities 116
Water Authority 115
Zoning Commission 115

FESHE Outcomes

This chapter provides information that addresses the outcomes for the Fire and Emergency Services Higher Education (FESHE) *Principles of Emergency Services* course.

4. List and describe the major organizations that provide emergency response service and illustrate how they interrelate.

6. Define the role of national, state, and local support organizations in fire and emergency services.

10. Identify the primary responsibilities of fire prevention personnel including code enforcement, public information, and public and private protection systems.

Learning Objectives

After reading this chapter, students will be able to:

1. Describe the functions of local agencies and organizations as they relate to the fire service. (Outcomes 4, 6, 10)

2. Identify state/provincial agencies and organizations that support the fire service. (Outcomes 4, 6)

3. Recognize the various federal organizations in the United States and Canada that support the fire service. (Outcomes 4, 6)

4. Identify North American trade, professional, and membership organizations. (Outcomes 4, 6)

5. Describe organizations that have common interests and interaction with the fire service. (Outcomes 4, 6)

Chapter 4
Roles of Public and Private Support Organizations

Courtesy of NFFF/USFA.

Case History

Cole was interested in becoming a firefighter. To learn more about the fire service, he decided to attend his local fire station's open house. While there, he met a firefighter, Emily, and asked her a few questions about how his local fire station works. Emily mentioned that this local fire station works with local organizations, such as law enforcement agencies, the city council, utilities, and businesses, to help her and her coworkers understand the community's needs. This knowledge, in turn, shapes what kinds of training her station emphasizes. On the state level, these firefighters work with state emergency management agencies, state police, and the fire marshal's office to understand safety issues that affect the entire state.

Cole learned that many different local, state, and federal organizations help with firefighter training, funding, and guidelines. Federal agencies also play a role in assisting local fire stations. The Department of Homeland Security (DHS) has several organizations that help firefighters, such as the Federal Emergency Management Agency, which provides fire fighting grants for training and equipment. Cole also learned that DHS has other organizations, such as the United States Fire Administration and the National Fire Academy, that provide training to firefighters from throughout the country.

Cole was introduced to the topic of nongovernmental organizations, and Emily discussed how her fire department works with one of these organizations. The National Fire Protection Association® (NFPA®) is a nonprofit organization that develops standards related to topics such as fire fighting training. State and local fire commissions then adopt these standards to use as local requirements for firefighters. Her station also relies on safety information from other nongovernmental organizations, such as the Underwriters Laboratories, Inc., which test different types of fire fighting equipment. This safety information helps firefighters understand hazards that they may encounter when responding to a fire.

Figure 4.1 Both city and county governments may have authority over fire and emergency services organizations.

Figure 4.2 Law enforcement officers and firefighters coordinate efforts at emergency scenes. *Courtesy of Ron Jeffers.*

Firefighters interact with a large number of government agencies and private organizations that have a function or provide a service in conjunction with the fire department. The following chapter examines the duties of numerous local, state, and federal agencies, as well as some prominent trade and professional organizations.

Local Agencies and Organizations

Firefighters interact with local outside agencies on a regular basis. Local government agencies as well as nongovernment organizations are often those most familiar with firefighters and will have day-to-day interactions with them. The following section provides an overview of the functions of some local agencies and organizations.

Local Government

Local government at the village, town, borough, township, city, county, or district level is comprised of elected officials responsible for enacting legislation to govern that jurisdiction **(Figure 4.1)**. Depending on its structure, this governing body may control the funding and/or operation of the fire department. Fire departments operating in this manner are usually funded through the collection of taxes or fees for service within the jurisdiction.

Local Law Enforcement

Law enforcement officers and firefighters have roles in public safety and work together at many types of incidents on a regular basis. The fire department relies on police officers to perform traffic control duties as well as scene safety and security at incidents ranging from fires to medical emergencies and auto accidents **(Figure 4.2)**. In some jurisdictions, the police provide the investigation, enforcement, and prosecution duties after fire cause has been determined. The two organizations should develop and maintain a high level of cooperation in order to operate most efficiently.

114 Chapter 4 • Roles of Public and Private Support Organizations

Figure 4.3 A plans examiner reviewing construction documents to ensure compliance with local building codes.

Building Department (Code Enforcement)

Local building departments enforce codes and regulations adopted by governing bodies in their jurisdictions. The **building department** is generally responsible for:

- Reviewing and approving all new construction and alterations to existing structures
- Conducting plans reviews
- Issuing permits related to buildings and their use
- Performing field inspections to verify that approved plans are followed in the construction and alteration process

A close working relationship between building officials and the fire department may improve safety in buildings and reduce damage from fire. In some jurisdictions, fire inspection and code enforcement responsibility may be assigned to the building department **(Figure 4.3)**.

Building Department — Local governmental agency responsible for enforcing various codes and regulations.

Water Department/Water Authority

Agencies that provide water for fire protection in a community may be established under a wide variety of organizational names. However, the efficiency in which the proper volume and pressure of water is made available for fire fighting may affect the overall success of any fire attack. The fire department and **water authority** should work together to determine and provide water main sizes as well as hydrant and pumping station locations. The water authority should also promptly notify the fire department about any system closures for repair or maintenance. This agency may also work to increase waterflow and pressure in certain areas of service during major fires.

Water Authority — Municipal authority responsible for the water supply system.

Zoning/Planning Commission

The planning and zoning functions may be placed under the jurisdiction of a single organization. The purpose of zoning within a community is to determine the layout regarding residences, commercial space, and industrial property. Once areas are established for each function, the planning commission may determine how to develop each function to best serve the community as a whole. These functions are important to the fire department because various types of development require specific fire protection coverage. Fire departments in each jurisdiction must locate apparatus and fire houses based on the needs of each district.

Zoning Commission — Local agency responsible for managing land use by dividing the jurisdiction into zones, in which only certain uses, such as residential, commercial, or manufacturing, are allowed.

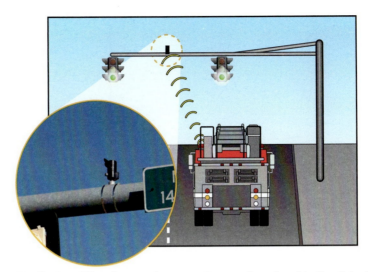

Figure 4.4 The Opticom™ emitter on the apparatus sends a signal to the detector on the traffic light support, which converts the optical signal to an electronic impulse. The system processes the signal then manipulates the controller to provide a green signal for the emergency vehicle(s) and red signals in all other directions.

Department of Public Works

The local public works department oversees the building and maintenance of public streets and throughways in a specified jurisdiction. Fire departments often seek to provide input when roads, bridges, and traffic flow patterns are designed so that they are adequate for large apparatus. Therefore, it is important for fire departments to maintain a good working relationship with the public works department. Street closures during repair projects will allow firefighters time to plan alternate routes in case of emergencies in the affected areas. Cooperation with the local public works department is also important, as it may be a vital source for heavy equipment during hazardous materials, technical rescue, snow removal from roads and hydrants, or other incidents.

In some jurisdictions, the public works department or similar agency may be responsible for traffic control devices. Local fire departments may work together with this agency to operate traffic signal preemption equipment that allows fire apparatus and other official vehicles to remotely control traffic lights in their travel path **(Figure 4.4)**.

Utilities

In order to provide a safe incident scene and reduce property loss, numerous public **utilities** must often be controlled during fire department operations. Hazards from flammable gases, electricity, and water are among the major utility issues that firefighters regularly face. In some instances, fire department personnel may mitigate these hazards by employing shutoff valves or switches that are easily accessible. However, many times the equipment and expertise of local utility workers are necessary to safely control hazards encountered during incidents. Local fire departments should build a relationship with utility providers and notify them promptly during operations involving their service and equipment.

Utilities — Services such as gas, electricity, and water that are provided to the public.

> **CAUTION**
> **Telephone and Cable Utility Lines**
> Cable and telephone lines are utility wires that generally are not thought of as posing a major hazard. However, these lines carry current and may pose a threat in certain hazardous environments. In addition, these wires are often run in close proximity to electrical lines, and under certain conditions they may become charged.

Judicial System

Firefighters may have occasion to have contact with the **judicial system** in a number of instances during the performance of duty. Members may be called upon to:

- Testify in court during arson or insurance claim cases.
- Recount what they saw and what actions they took during an incident.
- Seek a court order to force the correction of fire code violations. This action may be necessary when property owners fail to comply with the code after being notified of a violation.

Judicial System — System of courts set up to interpret and administer laws and regulations.

Office of Emergency Management

The Office of Emergency Management, or a similarly titled agency, handles emergency management of large natural or human-caused disasters in the local community. This agency is responsible for preplanning and response with all relevant local agencies and those at the state, county, and federal level. The fire department often assigns personnel to coordinate efforts through the Office of Emergency Management when planning for and during major incidents. Jurisdictions may employ an Emergency Management Director or task the fire or police chief with that responsibility **(Figure 4.5)**.

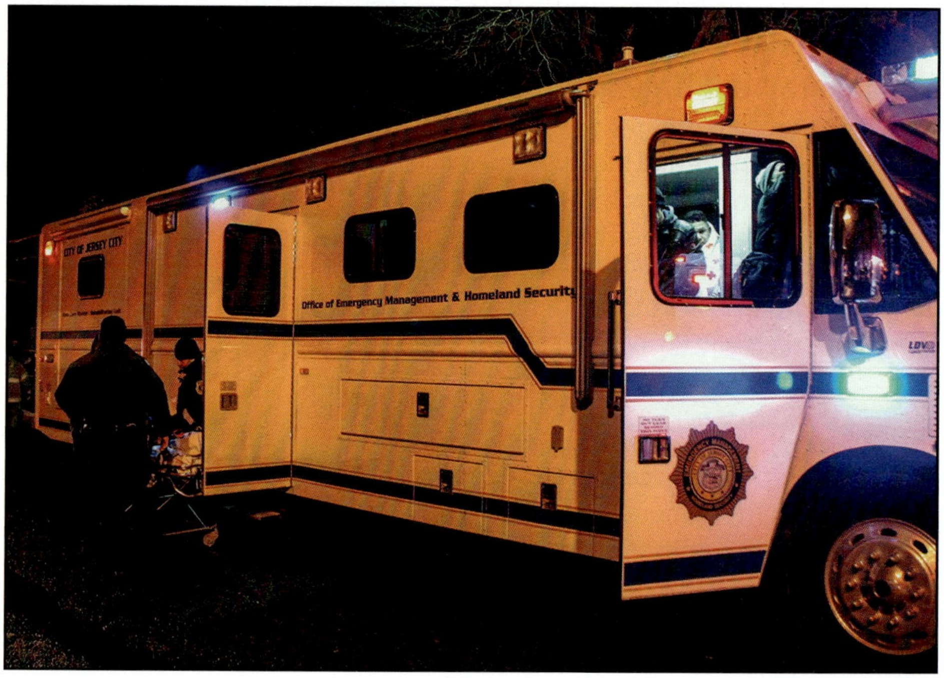

Figure 4.5 Emergency management personnel working with local firefighters during an incident. *Courtesy of Ron Jeffers*.

CERT (Community Emergency Response Team)

Citizens can be trained to be better prepared to respond to emergency situations in their communities through local Community Emergency Response Teams (CERT). Each CERT member completes 20 hours of training on disaster preparedness, basic disaster medical operations, fire safety, light search and rescue, and other essential topics.

CERT training also includes a disaster simulation in which participants practice skills they learned throughout the course. In the event of an emergency, CERT members can provide immediate assistance to victims, assist in organizing spontaneous volunteers at a disaster site, and provide critical support to first responders.

Figure 4.6 A local health department may help manage public health related emergencies.

Local Health Department

Health Department — Agency that focuses on public health issues in a given region.

Some jurisdictions operate a **health department** at the local level to inspect and enforce codes and standards at hospitals, schools, nursing homes, and other venues. This agency may be charged with determining if a venue may remain open after an incident or inspection. The local health department may also plan for and help manage any health-related emergencies, such as epidemics or hazardous material incidents **(Figure 4.6)**.

Nongovernment Organizations

Nonprofit and civic groups in the community are a valuable resource that may be used to spread fire and life safety messages. These groups contain interested citizens who may be instrumental in helping the fire department gain support in the community. Nonprofit and civic organizations may assist with community service projects, while other civic groups may provide a forum to speak to members and garner community support. The American Red Cross and the Salvation Army provide disaster relief services and support during emergencies.

State/ Provincial Agencies and Organizations

There are numerous organizations and agencies established to work in conjunction with firefighters on various issues of concern. Some of these groups are independent, while others operate under state or provincial authority. The purpose or responsibilities of similarly titled agencies may vary by state or province. A number of the more common organizations are briefly described in the following sections.

State Fire Marshal or Commissioner

The state fire marshal or commissioner is usually the principal authority on matters of fire protection and prevention in a state or province. A state or provincial legislative body usually delegates power to the fire marshal. These powers vary by jurisdiction, as do the organization of the agency. The fire marshal is an independent office in some states, while in others, it may be a function of the state police or Department of Public Safety. Typical responsibilities of the state fire marshal's office include:

- Reviewing and approving construction plans for fire safety
- Conducting fire safety and arson prevention activities
- Investigating and determining the origin and cause of fires
- Regulating the storage and use of combustibles, hazardous materials, and explosives
- Providing fire code requirements
- Developing and delivering fire service training programs (in some states)

State fire marshals may also inspect hospitals, day care facilities, nursing homes, and prisons, as well as state colleges and other facilities. Additionally, in some states, fire marshals may perform background checks of firefighter candidates, inspect carnival rides, and collect data from incident reports of fire departments throughout the state.

State/Provincial Fire Training

Most states and provinces have a firefighter training agency. Training organizations may have some full-time administrators and staff members as well as a team of instructors to accomplish training requirements. State level agencies may be associated with the state fire marshal's office or state colleges and vocational training centers. In some jurisdictions, state agencies may provide technical assistance to fire departments as well as training programs **(Figure 4.7)**.

Fire Commission

Some states have appointed fire commissions to establish professional qualification standards for the various positions within a fire department. The state funds these commissions to assist in training, testing, and administering the state fire service certification program. Most commissions have a director who coordinates with local, county, or state agencies that deliver the training.

State Law Enforcement

Many state police or patrol agencies work with fire department arson squads to investigate cases of suspected arson, especially when the numerous cases in several local jurisdictions appear to be linked. In some states,

Figure 4.7 Firefighters training on roof ventilation.

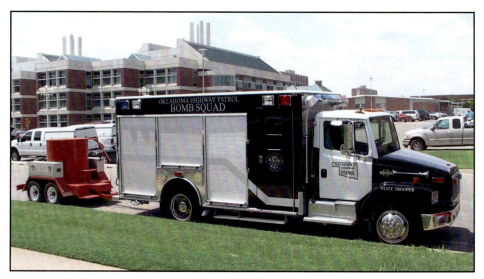

Figure 4.8 State law enforcement agencies may provide specialized personnel and equipment, such as a bomb squad.

the state fire marshal's office is a division of the state law enforcement. The fire marshal's office may also provide assistance with forensic lab services, hazardous materials, explosives, and incidents involving state property or highways **(Figure 4.8)**.

State Highway Department/ Department of Transportation/ Turnpike Commission

State departments responsible for roads, highways, bridges, and turnpikes should coordinate with local fire departments so that efficient access to all areas of a jurisdiction may be maintained. These agencies should notify fire departments whenever construction or repair projects may impede the travel of fire apparatus. State highway departments may also provide an excellent resource for heavy construction equipment, sand, or other materials that may be needed during hazardous material or other incidents on or near highways. Additionally, many state transportation agencies control highway message boards that can provide instructions or advisories during emergencies.

State Environmental Protection Agency

State agencies that monitor and protect the environment may become involved with the fire department during incidents that may involve hazardous materials contamination of the land, air, or water. This state agency may have an individual or team respond to monitor conditions and provide expertise to the fire department on how to lessen environmental impact.

State Occupational Safety and Health Administration

Based on legislative initiatives, a state occupational safety and health agency may produce and enforce programs to protect the safety and health of firefighters. The federal government requires states that manage their own programs to have regulations that are at least equal to those mandated by the federal Occupational Safety and Health Administration (OSHA) or the Environmental Protection Agency (EPA). Some state OSHA agencies may investigate firefighter injuries and deaths.

State Health Department

The state health department is responsible for coordinating the state response to numerous health concerns and public health emergencies, such as epidemics or other natural or human-caused disasters. This agency may also interact with the fire department after an incident, such as a fire involving establishments that provide food or lodging. The local health department generally regulates the certifications and licenses of the facility to ensure that it is following state guidelines.

State Forestry/Department of Natural Resources

State forestry departments generally work with local fire departments to prevent, contain, and extinguish wildland fires. These agencies have the expertise and often the specialized equipment necessary to fight large wildfires. In some states, forestry departments may be responsible for fighting wildland fires outside of city limits.

Office of Emergency Management

At the state level, the Office of Emergency Management generally interacts with the fire department in local emergency management offices to provide additional resources during natural or human-caused disasters. Resources can include personnel, vehicles, and communication equipment.

Special Task Forces

A special task force is group of people who have a specific skill set or equipment necessary to meet a specific challenge that are brought together to address a specialized need. Examples include:

- Members of law enforcement agencies may work with fire marshals to form an arson task force.
- A task force may be assigned to incidents involving the transportation of hazardous materials through multiple jurisdictions.
- Firefighters may be assigned to respond to urban search and rescue incidents.

State Firefighters Association

Many states feature a firefighters' group or association that lobby politicians, develop group insurance plans, promote fire and life safety, sponsor training, and advocate for causes that will help firefighters. These associations often hold regular meetings and conferences that invite guest speakers of interest to firefighters.

State Fire Chiefs Association

Numerous states also have an association of chief officers that is established to serve the interests of its members in the areas of political lobbying for fire protection causes, such as training, fire code issues, and budgetary requirements. These groups also hold meetings and various conferences that may provide opportunities for education and other professional development activities.

State Fire Marshals Association

As with other ranks and specialties in the fire service, fire marshals in many states have formed groups to promote the interests of its members. These organizations often sponsor opportunities for professional development, offer a forum for networking, and lobby for causes important to the membership.

Other State Agencies

Each state has public agencies that interact with the fire service to varying degrees. Firefighters should be aware of their state agencies that fire personnel often encounter and are the most helpful. Additionally, numerous membership organizations are available to firefighters of all ranks that provide education, training, and support.

Federal Organizations in the United States and Canada

Numerous federal agencies operate to safeguard public and environmental safety. Other agencies have a mission that directly relates to the fire service. The agencies listed in the following section have some interest or interaction with the fire service. While this list focuses primarily on the United States government, Canada often operates agencies that are similar in scope and purpose.

Congressional Fire Services Institute (CFSI)

In 1989, the Congressional Fire Services Institute (CFSI) was formed to educate members of Congress about fire and life safety issues. This institute enables the Congressional Fire Service Caucus to expand its membership within Congress and reach out to the fire service community. The CFSI tracks legislation, monitors hearings, develops policy, and identifies consensus priorities within the fire service. It is then able to report these findings to the Caucus and the general public.

- Army Corps of Engineers
- Bureau of Alcohol, Tobacco, Firearms, and Explosives (BATF)
- Chemical Safety and Hazard Investigation Board (CSB)
- Consumer Product Safety Commission (CPSC)
- Emergency Management Institute (EMI)
- Environmental Protection Agency (EPA)
- Federal Aviation Administration (FAA)
- Federal Emergency Management Agency (FEMA)
- National Fire Academy (NFA)
- National Guard/Civil Support Team
- National Highway Traffic Safety Administration (NHTSA)
- National Institute for Occupational Safety and Health (NIOSH)
- National Institute of Science and Technology (NIST)

- National Transportation Safety Board (NTSB)
- National Weather Service (NOAA)
- Nuclear Regulatory Commission (NRC)
- Occupational Safety and Health Administration (OSHA)
- Public Safety Canada
- Transport Canada
- United States Coast Guard (USCG)
- United States Department of Energy (DOE)
- United States Department of Homeland Security (DHS)
- United States Department of Interior (DOI)
- United States Department of Labor (DOL)
- United States Department of Transportation (USDOT)
- United States Fire Administration (USFA)
- United States Forest Service (USFS)

Army Corps of Engineers

As necessary, the Army Corps of Engineers may be deployed during natural or human-caused incidents to aid civilian agencies with expertise, heavy equipment, and urban search and rescue capabilities. The Corps supports immediate response priorities, provides critical commodities and emergency power, and works to restore critical infrastructure.

Bureau of Alcohol, Tobacco, Firearms, and Explosives (BATFE)

As a division of the United States Department of Justice, the BATFE assists local agencies in major arson investigations and regulates storage and transport of explosives. BATFE has an arson and explosives program as well as a fire research laboratory and deploys a response team for large-scale or specially selected fire investigations.

Chemical Safety Hazard Investigation Board (CSB)

This independent federal agency is charged with investigating industrial chemical incidents, including fires and explosions. The Clean Air Act Amendments of 1990 established this board, which became operational in 1998.

Consumer Product Safety Commission (CPSC)

The CPSC is an independent agency responsible for the recall of unsafe products and also conducts research on the injury and loss patterns of consumer goods. In addition, the commission regulates products such as cigarette lighters and enforces flammability standards for clothing, rugs, and furniture.

Emergency Management Institute (EMI)

The Emergency Management Institute (EMI) provides training to public sector managers to prepare for, mitigate, respond to, and recover from many types of emergencies. The EMI is part of the Federal Emergency Management Agency within the U.S. Department of Homeland Security.

Figure 4.9 Federal Emergency Management Agency (FEMA) personnel respond to large scale disaster incidents. *Courtesy of FEMA News Photos, photo by Dianna Gee.*

Environmental Protection Agency (EPA)

The EPA is an independent federal agency tasked with protecting human health and the environment. The EPA provides an on-scene coordinator at incidents involving hazardous substances. The coordinator has the authority to ensure that a proper cleanup is conducted to minimize environmental impact.

Federal Aviation Administration (FAA)

The Federal Aviation Administration, an agency within the U.S. Department of Transportation, interacts with the fire service concerning its responsibility over fire protection in all aspects of civil aviation. In addition, the FAA controls transport of all hazardous materials by air.

Federal Emergency Management Agency (FEMA)

FEMA was formed in 1979 by President Jimmy Carter. FEMA gives the President the capability to provide for national needs in preparing for and mitigating all types of emergencies at a federal level **(Figure 4.9)**.

As part of the U.S. Department of Homeland Security, FEMA is charged with aiding both citizens and first responders to build, sustain, and improve the capability to prepare for, protect against, respond to, and recover from natural as well as human-caused disasters.

FEMA programs include:

- Fire prevention and control
- Continuity of government
- Strategic stockpiles
- Civil defense
- Federal insurance plans
- Floodplain management and dam safety
- Hurricane preparedness
- Earthquake preparedness
- Radiological emergency preparedness

These programs strive to protect and maintain the continuity of civilian government while preserving life and property during emergencies.

National Fire Academy (NFA)

Since its establishment by the Federal Fire Prevention and Control Act of 1974 (Public Law 93-498), the NFA has worked to advance the professional development of fire service personnel and others engaged in fire prevention and control **(Figure 4.10)**.

The National Fire Academy, working within the United States Fire Administration, provides programs at its facility in Emmitsburg, Maryland, and in the field through state training agencies. Programs are offered in the following areas of study:

- Organizational and executive development
- Fire service education and public fire education
- Management technology
- Arson mitigation
- Hazardous materials
- Incident command
- Fire prevention

Figure 4.10 The National Fire Academy presents intensive courses on fire and emergency services topics at its Emmitsburg, MD campus. *Courtesy of the National Fire Academy.*

Resident courses are typically one or two weeks in duration, while field courses are typically presented in two-day weekend classes. The NFA also offers online programs and state weekend opportunities for training on campus. The NFA offers many classes that are reviewed and assigned recommended college credit hours by the American Council on Education.

Training Resources and Data Exchange (TRADE)

The Training Resources and Data Exchange (TRADE) program is a regionally based network designed to foster the exchange of fire-related training information and resources among federal, state, and local levels of government. The network is made up of the directors of each state's fire training systems as well as senior training personnel from the nation's largest fire departments. The objectives of TRADE are to:

- Identify fire, rescue, and emergency medical service training needs at the regional level.
- Identify and exchange training programs and resources within regions.
- Provide the NFA with an annual assessment of fire training needs within a region.
- Identify national trends with an impact on fire-related training and education.

National Guard/Civil Support Team

The governor of a state may deploy the National Guard to aid in the mitigation of an incident. Many of the National Guard's resources may be used to mitigate an incident, including search and rescue, hazardous materials response, fire fighting, heavy equipment, and transport services. The National Guard also operates the Civil Support Teams (CST), which may be called upon to assist civilian response agencies during suspected incidents involving weapons of mass destruction.

National Highway Traffic Safety Administration (NHTSA)

The purpose of the National Highway Traffic Safety Administration's (NHTSA) is to save lives, reduce injuries, and lessen economic impact from vehicle crashes. The NHTSA seeks to accomplish its mission through education, research, establishment of safety standards, and enforcement activity. The NHTSA is an agency of the U.S. Department of Transportation and the federal focal point for EMS.

Office of EMS Mission

The mission of the Office of EMS within the NHTSA is to reduce injuries and deaths by providing leadership and coordination in the EMS community. Programs include:

- National standard curricula for various levels of EMS providers
- National registration of EMT-Paramedics
- National EMS Advisory Council to provide recommendations to NHTSA relating to EMS issues
- Federal Interagency Committee on EMS to ensure coordination of federal agencies involved with EMS and 9-1-1 services
- National EMS information system
- In Canada, EMS works within the Canadian Medical Association

National Institute for Occupational Safety and Health (NIOSH)

As part of the Centers for Disease Control and Prevention (CDC), NIOSH conducts research and provides education in areas that affect the health and safety of workers. NIOSH also operates the Firefighter Fatality Investigation and Prevention Program and issues advisories on emerging issues related to firefighter safety.

National Institute of Standards and Technology (NIST)

The National Institute of Standards and Technology (NIST), formerly the National Bureau of Standards established in 1901, is part of the U.S. Department of Commerce. NIST's Building and Fire Research Laboratory (BFRL) conducts fire research and tests involving building technology, fire dynamics, and fire service operations **(Figure 4.11)**.

Figure 4.11 The National Institute of Standards and Technology (NIST) campus in Gaithersburg, MD. *Courtesy of NIST.*

Underwriters Laboratories and NIST

Underwriters Laboratories (UL) and NIST work in a cooperative effort to improve the understanding of fire behavior and enhance firefighter safety with a variety of projects, including wind-driven fire and flowpath control research.

National Transportation Safety Board (NTSB)

The National Transportation Safety Board (NTSB) investigates significant accidents involving civilian aircraft, railroads, maritime, and pipeline transportation. This agency also investigates the causes of accidents involving the release of hazardous materials in all forms of transport and issues recommendations aimed at preventing future accidents.

National Weather Service

The National Oceanic and Atmospheric Administration (NOAA) meteorologists work with federal, state, and local fire control agencies to help predict storms and aid responders in planning and executing suppression operations. NOAA has the capability to establish weather forecast offices on site in order to provide continuous meteorological support at an incident.

Nuclear Regulatory Commission (NRC)

The NRC operates primarily to assist in the training of emergency responders for responding to nuclear emergencies. The commission also develops and enforces standards on building and operating nuclear facilities and what do they do about fire protection issues in the nuclear industry for reactors and nuclear fuel cycle facilities.

Occupational Safety and Health Administration (OSHA)

Federal law established Occupational Safety and Health Act of 1970 to ensure safe and healthy conditions for the workers in the United States. As part of the U.S. Department of Labor, OSHA pursues its goals through the following:

- Enforcing regulations developed under the OSHA Act of 1970
- Assisting states in their efforts to ensure safe and healthy working conditions
- Providing research, information, education, and training in the field of occupational safety and health

Public Safety Canada (PS)

Created in 2003, Public Safety Canada was established to provide coordination of all federal departments and agencies responsible for safety and security in Canada. PS develops policies and standards for national response systems. Working with regional agencies and first responders, PS provides funds, tools, and training to protect Canada's infrastructure.

Figure 4.12 The U.S. Coast Guard is equipped to perform marine rescue and fire fighting. *US. Navy photo by Chief Mass Communication Specialist Bill Mesta.*

Transport Canada

Transport Canada develops regulations and policies as well as provides some services related to safety and security for the road, rail, marine, and aviation systems in Canada. Transport Canada also regulates the transportation of dangerous goods traveling by rail. It also operates the Canadian Transport Emergency Center that assists emergency responders during incidents involving dangerous goods.

United States Coast Guard (USCG)

The USCG provides marine rescue and limited fire fighting in waterways under its jurisdiction. It also provides on-scene coordination for spills or other emergencies on these waterways that may be a threat or that may threaten. USCG authority in these matters is similar to that of EPA coordinators **(Figure 4.12)**.

United States Department of Energy (DOE)

The U.S. Department of Energy, created in 1977, oversees U.S. policies on energy production, research, and conservation as well as the safe handling of nuclear materials.

United States Department of Homeland Security (DHS)

Created in the aftermath of the September 11, 2001, terror attacks, the DHS became the most significant transformation of government agency structure in over fifty years. Twenty-two agencies were brought under the control of DHS in 2003. Although created to strengthen security against terrorist attacks, the department also assists in preparation for and response to all types of disasters.

DHS is made up of seven major divisions:

- The Transportation Security Administration
- U.S. Customs and Border Patrol
- U.S. Citizenship and Immigration Services

- U.S. Immigration Customs Enforcement
- U.S. Secret Service
- U.S. Coast Guard
- Federal Emergency Management Agency (FEMA), which includes:
  — The U.S. Fire Administration
  — National Fire Academy
  — Emergency Management Institute

United States Department of the Interior (DOI)

The Department of the Interior provides fire protection services in some areas of the country. As part of the DOI, the Bureau of Land Management (BLM) provides protection against wildland fires on 545 million acres of public land. The BLM also supports the National Interagency Fire Center in Boise, Idaho. In addition, the National Park Service also has a major involvement in fire prevention, management, and suppression in the national park system.

United States Department of Labor (DOL)

The U.S. Department of Labor (DOL) is tasked with administering and enforcing the Occupational Safety and Health Act. The DOL compiles statistics on occupational injury and illness through the Bureau of Labor Statistics.

United States Department of Transportation (US DOT)

The DOT has an area of responsibility that includes the United States' highways, airways, railways, and coastal waterways, as well as interstate petroleum pipelines. In order to maintain public safety, the DOT has developed regulations that control the transportation of hazardous materials. Several subdivisions maintain responsibility for specific aspects of the transportation system:

- Pipeline and Hazardous Materials Safety Administration
- Federal Aviation Administration
- Federal Highway Administration
- Materials Transport Bureau
- National Highway Traffic Safety Administration.

United States Fire Administration (USFA)

The United States Fire Administration, now part of FEMA within the Department of Homeland Security, was created by the Federal Fire Prevention and Control Act of 1974 (Public Law 93-498). The mission of the USFA is to:

- Improve prevention and control efforts to reduce the nation's losses from fire
- Supplement existing research, training, and educational programs
- Encourage new state and local government programs and initiatives

Headquartered at the National Emergency Training Center in Emmitsburg, Maryland, the USFA maintains an extensive program to analyze fire data from around the nation. The USFA, in concert with the National Institute for Occupational Safety and Health, also administer a firefighter safety and health program.

United States Forest Service (USFS)

As part of the Department of Agriculture, the United States Forest Service provides fire protection to national forests, grasslands, and nearby private lands across the country. The USFS maintains the equipment, personnel, aircraft, and communications network to combat large-scale wildland fires. In addition, the Forest Service conducts wildland fire research and supplies the states with technical and financial assistance, as well as training and surplus equipment to fight wildfires in their jurisdictions.

North American Trade, Professional, and Membership Fire and Emergency Services Organizations

Trade and professional membership organizations generally rise out of a need for individuals with similar interests uniting to advance their common cause. In the fire service, these organizations serve vital roles in lobbying, training, research, and advocacy. The following section describes some of the membership organizations a firefighter may encounter in the United States and Canada. These groups include:

- American National Standards Institute (ANSI)
- Canadian Association of Fire Chiefs (CAFC)
- The Canadian Centre for Emergency Preparedness (CCEP)
- Canadian Fallen Firefighters Foundation (CFFF)
- Canadian Fire Alarm Association (CFAA)
- Canadian Fire Safety Association
- Canadian Standards Association
- Canadian Volunteer Fire Services Association (CVFSA)
- Council of Canadian Fire Marshals and Fire Commissioners
- FM Global Research
- Fire Prevention Canada
- Insurance Services Office, Inc. (ISO)
- International Association of Arson Investigators (IAAI)
- International Association of Black Professional Firefighters
- International Association of Fire Chiefs (IAFC)
- International Association of Women in the Fire and Emergency Services
- International City/County Management Association (ICMA)
- International Code Council (ICC)
- International Fire Marshals Association (IFMA)
- International Fire Service Training Association (IFSTA)
- International Municipal Signal Association (IMSA)
- International Society of Fire Service Instructors (ISFSI)
- National Association of Hispanic Firefighters (NAHF)
- National Association of State Fire Marshals

- National Fallen Firefighters Foundation (NFFF)
- National Fire Protection Association® (NFPA®)
- National Interagency Fire Center (NIFC)
- National Volunteer Fire Council (NVFC)
- National Wildfire Coordinating Group (NWCG)
- North American Fire Training Directors (NAFTD)
- Society of Fire Protection Engineers (SFPE)
- Underwriters Laboratories (UL)
- Underwriters Laboratories of Canada (ULC)

American National Standards Institute (ANSI)
ANSI is a not-for-profit organization that examines and accredits programs that assess conformance to standards and oversees the creation of new standards.

Canadian Association of Fire Chiefs (CAFC)
The Canadian Association of Fire Chiefs is headquartered in Ottawa, Ontario. It is an independent, nonprofit organization with a voluntary membership of over 1,200 individuals and organizations drawn from fire departments, health care facilities, and all levels of government, as well as colleges and universities. The CAFC is funded through sales of fire-related publications and membership dues.

The Canadian Centre for Emergency Preparedness (CCEP)
The Canadian Centre for Emergency Preparedness is a not-for-profit organization based in Burlington, Ontario, that seeks to increase the disaster preparedness of businesses, communities, and individual Canadians. This group strives to take individuals and groups from awareness to action in preparing for disasters.

Canadian Fallen Firefighters Foundation (CFFF)
The Canadian Fallen Firefighters Foundation (Foundation Canadienne des Pompiers Morts en Service in French) was established to memorialize firefighters killed in the line of duty, as well as provide assistance to their families.

Canadian Fire Alarm Association (CFAA)
The CFAA is engaged in efforts to "maximize the use and effectiveness of fire alarm systems in the protection of life and property." Established in 1973, the group is composed of Canadian fire alarm professionals.

Canadian Fire Safety Association (CFSA)
The Canadian Fire Safety Association was established in 1971 as a non-profit organization engaged in promoting fire safety through seminars, training programs, newsletters, scholarships, and regular meetings.

Canadian Standards Association (CSA)
Accredited by ANSI and OSHA, the CSA is a recognized testing laboratory in Canada and the United States. Products tested by the CSA are regularly sold in Canada and U.S.

The Canadian Volunteer Fire Services Association (CVFSA)

The CVFSA is a national organization that was established to the volunteer fire service communities across Canada. CVFSA provides training and education and organizational and administrative standards to the volunteer sector.

Council of Canadian Fire Marshals and Fire Commissioners

The Council of Canadian Fire Marshals and Fire Commissioners was formed to meet the following goals:

- Advise on and promote legislation, policies, and procedures pertaining to fire prevention
- Participate in the development of fire safety codes and standards
- Promote fire safety awareness and professional development in the fire service
- Compile, disseminate, and identify trends in fire loss statistics
- Provide a forum to discuss fire safety matters
- Provide advice regarding the testing and certification of fire protection equipment, materials, and services related to fire safety

FM Global Research

FM Global Research conducts research in property loss control by using data collected in testing, surveys, and studies. The data collected is primarily for use by FM Global, but is available to others. The FM Global Research labs also test and approve fire protection equipment, such as automatic sprinklers and sprinkler control valves.

Fire Prevention Canada

Fire Prevention Canada incorporated in 1976 as a nonprofit organization. Its primary mission is to increase the visibility and awareness of fire prevention practices at a national level. Awareness and visibility are accomplished through its patron Governor General of Canada and the efforts of the fire service with public education and prevention programs.

Home Fire Sprinkler Coalition

The Home Fire Sprinkler Coalition is an organization that informs the public about the livesaving value of home fire sprinkler protection. The coalition offers educational resources to fire departments, home builders, and consumers to encourage the installation of home fire sprinkler systems.

Insurance Services Office, Inc. (ISO)

Formed in 1971, the Insurance Services Office, Inc. evaluates and rates the fire protection in communities across the U.S. These ratings may directly affect the insurance premiums in each jurisdiction. As part of its evaluation of fire protection, the ISO produces the Commercial Fire Rating Schedule and the Fire Suppression Rating Schedule.

Figure 4.13 The International Association of Black Professional Fire Fighters is dedicated to supporting black career firefighters. *Courtesy of Ron Jeffers.*

International Association of Arson Investigators (IAAI)

The IAAI advances the training and standards of fire investigators. Active members of the IAAI must be engaged in suppression of arson activities for a private or governmental agency. The IAAI also maintains an associate membership for individuals with other professional backgrounds.

International Association of Black Professional Firefighters (IABPFF)

Since 1970, the International Association of Black Professional Firefighters has been dedicated to assisting black career firefighters in areas such as working conditions, advancement, and interracial programs. The IABPFF also serves as a liaison among black firefighters and includes the Black Chief Officers Committee **(Figure 4.13)**.

International Association of Fire Chiefs (IAFC)

The IAFC was formed in 1873 to further the professional advancement of the fire service. This organization is open to all chief officers of public, industrial, or government fire departments. Other individuals actively involved in fire fighting and administrative duties, such as fire marshals, commissioners, and directors, may become members as well. The IAFC holds an annual conference for educational advancement in a different North American city each year.

The IAFC has sections that focus on specific fire service interests. These sections include:

- Emergency Medical Services (EMS)
- Emergency Vehicle Management
- Federal and Military Fire Service
- Industrial Fire and Safety
- Metro Chiefs
- Safety, Health, and Survival
- Volunteer and Combination Officers

Figure 4.14 Female fire fighters during a training conference sponsored by iWOMEN.

International Association of Fire Fighters (IAFF)

Formed in 1918, the International Association of Fire Fighters is a labor union representing more than 300,000 career firefighters and emergency medical service personnel in the U.S. and Canada. Affiliated with the American Federation of Labor-Congress of Industrial Organizations (AFL-CIO), the IAFF is structured around chartered local unions, state associations, and joint councils. The association works to improve firefighter health and safety.

International Association of Women in Fire and Emergency Services (iWOMEN)

The International Association of Women in Fire and Emergency Services (iWOMEN) is a nonprofit group that provides a support network for women in the fire service as well as those looking to pursue fire-related careers. The organization also provides input to the National Fire Academy **(Figure 4.14).**

International City/County Management Association (ICMA)

Government administrators from around the world comprise the International City/County Management Association. This professional organization strives to strengthen government through educational programs that advance professional management skills. The ICMA sponsors training and provides publications to disseminate information.

International Code Council (ICC)

The nonprofit International Code Council, established in 1994, has developed a single set of comprehensive and coordinated national model construction codes. Established by the Building Officials and Code Administrators (BOCA), the ICC was able to consolidate the International Congress of Building Officials (ICBO) and the Southern Building Code Congress International (SBCCI) model codes into one version that may be used throughout the U.S.

International Fire Marshals Association (IFMA)

IFMA, organized as a section of the NFPA® in 1906, to unites those involved in fire prevention and investigation by correlating activities and exchanging information.

International Fire Service Training Association (IFSTA)

The International Fire Service Training Association was founded in 1934 to develop training manuals for the fire service. IFSTA conducts meetings each July and January at which time committees of subject-matter experts from across the U.S. and Canada review, revise, and validate selected manuals. Fire Protection Publications, an extension of Oklahoma State University, publishes these manuals. IFSTA publishes numerous manuals on a wide variety of fire service topics, including its most popular **Essentials of Fire Fighting**. These books are widely used throughout the U.S., Canada, and beyond North America **(Figure 4.15)**.

International Municipal Signal Association (IMSA)

Organized in 1896, the International Municipal Signal Association provides certification programs for the safe installation, operation, and maintenance of public safety systems. Through training programs, such as Fire Alarm Interior Levels I and II as well as Fire Alarm Municipal Levels I and II, the association provides technical knowledge on fire and police alarms and traffic control systems.

Figure 4.15 The International Fire Service Training Association (IFSTA) reviews and validates fire service training materials, which are then published by Fire Protection Publications.

International Society of Fire Service Instructors (ISFSI)

The society was established in 1960 with the goal of providing a medium for the exchange of ideas and training techniques to assist in the professional development of the fire service and its instructors.

National Association of Hispanic Firefighters (NAHF)

The National Association of Hispanic Firefighters, founded in 1995, is dedicated to the recruitment, professional development, and fellowship of Hispanic firefighters. The NAHF offers a self-study Spanish language program for firefighters called *Tactical Spanish for Firefighters and EMS.*

National Association of State Fire Marshals

The National Association of State Fire Marshals represents the highest fire official in each state and the District of Columbia. While the duties of the state fire marshal may vary by jurisdiction, they may include:

- Ability to adopt and enforce a code
- Perform fire investigation
- Report fire data
- Perform public education and prevention programs
- Advise the governor and state legislators regarding fire safety issues
- Supervise firefighter training
- Respond to hazardous material incidents
- Fight wildland fires
- Be aware of pipeline safety

National Fallen Firefighters Foundation (NFFF)

In 1992, Congress created the National Fallen Firefighters Foundation (NFFF). The foundation leads the nationwide effort to remember America's fallen firefighters and to assist family members and other firefighters who have been affected by a line-of-duty death. These efforts are funded through private donations from individuals, foundations, and corporations **(Figure 4.16)**.

Figure 4.16 The National Fallen Firefighters Memorial located on the campus of the National Fire Academy. *Courtesy of NFFF/USFA.*

The NFFF sponsors a tribute each October to honor those firefighters who have fallen in the line of duty the previous year. The foundation provides the travel, lodging, and meal expenses for the immediate survivors of the fallen firefighters and conducts support programs and awards scholarships to spouses and children.

In addition to offering resources to fire departments that have suffered a member killed in the line of duty, the NFFF also sponsors the Everyone Goes Home® program with the goal of preventing line-of-duty deaths. The 16 Firefighter Life Safety Initiatives of the Everyone Goes Home Program are listed in Appendix B of this manual.

National Fire Protection Association® (NFPA)®

Organized in 1896, the National Fire Protection Association® has developed more than 300 technical consensus standards. These standards include many that are used by the fire service for professional qualifications, health and safety, and personal protective equipment. Technical committees are comprised of people from many professional backgrounds with experience in public and private fire protection. NFPA® committee work is open for public review, comment, and appeal.

The NFPA® publishes *the Fire Protection Handbook*, sponsors Fire Prevention Week, and conducts seminars. Its 16 member sections include the following:

- Fire Service
- Building Fire Safety Systems
- Industrial Fire Protection
- Electrical
- Architects, Engineers, and Building Officials (AEBO)
- Wildland Fire Management
- Education
- Research
- Latin America
- Health Care
- Lodging Industry
- Fire Science and technology Educators
- Aviation
- Rail Transport
- International Fire Marshals Association
- Metropolitan Fire Chiefs

National Interagency Fire Center (NIFC)

Located in Boise, Idaho, the National Interagency Fire Center is the support and coordination center for wildland fire fighting. The NIFC was created in 1965 to reduce costs and avoid duplication of services as well as to coordinate fire planning and operations. The NIFC is comprised of eight agencies and organizations; these agencies include the U.S. Forest Service, Bureau of Land Management, National Park Service, The U.S. Fire Administration, and other stakeholders in wildland fire fighting operations.

National Propane Gas Association (NPGA)
The National Propane Gas Association works to advance safety in the propane industry. The NPGA is involved in codes and standards organizations, including the NFPA®. The association supports a propane emergencies program that many state fire training agencies adopted.

National Volunteer Fire Council (NVFC)
The National Volunteer Fire Council provides a unified voice as well as information and resources for volunteer fire and EMS organizations. The NVFC was established in 1976 and is comprised of representatives of forty-nine state fire associations.

National Wildfire Coordinating Group (NWCG)
The National Wildfire Coordinating Group is comprised of the U.S. Forest Service, the Bureau of Land Management, National Park Service, and the Bureau of Indian Affairs, as well as the Fish and Wildlife Service and the National Association of State Foresters. This group provides national leadership in the establishment, implementation, and communication of policies, standards, and guidelines for wildland fire management.

North American Fire Training Directors (NAFTD)
The NAFTD acts as a representative organization for state, provincial, and territorial fire training programs. The association acts to enhance state, provincial, and territorial fire training programs and their interests with agencies and organizations whose policies may affect them.

Society of Fire Protection Engineers (SFPE)
The Society of Fire Protection Engineers was established in 1950 as a section of the NFPA®. It incorporated as a separate entity in 1971 and acts to represent those individuals working in the field of fire protection engineering. With over four thousand members, the society supports the professional engineering (licensing) exam for **fire protection engineers**.

> **Fire Protection Engineer (FPE)** — Graduate of an accredited institution of higher education who has specialized in engineering science related to fire protection.

Underwriters Laboratories (UL)
Underwriters Laboratories, founded in 1894, promotes public safety through scientific investigation to determine the hazards posed by various materials. Subsequent to passing a rigorous testing process, a material is listed and labeled as having met UL standards. In addition, UL offers online fire service training **(Figure 4.17)**.

Figure 4.17 The UL label on this microwave oven shows that the product has been tested and passed vigorous testing by Underwriters Laboratories.

Underwriters Laboratories of Canada (ULC)
The nonprofit ULC was founded in 1920 and is an affiliate of Underwriters Laboratories, Inc. Its mission and methods relative to promoting public safety through product testing are the same.

Other Related Organizations
Numerous other organizations have some common interests and interaction with the fire service. While their major goals and mission may not be fire-service oriented, members of the fire department may have contact with some of these groups. Several organizations are listed in the following section.

American Association of Railroads Bureau of Explosives (BOE)
The railroad industry formed the Bureau of Explosives (BOE) in 1907 to aid carriers, shippers, and container manufacturers in their goal to improve hazardous material transportation safety. The BOE developed the first hazardous material safety rules that were later adopted and expanded by the federal government.

American Trucking Association (ATA)
The American Trucking Association has worked as an advocate for the trucking industry since 1933. The ATA develops and advocates policies to promote highway safety, security, environmental sustainability, and profitability for its members.

National Tank Truck Carriers (NTCC)
The NTTC works as an advocate for its members before Congress and numerous federal regulatory agencies. The organization's mission is to protect the safety of its members as well as the general public while ensuring the sustainability of tank truck transport. This mission is accomplished through cooperation with the government involving industry standards and enforcement measures.

Chapter Summary
Knowledge of available resources outside the fire service is crucial to the safety and efficiency of many operations. Establishing and maintaining a good working relationship with other government agencies, as well as private companies, is vital to ensure public safety, and the safety of responders during many incidents.

Fire departments encounter other government agencies and private organizations during routine as well as emergency activities. All levels of government, from local departments to federal agencies, may have the resources necessary to aid the fire department during an incident.

Many fire service groups, private trade associations, and government or private organizations have an interest in fire and life safety. A good working knowledge of these groups will aid members of the fire service in efforts to advance safety, training, and public education programs, as well as becoming informed of emerging fire service trends.

Review Questions

1. What are the ways that local government and local law enforcement interact with the fire service?
2. Describe the functions of at least three local agencies or organizations.
3. How can nonprofit and civic groups help the fire service?
4. List typical responsibilities of the state fire marshal's office.
5. Describe the functions of at least three state/provincial agencies or organizations.
6. Identify at least five federal organizations that have some interest or interaction with the fire service.
7. What agency gives the President the capability to provide for national needs in preparing for and mitigating all types of emergencies at a federal level?
8. What federal agency provides training programs for fire service personnel?
9. List at least five North American trade, professional, or membership fire and emergency services organizations.
10. What are the functions of the American Association of Railroads Bureau of Explosives (BOE), American Trucking Association (ATA), and the National Tank Truck Carriers (NTTC)?

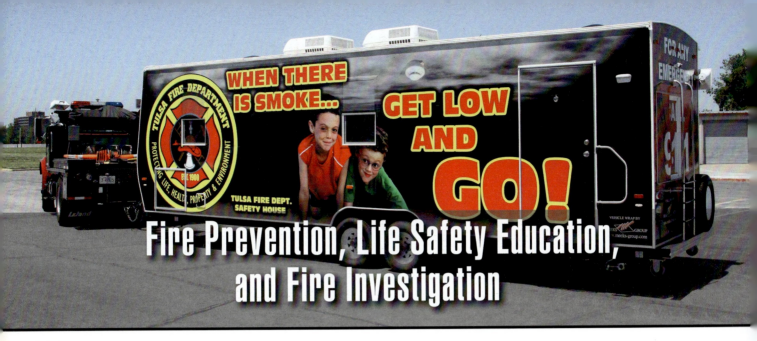

Fire Prevention, Life Safety Education, and Fire Investigation

Chapter Contents

Case History 145	Fire and Life Safety Education 165
Principles of Fire Prevention 145	Presenting Fire and Life Safety Information 165
Need for Fire Prevention 146	Fire and Life Safety Presentation Topics 167
Fire Hazards 147	Fire Station Tours 168
Fire Risk Analysis 152	**Fire Investigation** 168
Components of Fire Prevention 153	Role of the Investigator 169
Preincident Surveys 154	Observations Made by Emergency Responders 170
Facility Surveys 155	Conduct and Statements at the Scene 174
Inspection and Code Enforcement 159	Securing the Fire Scene 175
Codes and Standards 159	Legal Considerations 176
Consistent Codes and Standards 160	Protecting and Preserving Evidence 177
Effects on Construction 160	**Chapter Summary** 178
Role of the Fire Inspector 161	**Review Questions** 178
Fire Protection Engineer/Building Plans Examiner 162	
Permits 162	

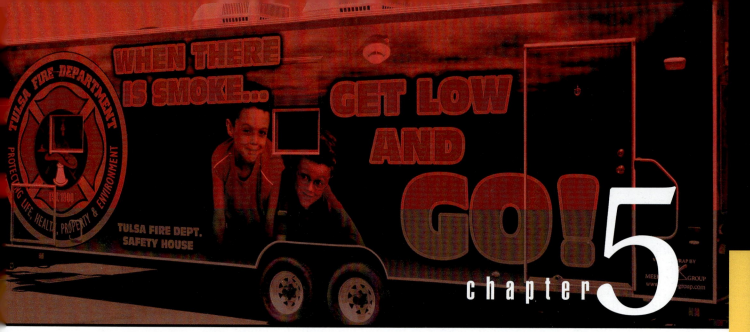

Chapter 5

Key Terms

Arson 169	Fire Prevention 155
Code 159	Overhaul 170
Fire Hazard 148	Preincident Planning 151
Fire Inspector 160	Preincident Survey 153
Fire Investigator 169	Standard 161
Fire Marshal 169	

FESHE Outcomes

This chapter provides information that addresses the outcomes for the Fire and Emergency Services Higher Education (FESHE) *Principles of Emergency Services* course.

5. Identify fire protection and emergency-service careers in both the public and private sector.

7. Discuss and describe the scope, purpose, and organizational structure of fire and emergency services.

10. Identify the primary responsibilities of fire prevention personnel including code enforcement, public information, and public and private protection systems.

Fire Prevention, Life Safety Education, and Fire Investigation

Learning Objectives

1. Explain the need for fire prevention. (Outcomes 7, 10)
2. Identify fire hazards. (Outcome 10)
3. Explain a fire risk analysis. (Outcome 10)
4. Identify the components of fire prevention. (Outcome 10)
5. Describe the role of preincident surveys in fire and life safety. (Outcome 10)
6. Explain the role of the fire inspector and fire protection/engineer/building plans examiner in inspection and code enforcement. (Outcomes 5, 10)
7. Identify how inspection and code enforcement help ensure fire and life safety. (Outcome 10)
8. Describe the purpose of permits. (Outcome 10)
9. Explain steps in presenting fire and life safety information. (Outcome 10)
10. Identify fire and life safety presentation topics. (Outcome 10)
11. Explain actions to take when giving a fire station tour. (Outcome 10)
12. Describe the various responsibilities of fire and emergency services personnel regarding fire investigation. (Outcome 10)

Chapter 5
Fire Prevention, Life Safety Education, and Fire Investigation

Case History

The McDonald family has a reunion every August. One of the family members attending the reunion was a fire and life safety educator who worked with a large fire department. At the reunion, her brother suggested that she teach his children basic fire safety for the home. She taught the children not to hide from a fire inside the home as well as the proper techniques to use for a stove fire. She also stressed to the children to "fall and crawl" for easier breathing, to "stop, drop, and roll" until the fire is out, and to meet at a family designated location outside the home.

Five years later, the McDonald's oldest daughter was helping her mother cook dinner when a pan of grease caught on fire. The mother reached for the pan handle and started to run with the pan to the back door. The daughter shouted to turn off the stove and to place a lid over the pan, which would smother the fire. The daughter told her mother that if she had run with the burning grease, she could have been severely burned or set something else on fire. These important fire prevention lessons had been taught earlier in the girl's life, and she applied them years later.

Principles of Fire Prevention

The prevention of fires begins with fire departments understanding the potential community risk or requirements of a community. Therefore, many fire departments invest financial and human resources in fire and life safety programs to help reduce community hazards and to prevent dangerous conditions or actions. A fire and life safety educator preplans for public awareness to be implemented in the community throughout the year. Public awareness includes educational presentations to children, group visits to senior citizens at residential living centers, and presentations to people who live with disabilities. The fire and life safety educator may also distribute fire safety literature to the schools in the community and issue press announcements to the media.

The last part of this chapter contains information on the responsibilities of the firefighter and the fire investigator. A firefighter's observations en route, upon arrival, and during and after the fire can assist in a fire investigation. The firefighter is often responsible for securing the fire scene and protecting the evidence. The firefighter's conduct at the scene and legal considerations are also summarized.

Three E's of Fire Prevention

President Harry S. Truman's 1947 conference on fire prevention coined the original "Three E's"— Education, Enforcement, and Engineering. The fire service began to use the terms shortly after they were introduced, often referred to as the "Three E's of Fire Prevention." The fire service usually lists the "Three E's" in the order mentioned. However, some of the related professional organizations have had a controversy over which term to list first. The important issue is that all three terms must be used to have an effective fire prevention program.

- *Education* influences citizens' behavior by raising their awareness concerning fire causes and proper actions for safety and prevention.
- *Engineering* involves changes in the physical environment by utilizing structure design and various types of safety features, such as fire sprinklers systems, smoke alarms, and notification devices.
- *Enforcement* includes the passing, revising, and enforcing of fire and building codes. It includes conducting periodic testing of fire and life safety systems to ensure that the systems are operational and the inspection of buildings meet all jurisdictional codes and regulations.

Need for Fire Prevention

The primary focus of the fire service is the protection of life and property. This goal can best be reached, with the least risk to fire service personnel, by preventing fires before they start. There are numerous safety, economic, and environmental benefits to a successful fire prevention and life safety program **(Figure 5.1)**.

Loss of Life, Property, and Environmental Concerns

In 2012, fires in the United States topped 1.3 million. These fires resulted in the deaths of almost 2,900 civilians or one death approximately every three hours. During the same year, 69 firefighters were killed in the line of duty. In addition, these fires resulted in a property loss of over $12.4 billion. The toll on the environment was equally severe, as 9.3 million acres of wildland burned, resulting in the loss of natural resources, habitats, and recreation. While even the most effective program will not prevent every fire, a reduction in loss will benefit all aspects of society and our environmental resources.

Monetary Loss

Fires of all types impact the monetary resources of local communities, citizens, and businesses. However, direct property losses are only one way in which fire affects the fiscal well-being of a community. Business owners lose revenue and employees lose their jobs when an establishment is closed due to a fire. If a business does not reopen after a fire, the jurisdiction will lose the tax revenue as well. When a loss is incurred at an insured premise, the insurance carrier pays a claim to the policyholder. Communities with higher fire losses

Figure 5.1 Fire prevention seeks to prevent losses from fires. *Courtesy of Chris Mickal.*

generally pay higher insurance premiums based on their exposure to risk. Not only is preventing fires in the best interest of the public and firefighter safety, it is financially prudent. In neighborhoods with high instances of incendiary fires, property values are often negatively affected, people may move away, and the quality of life in a community is negatively impacted.

Fire Hazards

In order for a fire to burn, four elements must be present: a fuel supply, sufficient heat, oxygen, and a self-sustained chemical reaction. Conditions necessary to create a fire may be prevented by eliminating one or more of these elements **(Figure 5.2, p. 148)**.

NOTE: See Chapter 6 on Scientific Terminology for additional information concerning fire behavior.

Generally, control of the fire hazards involving fuel supply and heat sources are the most manageable. If heat sources are isolated from a fuel supply, the condition will remain safe. Not all fuel supplies are easily ignited, but misuse of any fuel supply under extreme heat conditions may lead to a fire. Some common fuel and heat source hazards include the following:

- **Fuel hazards**:
  — Ordinary combustibles, such as wood, cloth, and paper
  — Flammable and combustible gases, such as natural gas, liquefied petroleum gas (LPG), and compressed natural gas (CNG)

Chapter 5 • Fire Prevention, Life Safety Education, and Fire Investigation 147

Figure 5.2 Bags of nitrates and chlorates represent a fuel hazard.

Figure 5.3 A common hazard is allowing combustible materials to build up.

— Flammable and combustible liquids, such as gasoline, oils, lacquers, and alcohol

— Chemicals, such as nitrates, oxides, and chlorates

— Dusts, such as grain, wood, metal, and coal

— Metals, such as magnesium, sodium, or potassium

— Plastics, resins, and cellulose

- **Heat sources**:

    — **Chemical heat energy** — Improperly stored materials may result in the production of chemical heat energy. These materials may come into contact with each other and react, or they may decompose and generate heat.

    — **Electrical heat energy** — Poorly maintained or defective electrical appliances, exposed wiring, or lighting sources may provide electrical heat sources.

    — **Mechanical heat energy** — Moving parts on a machine, such as belts and bearings, are sources of mechanical heating.

    — **Nuclear heat energy** — Heat that is created by fission.

Common Fire Hazards

The term *common* in this section refers to the probable frequency of a hazard being found and not the severity of the hazard. A common **fire hazard** is a condition that may be prevalent in many occupancies and may increase the

> **Fire Hazard** — Any material, condition, or act that contributes to the start of a fire or that increases the extent or severity of fire.

likelihood of a fire **(Figure 5.3)**. Firefighters must be alert to the following conditions and be aware of the dangers that they may pose to both occupants and firefighters:

- **Poor housekeeping and improper storage of combustible material.** It is difficult for occupants to evacuate in an emergency when an extreme amount of contents fill an occupancy and all floor space is covered with content piles of various heights. This hazard can also cause problems for firefighters attempting to search and maneuver hoselines through an occupancy. A large buildup of contents in a structure also increases the fire load and may increase the chance that a combustible material may contact an ignition source. An extreme amount of clutter may also conceal other fire hazards, such as electrical wiring issues.

- **Defective or improper use of lighting, heating, or other power equipment.** Improperly installed or malfunctioning heating, lighting, or other electrical equipment may also provide an ignition source for nearby combustibles.

- **Misuse of flammable or combustible liquids.** Cleaning chemicals or flammable liquids that are improperly used or stored around an occupancy may also provide an ignition source.

The attitude or behaviors of building occupants can also impact the risk of fires or other emergencies, such as an individual smoking in bed or leaving home with food cooking. Installing and testing smoke detectors and practicing a home escape plan can reduce the risk to an occupant if a fire occurs. Establishing a comprehensive fire and life safety education program can reduce the risk of fires in a community.

Special Considerations

A special fire hazard is one that arises as a result of the processes or operations that are characteristic to the individual occupancy. Commercial, manufacturing, and public assembly occupancies have their own special fire considerations **(Figure 5.4)**. Some of these considerations are listed in the following section:

Figure 5.4 Cans of flammable paints and lacquers represent a fire hazard in commercial properties.

- **Commercial Occupancies**:
    - Display or storage of large quantities of products
    - Mixed variety of contents in the building
    - Difficulty gaining access to occupancies when closed
    - Existence of party walls, common attics, cocklofts, or other voids in multiple occupancies
- **Manufacturing**:
    - High hazard processes using volatile substances or extreme temperatures
    - Flammable liquids in dip tanks, ovens, or dryers
    - Flammable liquids used in mixing, coating, spraying, or degreasing process
    - Storage of large quantities of combustible material
    - Operation of vehicles, fork trucks, or other powered equipment inside buildings (including use and storage of LPG or other fuels to power these vehicles) **(Figure 5.5)**
    - Large-scale use of flammable or combustible gases
- **Public Assembly**:
    - Large numbers of occupants, especially when posted limits are exceeded
    - Insufficient, locked, or blocked exits
    - Storage of material in the paths of egress
    - Highly combustible interior finishes
    - Improperly installed and maintained cooking facilities

Figure 5.5 Aircraft in a hangar poses a fire hazard.

Rhode Island Station Nightclub Fire

In 2003, the Station Nightclub fire in West Warwick, Rhode Island, killed 100 people and injured over 200. This fire illustrates the potential danger of combustible finishes and interior pyrotechnics. Pyrotechnics ignited foam insulation used for soundproofing, resulting in a fast-spreading fire that produced thick smoke that made it difficult for patrons to find fire exits.

Target Hazard Properties

A target hazard is viewed as a facility in a jurisdiction that has a potential for a great loss of life or property. Fire departments should give these properties special attention in terms of **preincident planning**, building familiarization, and fire prevention activities. Some examples include the following:

- Shopping centers
- Hospitals
- Theatres
- Nursing homes
- Schools
- Nightclubs and social clubs
- Lumberyards
- Bulk oil storage facilities
- Distribution centers
- Rows of wood frame tenements **(Figure 5.6)**
- "Big box" stores

Preincident Planning— Act of preparing to manage an incident at a particular location or a particular type of incident before an incident occurs.

Figure 5.6 Target fire hazards include rows of adjoining apartments. *Courtesy of Ron Jeffers.*

Multiple Hazards

Some facilities may present multiple hazards based on the process conducted at the location as well as the size and scope of the facility. For example, a lumber mill with a sales and storage yard may present a special hazard as the result of the milling operation and a target hazard due to the size and capacity of the sales and storage yard.

Fire Risk Analysis

Fire department members will review information from numerous sources when conducting a fire risk analysis **(Figure 5.7 a and b)**. The potential for loss of life or property from fire may be assessed based on trends emerging from past events, any changes in the jurisdiction that impact fire protection, and the potential for a significant fire based on the different types of hazards found in the community.

Trends

Agencies at various levels of government or private organizations with an interest in fire protection and prevention may study fire data to establish and address emerging trends. For example:

- A municipal fire department may establish that a particular area is experiencing an increasing number of wildland fires since visitors have begun to frequent a new campsite.

- At the state level, the fire marshal may note a trend emerging in different cities where an increasing number of vehicles are stolen and set ablaze.

Figure 5.7 a and b Firefighters conducting a preincident survey of a business.

- At the national level, the National Fire Information Council (NFIC) seeks to improve public safety through the collection and dissemination of fire-related response information.

This information, provided by fire departments through the National Fire Incident Reporting System (NFIRS), may aid in establishing trends in fire loss experiences in certain occupancies, vehicles, appliances, or other equipment. This data, in turn, may lead to the discovery of a faulty component or improved design features.

Demographics

The risk from fire is not the same for all members of the American population. Different segments of society may suffer more fire injuries and fatalities than others do. Although there have been general reductions in fires and fire casualties, the elderly, the young, and the poor are at a higher risk than others in the population. Males, African Americans, and Native Americans also share a heightened risk for death or injury by fire.

Components of Fire Prevention

Fire prevention contains numerous components to reduce hazardous conditions or prevent dangerous actions that may result in a fire. These facets of fire prevention work to address community risk, educate the public, and determine the cause and origin of fires so that similar incidents may be avoided in the future. The followings sections describe the components of fire prevention:

- Surveys, Inspection, and Code Enforcement
- Fire and Life Safety Education
- Fire Investigation

Surveys, Inspection, and Code Enforcement

Conducting **preincident surveys**, inspecting occupancies, and performing code enforcement functions are examples of proactive measures designed to either prevent or minimize property damage and loss of life during a fire. Preincident surveys promote firefighter safety and efficiency if a fire occurs. Inspection and code enforcement may prevent a fire, minimize property loss, or enable occupants to reach safety during a fire or other emergency **(Figure 5.8, p. 154)**.

Preincident surveys are primarily designed to familiarize firefighters with building features, systems, special hazards, and any other information deemed useful in developing an Incident Action Plan (IAP) during an emergency. Trained firefighters or fire inspectors may conduct code enforcement inspections based on jurisdictional policy, and concentrate on ensuring that an occupancy meets the requirements of the fire code.

Fire and Life Safety Education

When presenting the public with fire and life safety information, the fire department attempts to make members of the community aware of potentially dangerous activities or conditions that may result in fires or other emergencies that cause risk to lives or property damage. The information content and delivery may be formatted to appeal to a target audience with a particular risk profile.

> **Preincident Survey —** Assessment of a facility or location made before an emergency occurs, in order to prepare for an appropriate emergency response.

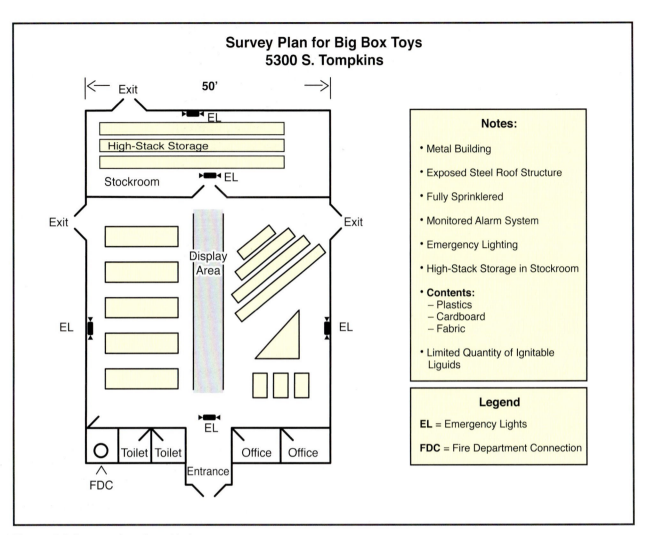

Figure 5.8 Survey plans for a big box toy store.

Fire Investigation
Fire investigation consists of determining the cause and origin of a fire with the goal of using that information to help prevent other fires of the same type from occurring. In the case of an incendiary fire, an investigation may lead to the arrest and prosecution of those individuals responsible for starting the fire.

Preincident Surveys

Firefighters regularly conduct preincident surveys of commercial, industrial, and target hazard occupancies in their districts. Firefighters conduct surveys to gain and maintain familiarity with the layout, building features, and fire protection attributes of occupancies to which they may respond. With a greater knowledge of the building, firefighters may create a preplan to operate more efficiently and be warned of any inherent hazards found in the structure.

At the property owner's request, some jurisdictions may conduct residential safety surveys for single and small multi-unit dwellings. The purpose of preincident surveys is to make owners and occupants aware of any hazardous conditions that may cause a fire to start or make egress during a fire more difficult for residents. The fire department may provide residential safety surveys in conjunction with a fire and life safety education program **(Figure 5.9)** The specifics of these programs will be explained later in this chapter.

Figure 5.9 A fire inspector conducting a fire safety survey in a residence.

Facility Surveys

When conducting surveys, firefighters should exhibit good communication skills and a willingness to answer questions posed by members of the public. Interaction with members of the community during routine business is an opportunity for firefighters to portray the fire service in a positive manner and communicate a message of fire and life safety to all segments of the population.

Personnel Requirements

The public expects firefighters to be qualified to discuss **fire prevention** and to be able to provide advice on correcting fire hazards. When conducting any public activity, firefighters must present a neat and professional appearance in order to help gain public respect and uphold the image of the fire service.

NFPA® 1001, *Standard for Fire Fighter Professional Qualifications*, contains job performance requirements for basic understanding of fire prevention and life safety education principles. Firefighters tasked with performing fire safety surveys are expected to recognize fire hazards and report them through the system that their jurisdiction provides. Although firefighters are looked upon to provide corrective advice, their expertise may be limited. Firefighters should refer complex issues to a qualified inspector or a fire and life safety educator.

Fire Prevention — (1) Part of the science of fire protection that deals with preventing the outbreak of fire by eliminating fire hazards through such activities as inspection, code enforcement, education, and investigation programs. (2) Division of a fire department responsible for conducting fire prevention programs of inspection, code enforcement, education, and investigation.

Scheduling the Preincident Survey

Time management may be a complicated task in the daily responsibilities of a fire company. Company and chief officers should make it a priority to set aside time to conduct preincident surveys. Before a visit, contact the party responsible for a selected property to be surveyed. When setting up an appointment to conduct the survey, the company officer should:

- Inform the owner or occupant about the scope and purpose of the survey.
- Attempt to schedule a day and time that will suit the schedule of the fire department and be less burdensome to the owner.
- Obtain permission from the appropriate party before beginning a survey.
- Be courteous and cooperative when scheduling and conducting preincident surveys. This attitude will create a favorable impression on the public and may foster greater cooperation with fire prevention efforts.

Figure 5.10 Firefighters doing a preincident survey inside a factory.

Conducting the Preincident Survey

When reporting to the premises to begin a survey, personnel should enter the building and meet with the owner or representative with whom their appointment has been made. Upon checking in with their contact, firefighters may begin their survey on the exterior of the building. They should make observations, notes, and photographs of the building's shape and size, as well as any features that may be points of concern during fire fighting operations.

Upon moving to the interior of the structure, personnel should sketch the floor layout and egress routes and note significant elements of manufacturing, storage, or occupancy. Firefighters should also document any hazards, improper practices, or unsafe conditions they encounter. Any hazards or unsafe conditions may be forwarded through department channels to fire prevention for follow-up action. In jurisdictions where fire companies carry out code enforcement duties, fire companies should conduct this task during a separate site visit **(Figure 5.10)**.

Firefighters should note or photograph the following:

- Lockbox locations
- Standpipe and fire department connection locations
- Fire alarm control panels
- Contact information for a responsible party
- Fire walls
- Sprinkler valves
- Utility shutoffs
- Location of combustible or flammable material storage
- Stairs
- Fire barriers
- Areas of refuge
- Contents of special value

Firefighters should be particularly observant of hazardous materials commonly used at the facility they are surveying. Conducting a preincident surveys allow firefighters the opportunity to document the location and the characteristics of many of the hazardous substances they may encounter when responding to emergencies at various occupancies.

When documenting information for a large or complicated occupancy, it may be necessary to make more than one visit to complete the survey. Many firefighters prefer to begin their survey on the roof as it is the highest point, and offers a vantage point over the property. From the roof, firefighters may work their way down through each floor, making a rough sketch or photographing the layout before moving to a lower floor. When visiting a previously planned structure, firefighters may simply note any changes to layout, occupancy, or hazards since the last survey.

Maps and Sketch Making

Maps that convey information concerning building construction, occupancy, fire loading, hazards, and built-in fire protection are valuable to the fire department. Some larger occupancies may already have maps available for the fire department with much of this information. For many buildings, however, firefighters must develop their own maps. When these maps detail the arrangement of the property with respect to streets and other local features, they are referred to a *plot plans*.

Computer-Assisted Preincident Planning

Many fire departments have access to preplanning software that allows a digital image to be inserted into an electronic file that contains the data required by local jurisdictions. Firefighters should always seek permission from the owner or occupant before taking pictures during any preincident survey.

Exit Interview

An exit interview gives firefighters the opportunity to thank the occupants or owners for their cooperation. This interview reinforces how the data will be used to increase safety and efficiency should an emergency arise **(Figure 5.11, p. 158)**. Firefighters can also ask or answer any questions that may have developed over the course of the survey. In addition, department members may use this opportunity to offer advice on how to improve safety conditions in the building.

Gathering Additional Information

Firefighters can obtain information that may be helpful by speaking with property or business owners or their representatives. Inquiring about the number of employees, shift schedule, and specific hazards of the manufacturing process can be invaluable during an incident. In order to ensure the entire department has access to this information, it should be shared with the fire prevention bureau.

Figure 5.11 Firefighters thanking the business owner after conducting an exit interview.

Residential Safety Surveys

Fire departments should take an active role in providing residential fire safety surveys in their jurisdiction. These surveys are voluntary in single or smaller multi-unit occupancies. When preparing to undertake a residential survey program, the fire department should clearly communicate the goals and objectives of the program to the community in order to gain acceptance. The fire department should make clear that the survey is an opportunity for firefighters to make family members aware of household fire hazards and that it is not a code enforce activity.

Generally, firefighters should work in pairs when conducting residential surveys. Based on jurisdictional policy, they may require written permission to conduct the survey. A homeowner should escort them while they conduct the survey, and they should not enter a bedroom without prior permission. For additional information, firefighters may refer to NFPA® 1452, *Guide for Training Fire Service Personnel to Conduct Community Risk Reduction*.

The main objectives of residential fire safety surveys include:

- Preventing accidental fires
- Improving life safety conditions
- Helping the owner or occupant to understand and improve existing conditions

Firefighters also gain valuable information as they conduct residential surveys. Some of the benefits include an improved knowledge of building construction, trends in local development, and occupancy conditions.

Use of a standard form when conducting a survey will serve as a guide and help summarize the firefighter's findings for review with the occupant. Some of the following items are common causes of residential fires:

- Heating appliances
- Cooking procedures
- Smoking materials
- Combustible or flammable liquids

Many fire departments also provide fire prevention literature, promote exit drills in the home (EDITH), and explain smoke alarm and residential sprinkler options **(Figure 5.12)**. Discussion of these topics is an important way for firefighters to help protect life and property. In addition, many departments provide fire and life safety education to the community beyond the scope of a fire safety survey. The safety survey is an excellent opportunity to help educate the public about the evolution of the fire department into an all-hazard response agency.

NOTE: More information concerning fire and life safety education follows in a subsequent section.

Figure 5.12 A fire inspector examines a smoke detector during a residential fire safety survey.

Smoke Alarm Giveaway Programs and Carbon Monoxide (CO) Detectors

Many fire departments provide free smoke detectors to residents that do not have them. Some agencies also provide replacement batteries as a public service. Local service clubs or citizen donations sometimes financially support these battery programs or they may be funded through a grant program. With the increasing prevalence of CO detectors, firefighters should make the presence and location of these devices part of the residential survey. Some departments have giveaway or loaner programs for these detectors as well.

Inspection and Code Enforcement

Inspection programs are established to protect lives and property from fires or other hazards. People engaged to conduct inspections may be employed privately or work for the fire department or another government agency. Depending on jurisdiction and specific duties, these individuals may be referred to as fire and life safety inspectors, code enforcement officers, or simply inspectors. In the private sector, they may work for insurance companies or the loss control division of a corporation. In the public sector, inspectors are commonly under the authority of the building official's office or fire department.

Codes and Standards

Inspectors must be familiar with the codes adopted for use in their jurisdiction. **Codes** used in fire safety may be model codes that third-party organizations have developed and jurisdictions may adopt as written, or with amendments.

Code — A collection of rules and regulations that has been enacted by law in a particular jurisdiction. Codes typically address a single subject area; examples include a mechanical, electrical, building, or fire code.

Building Codes

Common codes in use throughout North America include the International Building Code and International Fire Code, published by the International Code Council (ICC). The National Fire Protection Association® has developed NFPA® 1, *Fire Code*, as well as NFPA® 5000, *Building Construction and Safety Code*, in addition to NFPA® 101, *Life Safety Code*. A widely used code in Canada, published by the National Research Council of Canada, is a model code called the National Building Code of Canada.

Most model codes are revised every three to five years. The revised edition of the code does not automatically take effect in a jurisdiction until the authority having jurisdiction (AHJ) formally adopts it. It is not unusual for an older edition of a model code to continue to be enforced for a number of years after a new edition has been released.

Local amendments adopted with previous versions of a code may have lost their relevance with the passage of time or may be in conflict with the new version of the code. Inspectors must be well versed in the content of the codes in their jurisdiction and must work diligently to assist in the transition process between older and newly revised model codes.

Consistent Codes

Along with revisions of the same model code, an inspector must be aware of code provisions enacted at all levels of government. Having this knowledge prevents conflict between various codes. For example, fire and life safety codes must be consistent with other codes to avoid duplication of effort or attempting to enforce contradictory regulations. When conflicts between codes arise, an inspector should be aware of jurisdictional policy. The following are examples of other codes that may affect fire code enforcement:

- Housing codes
- Zoning ordinances
- Subdivision regulations
- Electrical, mechanical, and plumbing codes
- Health regulations
- Accessibility in accordance with the Americans with Disabilities Act (ADA)
- Emergency Planning and Community Right to Know Act (EPCRA)

Effects On Construction

A plans review and permitting process are established to ensure that new or remodeled structures meet the requirements of local building and fire codes. The building official, a **fire inspector**, or other representative from the fire department may have overseen this process. This process helps to ensure that all requirements are addressed before construction on the project actually begins.

> **Fire Inspector** — Fire personnel assigned to inspect property with the purpose of enforcing fire regulations.

Building Design

During the design process, applying model codes to new construction and renovations helps to address the issues of fire and life safety. Architects, specialists in building construction, site development, fire suppression and detection systems, and other fields, work together with inspectors to design a building that will suit the owner's desires and the requirements of the AHJ.

Local Safety Amendments

Jurisdictions that adopt model codes may feature additional safety amendments as necessary for local circumstances. Fire and life safety codes typically establish minimum requirements for each type of construction and occupancy in a variety of categories. These requirements may include:

- Number of exits
- Fire ratings for doors, walls, stairways, and areas of refuge
- Automatic sprinklers or other types of suppression systems
- Features to control air flow in the ventilation system during a fire

These safety considerations may apply to a building based on size, occupancy, special hazards, or any other safety consideration noted by the AHJ. A jurisdiction may choose to amend the recommendations of a model code. These amendments may include changes to minimum levels of suppression capabilities, construction features, or other specifications deemed important to local public safety needs.

Role of the Fire Inspector

An inspector must possess a high level of expertise to ensure that all hazards are identified and that corrective action is taken to mitigate any dangerous situation. This expertise is gained through experience and certification based on nationally recognized **standards**, such as NFPA® 1031, *Standard for Professional Qualifications for Fire Inspector and Plan Examiner*. Although responsibilities may vary by jurisdiction, a fire inspector may generally carry out the following duties:

- Enforce fire and life safety codes.
- Address citizen complaints relating to fire and life safety.
- Interpret and apply adopted codes and standards.
- Perform fire and life safety inspections of new and existing structures.
- Determine occupancy loads for commercial, industrial, and public assembly occupancies.
- Participate in legal proceedings for fire and life safety code issues.
- Verify water supply fire flow capabilities in order to determine the ability of water supply systems to provide the required level of protection.

> **Standard** — (1) A set of principles, protocols, or procedures that explain how to do something or provide a set of minimum standards to be followed. Adhering to a standard is not required by law, although standards may be incorporated in codes, which are legally enforceable.

Role in Fire Prevention

The fire inspector's ultimate goal is to maintain safety by preventing fires from occurring. Depending on local policy, inspectors may conduct fire and life safety inspections at local places of assembly, healthcare facilities, schools, chemical companies, or any target hazard in the community.

The inspector may perform inspections depending on the occupancy or potential hazard. Some typical inspections include the following:

- **Annual (routine)** — Places of assembly, healthcare, or education.
- **Issuance of a permit** — Instances where a property owner must obtain a permit for a special use at a property, such as a tent or other temporary change.
- **Response to a complaint** — Overcrowding, blocked exits, or other safety issue.
- **New construction** — New or renovated building to ensure code compliance.

- **Change in occupancy** — When a building undergoes an occupancy classification change code, compliance for the new use must be determined.
- **Owner/occupant request** — The result of an insurance company requirement or to satisfy the needs of another government agency.
- **Incident of fire** — A postincident inspection should be made in any public building.

Fire Protection Engineer/Building Plans Examiner

Professionals other than fire and life safety inspectors may have a major role in the fire safety in local communities. Fire protection engineers and plans examiners aid in compliance and review of buildings and fire prevention activities.

Role in Fire Prevention

Plans examiners may be uniformed or civilian fire department personnel, or they may work under the supervision of the building official's office. They ensure code compliance by reviewing architectural plans and fire protection system plans for new construction as well as renovations. They should meet the recommendations of NFPA® 1031, *Standard for Professional Qualifications for Fire Inspector and Plans Examiner*.

Fire protection engineers may also be uniformed or civilian members of a fire department. In some cases, a fire protection engineer may work for a private company contracted with the fire department for a specific project or time period to offer consulting services. Fire protection engineers review architectural and fire protection systems plans for proposed buildings to ensure compliance with local fire and life safety codes and ordinances. They may also consult fire department administrators in the areas of operations and fire prevention.

Permits

A permit is an official document that grants the holder permission to perform a specific activity. Permits may be issued for a single activity, such as a fireworks display, or a continuous operation, such as the manufacture of fireworks. The building, fire, or code enforcement departments of a jurisdiction may be empowered to issue and monitor various types of permits **(Figure 5.13)**.

Types

Each code defines the type of situation or activity requiring a permit. Permits are generally issued for operation or construction purposes. An operational permit is issued to conduct an operation or business for a specified period of time or until the permit is renewed or revoked. A construction permit is issued for the installation or alteration of a system or equipment. Typically, permits that relate directly to fire and life safety are used to control the following activities:

- Maintenance, storage, or use of hazardous products
- Hazardous operations or processes
- Installation/operation of equipment in connection with hazardous materials operations, maintenance, or storage

City of Edmond — Building Permit Application

Date: _____

Commercial

Applicant Name _____ Contact Name _____

Mailing Address _____ City _____ State _____ Zip _____

Office Phone _____ Other Phone _____ Pager _____

Plans by _____ Contact Name _____

Address _____ Phone Number _____

Please Check If: New Construction ☐   Alteration ☐   Addition ☐

Desc. (If Alteration or Addition) _____ Existing Square Footage _____

Project Address _____ Project Name _____

Section _____ Township _____ Range _____ Lot(s) _____ Block _____ Addition _____

Zoned _____ Urban District: Yes _____ No _____ Estimated Cost $ _____ of building without lot

Utilities   Edmond Water ☐   Well ☐

             Edmond Sewer ☐   Septic ☐   (If Septic, must provide Form #641-581 signed by DEQ)

             Edmond Electric ☐   OG&E ☐

Type of Construction (check one): 1A ____ 1B ____ 2A ____ 2B ____ 3A ____ 3B ____ 4 ____ 5A ____ 5B ____

Multi-Family: Yes ____ No ____   If yes, no. of dwelling units (per building*): _____   *One building per application

Type of Occupancy (per IBC): _____   Occupant Load _____   Fire Suppression System: Yes ____ No ____

Area of Project _____   No. of Floors _____   No. of Curb Cuts _____   No. of Toilets/Urinals _____

Size of Water Meter Kit: ½" ☐   1" ☐   1 ½" ☐   2" ☐

Irrigation System: Yes _____ No _____ Size of Irrigation Meter Kit: ½" ☐   1" ☐   1 ½" ☐   2" ☐

At anytime will there be any streets or alleys blocked by trucks or equipment?   Yes _____ No _____

Building Plans NEW Commercial: Six (6)* sets of plans with two (2) sets of specifications
Building Plans Alter/Add Commercial: Five (5)* sets of plans with two (2) sets of specifications
*Provide one additional set of plans for food service projects

Note: Each application shall be accompanied by proper plans drawn to scale. PLOT PLAN, FLOOR PLAN, ELEVATIONS, STRUCTURAL, ELECTRICAL, PLUMBING AND MECHANICAL.

BUILDING CANNOT BE OCCUPIED WITHOUT A CERTIFICATE OF OCCUPANCY ISSUED BY
<u>THE BUILDING DEPARTMENT</u>.

BY SIGNING THIS FORM, YOU GUARANTEE THE BUILDING PLANS SUBMITTED MEET ALL REQUIREMENTS SET
<u>FORTH BY THE EDMOND CITY COUNCIL</u>.

Signed: _____

As Owner/Agent for Owner

Figure 5.13 A sample building permit application.

- Open burning
- Large area tents
- Construction of a significant structure in the jurisdiction

Process

The permit process begins when the property owner or occupant recognizes the need to obtain a permit for an operation of condition that will exist on a particular property. Citizens frequently contact inspectors for advice concerning the permit process, which includes the following steps: application, review, issuance, and expiration of the permit.

Application

In many jurisdictions, an applicant must complete a specific form in order to begin the permit process. A person can usually obtain this form from a fire and life safety inspector or other municipal inspector at their office. Depending on the type of permit sought and jurisdictional requirements, the applicant may be required to submit additional documentation along with the application. This documentation may include:

- Shop drawings
- Construction documents
- Plot diagrams
- Safety data sheets (SDSs) or other chemical documentation

Review

When the applicant has submitted the necessary documentation, an inspector will begin the review process. Jurisdictions will often specify a time frame for the review process to help the applicant track the progress of the project. The inspector must verify that all the required documentation has been supplied and all forms are complete. If any paperwork is out of order, the application may be rejected entirely or the applicant may be asked for the missing documentation.

If the paperwork is in order, the inspector may then perform a site visit to verify and further investigate the request. If a site visit is warranted, the inspector should follow local policy regarding timeliness.

The formal review of the application includes an examination of the request and comparison to local fire code requirements. The inspector must be satisfied that the request does not violate the code or any other law or ordinance. If the inspector is satisfied that the request conforms to these requirements, a permit may be issued.

Issuance

Once issued, the permit must be kept at the location for which it applies and must be readily available for viewing by an inspector. The permit should explain the conditions under which it was issued as well as the actions it allows and the time frame for which it is in force. A copy of the permit and all pertinent information from the application should be included in the fire inspection file for that property.

Expiration

Permits may be reissued, renewed, or revoked depending on circumstances. In all cases, an inspector should review the work in progress to determine if the conditions of the initial issuance are still being met. If changes to the permit language are required, the owner may have to submit additional documentation in order for a permit to be reissued or renewed.

A fire and life safety inspector has the authority to revoke a permit if upon inspection it is noted the stipulations of issuance are not being followed. The permit may also be revoked if it is found that any documentation by which it was issued was a misrepresentation of actual conditions on the property in question. Depending on the problem, a permit may be revoked permanently or temporarily, pending corrective action.

Fire and Life Safety Education

Educating all ages to recognize potential hazards and taking appropriate corrective action is an important fire department function. Teaching survival techniques, such as **"***Stop, Drop, and Roll*" or "*Crawl Low Under Smoke*," can alter behavior and favorably impact life safety in a community. Fire department public education has grown to encompass the many aspects of personal safety that the NFPA® Risk Watch® program addresses:

- Motor vehicle safety
- Fire and burn prevention
- Choking, suffocation, and strangulation prevention
- Poisoning prevention
- Falls prevention
- Firearms injury prevention
- Bicycle and pedestrian safety
- Water safety

    Fire departments often provide education and assistance with the following:

- Community Emergency Response Team (CERT)
- CPR/first aid
- Carbon monoxide detectors
- Installation of child safety seats

Presenting Fire and Life Safety Information

When addressing an audience on any topic, the first step is preparing the audience to listen to your message. This approach involves gaining the attention of the audience and making them active participants by motivating with the importance of your message. Parents in a presentation on smoke detectors will be motivated by their desire to protect their children with an early warning in case of fire. A sense of personal involvement is a key to developing interest for any message.

The second step, known as the presentation step, is the actual transfer of ideas and explaining information. The life safety educator strives to bring the subject "to life" with the use of multimedia aids, props, and demonstrations.

Props, such as smoke detectors and fire alarm pull stations, and demonstrations of crawling under smoke, or stop, drop, and roll, may be used to achieve greater interest.

The third step involves the participants using the information they have been taught. This step, known as application, provides opportunities for audience involvement using the techniques or information provided by the life safety educator. Participants should be allowed to practice what they have learned whenever possible. This may involve practicing stop, drop, and roll or testing a smoke detector. The educator or firefighter should supervise this step closely, correcting any errors as they occur.

The final step in the process of instruction involves evaluation of the audiences understanding of the material and effectiveness of the instructor and/or teaching aids used in the presentation. Audience members may be asked to demonstrate a skill unassisted or asked prepared questions to gauge their ability to employ the material presented in the class. In addition, course and instructor evaluation forms may be used to verify the quality of a program or document a need for revision.

Role of the Fire and Life Safety Educator

The public expects a high level of customer service, including information and education delivered in a professional manner, with curriculum that includes current risks and emerging trends **(Figure 5.14)**. Firefighters who conduct fire and life safety education should be familiar with the recommendations of NFPA® 1035, *Standard on Fire and Life Safety Educator, Public Information Officer, Youth Firesetter Intervention Specialist and Youth Firesetter Program Manager Professional Qualifications* For additional information on fire and life safety education, refer to IFSTA's **Fire and Life Safety Educator** manual.

Figure 5.14 A Public Information Officer relays fire and life safety information to the public. *Courtesy of Ron Jeffers.*

Fire and Life Safety Presentation Topics

Firefighters are often asked to assist in or teach some basic fire and life safety information. Some topics a firefighter may be asked to present include the following:

- Stop, drop, and roll
- Home safety practices
- Placing, testing, and maintaining smoke detectors
- Information about sprinkler systems
- CPR/first aid
- Wildland/urban interface safety issues

The topics a fire department may present as part of fire and life safety presentations may be generated by the major safety concerns of the local community.

Stop, Drop, and Roll

Firefighters should do more than simply tell people what to do if their clothing catches on fire. Both adults and children need to be effectively educated with firefighters first demonstrating and then soliciting individuals to perform the action. Demonstrate that if their clothes catch on fire, they must immediately STOP moving, DROP to the ground (covering their face with both hands as they drop), and ROLL over and over until the flames are smothered.

Point out that if someone's clothes catch on fire, an observer may need to assist the person in dropping to the ground and smothering the flames. Coats, rags, blankets or other heavy cloth items that are close to the victim can be used to help smother the flames. Once the fire is out, cool the area with clear cold water (if possible). Summon emergency medical assistance immediately **(Figure 5.15)**.

Figure 5.15 Fire safety educators teach kids to stop, drop, and roll. *Courtesy of Dayna Hilton and Johnson County RFD#1.*

Smoke and Carbon Monoxide (CO) Alarms

Smoke and carbon monoxide (CO) alarms provide early warning and facilitate egress for occupants faced with a potential life-threatening emergency. This early warning is often credited with saving the lives of occupants and lessens the risk to firefighters who may be injured or killed during search and rescue operations. Firefighters should have a good working knowledge of these devices because a major part of the residential safety survey involves checking for working smoke and CO alarms. For additional information concerning smoke detectors, consult IFSTA's **Fire Protection, Detection, and Suppression Systems** manual

Fire Station Tours

Firefighters are frequently asked by citizens to give tours of the firehouse. These requests may be from schools, community organizers, or from citizens who walk in off the street. A tour should be more than an opportunity to show off the firehouse and apparatus. Firefighters should provide a relevant fire and life safety message and coordinate it with literature that citizens may take home to reinforce the verbal message. These tours demonstrate the knowledge and professionalism found in the fire service. They are also opportunities to present a consistent message to the public.

Figure 5.16 A firefighter gives a station tour to children.

Because visitors likely have limited exposure to the fire department, their impressions from the tour will be lasting. Members should dress and act professionally at all times, the firehouse should be orderly and in a state of readiness, and all activities should be productive.

Firefighters should attempt to answer all questions courteously and to the best of their ability. Members giving the tour should know department policy on safety in the firehouse. It may be necessary to remind visitors that they should not climb on apparatus or attempt to don any pieces of protective equipment.

Do not allow visitors to roam around the firehouse unescorted. Brief visitors on what steps to take to remain safe if the firefighters must respond to an alarm during the tour. Take special care to protect children around slide poles or repair shop areas **(Figure 5.16)**.

If equipment or apparatus is to be demonstrated, it should be done with extreme care. Place a firefighter at each corner of the apparatus during a demonstration to keep young children from approaching too close. Before any apparatus is moved, perform a visual check around the truck. Do not allow visitors on elevating platforms or aerial ladders. During station tours, do not sound sirens and air horns because of the high decibel levels they produce.

Fire station mascots (dogs, cats) can pose a safety and liability issue during tours. Excited animals may snap at or bite strangers. Many fire departments restrict the presence of animals in the firehouse for this and other reasons. If a mascot is present during a tour, consider keeping it separate from the tour group in order to reduce the chance of injury.

Fire Investigation

Every firefighter has a role to play in order to assist with a fire investigation. A qualified investigator(s) will lead the inquiry into cause and origin. However, as firefighters working in engine and ladder companies are often the first representatives of the fire department to arrive at an incident, their observations may provide valuable insight during the investigative process.

Role of the Investigator

Typically, the fire chief or **fire marshal** in a jurisdiction has the legal responsibility for origin and cause determination. However, the actual investigative process is often conducted under their authority by individuals specially trained in this area of expertise. **Fire investigators** have received training in fire pattern recognition, evidence collection and preservation, interview techniques, and legal procedures and testimony. The International Association of Arson Investigators® (IAAI®) and the National Association of Fire Investigators (NAFI) offer training and testing to become a certified fire and explosion investigator. NFPA® 1033, *Standard for Professional Qualifications for Fire Investigator,* may be obtained through accredited organizations.

NOTE: Refer to NFPA 921®, *Guide for Fire and Explosion Investigations*, for further information on conducting fire investigations.

Some fire departments operate fire investigation or **arson** squads, while other jurisdictions use a joint fire department and law enforcement unit. In some locations, the police department has the sole responsibility for suspected incendiary fire investigations **(Figure 5.17)**. In addition, some state governments retain the authority to conduct cause-and-origin investigations. There are also private companies that conduct separate fire investigations on behalf of an insurance company or corporation that has suffered a loss from fire.

> **Fire Marshal** — Highest fire prevention officer of a state, province, county, or municipality. In Canada, this officer is *also known as* the Fire Commissioner.
>
> **Fire Investigator** — Public or private sector individual tasked with discovering the origin and cause of a fire, as well as who may be responsible or liable for a fire.
>
> **Arson** — Crime of willfully, maliciously, and intentionally starting an incendiary fire or causing an explosion to destroy one's property or the property of another. Precise legal definitions vary among jurisdictions, wherein it is defined by statutes and judicial decisions.

Figure 5.17 Local law enforcement may work with the fire department on arson investigations. *Courtesy of Ron Jeffers.*

Three Factors Needed to Start a Fire
- Fuel that ignites
- Form and source of the heat of ignition
- Act or omission that helped bring these two factors together

All fires should be investigated in order to determine if they were caused accidentally or are incendiary in nature. The investigator relies heavily on the fire officers and firefighters arriving first on the scene to assist in describing conditions, protecting potential evidence, and indicating any circumstances that were out of the ordinary at the fire scene.

Observations Made by Emergency Responders

While firefighters often have the first trained observations at a fire scene, important information may be observed throughout the fire fighting operation. For example, while checking for and extinguishing hidden fire during **overhaul** operations, personnel may find indications of fire patterns or unusual fire behavior not previously noticeable. In addition, firefighters working in the building may come across flammable liquid containers or signs of forced entry not visible during heavy smoke conditions earlier in the fire.

While the fire department must respond to and extinguish a fire as quickly and efficiently as possible, firefighters must understand that their actions may affect the investigation to determine the cause and origin after the fire has been extinguished. Members should be judicious in their overhaul practices as they may need to preserve evidence that may prove valuable in the investigation of accidental or incendiary fires. Because investigators seldom arrive with the first fire companies, firefighters should note occupants' or witnesses' statements and strive to protect any potential evidence until the arrival of a qualified investigator.

NOTE: For additional information concerning fire investigations, refer to IFSTA's **Introduction to Fire Origin and Cause** manual as well as the **Fire Investigator manual**.

Observations En Route

Upon receipt of an alarm, firefighters should begin to gather information on the following:

- **Time of day** — Are the people and circumstances at the scene as one would expect for the particular occupancy and response area? For example, in a residential neighborhood at 3 a.m., residents would likely be dressed in night clothes rather than street clothes. Firefighters would not normally expect a business owner to be at his or her factory in the middle of the night, whereas a custodian or maintenance worker would likely arouse little suspicion at the same hour.

- **Visible flames and smoke** — Firefighters should notice and remember any visible smoke or fire conditions on arrival. Firefighters should be able to recall the floor and section of the building as well as be able to describe the volume, color, and spread of the smoke and fire.

Overhaul — Operations conducted once the main body of fire has been extinguished; consists of searching for and extinguishing hidden or remaining fire, placing the building and its contents in a safe condition, determining the cause of the fire, and recognizing and preserving evidence of arson.

- **Weather and natural hazards** — Do current weather conditions match what is found at the occupancy? Is the furnace running on a hot day? Are windows open on a cold night? Arsonists sometimes open windows to help the wind feed a fire, or set fires during inclement weather expecting a longer response time by the fire department.

- **Barriers** — Note any unusual impediments to fire fighting activities, as they may have been placed to purposely delay their efforts. For example, common ways an arsonist can complicate fire department operations are to block streets with trash containers, use vehicles to block hydrants, and sabotage sprinkler and standpipe connections.

- **People leaving the scene** — Under most circumstances, bystanders tend to linger around and watch the action of a fire scene. However, people seen leaving, especially hurriedly, may be suspicious. Firefighters should note a physical description of any people who leave abruptly or act strangely at the fire scene. If they leave by car, firefighters should record its make, model, color, and license number.

Observations Upon Arrival

Upon arrival at the scene, and while engaging in initial operations, firefighters should gather as much of the following information as possible:

Figure 5.18 Noticing the location of the fire helps to extinguish it quicker. *Courtesy of Ron Jeffers.*

- **Extent of fire on arrival** — Note the location and extent of visible fire and smoke. If the fire is self-vented, is it showing through a window (horizontal) or through the roof (vertical)? **(Figure 5.18)** If the person who reported the fire is available, inquire about the extent of the fire when it was discovered and ask that person to remain at the scene for further questioning.

- **Wind direction and velocity** — Note the wind direction and velocity as these conditions may greatly impact the path, speed, and intensity of the fire.

- **Doors or windows locked or unlocked** — Before entry is made, note the position and condition of doors and windows. Note any prior forcible entry, unlocked doors, or articles placed over windows to delay the discovery of the fire.

- **Position of electrical switches** — If possible, determine if a light or appliance that may have been involved in the fire had a switch in the open or closed position. Do not change the position of the switch.

- **Location of the fire** — This may help to determine the origin of the main body of fire. In addition, note if there appears to be more than one seemingly unconnected fire. An arsonist may have set multiple fires to guarantee as much damage as possible.

- **Containers or cans** — Note any metal cans or plastic containers inside or in the immediate vicinity of the building. They may have been used to transport accelerants and could yield valuable evidence during the investigation.

- **Burglary tools** — Take note of pry bars and screwdrivers found in unusual places in the structure. An arsonist may have used them to force entry.

- **Familiar faces** — Scan the crowd of bystanders for familiar faces. These known individuals may be people who like to watch fires or may be habitual firesetters.

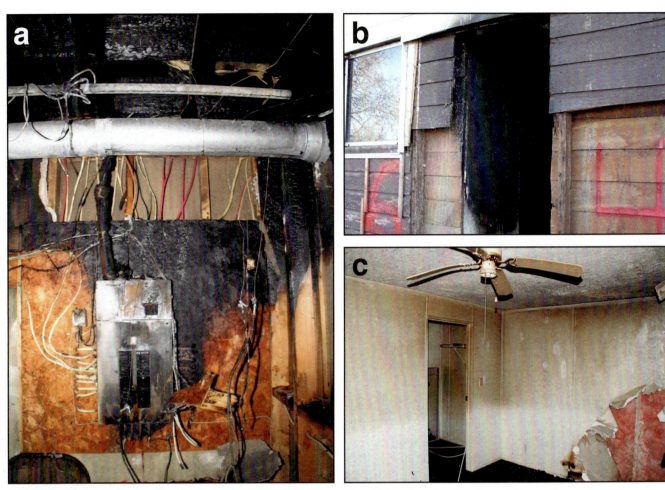

Figure 5.19 a, b, c Different fire patterns. *a-Courtesy of Wayne Chapdlaine, b and c Courtesy of Donny Howard*

Firefighter Observations

Firefighters should note numerous details as they continue the fire fight. These observations may include:

- **Fire patterns** — Note areas of irregular burning or locally heavy charring in areas of little fuel. While personnel are engaged in fire fighting, observe and note the spread and intensity of the fire. This observation may aid in determining the origin and fuel(s) involved **(Figure 5.19 a, b, c)**.

- **Unusual odors** — Firefighters must wear SCBA during fire fighting operations until the atmosphere has been tested and determined safe. However, note and report to investigators any odor that may linger outside or in the vicinity of the building.

- **Abnormal behavior of fire when water is applied** — Re-ignition of fire, or an increase in the intensity of the fire after water has been applied, may indicate the use of an accelerant or improper storage of flammable liquids. Water applied to a flammable liquid or accelerant may cause it to splatter, allowing flame intensity to increase and the fire to spread in several directions.

- **Color of flames and smoke** — Note the color and intensity of flames and smoke. This information may aid investigators in determining what material was burning **(Figure 5.20)**.

- **Obstacles that hinder fire fighting** — Note obstacles that seem to be placed intentionally in the way to hamper fire fighting efforts. For example, furniture that blocks doors, stairs with missing treads, or a floor with a hole cut in it to slow efforts, injure firefighters, or aid in the spread of the fire.

- **Incendiary devices** — Fragments of bottles, metal parts, and electrical wiring may be components of an incendiary device that has been used to start or spread a fire. There may be numerous devices, and firefighters may find at least one that has not functioned properly and can be thoroughly examined by investigators.

- **Trailers** — Combustible materials, such as rolled newspapers, rags, blankets, and ignitable liquids, may be used as trailers to spread fire from one point to another. These articles often leave char or fire patterns and may be used in conjunction with incendiary ignition devices.

- **Demolition, construction, and structural alterations** — When a structure undergoes renovation or demolition, plaster or drywall is often removed, holes are made in ceilings, and walls and fire doors are secured in an open position. These changes, as well as any damage created to purposely cause fire spread, may help to allow fire to quickly spread between floors or rooms. These alterations should be brought to the attention of the fire investigator.

- **Heat intensity** — Observe indications of high heat intensity, especially in relation to other areas of a room. This condition may indicate a different fire load or the use of an accelerant.

- **Fire detection and suppression systems** — Note and relay the presence and operating condition of any suppression or detection systems to investigators.

- **Location of fire** — Note possible sources of ignition as well as any fires remote from likely sources of ignition. Fires that appear to originate in closets, bathtubs, file drawers, or the center of a room may indicate suspicious activity.

- **Personal possessions/equipment and inventory** — While not all people possess the same amount of material goods, firefighters should be aware of occupancies that look conspicuously empty. For example, a lack of basic furnishings, pictures missing from the walls, and the absence of clothing or appliances, may raise suspicion. Commercial or retail occupancies follow similar scrutiny. A store with inappropriately damaged stock or an office with empty files and no computer equipment may indicate the fire was a planned event.

Figure 5.20 Firefighters must note the color and intensity of flames and smoke to help determine the type of fire they are fighting. *Courtesy of Ron Jeffers.*

Figure 5.21 Firefighters discussing observations after the incident. *Courtesy of Ron Jeffers.*

Responsibilities After the Fire

Firefighters should report any of the aforementioned facts or conditions to the officer in charge as soon as possible. The firefighter or company officer may be asked to write a statement or provide an interview that details the important facts that he or she has observed. Even information that was relayed from a third party, such as a bystander or occupant, should be reported. Although this information may be hearsay, it can give the investigator a possibly valuable lead to follow **(Figure 5.21)**.

Written Account

Legal cases may come to trial several years after an incident takes place. Firefighters with key observations should maintain a written account of their observation so as not to rely on memory alone while providing testimony. Lawyers may use these written accounts while representing other parties in a lawsuit. Any notes must be written in a professional manner.

Conduct and Statements at the Scene

Although firefighters should be observant while at the scene, they should not attempt to interrogate any person they consider a potential suspect. Firefighters should relay their observations to an investigator. A trained investigator should conduct an interview with any potential suspect or involved party. If an occupant or property owner should provide any information, a firefighter should listen closely and relay the material to an investigator as soon as possible.

Firefighters should refrain from making any statements or accusations or providing personal opinions concerning a fire to anyone. The property owner, members of the news media, or other bystanders may hear these statements and opinions and mistake them as facts. Careless joking or conjecture may prove embarrassing to the fire department and strain community relations. A sufficient reply to any question regarding origin and cause is to state that the "fire is under investigation" and refer the person to an investigator or chief officer.

Media Inquiries

Only personnel authorized by the fire department should speak to the media. Any inquiries with firefighters should be referred to the Public Information Officer (PIO), the Incident Commander (IC), or lead fire investigator.

Securing the Fire Scene

The fire scene must remain secure until investigators have concluded their work and are ready to release the premise back to the property owner. During salvage and overhaul operations, firefighters should take care not to contaminate the scene while operating power tools, hoselines, or other equipment.

If an investigator is not immediately able to respond, the building and any surrounding property of interest should be secured and guarded until personnel trained in evidence collection procedures can mark, tag, or photograph potential evidence **(Figure 5.22)**. This duty may be given to members of law enforcement or fire department members who are appropriately trained.

Figure 5.22 Firefighters covering furniture with a salvage cover to prevent water damage during fire fighting operations.

Chapter 5 • Fire Prevention, Life Safety Education, and Fire Investigation **175**

Figure 5.23 Firefighters boarded up the windows of this building to prevent internal weather damage.

The fire department has the authority to deny access to a building during fire fighting operations and for a reasonable time after suppression efforts have terminated. Department authority ends when the last fire department official leaves the scene. Future access may require either the owner's written permission or a search warrant.

While safeguarding the scene, firefighters should not allow anyone to enter without an investigator's permission. Any individuals granted entry should be escorted by a member of the fire department for the visitor's safety and in order to maintain scene security. The firefighter monitoring entry should maintain a written log that includes the person's name, times of entry and departure, and a brief listing of any items removed from the building.

A building may be secured in a variety of ways depending on incident circumstances. If exterior doors are generally intact, they may be locked and a firefighter positioned at one entrance that is to serve as a monitored access point. Closing or boarding doors and windows and covering any other openings will also help to prevent further damage to building contents or potential evidence **(Figure 5.23)**. This type of salvage work also reflects in a positive way on the fire department and is a function of good customer service.

Sometimes, an area around the building's exterior must be controlled for safety or investigation purposes. Cordoning off an area with rope or fire line tape is a quick and effective means of establishing a boundary that unauthorized people are not allowed to cross. Once a cordon is in place, law enforcement or fire department personnel must enforce the boundaries in order to ensure no one ventures beyond the established line.

Legal Considerations

As indicated in a previous section, firefighters may remain in control of a fire building or scene for as long as reasonably necessary. However, once they leave, a search warrant may be required in order for investigators to regain access.

Case of Michigan v. Tyler

This requirement for a search warrant is based on the case of Michigan v. Tyler (436 U. S. 499, 56 L. Ed. 2d 486 [1978]). The United States Supreme Court held in that case that "once in a building [to extinguish a fire], firefighters may seize [without a warrant] evidence of arson that is in plain view... [and] officials need no warrant to remain in a building for a reasonable time to investigate the cause of a blaze after it has been extinguished."

The court agreed, with modification, with the Michigan State Supreme Court's statement that "[if] there has been a fire, the blaze extinguished and the firefighters have left the premises, a warrant is required to re-enter and search the premises, unless there is consent..."

These legal decisions mean that if there is evidence that may be gained from a property, the fire department must maintain at least one member at the scene until an investigator arrives to process anything of value in determining origin and cause. To return to the scene to search for additional evidence, even with the owner's consent or a search warrant, may complicate investigation or prosecution efforts.

Each fire department should develop a standard operating procedure or guideline (SOP/G) based on legal interpretations that affect its jurisdiction. This information may be obtained from the office of the district attorney or state attorney general.

Protecting and Preserving Evidence

The need to protect and preserve potential evidence at a fire scene is important for accidental as well as incendiary fires since accidental fires may be the subject of civil litigation and/or criminal prosecution **(Figure 5.24)**. Any material that may help determine the origin and cause of a fire may be considered evidence. Examples of such materials include the following:

- Burned or scorched appliances that may have been involved in the ignition of the fire
- Burned, scorched, or melted electrical cords, extension cords, and outlets
- Areas of concentrated fire seen early in suppression operations
- Portable space heaters
- Smoke alarms

Firefighters should consider that excess use of water, trampling unnecessarily through debris, or displacing the contents of a fire building may inhibit the collection or use of important evidence. Firefighters should safeguard partially burned papers found in a furnace, fireplace, or stove until an investigator can collect them. Footprints or tire marks outside a structure may yield important information and may be protected with a box

Figure 5.24 A fire investigator examining the aftermath of a fire scene.

until they can be photographed or plaster casts made. After receiving the approval of investigators, firefighters may overhaul the structure and its contents to prevent rekindle of the fire.

Chapter Summary

The fire service responds to many types of emergencies in addition to fires. In order to increase the safety of firefighters and civilians alike, fire departments must work to prevent fire and other emergencies by educating the public concerning hazard recognition and effective actions during a fire or other emergency. Firefighters need to educate the public through a variety of media using active public relations programs.

These efforts should be integrated with residential fire and life safety surveys and preincident surveys of businesses, places of assembly, and other occupancies. Fire code inspectors and other professionals work to ensure buildings are built and renovated in accordance with jurisdictional codes and industrial processes., They also ensure occupancy use is safe and correct for existing conditions. Thorough knowledge of fire and life safety codes that a jurisdiction adopts is necessary in working with building owners, developers, and occupants when new construction or a change of occupancy is planned.

When a fire does occur, the fire department must not only extinguish it as safely and efficiently as possible, but also determine the origin and cause of the fire. Firefighters have the responsibility of observing a wide variety of conditions as they arrive first at the scene of an incident. When any potential evidence is discovered, it must be safeguarded until an investigator can properly document and preserve it for use in an origin-and-cause investigation.

Review Questions

1. What is the primary focus of the fire service?
2. List common fuel and heat source hazards.
3. What are some common fire hazards?
4. What are special fire hazards and target hazards?
5. Name some of the information needed to conduct a fire risk analysis.
6. List the components of fire prevention.
7. What is the purpose of a preincident survey?
8. What are some actions that should be taken when conducting facility surveys?
9. What are the main objectives of residential fire safety surveys?
10. What are some other codes that may affect fire code enforcement?
11. What are some duties of the fire inspector?
12. List steps in the permit process.
13. What are the steps in presenting fire and life safety information?
14. What are some fire and life safety presentation topics?
15. What precautions should be taken when giving fire station tours?
16. What observations should be made by emergency responders throughout the fire fighting operation?
17. What are some actions taken when securing the fire scene?
18. Why is protecting and preserving evidence important?

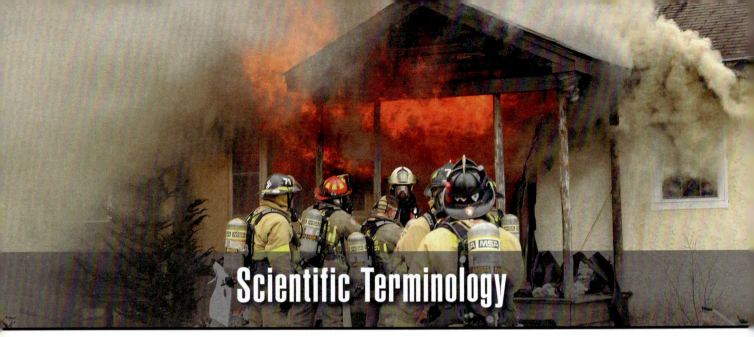

Scientific Terminology

Courtesy of Bob Esposito.

Chapter Contents

Case History 183	
Property of Matter **184**	
Physical States of Matter 184	
Specific Gravity and Vapor Density 184	
Physical and Chemical Changes 185	
Mass and Energy 185	
Combustion **186**	
Fire Tetrahedron 187	
Stages of Fire Development 196	
Special Considerations **202**	
Rollover 202	
Thermal Layering of Gases 203	
Backdraft 203	
Smoke Explosion 205	
Products of Combustion 205	

Classification of Fires **205**
    Class A Fires 206
    Class B Fires 206
    Class C Fires 206
    Class D Fires 207
    Class K Fires 207

Fire Extinguishment Theory **207**
    Temperature Reduction 208
    Fuel Removal 208
    Oxygen Exclusion 209
    Chemical Flame Inhibition 209

Chapter Summary **209**
Review Questions **209**

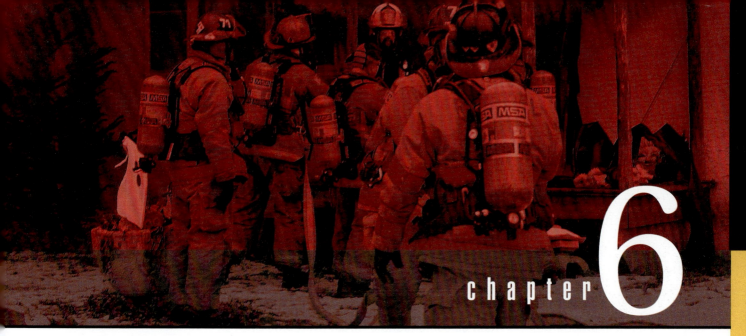

Chapter 6

Key Terms

Atmospheric Pressure 184	Law of Heat Flow 190
Backdraft 204	Lower Flammable (Explosive) Limit (LFL) 195
Chemical Reaction 186	Oxidizing Agent 187
Class A Fire 206	Oxygen-Enriched Atmosphere 196
Class B Fire 206	Potential Energy 185
Class C Fire 206	Products of Combustion 191
Class D Fire 207	Pyrolysis 193
Class K Fire 207	Rollover 202
Chemical Reaction 186	Smoke Explosion 205
Combustion 186	Specific Gravity 185
Endothermic Reaction 185	Surface-to-Mass Ratio 194
Exothermic Reaction 185	Thermal Layering of Gases 203
Flammable/Explosive Range 195	Upper Flammable Limit 195
Flashover 199	Vapor Density 185
Fuel 185	Ventilation-Controlled 201
Fuel Controlled 198	Volatility 195
Halon 196	

FESHE Outcomes

This chapter provides information that addresses the outcomes for the Fire and Emergency Services Higher Education (FESHE) *Principles of Emergency Services* course.

2. Analyze the basic components of fire as a chemical chain reaction, the major phases of fire, and examine the main factors that influence fire spread and fire behavior.

Chapter 6 • Scientific Terminology   **181**

Scientific Terminology

Learning Objectives

1. Describe properties of matter. (Outcome 2)
2. Explain combustion, the fire tetrahedron, and stages of fire development. (Outcome 2)
3. Describe hazardous conditions involving fire that have a drastic impact on incident safety. (Outcome 2)
4. Distinguish among the five main classifications of fires. (Outcome 2)
5. Describe the four aspects of fire extinguishment theory. (Outcome 2)

Chapter 6
Scientific Terminology

Case History

Greg was riding with Engine Four when it arrived at a house fire that was burning furiously. He was in his personal protective equipment (PPE). He took the nozzle, and two other firefighters backed him up on the line. As they approached the front door, the firefighters noticed that the smoke was being drawn in and then coming out the door in puffs. They could not see any flame but felt intense heat. The firefighter immediately behind Greg told him, "We have got to get out of here now. There is going to be a backdraft." Greg knew that a backdraft meant that there was going to be an instantaneous explosion if they opened the door. They quickly backed away, and within seconds, the front door was blown over their heads. They credited their quick size-up and action in recognizing that the structure had consumed most of the oxygen and was seeking additional oxygen to continue burning. Opening the door would have supplied this oxygen, causing the superheated gases to immediately increase in intensity.

This chapter introduces several basic concepts from physical science that describe the ignition and development of a fire. Firefighters can use the information in this chapter to interpret what they see on the fireground and develop methods to prevent, extinguish, and investigate fires. An understanding of fire behavior and the phases a fire passes through as it grows will help firefighters select the proper tactics to attack and extinguish fires. This knowledge also helps them recognize the potential hazards to themselves and others while they work on the fireground.

Firefighters responding to a fire may have to cope rapidly with a variety of conditions. The fire may be exposing (endangering) another structure or groups of structures, as in wildland/urban interface fires. The smoke and flames may be creating a life hazard (danger to survival) to occupants. The room of fire origin may be close to flashover (simultaneous ignition of room contents). If a building is not ventilated, there may be a backdraft (fire explosion) potential. All of these conditions result from fire and the way it behaves.

NOTE: Many of the concepts discussed in this chapter hold true for wildland fires, but a number of additional factors (not addressed in this manual) must be addressed for those incidents.

Properties of Matter

Physical material is called *matter*, which is anything that occupies space and has mass (weight). Matter is capable of undergoing many types of physical and chemical changes. The following section focuses on these changes as they relate to how matter reacts to fire.

Physical States of Matter

Matter possesses properties that are observable, such as its physical state as a solid, liquid, or gas, or its color or smell. An example of the three states of matter is easily found in the observable states of water. At normal **atmospheric pressure** and temperatures above 32°F (0° Celsius), water is found as a liquid. When the temperature of water falls below 32°F (0° Celsius) and the atmospheric pressure is the same, water changes its state to a solid and becomes ice. At temperatures above its boiling point, water changes to its gas form and is called *steam*.

Atmospheric Pressure — Force exerted by the atmosphere at the surface of the earth due to the weight of air. Atmospheric pressure at sea level is about 14.7 psi (101 kPa) and is measured as 760 mm of mercury on a barometer. Atmospheric pressure increases as elevation is decreased below sea level and decreases as elevation increases above sea level.

Pressure Increases

Pressure is the other factor that determines when a change of state occurs. As the pressure on the surface of a substance decreases, so does the temperature at which it boils. The opposite is also true. As the pressure increases on the surface, the temperature at which the substance boils will increase as well. This principle is evident in the operation of pressure cookers, as they heat water beyond its boiling point in the outside atmosphere. The boiling point of the liquid increases as the pressure increases. This result is due to when water (at 80° F) (25° Celsius) turns into steam (212°F) (100° Celsius), the steam expands 1,667 times from the volume of water.

Specific Gravity and Vapor Density

Matter can also be described using terms derived from its physical properties of mass and volume. Density is the measure of how tightly the molecules of a solid substance are packed together.

Mass and Weight

The terms mass and weight are often confused. The weight of an object is a measure of the force on that object exerted by gravity. The mass of an object is the measure of its amount of matter. For example, the weight of a block of wood on the moon would be approximately one-sixth that of the same block on earth due to the moon's weaker gravity. However, its mass would be the same on the moon as on earth.

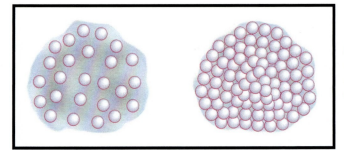

Figure 6.1 The molecules on the right are denser than the molecules on the left.

A common descriptive ratio of the weight of a liquid is its specific gravity. **Specific gravity** is the ratio of the mass of a given volume of liquid compared with the mass of an equal volume of water. As a result, water has a specific gravity of 1. Liquids with a specific gravity less than 1 are lighter than water, such as gasoline and most flammable liquids. Meanwhile, those liquids with a specific gravity greater than 1 are heavier **(Figure 6.1)**.

Vapor density is defined as the density of a gas or vapor in relation to air. Since air is used as the comparison, it has a vapor density of 1. Gases, such as methane, with a vapor density less than 1 will rise. Gases, such as propane, with a vapor density greater than 1 will sink.

Physical and Chemical Changes

A physical change occurs when a substance changes one of its observable properties, such as size, shape, or appearance, but remains chemically the same. Common examples of physical changes are water freezing (liquid to solid) or boiling (liquid to gas).

A chemical change occurs when a substance changes from one type of matter into another. A chemical change often involves the reaction of two or more substances to form other types of compounds. Oxidation is a chemical reaction involving the combination of oxygen with other materials. This process can be slow, such as the combination of oxygen with iron to form rust or rapid, as in the combination of methane.

Chemical and physical changes almost always involve an exchange of energy. The **potential energy** of a **fuel** is released during **combustion** and converted to kinetic energy. Reactions that give off energy as they occur are known as exothermic, while those reactions that absorb energy are called endothermic. Fire is an **exothermic reaction** called combustion that releases energy in the form of heat and sometimes light. Converting water from a liquid to a gas (steam) requires the input of energy and is an **endothermic** reaction. This process is an important part of fire suppression.

Mass and Energy

As a fire consumes fuel, its mass of the fuel is reduced. Because mass and energy can neither be created nor destroyed, the reduction in the mass of the fuel results in energy in the form of heat and light.

Firefighters should be aware that the more fuel there is available to burn, the more potential there is for greater amounts of energy being released as heat during a fire. The more energy (heat) that is released, the more extinguishing agent (usually water) that is needed to extinguish the fire.

Specific Gravity — Mass (weight) of a substance compared to the weight of an equal volume of water at a given temperature. A specific gravity less than 1 indicates a substance lighter than water; a specific gravity greater than 1 indicates a substance heavier than water.

Vapor Density — Weight of a given volume of pure vapor or gas compared to the weight of an equal volume of dry air at the same temperature and pressure. A vapor density less than 1 indicates a vapor lighter than air; a vapor density greater than 1 indicates a vapor density heavier than air.

Potential Energy — Stored energy possessed by an object that can be released in the future to perform work once released.

Fuel — (1) Flammable and combustible substances available for a fire to consume. (2) Material that will maintain combustion under specified environmental conditions. (NFPA® 921)

Exothermic Reaction — Chemical reaction between two or more materials that changes the materials and produces heat.

Endothermic Reaction — Chemical reaction in which a substance absorbs heat energy.

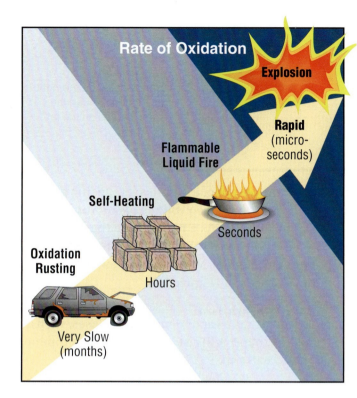

Figure 6.2 Combustion, a self-sustaining chemical reaction, may be slow (rusting) or fast (explosion).

Combustion

Combustion is a rapid, self-sustaining chemical chain reaction that releases heat energy in the form of heat, light, and by-products that can cause further reactions. Modes of combustion are differentiated by where the reaction occurs. In flaming combustion, the oxidation involves fuel in the gas phase. Heat is required to convert liquid or solid fuels into gases. When heated, both types of fuels will give off vapors that can mix with oxygen and burn, producing flames. Some solid fuels, especially those that are porous and can char, may undergo oxidation at the surface level of the fuel. This process is known as non-flaming or smoldering combustion, such as burning charcoal, smoldering fabric, and upholstery.

The time it takes for a **chemical reaction** to occur determines the type of reaction that may be observed. As mentioned previously, oxidation at the slow end of the time spectrum is rust. Meanwhile, at the faster end of the spectrum, an individual may observe explosions that result from the rapid reaction of fuel and an oxidizer. These reactions release a large amount of energy over a short period of time.

While fire may take a variety of forms, all fires involve a heat-producing chemical reaction between some type of fuel and oxygen or similar substance. When any fuel burns, heat is generated faster than it can be dissipated, causing an increase in temperature **(Figure 6.2)**. The following sections explain topics that are of importance to firefighters concerning the combustion process.

> **Combustion** — A chemical process of oxidation that occurs at a rate fast enough to produce heat and usually light in the form of either a glow or flame. Reproduced with permission from NFPA® 921, *Guide for Fire and Explosion Investigations*, Copyright© 2011, National Fire Protection Association®.

> **Chemical Reaction** — Change in the composition of matter that involves a conversion of one substance into another.

Fire Tetrahedron

In order for combustion to occur, four components are necessary:

- Heat
- Fuel (reducing agent)
- Oxygen (**oxidizing agent**)
- Self-sustained chemical chain reaction

These components may be represented graphically as the *fire tetrahedron*. The fire tetrahedron represents the flaming mode of combustion. Each component of the tetrahedron must be in place in order for combustion to occur. If ignition has already occurred and one of the components is removed, the fire will be extinguished. Firefighters should have a thorough understanding of how the principles of the tetrahedron work. The following sections explain each component **(Figure 6.3)**.

Oxidizing Agent — Substance that oxidizes another substance; can cause other materials to combust more readily or make fires burn more strongly. *Also known as* Oxidizer.

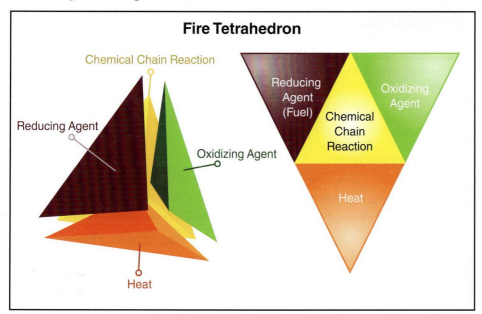

Figure 6.3 The four components of combustion, commonly referred to as a fire tetrahedron.

Fire Triangle

For many years, the fire triangle, consisting of heat, fuel, and oxygen, was used to explain combustion. This model does not provide a complete explanation of the combustion process. The fire tetrahedron, with the self-sustained chemical chain reaction added as a component, is a more accurate description of flaming combustion. **(Figure 6.4, p. 188)**

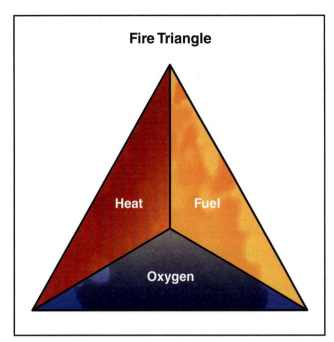

Figure 6.4 The fire triangle represented only three components of combustion.

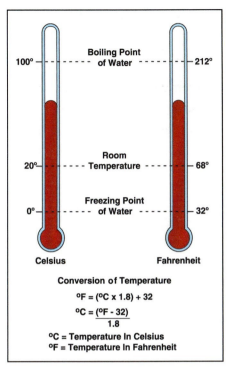

Figure 6.5 A comparison of the Celsius and Fahrenheit scales.

Heat and Temperature

In order to have a working knowledge of fire behavior, an individual must understand the terms *heat* and *temperature*. These terms are sometimes used interchangeably; however, each term has a distinct meaning.

Energy exists in two states: potential and kinetic. Potential energy is the energy possessed by an object that may be released in the future, and kinetic energy is the energy possessed by a moving object. Heat is kinetic energy associated with the movement of atoms and molecules that comprise matter. Before ignition, a fuel has potential chemical energy. When that fuel burns, the chemical energy is converted into kinetic energy in the form of heat and light. Temperature is a measurement of kinetic energy. Heat will move from objects of higher temperature to objects of lower temperature. This concept is important in understanding both the development of fire and the tactics used to control it.

Energy is the capacity to perform work. Work occurs when a force is applied to an object over a distance or when a chemical, biological, or physical transformation is made in a substance. Although energy is not measured directly, the work that it accomplishes is measurable. In the case of heat, work refers to the increase in the temperature of the substance.

In the International System of Units (SI), the measure for heat energy is expressed in joules. A joule is equivalent to one newton over a distance of one meter. In the Customary System, the unit used to measure heat energy is the British thermal unit (BTU), which is the amount of heat required to raise the temperature of 1 pound of water 1 degree. While not prevalent in scientific texts, the BTU is frequently used in the fire service **(Figure 6.5)**.

Celsius and Fahrenheit are the most commonly used scales to measure temperature in the fire service. The Celsius scale is used in the SI system, while Fahrenheit is the measurement used in the Customary System. When comparing these two scales, the freezing point of water is often used as a reference. This temperature is recorded as zero in Celsius and 32 degrees in Fahrenheit.

Energy exists in many forms and may change from one form to another. In the study of fire behavior, the conversion of energy into heat is important because heat is the energy component of the fire tetrahedron. When a fuel is heated, its temperature increases. Applying additional heat causes pyrolysis (the chemical decomposition of a substance through the action of heat) in solid fuels and vaporization of liquid fuels, releasing ignitable vapors. A spark or another external source may provide the energy required for ignition, or the fuel may be heated until it ignites without a spark or another source. Once ignited, the process continues the production and ignition of fuel vapors or gases so that the combustion reaction is sustained.

There are two forms of ignition: piloted ignition and autoignition. Piloted ignition occurs when a mixture of fuel and oxygen encounter an external heat (ignition) source with sufficient heat energy to start the combustion reaction. Autoignition occurs without any external flame or spark to ignite the fuel gases or vapors. In this case, the fuel surface is chemically heated to the point at which a combustion reaction occurs. Autoignition temperature is the temperature to which the surface of a substance must be heated in order for ignition and self-sustained combustion to occur. The autoignition temperature of a substance is always higher than its piloted ignition temperature. While both types of ignition may occur under fire conditions, piloted ignition is the most common.

Sources of Heat Energy

Chemical, mechanical, electrical, light, nuclear, and sound energy may cause a substance to heat by increasing the speed at which its molecules move. The most common sources of heat that result in the ignition of a fuel are chemical, electrical, and mechanical. Each of these sources is described in the following section.

- Chemical heat energy — The most common source of heat in combustion reactions. When a combustible is in contact with oxygen, oxidation occurs. This process almost always results in the production of heat. Self-heating or spontaneous heating is another form of *chemical heat energy* that occurs when a material increases in temperature without the addition of an external heat source. Normally, oxidation slowly produces heat, and heat is lost to the surroundings almost as fast as it is generated. An external heat source, such as the sun, can accelerate this process. In order for self-heating to progress to spontaneous ignition, the material must be heated to its autoignition temperature. For spontaneous ignition to occur, the following factors are required:

  — The insulation properties of the material immediately surrounding the fuel must be such that the heat cannot dissipate as fast as it is generated.

  — The rate of heat production must be great enough to raise the temperature of the material to its ignition temperature.

- The available air supply (ventilation) in and around the material must be adequate to support combustion.
- Spontaneous ignition may occur in rags soaked in linseed oil, rolled into a ball, and thrown into a corner. The temperature of the rags may eventually increase enough to cause ignition if the heat generated by the natural oxidation of the oil and the cloth is not allowed to dissipate.
- The rate of the oxidation reaction, and thus the heat production, increases as more heat is generated and held by the materials insulating the fuel.
- The more heat that is generated and absorbed by the fuel, the faster the reaction will cause the heat generation.

• Electrical heat energy — Generates temperatures high enough to ignite any combustible materials in the vicinity of the heated area. Electrical heating may occur in several ways, including the following:

- Resistance heating — When electric current flows through a conductor, heat is produced. Some electrical appliances, such as ranges, ovens, or portable heaters, are designed to make use of resistance heating. Other electrical equipment is designed to limit resistance heating under normal operating conditions.
- Overcurrent or overload — When current flowing through a conductor exceeds design limits, it may overheat and present an ignition hazard. Overcurrent or overload is an unintended form of resistance heating.
- Arcing — In general, an arc is a high-temperature luminous electrical discharge across a gap or through a medium, such as charged insulation. An arc may be generated when a conductor is separated (as in an electric motor or switch) or by high voltage, static electricity, or lightning.
- Sparking — When an electrical arc occurs, luminous particles may be formed and spatter away from the point of arcing.

• Mechanical heat energy — The movement of two surfaces against each other creates the heat of friction. Friction or compression creates mechanical heat energy. Heat of compression is generated when a gas is compressed. Diesel engines use this principle to ignite fuel vapor without a spark plug. This principle is also the reason SCBA cylinders feel warm to the touch after being filled.

Transmission of Heat

The transfer of heat from one point to another controls the growth of any fire from the initial fuel package to any other fuel beyond the area of fire origin. Firefighters use their knowledge of heat transfer to help estimate the size of a fire and evaluate the effectiveness of fire suppression operations. In order for heat transfer to take place, two objects must be at different temperatures. Heat moves from warmer to cooler objects. The greater the difference in temperatures between two objects, the greater the transfer rate will be. This principle is called the **Law of Heat Flow**.

Law of Heat Flow — Natural law that specifies that heat tends to flow from hot substances to cold substances. This phenomenon is based on the supposition that one substance can absorb heat from another.

Figure 6.6 Direct contact with pipes or ducts can conduct heat through walls.

Heat may be transferred between two bodies by means of conduction, convection, or radiation. Each mechanism is detailed in the following sections:

Conduction. When a piece of metal rod is heated at one end with a flame, the heat travels throughout the rod. This transfer of energy is due to the increased activity of the atoms within the rod. As heat is applied to one end of the rod, atoms in that area begin to move faster than other areas. This activity causes an increase in the collision of atoms. Collisions transfer energy to the atom being hit and cause that energy in the form of heat to be transferred throughout the rod **(Figure 6.6)**.

This heat transfer, known as conduction, is the point-to-point transmission of heat energy. Conduction occurs when an object is heated as a result of direct contact with a heat source. This method of heat transfer is common in the development of many fires. As the fire grows, hot gases begin to flow over objects at a greater distance from the point of ignition, causing the fire to spread by allowing the heat from the fire gases to directly contact fuel packages in other areas farther away from the origin of the fire. Good insulators are materials that do not conduct heat well. Because of their physical makeup, they disrupt the point-to-point transfer of heat or thermal energy. The best insulators used in building construction are those made of fine particles or fibers, with void spaces between them filled with a gas or air. Gases do not conduct heat very well because their molecules are far apart.

Convection. As a fire begins to grow, the air in the surrounding area is heated by the spread of hot gases and **products of combustion**. Convection is the transfer of heat energy by the movement of heated liquids or gasses. When heat is transferred by convection, there is a movement or circulation of a liquid or gas from a warmer area to a cooler area **(Figure 6.7, p. 192)**.

Radiation. Radiation is the transmission of energy as an electromagnetic wave without an intervening medium. This energy travels in a straight line at the speed of light. The heat that the sun provides to warm the surface of the

Products of Combustion — Materials produced and released during burning.

Chapter 6 • Scientific Terminology **191**

Figure 6.7 Convection is the transfer of heat energy by the movement of heated liquids or gases.

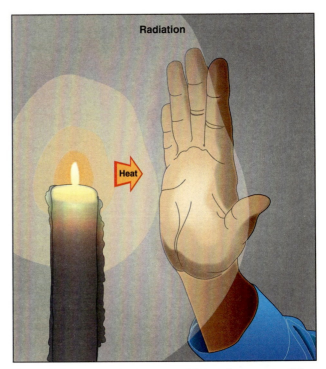

Figure 6.8 Radiated heat is one of the major causes of fire spread to exposures.

earth is a common example of this method of heat transmission. Radiant heat causes many exposure fires (fires ignited in a fuel package or structure remote from the building of origin) **(Figure 6.8)**.

As a fire grows, it radiates an increasing amount of energy in the form of heat. This heat may be capable of traveling through substantial air spaces that would disrupt convection or conduction. Materials that reflect radiated heat energy will disrupt this transmission of heat.

Firefighters benefit from understanding the various ways in which heat is transmitted. The ability to determine various ways in which a fire is likely to spread will aid personnel in the determination of safe and efficient tactics. Deploying a hoseline for exposure protection (radiation), cooling of fuel by applying water (conduction) or performing ventilation (convection) may be necessary at a scene.

NOTE: Several methods of heat transfer may occur simultaneously, such as a stove top fire spreading to kitchen cabinets (radiation and conduction) and vertically through an exhaust hood (convection).

Fuel

Fuel is the material or substance being oxidized or burned in the combustion process. This material is also known as the reducing agent because its mass is being reduced as the fire consumes it.

Fuels may be organic or inorganic. Inorganic fuels, such as magnesium, do not contain carbon. Most common fuels are organic, containing carbon and other elements. These fuels may be further divided into hydrocarbon-based fuels, such as gasoline, fuel oil, and plastics, and cellulose-based materials, such as wood and paper.

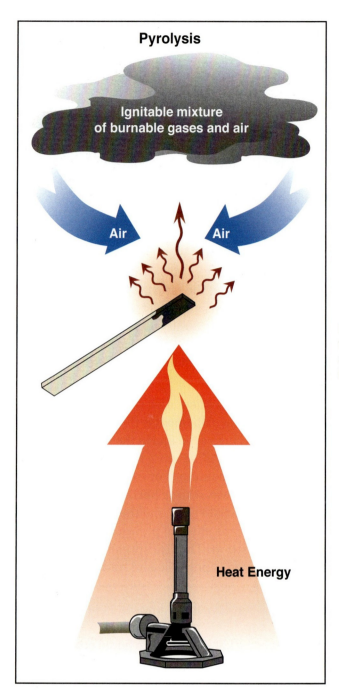

Figure 6.9 Heat causes the pyrolysis of solid fuels and the production of ignitable vapors or gases.

A fuel may be found in any of three states of matter: solid, liquid, or gas. In order to burn, fuels must be in the gaseous state. Heat energy is required to change solids and liquids into gas. Vapor is the common term used to describe the gaseous state of a fuel that would normally exist as a liquid or solid at standard temperature and pressure.

Vapors are evolved from solid fuels by **pyrolysis**, which is the chemical decomposition of a substance through the action of heat. During this process, as solid fuels are heated, combustible vapors are driven from the substance. If there is sufficient fuel and heat, pyrolysis may generate sufficient quantities of burnable vapors, which can ignite and sustain combustion if the other elements of the fire tetrahedron are present **(Figure 6.9)**.

Pyrolysis — Thermal or chemical decomposition of a solid material by heating, generally resulting in the lowered ignition temperature of the material; the pre-ignition combustion phase of burning during which heat energy is absorbed by the fuel, which in turn gives off flammable tars, pitches, and gases; often precedes combustion. Pyrolysis of wood releases combustible gases and leaves a charred surface.

Surface-To-Mass Ratio — Ratio of the surface area of the fuel to the mass of the fuel.

Because of their nature, solid fuels have a definite shape and size that significantly affects their ease of ignition. The **surface-to-mass ratio** of the fuel is a major factor in ignitability. Surface-to-mass ratio is the area of fuel in proportion to mass.

An example of surface-to-mass affecting ignitability may be found in the cutting of wood. In order to produce usable wood, a tree must be cut into a log. The mass of the log is very high, but the surface area is relatively low, making the surface-to-mass ratio low. The log may be milled into boards, reducing the mass of the individual boards compared with the whole log. However, the surface area is increased, therefore increasing the surface-to-mass ratio. As a result of the milling process, saw dust is created. This saw dust has an even higher surface-to-mass ratio. As the fuel particles become smaller, the ratio increases (more finely divided, as in saw dust as opposed to logs). This process also increases the ignitability of the material. With a greater surface area, more material may be exposed to heat, generating more burnable vapors due to pyrolysis **(Figure 6.10)**.

The proximity and orientation of a solid fuel relative to the source of heat also impacts the way it may burn. Fire spread involving a solid fuel in the vertical position (standing on edge) will be more rapid than if the fuel were arranged horizontally. For example, if a sheet of plywood paneling was resting horizontally across two sawhorses, a fire would consume the wood at a slower rate than if the same sheet of plywood were standing in a vertical position. The fire spreads more rapidly because the heated vapors rise and transfer more heat to the paneling.

Liquid fuels have a volume and mass, but since they assume the shape of their container, they have no fixed shape other than a flat surface. With liquid products, fuel vapors are generated by a process called *vaporization*. This process is achieved when liquid is transformed into its vapor or gaseous state.

This transformation from liquid to vapor occurs as the molecules of the substance escape from the liquids surface into the surrounding atmosphere. In order for these molecules to break free of the liquid's surface, there must be some energy input. In most cases, the energy is in the form of heat. For example, water left in a pan will eventually evaporate. The energy required to accomplish this process comes from the sun or surrounding environment. Water placed on a stove and heated to boiling will evaporate more quickly due to the amount of energy applied to the pan. The rate of vaporization may be determined by the substance, the amount of energy applied, and the surface area that is exposed.

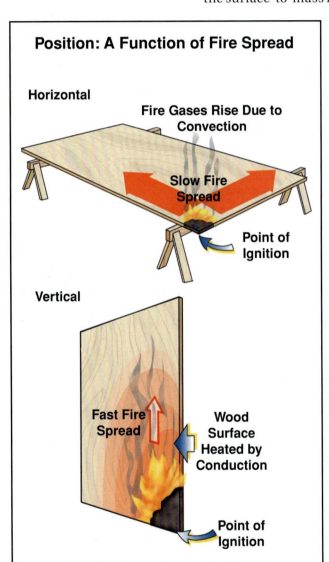

Figure 6.10 The actual position of a solid fuel affects the way it burns. Fuel in a vertical position will burn much more quickly due to increased heat transfer.

Vaporization of liquid generally requires less energy input than pyrolysis does for solid fuels. This scenario is primarily due to the different densities of substances in liquid or solid states and by the fact that molecules of a substance in liquid state have more energy than when they are in a solid state. Solids also tend to absorb more energy because of their mass. The **volatility** with which a liquid gives off vapor influences its ignitability. All liquids give off vapors to a certain degree in the form of simple evaporation. Liquids that give off flammable vapors at lower temperatures are more dangerous than those that require higher temperature to vaporize.

Like surface-to-mass ratio for solid fuels, the surface-to-volume ratio of liquids is important to their ignitability. When a spill or leak of liquid occurs, the product will flow and accumulate in low areas. When contained, the specific volume of a liquid may have a relatively low surface-to-volume ratio. However, upon release, this ratio may increase significantly along with the amount of fuel vaporized from the surface.

Gases have mass, but no definite shape or volume. For combustion to occur involving fuel in a gaseous state, it must be mixed with air (oxidizer) at the proper ratio. The range of concentrations of the fuel vapor and air is known as the **flammable/explosive range**. This range is measured using the percentage by volume of vapor in air for the **lower flammable limit (LFL)** and for the **upper flammable limit (UFL)**. The LFL identifies the minimum concentration of fuel vapor and air will support combustion. A concentration below the LFL is said to be too lean to burn. Accordingly, the UFL identifies the concentration above which combustion cannot take place. Concentrations above the UFL are said to be too rich to burn.

> **Volatility** — Ability of a substance to vaporize easily at a relatively low temperature.
>
> **Flammable/Explosive Range** — Percentage of a gas vapor concentration in the air that will burn if ignited.
>
> **Lower Flammable (Explosive) Limit (LFL)** — Lower limit at which a flammable gas or vapor will ignite and support combustion; below this limit the gas or vapor is too lean or thin to burn (too much oxygen and not enough gas, so lacks the proper quantity of fuel).
>
> **Upper Flammable Limit (UFL)** — Upper limit at which a flammable gas or vapor will ignite; above this limit the gas or vapor is too *rich* to burn (lacks the proper quantity of oxygen).

Gaseous Fuels
Gaseous fuels are already in the natural state required for ignition. No pyrolysis or vaporization is required to ready the fuel, and less energy is required for ignition.

Oxygen (Oxidizing Agent)

Oxidizing agents are materials that release oxygen or other oxidizing gases during the course of a chemical reaction. Oxidizers are not combustible by themselves, but support combustion when combined with a fuel. While oxygen is the most common **oxidizer**, other substances in the same category are bromates, chlorates, nitrates, and peroxides.

With oxygen comprising approximately 21 percent of our air, it is considered the primary oxidizing agent. At a room temperature of 70°F (20° Celsius), combustion may be supported at oxygen concentrations as low as 14 percent. However, as temperatures in a fire compartment increase, lower concentrations of oxygen will support flaming combustion.

NOTE: Air is composed of 78 percent nitrogen, 21 percent oxygen, and small amounts of several other gases, including argon, carbon dioxide, neon, and helium.

> **Oxidizer** — Any material that readily yields oxygen or other oxidizing gas, or that readily reacts to promote or initiate combustion of combustible materials. (Reproduced with permission from NFPA® 400-2010, Hazardous Materials Code, Copyright©2010, National Fire Protection Association®)

Oxygen-Enriched Atmosphere — Area in which the concentration of oxygen is in excess of 21 percent by volume or 21.3 kPa; typically 23.5 percent for confined spaces, as defined by the Occupational Safety and Health Administration (OSHA).

When oxygen concentrations exceed 21 percent, the atmosphere is said to be oxygen-enriched. Materials that burn at ordinary oxygen levels may ignite more easily and burn with greater intensity in an **oxygen-enriched atmosphere**. These conditions may be encountered in healthcare facilities, industrial occupancies, and even private homes where oxygen breathing equipment is in use.

Some petroleum-based products may autoignite in oxygen-enriched atmospheres. Some materials that do not ordinarily burn at normal oxygen levels will burn readily in oxygen-enriched environments. These materials include Nomex®, the fire-resistant material from which many articles of firefighter protective clothing is made. Nomex® will not burn at normal oxygen levels. However, in an enriched atmosphere of 31 percent oxygen, Nomex® ignites and burns vigorously.

Self-Sustained Chemical Chain Reaction

The self-sustained chemical reaction involving methane and oxygen is an example of flaming combustion. Complete oxidation of methane results in the production of carbon dioxide and water as well as the production of heat and light. This process is somewhat complex. During combustion, the molecules of methane and oxygen break apart to form free radicals (electrically charged, highly reactive parts of molecules). Free radicals combine with oxygen or with the elements that form the fuel material (in the case of methane carbon and hydrogen), producing intermediate combustion products (new substances), even more free radicals, and increasing the speed of the oxidation reaction **(Figure 6.11)**.

At various points in the combustion of methane, this process results in production of carbon monoxide and formaldehyde, which are both flammable and toxic. When more chemically complex fuels burn, different types of free radicals and intermediate combustion products are created, many of which are also flammable and toxic.

Flaming combustion is an example of a chemical chain reaction. Sufficient heat will cause fuel and oxygen to form free radicals and initiate self-sustained chemical reaction. The fire will continue to burn until fuel or oxygen is exhausted or an extinguishing agent is applied in sufficient quantity to interfere with the ongoing reaction. In some cases, extinguishing agents deprive the combustion process of fuel, oxygen, or sufficient heat to sustain the reaction. Chemical flame inhibition is when a **Halon**-replacement agent interferes with this chemical reaction, forms a stable product, and terminates the combustion reaction.

Halon — Halogenated agent; extinguishes fire by inhibiting the chemical reaction between fuel and oxygen.

Stages of Fire Development

Both confined and unconfined fires will progress through stages or phases. In laboratory simulations, these stages have shown to be distinct when conditions in the room or compartment remain the same. These stages include incipient growth, fully developed, and decay. Fires outside the laboratory may not develop through each of the stages in this exact sequence.

Actual conditions in a building made up of multiple compartments (rooms) may vary widely. For example, in the compartment of origin, the fire may be in the fully developed stage. Meanwhile, in adjacent compartments, the fire

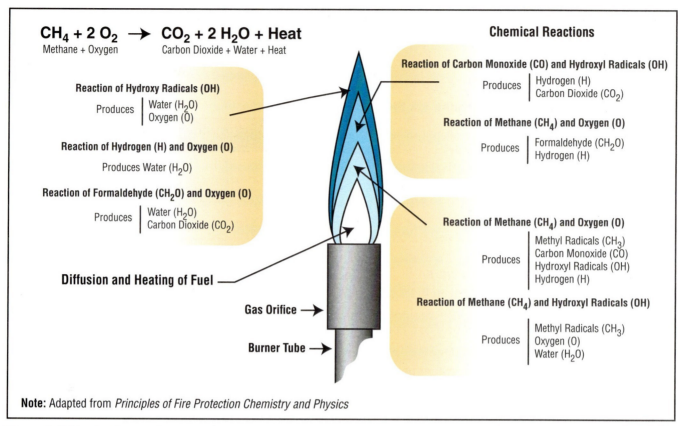

Figure 6.11 Combustion produces a variety of new substances.

may be in the growth stage. At the fire scene, use the stages of fire development as a guide to what may occur during the fire, not an exact pattern of what will occur at every fire. During fireground operations, firefighters should assess changing hazards and fire conditions at the incident scene rather than assume the fire will follow the same pattern as laboratory tests.

Study Results

Several organizations, including National Institute of Standards and Technology (NIST) and Underwriters Laboratories (UL), have conducted extensive analysis of fire behavior during a variety of conditions. The results of these studies help determine the tactics employed in modern fire fighting.

Factors that Affect Fire Development

A number of factors influence fire development within a compartment, including:

- Fuel type
- Availability and location of additional fuels
- Compartment volume and ceiling height
- Ventilation
- Thermal properties of the compartment
- Ambient conditions
- Fuel load **(Figure 6.12, p. 198)**

Chapter 6 • Scientific Terminology **197**

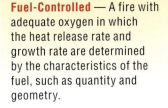

Figure 6.12 Fuel characteristics or the availability of an air supply may limit fire development.

Incipient

The incipient stage starts with ignition as the three elements of the fire triangle come together and the combustion process begins. At this point, the fire is small and confined to the material first ignited **(Figure 6.13)**.

Once combustion begins, development of an incipient fire is largely dependent on the characteristics and configuration of the fuel involved. This is a **fuel-controlled** fire. Air in the compartment may provide adequate oxygen to continue fire development. During this phase of development, radiant heat warms the adjacent fuel material and continues the process of pyrolysis.

A plume of hot gases from the fire begins to rise and mix with the cooler air in the room. As the plume reaches the ceiling, these gases begin to spread horizontally in what is referred to as a *ceiling jet*. The hot gases in the room contact the surfaces of the compartment and the contents within, transferring heat to these objects.

In this stage of development, the fire has not yet significantly influenced the environment of the compartment. The temperature is only slightly above normal, and the concentration of products of combustion is low. During this stage, occupants may be able to escape easily and the fire could be suppressed with a fire extinguisher. However, depending on the type and configuration of fuel, the fire may transition rapidly from the incipient to the growth stage.

Fuel-Controlled — A fire with adequate oxygen in which the heat release rate and growth rate are determined by the characteristics of the fuel, such as quantity and geometry.

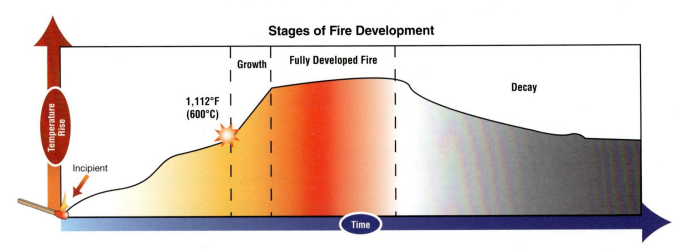

Figure 6.13 The stages of fire development in a compartment.

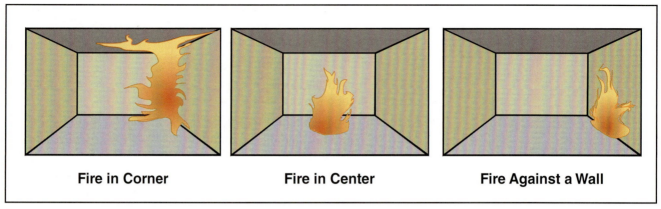

Figure 6.14 The location of a fire inside a compartment can affect its rate of growth.

Growth

As the fire transitions from the incipient to the growth stage, it begins to influence the environment in the compartment and has grown large enough for the compartment configuration and amount of ventilation to influence it. The first effect is the amount of air that is entrained into the plume. In a compartment fire, the location of the fuel package in relation to the compartment walls affects the amount of air that is entrained and thus the amount of cooling that takes place. Unconfined fires draw air from all sides, and the entrainment of this air cools the plumes of hot gasses, reducing the flame length and vertical extension.

When a fuel package is not in the middle of a room, the combustion zone (area where sufficient air is available to feed the fire) expands vertically and a higher plume results. A higher plume increases the temperatures in the developing layer of hot gases at the ceiling, increasing fire development. In addition, heated surfaces around the fire radiate back heat toward the burning fuel, further increasing the speed of fire development **(Figure 6.14)**.

Flashover

Flashover typically occurs in the growth stage of a fire, but it may come in the fully developed stage. When flashover occurs, the combustible materials in the compartment and the gases produced by pyrolysis ignite almost simultaneously. This results in full-room involvement.

During a flashover, the environment in a compartment is changing from two layers (hot on top and cooler on the bottom) to a single, well mixed hot gas condition from floor to ceiling that is untenable, even for fully protected firefighters. The transition period from preflashover to postflashover may occur rapidly. During a flashover, the volume of fire may increase from approximately one-fourth to one-half of the compartment's upper volume, filling the compartment and possibly extending out of any openings. When a flashover occurs, burning gases may push out of openings with considerable velocity.

Flashover — (1) Stage of a fire at which all surfaces and objects within a space have been heated to their ignition temperature, and flame breaks out almost at once over the surface of all objects in the space. (2) Rapid transition from the growth stage to the fully developed stage

Common Elements of Flashover

- **Transition in fire development** —Represents a transition from the growth stage to the fully developed stage.
- **Rapidity** — Although not instantaneous, flashover happens rapidly, often in a matter of seconds, spreading complete fire involvement within the compartment.
- **Compartment** — There must be an enclosed space, such as a room or enclosure.
- **Ignition of all exposed surfaces** —All combustible surfaces in the compartment becomes ignited.

Two interrelated factors determine whether a fire within a compartment will progress to flashover. First, there must be sufficient fuel, and the heat-release rate must be sufficient for flashover conditions to develop. For example, a couch on fire in a room will have a much greater likelihood of progressing to flashover than a small metal waste basket on fire in the same room.

Ventilation is the second factor. Regardless of the fuel package, heat release is dependent of oxygen. A developing fire must have sufficient oxygen to reach flashover. Heat release is limited by available air supply. If there is insufficient natural ventilation, a fire may enter the growth stage but not reach the heat release rate to transition through flashover to a fully developed fire.

Survival rates for firefighters in flashover are low **(Figure 6.15)**. While an exact temperature is not associated with a flashover, it may typically occur at 1,100°F (590° Celsius) ceiling temperature. Firefighters should be aware of the following flashover indicators:

- **Building indicators** — Flashover may occur in any building; interior configuration, fuel load, thermal properties, and ventilation will determine how rapidly it can occur
- **Smoke indicators** — Rapidly increasing volume, turbulence, darkening color, optical density, and lowering of the hot gas level
- **Air flow indicators** — High velocity and turbulence, bi-directional movement with smoke exiting the top of a doorway and fresh air moving in at the bottom, or pulsing air movement
- **Heat indicators** — Rapidly increasing temperature in the compartment, pyrolysis of contents located away from the fire, darkened windows, or hot surfaces
- **Flame indicators** — Isolated flames in the hot gas layers or near the ceiling

Fully Developed

The fully developed stage occurs when all combustible materials in the compartment are burning. During this stage:

- Burning fuels in the compartment are releasing the maximum amount of heat possible for the available oxygen and fuel, producing large volumes of fire gasses.

Figrue 6.15 Flashover can happen almost instantaneously. *Courtesy of NIST.*

- The fire is **ventilation-controlled** because the heat release is dependent on the compartment openings. These openings provide oxygen, which supports ongoing combustion and releases products of combustion. If the available air supply is increased, a higher heat release rate will result.

- Flammable products of combustion are likely to flow from the compartment of origin into adjacent compartments or out through openings to the building's exterior. Flames will extend out of compartment openings because there may be insufficient oxygen in the compartment for complete combustion.

Ventilation-Controlled — Fire with limited ventilation in which the heat release rate or growth is limited by the amount of oxygen available to the fire. (NFPA® 921, Guide for Fire and Explosion Investigations).

Decay

A compartment fire will decay as the fuel is consumed or if the oxygen concentration falls to the point that flaming combustion is diminished. Both of these situations can result in the combustion reaction coming to a stop. However, decay due to reduced oxygen concentration can follow a considerably different path if the ventilation of the compartment changes before combustion ceases and the temperature in the compartment lowers.

Rapid Fire Development

The fire service has used numerous terms to describe various fire events that result in rapid fire development. Among the more common terms are:

- Flashover
- Backdraft
- Smoke explosion

Chapter 6 • Scientific Terminology

Figure 6.16 Backdraft is a dangerous stage in a fire during which highly concentrated flammable gases are introduced to oxygen from an entry point below the gases. *Courtesy of Bob Esposito.*

Rapid fire development due to events such as those previously mentioned has been responsible for numerous firefighter deaths and injuries. Firefighters should be able to recognize the indicators of rapid fire development and determine the best course of action to take before they occur. Flashover was discussed in the previous section concerning the stages of fire. Backdraft and smoke explosions will be detailed in the subsequent section that explains special fire considerations **(Figure 6.16)**.

Special Considerations

Several hazardous conditions may occur on the fireground that have a drastic impact on incident safety. Firefighters should be aware of the potential for hazardous conditions and take actions to avoid or manage their effects. These hazardous conditions include rollover, thermal layering of gases, backdraft, and the effects of the products of combustion.

Rollover

Rollover describes a condition where the unburned gases that have accumulated at the top of a compartment ignite and flames spread through the hot gas layer or across the ceiling. Rollover may occur during the growth stage as the hot gas layer forms at the ceiling of a compartment. Flames may be observed in the layer when the combustible gases reach their ignition temperature. While the flames add to the total heat generated in the compartment, this condition is not flashover. Rollover will generally precede flashover, but it will not always result in flashover. Rollover contributes to flashover conditions because the burning gases at the upper level of the room generate a tremendous amount of radiant heat, which transfers to the fuel packages in the room. The new fuels begin pyrolysis and release additional gases that contribute to flashover.

Rollover — (1) Condition in which the unburned fire gases that have accumulated at the top of a compartment ignite and flames propagate through the hot-gas layer or across the ceiling. These superheated gases are pushed, under pressure, away from the fire area and into uninvolved areas where they mix with oxygen. When their flammable range is reached and additional oxygen is supplied by opening doors and/or applying fog streams, they ignite and a fire front develops, expanding very rapidly in a rolling action across the ceiling. See Backdraft, Flashover, and Incipient Phase. (2) Involves a vehicle rolling sideways onto its side and possibly continuing onto its top, then the opposite side.

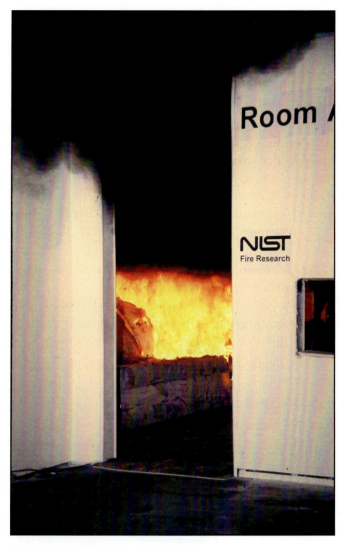

Figure 6.17 Thermal layering is also described as thermal balance because the fire gases form into layers according to their temperatures. *Courtesy of NIST.*

Thermal Layering of Gases

The **thermal layering of gases**, sometimes referred to a heat stratification or thermal balance, is the tendency of gases to form into layers according to temperature. Generally, the hottest gases will be in the upper layer, while cooler gases form the lower layers. Changes in ventilation and flow path can significantly affect thermal layering **(Figure 6.17)**.

As the volume and temperature of the hot gas layer increases, so does the pressure. Higher pressure in this layer causes the hot gases to spread downward within the compartment and laterally through any openings. The pressure of a cooler gas layer is lower, resulting in the inward movement of air from outside the compartment at the bottom of an opening as hotter gases exit through the top. The interface of the hot and cooler gas layers at the opening is called the neutral plane. This term refers to the neutral pressure where the layers meet. The neutral plane only exists at openings where hot gases are exiting and cooler air is entering a compartment.

Backdraft

A ventilation-controlled compartment fire may produce a large volume of flammable smoke and other gases due to incomplete combustion. While the heat release rate from a ventilation-controlled fire is limited, elevated temperatures

Thermal Layering of Gases — Outcome of combustion in a confined space in which gases tend to form into layers, according to temperature, with the hottest gases found at the ceiling and the coolest gases at floor level.

are usually present within the compartment. Any low level ventilation (such as opening a door or window) without prior upper level ventilation (such as vertical ventilation on the roof) may result in explosive rapid combustion of the flammable gases called a **backdraft**. Backdraft occurs in the decay stage of a fire, in a space containing a high concentration of heated flammable gases that lack sufficient oxygen for flaming combustion **(Figure 6.18)**.

When potential backdraft conditions exist in a compartment, the introduction of a new source of oxygen will return the fire to a fully involved state rapidly, often explosively. A backdraft can be created with a horizontal or vertical opening when hot fuel-rich smoke is mixed with air. Backdraft conditions may develop in a room, void space, or an entire building. Whenever a compartment contains hot products of combustion, the potential for backdraft should be considered before creating any openings in the compartment. Backdraft indicators may include:

- **Building indicators** — Fire confined to a single compartment or void space; building contents have a high heat release rate.
- **Smoke indicators** — Optically dense smoke, light colored or black becoming dense gray-yellow (smoke color alone is not a reliable indicator). Neutral plane rising and lowering similar to pulsing or breathing movement.
- **Air flow indicators** — High velocity, turbulent smoke discharge, sometimes appearing to pulse or breathe.
- **Heat indicators** — High heat, smoke-stained windows.
- **Flame indicators** — Little or no visible flame.

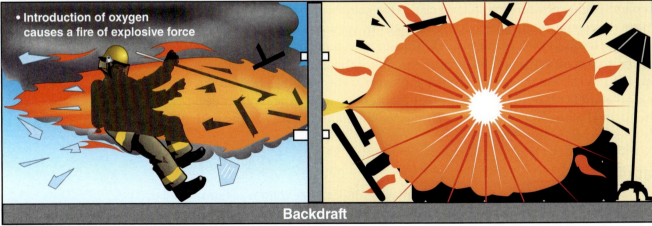

Figure 6.18 Improper ventilation during fire fighting operations may result in a backdraft.

Smoke Explosion

A **smoke explosion** may occur before or after the decay stage. It occurs as unburned fuel gases come in contact with an ignition source. When smoke travels from a fire, it can cool and accumulate in other areas and mix with air. The smoke within its flammable range contacts an ignition source and results in explosively rapid combustion. Smoke explosions are violent because they involve premixed fuel and oxygen. This process is similar to the ignition of propane and air within its flammable range. The smoke is generally cool, less than 1,112°F (600°C) and located in void spaces connected to the fire or in uninvolved areas remote to the fire.

> **Smoke Explosion** — Form of fire gas ignition; the ignition of accumulated flammable products of combustion and air that are within their flammable range.

Products of Combustion

As any fuel burns, its chemical composition changes. This change results in the production of new substances and the release of energy. In a structure fire, multiple fuels are involved with a limited air supply. This mixture results in incomplete combustion, which produces complex chemical reactions that cause toxic and flammable gases, vapors, and particulates.

While the heat generated by a fire is a danger to anyone directly exposed to it, smoke and fire gases cause the majority of deaths in fires. The materials comprising smoke vary based on the fuel involved. However, all smoke may be considered toxic. The smoke generated by a fire contains asphyxiant gases and irritants. Asphyxiant gases are products of combustion that cause central nervous system depression, resulting in reduced awareness, loss of consciousness, and death. Carbon monoxide (CO), hydrogen cyanide (HCN), and carbon dioxide (CO_2) are among the most common gases found in smoke. The reduction in oxygen levels as a result of a fire in a compartment will also cause a narcotic effect in humans. Irritants in smoke are those substances that cause breathing discomfort (pulmonary irritants) and inflammation of the eyes, respiratory tract, and skin (sensory irritants). Depending on the fuels involved, smoke contains numerous substances that may be considered irritants. **(Table 6.1, p. 206)**.

Carbon monoxide (CO) is one of the most common hazardous substances contained in smoke. Although not the most dangerous material usually associated with smoke, it is almost always present when combustion occurs. CO may be detected in the atmosphere with monitoring devices, and is often reported as being found in the blood of fire victims. CO displaces oxygen in the blood, creating carboxyhemoglobin (COHb), which starves the cells of oxygen.

NOTE: Because the substances found in smoke are deadly (either alone or in combination), firefighters must always use SCBA for protection when operating in smoke.

Classification of Fires

The classification of a fire is important to a firefighter in determining the safest, most efficient method of extinguishment. Each class of fire has unique properties and extinguishing requirements. The five classes of fire and routine methods of extinguishment are described in the following sections.

Table 6.1
Common Products of Combustion and Their Toxic Effects

Asbestos	A magnesium silicate mineral that occurs as slender, strong, flexible fibers. Breathing asbestos dust causes asbestosis and lung cancer.
Carbon Monoxide	Colorless, odorless gas. Inhalation of carbon monoxide causes headache, dizziness, weakness, confusion, nausea, unconsciousness, and death. Exposure to as little as 0.2% carbon monoxide can result in unconsciousness within 30 minutes. Inhalation of high concentration can result in immediate collapse and unconsciousness.
Formaldehyde	Colorless gas with a pungent odor that is highly irritating to the nose. 50-100 ppm can cause serious injury. Exposure to high concentrations can cause injury to the skin. Formaldehyde is a suspected carcinogen.
Nitrogen Dioxide	Reddish-brown gas or yellowish-brown liquid which is highly toxic and corrosive.
Particulates	Small particles that can be inhaled and deposited in the mouth, trachea, or the lungs. Exposure to particulates can cause eye irritation, respiratory distress, in addition to health hazards specifically related to the particular substances involved.
Sulfur Dioxide	Colorless gas with a choking or suffocating odor. Sulfur dioxide is toxic and corrosive, and can irritate the eyes and mucous membranes.

Source: *Computer Aided Management of Emergency Operations (CAMEO) and Toxicological Profile for Polycyclic Aromatic Hydrocarbons.*

Class A Fires

Class A Fire — Fires involving ordinary combustibles such as wood, paper, cloth, and similar materials.

Fires involving ordinary combustibles, such as wood, cloth, paper, rubber, and many plastics, are considered **Class A fires**. Water may be used to cool the burning material below its ignition temperature. The addition of a Class A foam agent reduces the surface tension of the water, allowing it to penetrate more easily into piles of material. Class A fires may be difficult to extinguish using oxygen exclusion methods, such as CO2, because this product does not provide the cooling effect required for total extinguishment.

Class B Fires

Class B Fire — Fires of flammable and combustible liquids and gases such as gasoline, kerosene, and propane.

Class B fires involve flammable and combustible liquids and gases, such as gasoline, oil, lacquer, paint, and alcohol. The smothering or blanketing effect of oxygen exclusion is most effective for extinguishment while helping to reduce the production of additional vapors. Other extinguishing methods may involve removal of fuel, temperature reduction, or the interruption of the chain reaction with dry chemical agents **(Table 6.2)**.

Class C Fires

Class C Fire — Fires involving energized electrical equipment.

Fires involving energized electrical equipment are classified as **Class C** fires. Household appliances, computers, transformers, and electrical transmission lines are examples of materials that may support an electrical fire. These fires may be extinguished by using a nonconducting agent, such as carbon dioxide or dry chemical. These fires may be controlled by de-energizing the equipment, then using the appropriate agent based on the fuel involved.

Table 6.2
Flammable Ranges of Common Flammable Gases and Liquids (Vapor)

Substance	Flammable Range
Methane	5%–15%
Propane	2.1%–9.5%
Carbon Monoxide	12%–75%
Gasoline	1.4%–7.4%
Diesel	1.3%–6%
Ethanol	3.3%–19%
Methanol	6%–35.5%

Source: *Computer Aided Management of Emergency Operations* (CAMEO)

Class D Fires

Class D fires involve combustible metals, such as magnesium, aluminum, titanium, zirconium, potassium, and sodium. Because certain airborne concentrations of metal dusts can cause explosions when a suitable ignition source is present, these metals are particularly hazardous as powders. The high temperature of some burning metals makes water an ineffective extinguishing agent. Many Class D materials, such as magnesium, may react violently when exposed to water.

No single agent effectively extinguishes fire involving all combustible metals. Agents are labeled to indicate the types of metal fires they may effectively control. These agents act to cover and exclude oxygen from burning material.

> **Class D Fire** — Fires of combustible metals such as magnesium, sodium, and titanium.

Class K Fires

Fires involving combustible cooking media, such as vegetable or animal oils and fats, are categorized as **Class K**. The widespread use of high efficiency cooking equipment that is highly insulated and slow to cool has prompted the fire service to introduce the use of wet-chemical extinguishers that contain a special potassium acetate-based, low pH agent that is dispensed in a fine mist. For additional guidance concerning Class K fire suppression, consult NFPA® 96, *Standard for Ventilation Control and Fire Protection of Commercial Cooking Operations*.

> **Class K Fire** — Fires in cooking appliances that involve combustible cooking media, such as vegetable or animal oils and fats; commonly occurring in commercial cooking facilities such as restaurants and institutional kitchens.

Fire Extinguishment Theory

Fire is extinguished by limiting or interrupting one or more of the essential elements of the combustion process (fire tetrahedron). This process may be accomplished in the following ways:

- Reducing its temperature
- Eliminating available fuel
- Eliminating available oxygen
- Stopping the self-sustained chemical chain reaction

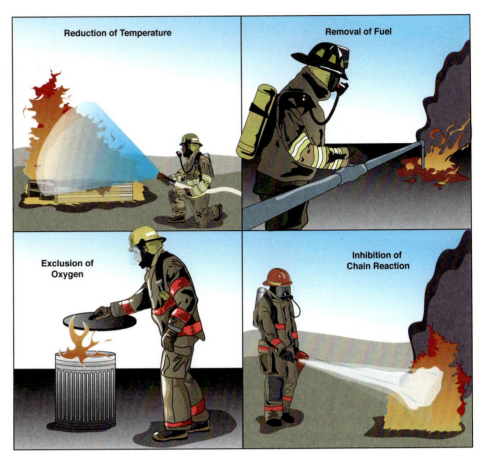

Figure 6.19 To extinguish a fire, remove the fuel, exclude the oxygen, or inhibit the chemical chain reaction.

Temperature Reduction

Cooling with water is one of the most common methods of fire extinguishment. This process causes a reduction in temperature of a fuel to the point where it will not produce sufficient vapor to burn. Cooling may extinguish solid fuels with high ignition temperatures or liquid fuels with high flash points. The use of cooling is also effective for smoldering fires. However, cooling with water cannot sufficiently reduce vapor production to extinguish fires involving low flash point flammable liquids or gases. **(Figure 6.19)**.

In addition to cooling fuels, water may also be used to control burning gases and reduce the temperature of hot products of combustion in the upper layer. This application slows the pyrolysis of combustible material, reduces radiant heat from the upper layer, and reduces the potential for flashover.

Fuel Removal

Removing the fuel sources effectively extinguishes some fires. A fuel source may be removed by stopping the flow of a liquid or gaseous fuel or by removing the solid fuel in the path of a fire. Another method that may be appropriate depending on the fuel package involves allowing the fire to burn until all the fuel is consumed.

Oxygen Exclusion
Reducing the oxygen available to the combustion process reduces the growth and may extinguish a fire over time. A simple example of oxygen exclusion is the extinguishment of a stove top fire by placing a cover over a pan of burning food. Oxygen content in a compartment may be reduced by flooding an area with an inert gas, such as carbon dioxide. This action displaces the oxygen and disrupts the combustion process. Oxygen may also be separated from fuel by blanketing the fuel with fire suppressant foam.

Chemical Flame Inhibition
Extinguishing agents, such as dry chemical, Halon, and Halon-replacement agents, interrupt the combustion reaction and stop flame production. This method of extinguishment is effective on gas and liquid fuels because they must produce flaming combustion to burn. These agents do not work effectively to extinguish non-flaming fires because there is no chemical chain reaction to inhibit. The high agent concentrations and extended time necessary to extinguish smoldering fires in this manner make these agents impractical for application on such a fire.

Chapter Summary
Firefighters should understand fire behavior, the stages of fire growth, and the dangers of products of combustion. They should also be familiar with the classifications of fires to operate safely and efficiently. This knowledge is of the utmost importance on the fireground as emergency responders continually evaluate the fire and determine how to attack the fire itself, protect exposures, and ensure the safety of everyone at the scene. Once an understanding of the factors that affect fire spread has been gained and the type of fire determined, the proper course of action for extinguishment may be developed.

Review Questions
1. What are the three states of matter?
2. Define specific gravity and vapor density.
3. What is the difference between a physical change and a chemical change?
4. What is the difference between flaming combustion and smoldering combustion?
5. List the four components of the fire tetrahedron.
6. What are the stages of fire development?
7. List factors that affect fire development.
8. What are the indicators of flashover?
9. What are types of hazardous conditions on the fireground that impact incident safety?
10. List indicators of backdraft.
11. Why are products of combustion dangerous?
12. What are the five classifications of fires?
13. What are the four ways of extinguishing a fire?

Building Construction

Courtesy of Bob Esposito.

Chapter Contents

Case History 213	Firefighter Hazards Related to Building Construction 225
Types of Building Construction 214	Dangerous Building Conditions 225
Type I Construction 214	Lightweight and Truss Construction Hazards 229
Type II Construction 217	New Building Construction Technologies 231
Type III Construction 217	Construction, Renovation, and Demolition Hazards 231
Type IV Construction 218	**Chapter Summary** 232
Type V Construction 219	**Review Questions** 233
Effects of Fire on Common Building Materials 220	
Wood 220	
Masonry 222	
Cast Iron 223	
Steel 223	
Reinforced Concrete 224	
Gypsum 224	
Glass/Fiberglass 224	

chapter 7

Key Terms

Building Code 214	Load-Bearing-Wall 220
Fire Wall 222	Masonry 222
Fire Resistance 214	Non-Load-Bearing Wall 220
Fire Retardant 221	Spalling 224
Gypsum Board 224	Truss 218
Lightweight Steel Truss 229	Veneer Walls 222
Lightweight Wood Truss 229	

FESHE Outcomes

This chapter provides information that addresses the outcomes for the Fire and Emergency Services Higher Education (FESHE) *Principles of Emergency Services* course.

2. Analyze the basic components of fire as a chemical chain reaction, the major phases of fire, and examine the main factors that influence fire spread and fire behavior.

10. Identify the primary responsibilities of fire prevention personnel including code enforcement, public information, and public and private protection systems.

Building Construction

Learning Objectives

1. Distinguish among types of building construction. (Outcomes 2, 10)
2. Explain the effects of fire on common building materials. (Outcomes 2, 10)
3. Describe firefighter hazards related to building construction. (Outcomes 2, 10)

Chapter 7
Building Construction

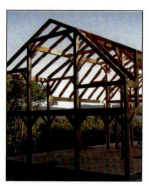

Case History

Jack, who had interest in becoming a firefighter, would frequently visit the firehouse. Most of the crews at the firehouse knew Jack well and were glad to share information and teach him about fire fighting. One afternoon, he arrived in time to greet the crews as they returned from a fire. After cleaning up, most of the firefighters gathered in the kitchen to talk about the incident. As Jack listened, he heard terms, such as *balloon frame*, *wood frame*, and *dimensional lumber,* which were unfamiliar to him. These terms were used to describe the construction of the building in question and in comparing it to other buildings. As he asked questions of the firefighters, even more questions arose in his mind. He thought it was curious that a fire that began in a basement could travel to the attic without doing much damage to the floors in between. He also wondered how the construction of a roof would make it more susceptible to collapse. He began to wonder how to look at a building in a way that would provide important information regarding potential fire behavior within the building and behavior of the building during a fire. Buildings began to look a little more complex and much more interesting to him with this new perspective. He made a mental note to begin studying building construction from a firefighter's perspective.

All firefighters should have a basic knowledge of building principles and construction in order to understand how a structure may react to fire and fire fighting activity. An understanding of the strengths and weaknesses of various types of construction will aid in planning a safe and effective fire attack.

New technology and building designs are introduced on a regular basis; therefore, firefighters must remain a student of building construction throughout their careers. The following chapter is an introduction to basic building types and their fire protection characteristics. More in-depth information on this subject may be found in the **IFSTA Building Construction Related to the Fire Service** manual.

Fire Resistance — The ability of a structural assembly to maintain its load-bearing ability under fire conditions.

Building Code — Body of local law, adopted by states/provinces, counties, cities, or other governmental bodies to regulate the construction, renovation, and maintenance of buildings. *See Code and Occupancy Classification.*

Types of Building Construction

The architect, structural engineer, or contractor determines the type of building construction used in a structure. In general, construction classifications are based on the type of materials used in the construction and the **fire-resistance** rating requirements of certain structural components. The following sections describe the general characteristics of each construction type specified in the model **building codes** for the United States and Canada.

The five types of building construction are as follows:

- Type I construction (also known as *fire-resistive construction*)
- Type II construction (also known as *noncombustible* or *limited-combustible construction*)
- Type III construction (also known as *ordinary construction*)
- Type IV construction (also known as *heavy timber* or *mill construction*)
- Type V construction (also known as *wood-frame construction*)**(Table 7.1)**

Construction Classifications and Subclassifications

NFPA® 220, *Standard on Types of Building Construction*, details each classification and subclassification using a three-digit code. For example, Type I construction can either be 4-4-2 or 3-3-2. The digits have the following meaning:

- The first digit refers to the fire-resistance rating (in hours) of exterior bearing walls.
- The second digit refers to the fire-resistance rating of structural frames or columns and girders that support loads of more than one floor.
- The third digit indicates the fire-resistance rating of the floor construction.

In Type IV construction, the designation 2HH is used. The structural members indicated by this designation are of heavy timber with minimum dimensions greater than those used in Type III or IV construction. The highest requirements for fire resistance are found in Type I construction.

NOTE: The International Building Code (IBC) makes use of construction classifications similar to NFPA 220®, although the requirements for individual structural members differ.

Type I Construction

Type I construction (also known as *fire-resistive construction*) provides the highest level of protection from fire development and spread as well as collapse. All structural members are composed of noncombustible or limited combustible materials with a high fire-resistance rating. Components such as walls, floors, and ceilings must be able to resist fire for a period of three to four hours depending on the component. **(Figure 7.1, p. 216).**

Table 7.1
Fire-Resistance Rating Requirements for Building Elements (Hours)

Building Element	Type I A	Type I B	Type II A[e]	Type II B	Type III A[e]	Type III B	Type IV HT	Type V A[e]	Type V B
Structural Frame[a]	3[b]	2[b]	1	0	1	0	HT	1	0
Bearing Walls 　Exterior[g] 　Interior	3 3[b]	2 2[b]	1 1	0 0	2 1	2 0	2 1/HT	1 1	0 0
Nonbearing Walls and Partitions 　Interior[f]	0	0	0	0	0	0	See Section 602.4.6*	0	0
Floor Construction 　Including Supporting Beams and Joists	2	2	1	0	1	0	HT	1	0
Floor Construction 　Including Supporting Beams and Joists	1½[c]	1[c,d]	1[c,d]	0[c,d]	1[c,d]	0[c,d]	HT	1[c,d]	0

For SI: 1 foot = 304.8 mm
HT = Heavy Timber

a. The structural frame shall be considered to be the columns and the girders, beams, trusses, and spandrels having direct connections to the columns and bracing members designed to carry gravity loads. The members of the floor panel or roof panels which have no connection to the columns shall be considered secondary members and not a part of the structural frame.

b. Roof supports: Fire-resistance ratings of structural frame bearing walls are permitted to be reduced by 1 hour where supporting a roof only.

c. Except in Group F-1, H, M, and S-1 occupancies, fire protection of structural members shall not be required, including protection of a roof framing and decking where every part of the roof construction is 20 feet or more above any floor immediately below. Fire-retardant-treated wood members shall be allowed to be used for such unprotected members.

d. In all occupancies, heavy timber shall be allowed where a 1-hour or less fire-resistance rating is required.

e. An approved automatic sprinkler system in accordance with Section 903.3.1.1* shall be allowed to be substituted for 1-hour fire-resistance-rated construction, provided such system is not otherwise required by other provisions of the code or used for an allowable area increase in accordance with Section 504.2*. The 1-hour substitution for the fire-resistance exterior of walls shall not be permitted.

f. Not less than the fire-resistance rating required by other sections* of this code.

g. Not less than the fire-resistance rating based on fire separation distance.

* Section numbers refer to sections in the *2006 International Building Code®*.

Courtesy of the International Code Council®, International Building Code®, 2006, Table 601.

Figure 7.1 Type I fire-resistive construction is engineered to resist fire for 3-4 hours. *Courtesy of Ron Moore and McKinney (TX) FD.*

Figure 7.2 Type I construction has structural members made of non-combustible or limited combustible materials. *Courtesy of McKinney (TX) Fire Department.*

Steel used in fire-resistive buildings must be protected from the heat of a fire. In contemporary practice, the most commonly used insulating materials are gypsum, spray-applied materials, intumescent coating (a paint-like product that expands when heated to create an insulating barrier), and lightweight concrete due to its durability.

The fire-resistive compartmentation created by doors, partitions, and floors is designed to retard the spread of fire and products of combustion through the building. These features are intended to provide an opportunity for occupant evacuation and may benefit fire fighting operations. Due to the limited combustibility of the construction materials, the primary fire hazards in these structures are comprised of the contents of the building. However, openings made in partitions during renovations or improperly designed and maintained heating, ventilation, and air conditioning (HVAC) systems may compromise the ability of a Type I occupancy to confine the fire to a compartment **(Figure 7.2)**.

Common Types of Openings in a Structure

Openings in partition walls, floors, and ceilings may include those used for the following:

- Electrical service
- Plumbing
- Television, cable, or data lines
- HVAC
- Laundry chutes
- Crawl spaces/attics
- Trash chutes

216   Chapter 7 • Building Construction

Figure 7.3 Buildings of Type II construction are permitted to have materials that are slightly less fire-resistive than buildings of Type I construction. *Courtesy of McKinney (TX) Fire Department.*

Figure 7.4 Type II construction is engineered to limit the spread of fire. *Courtesy of McKinney (TX) Fire Department.*

Type II Construction

Buildings classified as Type II (also known as *noncombustible* or *limited combustible*) are similar to Type I, but they offer lower fire resistance and lack the compartmentation of Type I construction. Materials with no fire-resistance ratings, such as untreated wood, may be used in limited quantities. Steel components used in Type II construction do not need to meet the fire-resistance rating of Type I buildings **(Figure 7.3)**.

Heat buildup from a fire inside the building may cause structural supports, even those made of steel, to fail. In addition, Type II buildings often feature flat, built-up roofs that may contain combustible felt, insulation, roof tar, or synthetic membrane. Fire extension to the roof may cause the roof to become involved and fail **(Figure 7.4)**.

Type III Construction

Type III (also known as *ordinary*) construction may be found in older schools, mercantile buildings, and some residential structures. This type of construction features exterior walls and structural members that are constructed of noncombustible materials. Interior walls, columns, beams, floors, and roofs are partially or completely constructed of wood **(Figure 7.5, p.218)**. These buildings may contain numerous conditions that affect their behavior during a fire, including the following:

- Voids between wooden channels created by roof and truss systems as well as wall studs. Unless fire stops are installed, voids may allow unimpeded fire spread in these areas.

- Older Type III buildings may have undergone numerous renovations, including the creation of large hidden voids above ceilings or below floors that may contribute to undetected fire spread.

- Structural components may have been removed during renovations in order to open up floor space or change the buildings configuration. These alterations may reduce the load carrying capacity of a structural member **(Figure 7.6, p.218)**.

Figure 7.5 Type III construction is commonly found in older schools, businesses, and residential structures.

Figure 7.6 Type III construction features interior structural members constructed completely or partially of wood, while exterior walls and structural members are noncombustible or limited combustible materials. *Courtesy of McKinney (TX) Fire Department.*

Truss — (1) Structural member used to form a roof or floor framework; trusses form triangles or combinations of triangles to provide maximum load-bearing capacity with a minimum amount of material. Often rendered dangerous by exposure to intense heat, which weakens gusset plate attachment. (2) Beams consisting of one tensile chord, one compression chord, and truss blocks or spaces between the two.

- Change in building occupancy may result in additional loads that the building was not designed to accommodate.
- Prefabricated wood **truss** systems, which may fail quickly when exposed to fire, may also be found in some new or renovated Type III structures.

Type IV Construction

Type IV construction (also known as *heavy timber/mill construction*) is characterized by the use of wooden structural members that are generally greater than 8 inches (.20 meters) in dimension, with a fire-resistance rating of at least two hours. Any other materials used in construction and not composed of wood must have a fire-resistance rating of at least one hour.

Originally constructed with a lack of void spaces to help prevent fire travel, Type IV buildings are stable and resistant to collapse due to the sheer mass of their structural members. When involved in a fire, the heavy timber structural elements form an insulating effect derived from the timber's own char that reduces heat penetration inside the beam **(Figure 7.7)**.

In the nineteenth century and early twentieth century, Type IV construction was used extensively in factories, mills, and warehouses. It is not commonly used in new construction

Figure 7.7 Type IV construction features exterior and interior walls and their associated structural members made of noncombustible or limited combustible materials.

Figure 7.8 Type V construction may appear to be Type III because of a veneer, but the actual fire-resistance rating is not comparable.

for multistory buildings, although many buildings of this type remain in use. Many old Type IV warehouse and industrial buildings have been converted to residential use. Today, heavy-timber wood frame construction is encountered primarily where it is desired for appearance.

The primary fire hazard associated with Type IV construction is the massive amount of combustible contents found in the building as well as the heavy timbers used in construction. These timbers give off a tremendous amount of heat and often pose a serious threat to any other buildings in the vicinity.

Type V Construction

Type V construction (also known as *wood frame*) consists of exterior walls, bearing walls, floors, roofs, and supports made completely or partially of wood or other approved materials of smaller dimensions than those used in Type IV construction. Type V construction is the most commonly used method for construction of single-family residences. This method presents almost unlimited potential for fire spread within the building of origin or similarly constructed nearby structures **(Figure 7.8)**.

Modern Type V construction often includes the use of prefabricated wood truss systems for roof and floor construction. Roof systems may include 2 x 4 inch (50 x 100 mm) dimensional lumber manufactured into trusses using gusset plates for connectors. Floor joists may be comprised of wooden I-beams constructed of thin plywood or wood composite and glued to two 2 x 4 inch (50 x 100 mm) pieces that form a top and bottom of the truss. Under fire conditions, truss systems fail and burn more rapidly than solid lumber.

Balloon-Frame Construction

In balloon-frame construction, the exterior wall studs are continuous from the foundation to the roof. Ribbon boards that are recessed into the vertical stud are used to support the joists, which, in turn, support the second floor. The term *balloon frame* originated in the fragile appearance of the thin, closely spaced studs as opposed to the more substantial members used in early timber construction. They were said to be as fragile as a balloon.

The vertical combustible spaces between the studs in balloon-frame construction provide a channel for the rapid communication of fire from floor to floor. Unlike timber framing, light wood framing is usually not left exposed. The framing is covered with an interior of plaster, which will act to slow the spread of fire into the stud space. However, a fire may originate in or spread to this area through penetrations for plumbing, heating, or electrical service.

Once fire is in the stud space, it may readily extend from floor to floor or even foundation to attic in the walls. A fire in a balloon-frame building may be difficult to contain and may give arriving firefighters the impression that the fire originated in the attic when it may have started on a lower floor and communicated through the stud wall. Although this method of construction has not been widely used since the 1920s, many balloon-frame buildings remain.

Platform Framing

In platform framing, the exterior vertical wall studs are not continuous to the second floor. The first floor is constructed as a platform upon which the exterior vertical studs are erected on the second floor. A wood sill is attached to the foundation upon which a header and floor joists or trusses are attached. Subflooring is then added to form a floor deck with the first floor wall framing attached to a top and bottom plate. Once the first-floor framing is in position and braced, the second floor joists are erected on the top plates of the first floor walls.

Effects of Fire on Common Building Materials

Materials react differently when exposed to heat or fire. Firefighters should have a working knowledge of how common building materials can be expected to react during fire fighting operations in order to employ the safest and most efficient suppression tactics. The following section reviews several common materials and explains how they may react to fire involvement.

Wood

Wood is widely used in numerous structural support systems, including **load-bearing** (those that support structural weight) or **non-load-bearing walls**. In addition, wood products may be used for roof decking, exterior wall sheathing, and a variety of interior finishes.

The reaction of wood to fire conditions depends mainly on two factors: dimension and moisture content. The smaller the wood size, the more likely it is to quickly lose structural integrity. Larger dimension lumber, such as those used in Type IV construction, retains much of its original structural integrity even after extensive exposure to fire. Gypsum may protect smaller dimensional lumber in the form of drywall or plaster to increase its resistance to heat and fire.

The moisture content of wood also has a bearing on the rate at which it will burn. Wood with a high moisture content (sometimes called *green wood*) does not burn as fast as wood that has been dried or cured. In some cases, fire retardants may be added to wood in an attempt to reduce the speed at which it ignites or burns (**Figure 7.9**).

Load-Bearing Wall — Wall that supports itself, the weight of the roof, and/ or other internal structural framing components, such as the floor beams and trusses above it; used for structural support.

Non-Load-Bearing Wall — Wall, usually interior, that supports only its own weight. These walls can be breached or removed without compromising the structural integrity of the building.

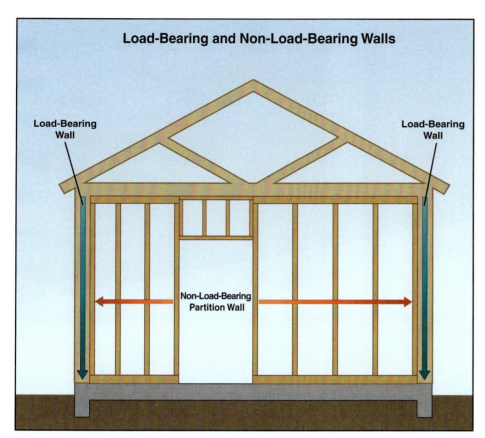

Figure 7.9 Knowing the difference between the load-bearing and non-load bearing walls allows firefighters to make sound decisions during fire evolutions.

The moisture content found in wood also affects its strength. In a living tree, wood has high moisture content. When a tree is cut, the water begins to evaporate. As the water content diminishes naturally or through curing, the wood begins to shrink in size and increase in strength. Most structural lumber has a moisture content of 19 percent or less.

Wood may be treated to reduce its combustibility. Building codes may permit the use of fire-retardant-treated wood in certain building applications. Although this wood resists ignition and has increased fire endurance compared to non-treated wood, it should not be confused with materials that are fire resistive.

The two methods of fire-retardant treatment are pressure impregnation and surface coating. Pressure impregnation consists of placing wood in a cylinder where a vacuum process draws the air out of the cells of the wood. A **fire retardant** chemical is introduced to the cylinder, which is then pressurized, forcing the chemical into the cells of the wood. This process produces a treatment that is permanent when the wood is used under the correct conditions. Surface coating is used primarily to reduce the surface burning of wood.

Water used during fire fighting operations has no substantial negative effect on the strength of wood construction materials. Application of water to burning wood minimizes damage by stopping the charring process. Firefighters should be aware that charring weakens the strength of wood and should inspect structural members for charring to ascertain the structural integrity of a building.

Fire Retardant — Any substance, except plain water, that when applied to another material or substance will reduce the flammability of fuels or slow their rate of combustion by chemical or physical action.

Chapter 7 • Building Construction **221**

Figure 7.10a and b A structure with a masonry exterior of brick and concrete. A structure with a masonry exterior of composite masonry units (cinder blocks) and concrete.

Newer construction often features lightweight building components that are made from wood fibers, plastics, and other substances joined by glue or resin binders. Such materials include plywood, particleboard, fiberboard, and paneling. While these materials may be strong, they may contain products that are highly combustible that deteriorate rapidly under fire conditions and emit significant toxic gases.

Masonry

Masonry includes brick, stones, and concrete products, such as concrete masonry units (CMUs) also called *cinder blocks*. Masonry may be used for fire wall assemblies to meet the requirements of a specified fire-resistance. The components of the assembly include the wall structure, doors, windows, and any other opening protection necessary to meet the protection-rating criteria **(Figure 7.10)**.

Firewall assemblies may be used to separate two connected structures (such as townhouses or a strip mall) and prevent the spread of fire from one structure to the next. Firewall assemblies may also divide large buildings into smaller portions in order to confine a potential fire to one area of the structure. Cantilever firewalls are freestanding walls that may be found in large churches and shopping centers. While block walls are often used as load-bearing walls, many brick and stone walls are **veneer walls**, which are decorative and usually attached to another frame structure that is load-bearing.

Fire and exposure to high temperatures have minimal effect on masonry. Bricks rarely show signs of serious deterioration, although stones may spall or lose small portions of their surface when heated. Blocks may crack, but usually retain most of their structural stability. However, the mortar between bricks, blocks, and stone may be subject to more aggressive deterioration and should be inspected for signs of weakening **(Figure 7.11)**.

> **Masonry** — Bricks, blocks, stones, and unreinforced and reinforced concrete products.
>
> **Firewall** — Fire-rated wall with a specified degree of fire resistance, built of fire-resistive materials and usually extends from the foundation up to and through the roof of a building that is designed to limit the spread of a fire within a structure or between adjacent structures.
>
> **Veneer Walls** — Walls with a surface layer of attractive material laid over a base of a common material.

Rapid cooling with water, as may occur during fire fighting operations, may cause bricks, blocks, or stone to spall and crack. This result is a common problem when water is used to extinguish a chimney fire. All masonry products should be inspected for this damage after extinguishment has been completed.

Cast Iron

Cast iron is rarely used in modern construction, but may be found in some older buildings. It was once commonly used as an exterior surface covering (veneer) and for columns. These large veneer walls were often fastened to masonry on the front of a building for decorative purposes. Cast iron tolerates high heat and fire conditions, but may crack and shatter when cooled with water. A major concern during an incident is that the bolts used to connect the cast iron to the building may fail, causing heavy sections to fall to the ground **(Figure 7.12)**.

Steel

Steel is the primary material used for structural support in the construction of large modern buildings. Firefighters should be aware that steel structural members elongate when heated. If a beam is restrained from movement as when it is part of a roof assembly, it may buckle and fail near the middle or push out load-bearing walls, causing a collapse **(Figure 7.13)**. For practical purposes, steel can be expected to fail at temperatures in the vicinity of 1,000° F (540° C). The temperature at which a specific steel member may fail depends on numerous variables, including:

- Size of the member
- The load it is under
- Composition of the steel
- Geometry of the steel
- If it is protected from direct exposure to fire **(Figure 7.14)**

Firefighters should be aware of the type of steel members that are in a building and the length of time they may have been exposed to fire. Lightweight steel construction may be expected to fail quickly when exposed to fire. Water may be applied to structural steel members in an effort to cool them and reduce the risk of failure **(Figure 7.15, p. 224)**.

Figure 7.11 A cantilever fire wall is a freestanding firewall.

Figure 7.12 The darker components of this structure are made of cast iron.

Figure 7.13 Steel serves as a structural component in this building. *Courtesy of McKinney (TX) Fire Department.*

Figure 7.14 Example of lightweight steel trusses. *Courtesy of McKinney (TX) Fire Department.*

Figure 7.15 Examples of steel I-beams used in constructing a high-rise structure.

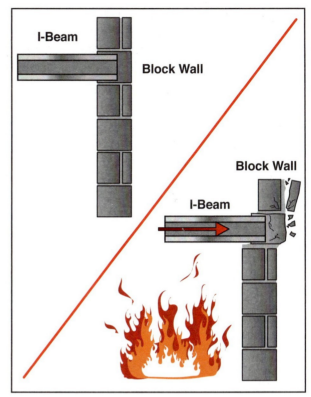

Figure 7.16 Steel beams expand when heated and can push a wall outward causing a collapse.

> **Spalling** — The expansion of excess moisture within masonry materials due to the exposure to the heat of a fire, resulting in tensile forces within the material and causing it to break apart. The expansion causes sections of the material's surface to violently disintegrate, resulting in explosive pitting or chipping of the material's surface.

> **Gypsum Board** — Widely used interior finish material; consists of a core of calcined gypsum, starch, water, and other additives that are sandwiched between two paper faces.

Reinforced Concrete

Reinforced concrete is internally reinforced with steel bars or mesh in order to give the material the tensile strength of steel in addition to the compressive strength of concrete. Reinforced concrete may spall and lose its strength under severe fire conditions. Heating may cause a failure of the bond between the concrete and steel reinforcement. When concrete surfaces have been exposed to heat or fire conditions, firefighters should look for cracks and **spalling**, indicating that the strength of the concrete may be compromised (**Figure 7.16**).

Gypsum

Gypsum is an inorganic product from which **gypsum board**, plaster, and plasterboard are constructed. Gypsum is unique because of its high water content, which requires a great deal of heat to evaporate, giving this material excellent heat and fire-resistant properties. Gypsum is commonly used to provide insulation to steel and wood structural members that may be exposed to fire conditions.

Glass/Fiberglass

Glass is typically used for doors and windows rather than structural support and is not an effective barrier to fire extension. Heated glass may crack and shatter when struck by a cold fire stream. Fiberglass is typically used to provide insulation. While the glass component is not a significant fuel, the materials used to bind the fiberglass may be combustible and prove difficult to extinguish.

Figure 7.17 Combustible furnishings burn readily and emit many toxic gases.

Firefighter Hazards Related to Building Construction

Firefighters should understand building construction methods in order to better plan fire fighting operations and to recognize conditions that are inherently dangerous due to building design and fire conditions. It is every firefighter's responsibility to constantly monitor the fireground for hazardous conditions. The following sections describe some of the critical issues relating to building construction that may impact firefighter safety.

Dangerous Building Conditions

Firefighters should be aware of dangerous conditions caused by the effect of the fire on a structure, as well as the dangerous conditions that may result from actions taken during fire suppression efforts. In order to effectively fight fires, ventilation holes may be cut in the roof, floors, ceilings, or walls to search for and extinguish hidden fire. Ventilation holes could result in a large volume of water to collect in a structure. The effect these actions may have on a fire-weakened structure must be constantly evaluated.

There are two primary types of danger that a building may pose during fire fighting efforts:

- Conditions, such as large quantities of combustible materials, that may contribute to fire spread (**Figure 7.17**).
- Conditions such as type of construction, renovation, demolition, or poor maintenance, that make a building susceptible to collapse.

Fire Loading

Fire load is the amount of fuel within a compartment expressed in pounds per square foot. It is used as a measure of the potential heat release of a fire within a compartment. Heavy fire loading is the presence of large amounts

Figure 7.18 Interior furnishings and materials contribute to the fuel load and smoke generation capability of an occupancy. *Courtesy of Eddie Avila.*

of combustible material in a building. The quantity and arrangement of the building's contents should be considered when determining the possible duration and intensity of a fire.

Heavy content loading is a critical hazard in commercial and warehouse occupancies because a fire may quickly overwhelm the capabilities of a fire sprinkler system if material storage is excessive or poorly arranged. Buildings overloaded with contents may also hamper firefighters in gaining access or hinder civilians making egress during a fire. Vigilant code enforcement and preincident surveys may minimize the occurrence and effect of heavy fire load occupancies.

Combustible Furnishings and Finishes

The furnishings and cosmetic finishes in some occupancies may contribute to fire spread and production of toxic smoke and fire gases. Rapid fire spread and toxic smoke have been identified as major factors in the loss of many lives in fires. Some codes regulate furnishings found in occupancies, such as hospitals, theaters, and detention facilities. Code enforcement activities in these occupancies may aid in the recognition and remedy of many of these hazards (**Figure 7.18**).

Roof Coverings

The final outside layer placed on a roof assembly is considered the covering. Common materials for roof coverings include wood, composite shingles, tile, slate, tin, asphalt, tar paper, and synthetic membrane (**Figure 7.19**). The combustibility of a roof surface affects surrounding buildings that may become exposures to a large primary fire. Wood shake shingles, even when treated with fire retardant, may contribute significantly to fire spread and the possibility of a conflagration during wildland/urban interface fires. Firefighters must be prepared to employ aggressive exposure protection tactics when presented with fires involving wood shake shingles (**Figure 7.20**).

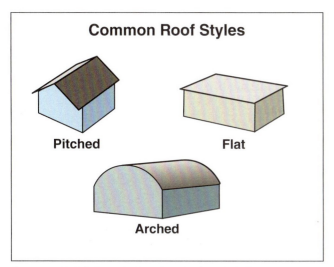

Figure 7.19 Pitched, flat, and arched roof styles are prevalent in most jurisdictions.

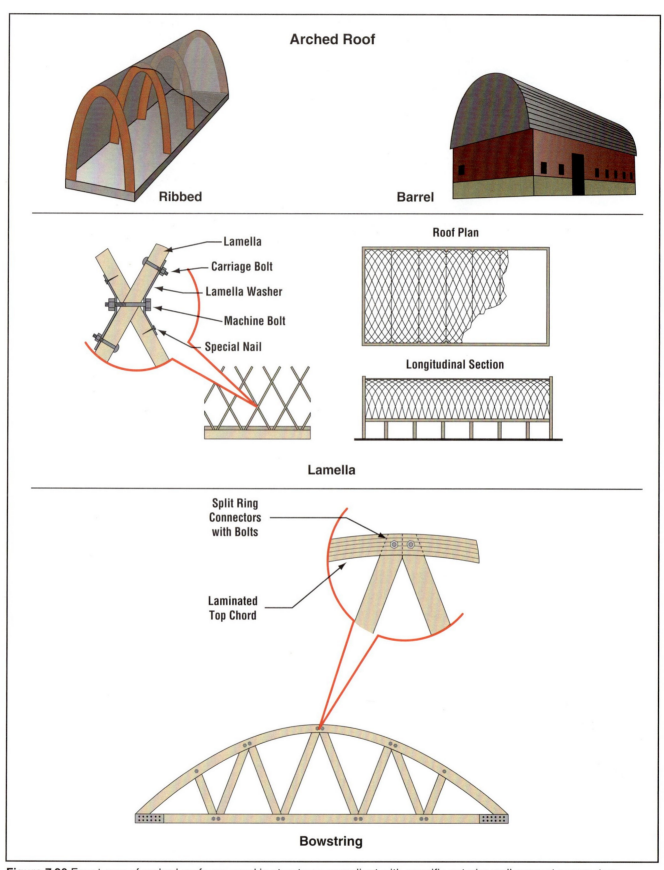

Figure 7.20 Four types of arched roofs are used in structures compliant with specific exterior wall support parameters.

Figure 7.21 The large open space in this structure could contribute to rapid fire spread.

Figure 7.22 Older, poorly maintained buildings are likely to collapse more quickly than newer buildings.

Wooden Floors and Ceilings

Combustible structural components, such as framing, floors, and ceiling made primarily of wood, also contribute to the fire loading in a structure. Prolonged exposure to fire may weaken these elements and increase the risk of collapse.

Large, Open Spaces

Large, open spaces may commonly be found in retail stores, enclosed malls, warehouses, churches, common attics, theaters, and other occupancies found in local jurisdictions. Fire fighting operations in these buildings should include appropriate vertical ventilation in order to channel products of combustion and slow the spread of the fire. Firefighters should be aware that advanced fire conditions may exist at the ceiling with drastically less heat and smoke at floor level (**Figure 7.21**).

Building Collapse

Structural collapses at an incident have killed or injured many firefighters. Collapse often occurs as a result of damage to the building's structural system caused by the progression of fire, fire fighting operations, or a combination of both. Fire begins to weaken the structural support systems of a building until it becomes incapable of holding its own weight in addition to the weight of any contents. The longer a fire burns, the more likely the building will collapse. The timeframe for collapse varies with the severity of the fire, the fire loading, type of construction, and the overall condition of the building (**Figure 7.22**).

Some buildings are more inclined to collapse than others. Those structures featuring lightweight truss construction will fail more quickly than a heavy timber structure. Older, poorly maintained buildings that have had severe weather exposure will be expected to fail before a newer well-maintained building. Information concerning building construction and age should be obtained during a preincident survey.

The following conditions may indicate the potential for building collapse:

- Cracks or separations in walls, floors, or ceilings (**Figure 7.23**)
- Evidence of preexisting instability, such as the presence of tie rods and bearing plates to hold walls together

Figure 7.23 Cracks in walls are a warning of building instability.

Figure 7.24 Loose or missing bricks indicate a potentially dangerous situation.

- Loose bricks or blocks falling from the building (**Figure 7.24**)
- Deteriorated mortar between the masonry
- Walls that appear to be leaning
- Structural members that appear to be distorted
- Fires beneath floors or roofs that support heavy machinery, HVAC units, or other heavy loads
- Prolonged fire exposure to structural members
- Unusual creaks or cracking noises
- Structural members pulling away from walls
- Excessive weight of building contents

Fire fighting operations may also increase the risk of building collapse. Improper vertical ventilation techniques may result in cutting structural supports that may weaken the roof system. The water used to extinguish the fire adds extra weight to floors if measures are not taken to drain it off. A hoseline flowing 250 gallons per minute (1 000 L/min) can add one ton (0.9 T) of water to a building every minute. Water that is not drained or converted to steam by the fire will add to the collapse potential of an already weakened structure.

Lightweight and Truss Construction Hazards

The use of lightweight and truss-supported building systems have become more common in many types of buildings, including houses, apartments, and commercial buildings. The two most popular types are **lightweight steel trusses** and **lightweight wood trusses**. Steel trusses are constructed from a long steel bar that is bent at 90-degree angles with flat or angular pieces welded to the top and bottom. Lightweight wood trusses are made of 2 x 3 or 2 x 4-inch (50 x 75 mm or 50 x 100 mm) dimensional lumber connected by small plates with ⅜ inch (metrics) teeth called *gusset plates*. Some newer wooden trusses are manufactured using finger joint bonding (glue) and have no gusset plate connectors (**Figure 7.25, p. 230**).

Lightweight Steel Truss — Structural support made from a long steel bar that is bent at a 90-degree angle with flat or angular pieces welded to the top and bottom.

Lightweight Wood Truss — Structural supports constructed of 2 x 3-inch or 2 x 4-inch (50 x 75 mm or 50 x 100 mm) members that are connected by gusset plates.

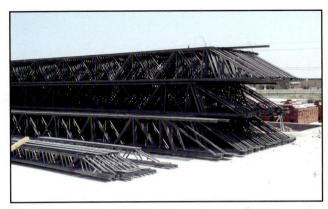

Figure 7.25 Lightweight steel trusses are especially vulnerable to collapse during a fiew.

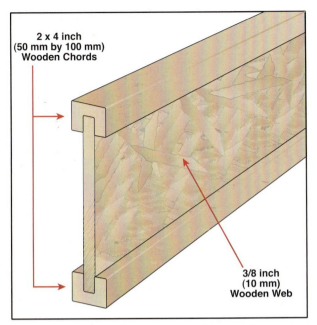

Figure 7.26 Wooden I-beams have a large surface area for combustion, so they can burn through and weaken quickly.

Lightweight trusses, whether wood or steel, will fail quickly under fire conditions. Steel trusses fail at approximately 1,000° F (540° C), and gusset plates in wood trusses quickly warp and fall out upon exposure to heat or fire. Trusses may be protected with fire-retardant material to increase their survivability, but most are not protected in this manner.

Other types of trusses may be found in various buildings in virtually every community. One such design, the bowstring truss, is often used in structures that require large open spaces, such as car dealerships, factories, bowling alleys, and supermarkets. Bowstrings are often identified by their rounded appearance, although they may be hidden behind a parapet wall from above and a suspended ceiling from below. In those instances, it may only be possible to identify the type of roof system by performing a visual inspection of the rooftop.

Wooden I-beams are also used in lightweight construction. These members have characteristics similar to wood trusses, and similar precautions should be taken when these beams have been exposed to heat or fire conditions (**Figure 7.26**).

> **WARNING!**
> Truss constructed buildings have been known to fail in as little as 5 to 10 minutes of fire exposure. Firefighters should exercise extreme caution when operating in or on any building of lightweight construction in which the structural elements have been exposed to heat or fire.

New Building Construction Technologies

There are many new building construction technologies that firefighters may encounter. Some of these construction features include:

- Alternative energy sources, such as solar panels mounted to the roof of a building (**Figure 7.27**).
- Styrofoam foundation forms
- Energy efficient windows
- Hurricane or tornado-resistant windows
- Building access and security features, such as electronic magnetic locking doors and window systems, that may hamper access or egress by civilians and firefighters

Figure 7.27 Solar energy panels produce clean and reliable energy.

Each of the previously mentioned features may impact firefighters during suppression operations. Solar panels mounted on a roof add weight to the roof system, which may already be weakened by a fire. In addition, these panels may hinder vertical ventilation by obstructing a portion of the roof that is most advantageous to cut. The panels and associated wiring may also be energized as long as the panels are receiving sunlight. In addition, light towers or other lighting used during incident operations may energize these panels. Security access systems and window and door bars may also delay entry and egress as well as complicate forcible entry operations.

Figure 7.28 A building in this early stage of construction does not yet have any fire resistant structural elements in place. *Courtesy of McKinney (TX) Fire Department.*

Construction, Renovation, and Demolition Hazards

When buildings are under construction, renovation, or demolition, the risk of fire or other emergencies generally increases. The increased fire load associated with equipment and raw materials, as well as the potential for ignition sources from welding, cutting, and grinding, pose a serious threat. While under construction, a building may be subject to rapid fire spread as protective features, such as drywall, firewalls, fire doors, and automatic sprinkler systems may not yet be in place (**Figure 7.28**).

Figure 7.29 Buildings in the process of demolition may be subject to high risk of rapid fire growth.

Buildings under renovation or demolition may have similar issues with fire spread as breached walls, open stairwells, missing doors, and disabled fire protection systems may also increase hazards to firefighters operating in these structures (**Figure 7.29**). Additional hazards at construction sites may include the following:

- Stairs
- Hallways
- Standpipe connections blocked by raw material or debris
- Poor housekeeping conditions
- Vehicles or dumpsters blocking access points or fire department connections (FDC)

During renovations in some buildings, occupants may be allowed to continue to inhabit certain areas throughout the project. This additional fire load and life hazard should be considered by firefighters during a preincident survey.

Chapter Summary

In order to operate safely and efficiently on the fireground, firefighters should have knowledge of building construction materials and practices. Understanding how various types of construction will react under fire conditions is essential in planning and executing fire fighting operations. The importance of knowing the five types of building construction and the benefits and inherent shortcomings of each will aid firefighters in sizing up an incident to determine whether an interior fire attack is advisable. Much of the information about specific buildings in a jurisdiction may be learned by close observation of buildings under construction as well as comprehensive preincident surveys. Firefighters should understand the methods and dangers of lightweight construction that has been exposed to heat or fire and should be aware of the hazards associated with the construction, renovation, or demolition of a building.

Review Questions

1. List the five types of building construction.
2. Which type of construction provides the highest level of protection from fire development and spread?
3. What is the main characteristic of Type IV construction?
4. What type of construction is the most commonly used method for construction of single-family residences?
5. What two factors determine the reaction of wood to fire conditions?
6. What effect does fire and exposure to high temperatures have on masonry?
7. What factors determine the temperature at which a specific steel member may fail?
8. Why does gypsum have excellent heat and fire-resistant properties?
9. What are the two primary types of danger that a building may pose during fire fighting efforts?
10. Define fire load.
11. Why might the furnishings and cosmetic finishes in some occupancies create dangerous conditions during a fire?
12. What conditions may indicate the potential for building collapse?
13. Why are lightweight trusses a hazard under fire conditions?
14. List new building construction technologies that firefighters may encounter.
15. What hazards are present in buildings under construction, undergoing renovation, or undergoing demolition?

Fire Detection, Alarm, and Suppression Systems

Chapter Contents

Courtesy of Bob Esposito.

Case History 237
Reasons for Installing Fire Detection Alarm and Suppression Systems 237
Types of Alarm Systems 238
    Heat Detectors 239
    Smoke Detectors and Smoke Alarms 240
    Automatic Alarm Systems 243
    Auxiliary Services 244
Automatic Sprinkler Systems 244
    Sprinkler Systems and Their Effects on Life Safety 246
    Sprinkler Position 246
    Applications of Sprinkler Systems 250
    Classes of Standpipe Systems 251
    Types of Standpipe Systems 253

Chapter Summary 254
Review Questions 254

chapter 8

Key Terms

Ambient Temperature.............................239	Flame Detector...242
Automatic Sprinkler System....................244	Ionization Detector..................................241
Carbon Monoxide Detector......................242	Rate-of-Rise Heat Detector......................239
Combination Detector.............................242	Residential Sprinkler System..................251
Fire Department Connection (FDC).......251	Sprinkler..245
Fixed-Temperature Heat Detector..........239	Wet-Pipe Sprinkler System......................250

FESHE Outcomes

This chapter provides information that addresses the outcomes for the Fire and Emergency Services Higher Education (FESHE) *Principles of Emergency Services* course.

10. Identify the primary responsibilities of fire prevention personnel including code enforcement, public information, and public and private protection systems.

Fire Detection, Alarm, and Suppression Systems

Learning Objectives

1. Summarize the reasons for installing fire detection, alarm, and suppression systems. (Outcome 10)
2. Identify characteristics of various types of alarm systems. (Outcome 10)
3. Describe automatic sprinkler systems. (Outcome 10)
4. Distinguish among applications of sprinkler systems. (Outcome 10)
5. Describe standpipe systems. (Outcome 10)

Chapter 8
Fire Detection, Alarm, and Suppression Systems

Case History

Being a fire chief in a small town poses many challenges. In addition to being responsible for fire suppression and public education, a fire chief is often responsible for code enforcement and plan review for the town. A small city in Oklahoma had an active program to revitalize the downtown. Many of the structures, some dating before 1900, fit perfectly with having apartments over first-floor commercial occupancies. The first of many "apartments over" required an alarm system per the existing fire code. After much consultation, the owner and developer agreed to install the code-mandated alarm system in the building.

The fire department received an automatic alarm from this system early one morning. There was heavy fire in the building when the first engine arrived. Because of the initial involvement and construction, the building was a total loss. The occupants in the building to the east were alerted early by the alarm system and escaped unharmed. One of the occupants stated in the local newspaper "that the fire alarm system saved his life."

Firefighters encounter fire alarm detection and suppression systems on a routine basis during preplanning, inspection, and fire fighting operations. Personnel should understand the capabilities and limitations of both detection and suppression systems found in local jurisdictions in order to plan an appropriate response strategy.

Reasons for Installing Fire Detection, Alarm, and Suppression Systems

There are a number of reasons for installing fire detection, alarm, and suppression systems in residential buildings and other properties. In some jurisdictions, codes require such systems to be installed in various properties based on code requirements or for insurance purposes. The systems installed in these occupancies may fulfill one or more of the following functions:

- Notify occupants of a building to take necessary action to escape a fire
- Summon organized assistance and/or assist in fire control actions

Figure 8.1 A manual pull station.

- Supervise fire control systems to ensure that operational status is maintained
- Initiate required auxiliary functions involving environmental, utility, and building system controls (elevators, HVAC systems). Fire detection, alarm, and suppression systems may incorporate one or all of these features. Such systems may operate mechanically, hydraulically, pneumatically, or electrically (**Figure 8.1**).

Automatic sprinkler systems remain the most reliable form of fire suppression systems for commercial, industrial, institutional, or residential occupancies. Fires in buildings equipped with automatic sprinklers generally result in less water damage than those extinguished by traditional fire attack methods. A majority of fires in sprinklered buildings are controlled by the activation of five or less sprinkler heads.

Types of Alarm Systems

The most basic alarm system is designed to only be initiated manually by pulling a handle. While these systems are properly termed *protected premises alarm systems*, they are more commonly called *local alarm systems*. These systems are installed in some school buildings and other public properties. In these systems, the signal only alerts building occupants of the need to evacuate the premises; it does *not* notify the fire department. Therefore, when a local alarm system is activated, it still necessary for someone to dial 9-1-1 to alert the fire department.

A protected premises fire alarm, which is the most basic type of alarm system and is also known as a local fire alarm, may be installed in smaller buildings with various types of occupancies. When activated, these systems only alert the occupants of the premises to evacuate; they do not notify the fire department. As a result, when a local alarm system is activated, it is still necessary for someone to alert the fire department. **(Figure 8.2 a, b, and c)**.

There are several basic types of automatic alarm-initiative devices. They are designed to detect heat, carbon monoxide (CO), smoke, fire gases, or flame. The sections that follow describe the most common types of devices in use.

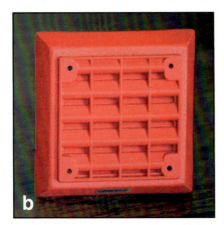

Figure 8.2 a A fire alarm bell. **b.** A fire alarm speaker. **c.** A combination fire alarm speaker and strobe light.

Heat Detectors

Heat detectors initiate an alarm when the ambient temperature near the detector reaches a predetermined level. There are numerous designs of heat detection devices, but all are either fixed temperature devices or rate-of-rise detectors (**Figure 8.3**).

Fixed-Temperature Heat Detectors

Fire detection and alarm systems using heat-detection devices are among the oldest still in service. **Fixed-temperature heat detectors** are relatively inexpensive compared to other types of systems and are the least prone to false activations. While these devices are reliable, they are typically the slowest to activate under fire conditions. These devices also are not typically resettable and must be replaced after activation.

A fixed temperature heat detector activates when it is heated to the temperature at which it is rated. If the **ambient temperature** in a room is low to begin with, a fire would burn undetected until it raised the temperature of the heat detector to its activation point. Depending on the size of the room and fuel load, a fire may burn for quite some time before activating a fixed-temperature heat detector.

Because heat rises, heat detectors are installed at the highest portion of a room, usually the ceiling. Heat detectors should have an activation temperature slightly higher than what may normally be expected in that space.

The various types of fixed devices described in this section activate by one or more of the following mechanisms:

- Expansion of heated material
- Melting of heated material
- Changes in electrical resistance of heated material

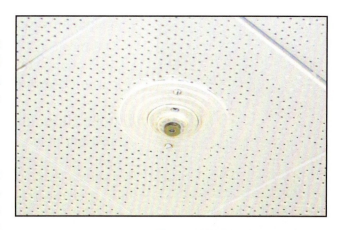

Figure 8.3 Heat detectors should have an activation temperature slightly above the normally expected ceiling temperature.

Fixed-Temperature Heat Detector — Temperature-sensitive device that senses temperature changes and sounds an alarm at a specific point, usually 135°F (57°C) or higher.

Ambient Temperature — Temperature of the surrounding environment.

Rate-of-Rise Heat Detectors

A **rate-of-rise heat detector** operates on the principle that fires rapidly increase the temperature in a given area. Typically, rate-of-rise heat detectors are designed to send an alarm when the rise in temperature exceeds 12°F to 15°F (7°C to 8°C) per minute. Because a sudden rise in temperature initiates an alarm regardless of the original temperature, an alarm may be activated at a room temperature far below what is required for initiating a fixed temperature device.

If properly installed, rate-of-rise heat detectors are reliable and not subject to false activations. However, if these devices are placed in a location subject to rapid increases in temperature, such as a small enclosed kitchen, a false activation may occur due to cooking.

There are several types of rate-of-rise heat detectors in common use, including:

- Pneumatic rate-of-rise line detector — Monitors large areas of a building
- Pneumatic rate-of-rise spot detectors — Monitors a small area surrounding the device
- Rate-compensated detector — Used in areas normally subject to regular

Rate-of-Rise Heat Detector — Temperature-sensitive device that sounds an alarm when the temperature changes at a preset value, such as 12°F to 15°F (7°C to 8°C) per minute.

temperature changes, which are slower than those under fire condition
- Thermoelectric detector — Operates on the principle that when two wires of dissimilar metals are twisted together and heated at one end, an electrical current is generated at the other end

Smoke Detectors and Smoke Alarms

Smoke detectors and smoke alarms are actually two distinct types of devices. A smoke alarm detects the presence of smoke and sounds an alarm to alert occupants. A smoke detector also detects the presence of smoke; however, it must transmit a signal to another device to sound the alarm (**Figure 8.4**). In many cases, the devices in large residential, commercial, or industrial buildings are smoke detectors. Meanwhile, those devices in single-family or small multifamily dwellings are smoke alarms (**Figure 8.5**).

Figure 8.4 A smoke detector in an industrial facility.

Figure 8.5 A residential smoke alarm.

Photoelectric Smoke Detectors

A photoelectric smoke detector uses a photoelectric cell coupled with a tiny light source. The photoelectric cell functions in one of two ways to detect smoke: beam application and refractory application.

A beam application smoke detector employs a beam of light that focuses across an area being monitored and on to a photoelectric cell. The cell constantly converts the beam into electrical current, which keeps a switch open. When smoke obscures the path of the light beam, the required amount of current is no longer produced and an alarm is initiated. Photoelectric smoke detectors work satisfactorily to detect all types of fires and are generally more sensitive to smoldering fires than the ionization detectors that will be discussed in the next section (**Figure 8.6**).

A refractory application photocell uses a light beam that passes through a small chamber to a point away from the light source. Under normal conditions, the light will not strike the photocell, no current is produced, and the switch will remain open. When smoke enters the chamber, it causes the light beam to be refracted (scattered) in all directions. A portion of this light will strike

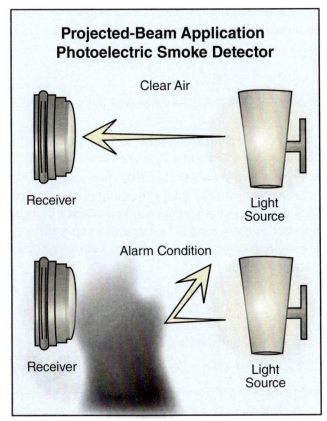

Figure 8.6 The projected beam application photoelectric smoke detector is activated when smoke enters the light beam.

240 Chapter 8 • Fire Detection, Alarm, and Suppression Systems

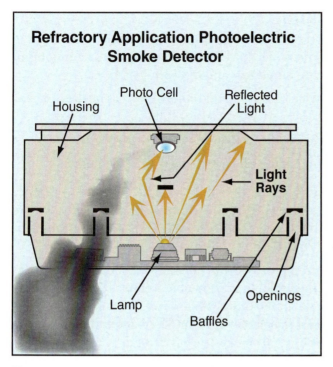

Figure 8.7 When smoke causes the light beam in a refractory application photoelectric smoke detector to scatter, light reflects onto the photosensitive device and an alarm sounds.

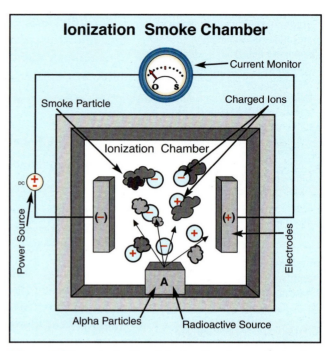

Figure 8.8 In an ionization detector, the alarm sounds when products of combustion reduce the normal current between the charged plates. This type of detector usually responds best to flaming fires.

the photocell, which, in turn, causes current to flow and the switch to close. Occasionally, when foreign objects such as dust and spiders enter the detector, they have been known to initiate false activations of these devices. (**Figure 8.7**).

Ionization Smoke Alarms

During the process of combustion, minute particles and aerosols too small for the naked eye to see are produced. **Ionization detectors** use a tiny amount of radioactive material to ionize air molecules as they enter a chamber within the detector. The ionized particles allow an electrical current to flow between negative and positive plates in the chamber. When particulates from the combustion process enter the chamber, they attach themselves to electrically charged molecules of air, making this air within the chamber less conductive. The decrease in current flowing between the plates transmits an alarm initiating signal. Although ionization detectors respond satisfactorily to most fires, they generally respond faster to flaming fires than smoldering ones (**Figure 8.8**).

Ionization Detector — Type of smoke detector that uses a small amount of radioactive material to make the air within a sensing chamber conduct electricity.

Smoke Alarm Legislation

Firefighters should be aware of jurisdictional laws that address minimum requirements for smoke alarms. Such laws may address the occupancies where such devices are required as well as the power source (battery or hardwired) that may be necessary.

Chapter 8 • Fire Detection, Alarm, and Suppression Systems

Flame Detector — Detection and alarm device used in some fire detection systems (generally in high-hazard areas) that detect light/flames in the ultraviolet wave spectrum (UV detectors) or detects light in the infrared wave spectrum (IR detectors).

Combination Detector — Alarm-initiating device that is capable of detecting an abnormal condition by more than one means. The most common combination detector is the fixed-temperature/rate-of-rise heat detector.

Carbon Monoxide Detector — Device designed to detect the presence of carbon monoxide (CO) and sound an alarm to prevent personnel within the structure from being poisoned by the colorless and odorless gas.

Other Detectors

The following section provides basic information concerning other types of detectors a firefighter may encounter, including **flame detectors, combination detectors, carbon monoxide detectors**, and indicating devices.

The following are the three basic types of flame detectors (also known as *light detectors*):

1. Those that detect light in the ultraviolet wave spectrum (UV detectors)
2. Those that detect light in the infrared wave spectrum (IR detectors)
3. Those that detect both types of light

These types of detectors are sensitive and usually positioned in areas where other light sources are unlikely since sunlight, welding, or other bright light sources may cause activation.

Depending on the requirements of the occupancy, various combinations of detection equipment are available to protect a premise. Some detectors may combine two types of detection into a single device. Some combinations include fixed-temperature/rate-of-rise heat detectors, combination heat/smoke detectors, and combination smoke/fire gas detectors. This versatility makes detectors more responsive to a variety of fire conditions (**Figure 8.9**).

Carbon monoxide (CO) detectors are often the cause for fire department response. These devices may have a single purpose to detect the presence of carbon monoxide or combined with smoke detection capability in a single unit (**Figure 8.10**).

Figure 8.9 A combination ultraviolet (UV) detector and infrared (IR) detector.

Figure 8.10 A carbon monoxide (CO) detector.

CO detectors monitor an area for the presence of the colorless, odorless gas that is both toxic and combustible. Formed as a result of incomplete combustion of carbon, the source for these alarms is often poorly designed or maintained heating equipment, generators, grills used indoors, or idling automobiles.

Many jurisdictions require the installation of CO detectors as well as smoke alarms in residential occupancies. When a CO detector alerts occupants of a building, they typically call the fire department to investigate. Each jurisdiction may establish its own policy for responding to such alarms, but typical actions include evacuating occupants and contacting the local gas utility provider.

There are a variety of audible and visual alarms indicating devices in use to warn occupants of fire alarm activation. Many indicating devices combine an audible warning, such as a horn or chime, with a visual indication, usually a strobe light. Depending on the occupancy, alarm indicating devices may use a pre-recorded message with evacuation instructions or an extra loud signal in high noise work areas.

Tamper-Proof Smoke Alarms
Some jurisdictions require smoke alarms that feature a tamper-proof, non-removable battery to be installed. This requirement addresses the possibility of the occupant removing smoke alarm batteries during a false alarm or if a battery is needed for another household device. These devices typically have a 10-year battery life.

Automatic Alarm Systems

Some alarm systems are designed to transmit a signal to an off-site location that will monitor the system and contact the fire department upon receipt of a signal. There are various brands available that operate using dedicated wire pairs, leased telephone lines, fiber optic cable, or wireless communication links. Types of alarm systems include the following:

- Auxiliary systems
- Remote receiving stations
- Proprietary systems
- Central station systems

The following section provides an outline of the basic features of these systems.

The protected property owns and operates the proprietary systems. These systems are often used to protect large industrial buildings, high-rises, and groups of commonly owned buildings, such as a college campus or industrial complex. Each building has its own system that is wired into a common receiving point at the facility. A trained staff on duty twenty-four hours a day monitors the system for alarms and notifies the fire department upon activation (**Figure 8.11, p. 244**).

Central station systems have some similarities to proprietary systems. However, the major difference is that instead of having a receiving point on site that employees of the protected property staff, the receiving point is an

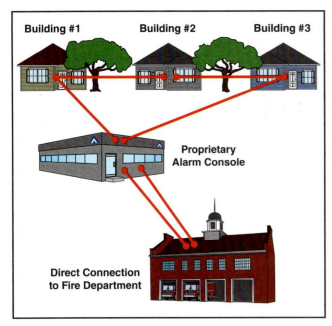

Figure 8.11 A proprietary system is used to protect commonly owned buildings that are close together.

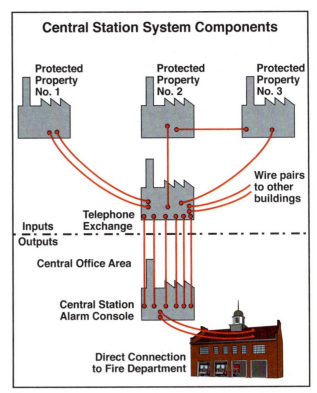

Figure 8.12 In a central station system, the receiving point for the alarm is located at an outside, contracted service point.

off-site facility that maintains contracts to monitor alarms for numerous customers. Upon receipt of an alarm at the central station, an employee contacts the appropriate response agency and may also contact a responsible party for the protected property **(Figure 8.12)**.

Auxiliary Services

Emergency signaling systems in some major occupancies feature the ability to control environmental services (HVAC), security cameras, property access controls, elevators, and other features. This capability may allow remote control of fire doors and dampers, pressurization of stairwells to exclude smoke, and control of access to areas of hazardous material storage.

Automatic Sprinkler Systems

Automatic sprinkler systems consist of a series of sprinklers (also called heads) arranged so that the system will distribute sufficient quantities of water directly onto a fire. This purpose is to contain a fire until the fire department can arrive for final extinguishment. Water is supplied to individual heads through a system of piping. Sprinkler heads may extend from exposed pipes or protrude from a ceiling or wall from hidden pipes **(Figure 8.13)**.

> **Automatic Sprinkler System** — System of water pipes, discharge nozzles, and control valves designed to activate during fires by automatically discharging enough water to control or extinguish a fire.

244 Chapter 8 • Fire Detection, Alarm, and Suppression Systems

Figure 8.13 An operating sprinkler system.

Early Sprinkler Systems

During the industrial revolution in the United States, various forms of sprinkler systems protected mills and factories throughout New England. Due to many fires with a large loss of life, sprinklers began to be required in many public occupancies in North America. Sprinklers are currently required in many types of occupancies, including single family residences in some jurisdictions.

Sprinkler coverage may protect all areas of a building or only certain areas, such as high hazard sections or exit routes. The authority having jurisdiction (AHJ) can mandate through fire codes and/or local ordinances what level of protection is required for various occupancies.

In order to ensure reliability, the sprinkler and its components should be listed with a nationally recognized testing laboratory, such as Underwriters Laboratories Inc., Underwriters Laboratories of Canada, or FM Global. Automatic sprinkler systems are reliable fire protection devices whose basic operation of piping and valves, as well as the various applications and life safety benefits, should be understood by all firefighters.

> **Sprinkler** — Water flow discharge device in a sprinkler system; consists of a threaded intake nipple, a discharge orifice, a heat-actuated plug, and a deflector that creates an effective fire stream pattern that is suitable for fire control.

Common Causes of Sprinkler System Failure or Incomplete Operation

Automatic sprinkler systems are reliable. When failures are reported, the reason is rarely the sprinkler system, but likely one of the following circumstances:

- Partially or completely closed water main valve
- Interruption of the municipal water supply
- Damaged or painted-over sprinkler heads
- Frozen or broken pipes
- Excess debris or sediment in the pipes
- Failure of a secondary water supply
- Tampering or vandalism

Sprinkler Systems and Their Effect on Life Safety

Sprinkler systems help enhance the life safety of building occupants as they discharge water directly onto a fire while it is still relatively small. With the fire contained in the early growth stages, the products of combustion are limited. Sprinklers are also effective in the following situations:

- Preventing fire spread upwards in multistory buildings
- Protecting the lives of occupants in other parts of the building by confining the fire to the area of origin
- Protecting means of egress

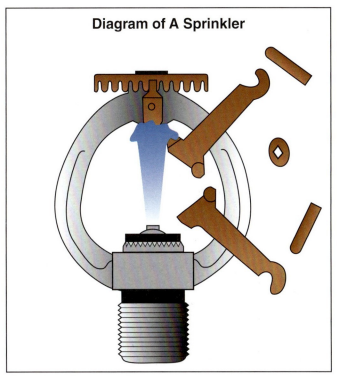

Figure 8.14 Diagram of a sprinkler.

Sprinklers

Sprinklers begin to discharge water with the release of a cap or plug that is activated by a heat-responsive element. The sprinkler head may be viewed as a fixed-spray nozzle operated individually by a thermal detector. There are numerous types and designs of sprinklers for use with various occupancies and installation requirements. An exterior connection, known as a fire department connection (FDC), allows a fire department pumper to augment the pressure and water supply of the sprinkler system for more effective operation (**Figure 8.14**).

A sprinkler head is commonly identified by the temperature at which a sprinkler is designed to operate. There are several ways in which a sprinkler may be marked to identify this temperature, including color coding of the frame arms, different colored liquid in bulb-type sprinklers, or stamping the temperature on the sprinkler itself. Three of the most common release mechanisms for sprinklers include fusible links, frangible bulbs, and chemical pellets. All of these mechanisms fuse (melt) or open in response to heat (**Table 8.1**). Additional information regarding sprinkler systems may be found in IFSTA's **Fire Protection, Detection and Suppression Systems** manual.

Sprinkler Position

Pendent, upright, and sidewall are three basic mounting positions for sprinklers. These types are not interchangeable, as they are each designed with a specific spray pattern based on their intended installation. In addition to these common types, there is various special purpose sprinklers used for specific applications (**Figure 8.15 a, b and c**).

Control Valves

A sprinkler system must be equipped with a main water control valve. Control valves within the system may be used to shut off the water supply in order to replace sprinkler heads, perform maintenance, or discontinue operations. The main control valve is located between the source of the water supply and the sprinkler system. The control valve is usually located under the sprinkler

Table 8.1
Sprinkler Temperature Ratings, Classifications, and Color Codings

Max. Ceiling Temp. °F	°C	Temperature Rating °F	°C	Temperature Classification	Color Code	Glass Bulb Colors
100	38	135 to 170	57 to 77	Ordinary	Uncolored or black	Orange or red
150	66	175 to 225	79 to 107	Intermediate	White	Yellow or green
225	107	250 to 300	121 to 149	High	Blue	Blue
300	149	325 to 375	163 to 191	Extra high	Red	Purple
375	191	400 to 475	204 to 246	Very extra high	Green	Black
475	246	500 to 575	260 to 302	Ultra high	Orange	Black
625	329	650	343	Ultra high	Orange	Black

Figure 8.15 These sprinklers react to heat: **(a)** fusible link sprinkler **(b)** frangible bulb sprinkler **(c)** chemical pellet sprinkler.

alarm valve, the dry pipe, or deluge valve, or outside the building near the sprinkler system that it controls. The control valve should always be returned to the open position after any maintenance is complete. These valves should be secured in the open position and monitored to be sure they are not tampered with (**Figure 8.16, p. 248**).

Main water control valves are manually operated indicating valves. An indicating control valve shows at a glance whether it is open or shut. The following are the four common types of indicating valves used in sprinkler systems:

- **Outside stem and yoke (OS&Y) valve** — Has a yoke on the outside with a threaded stem that controls the opening and closing of the gate. The threaded portion of the stem is out of the yoke when the valve is open and inside the yoke when the valve is closed (**Figure 8.17, p. 248**).

- **Post indicator valve (PIV)** — Hollow metal post that is attached to the valve housing. The valve stem is inside this post, and a moveable target is located on the stem with the words "Open" and "Shut" to indicate the valves status (**Figure 8.18, p. 248**).

Figure 8.16 Sprinkler system control valves are located between the source of the water supply and the sprinkler system.

Figure 8.17 The outside stem and yoke valve (OS&Y) is a type of indicating valve.

Figure 8.18 On a post indicator valve (PIV), the words "OPEN" and "SHUT" indicate the position of the valve.

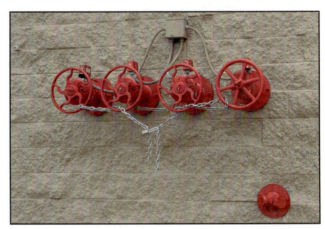

Figure 8.19 A wall post indicator valve (WPIV) extends through the wall with the target and valve operating nut on the outside of the building.

- **Wall post indicator valve (WPIV)** — Similar to a PIV, except that it extends through a wall with the target and nut on the outside of the building (**Figure 8.19**).
- **Post indicator valve assembly (PIVA)** — Does not use a target with words to indicate valve status, but has a sight area that is open when the valve is open and shut when the valve is shut (**Figure 8.20**).

Waterflow Alarms

A waterflow alarm operates to indicate that water is flowing through a sprinkler system. These alarms may be either hydraulically (operated by the movement of water) or electrically operated. The hydraulic alarm may alert an occupant or passerby to waterflow, while an electric waterflow alarm may also be configured to notify the fire department (**Figure 8.21**).

NOTE: Hydraulic waterflow alarms are sometimes referred to as water motor gongs. This type of alarm uses water from the sprinkler system to drive a water motor and sound a local alarm gong. These devices have a distinctive sound as they clang rapidly due to the velocity of the water flowing through the motor.

Figure 8.20 The post indicator valve assembly (PIVA) has a sight area that is open when the valve is open and closed when the valve is closed.

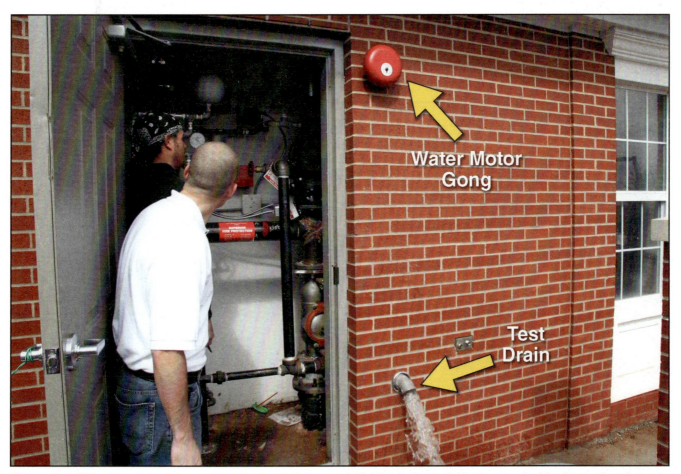

Figure 8.21 Waterflow alarm systems alert occupants that water is flowing in the sprinkler system. In this picture, the alarm at the top is ringing while the system is tested.

Applications of Sprinkler Systems

The following sections highlight the major applications of sprinkler systems. Firefighters should have an understanding of each type of system:

- Wet pipe
- Dry pipe
- Preaction
- Deluge
- Residential

Wet-Pipe System

A **wet-pipe sprinkler system** is used in locations that will not be subjected to temperatures below 40° F (4°C). This is the simplest type of sprinkler system and generally requires little maintenance. This system contains water under pressure at all times and is connected to a public or private water supply. When a sprinkler head is activated, water discharges and an alarm is actuated.

Dry-Pipe System

Dry-pipe systems are used in locations where the piping may be subjected to temperatures below 40°F (4°C). In this system, air under pressure replaces water in the piping. When a sprinkler is activated, the pressurized air escapes first. Once the pressure is released, the valve will open, allowing water to flow into the piping system and discharge through any heads that may have fused (**Figure 8.22**).

> **Wet-Pipe Sprinkler System** — Fire-suppression system that is built into a structure or site; piping contains either water or foam solution continuously; activation of a sprinkler causes the extinguishing agent to flow from the open sprinkler.

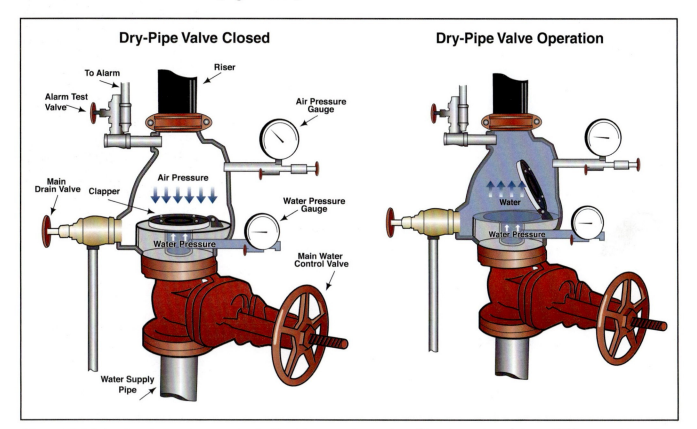

Figure 8.22 A dry pipe system is pressurized with air to prevent water from filling the system piping until it is needed.

Preaction System

A preaction sprinkler system is a dry system that employs a deluge-type valve (see following section), fire detection devices, and closed sprinklers. This type of system is installed in occupancies where preventing water damage is critical. The sprinkler pipes remain dry until a heat or smoke detector is actuated. In response to detection device activation, the sprinkler pipes will be filled with water and any fused heads will discharge onto a fire.

Deluge System

In a deluge system, water flows from all sprinklers in a designated area where the system has been activated. Along with open sprinklers, this system is generally equipped with a deluge valve. When the deluge valve is activated, water flows from each open sprinkler that is controlled by the specific valve that was activated. A deluge system is often used to protect extra hazardous occupancies, such as aircraft hangars. Such systems may also be used to supply other fire suppression products, such as foam.

Residential Systems

Residential sprinkler systems may be found in some single- or two-family dwellings. This type of system is designed to prevent total fire involvement in the room of origin, giving occupants an opportunity to escape. Constructed as either wet or dry systems, the recommendations for these installations may be found in NFPA® 13D, *Standard for Installation of Sprinkler Systems in One and Two Family Dwellings and Manufactured Homes* **(Figure 8.23)**.

Residential Sprinkler System — Wet- or dry-pipe fire suppression system that is built into a residential structure; activation of a sprinkler causes the extinguishing agent to flow from the open sprinkler.

The Fire Sprinkler Initiative

Some jurisdictions have adopted residential sprinkler systems. The Fire Sprinkler Initiative has information regarding this initiative.

A standpipe system is an installation of wet or dry pipes used in large single-story or multistory buildings to provide water supply for fire fighting. An exterior fire department connection is used to allow fire department pumpers to supply water or augment the pressure and volume of the system to enhance fire fighting efforts. Hose connections are located at specific points within the building in a cabinet, hose station, or stairwell **(Figure 8.24 p. 252)**.

Figure 8.23 A residential sprinkler during construction of a house.

Depending on the needs of occupancy, a standpipe system may be simple, consisting of a vertical pipe (riser) with hose connections and an Fire Department Connection (FDC). In other instances, a system may be quite complex, consisting of many risers, multiple fire pumps, and reservoirs located on upper floors for water storage. Standpipes may be designed to supply various diameters of hose based on the classifications that are discussed in the following section.

Fire Department Connection (FDC) — Point at which the fire department can connect into a sprinkler or standpipe system to boost the water pressure and flow in the system. This connection consists of a clappered siamese with two or more 2½-inch (64 mm) intakes or one large-diameter (4-inch [102 mm] or larger) intake

Classes of Standpipe Systems

NFPA® 14, *Standard for the Installation of Standpipe and Hose Systems*, is often used for the design and installation of standpipe systems. This standard recognizes three classes of standpipe system: Class I, Class II, and Class III.

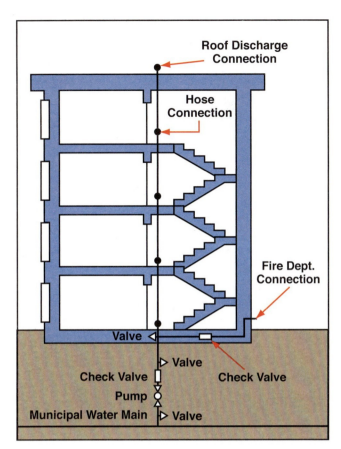

Figure 8.24 A standpipe system is designed to enable building occupants or firefighters to put water on a fire from different levels in the building.

Figure 8.25 A Class I standpipe system is designed to be used by firefighters to supply water to large hoselines.

Class I Standpipe Systems

Firefighters trained in the operation of large 2 ½-inch diameter handlines are the primary users of a Class I standpipe system. These systems are designed to supply effective fire streams for use during more advanced fires. Class I systems feature 2-½ inch hose connection attached to the standpipe riser (**Figure 8.25**).

Class II Standpipe Systems

Building occupants with no specialized fire fighting training are the intended users of a Class II standpipe system. These systems often feature a hose cabinet with 1 ½-inch single jacket fire hose and a lightweight twist type shutoff nozzle. The hose connection on the standpipe riser is 1 ½ inches in diameter. These installations are sometimes referred to as house lines. Class II systems are not as prevalent as they once were, however, firefighters may still encounter them (**Figure 8.26**).

Class III Standpipe Systems

A Class III standpipe system combines the features of Class I and Class II systems. These installations provide both a 2 ½-inch connection and 1 ½-inch connection. The design of the system should allow both to be used simultaneously (**Figure 8.27**).

Figure 8.26 A Class II standpipe system is designed for use by building occupants.

Figure 8.27 Class III standpipe systems have connections for both small and large hoselines.

Types of Standpipe Systems

In addition to the classes of standpipes, there are two basic types of systems:

- A wet standpipe system maintains water in the risers at all times. It is capable of automatically supplying water when a valve at the riser connection is opened.

- A dry system does not have a permanent water supply. The fire department at the FDC must supply water.

- A wet system with an automatic water supply is desirable, as it provides the constant availability of water at a standpipe connection. However, a wet system is more costly to install and maintain, and it cannot be used for applications where it will be exposed to temperatures that are below freezing.

Pressure Regulating Devices

When the pressure at the discharge opening of a standpipe connection exceeds 100 psi, a pressure regulating device may be installed to ensure a hoseline operated from that outlet is manageable.

There are several types of pressure regulating devices in common use. One design consists of a restricting orifice inserted into the waterway, while another design features vanes that may be rotated to change the cross sectional area of the waterway. A pressure reducing valve is another type of pressure regulating device.

All pressure regulating devices should be installed and adjusted to meet the pressure and flow requirements specific to the location in which they are to be utilized. If a pressure regulating device is not properly installed or adjusted, the available flow may be greatly reduced, endangering firefighters and hampering suppression efforts.

Chapter Summary

Fire detection, alarm, and suppression systems play a vital role in detecting a fire in its early stages, warning occupants, and notifying the fire department. In addition, automatic sprinklers may further enhance the life safety of civilians and firefighters by helping to contain a fire while it is still relatively small.

Firefighters must understand the capabilities of modern detection and suppression systems in order to utilize the advantages of early warning and the detailed amount of information and systems control options that may be provided by some systems. Furthermore, firefighters should become familiar with the specific types of fire detection and suppression systems found in their jurisdiction and preplan ways to best use and support the features of these systems.

Review Questions

1. What are some reasons for installing fire detection, alarm, and suppression systems?
2. How do fixed-temperature heat detectors and rate-of-rise heat detectors operate?
3. What are some of the different types of smoke detectors and smoke alarms?
4. What are some types of automatic alarm systems?
5. Describe the four common types of indicating valves used in sprinkler systems.
6. Explain wet pipe and dry pipe applications of sprinkler systems.
7. What are the characteristics of preaction systems and deluge systems?
8. What is a residential sprinkler system designed to prevent?
9. What are the three classes of standpipe systems?
10. What are the two basic types of standpipe systems?

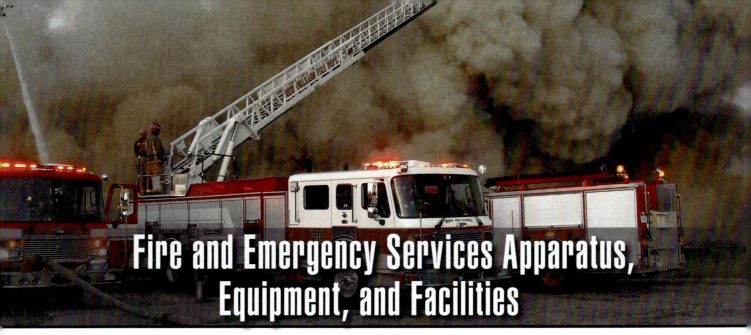

Fire and Emergency Services Apparatus, Equipment, and Facilities

Courtesy of Chris Mickal.

Chapter Contents

Case History 259
Fire Department Apparatus 260
　The Engine (Pumper) 260
　Smaller Fire Apparatus 263
　Mobile Water Supply Apparatus 264
　Wildland Fire Apparatus 264
　Aerial Apparatus 265
　Quintuple Aerial Apparatus (Quint) 269
　Rescue Apparatus 269
　Fire Service Ambulances 271
　Aircraft Rescue and Fire Fighting Apparatus
　　(ARFF) 273
　Hazardous Materials Response Unit 274
　Mobile Air Supply Unit 275
　Fireboats and Search and Rescue Boats 276
　Power and Light Unit 277
　Mobile Fire Investigation Unit 277
　Fire Fighting Aircraft 278
　Other Special Units 278
**Uniforms and Personal
　Protective Clothing** 279
　Uniforms 279
　Personal Protective Clothing 280

Breathing Apparatus 280
　Air Purifying Respirators (APRS) 280
　Supplied Air Respirators (SARS) 280
　Self-Contained Breathing Apparatus (SCBA) 280
Tools and Equipment 281
　Equipment 281
　Power Tools 281
　Hand Tools 281
**Ropes, Webbing, Related Hardware,
　and Harnesses** 281
Ground Ladders 282
**Fire Hose, Nozzles, and Hose Appliances
　and Tools** 282
Fire Department Facilities 287
　Fire Stations 287
　Administrative Offices and Buildings 288
　Telecommunication Centers 289
　Training Centers 289
Maintenance Facilities 292
Chapter Summary 293
Review Questions 293

chapter 9

Key Terms

Aerial Apparatus....................................265	Open-Circuit Self-Contained Breathing...................280
Apparatus..260	Personal Protective Clothing 280
Burn Building...290	Personal Protective Equipment (PPE)..280
Closed-Circuit Self-Contained Breathing Apparatus 280	Search and Rescue Boat 276
Drill Tower 290	Smokehouse 290
Engine..260	Tender..264
Fireboat 276	
Fire Station..287	
Mobile Water Supply Apparatus 264	

FESHE Outcomes

This chapter provides information that addresses the outcomes for the Fire and Emergency Services Higher Education (FESHE) *Principles of Emergency Services* course.

8. Describe the common types of fire and emergency service facilities, equipment, and apperatus..

Fire and Emergency Services Apparatus, Equipment, and Facilities

Learning Objectives

1. Identify fire department apparatus. (Outcome 8)
2. Recognize the uses for uniforms and personal protective clothing. (Outcome 8)
3. Distinguish among basic types of breathing apparatus. (Outcome 8)
4. Describe how tools and equipment can be used in the fire and emergency services. (Outcome 8)
5. Identify uses for ropes, webbing, related hardware, and harnesses in the fire and emergency services. (Outcome 8)
6. Describe why ground ladders are used in fire and emergency services. (Outcome 8)
7. Explain the basic functions of fire hose, nozzles, and hose appliances and tools. (Outcome 8)
8. Recognize uses for various fire department facilities. (Outcome 8)

Chapter 9
Fire and Emergency Services Apparatus, Equipment and Facilities

Case History

The chief reflected on the recent approval for the new aerial. The new firefighters sitting in the dayroom listened carefully as he spoke. At 105 feet and $1.1 million, it was to be the largest and most expensive piece the department had ever purchased. As the chief said this, he recalled that when he was appointed to the department 35 years ago, the truck company was operating a 1965 Maxim straight ladder aerial. The purchase price for the truck was under $100,000. The aerial truck was equipped with a straight 95-foot ladder and only had a two-man cab with a department-installed, two-way radio. The truck was an open cab with windshield wipers on both the outside and the inside of the windscreen. Air packs, added in the 1970s, were carried in boxes in a compartment.

The new apparatus would be delivered with a six-person, enclosed cab, air bags for personnel protection in the event of an accident, anti-rollover protection, an integral waterway to allow water tower operations, and the latest in radio and computer devices. The truck would carry rescue, ventilation, and fire suppression tools that were not invented when the chief began his career. Each seating position was equipped with safety belts integral with the ignition system to increase the safety of the firefighters riding there. The air packs were mounted in each seat of the five riding positions. Ventilation fans, rope rappelling harnesses, powered tools, generators, and lighting equipment were to be included in the new aerial.

The chief and the firefighters talked about how much has changed, and one firefighter said that there can't be much left to invent. Fire fighting is always the same now. The Chief chuckled and said, "Don't be so sure. In 35 years, you will be having this same conversation with the latest recruits!"

The operations of a fire department are centered around its personnel, apparatus, equipment, and facilities. Familiarity with available resources is important if a firefighter is to work safely and efficiently. While all fire departments may not possess every piece of apparatus or equipment described in the following chapter, firefighters should be aware of the wide variety of resources used in the fire service throughout North America.

NFPA Standards

The NFPA® produces a number of standards that provide minimum specifications regarding the manufacturing or construction of fire and emergency services apparatus, tools, equipment, and facilities. When an NFPA® standard is updated, it does not mean that equipment in use is automatically unsafe or unusable. All newly purchased apparatus, equipment, and facilities will meet the current edition of the appropriate standard at the time of manufacture.

Apparatus — Motor-driven vehicle or vehicles designed for fighting fires; different types include engines, water tenders, and ladder trucks.

Engine — Ground vehicle providing specified levels of pumping, water, and hose capacity and staffed with a minimum number of personnel. *Also known as* Fire Department Pumper.

Fire Department Apparatus

Fire **apparatus** are designed to perform specific functions at emergency incidents. Some apparatus perform only one major function, while others are designed to be multifunctional. All fire departments may not operate every type of apparatus listed in the sections that follow. However, the apparatus described in this chapter are among the most commonly used throughout the United States and Canada.

Figure 9.1 A modern fire department pumper carries a variety of portable equipment. *Courtesy of Ron Jeffers.*

The Engine (Pumper)

The **engine** is the most common of all fire apparatus. Sometimes called a *pumper* in reference to its main feature, the primary function of this truck is to confine and extinguish fires with water delivered through hoselines under pressure from its fire pump **(Figure 9.1)**.

Pumpers should be designed to meet the recommendations set forth in NFPA®1901, *Standard for Automotive Fire Apparatus*. The following rated capacities are among those most encountered on standard fire department pumpers:

- 750 gpm (gallons per minute) (3 000 L/min)
- 1,000 gpm (4 000 L/min) (4 000 L/min
- 1,250 gpm (5 000 L/min)
- 1,500 gpm (6 000 L/min)
- 1,750 gpm (7 000 L/min)
- 2,000 gpm (8 000 L/min)

Gauges and controls are provided to aid the driver/operator in controlling the pressure and volume of water supplied to an emergency scene **(Figure 9.2)**. Pumpers usually have an onboard water tank, sometimes referred to as a *booster tank*, with capacities generally ranging from 500 to 1,500 gallons (2 000 L to 6 000 L). This tank may be used to begin a fire attack before the pumper is connected to an external water supply, such as a fire hydrant or body of water. Some pumpers feature an additional tank for special extinguishing agents, such as foam concentrate. This smaller tank may be connected to a proportioning device that, along with a pump, will mix foam concentrate with water to produce a foam fire stream

Pumpers are equipped with various arrangements of hose beds in different places on the apparatus. These beds carry different sizes of hose used for water supply of fire attack purposes. Many of the hoselines used for fire attack may be pre-connected to a pump discharge outlet and are equipped with a pre-attached nozzle to aid in rapid deployment and operation of the hoseline **(Figure 9.3)**.

Figure 9.2 Pump and flow of water are controlled at the pump panel.

NFPA® 1901 recommends that an engine carry a minimum of 400 feet (120 m) of small attack hose, ranging in size from 1 ½ to 2 inches (38 mm to 50 mm) in diameter. The standard also recommends a minimum of 800 feet (240 m) of hose that is 2 ½ inches or larger in diameter for water supply purposes. Some apparatus may be equipped with a reel of smaller (1 inch [25 mm] or less) noncollapsible hose. This hoseline, often called a *booster reel*, may be used for extinguishing small exterior fires, but it is no longer part of the NFPA®1901 standard **(Figure 9.4)**.

In addition to hose and water, engines may carry a wide variety of standard equipment and specialized tools and appliances based on jurisdictional requirements. Many fire department pumpers carry the following standard equipment:

- Ground ladders
- Forcible entry tools
- Nozzles
- Hose adapters and appliances
- Hose tools

Figure 9.3 Firefighters using hose lines to extinguish a fire during a training exercise.

Figure 9.4 A booster hose is used on smaller exterior fires.

Chapter 9 • Fire and Emergency Services Apparatus, Equipment, and Facilities 261

Figure 9.5 Portable water tanks are used for water shuttle operations.

- Self-contained breathing apparatus (SCBA)
- Portable fire extinguishers
- Pike poles
- Salvage covers
- Emergency medical equipment

In some jurisdictions, an engine company may perform additional functions and require specialized equipment, including the following:

- Rescue and extrication tools
- Power saws
- Ventilation fans
- Portable water tanks for water shuttle operations **(Figure 9. 5)**

Commercial and Custom Fire Apparatus

A commercial fire apparatus consists of a body (compartments, pump, hose bed) that has been designed, built, and placed on a cab and chassis that has been built by and acquired from a commercial truck builder. That particular cab and chassis may be specified for use in a variety of commercial vehicles of size and weight in addition to fire apparatus. Generally, the largest advantage with a commercial apparatus is that the initial cost may be lower than a custom apparatus. In addition, parts and service centers may be found locally. A disadvantage to commercial cabs is that they may not have the seating capacity or in-cab storage that some fire departments require.

A fire apparatus manufacturer engineers and manufactures custom fire apparatus, which includes the cab, chassis, compartments, and other components. A custom apparatus may provide a convenient seating arrangement for firefighters as well as cab space configured to store equipment safely and to allow for the use of communications and other computer equipment.

The purchaser has considerable latitude in the inclusion and layout of many options for the apparatus. Manufacturers are essentially able to meet all the fire department's expectations by designing the apparatus to fit its specifications.

Smaller Fire Apparatus

Fire apparatus may be purchased to fill the needs of the jurisdiction, and, in some cases, a smaller unit is specified to provide a small, maneuverable apparatus that is operable with fewer personnel. A smaller vehicle may serve different functions based on the region in which it operates and may be known by different names that are regionally accepted. Some of the names for smaller apparatus include:

- Mini-pumper
- Midi-pumper
- Quick attack apparatus
- Rapid intervention vehicle
- Quick response vehicle

Smaller apparatus may be used to handle small fires that do not require the equipment or personnel of a large pumper **(Figure 9.6)**. Additionally, these smaller apparatus may be used to begin an initial attack on a larger fire. Once the larger apparatus arrives on the scene, additional personnel can provide support. A jurisdiction may employ this concept to reduce fuel expenses and wear on larger, more expensive apparatus.

An initial attack apparatus may operate with a crew of two to five firefighters, depending on the size of the crew cab to accommodate riders. These vehicles are often pickup truck chassis with custom bodies and pumps of at least 250 gpm (1 000 L) and booster tanks of at least 200 gallons (800 L). They carry an equipment inventory similar to a standard engine, but fewer pieces of each type of equipment. Often these vehicles carry emergency medical equipment and are used in a first responder role on medical incidents. This size attack apparatus may also be deployed for vehicle fires inside parking structures. Some vehicles may feature a master stream appliance that can be positioned easily due to the size of the apparatus and supplied with water from a larger pumper that would not be able to maneuver into an effective attack position **(Figure 9.7)**.

Figure 9.6 An example of a smaller pumper truck.

Figure 9.7 A rapid intervention truck. *Courtesy of James Nilo.*

Figure 9.8 A water tender truck is used to transport water. *Courtesy of Ron Jeffers.*

Figure 9.9 A mobile water supply truck. *Courtesy of Ron Jeffers.*

Mobile Water Supply Apparatus

The **mobile water supply apparatus**, or water **tender**, is used to transport water to areas beyond a pressurized water system or where water supply is inadequate. Mobile water supply apparatus feature onboard tanks of at least 1,000 gallons (3 785 L) that may be used to augment the smaller tanks found on many pumpers **(Figure 9.8)**.

The size of a water tank on a mobile water supply apparatus depends on a variety of factors:

- **Terrain** — Apparatus must be capable of traversing the terrain, such as narrow winding roads and steep hills. Large, heavy apparatus may not be able to negotiate these road conditions.
- **Bridge weight and height limits** — Bridges in some jurisdictions may not bear the weight or accommodate the height of a large tender. Alternate routes must be pre-selected, which may divert response a significant distance from an incident.
- **Monetary constraints** — A jurisdiction may not have available funds to purchase a large piece of apparatus.
- **Size of other mobile water supply apparatus in the area** — Mobile water supply shuttles operate more efficiently when apparatus of similar capacity are used.

NFPA® 1901 recommendations should guide the purchaser and manufacturer when designing a mobile water supply apparatus to produce the safest, most efficient vehicle available. These recommendations include limiting single rear axle apparatus to water tanks of 1,500 gallons (6 000 L) or less. If a greater capacity is required, a dual rear axle of tractor-trailer design should be considered. Straight-chassis vehicle may have up to a 4,000 gallon (16 000 L) tank capacity, with larger tanks requiring a tractor-trailer configuration. Some tenders feature a fire pump similar in size to those found in traditional pumping apparatus. These units may operate as an attack pumper or a mobile water supply apparatus depending on incident requirements **(Figure 9.9)**.

Wildland Fire Apparatus

Apparatus that can traverse rugged off-road terrain are often required to control wildfires **(Figure 9.10)**. These vehicles must be lightweight and highly maneuverable. They should be designed and manufactured in accordance with the recommendations provided in NFPA® 1906, *Standard for Wildland*

Mobile Water Supply Apparatus — Fire apparatus with a water tank of 1,000 gallons (3 785 L) or larger whose primary purpose is transporting water; may also carry a pump, some hose, and other equipment. *Also known as* Tanker or Tender.

Tender — Term used within the Incident Command System for a mobile piece of apparatus that has the primary function of supporting another operation; examples include a water tender that supplies water to pumpers, or a fuel tender that supplies fuel to other vehicles.

Figure 9.10 Wildland fire trucks must be able to traverse rugged off-road terrain. *Courtesy of Bob Esposito.*

Figure 9.11 Brush trucks are wildland apparatus with pumping capabilities. *Courtesy of Steve Lofton.*

Fire Apparatus. These vehicles are typically built on utility bodies and feature all-wheel drive and onboard tanks ranging from 50 to 1,000 gallons (4 000 L) depending on the size of the apparatus.

NOTE: Due to the improper installation or the absence of baffles, mobile water supply apparatus have experienced rollovers. Onboard water tanks on fire apparatus should include the installation of proper baffles (partial walls in the tank) to lessen the effect of water shifting in the tank as the apparatus makes turns or traverses uphill or downhill grades. See IFSTA's **Pumping Apparatus Driver/Operator** Handbook for additional information.

Wildland apparatus, also known as brush engines in some regions, are often equipped to pump and roll with an auxiliary-engine-driven pump **(Figure 9.11)**. Some vehicles may include a traditional power take-off pump for stationary operation in addition to or instead of pump-and-roll capability. Some wildland apparatus carry a portable pump that may be taken from the apparatus and placed in a static water source to provide water to the fire scene. In addition, wildland fire apparatus may have pre-piped water nozzles mounted to the vehicle to protect it from hot spots or to extinguish fires using pump-and-roll techniques.

Depending on the size of the apparatus, wildland vehicles may carry non-collapsible hose reels or 1 ½-inch (38 mm) diameter single-jacket forestry hose with adjustable flow nozzles. A variety of rakes, shovels, axes, backpack water tanks, and other equipment also may be stored on the apparatus.

Aerial Apparatus

An **aerial apparatus** is equipped with a powered ladder that provides firefighters access to upper floors and roofs during incident operations. These pieces of apparatus should be designed and manufactured in accordance with the recommendations of the NFPA® 1901 standard.

Aerial apparatus may be divided into the following categories:

- Aerial ladder apparatus
- Elevating platform apparatus, including three subcategories:
    — Aerial ladder platforms
    — Telescoping aerial platforms
    — Articulating aerial platforms
- Water towers

Aerial Apparatus — Fire fighting vehicle equipped with a hydraulically operated ladder, elevating platform, or other similar device for the purpose of placing personnel and/or water streams in elevated positions.

Figure 9.12 An aerial ladder truck.

Figure 9.13 Hydraulic pumps are used to power the hydraulic arms that lift and extend aerial devices.

Aerial Ladder Apparatus

An aerial ladder is a powered ladder mounted to a fire apparatus chassis. Its working height is measured from the ground to the highest ladder rung with the ladder at maximum elevation and extension **(Figure 9.12)**. The working height of North American aerial ladders is generally from 50 to 135 feet (15 to 40 m). Models manufactured in other countries may exceed these heights. The main uses for aerial ladders are as follows:

- Rescue
- Ventilation
- Elevated master stream application
- Gaining access to upper levels

Hydraulic pumps and motors are used to operate most aerial ladders **(Figure 9.13)**. In order for the apparatus to meet the recommendations of NFPA® 1901, an electric or mechanical backup system must be provided. Apparatus may consist of a two- or three-axle single chassis vehicle or a three-axle tractor trailer-vehicle. Tractor-drawn aerial apparatus, also known as tiller trucks, are longer than single-chassis apparatus. The tiller operator, riding in a position at the rear of the vehicle, controls the steerable rear wheels of the tractor-drawn aerial apparatus. The ability to steer the rear wheels makes the tiller truck more maneuverable than single-chassis apparatus. The tiller truck allows access to narrow streets and maneuverability in heavy traffic.

Aerial Ladder Platform Apparatus

An aerial ladder platform apparatus is similar to a two- or three-axle single chassis aerial ladder apparatus, except that a work platform is mounted to the tip of the aerial ladder. The aerial may be mounted at the rear or midpoint on the apparatus combined with the features of the safe work area of a platform **(Figure 9.14)**. NFPA® 1901 requires that the platform be constructed of metal, usually steel or an aluminum alloy. The working height of an elevated platform is measured from the ground to the top surface of the highest platform handrail with the aerial device at maximum extension and elevation. Aerial platforms generally range in height from 85 to 110 feet (25 to 30 m) **(Figure 9.15)**.

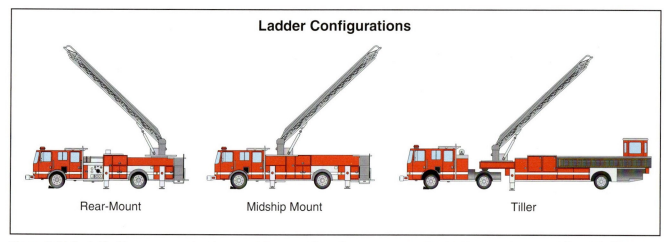

Figure 9.14 Aerial ladders are mounted in one of three configurations: rear-mount, midship mount, or on a tillered trailer.

Many aerial platform apparatus feature at least one (sometimes two) permanently mounted master stream appliances on the work platform **(Figure 9.16)**. In addition, a water spray nozzle is located under the platform that may be activated to protect the firefighters who are inadvertently exposed to high heat conditions during fire fighting operations. Electrical, breathing air, and sometimes hydraulic connections are provided in the platform. Other equipment may be mounted in the platform, such as spotlights, forcible entry tools, and appliances for connecting handlines. Recommended design components, such as a backup control system and two control stations, can typically be found in the platform and at the turntable.

Aerial platforms equipped with breathing air connections typically allow members operating in the platform to wear an SCBA mask attached to a hose that is fed a supply of air from tanks mounted near the turntable of the apparatus. Many of these hoses are long enough to allow firefighters to leave the platform and work nearby while connected to the apparatus breathing air system **(Figure 9.17, p. 268)**.

Figure 9.15 With the ladder at maximum elevation and extension, measure from the ground to the highest ladder rung to determine its working height.

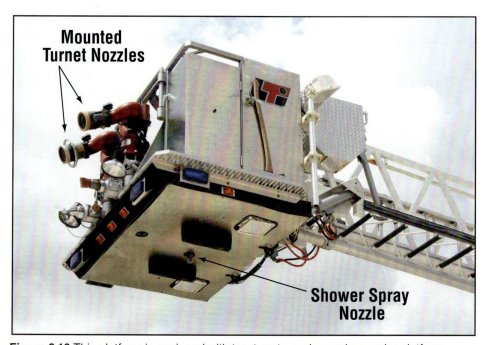

Figure 9.16 This platform is equipped with two turret nozzles and an under platform shower spray nozzle.

Chapter 9 • Fire and Emergency Services Apparatus, Equipment, and Facilities **267**

Telescoping Aerial Platform Apparatus

Although NFPA® places the aerial ladder platform and the telescoping aerial platform under the same definition, each apparatus features different capabilities. The primary difference between the two vehicles is that the aerial ladder platform has a large ladder that allows firefighters to climb up and down in the performance of fire fighting tasks. A telescoping aerial platform features a small ladder attached to a telescoping boom that is designed for use in emergencies to evacuate the platform.

A telescoping aerial platform consists of two or more sections of either box beam or tubular truss beam construction. In box-beam construction, four sides are welded together to form a box shape with a hollow center that sometimes contains the hydraulic, breathing air lines, and waterway required at the platform. Tubular truss construction, similar to aerial ladders, consists of tubular steel welded to form a box shape using a cantilever or triangular truss design.

Figure 9.17 A firefighter wearing a breathing mask connected to the aerial device's built-in breathing air system.

Articulating Aerial Platform Apparatus

Articulating aerial platform apparatus differ from telescoping platform apparatus in that they consist of two or more boom sections that are connected by a hinge and fold like an elbow. The construction of the boom is similar to that of the telescoping platform **(Figure 9.18)**. This type of apparatus can perform many of the same functions, such as rescue, ventilation, master stream application, or accessing upper floors, as other aerial apparatus.

NFPA®1901 recommends that articulating platforms vehicles, such as platform apparatus, have at least one permanently mounted master-stream appliance. Many articulating platforms also feature breathing air and electric power available in the platform to facilitate operations.

Water Towers

Some fire departments outfit engine company apparatus with hydraulically operated water towers **(Figure 9.19)**. These telescoping or articulating devices are designed to deploy an elevated master stream appliance. The driver/operator may control the tower and nozzle mounted at the tip from ground level. Most

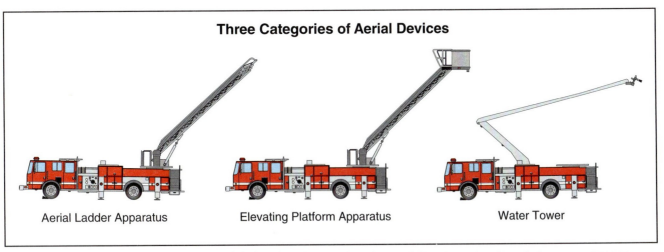

Figure 9.18 NFPA® 1901 categorizes aerial devices into three categories: aerial ladders, elevating platforms, and water towers.

Figure 9.19 This apparatus is equipped with an articulating water tower. *Courtesy of Las Vegas (NV) Fire and Rescue.*

Figure 9.20 Quints have a fire pump, water tank, supply and attack hoses, ground ladders, and an aerial device or elevated platform. *Courtesy of Ted Boothroyd.*

water towers are designed to allow for deployment from a few degrees below horizontal to almost 90 degrees from the ground. These devices may range from 35 to 130 feet (10.5 to 40 m) and flow 1,000 to 5,000 gpm (4 000 to 2 0000 L).

Quintuple Apparatus (Quint)

The term *quint* is used to describe a fire apparatus that has five major features. NFPA® 1901 establishes the qualifying features of a quint apparatus **(Figure 9.20)**. The criteria includes:

- Size of the pump
- Size of the water tank
- Amount of supply and attack hose
- Types and length of ground ladders
- Pre-piped waterway on the aerial device

Some fire departments operate a quint in addition to (or rather than) standard aerial apparatus. A quint can be operated as follows:

- May be used as a ladder company, engine company, or both (depending on the nature of the call).
- Supplies its own elevated master stream appliance.
- Extinguishes fires encountered when an engine is not present.
- Extinguishes a fire should it arrive first on the scene. Some departments have replaced their traditional engine and ladder companies with quint apparatus.

Rescue Apparatus

Rescue apparatus are used to carry the tools and equipment necessary to rescue people from positions of danger, such as fires and motor vehicle, industrial, or agricultural accidents. These vehicles may also carry the specialized equipment necessary to operate at technical rescue incidents, such as trench cave-ins or structural collapses. Rescue apparatus are staffed with personnel

Figure 9.21 Rescue trucks carry specialized tools and equipment necessary to perform technical rescue incidents. *Courtesy of Ron Jeffers.*

Figure 9.22 A medium rescue truck. *Courtesy of Ron Jeffers.*

based on jurisdictional requirements and are categorized into three general types of apparatus: light, medium, and heavy, the requirements of which vary by geographic region **(Figure 9.21)**.

Light Rescue Vehicle

Often functioning as a first response vehicle, the light rescue vehicle is designed to carry basic hand tools, such as saws and jacks, power equipment, a limited number of hydraulic rescue tools, and emergency medical supplies. Many of these vehicles are constructed on one ton or 1½ ton chassis with a multiple-compartment utility style body. A crew responding in a light rescue vehicle may handle a small incident or work to stabilize the scene until additional support arrives.

Medium Rescue Vehicles

Larger in size and capacity, the medium rescue vehicle may carry a variety of hand tools similar to the light rescue vehicle as well as the following additional equipment **(Figure 9.22)**:

- Large assortment of hydraulic rescue tools
- Air bag lifting systems
- Several types of power saws
- Acetylene cutting equipment
- Ropes and rigging equipment

In many jurisdictions, a medium-duty rescue apparatus is capable of handling the majority of rescue incidents. This vehicle often carries a variety of fire fighting equipment, as well as specialized rescue equipment to support incident operations.

Heavy Rescue Vehicle

A heavy rescue vehicle is a large apparatus outfitted with equipment capable of mitigating almost any type of entrapment issue. In addition to the fundamental equipment carried on smaller rescue vehicles, a heavy rescue vehicle may feature the following specialized tools and equipment:

- A-frames or gin poles
- Cascade system
- Large power plant (generator)
- Trenching and shoring equipment
- Small pumps
- Winches
- Hydraulic booms
- Large amounts of rope and rigging equipment
- Air compressor
- Ground ladders
- Large lighting equipment

Figure 9.23 A heavy rescue truck. *Courtesy of Ron Jeffers.*

The heavy rescue vehicle, which has more capacity for storage, may also carry additional fire fighting equipment or other specialized tools that may be required in a particular jurisdiction **(Figure 9.23)**.

Fire Service Ambulances

Many fire departments provide emergency medical care, which includes the transport of sick or injured people to the hospital. Ambulances are classified by their gross vehicle weight rating (GVWR) and their construction. The classifications are as follows:

- **Type I Ambulance** (10,001 – 14,000 GVWR) — Vehicle has a cab and chassis furnished with a modular body **(Figure 9.24, p. 272).**

- **Type I - AD (Additional Duty) Ambulance** (14,001 GVWR or more) — Vehicle has a cab and chassis with a modular body and increased GVWR, storage, and payload **(Figure 9.25, p. 272)**

- **Type II Ambulance** (9,201 – 10,000 GVWR) — Long wheel base van with integral cab-body **(Figure 9.26, p. 272)**

- **Type III Ambulance** (10,001 – 14,000 GVWR) — Vehicle is a cutaway van with integrated modular ambulance body **(Figure 9.27, p. 272)**

- **Type III - AD (Additional Duty) Ambulance** (14,001 GVWR or more) - Vehicle is a cutaway van with an integrated modular body and increased GVWR, storage, and payload **(Figure 9.28, p. 272)**

NFPA® 1917

NFPA® 1917, *Standard for Automotive Ambulances*, was developed with consideration of the Federal Specification KKK-A-1822 for ambulances as well as NFPA® 1901, *Standard for Automotive Fire Apparatus*. The NFPA® 1917 standard addresses the design, performance, and testing of ambulances.

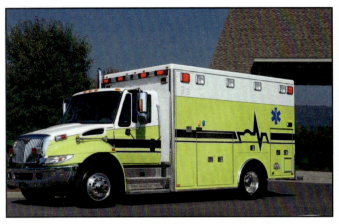

Figure 9.24 Type I Ambulance. *Courtesy of American Emergency Vehicles (AEV).*

Figure 9.25 Type I-AD (Additional Duty) Ambulance. *Courtesy of American Emergency Vehicles (AEV).*

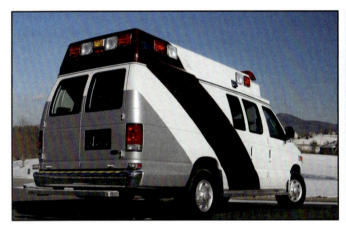

Figure 9.26 Type II Ambulance. *Courtesy of American Emergency Vehicles (AEV).*

Figure 9.27 Type III Ambulance *Courtesy of American Emergency Vehicles (AEV).*

Figure 9.28 Type III D (Additional Duty) Ambulance. *Courtesy of American Emergency Vehicles (AEV).*

Figure 9.29 An aircraft firefighting apparatus.

Figure 9.30 An ARFF apparatus demonstrating a pump-and-roll capability. *Courtesy of Edwin A. Jones, USAR*

Figure 9.31 A large aircraft fire fighting apparatus.

Aircraft Rescue and Fire Fighting Apparatus (ARFF)

Aircraft rescue fire fighting apparatus (ARFF) are specially designed and built to suppress aircraft fires. Major fire fighting vehicles are the largest of all ARFF apparatus. These vehicles provide the bulk of the extinguishing capabilities of the apparatus responding to an aircraft emergency **(Figure 9.29)**. Based on the water tank size of the vehicle, the FAA divides ARFF vehicles into the following four classes:

- **Class 1** — Minimum water tank capacity 1,000 gallons (4 000 L)
- **Class 2** — Minimum water tank capacity 1,500 gallons (6 000 L)
- **Class 3** — Minimum water tank capacity 2,500 gallons (10 000 L)
- **Class 4** — Minimum water tank capacity 3,000+ gallons (12 000+L)

Due to the large quantity of fuel that may be present during aircraft fires, a large amount of extinguishing agents may be needed to rapidly apply to protect the occupants. Specially outfitted aircraft fire fighting vehicles provide ARFF personnel with the ability to operate turrets, handlines, ground sweeps, and extendable booms. Additionally, ARFF apparatus often carry medical equipment, ladders, and extrication tools to provide rescue and treatment of injured people.

ARFF vehicles have the ability to discharge fire streams while moving (pump-and-roll capability). The large master stream nozzles on the roof of the apparatus are controlled from inside the cab of the apparatus. Many ARFF apparatus also feature other suppression agents that are foam compatible, such as carbon dioxide or dry chemical. These agents are often able to be discharged simultaneously with foam application or independently if needed **(Figure 9.30)**.

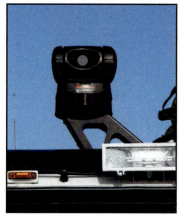
Figure 9.32 A forward looking infrared camera mounted on the roof of an ARFF apparatus.

Some incidents may occur off the runway or other paved surface necessitating that ARFF vehicles are built on special chassis with power to all wheels **(Figure 9.31)**. Other design features of many ARFF vehicles include:

- Antilock brakes
- Central inflation/deflation tire systems
- Driver-enhanced vision systems **(Figure 9.32)**
- High mobility (independent) suspension systems

Chapter 9 • Fire and Emergency Services Apparatus, Equipment, and Facilities **273**

Figure 9.33 Hazardous materials apparatus carry decontamination equipment, such as shown here.

Some airports may operate structural fire apparatus that serve a dual role in protecting airport structures as well as supporting ARFF apparatus during aircraft emergencies.

Hazardous Materials Response Unit

Many fire departments are responsible for hazardous materials incident responses that require specialized training, tools, and equipment. Often an apparatus is specially designed and outfitted for this purpose. In addition to transporting trained haz mat team members, the haz mat response unit must be capable of storing a wide variety of tools to mitigate different types of hazardous materials incidents **(Figure 9.33)**. These units may carry the following items:

- Standard hand tools, wrenches screwdrivers, saws
- Nonsparking (copper-beryllium) tools
- Patches, plugs, and duct tape
- Patch kits for specific types of drums or cylinders
- Specialized protective clothing
- SCBA
- Monitors, gas metering devices, and radiological monitors
- Reference books (including the *North American Emergency Response Guidebook*)

- Computers and printers with Internet access
- Cellular and satellite phones and fax machines
- Radio equipment with multiagency capability
- Weather monitoring equipment
- MSDS sheets for products known to be in the response area
- Decontamination equipment

Mobile Air Supply Unit

A mobile air supply unit's primary purpose is to fill or resupply exhausted SCBA cylinders at the scene of an incident. Some units simply carry a large quantity of replacement cylinders. The following is a variety of ways to fill or resupply SCBA cylinders from a mobile air supply unit:

- Fill bottles by other apparatus that may feature a single or multiple cascade system containing three or more large cylinders.
- Resupply individual SCBA cylinders to provide a continuous supply of air. Other apparatus may be outfitted with a compressor fitted with a purification system that is capable of filling storage cylinders.
- Make simple field repairs to SCBA harnesses and cylinders. Some air supply units may carry tools and common replacement parts to help with simple field repairs to SCBA harnesses and cylinders.
- These vehicles range from pickup trucks with utility bodies to large vans or large custom-designed apparatus that may also provide lighting or support equipment. Some fire departments may operate a trailer towed by a service vehicle as an air supply unit **(Figure 9.34)**.

Figure 9.34 A mobile air supply apparatus. *Courtesy of McKinney (TX) Fire Department.*

Figure 9.35 A mobile command post is equipped with necessary equipment and communication materials directly to the emergency scene. *Courtesy of Ron Jeffers.*

Figure 9.36 Mobile command units respond to large-scale operations that involve many agencies or jurisdictions and may last a long period of time.

Mobile Command Post

The Incident Commander and other Command Staff members use a Mobile Command Post to provide communication, reference, and other support capabilities necessary to operate at an emergency scene **(Figure 9.35)**. For most incidents, a sport utility vehicle or pickup truck outfitted as a command vehicle will provide sufficient space to carry radio equipment, computer hardware, unit status board, and reference material detailing water supply, preincident plans, hazardous material data, and other important information.

During long durations or large-scale incidents, a unit capable of functioning as a field dispatch center and temporary headquarters may be required. A larger vehicle, such as a converted bus, trailer, motor home, or custom-designed vehicle, may be necessary to provide the level of communication, conference, computer support, and television and video equipment required to manage the incident **(Figure 9.36)**.

Fire Boats and Search and Rescue Boats

Many cities with a waterfront operate **fireboats** to protect docks, wharves, piers, and other maritime facilities. Fireboats may be small, high speed, shallow draft vessels or larger boats with high capacity pumps capable of delivering as much as 26,000 gpm (104 000 l/min) through numerous large master stream appliances. Staffing requirements vary based on the size and capabilities of the boat. Some fireboats have permanently assigned crews, while firefighters from nearby land-based fire companies may staff other fireboats when a marine response is needed **(Figure 9.37)**.

Fireboats may also be used to provide water supply to land-based fire companies for large-scale operations near the waterfront or when traditional water supply delivery systems have been compromised, such as after an earthquake. Fire departments often employ fireboats, or specially constructed **search and rescue boats,** for use in operations in the marine environment. Many of these boats are built to support dive teams or other search and rescue missions **(Figure 9.38)**.

Fireboat — Vessel or watercraft designed and constructed for the purpose of fighting fires; provides a specified level of pumping capacity and personnel for the extinguishment of fires in the marine environment.

Search and Rescue Boat — Watercraft designed and equipped to carry personnel during search and rescue operations such as boating accidents, flood evacuations, and dive rescues.

Figure 9.37 Some fire departments use fire boats for marine fire fighting operations.

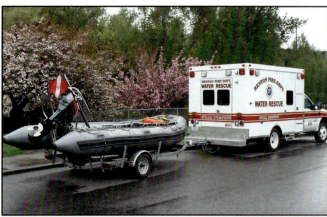

Figure 9.38 Search and rescue boats serve as mobile platforms for search and rescue operations on lakes and rivers.

Power and Light Unit

Some fire departments use special apparatus to provide additional scene lighting and electrical power to incident scenes. Large capacity generators on these apparatus may be used to power all manner of electrical equipment and scene lighting as well as standby power to buildings. These vehicles are often equipped with banks of floodlights and telescoping light towers as well as portable floodlights and a variety of electrical cords and adapters. In some jurisdictions, this capability is combined on the same vehicle with mobile air supply capability to create an air/power/light unit **(Figure 9.39)**.

Figure 9.39 A lighting and power apparatus. *Courtesy of McKinney (TX) Fire Department.*

Mobile Fire Investigation Unit

A mobile fire investigation unit may be equipped with the materials required to determine fire cause and origin. Equipment for the collection, preservation, and preliminary analysis of evidence may be stored on this vehicle. Such equipment may include:

- Flammable liquids detectors
- Gas chromatographs
- Magnifying lenses
- Common hand tools
- Lighting equipment
- Cameras
- Audio and video recorders
- Fingerprint kits
- Sifting screens
- Materials for making plaster casts
- Materials for making fire scene sketches and written notes
- Containers for storing and protecting evidence, plastic bags, steel cans with lids, and boxes

Figure 9.40 A fixed-wing air tanker conducting a slurry drop on a wildland fire.

Figure 9.41 A helicopter tanker conducting a slurry drop.

Fire Fighting Aircraft

To supplement ground units for wildland fire fighting, fixed wing aircraft and helicopters may be used. Airplanes may be used to drop fire retardant chemicals as well as deploy hand tool crews via parachute. A wide variety of fixed-wing aircraft are used in wildfire operations across North America. Based on their size and design, some planes may carry from 120 to 29,000 gallons (480 L to 116 000 L) of fire retardant chemical or water. By opening certain bay doors, the pilot may be able to choose the amount of product released. This capability will provide differing characteristics of penetrating power as well as length and width of the drop area **(Figure 9.40)**.

Helicopters, like fixed-wing aircraft, may drop fire retardant or water on wildfires with the use of a tank attached to the underside of the helicopter or from a bucket suspended underneath. Although limited in capacity, the helicopter offers the advantage of greater accuracy **(Figure 9.41)**.

Helicopters

Helicopters may also be used for airborne command posts, aerial photography, and fire area mapping. In addition to fire suppression duties, a helicopter may be employed to affect rescues in areas of rough terrain. A winch in the helicopter may be used to lower rescuers and raise victims in a basket stretcher. Additionally, helicopters may be used for medical transport from incident scenes to a medical facility.

Other Special Units

In order to fulfill particular jurisdictional needs, fire departments may operate other special units. These units may serve a wide variety of specialized functions that may include high-expansion foam units, dry chemical application vehicles, large caliber master stream apparatus, and smoke removal and control trucks. Other vehicles may not directly serve fire fighting functions but act in a support role for fire department activities **(Figure 9.42)**. These vehicles may include:

Figure 9.42 Fire departments may use special operations vehicles that serve a variety of purposes. *Courtesy of Ron Jeffers.*

Figure 9.43 A fire department refueling truck can deliver fuel to several fire apparatus at long duration incidents.

- Fuel re-supply trucks **(Figure 9.43)**
- Mechanical service trucks
- Tow trucks
- Thawing apparatus
- Rehab units **(Figure 9.44)**

NOTE: In some fire departments, a single vehicle may perform several support functions.

Uniforms and Personal Protective Clothing

Members of the fire and emergency services wear a variety of uniforms and protective clothing. As mentioned in Chapter 1, uniforms identify the wearer as a member of the fire service. Personnel assigned to a particular function wear uniforms most suitable to their specific duties. Different types of personal protective clothing may be designed to protect firefighters from a variety of hazardous conditions to which they will likely be exposed. The following section will describe common uniforms and personal protective clothing.

Figure 9.44 Many fire departments use rehabilitation vehicles to allow firefighters to rest and rehydrate during fire and emergency operations. *Courtesy of Ron Jeffers.*

Uniforms

Fire service uniforms may range from station/work uniforms to Class A dress uniforms. Casual uniforms usually consist of polo shirts, trousers, and steel-toed shoes or boots. Station/work uniforms consist of button down shirts, work trousers, and steel-toed shoes or boots. Full dress or Class A uniforms include a department cap, formal coat, dress shirt, tie, pants, and shoes. Some fire departments may specify a uniform for physical fitness training that may include shorts or sweat pants and a t-shirt as well as athletic footwear.

All station/work uniforms should meet the recommendations of NFPA® 1975, *Standard on Station/Work Uniforms for Emergency Services.* The purpose of the standard is to provide a recommendation for work wear that is functional and will not contribute to firefighter injury and will not reduce the effectiveness of outer protective clothing.

Personal Protective Clothing — Garments emergency responders must wear to protect themselves while fighting fires, mitigating hazardous materials incidents, performing rescues, and delivering emergency medical services.

Personal Protective Equipment (PPE) — General term for the equipment worn by fire and emergency services responders; includes helmets, coats, trousers, boots, eye protection, hearing protection, protective gloves, protective hoods, self-contained breathing apparatus (SCBA), personal alert safety system (PASS) devices, and chemical protective clothing. When working with hazardous materials, bands or tape are added around the legs, arms, and waist.

Open-Circuit Self-Contained Breathing Apparatus — SCBA that allows exhaled air to be discharged or vented into the atmosphere.

Closed-Circuit Self-Contained Breathing Apparatus — SCBA that recycles exhaled air; removes carbon dioxide and restores compressed, chemical, or liquid oxygen. Not approved for fire fighting operations.

Personal Protective Clothing

Firefighters rely on **personal protective clothing (PPC)** and **personal protective equipment (PPE)** to protect themselves from hazardous conditions. Some common types of protective clothing include the following **(Table 9.1, p. 294)**:

- Structural
- Wildland
- Proximity
- Hazardous materials

Each type of personal protective clothing is composed of components that will protect the wearer from specific hazards for which the ensemble was designed. When selecting protective clothing, consult NFPA® recommendations for specific levels of protection.

Breathing Apparatus

Firefighters must use breathing apparatus to perform their duties safely when working in environments that are oxygen deficient (less than 19 percent oxygen) or those environments contaminated by chemical releases or products of combustion. Several of the most widely used types of breathing apparatus are explained in the following sections.

Air Purifying Respirators (APRs)

Air purifying respirators (APR) use filters attached to a face mask to protect the wearer from harmful particles, vapors, or gases. An APR mask may be partial or full-face covering. Some of these respirators feature powered fans to create positive pressure in the facepiece. The higher pressure in the facepiece helps to keep contaminants out if the seal between the wearer's face and the APR facepiece becomes slightly dislodged **(Table 9.2, p. 297)**.

Supplied Air Respirators (SARs)

A supplied air respirator (SAR) consists of a source of breathing air (air cylinder or compressor), a facepiece with a regulator, and up to 300 feet (91 m) of air hose. The system also features an emergency breathing air cylinder should the primary air supply fail. These respirators allow emergency responders to operate in hazardous environments without having to carry a large air cylinder **(Table 9.3, p. 298)**.

Self-Contained Breathing Apparatus (SCBA)

A self-contained breathing apparatus (SCBA) supplies breathable air from a supply that the wearer carries. The fire service uses two basic types of SCBA. **Open-circuit SCBAs,** commonly used by firefighters during fire suppression and other emergency operations, consist of a compressed air cylinder that supplies breathable air to the wearer **(Table 9.4, p. 299)**. Exhaled air is expelled to the outside atmosphere through special vents in the facepiece. In contrast, **closed-circuit SCBA** reuse exhaled air by filtering out carbon dioxide and supplementing the air with oxygen from a source within the breathing apparatus **(Table 9.5, p. 300)**. Closed circuit SCBA generally provides a longer duration of air supply than open circuit SCBA provides. Hazardous material response teams sometimes use closed circuit SCBA.

Breathing apparatus used by the fire service should meet NFPA® standards and be tested and approved by the National Institute for Occupational Safety and Health (NIOSH). Tables 9.2, 9.3, and 9.4 describe each type of breathing apparatus, its use, and basic components.

Tools and Equipment

Fire department operations depend on the safe and efficient use of a wide variety of tools and equipment. Some of these items, such as saws and pry bars, are also used in trades outside the fire service, while other types of equipment are unique to emergency responders. Firefighters must learn the purpose and operation of each tool and piece of equipment they may be required to use. Safe operation and proper maintenance are priorities for all fire service tools and equipment. Tools and equipment in the fire service generally fall into one of two categories: hand tools (nonpowered) or power tools. Each category will be described in the following sections.

Hand Tools

Hand tools can be identified by their specific function, such as striking, cutting, or prying. Many different types of hand tools perform similar fireground functions. Some jurisdictions may prefer one style over another based on prevalent conditions in specific districts **(Table 9.6, p. 301)**.

Power Tools

The fire service employs many different types of power tools. These tools may be described by their source of power and their purpose. Power tools may be used for tasks such as vehicle extrication, ventilation, forcible entry, and technical rescue **(Table 9.7, p. 311)**.

Equipment

Fire service equipment includes any number of articles used to aid in accomplishing an objective at an incident. The types of equipment that individual fire departments use may vary according to the needs of the jurisdiction. However, typical pieces of fire service equipment include items such as salvage covers, smoke ejectors, cribbing blocks, basket stretchers, and electrical cords and adapters.

NOTE: This section does not describe the skills required to use, inspect, or maintain fire service tools and equipment. For more detailed information, consult IFSTA's **Essentials of Fire Fighting** manual.

Ropes, Webbing, Related Hardware, and Harnesses

Firefighters and other emergency responders may use ropes to raise or lower tools, equipment, personnel, and victims. In addition, a rope may be used to prevent firefighters from becoming lost during fireground search operations.

Rope is classified into two categories of use: life safety or utility. Ropes made of synthetic fibers may be classified for use as life safety ropes, while natural fiber (hemp or cotton) ropes are used for utility purposes.

Firefighters may employ either flat or tubular webbing made from the same synthetic materials as life safety rope. Other types of hardware serving specific functions may comprise rope systems along with webbing. Carabiners, rescue harnesses, and other devices to aid in the raising and lowering of people or equipment are often integrated with rope to perform technical rescue operations **(Table 9.8, p. 315)**.

Ground Ladders

Firefighters often use ground ladders to access points above or below ground level. Ground ladders are not permanently mounted on an apparatus and may be carried to the location where they will be raised. Ground ladders are manually raised and positioned and provide relatively quick access to windows, balconies, rooftops, and other points above or below grade **(Figure 9.45)**.

To perform specific fireground tasks, ground ladders are manufactured as single ladders or extension ladders of various lengths. Ground ladders may be constructed of wood, fiberglass, or aluminium. Firefighters must become familiar with the various components of each ladder and be able to position and raise each type safely and efficiently **(Figure 9.46 a and b, 284-285)**.

Fire Hose, Nozzles, and Hose Appliances and Tools

Fire hose is manufactured in a variety of diameters and lengths for use in different fireground operations. Although it is commonly found in 50 or 100 foot (15 m or 30 m) sections, it may be specified in shorter or longer lengths. Supply hose is generally used to carry water from a source to a fire apparatus or portable pump, while attack hose is used to deliver water from a pumper or portable pump to suppress a fire. Fire hose is manufactured in diameters ranging from ¾-inch to six inches (20 to 150 mm) and is commonly composed of a rubber interior liner with a rubber or woven fiber outer covering. Either threaded or unthreaded couplings are used to connect multiple lengths of hose or attach sections to pump fittings or nozzles **(Figure 9.47, p. 286)**.

Nozzles are used to create fire streams appropriate for a particular fire fighting operation. Nozzles feature a shut-off mechanism to control the flow of water and may belong to one of two broad categories: smooth bore or fog. A smooth-bore nozzle features an open waterway that allows water to travel directly through the nozzle in a solid stream. Fog nozzles contain mechanisms to deflect water into a pattern of droplets that may be delivered out of the nozzle tip in a straight stream pattern of one of a variety of narrow or wide fog pattern settings **(Figure 9.48, p. 286)**.

Hose appliances consist of hardware, such as valves and fittings, through which water flows. These appliances may serve to split or connect hoselines, control water flow, or connect hoselines of dissimilar thread pattern or diameter **(Figure 9.49, p. 287)**.

Firefighters also employ various tools to loosen or tighten hose couplings, secure hose to an object, or protect hose from damage at an incident scene. Tools differ from appliances in that water does not flow through them.

Figure 9.45 Types of ground ladders used in fire and emergency services.

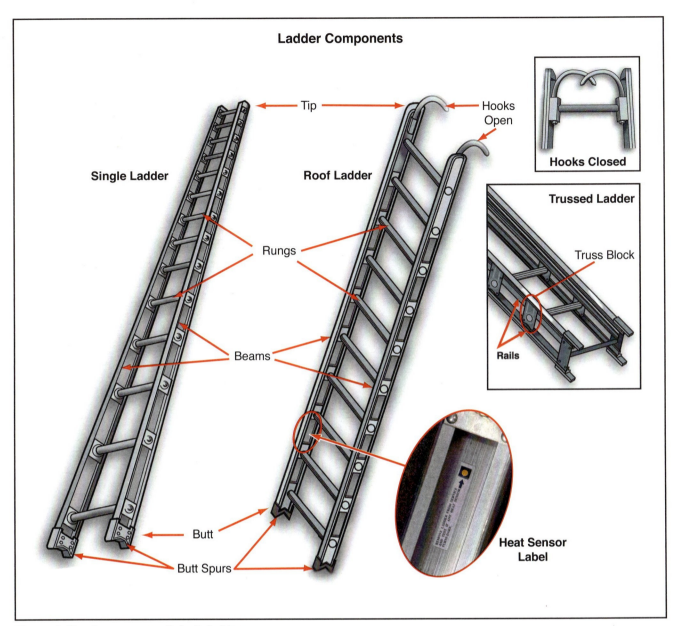

Figure 9.46 a Components of single ladders.

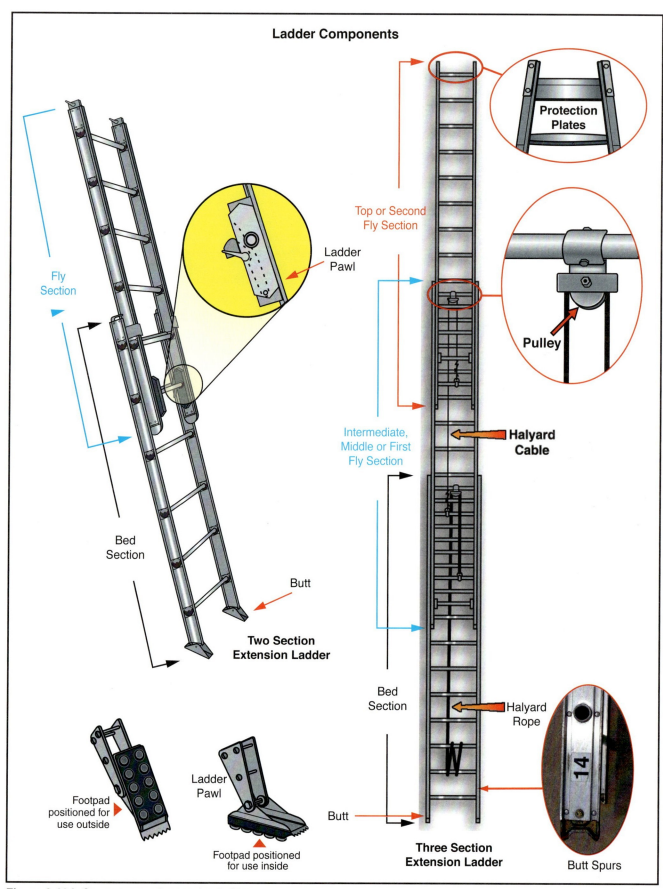

Figure 9.46 b Components of extension ladders.

Chapter 9 • Fire and Emergency Services Apparatus, Equipment, and Facilities **285**

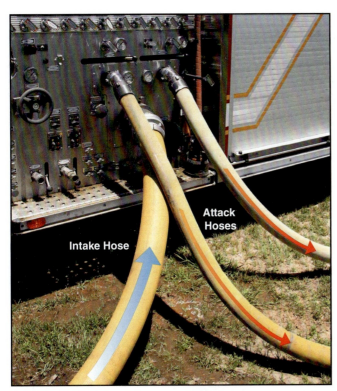

Figure 9.47 Supply hose carries water from a pumper while attack hoses carry water from the pumper to the fire.

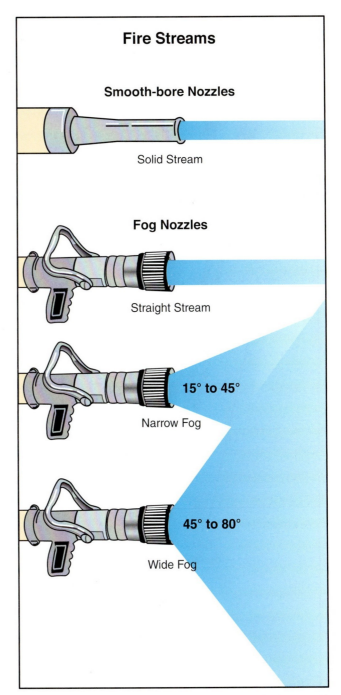

Figure 9.48 Types of fire streams created by solid-bore and fog nozzles.

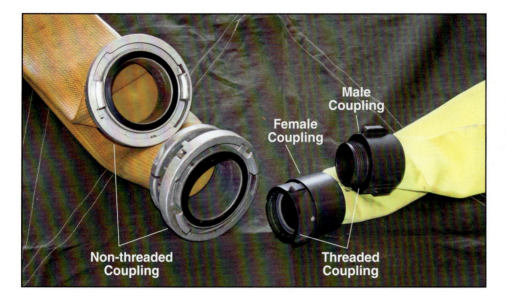

Figure 9.49 Examples of non-threaded hose couplings (left) and threaded hose couplings (right).

Fire Department Facilities

Fire departments may operate various facilities to service and support daily functions. Depending on the size of the department, some or all of these facilities may be found at one location or dispersed throughout the jurisdiction. Some of these facilities include:

- Fire stations
- Administrative offices and buildings
- Telecommunications centers
- Training centers
- Maintenance and repair shops

Fire Stations

Fire stations, known in some regions as *firehouses* or *fire halls*, are used to house apparatus, personnel, and equipment. The required number and size of fire stations vary with the needs of the jurisdiction. Small communities may utilize one small building to house all of its apparatus, while a large city may operate many stations of various sizes to strategically locate apparatus and personnel throughout response areas **(Figure 9.50, p. 288)**.

Although the size and layout may vary widely, some components are present in almost all fire stations. A garage or apparatus bay for housing fire trucks may contain the following equipment and utilities to ensure safe and efficient operation of the apparatus:

- Water piping and outlets for refilling the apparatus water tank
- Floor drains under the apparatus
- An electrical shore line system for maintaining the apparatus battery
- Air compressor for filling tires or air brake systems
- Apparatus exhaust removal system

An area should be designated in the fire stations for storage of tools and equipment used for apparatus and station maintenance. The tool inventory

> **Fire Station** — (1) Building in which fire suppression forces are housed. *Also known as* Fire Hall or Fire House. (2) Location on a vessel with fire fighting water outlet (fire hydrant), valve, fire hose, nozzles, and associated equipment.

Figure 9.50 Fire stations vary in size with the needs of the jurisdiction.

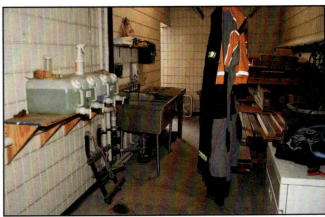

Figure 9.51 A section in a fire station for storing and maintaining tools and equipment.

Figure 9.52 The personal protective equipment (PPE) storage area in a fire station.

is usually based on how much maintenance work is expected for firefighters to perform **(Figure 9.51)**.

Fire stations should also be equipped with an appropriate area for the storage of personal protective equipment (PPE). Often this arrangement will include racks, lockers, or cubicles for each firefighter to store PPE when he or she is off shift. This facility should have adequate ventilation to aid in the drying of wet PPE **(Figure 9.52)**.

Some stations are equipped with an air cascade system or breathing air compressor for refilling SCBA cylinders. Hose racks and hose dryers may also be found in many fire stations to help maintain and store hose that is not being used on the apparatus.

Other than the basic utility features of a fire station, the amenities provided for firefighters working in the station vary widely between jurisdictions. Fire stations that quarter personnel working a shift schedule will feature kitchen and sleeping facilities as well as some type of space for relaxation and recreation. Often meeting or training rooms are part of fire station design. These spaces may serve dual purposes as meeting rooms and recreation areas in some stations. Generally, fire stations contain office space for company or chief officers housed in the building, and some stations may also provide halls for fundraising or banquet accommodations.

Administrative Offices and Buildings

Office space for administrative purposes may be attached to a fire station, or it may be in a separate building. Typically, these offices house the administrative chief officers of the fire department as well as staff for other support functions and clerical personnel. Space may be allocated for storage of records, conference rooms, and offices for other divisions within the fire department, such as inspection, training, and communications.

Figure 9.53 An example of a 9-1-1 telecommunications center.

Figure 9.54 An example of mobile communications center.

Telecommunication Centers

The telecommunication center is the focal point for fire department communications. Depending on jurisdictional requirements, it may be located at the fire department's administrative offices, a fire station, or a separate facility. In some jurisdictions, the telecommunications center is a joint facility that dispatches many different fire departments or public safety agencies in a region **(Figure 9.53)**.

The telecommunications center monitors all companies operating in the field. It is equipped with an array of radios and telephone equipment to send and receive information as required during incidents. Other equipment in the telecommunication center includes recording devices for radio and phone calls, as well as computer-aided dispatch systems that provide information on locations and units assigned for response. Apparatus equipped with mobile data terminals (MDTs) or laptops may retrieve dispatch information, including street maps, pre-incident plans, and detailed occupancy information **(Figure 9.54)**. Some telecommunication centers are equipped with Enhanced 9-1-1 systems that are able to display the caller's address or location using cellular signals.

Training Centers

Some fire departments have dedicated facilities for education and training activities. Some of these facilities may be shared by more than one jurisdiction. Training centers often feature an administration building that may contain any of the following:

- Administrative offices
- Classrooms **(Figure 9.55, p. 290)**
- Auditorium
- Cafeteria
- Exercise equipment
- Locker, shower, and dressing areas
- Storage for apparatus, equipment, and supplies

Training centers also feature other buildings and facilities designed for specific educational purposes **(Figure 9.56, p. 290)**. Facilities to develop and maintain other fire service skills may be built into a smoke house or be

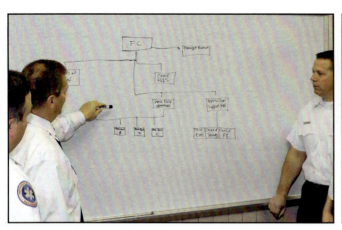

Figure 9.55 A chief officer training firefighters in a station classroom.

Figure 9.56 Foam application training is practiced with flammable/combustible liquid fires.

contained in a separate structure. These skills may include props to practice forcible entry, ventilation, and various types of hose stretches. The following sections explain these buildings and facilities.

Burn Building

A **burn building** is designed and constructed to allow firefighters to practice interior structural fire attacks repeatedly under controlled conditions. Typically, burn buildings are designed to burn either piped-in natural gas or propane or straw and wood products **(Figure 9.57)**. Buildings that use natural gas or propane allow for greater control over fire extinguishment should a problem arise. Live burn training evolutions must comply with NFPA® 1403, *Standard on Live Fire Training Evolutions.*

NOTE: The judicial system has used NFPA® 1403 in several jurisdictions as the defining minimum industry standard for live fire training. Fire service instructors have been prosecuted or sued for noncompliance to this standard.

Drill Tower

Drill towers are usually between three and seven floors in height and are equipped with interior stairs that allow access to upper floors **(Figure 9.58)**. These structures may be used to train with ladders, ropes, aerial devices, or high-rise tactics. Some drill towers are equipped with safety nets around the outside of the tower to catch a firefighter in case of a fall during training.

Smokehouse

A **smokehouse** provides a simulated smoke condition for firefighter training. This type of environment helps firefighters develop confidence in their SCBA and build proficiency in search and rescue techniques. Some smoke houses are designed to replicate the interior of a house, while others are designed with obstacles or movable partitions to change the interior layout. A machine that uses a water-based or vegetable oil fluid to produce a nontoxic smoke product supplies the smoke in these structures. Construction of smoke houses should include a venting system to quickly clear smoke and adequate exit points from which firefighters may be removed during an actual emergency **(Figure 9.59)**.

Burn Building — Structure designed to contain live fires for the purpose of fire suppression training.

Drill Tower — A tall training structure, typically more than three stories high, used to create realistic training situations, especially ladder and rope evolutions.

Smokehouse — (1) Specially designed fire training building that is filled with smoke to simulate working under live fire conditions; used for SCBA and search and rescue training.

Figure 9.57 Burn buildings are designed for live fire training.

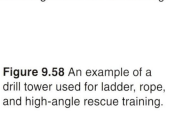

Figure 9.58 An example of a drill tower used for ladder, rope, and high-angle rescue training.

Figure 9.59 A smoke house provides simulated conditions for firefighter training.

Figure 9.60 Training pads are used for teaching students the skills they will need to know in similar situations.

Training Pads

Training centers may feature one or more pads or outside areas designed for a specific purpose or training prop **(Figure 9.60)**. These purposes may include the following:

- Flammable and combustible liquid and gas fire fighting
- Driver/operator training
- Pump testing and drafting facilities
- Vehicle extrication
- Hazardous materials props
- Trench and confined space rescue props

Maintenance Facilities

Proper maintenance of fire apparatus and equipment requires special facilities and trained personnel. Some fire departments have maintenance facilities located at a fire station, while other jurisdictions may operate a separate building to house these support activities. Apparatus maintenance and repair shops are equipped much like a commercial truck repair facility.

Maintenance of other fire fighting equipment, such as SCBA, hose, ladders, and saws, may be conducted in an area of the apparatus maintenance shop or at another location. However, many fire departments conduct all maintenance and repair functions in a central location to consolidate tools and other specialized equipment.

Other Maintenance Providers

Some fire departments use municipal motor pools, public works garages, or commercial repair centers for maintenance and repair of apparatus. Likewise, they may contract with equipment and tool vendors to maintain other fire department equipment.

Chapter Summary

Firefighters need to operate a wide variety of specialized equipment in order to safely and efficiently mitigate the many types of incidents for which they are called to respond. Fire apparatus are highly specialized vehicles, and each type of apparatus serves a designated purpose during fire and other emergency incidents. Firefighters' knowledge of their fire departments' equipment and apparatus is essential for responding to incidents safely.

Training facilities provide realistic environments by using many props that simulate actual incident scenarios. These facilities teach firefighters skills that can be learned and sharpened under controlled conditions using many props that simulate actual incident scenarios. Fire apparatus and equipment must be maintained in proper working order at all times. This work may be conducted at a facility located in a fire station or a separate maintenance building. Some jurisdictions use a municipal maintenance center or contract with a commercial repair or service center for maintenance or repair requirements.

Review Questions

1. Describe four types of fire department apparatus.
2. What are the characteristics that work wear should meet?
3. What is the purpose of personal protective clothing?
4. Explain the differences between an air purifying respirator (APR) and a supplied air respirator (SAR).
5. List the basic types of self-contained breathing apparatus (SCBA) used in the fire service.
6. List the two categories of tools used in the fire service.
7. What are the two categories of rope used in the fire service?
8. What is the function of ground ladders in the fire service?
9. Explain the difference between supply hose and attack hose.
10. What are nozzles used to create?
11. What are some purposes of hose appliances?
12. Describe three types of fire department facilities.

Table 9.1
Personal Protective Clothing
Structural, Wildland, and Proximity Protective Clothing

Type	Uses
Structural (also called *turnouts*, *bunker gear*, or *bunkers*)	To protect emergency responders from: - Excessive heat and thermal injury encountered during structural fire fighting operations. - Physical injury during structural fire fighting, vehicle extrication, and search and rescue operations.

Type	Uses
Wildland (also called *brush gear*)	To protect emergency responders from excessive heat and thermal injury encountered during wildland/urban interface fire fighting operations.

Type	Uses
Proximity (also called *silvers*)	To protect emergency responders from: - Excessive heat and thermal injury encountered during aircraft rescue and fire fighting and industrial fire fighting operations. - Physical injury aircraft rescue and fire fighting and industrial fire fighting operations.

Table 9.1
Personal Protective Clothing (continued)
Hazardous Materials Protective Clothing
(Based upon U.S. Environmental Protection Agency [EPA] Levels of Protection)

Type	Uses
Level A (also called *vapor-protective*)	To protect emergency responders at hazardous materials incidents when: - Unknown or unidentified chemical hazards. - Identified chemical(s) are highly hazardous to respiratory system, skin and eyes. - A high potential for splash, immersion, or exposure to unexpected vapors, gases, or particulates of material that are harmful to skin or capable of being absorbed through the intact skin. - Substances are present with known or suspected skin toxicity or carcinogenicity. - Operations are conducted in confined or poorly ventilated areas.

Type	Uses
Level B (also called *liquid-splash protective*)	To protect emergency responders at hazardous materials incidents when: - Substances identified and require a high level of respiratory protection but less skin protection. - Atmosphere contains less than 19.5 percent oxygen or more than 23.5 percent oxygen. - Presence of incompletely identified vapors or gases is indicated by a direct-reading organic vapor detection instrument, but the vapors and gases are known not to contain high levels of chemicals harmful to skin or capable of being absorbed through intact skin. - Presence of liquids or particulates is indicated, but they are known not to contain high levels of chemicals harmful to skin or capable of being absorbed through intact skin.

Table 9.1
Personal Protective Clothing (continued)
Hazardous Materials Protective Clothing
(Based upon U.S. Environmental Protection Agency [EPA] Levels of Protection)

Type	Uses
Level C	To protect emergency responders at hazardous materials incidents when: - Atmospheric contaminants, liquid splashes, or other direct contact will not adversely affect exposed skin or be absorbed through any exposed skin. - Types of air contaminants have been identified, concentrations have been measured, and an APR is available that can remove the contaminants. - All criteria for the use of APRs are met. - Atmospheric concentration of chemicals does not exceed IDLH levels. The atmosphere must contain between 19.5 and 23.5 percent oxygen.

Type	Uses
Level D (work uniforms, street clothing, coveralls, or firefighter structural protective clothing)	To protect emergency responders at hazardous materials incidents when: - Atmosphere contains no hazard. - Work functions preclude splashes, immersion, or the potential for unexpected inhalation of or contact with hazardous levels of any chemicals. **NOTE:** May not be worn in the hot zone and are not acceptable for haz mat emergency response above the Awareness Level.

Table 9.2
Breathing Apparatus
Air Purifying Respirators (APRs)

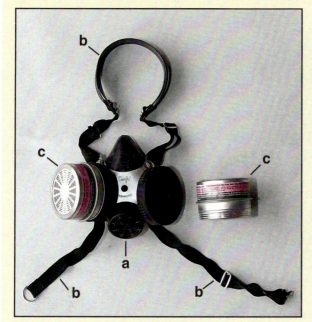

Partial-face Covering APR

Full-face Covering APR

Type	Uses	Components
Partial-face covering APR	Protects firefighters and other emergency response personnel by filtering out hazardous particles, hazardous or toxic chemical vapors and gases, or a combination of particles and vapors and gases. The level and type of protection provided is dependent upon the types of filters used with the APR.	a. Partial facepiece b. Straps c. Filters
Full-face covering APR		a. Full Facepiece b. Straps c. Filters

Chapter 9 • Fire and Emergency Services Apparatus, Equipment, and Facilities **297**

Table 9.3
Breathing Apparatus
Supplied Air Respirators (SARs)

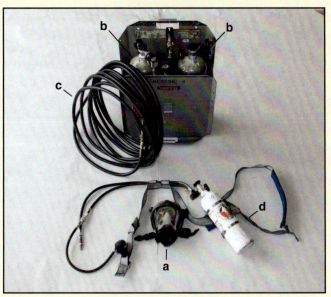

Supplied Air Respirators (SARs)

Type	Uses	Components
Supplied Air Respirators (SARs)	To protect firefighters and other emergency response personnel during operations involving toxic, potentially toxic, or oxygen deficient atmospheres. The unit is designed to vent exhaled air to the outside atmosphere.	a. Facepiece b. Air supply c. Air hose d. Emergency breathing system

Table 9.4
Breathing Apparatus
Self-Contained Breathing Apparatus (SCBA)

Open Circuit SCBAs

Type	Uses	Components
Open Circuit SCBAs	To protect firefighters and other emergency response personnel during operations involving toxic, potentially toxic, or oxygen deficient atmospheres. The unit is designed to vent exhaled air to the outside atmosphere.	a. Facepiece b. Hose and regulator c. Air cylinder d. Backplate e. Straps f. Pressure gauge g. Low pressure alarm h. Personal Alert Safety System (PASS) device

Table 9.5
Breathing Apparatus
Self-Contained Breathing Apparatus (SCBA)

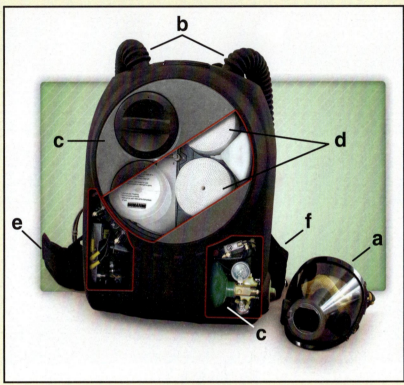

Closed Circuit SCBAs

Type	Uses	Components
Closed Circuit SCBAs	To protect firefighters and other emergency response personnel during operations involving toxic, potentially toxic, or oxygen deficient atmospheres by providing a greater duration of breathable air. The unit is designed to re-use exhaled air by scrubbing out carbon dioxide and adding fresh oxygen from a cylinder to the breathing mixture.	a. Facepiece b. Hoses c. Oxygen cylinder d. CO_2 filtration system e. Housing assembly f. Waist straps *Not Shown:* • Shoulder straps Pressure gauge • Low pressure alarm • Personal Alert Safety System (PASS) device

Courtesy of Biomarine, Incorporated.

Table 9.6a
Hand Tools
Striking Tools

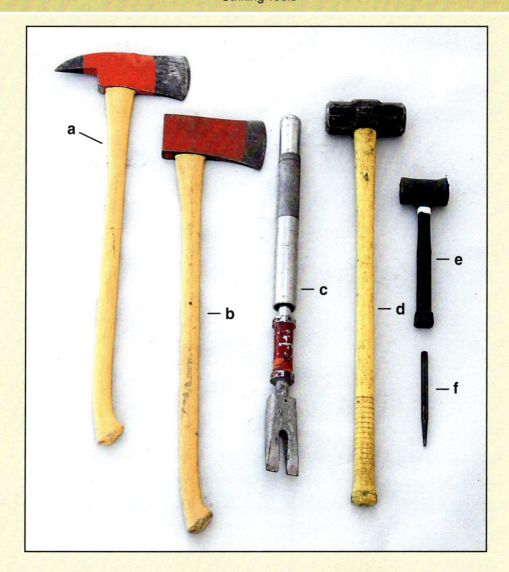

Function

Used to strike or penetrate an object (wall, roof, floor, etc.) or to force another tool to do so.

Examples

a. Pick head axe
b. Flat head axe
c. Ram bar
d. Sledge hammer
e. Mallet
f. Punch

Table 9.6b
Hand Tools
Prying Tools

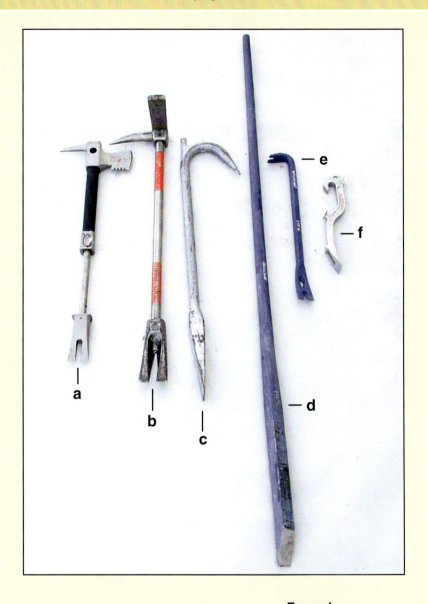

Function

Used to gain a mechanical advantage or leverage in order to move or remove an object.

Examples

a. Pry-axe
b. Halligan bar
c. Claw tool
d. Pry bar
e. Crowbar
f. Spanner wrench

Table 9.6c
Hand Tools
Cutting Tools

Chopping

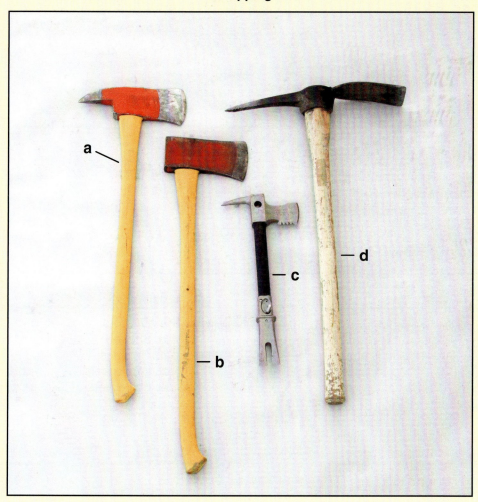

Function

Used to cut away or into an object, structure, or vehicle in order to gain access such as forcible entry or extrication, to remove an object trapping a victim, or to provide an avenue for smoke, heated products of combustion, or hazardous atmospheres to ventilate. Subdivided into Chopping Tools, Snipping Tools, Handsaws, and Knives

* May also belong in another category of tools.

Examples

a. Flat-head axe*
b. Pick-head axe*
c. Pry-axe
d. Pick

Table 9.6c
Hand Tools
Cutting Tools

Snipping

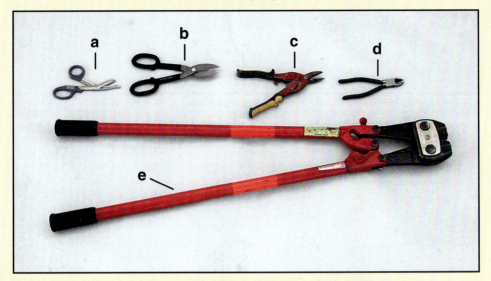

Function

Used to cut away or into an object, structure, or vehicle in order to gain access, such as forcible entry or extrication to remove an object trapping a victim.

Examples

a. Scissors
b. Shears
c. Tin snips
d. Wire cutters
e. Bolt cutters

**Table 9.6c
Hand Tools**
Cutting Tools

Handsaws

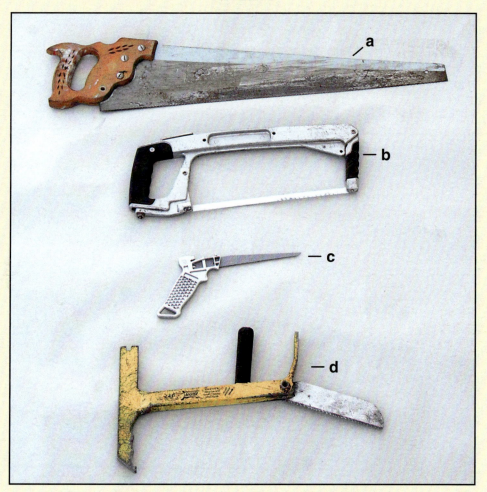

Function

Used to cut away or into an object, structure, or vehicle in order to gain access such as forcible entry or extrication, to remove an object trapping a victim, or to provide an avenue for smoke, heated products of combustion, or hazardous atmospheres to ventilate.

*May also belong in another category of tools.

Examples

a. Carpenter's saw
b. Hacksaw
c. Keyhole saw
d. Windshield Cutter (Glass saws)*

Table 9.6c
Hand Tools
Cutting Tools

Knives

Function	Examples
Used to cut away or into an object, structure, or vehicle in order to gain access such as forcible entry or extrication, to remove an object trapping a victim, or to provide an avenue for smoke, heated products of combustion, or hazardous atmospheres to ventilate.	a. Pocket b. Linoleum c. Utility d. V-blade

**Table 9.6d
Hand Tools**
Lifting Tools

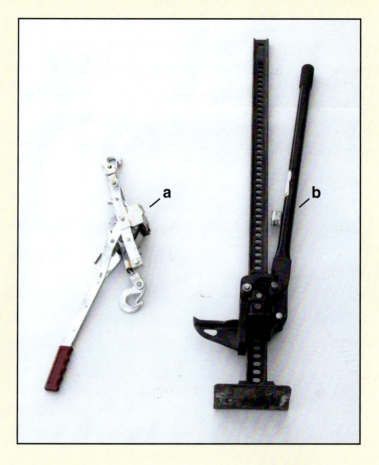

Function

Used to lift an object, part of a structure, or a vehicle.

Examples

a. Come-along
b. Ratchet (Hi-Lift ®) jack

Table 9.6e
Hand Tools
Stabilizing Tools

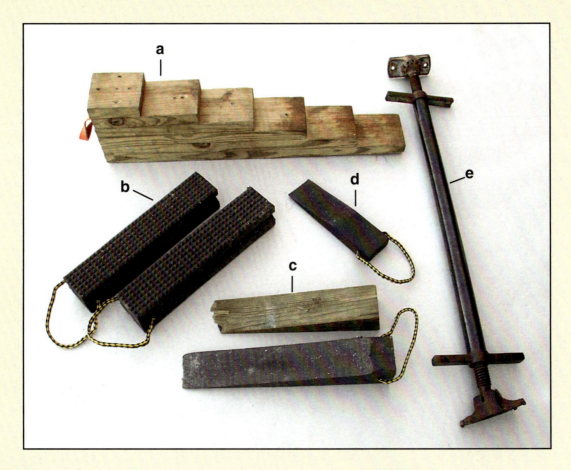

Function

Used to stabilize an object, part of a structure, or a vehicle during lifting operations.

Examples

a. Step chock
b. Chocks
c. Wedges
d. Shim
e. Strut

Table 9.6f
Other Hand Tools
Specialized Hand Tools

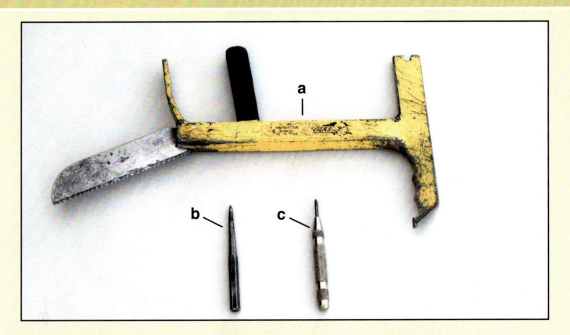

Function

Used to break or shatter glass during:
- forcible entry into a structure
- structural ventilation operations
- vehicle extrication operations

* May also be classified under another type of tool.

Examples

a. Glass saw*
b. Standard center punch
c. Spring-loaded center punch

Table 9.6e
Other Hand Tools
Mechanic's Tools

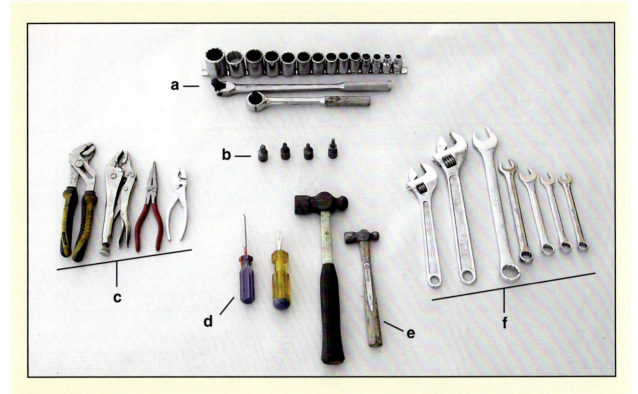

Function

Used to manipulate bolts, nuts, screws, and other fasteners during victim extrication from structures, vehicle, and industrial equipment.

Examples

a. Socket set
b. Torx® drivers (Star drivers)
c. Pliers
d. Screw drivers
e. Ball peen hammers
f. Wrenches

Table 9.7a
Power Tools

Electric Powered

Function

Some power tools are used to cut or remove an object, part of a structure, or vehicle. Others are used to move air in order to ventilate a structure.

Examples

a. Smoke blower
b. Rotary saw
c. Drill or driver
d. Reciprocating saw

Table 9.7b
Power Tools

Pneumatic Powered

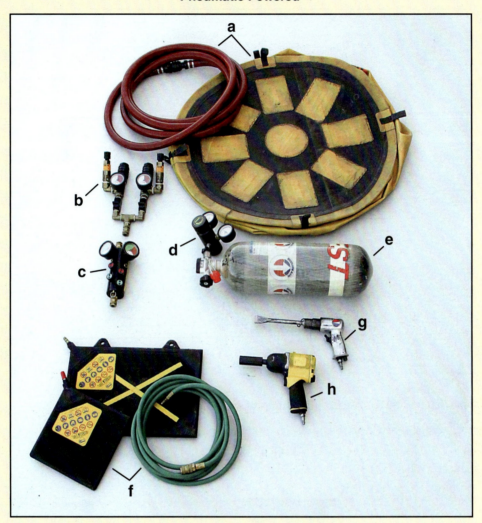

Function

Some power tools are used to cut, spread, lift, move, or remove an object, part of a structure, or vehicle.

Examples

a. Low pressure hose and air bag
b. Low/medium pressure air manifold
c. High pressure air manifold
d. Air pressure regulator
e. Air cylinder
f. High pressure air bags and hose
g. Air chisel
h. Air wrench

**Table 9.7c
Power Tools**

Hydraulic – Manually Operated

Function

Some power tools are used to spread, lift, move remove an object, part of a structure, or vehicle.

Examples

a. Porta-Power®
b. Hydraulic jack

Table 9.7c
Power Tools

Hydraulic – Power Driven

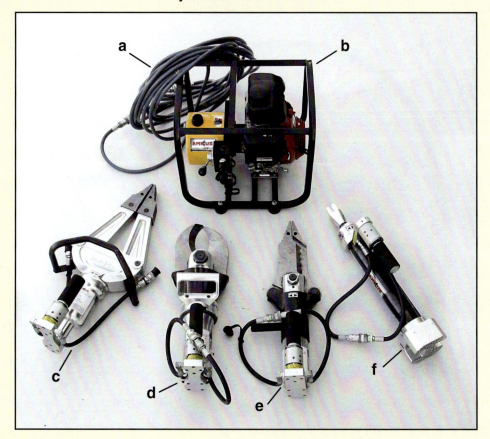

Function	Examples
Some power tools are used to cut, spread, lift, move, or remove an object, part of a structure, or vehicle.	a. Hydraulic hose lines b. Power unit c. Spreaders d. Sheers e. Combination spreaders/sheers f. Ram

Table 9.8
Ropes, Webbing, Hardware, and Harnesses

Natural and Synthetic Ropes

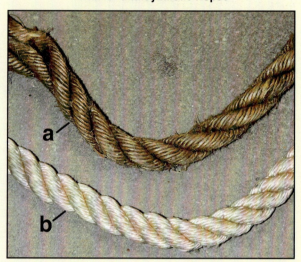

Uses	Types
To raise and lower tools, equipment, and personnel. May also serve as a guide line during structural search operations. **NOTE:** Natural fiber ropes are no longer used for life-safety purposes.	**a.** Natural fiber rope **b.** Synthetic rope

Webbing

Uses	Types
To raise and lower tools, equipment, and personnel. May also be used to anchor an item to prevent movement.	**a.** Flat webbing **b.** Tubular webbing

**Table 9.8 (continued)
Ropes, Webbing, Hardware, and Harnesses**

Hardware

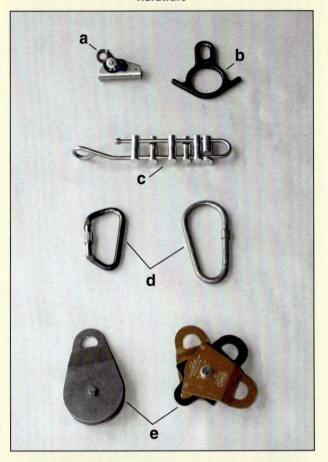

Uses	Types
To connect sections of a rope system together (carabiner), act as a brake during rappelling (figure-eight plate or brake-bar rack), to ascend a vertical rope (ascender), or change the direction of pull or create mechanical advantage (pulley).	**a.** Ascender **b.** Figure-eight plate **c.** Brake-bar rack (descender) **d.** Carabiner **e.** Pulley

Table 9.8 (continued)
Ropes, Webbing, Hardware, and Harnesses

Rescue Harnesses

a. b.

Uses	Types
To help protect rescuers and victims as they move and/or work in elevated positions during rope rescue operations.	**a.** Class I (loads up to 300 lb [1.33 k/N]) and Class II (loads up to 600 lb [2.67 k/N]) **b.** Class III (full body harness)

Chapter 9 • Fire and Emergency Services Apparatus, Equipment, and Facilities

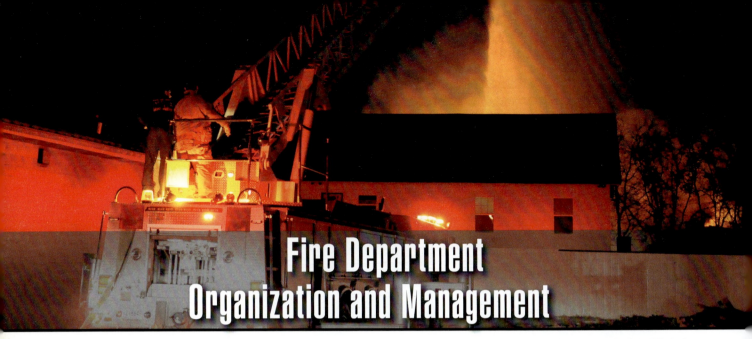

Courtesy of Chris Mickal.

Fire Department Organization and Management

Chapter Contents

Case History	321
Principles of Organization	**321**
Unity of Command	322
Span of Control	324
Division of Labor	324
Discipline	325
Local Government Structure	**325**
Commission	325
Council (Board) Manager	325
Mayor/Council	325
Fire District	326
Response Considerations	**326**
Automatic Aid	326
Mutual Aid	326
Fire Department Funding	**328**
Tax Revenues	328
Trust Funds	328
Enterprise Funds	329
Bond Sales	329
Grants/Gifts	329
Fundraisers	329
Subscription Fees	330
Current Challenges Facing Fire Protection	**330**
Funding/Budget Constraints	330
Outdated Equipment	330
Recruiting Personnel	321
Retaining Personnel	321
Adequate Water Sources	332
Policies and Procedures	**333**
Standard Operating Procedures (SOPs) Guidelines (SOGs)	333
Disciplinary Procedures	334
Formal Communications	334
Incident Command System	**335**
Functional Areas of ICS	336
NIMS Terms	337
NIMS-ICS Training	341
Emergency Operations	341
Personnel Accountability Systems	342
Rapid Intervention Crews (RIC)	344
Critical Incident Stress Management/Exposure to Potentially Traumatic Events	344
Chapter Summary	**345**
Review Questions	**346**

Chapter 10

Key Terms

Automatic Aid 326
Chain of Command...321
Command 323
Critical Incident Stress Management
   (CISM)..345
Discipline...325
Division of Labor 324
Finance/Administrative Unit 337
Incident Commander (IC).........................336
Incident Command System (ICS)...........333
Logistics Section...337
Mutual Aid 326
National Incident Management
   System - Incident Command System
   (NIMS-ICS)..335

Operations Section 337
Operations Section Chief 337
Outside Aid 326
Planning Section 337
Personnel Accountability System..........342
Rapid Intervention Crew or Team
   (RIC/RIT)...344
Span of Control...324
Standard Operating Procedures
   (SOPs)..325
Unified Command (UC).............................326
Unity of Command......................................323

FESHE Outcomes

This chapter provides information that addresses the outcomes for the Fire and Emergency Services Higher Education (FESHE) *Principles of Emergency Services* course.

1. Illustrate and explain the history and culture of the fire service.

7. Discuss and describe the scope, purpose, and organizational structure of fire and emergency services.

9. Compare and contrast effective management concepts for various emergency situations.

Chapter 10 • Fire Department Organization and Management **319**

Fire Department Organization and Management

Learning Objectives

1. Describe the basic principles of organization. (Outcome 7)
2. Differentiate among forms of local government. (Outcome 7)
3. Explain automatic aid and mutual aid. (Outcome 7)
4. Differentiate among the different ways a fire department can be funded. (Outcome 7)
5. Explain current challenges facing fire protection. (Outcome 1)
6. Recognize the importance of policies and procedures in the fire service. (Outcome 9)
7. Describe how the Incident Command System is used in the fire and emergency services field. (Outcome 9)
8. Identify terms used in the National Incident Management System. (Outcome 9)
9. Describe types of personnel accountability systems. (Outcome 9)
10. Explain the purpose of a Rapid Intervention Crew. (Outcome 9)
11. Recognize the need for critical Incident Stress Management. (Outcome 9)

Chapter 10
Fire Department Organization and Management

Case History

A fire department in a small town was hosting a pancake fundraiser for a new piece of equipment. As anticipated, many citizens attended the pancake breakfast on a Saturday morning. Public safety was important to many of the individuals who attended, and it made many of them consider where fire departments routinely receive funding. While enjoying their pancakes, the citizens visited and asked each other, "Did fire departments always need to raise money for equipment, or did the taxes from the communities pay for some level of protection?" The citizens contemplated the challenges that the local departments face when using obsolete and outdated equipment or when recruiting new members. Several people mentioned that they often drove by the local station but admitted that they did not give much thought to how it was organized and managed or how it responded to emergencies. The citizens always thought that if the community needed their fire department — the firefighters and the apparatus would come. During this fundraiser, the citizens realized that they had so much more to learn about the fire service and their fire department as well.

No two agencies are organized or managed in exactly the same manner because of the diverse nature of the jurisdictions that they protect and the composition of each fire department. However, there are similarities among many fire departments that may be explained in general terms. This chapter will introduce the following concepts:

- Principles of organization
- Response considerations
- Fire department funding
- Policies and procedures
- Incident management

Principles of Organization

In order to operate safely and efficiently, fire department members should establish and follow policies, procedures, job descriptions, and organizational charts that outline areas of responsibility for department members. An organizational chart illustrates fire department's structure, including its **chain of command**.

Chain of Command — (1) Order of rank and authority in the fire and emergency services. (2) The proper sequence of information and command flow as described in the National Incident Management System - Incident Command System (NIMS-ICS).

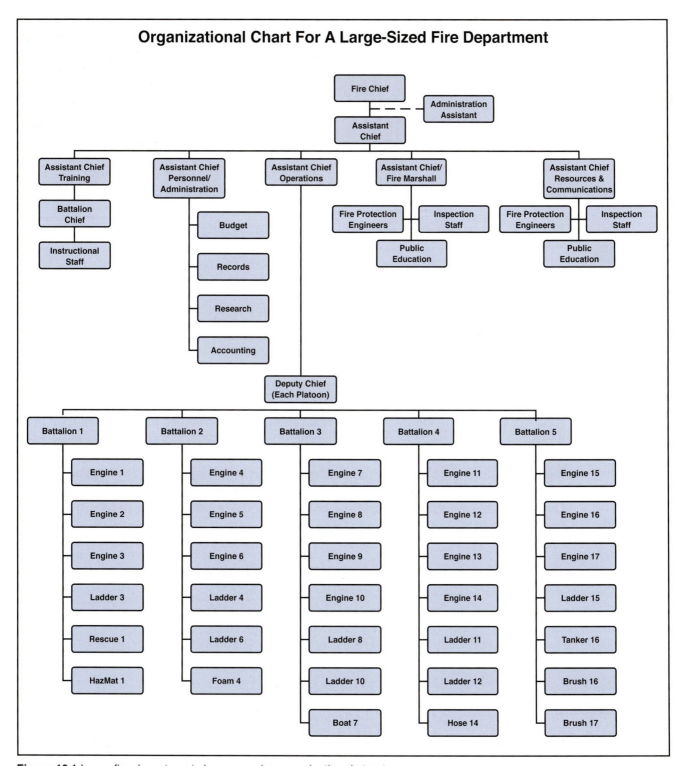

Figure 10.1 Large fire departments have complex organizational structures.

Fire departments with fewer members, fire stations, and fire companies will often have a simple chain of command. Departments consisting of numerous stations and personnel generally have a more complex chain of command with many levels of **Command** responsibility **(Figure 10.1)**. In order to operate effectively as part of a team, firefighters should be familiar with the following organizational principles:

- Unity of Command
- Span of Control
- Division of Labor
- Discipline

Unity of Command

Unity of command is the organizational concept that states a person can only report to one supervisor. Each subordinate reports to one superior up through the organizational structure. The chain of command is the organizational structure by which a pathway of responsibility is followed from the highest to the lowest level of the department.

The fire chief may issue an order that filters through the chain of command until it turns into an assignment at the appropriate level. Unity of command ensures that all personnel within the chain of command are aware of an order **(Figure 10.2)**. In this way, orders may be given for specific assignments without loss of accountability or control.

In cases where unity of command is not followed, firefighters may be placed into a situation that requires them to report to more than one supervisor. Reporting to more than one supervisor may be potentially dangerous for the firefighter on the fireground, and it may result in other difficult situations for the firefighter.

> **Command** — (1) Act of directing, ordering, and/or controlling resources by virtue of explicit legal, agency, or delegated authority. (2) Term used on the radio to designate the Incident Commander. (3) Function of NIMS-ICS that determines the overall strategy for the incident, with input from throughout the ICS structure.

> **Unity of Command** — Organizational principle in which workers report to only one supervisor in order to eliminate conflicting orders.

Figure 10.2 An officer issuing orders to a firefighter at an emergency scene. *Courtesy of Ron Jeffers.*

Figure 10.3 A five-to-one span of control is ideal for fire departments.

Span of Control

Span of Control — Maximum number of subordinates that one individual can effectively supervise; ranges from three to seven individuals or functions, with five generally established as optimum.

Span of control refers to the number of personnel one individual can effectively manage. There are situations in which span of control may vary slightly based on the complexity of the assignment and level of training and experience of the supervisor and subordinate **(Figure 10.3)**.

General Guidelines for Span of Control

A general guideline for the fire service is that an officer can effectively supervise three to seven subordinates during emergency operations. However, officers may have responsibility for an additional number of subordinates in administrative applications.

Division of Labor

Division of Labor — Subdividing an assignment into its constituent parts in order to equalize the workload and increase efficiency.

Division of labor is the practice of subdividing large jobs into smaller areas of responsibility that may be assigned to specific groups or individuals. The division of labor concept is used in the fire service for the following reasons:

- To assign responsibility
- To prevent duplication of effort
- To assign specific goal-oriented tasks
- To provide expertise regarding an assigned task

Discipline

Discipline is often understood to refer to corrective action or punishment. However, in this context, the term *discipline* involves an organization's responsibility to provide the direction necessary to achieve identified goals and objectives. This direction is often supplied by **standard operating procedures**, guidelines, and/or regulations. Whatever terminology is used, these documents should be clearly written and available for review by all personnel.

> **Discipline** — To maintain order through training and/or the threat or imposition of sanctions; setting and enforcing the limits or boundaries for expected performance.

Local Government Structure

The structure of local governments may vary. Regardless of the organizational structure, taxpayers remain at the top of the organizational chart. Funding for the department is derived from the taxpayers of the community. Governing bodies make decisions that often have an impact on the operations and services of the fire department in that jurisdiction. Forms of local government include the following:

- Commission
- Council (Board) Manager
- Mayor/Council
- Fire Districts

> **Standard Operating Procedures (SOPs)** — Standard methods or rules in which an organization or fire department operates to carry out a routine function. Usually these procedures are written in a policies and procedures handbook and all firefighters should be well versed in their content.

Commission

In a Commission form of government, voters elect a Board of Commissioners who conduct the legislative and executive functions of government. This Board may appoint officials to oversee various governmental needs. The Commission may also select one of its own to serve as the Chair. Department heads within this organization are often elected positions that are not appointed by the commissioners.

Council (Board) Manager

In the Council (Board) Manager form of government, voters elect a council or board of officials to conduct legislative responsibilities. These officials hire a professional manager to oversee executive (administrative) matters. The manager appoints department heads within each area of governmental responsibility.

Mayor/Council

Voters in the Mayor/Council form of government elect both a mayor and members of a council to govern the community. The mayor acts as the chief executive officer, and the council performs legislative duties. In the capacity of chief executive, the mayor appoints department heads to manage government responsibilities **(Figure 10.4)**.

Figure 10.4 A chief officer talking to a mayor.

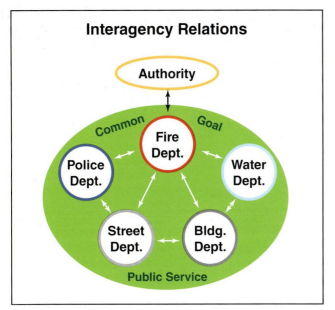

Figure 10.5 Organizations within a locality share a common authority, common goals, and dedication to serving their communities.

Outside Aid — Assistance from agencies, industries, or fire departments that are not part of the agency having jurisdiction over the incident.

Automatic Aid — Written agreement between two or more agencies to automatically dispatch predetermined resources to any fire or other emergency reported in the geographic area covered by the agreement. These areas are generally located near jurisdictional boundaries or in jurisdictional "islands."

Mutual Aid — Reciprocal assistance from one fire and emergency services agency to another during an emergency, based upon a prearranged agreement; generally made upon the request of the receiving agency.

Fire Districts

A fire district is not directly tied to a municipal or county jurisdiction. The district is a state-authorized governing body established to provide fire protection and other emergency services within a specific area that may overlap other jurisdictional boundaries. Generally, a fire district is governed by an elected Board of Commissioners or trustees who represent the residents of the district.

Response Considerations

Regardless of governmental structure, the fiscal limitations under which the department must operate are a primary consideration affecting the ability of a fire department to provide services to the community. Availability of funds typically results in creating a balance of overall services to meet jurisdictional needs. The fire department leaders' main goals are to use the available resources most efficiently to provide effective fire protection for residents and the safest operating conditions for firefighters. Many fire departments employ a strategy that consists of the following parts:

- Deploy resources that are funded.
- Reach contractual agreements with other agencies to supplement local resources for incidents that are beyond the agency's ability to control.

Outside aid agreements may be established with public agencies or private contractors that perform a specific, specialized service beyond the capabilities of the jurisdiction in which an incident may occur. These specialized services may include hazardous materials incident response, technical rescue, or other specialized discipline.

Automatic Aid

Automatic aid is a formal, written response agreement between fire departments as well as specific facilities (such as airports, refineries, or chemical plants) initiated under predetermined conditions. For example, automatic aid between departments may be dispatched whenever an emergency is reported in a predetermined response area or when specific resources are required that the requesting department does not have available **(Figure 10.5)**. This type of agreement may also provide for automatic aid upon the transmission of a given number of alarms at an incident.

Mutual Aid

Mutual aid is a reciprocal agreement between two or more fire protection agencies. These agreements may be local, regional, statewide, or interstate, meaning that the agencies may not have shared boundaries. The agreement defines how each agency will provide resources in various situations and how the shared resources will be monitored and directed during operations **(Figure 10.6)**. Responses under a mutual aid agreement usually are initiated only upon request of an agency that is a party to the agreement. This typically occurs

326 Chapter 10 • Fire Department Organization and Management

Figure 10.6 Adjoining fire departments or districts sometimes work together because of the size or nature of an emergency. *Courtesy of Ron Jeffers.*

Figure 10.7 A second fire department may provide backup for emergencies if the primary department's resources are already deployed. *Courtesy of Ron Jeffers.*

when a large incident or a number of smaller incidents deplete the resources of an agency. The response for such a request may be at the discretion of the providing agency, subject to provisions of an agreement or state law.

Fire departments enter into automatic aid and/or mutual-aid agreements for a variety of reasons **(Figure 10.7)**. Mutual aid agreements allow fire departments to:

- Allow sharing of limited or specialized resources by jurisdictions within a region.
- Address the need for neighboring fire protection agencies to assist one another when a response requirement exceeds the primary jurisdiction's capabilities.
- Address the National Fire Protection Association® (NFPA®) standards or Insurance Services Office (ISO) recommendations and other possible requirements for staffing, apparatus, or response times.
- Provide quicker response times to an incident when other departments' resources may be closer to an incident than those of the primary jurisdiction.
- Define a response model for agencies within a jurisdiction, such as a military base or industrial facility, with its own fire department that may necessitate automatic aid.
- Define response models for areas that may lie between neighboring jurisdictions.

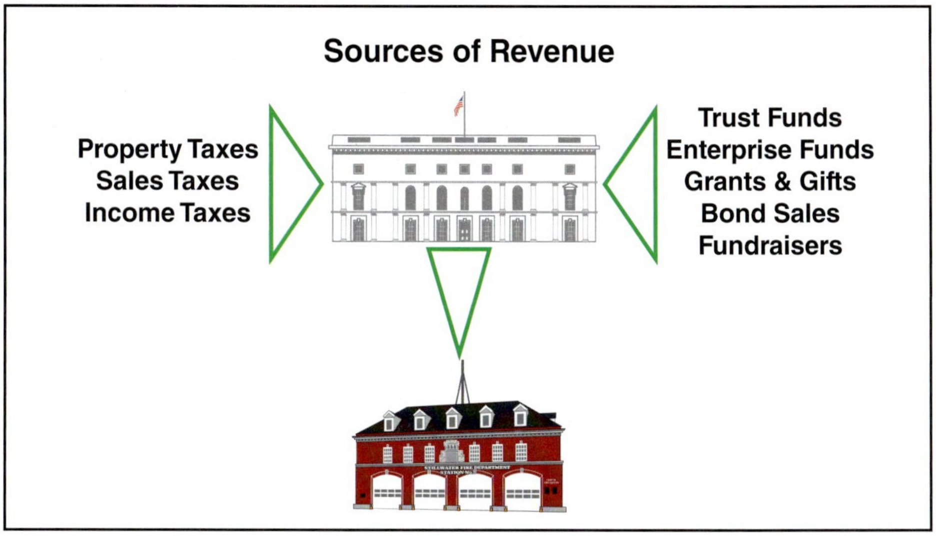

Figure 10.8 Illustrating how revenue makes its way to a fire and emergency services department.

Automatic and mutual aid agreements should undergo periodic review to ensure that they continue to efficiently serve the needs of all communities involved. Because these agreements will result in personnel from different agencies working together at incidents, joint training exercises to review policies and equipment should be conducted on a regular basis in order to ensure compatibility between agencies.

Fire Department Funding

In order to conduct operations, all fire departments must have a reliable source(s) of revenue. Fire departments may be funded by a variety of means depending on their composition and the jurisdiction they protect **(Figure 10.8)**. The following sections highlight some methods of revenue for different types of fire department organizational structures.

Tax Revenues

A portion of funds collected from sales or property taxes may be allocated to finance the operation of a fire department. These departments may be under the authority of a municipal, county, or other form of government authority. A fire department may also operate as a fire district that collects a tax specifically for the purpose of fire protection. An Emergency Services District is a similar organization created by voters for the purpose of funding fire protection, EMS, or both services. In many cases, the governing body for the jurisdiction maintains ultimate control over the fire department budget.

Trust Funds

Establishment of a trust fund within a fire department's budgetary system may serve several purposes, depending on the nature of the revenue and purpose of the trust. Some jurisdictions attach a tax within their boundaries to be held in a trust fund specifically for capital improvements for the fire service or other public safety entities. In other systems, a trust may be funded by private donations or charities to be used at the discretion of the fire department or for very specific purposes, such as maintenance of historic fire houses, purchase of medical equipment, or other earmarks.

Enterprise Funds

An enterprise fund may be established for a fee-charging government service. Some fire departments may choose this system for use with a fee-for-service component such as EMS transportation. In this way, a separate accounting of cost and revenue is maintained for that aspect of service. The establishment of separate funding may be intended to direct the program for which it was created to operate in the same manner as a private business earning revenue to justify continued operation.

Bond Sales

A government may use bonds to raise funds to support projects. The government agency will pay back the money it has borrowed with interest. Bonds are usually issued to finance long-life projects such as buildings or vehicles. The assets purchased with the bond money should have the same or longer useful life than the time frame of the bond. Because bonds normally have lower interest rates than other types of financing, they are an attractive way for a government agency to finance a capital expense. Bonds cannot be used to finance operational expenses such as salaries.

Grants/Gifts

Grants are issued to fire service organizations through government agencies, such as the Federal Emergency Management Agency as well as many private foundations that seek to provide assistance to agencies in local communities. Some private sources for grants may be a non-profit group, while others may be affiliated with a for-profit corporation. Many types of grants exist that may be awarded to fund specific programs or equipment. The entity that is sponsoring the grant usually has an application process that must be completed for review by a committee or board. This review ensures the legitimacy of the request and pairs it with the guidelines established for an award. Funding sources have diverse backgrounds and may have a variety of requirements to be satisfied before funds are awarded. An agency should attempt to match their needs with a funding organization that closely matches their goals.

Some charitable and for-profit corporations also give monetary awards to community organizations that serve the public. These are often one-time donations to aid with a special project or current need.

Fundraisers

The fire service, in particular the volunteer sector, has a long tradition of holding a wide variety of fundraisers to increase revenue for its operations. Events ranging from raffles to car washes to bingo games have helped many local departments purchase new apparatus and protective gear or meet operating expenses. The fire department membership plans and operates many events so that the maximum amount of proceeds goes into the coffers of the organization. The limitations on fundraising efforts are local laws and ordinances and the by-laws of the fire department. Fire service organizations should have an understanding of Internal Revenue Service laws before embarking on any fundraising efforts.

Subscriptions/Fees
Some jurisdictions may implement a subscription fee for fire protection or emergency medical response. In addition, departments may charge a fee for:

- Providing emergency services to interstate highways
- Mitigating a hazardous materials incident
- Providing technical rescue

Current Challenges Facing Fire Protection
In addition to the expected hardships found on the fireground and at other emergency operations, the fire service has always faced a multitude of challenges from sources other than incident response. The constraints of funding, maintenance of equipment, inadequate infrastructure, and personnel issues plague many fire departments throughout North America.

Funding/Budget Constraints
Many fire departments face serious budget issues that threaten to affect the level of service provided as well as the safety of firefighters and civilians in the jurisdiction **(Figure 10.9)**. The financial health of some cities has led to drastic cuts in municipal department budgets, including those of public safety agencies. Firefighters have been laid off, when on-duty personnel levels fall beyond a certain point. In some cases, firehouses have been permanently closed and companies disbanded for lack of funding.

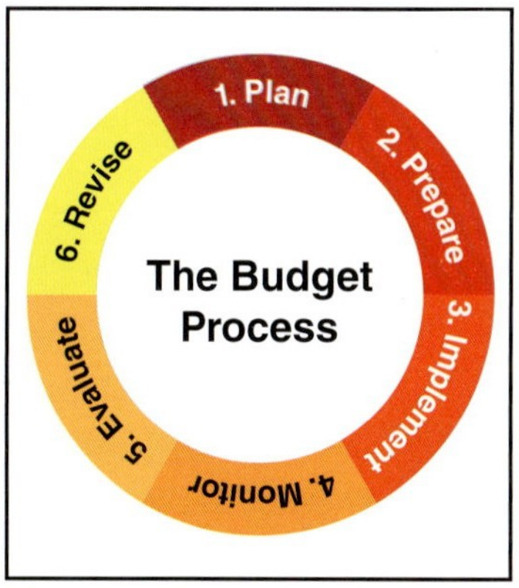

The federal government has attempted to provide some relief in the form of a FEMA Staffing for Adequate Fire and Emergency Response (SAFER) grant. This grant provides some amount of funding for firefighter positions with specific jurisdictional commitments in regard to continuation of funds beyond the life of the grant.

In an effort to maximize efficiency, fire departments sometimes cooperate to acquire selected apparatus or other specialized resources that may be too costly for a single jurisdiction to purchase or maintain. However, by funding the expenditure jointly and sharing limited resources, each department may benefit from its utilization during an incident.

Outdated Equipment
With the rapid pace of technological advancement and changing safety concerns in the fire service, many fire departments are facing challenges in keeping apparatus, protective equipment, and other tools and appliances up to date with safety standards. The Assistance to Firefighters Grant (AFG) program, administered by FEMA, was established to provide funding to some of the nation's fire departments that are deemed to have a great need for equipment upgrades, facility improvements, or training requirements. While helpful, this funding is not currently sufficient to address the needs of every fire department struggling with budget issues. Grants may not allow for the maintenance and continued certification of equipment subsequent to purchase, limiting the usability of the equipment **(Figure 10.10)**.

Some departments have purchased new equipment or received used equipment as a donation from other jurisdictions. While this used equipment may still be serviceable, it seldom features the most recent technology or safety innovations.

Figure 10.9 The steps of the budgeting process.

Figure 10.10 Adequate financial resources are needed to replace outdated apparatus. *Courtesy of Chris Mickal.*

Figure 10.11 Chief officers in volunteer departments should describe the benefits of becoming a volunteer fire fighter.

Recruiting Personnel

In the career fire service, most jurisdictions have little difficulty in recruiting applicants for firefighter positions. During turbulent economic conditions, public sector careers have typically been popular choices among applicants because of their relative security and competitive compensation with similarly skilled private sector employment.

In the volunteer fire service, many departments are facing an acute membership shortage. In many communities around the country, conditions have changed since the last generation of volunteer firefighters began their service. Citizens who once worked close to where they lived now often commute miles to their workplace. Citizens commuting to work leaves few available people in so-called "bedroom communities" during work hours. In some cases, those who are employed in the community may be unable to leave their jobs to respond to emergencies due to company policies.

Compared to life a generation ago, people generally have less time due to family or social responsibilities and commitments. In some cases, adults in a household work two jobs, leaving little time for volunteer efforts. Combining the time constraints of modern life with the increased training required of today's volunteer firefighter places a major burden on some communities to assemble an adequate fire fighting force. In some volunteer departments with budget shortfalls, the cost of mandated training classes may be placed on the firefighters, creating a financial hardship **(Figure 10.11)**.

Retaining Personnel

Retention of personnel is important to both the career and volunteer fire service. When firefighters leave the department, they take with them their training, experience, and knowledge of department operations. A significant amount of money is often invested in training programs, and the value of experience and organizational knowledge is an intangible, yet critical factor in daily fire department operations.

While some career firefighters leave to pursue other career paths or other fire service organizations, often the most dramatic effect on the personnel of a department occurs when a large number of veteran firefighters and of-

Figure 10.12 Volunteer firefighters in training. *Courtesy of Chris Mickal.*

Figure 10.13 Effective fire fighting requires an adequate water source or sources. *Courtesy of Chris Mickal.*

ficers retire within a short period of time. Personnel retirement may leave a fire department scrambling to fill leadership positions and cause the overall experience level of company officers to decline until newly promoted members gain experience in their positions.

Volunteer fire departments have a high turnover rate due to a number of factors that affect people's ability and desire to volunteer. The turnover rate is sometimes greater. Many times, young people will become volunteer firefighters in order to determine if a position as a career firefighter might be of interest. These members sometimes leave to take jobs with career departments. In other cases, volunteers may move away, change jobs, or need to take a second job, limiting their availability to be firefighters. Some volunteer departments have turned to various forms of incentives to retain trained personnel. These programs may offer a stipend for the number of calls or training sessions they attend. Other departments may pay for their volunteers to attend advanced fire fighting or emergency medical training programs **(Figure 10.12)**.

Adequate Water Supply

Responding with a sufficient number of highly trained personnel onboard a modern fire apparatus will be of little use if an adequate, sustained water supply is not available for fire fighting efforts. The potential for a deficient water supply may exist in rural, suburban areas, or even cities fully served by fire hydrants **(Figure 10.13)**. Some common causes for lack of sufficient water supply in pressurized (hydrant) systems include:

- Sediment buildup in water mains
- Small diameter mains in older neighborhoods
- Drawing from a number of hydrants simultaneously in the same area

In areas not served by fire hydrants, the fire department must draft water from a static source, such as a lake, pond, cistern, or a neighborhood swimming pool. If remote from the fire location, water tenders (mobile water supply apparatus) must deliver the water supply to the scene.

Factors affecting this supply source may include a scarcity of water due to weather conditions, the distance from the source to the fire, as well as the accessibility of the water source. Fire departments in jurisdictions using static supply sources must have an adequate number of mobile water supply apparatus available to ensure efficient delivery of water from a remote source to the incident location. This may be a labor-intensive and time-consuming process. Preplanning and frequent training will enable firefighters to quickly and efficiently locate water sources and fulfill water supply objectives.

Figure 10.14 A chief officer discussing policies with firefighters.

Policies and Procedures

Firefighters should become familiar with their department's policies and procedures to help ensure safe and efficient operations at emergencies and during routine functions. In order to understand the purpose of policies and procedures, it is important to identify their similarities and differences, as well as how they are used in the fire service.

A policy is a guide to decision-making within an organization. Policy originates with the top management of the fire department and outlines the types of decisions that must be made by officers or other management personnel in specific situations **(Figure 10.14)**.

Whereas a policy is a guide to decision-making, a procedure is a formal communication that is a detailed guide to action. A procedure describes in writing the steps to be followed in carrying out department policy for a specific situation or condition.

Standard Operating Procedures (SOPs)/ Guidelines (SOGs)

Many fire departments have predetermined written procedures that outline actions in response to routine or emergency situations that may occur. In some organizations, the procedure is fairly rigid and is expected to be followed unless extenuating circumstances preclude its viability. Other organizations may choose to adopt standard operating guidelines (SOGs), which are generally less rigid in their application and may give firefighters more latitude in choosing a method of implementation **(Figure 10.15)**.

In either case, the operating policy or guideline may vary by jurisdiction, but the principle is usually the same. During fire fighting operations, the **Incident Command System** is a means of coordinating a fire attack. However, policies do not attempt to replace size-up, professional judgment, or Command decisions. SOPs should be established to follow the most commonly accepted order of fireground priorities:

10.15 A chief officer posting station safety procedures.

Incident Command System (ICS) — Standardized approach to incident management that facilitates interaction between cooperating agencies; adaptable to incidents of any size or type.

- Life safety
- Incident stabilization
- Property conservation

The primary concern during all fire department operations is the life safety of firefighters and civilians. In conjunction with the life safety goal, the SOPs should also address how to bring a fire under control or mitigate another type of emergency incident. In addition, firefighters should use appropriate tactics and loss control techniques to minimize damage to property.

In addition to their role at an emergency scene, standard operating procedures are often used to conduct routine administrative functions in the fire service. The issues addressed by administrative SOPs may include uniforms, conduct, station duties, and other routine matters.

Disciplinary Procedures

Fire departments should adopt and enforce written disciplinary procedures for members who may neglect established policy. Disciplinary measures enacted against an individual may vary depending on several factors:

- Severity of the offense
- Number of occurrences for a specific offense
- Previous record of the individual
- Precedence on similar occurrences
- Subject to the provisions of a collective bargaining agreement in some jurisdictions
- Based on jurisdictional rules and regulations

Disciplinary procedure may vary based on the department's organization. Many career departments have specified levels of action that may start with a verbal warning and proceed to time off without pay, demotions, and termination of employment. Volunteer departments may warn members, suspend them for a period of time, demote them, or terminate their membership.

Formal Communications

Communication is the exchange of ideas and information that conveys an intended meaning in a form that is understood. The fire service operates under a paramilitary chain of command, or scalar structure, having an uninterrupted series of steps. Authority is centralized at the top of a pyramid type of organization. Major decisions are directed from the top of the pyramid, down through intermediate levels, and to the base of the organization. Likewise information may be transmitted to the top of the chain of command using the same channels of communication.

Each level through which information must pass may act as an unintended filter, reducing the quality and intent of the original message. For this reason, verbal communication is not the most effective means for transmitting information on a large scale throughout the fire department.

Verbal orders will be necessary during emergency incidents and for some routine duties. However, to create and maintain efficient and safe operations, fire service organizations should release policies, procedures, and other important information as formal written communication. Fire service organizations should use a system of two-way written communication to maintain a smooth flow of information throughout the chain of command.

Incident Command System

The **National Incident Management System - Incident Command System (NIMS-ICS)** is designed to safely and efficiently manage the operations of a variety of incidents of all sizes. Fire departments and other response agencies are mandated by Homeland Security Presidential Directive/HSPD-5 to adopt the National Incident Management System in order to be eligible for federal funds **(Figure 10.16)**. Emergency response organizations operating under NIMS use common terminology and command structures to ensure the ability to interface with "outside" organizations during an emergency **(Figure 10.17)**.

Figure 10.16 A mobile incident command post uses an organization board to track firefighter locations. *Courtesy of Ron Jeffers.*

In 2004, the U.S. government officially adopted the Incident Command System (ICS) as the national model management system for coordinating equipment, personnel, and communications within a common organizational structure. NIMS-ICS combines command strategy with organizational procedures **(Figure 10.18, p. 336)**. This system may be used during single-agency or multi-agency incidents and is applicable to emergency and nonemergency events alike.

National Incident Management System - Incident Command System (NIMS-ICS) — The U.S. mandated Incident Management System that defines the roles, responsibilities, and standard operating procedures used to manage emergency operations; creates a unified incident response structure for federal, state, and local governments.

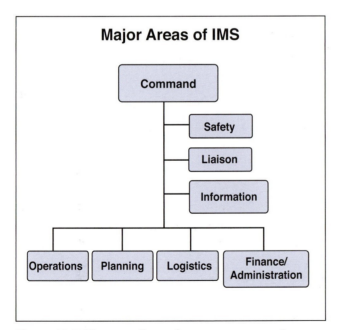

Figure 10.17 There are five major areas - command, operations, planning, logistics, and finance administration – within the Incident Management System.

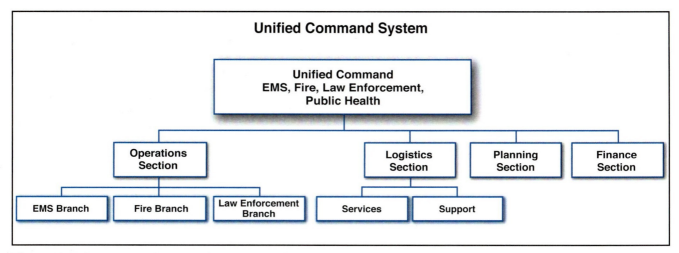

Figure 10.18 An example of an organizational chart for a unified command system.

> **Unified Command (UC)** — In the Incident Command System, a shared command role in which all agencies with geographical or functional responsibility establish a common set of incident objectives and strategies. In Unified Command there is a single Incident Command post and a Single Operations Chief at any given time.

The following components function interactively to provide an effective operational platform:

- Common terminology
- Modular organization
- Integrated communications
- **Unified Command** structure
- Consolidated action plans
- Manageable span of control
- Predesignated incident facilities
- Comprehensive resource management

Functional Areas of ICS

The five major organizational positions are identified as sections within NIMS-ICS. Each section is responsible for a major function as it pertains to management of the overall incident. These components will be described in the following paragraphs:

- Command
- Operations
- Planning
- Logistics
- Finance/Administration

Command

The person in overall command of the incident is designated the **Incident Commander (IC)**. The IC is ultimately responsible for all incident activities, including the development and implementation of a strategic plan. The IC has the authority to request additional resources to the scene and release personnel and equipment that are no longer necessary. Depending on the complexity of an incident, the IC may form a Command Staff with whom authority for decision-making may be delegated. Positions within the Command Staff may include the Safety Officer, Liaison Officer, and Public Information Officer.

> **Incident Commander (IC)** — Person in charge of the Incident Command System and responsible for the management of all incident operations during an emergency.

Operations

The **Operations Section** is responsible for direct management of all tactical activities that directly affect mitigation of the incident. The **Operations Section Chief** implements the Incident Action Plan as outlined by the Incident Commander. The Operations Section may be divided into branches in order to maintain a manageable span of control over complex or large scale operations.

Planning

The **Planning Section** is responsible for gathering, evaluating, processing, and disseminating information needed for effective decision-making during an incident as well as preparing and disseminating the Incident Action Plan (IAP). Planning, under the direction of the IC, is also responsible for tracking incident status as well as the resources assigned to the incident. The Planning Section Chief is responsible for identifying potential incident requirements in concert with the Logistics Section and preparing to obtain resources to meet those needs. The Planning Section also serves to streamline information sources for the Incident Commander coordinating and clarifying pertinent data for command review.

Logistics

The **Logistics Section** serves as the support mechanism for incident operations. The responsibility of logistics consists of a support branch that includes supply, ground support and facilities, and a service branch that includes communications, medical, and food. Long-term or complex incidents often rely heavily on an efficient Logistics Section to maintain continued operations.

Finance/Administrative

Finance/Administrative Unit has the responsibility for tracking and documenting all costs and financial aspects of an incident. This section is usually activated only during large-scale, long-term incidents. Ordinary incidents that are of limited scope and duration do not usually require financial tracking as expenditures may be limited and considered a matter of routine operating costs. The Finance/Administration unit is also responsible for addressing legal issues that may result from incident activities.

NIMS-ICS may add the Intelligence Section to gather information related to an incident. However, in many cases, the Planning Section will be responsible for gathering information at an incident as well. In mitigating incidents where the cause determination may require extensive investigative efforts, this section may be assigned within NIMS-ICS to fulfill that responsibility.

NIMS Terms

The National Incident Management System (NIMS) uses a number of terms with which firefighters should be familiar. The terms explained in the following section are widely used in the fire service. See Appendix E for the ICS organizational chart.

Operations Section — Incident Command System section responsible for all tactical operations at the incident. The Operations Section includes branches, divisions and/or groups, task forces, strike teams, single resources, and staging areas.

Operations Section Chief — Person responsible to the Incident Commander for managing all tactical operations directly applicable to accomplishing the incident objectives.

Planning Section — Incident Command System section responsible for collection, evaluation, dissemination, and use of information about the development of the incident and the status of resources; includes the situation status, resource status, documentation, and demobilization units, as well as technical specialists.

Logistics Section — Section responsible for providing facilities, services, and materials for the incident; includes the Communications unit, Medical Unit, and Food Unit within the Service Branch and the Supply Unit, Facilities Unit, and Ground Support Unit within the Support Branch.

Finance/Administrative Unit — Component of an Incident Management System responsible for documenting the ongoing financial impact of the incident; oversees the documentation of time and costs associated with personnel, as well as the documentation of private resources used throughout the incident. Includes the time unit, procurement unit, compensation/claims unit, and the cost unit.

Figure 10.19 An Incident Commander at a large-scale emergency scene. *Courtesy of Chris Mickal.*

Incident Commander (IC)

The Incident Commander is the individual at the top of the incident chain of command and is in charge of the overall incident **(Figure 10.19)**. The primary responsibilities of the IC include formulating the Incident Action Plan (IAP), coordinating resources to implement the plan, and ensuring that the goals and objectives are met in a safe and efficient manner.

Command Staff

Command staff positions may be established to assume responsibility for key functions that may develop based on the size and scope of an incident or the requirements established by the IC. These positions allow for a manageable span of control at the command level. Specific staff functions include that of the Public Information Officer, Safety Officer, and Liaison Officer **(Figure 10.20)**.

Figure 10.20 A safety officer overseeing an emergency incident. *Courtesy of Ron Jeffers.*

Incident Action Plan (IAP)
An Incident Action Plan should be formulated for each incident. The plan should identify the strategic goals and tactical objectives required to mitigate the incident. Incidents that are small-scale, short-duration events may not require a written IAP. However, complex or long-duration emergencies require the creation and maintenance of a written plan to guide each operational period.

Section
A Section is defined as the organizational level having responsibility for a major functional area of Incident Management. The Section is maintained organizationally between a Branch and Incident Command. Sections include: Operations, Planning, Logistics, Finance/Administration, and Intelligence.

Section Chief
A Section Chief is responsible for overseeing the function of a Section within the ICS. The individual in this position reports to the Command authority for all established Branches under the functional responsibility of that Section.

Branch
A Branch is the organizational level with functional or geographic responsibility for major parts of the operations or logistics functions. The Branch level is placed between Sections and Divisions or Groups in the Operations Section, and between the Section and Units in the Logistics Section.

Branch Director
Branch Directors are responsible for the functions of the Branch to which they are assigned. This may include supervision of several Groups or Divisions with geographic and/or functional assignments or units serving Logistics needs. A Branch Director may also be established for each agency with a particular specialty operating at a multijurisdictional incident.

Division
A Division is a geographic designation that assigns responsibility for all operations in a particular area. Divisions are assigned clockwise around an outdoor incident with Division A at the front (address side) of the incident. During operations in multistory buildings, Divisions are often identified by the floor to which they are assigned (first floor Division 1, second floor Division 2). In a single-story structure, the entire floor may be assigned as a Division (Interior Division). Organizationally, the Division level is between Single Resources, Strike Team or Task Force, and a Branch (if functioning) or Section.

Division Supervisor
The Division Supervisor is responsible for the activities of resources assigned to the geographic area in which the Division is located. The Division Supervisor reports to the Branch Director if that position has been established. In cases where a Branch or Section designation has not been established, the Division Supervisor may report to a Section Chief or the Incident Commander.

Group

Groups are an organizational level responsible for a specific functional assignment. Examples of Group assignments include Salvage Group, Search Group, and Ventilation Group. The Group level is organizationally placed between Single Resources, Task Forces, or Strike Teams, and the Branch, or Operation, or Command level, depending on the establishment of NIMS resources.

Group Supervisor

The Group Supervisor is parallel to the Division Supervisor in the Incident Command System. However, the Group Supervisor is responsible for oversight of a specific function, as opposed to a geographical area. Close coordination between Division Supervisors and the Supervisors of Groups operating in their location is essential to maintain safety and the continuity of the Incident Action Plan.

Unit

A Unit is an organizational element having responsibility for a specific function in the Planning, Logistics or Finance/Administration Section. A Unit may serve to address communications, medical, food, or supply needs, as well as other responsibilities deemed necessary by the IC.

Unit Leader

A Unit leader is the individual responsible for managing a particular activity in the Section to which the Unit is assigned. Some of these activities may include Triage in the Medical Group of the Operations Section, Supplies in the Logistics Section, or Cost Analysis in the Planning/Administration Section.

Strike Team/Task Force

A Strike Team consists of a specified combination of the same kind and type of resources assembled for a particular purpose. For example, in some jurisdictions, a Strike Team of water tenders (mobile water supply apparatus) may be used to meet the water supply needs of an incident. A Task Force is comprised of single resources assembled for a tactical need. In some jurisdictions, a Task Force consisting of a specified number of engine and ladder companies may be assembled to respond together for mutual aid to a neighboring city for a major incident **(Figure 10.21)**.

Strike Team/Task Force Leader

The leader of either a Strike Team or Task Force must have communications established with the members of the Strike Team or Task Force as well as their immediate Supervisor. If multiple jurisdictions are involved, the leader may be a member of any of those involved. However, it is important that all personnel are familiar with the strategy and tactics that will be employed at the incident.

Resources

Resources are considered personnel and major items of equipment, supplies, and facilities that are maintained for assignment to incident operations **(Figure 10.22)**. Resources are described by the kind and type and may be used in operational support or supervisory capacities.

Figure 10.21 A strike team assembles to attack a fire scene. *Courtesy of Ron Jeffers.*

Figure 10.22 A thermal camera is an example of modern technology that is used to check for heat signatures through heavy smoke. *Courtesy of Ron Jeffers.*

NIMS-ICS Training

NIMS-ICS training is available from the National Fire Academy, the Federal Emergency Management Agency, or other state or local agencies. It is important for firefighters to maintain up-to-date training of this material to maintain proficiency.

Emergency Operations

In order to properly manage firefighter safety during emergency operations, Incident Commanders should follow departmental procedures and perform the following:

- Conduct an initial size-up and risk assessment of an incident prior to committing to operations.

- Maintain accountability of personnel operating at the scene by location and function.

- Establish Rapid Intervention Crew(s) (RIC) and position them for immediate response to a firefighter(s) emergency.

NFPA® 1500 Standard

NFPA® 1500, *Standard on Fire Department Occupational Safety and Health Program*, contains information regarding the organization of initial attack operations in order to ensure the safety of firefighters operating in hazardous areas. The standard advises establishing personnel to remain outside the hazardous area to be available for assistance or rescue while other firefighters are working in a hazardous area. In addition, NFPA® 1500 addresses options for initial attack operations where imminently life-threatening conditions exist and immediate action may prevent a loss of life or serious injury.

Chapter 10 • Fire Department Organization and Management

Figure 10.23 Personnel accountability reports are used to track personnel through an extended incident.

Figure 10.24 An Accountability Officer tracks the location of all firefighters on an emergency scene. *Courtesy of Ron Jeffers.*

Personnel Accountability Systems

Personnel Accountability System — Method for identifying which emergency responders are working on an incident scene.

Personnel accountability systems are critical to the safety of firefighters. These systems identify and track the assignments and location of all personnel working at an incident. The system should be standardized throughout the department, allow for integration of mutual aid from other jurisdictions, and used at every incident **(Figure 10.23)**. Personnel accountability must also include individuals who may respond to an incident in vehicles other than fire apparatus. If the IC or Accountability Officer does not know who is on the fireground and where they are located, it is impossible to determine who and how many may be trapped inside or injured **(Figure 10.24)**.

Tag System

Based on department policy, some fire departments use a tag system to provide accountability for members operating at an incident. Firefighters are issued a personal identification tag they leave at a designated location or with a member who is responsible for accountability at a designated location. Some departments collect firefighters' tags from a board at the Command Post or in an apparatus. The Incident Commander or other officer may consult the board to determine who is on the fireground. Firefighter tags may be arranged on the board according to assignment in order to provide functional accountability. Firefighters retrieve their tags upon relief from the fireground **(Figure 10.25)**.

SCBA Tag System

An SCBA tag system provides for closer accountability for members working inside a hazardous atmosphere, such as a fire or hazardous materials incident. All personnel entering a hazardous atmosphere must wear appropriate per-

Figure 10.25 Firefighters give their personal identification tags to an accountability officer at an incident. *Courtesy of Ron Jeffers.*

Figure 10.26 A digital personnel accountability system offers responders another layer of safety in dangerous environments.

sonal protective equipment (PPE), including SCBA. A tag that is independent of any other accountability system is provided with the air pressure in the SCBA and the user's name.

As the firefighter enters the hazardous environment, a designated entry control person takes the firefighter's tag. At this point, entry control person may complete a safety check to ensure that the firefighter has donned the appropriate equipment and it is functioning properly. Monitoring the estimated usage of air supply, relief crews may be sent into the hazardous area well before the current team runs low on air. Upon leaving the hazardous area, firefighters retrieve their tags, maintaining an accurate account of personnel still engaged in operations.

Computer-Based Electronic Accountability

Wireless computer-based tracking systems are widely used to maintain accountability of personnel working at emergency incidents **(Figure 10.26)**. Systems may use barcode technology with scanners and readers and GPS transmitters placed on a firefighter's PPE for tracking purposes. These systems allow for detection and rapid notification of members who are immobile or call for assistance. In addition to the capability of firefighters to transmit a mayday alarm through these systems, the Incident Commander can send an evacuation alarm directly to firefighters at the scene. Although electronic accountability systems offer enhanced firefighter safety, manual systems, such as personnel lists, face-to-face or radio roll calls, and other methods of incident accountability, should be maintained and supplemented by the additional coverage of an electronic system.

Figure 10.27 A rapid intervention crew preparing to enter a structure to rescue trapped firefighters.

Rapid Intervention Crews (RIC/RIT)

The purpose of a **rapid intervention crew (RIC/RIT)** is to provide a dedicated team of firefighters standing by to rescue other firefighters who may become lost, trapped, or injured while engaged in interior operations **(Figure 10.27)**. The RIC concept provides for immediate deployment of a rescue team while allowing other firefighters to remain engaged in fire fighting efforts that may result in aiding the success of the RIC mission.

Rapid Intervention Crew or Team (RIC/RIT) — Two or more firefighters designated to perform firefighter rescue; they are stationed outside the hazard and must be standing by throughout the incident.

Ideally, a rapid intervention crew should respond with first alarm units. All crew members should be fully equipped with PPE and SCBA as well as lights, radios, forcible entry tools, search ropes, and any incident specific equipment they may need to perform a rescue. The officer of the RIC should report directly to the IC and pre-position the team nearby. The minimum number of personnel for an initial RIC is two firefighters. However, the actual size and number of RIC teams required at an incident depends on its size and complexity. Numerous studies have been conducted to illustrate the need for a relatively large number of firefighters to complete a successful rescue of a single member trapped in a structure.

NOTE: See FPP's *Rapid Intervention Teams* manual for additional information

Critical Incident Stress Management/Exposure to Potentially Traumatic Events

Responders who have participated in any incident, or a number of incidents over a period of time, may become victims of critical incident stress. Firefighters respond to emergencies that can have a severe impact on people, including traumatic injuries and death, and emergency responders may be physically and emotionally affected **(Figure 10.28)**.

Figure 10.28 EMS personnel assist injured firefighters. *Courtesy of Ron Jeffers.*

Because individuals react to stress in different ways, and the effects of unresolved stresses tend to accumulate, firefighters should seek out counselors who specialize in **critical incident stress management** to assist after incidents involving mass casualties, loss of a co-worker, loss of a child, or other serious incident. Not everyone needs to seek professional help after an incident, but every firefighter should know help is available and the proper avenue to seek assistance.

If firefighters are called to respond to a major incident where conditions exist that are likely to produce emotional stress for those involved, a briefing may be conducted prior to deployment to prepare members for what they may encounter. This may help firefighters mentally prepare themselves for a difficult assignment.

Critical Incident Stress Management (CISM) — Comprehensive crisis intervention system composed of seven elements: pre-crisis preparation, a disaster or large scale incident, defusing, critical incident stress debriefing, one-on-one crisis intervention/counseling, family/organizational crisis intervention, and follow-up/referral mechanisms.

Critical Stress Management Research

Research is conducted on a regular basis concerning the possible effects on firefighters who are exposed to potentially traumatic events. Several national organizations, including the International Association of Firefighters, International Association of Fire Chiefs, National Volunteer Fire Council, and National Fallen Firefighters Foundation, may be able to provide additional resources. Recent studies involving critical incident stress management include stress first aid, debriefing, and expanded employee assistance programs.

Chapter Summary

Fire departments operate under a variety of governing bodies depending on the jurisdiction in which they are located. Depending on the method with which a fire department is operated, it may obtain revenue through a variety of sources, including property taxes, fundraising efforts, and grant sources.

Although there are various organizational differences, fire department structure is generally paramilitary in nature, with a span of control that allows for adequate supervision of resources. These resources are sometimes drawn from outside the borders where the incident occurs. Responses featuring mutual aid or a predetermined level of automatic aid are common in some areas.

In order to maintain safe and efficient operations, fire service organizations should establish standard policies and procedures for incident response and routine operations. These policies should be part of a formal written communication process that may include disciplinary procedures as well as administrative correspondence.

With NIMS-ICS, all emergency responders can work together more effectively using its modular structure, common terminology, and coordination. Firefighters should learn and train periodically using the Incident Management System to ensure they are familiar with its structure and terminology. This effort will help ensure that officers and firefighters conduct operations according to policy and manage incidents with firefighter safety as a priority.

Another aspect of firefighter safety involves maintaining accountability for the actions and location of all personnel operating at an incident. Various systems, including radio or face-to-face reports, tags, and electronic devices, aid the IC in this task. Knowing the location and number of firefighters who may be in need of assistance is an important concern when a rapid intervention crew must be deployed at an incident. Having a RIC standing by is necessary to provide members who may need assistance the best possible chance of rescue.

Review Questions

1. Define Unity of Command, Span of Control, and Division of Labor.
2. Describe the concept of discipline in the fire and emergency services.
3. Describe at least two forms of local government.
4. Differentiate between automatic aid and mutual aid.
5. List two reasons fire departments enter into automatic or mutual aid agreements.
6. What are the different means of funding fire departments?
7. What are some of the current challenges facing fire protection?
8. What is the difference between a policy and a procedure?
9. Define at least three terms used in the National Incident Management System.
10. What is the tag system of personnel accountability?
11. What is a Rapid Intervention Crew?
12. Why is critical incident stress management needed in the fire service?

Courtesy of Ron Jeffers.

Contents

Appendix A
FESHE Course Outcomes and NFPA® 1001 JPRs With Page References 349

Appendix B
NFPA® Professional Qualification Standards 350

Appendix C
Firefighter Life Safety Initiatives 351

Appendix D
Incident Command System (ICS) Organizational Chart 352

Appendix A
Chapter and Page Correlation to FESHE Requirements

FESHE Course Outcomes	Chapter Reference	Page Reference
1	1	12-41
1	3	81-108
1	10	321-326, 333-344
2	6	186-209
3	1	28-33
4	1	13-21
4	4	114-139
4	10	326-328, 335-345
5	1	18-21
5	4	118-130
6	4	114-130
7	1	12-21
7	2	51-72
7	5	165-177
7	10	321-326, 335-344
8	9	260-289
9	10	321-326, 333-344
10	2	47-72
10	4	115-116
10	5	145-168
10	7	15
10	8	237-253
11	1	28-39, 40-41
12	1	27, 38

Appendix B
NFPA® Professional Qualification Standards

The National Board on Fire Service Professional Qualifications (NBFSPQ or Pro-Board) and the International Fire Service Accreditation Congress (IFSAC) set minimum competency standards for a wide variety of fire fighting and related activities. The following are the required standards:

- NFPA® 1001, *Standard for Fire Fighter Professional Qualifications*

- NFPA® 1002, *Standard for Fire Apparatus Driver/Operator Professional Qualifications*

- NFPA® 1003, *Standard for Airport Fire Fighter Professional Qualifications*

- NFPA® 1005, *Standard for Professional Qualifications for Marine Fire Fighting for Land-Based Fire Fighters*

- NFPA® 1006, *Standard for Technical Rescuer Professional Qualifications*

- NFPA® 1021, *Standard for Fire Officer Professional Qualifications*

- NFPA® 1026, *Standard for Incident Management Personnel Professional Qualifications*

- NFPA® 1031, *Standard for Professional Qualifications for Fire Inspector and Plan Examiner*

- NFPA® 1033, *Standard for Professional Qualifications for Fire Investigator*

- NFPA® 1035, *Standard on Fire and Life Safety Educator, Public Information Officer, Youth Firesetter Intervention Specialist and Youth Firesetter Program Manager Professional Qualifications*

- NFPA® 1037, *Standard for Professional Qualifications for Fire Marshal*

- NFPA® 1041, *Standard for Fire Service Instructor Professional Qualifications*

- NFPA® 1051, *Standard for Wildland Fire Fighter Professional Qualifications*

- NFPA® 1061, *Professional Qualifications for Public Safety Telecommunications Personnel*

- NFPA® 1071, *Standard for Emergency Vehicle Technician Professional Qualifications*

- NFPA® 1072, *Standard for Hazardous Materials/Weapons of Mass Destruction Emergency Response Personnel Professional Qualifications*

- NFPA® 1081, *Standard for Industrial Fire Brigade Member Professional Qualifications*

- NFPA® 1091, *Standard for Traffic Control Incident Management Professional Qualifications*

- NFPA® 1521, *Standard for Fire Department Safety Officer Professional Qualifications*

Appendix C
Firefighter Life Safety Initiatives

The Firefighter Life Safety Summit held in Tampa, Florida, in March 2004, produced 16 major initiatives that will give the fire service a blueprint for making changes.

1. Define and advocate the need for a cultural change within the fire service relating to safety, incorporating leadership, management, supervision, accountability, and personal responsibility.

2. Enhance the personal and organizational accountability for health and safety throughout the fire service.

3. Focus greater attention on the integration of risk management with incident management at all levels, including strategic, tactical, and planning responsibilities.

4. Empower all firefighters to stop unsafe practices.

5. Develop and implement national standards for training, qualifications, and certification (including regular recertification) that are equally applicable to all firefighters, based on the duties they are expected to perform.

6. Develop and implement national medical and physical fitness standards that are equally applicable to all firefighters, based on the duties they are expected to perform.

7. Create a national research agenda and data collection system that relate to the initiatives.

8. Utilize available technology wherever it can produce higher levels of health and safety.

9. Thoroughly investigate all firefighter fatalities, injuries, and near misses.

10. Ensure grant programs support the implementation of safe practices and/or mandate safe practices as an eligibility requirement.

11. Develop and champion national standards for emergency response policies and procedures.

12. Develop and champion national protocols for response to violent incidents.

13. Provide firefighters and their families access to counseling and psychological support.

14. Provide public education more resources and champion it as a critical fire and life safety program.

15. Strengthen advocacy for the enforcement of codes and the installation of home fire sprinklers.

16. Make safety a primary consideration in the design of apparatus and equipment.

National Fallen Firefighters Foundation
www.firehero.org

Appendix D
Incident Command System (ICS) Organizational Chart

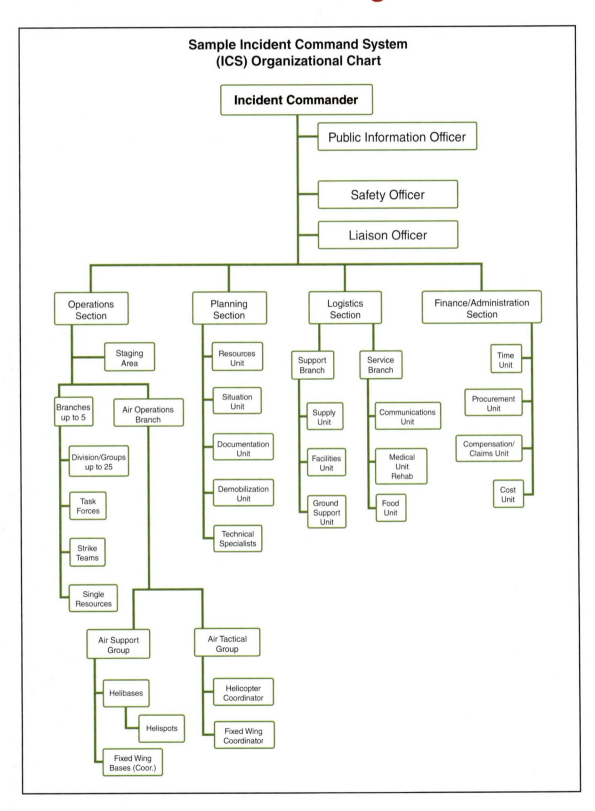

Glossary

Courtesy of Ron Jeffers.

Glossary

This glossary contains an extensive list of fire service terms and their definitions. Only the fire service definitions are given for the provided terms. In many cases, certain terms may have nonfire-service applications that are not covered here. Also, the spellings and definitions are consistent with IFSTA and Fire Protection Publications policy and may differ slightly from those used by other fire service organizations. Example: IFSTA uses one word for "firefighter," while the NFPA® uses two words for "fire fighter."

Entries have been alphabetized as though spaces and hyphens within the terms were not present. For example, the entry **Fireproof** comes before **Fire-Protection System**, which comes before **Fire Wall**. Numerical entries (**9-1-1**, **25 Percent Drain**) have been alphabetized by the first letter of their first numeral.

A

AAAE — *See* American Association of Airport Executives.

AAIB — *See* Air Accident Investigations Branch.

Abandonment — Termination of a first responder/patient relationship by the first responder, without consent of the patient and without care to the patient by qualified medical volunteers.

ABC — *See* American Board of Criminalistics.

A:B:C Extinguisher — *See* Multipurpose Fire Extinguisher.

ABCs — Airway, breathing, and circulation, the first three steps in the basic life support examination of any patient. Accompanied by control of severe bleeding, if necessary.

Abdominal Cavity — Large body cavity below the diaphragm and above the pelvis; contains the stomach with lower portion of the esophagus, small and large intestines (except sigmoid colon and rectum), liver, gallbladder, spleen, pancreas, kidney, and ureter.

Abeam — Directly off the side of a vessel; in a direction at right angles to the middle of the vessel's length. An object is said to be abeam when it is to the side of a vessel.

ABET — *See* Accreditation Board of Engineering and Technology.

Aboard — In or on a vessel; opposite of ashore.

Abort — (1) The act of terminating a planned aircraft maneuver, such as the takeoff or landing. (2) To terminate prematurely, or in the early stages.

Abrasion — Injury consisting of the loss of a partial thickness of skin from rubbing or scraping on a hard, rough surface. *Also known as* Brush Burn or Friction Burn.

Absolute Pressure — Gauge pressure plus atmospheric pressure.

Absorbent — Inert material or substance with no active properties that allow another substance to penetrate into the interior of its structure, and that can be used to pick up a liquid contaminant. An absorbent material is commonly used in the abatement of hazardous materials spills. Some examples of such absorbents are soil, diatomaceous earth, vermiculite, sand, and other commercially available products. *See* Contaminant.

Absorption — (1) Penetration of one substance into the structure of another, such as the process of picking up a liquid contaminant with an absorbent. *See* Absorbent and Contaminant. (2) Passage of materials (such as toxins) through some bodily surface into body fluids and tissue. *See* Routes of Entry and Toxin.

Abuse — Harmful behaviors and/or actions, as defined by local law, that place an individual at risk and require reporting.

AC Circuit — *See* Alternating Current Circuit.

Academic Misconduct — Any unethical behavior in which students present another student's work as their own, or gain an unfair advantage on test by bringing answers into the testing area, copying another students' answers, or acquiring a test questions in advance.

Academy — Training school; a place to train, learn, study, and achieve.

Accelerant — Material, usually a flammable or combustible liquid, that is used to initiate or increase the speed of a fire.

Accelerator — (1) Device attached to a dry-pipe sprinkler system for rapid removal of air in the system when a sprinkler operates. (2) Device, usually in the form of a foot pedal, used to control the speed of a vehicle by regulating the fuel supply.

Acceptance Test — Preservice test on fire detection and/or suppression systems after installation to ensure that the system operates as intended. *Also known as* Proof Test.

Acceptance Testing — Preservice tests on fire apparatus or equipment, performed at the factory or after delivery, to assure the purchaser that the apparatus or equipment meets bid specifications.

Access — (1) Place or means of entering a structure. (2) Roadways allowing fire apparatus to travel to an emergency. *See* Egress.

Access Hole — (1) Starter hole into which a cutting tool may be inserted to continue cutting a piece of sheet metal. (2) Space made in a door crack with a manual prying tool to facilitate the placement of a spreading tool.

Accessibility — Ability of fire apparatus to get close enough to a building, structure, site, or emergency scene to conduct emergency operations. *See* Access.

Accident — (1) Unplanned, uncontrolled event (or sequence of events) resulting from unsafe acts and/or occupational conditions; can result in injury, death, or property damage. Typically caused by persons who are either unaware or uninformed of potential hazards, ignorant of safety policies, or who fail to follow safety procedures.

Accident Investigation — Fact-finding rather than fault-finding procedures that look for causes of accidents; leads to analyzing causes in order to prevent similar accidents.

Accident Review Board (ARB) — Committee of department personnel and stakeholders whose responsibility is to review vehicular accident investigations.

Accidental Fire Cause — Cause classification for a fire that does not involve a deliberate human act to ignite or spread the fire into an area where the fire should not be.

Accommodation Ladder — Vessel's own gangway (usually one on each side) fitted with means of raising and lowering; also a set of steps or ladder used for getting from one deck to another.

Accommodation Spaces — Areas of a vessel designed for living; subdivided into officer, crew, and passenger accommodations. *Also known as* Cabins.

Accordion Fold — Method of folding a salvage cover; when completed, resembles the bellows of an accordion.

Accordion Load — Arrangement of fire hose in a hose bed or compartment in which the hose lies on edge with the folds adjacent to each other.

Accreditation Board of Engineering and Technology (ABET) — Organization that provides accreditation, promotion, and advancement of education in applied science, computing, engineering, and technology.

Acetylene (C_2H_2) — Colorless gas that has an explosive range from 2.5 percent to 100 percent; used as a fuel gas for cutting and welding operations.

ACGIH® — *See* American Conference of Governmental Industrial Hygienists®.

Acid — Compound containing hydrogen that reacts with water to produce hydrogen ions; a proton donor; a liquid compound with a pH less than 7. Acidic chemicals are corrosive. *See* Base, Corrosive, and pH.

Acoustic Search Device — Sensitive equipment used to listen for victims' responses in a collapsed structure.

Acquired Immune Deficiency Syndrome (AIDS) — Fatal viral disease that is spread through direct contact with bodily fluids from a previously infected individual.

Acquired Structure — Structure acquired by the authority having jurisdiction from a property owner for the purpose of conducting live fire training or rescue training evolutions. *Also known as* Acquired Building.

Acrolein ($CH_2 = CHCHO$) — Toxic gas produced by the burning of wood, paper, cotton, plastic materials, oils, or fats. When inhaled, acrolein can cause nose and throat irritation, nausea, shortness of breath, pulmonary edema, lung damage, or death.

Action Plan — Written plan of how objectives are to be achieved. *See* Incident Action Plan.

Activation Energy — minimum energy that starts a chemical reaction when added to an atomic or molecular system.

Active Listening — Method of listening characterized by maintaining eye contact with the message sender, imagining the sender's upcoming points, taking notes or mentally summarizing key points, paraphrasing especially important points, nodding the head, and thinking or saying something such as, "I understand."

Actual Mechanical Advantage — The rate of mechanical advantage a system offers in actuality; often a lower rate than the theoretical mechanical advantage rate due to the friction in a system.

Actuate — To set into operation; this term is often used to refer to an installed fire protection system or its components.

Actuator Valve — Valve that controls the flow of hydraulic oil from an aerial apparatus hydraulic system to the hydraulic cylinders.

Acute — Characterized by sharpness or severity; having rapid onset and a relatively short duration. *See* Chronic.

Acute Exposure — Single exposure (dose) or several repeated exposures to a substance within a short time period.

Acute Exposure Guideline Levels — Airborne concentration of a substance at or above which it is predicted that the general population, including "susceptible" but excluding "hypersusceptible" individuals, could experience notable discomfort. Established by the Environmental Protection Agency (EPA).

Acute Health Effects — Health effects that occur or develop rapidly after exposure to a hazardous substance. *See* Chronic Health Effects.

Acute Myocardial Infarction (AMI) — Critical phase of a heart attack where blockage of a coronary artery produces chest pain, nausea, heavy sweating, anxiety, and pallor. *Also known as* Heart Attack.

Acute Radiation Syndrome (ARS) — Serious illness that occurs when the entire body (or most of it) receives a high dose of radiation, usually over a short period of time.

ADA — *See Americans with Disabilities Act of 1990 - Public Law 101-336*.

Adapter — Fitting for connecting hose couplings that have dissimilar threads but the same inside diameter. *See* Fitting, Increaser, and Reducer.

Addiction — State of being strongly dependent upon some agent, such as drugs, tobacco, or alcohol.

Adhesion — Act of binding together substances of unlike compositions.

Adiabatic — Process of thermodynamic change of state in which no heat is added or subtracted from a system; adiabatic compression always results in warming, while adiabatic expansion always results in cooling.

Adjudication — To hear or settle a case by judicial procedure.

Adjunct — An accessory or auxiliary agent, such as an oral airway.

Adjustable Flow Nozzle — Nozzle designed so that the amount of water flowing through the nozzle can be increased or decreased at the nozzle; usually accomplished by adjusting the pattern of the stream.

Adjustable Fog Nozzle — Nozzle designed to allow the discharge pattern to be adjusted from straight stream to full fan fog; suitable for applying water, wet water, or foam solution. Some adjustable fog nozzles allow the rate of flow to be adjusted as well.

Adjutant — Firefighter assigned to drive and assist a chief officer. *Also known as* Chief's Aide.

Administration — Government agency having authority over port operations.

Administration Classification — Test classification based on how a test is administered.

Administrative Law — Body of law created by an administrative agency in the form of rules, regulations, orders, and decisions to carry out the regulatory powers and duties of the agency. *See* Law.

Administrative Search Warrant — Court order that allows investigators the right to enter a scene after the fire department has left the scene, or to reenter the scene if the investigator arrived while the fire department had control of the scene.

Admission — Statement that implicates the speaker in the commission of a crime.

Admission Valve — Pressure regulator valve that lets the air flow to the user.

Admixture — Ingredients or chemicals added to concrete mix to produce concrete with specific characteristics.

Adrenaline — Chemical released by the body that causes the breathing rate to increase and the body to prepare for "fight or flight."

Adsorbent — Material, such as activated carbon, that has the ability to condense or hold molecules of other substances on its surface.

Adsorption — Adherence of a substance in a liquid or gas to a solid. This process occurs on the surface of the adsorbent material.

Advanced Exterior Fire Fighting — Offensive fire fighting requiring the use of personal protective equipment, including self-contained breathing apparatus (SCBA), and performed outside a structure when the fire has progressed beyond the incipient phase.

Advanced Life Support (ALS) — Advanced medical skills performed by trained medical personnel, such as the administration of medications, or airway management procedures to save a patient's life.

Advancing Line — Line of fire hose that is moved forward.

Adverse Weather Condition — Any atmospheric condition, such as rain, snow, or cold, that creates additional problems or considerations for emergency personnel.

Adze — (1) Chopping tool with a thin, arched blade set at a right angle to the handle. (2) A wedge-shaped blade attached at right angles to the handle of the tool. *Also spelled* Adz.

AED — *See* Automated External Defibrillator.

Aerate — To mix with air.

Aeration — Introduction of air into a foam solution to create bubbles that result in finished foam.

Aerator — Device for introducing air into dry bulk solids to improve flow ability.

Aerial Apparatus — Fire fighting vehicle equipped with a hydraulically operated ladder, elevating platform, or other similar device for the purpose of placing personnel and/or water streams in elevated positions.

Aerial Attack — Use of aircraft to apply extinguishing agents to wildland fires. *See* Attack Methods (2).

Aerial Device — General term used to describe the hydraulically operated ladder or elevating platform attached to a specially designed fire apparatus.

Aerial Device Certification Testing — Pre-service testing, usually performed by a third-party testing agency, designed to give an unbiased opinion as to whether or not a piece of apparatus meets its design specifications and is worthy of being placed in service.

Aerial Fuels — Standing and supported live and dead combustibles that are not in direct contact with the ground; consists mainly of foliage, twigs, branches, stems, cones, bark, and vines.

Aerial Ignition — Use of an airborne incendiary device to assist in backfiring, burning out, or prescribed fires. Devices are normally carried in or suspended from helicopters.

Aerial Ladder — Power-operated ladder, usually employing hydraulics, that is mounted on a special truck chassis.

Aerial Ladder Platform — Power-operated ladder, usually employing hydraulics, with a passenger-carrying device attached to the end of the ladder.

Aerial Ladder Truss — Assembly of bracing bars or rods in triangular shapes that form a rigid framework for the aerial device.

Aerobic Capacity — Measure of cardiovascular fitness that takes into account oxygen capacity and efficiency of the lungs and blood in the cardiovascular and respiratory systems.

Aerodrome — *See* Airport.

Aerosol — Mist characterized by highly respirable, minute liquid particles. *See* Mist.

Aerosolize — To produce a fine mist or spray characterized by highly respirable, minute liquid particles.

Aesthetics — Branch of philosophy dealing with the nature of beauty, art, and taste.

AFA-CWA — *See* Association of Flight Attendants-CWA.

AFCI — *See* Arc-Fault Circuit Interrupter.

Affective — Descriptive of a person's attitudes, values, and habits.

Affective Learning Domain — Learning domain that involves emotions, feelings, attitudes, values, and habits. *See* Learning Domain.

AFFF — *See* Aqueous Film Forming Foam.

Affidavit — Written statement made under oath; must be performed before a notary or other officer with the authority to administer oaths.

Affiliation — Socio-psychological concept that people tend to act as a group, even with people they do not know very well; explains why no one typically leaves in an emergency until everyone leaves together.

Affirmative — Clear text radio term for "yes."

Affirmative Action — Administrative law adopted by the equal employment opportunity commission to implement the requirements of Title VII of the *Civil Rights Act of 1964*.

Affirmative Action Programs — Employment programs designed to make a special effort to identify, hire, and promote special populations where the current labor force in a jurisdiction or labor market is not representative of the overall population.

A-Frame — (1) Vertical lifting device that can be attached to the front or rear of the apparatus; consists of two poles attached several feet (meters) apart on the apparatus and whose working ends are connected to form the letter A. A pulley or block and tackle through which a rope or cable is passed is attached to the end of the frame. (2) A type of building construction in which a steep, gabled roof forms the major structural supports for the entire building.

A-Frame Ladder — Type of ladder that is hinged in the middle and can be used as a stepladder or a short extension ladder.

A-Frame Stabilizer — Stabilizing device that extends at an angle down and away from the chassis of an aerial fire apparatus.

AFSA — *See* American Fire Sprinkler Association.

Aft — Direction toward the back end or stern of a vessel, such as a ship or aircraft; term used relative to some other part of a vessel indicating the direction toward the stern. *Also known as* After.

After Action Reviews — Learning tools used to evaluate a project or incident to identify and encourage organizational and operational strengths and to identify and correct weaknesses.

Aftercooler — Air compressor component that cools the air that has been heated during compression.

Afterpeak — Area in the hull at the extreme rear end of a vessel; usually used for storage. *See* Forepeak.

Agar — Gelatinous or jelly-like substance used to grow bacterial cultures.

Age of Accountability — The minimum age at which State Courts have ruled that a child is intellectually capable of understanding right from wrong and the consequences associated with inappropriate behavior.

Agency for Toxic Substances and Disease Registry (ATSDR) — Lead U.S. public health agency responsible for implementing the health-related provisions of the Comprehensive Environmental Response, Compensation and Liability Act (CERCLA); charged with assessing health hazards at specific hazardous waste sites, helping to prevent or reduce exposure and the illnesses that result, and increasing knowledge and understanding of the health effects that may result from exposure to hazardous substances.

Agency Representative — Individual from an assisting or cooperating agency who has been assigned to an incident and has full authority to make decisions on all matters affecting that agency's participation at the incident. Agency representatives report to the incident liaison officer.

Agenda-Based Process — Classroom discussion format in which an agenda of topics or key points is provided to students for them to research, report on, and discuss as a group.

Agent — Generic term used for materials that are used to extinguish fires.

Aggravate — To worsen; to make worse.

Aggregate — (1) Gravel, stone, sand, or other inert materials used in concrete. These materials may be fine or coarse. (2) Term used in fire prevention and building codes to describe the sum total of individual parts or components of an assembly or feature, such as in units of exit.

Agroterrorism — Terrorist attack directed against agriculture, such as food supplies or livestock. *Also known as* Agricultural Terrorism.

Aground — Vessel resting wholly or partly on the ground instead of being entirely supported by the water. If done intentionally, a vessel is said to "take the ground"; if by accident, it is said to have "run aground."

Ahead — In front of a vessel; may indicate direction (an object may lie ahead) or movement (proceed at "full speed ahead").

AHJ — *See* Authority Having Jurisdiction.

AIDS — *See* Acquired Immune Deficiency Syndrome.

Aileron — Movable hinged rear portion of an airplane wing. The primary function of the ailerons is to roll or bank the aircraft in flight.

Air — Gaseous mixture that composes the earth's atmosphere; composed of approximately 21 percent oxygen, 79 percent nitrogen, plus trace gases.

Air Accident Investigations Branch (AAIB) — Part of the United Kingdom's Department of Transport that investigates civil aircraft accidents and other serious incidents.

Air Attack — (1) Using fixed-wing aircraft or helicopters to apply fire retardants or extinguishing agents on a wildland fire. Aircraft can also be used to transport crews, supplies, and equipment, or provide medical evacuation and reconnaissance. (2) Incident Management System term for the air attack coordinator.

Air Bag — (1) Inflatable bag built into the steering wheel, dashboard, or doors of an automobile that inflates immediately when the vehicle is impacted. (2) Large inflatable bag onto which persons can leap to escape danger. (3) *See* Air Lifting Bag.

Air Bank — *See* Cascade System.

Air Bill — Shipping document prepared from a bill of lading that accompanies each piece or each lot of air cargo. *See* Bill of Lading and Shipping Papers.

Air Bottle — *See* Air Cylinder.

Air Brake Test — A series of tests used to ensure the serviceability of an air braking system. Tests include air loss, air compressor buildup, air warning, and emergency parking brake activation.

Air Cascade System — *See* Cascade System.

Air Chamber — Chamber filled with air that eliminates pulsations caused by the operation of piston or rotary-gear pumps.

Air Chisel — *See* Pneumatic Chisel.

Air Cylinder — Metal or composite cylinder or tank that contains the supply of compressed air for the breathing apparatus. *Also known as* Air Bottle or Air Tank.

Air Drop — Process of dropping water, short-term fire retardant, or long-term fire retardant from an air tanker or helicopter onto a wildland fire.

Air Flow — The movement of air toward burning fuel and the movement of smoke out of the compartment or structure.

Glossary **359**

Air Lift Axle — Single air-operated axle that, when lowered, will convert a vehicle into a multiaxle unit, providing the vehicle with a greater load carrying capacity.

Air Lifting Bag — Inflatable, envelope-type device that can be placed between the ground and an object and then inflated to lift the object. It can also be used to separate objects. Depending on the size of the bag, it may have lifting capabilities in excess of 75 tons (68 000 kg).

Air Line Connection — *See* Chuck.

Air Line Pilots Association, International (ALPA) — Largest airline pilot union in the world, representing 61,000 pilots who fly for 40 U.S. and Canadian airlines.

Air Lock — (1) Intermediate chamber between places of unequal atmospheric pressure or temperature. (2) Situation that can develop in a centrifugal pump that has not been properly primed; rapid revolution of the impeller may create an air lock, which prevents priming the pump.

Air Mask — *See* Self-Contained Breathing Apparatus.

Air Mass — Extensive body of air, usually 1,000 miles (1 609 km) or more across, having the same properties of temperature and moisture in a horizontal plane.

Air Operations Area (AOA) — Area of an airport where aircraft are expected to operate, such as taxiways, runways, and ramps.

Air Pack — *See* Self-Contained Breathing Apparatus.

Air Pocket — (1) Condition that occurs during drafting when a portion of hard suction hose is elevated higher than the intake of the pump. (2) Void created by a cave-in. (3) Confined space where air is trapped in the top of a vehicle that has sunk beneath the water. (4) Condition of the atmosphere (as a local down current) that causes an airplane to drop suddenly.

Air Purification System — System designed to produce compressed breathing air for use in respiratory protection equipment.

Air Scoop — Hood or open end of an air duct that introduces air into an automobile, aircraft, or engine for combustion, cooling, or ventilation.

Air Spring — Flexible, air-inflated chamber on a truck or trailer in which the air pressure is controlled and varied to support the load and absorb road shocks.

Air Support Group Supervisor — Individual responsible to the air operations branch director for logistical support and management of helibase and helispot operations, and maintenance of a liaison with fixed-wing aircraft bases.

Air Surface Detection Equipment (ASDE) — Short-range radar displaying the airport surface, used to track and guide surface traffic in low-visibility weather conditions. ASDE may be used to direct radio-equipped emergency vehicles to known accident sites.

Air Tactical Group Supervisor — Individual responsible to the air operations branch director for the coordination of fixed-wing and/or rotary-wing aircraft operations over an incident.

Air Tank — *See* Air Cylinder.

Air Tanker — Any fixed wing aircraft certified by FAA as being capable of transport and delivery of fire retardant solutions.

Air Traffic Control (ATC) — Federal Aviation Administration (FAA) division that operates control towers at major airports.

Air-Aspirating Foam Nozzle — Foam nozzle designed to provide the aeration required to make the highest quality foam possible; most effective appliance for the generation of low-expansion foam.

Airborne Pathogens — Disease-causing microorganisms (viruses, bacteria, or fungi) that are suspended in the air.

Airco — One of several names for the plane carrying the air attack coordinator.

Aircraft Accident — Occurrence during the operation of an aircraft in which any person suffers death or serious injury, or in which the aircraft receives damage.

Aircraft Attitude — Angle of the aircraft front to rear while in flight. *Also known as* Attitude.

Aircraft Classes — Classification of aircraft by weight for various purposes. (Canadian terms are in parentheses.)

Aircraft Familiarization — Process of teaching personnel to become familiar with the various aircraft operated in an airport; includes familiarization with fuel capacity, fuel tank locations, emergency exit locations, operation of emergency exits, and passenger seating capacity.

Aircraft Fire Apparatus — Fire apparatus specifically designed for aircraft crash fire fighting/rescue operations.

Aircraft Hangar, Group I — Classification of aircraft hangar that has a single fire area in excess of 40,000 feet (12 192 m), has an access door height in excess of 28 feet (8.5 m), houses aircraft with a tail height in excess of 28 feet (8.5 m), and/or houses strategically important military aircraft.

Aircraft Hangar, Group II — Classification of aircraft hangar that has a single fire area that is less than 40,000 feet (12 192 m) and an access door height that is less than 28 feet (8.5 m). Construction type and fixed fire suppression systems also are used to determine Group II qualifications.

Aircraft Hangar, Group III — Classification of aircraft hangar that has aircraft access doors that are less than 28 feet (8.5 m) in height and a single fire area that is less than those given for the various types of building construction found in NFPA® 409, *Standard on Aircraft Hangars*.

Aircraft Incident — Non-accidental occurrence associated with the operation of an aircraft that affects or could affect continued safe operation if not corrected.

Aircraft Rescue and Fire Fighting (ARFF) — Term used to describe actions performed by rescue and fire fighting personnel to handle aircraft incidents and accidents.

Aircraft Rescue and Fire Fighting (ARFF) Apparatus — Motor-driven vehicle designed for aircraft rescue and fighting fires; capable of delivering Class B foam and providing specified levels of pumping, water, hose, rescue capacity, and personnel.

Aircraft Rescue and Fire Fighting Working Group (ARFFWG) — Non-profit international organization dedicated to the sharing of aircraft rescue and fire fighting information between airport firefighters, municipal fire departments, and all others concerned with aircraft fire fighting.

Aircraft Tug — Special, low-profile vehicle designed to tow aircraft on the airport ramp or push aircraft backwards away from an airport gate. *Also known as* Pushback Tractor or Tug.

Aircraft Velocity — Speed of an aircraft relative to its surrounding air mass. *Also known as* Airspeed.

Air-Entrained Concrete — Concrete with air entrapped in its structure to improve its resistance to freezing.

Airfield — *See* Airport.

Airfoil — Any surface, such as an airplane wing, aileron, elevator, rudder, or helicopter rotor, designed to obtain reaction from the air through which it travels. This reaction keeps the aircraft aloft and controls its flight attitude and direction.

Airframe — (1) Major components of an aircraft, such as the fuselage, wings, stabilizers, and flight control surfaces, that are necessary for flight. (2) Basic model of an aircraft; for example, the Boeing 707 airframe has both civilian and military applications in a variety of configurations.

Air-Handling System — *See* Heating, Ventilating, and Air-Conditioning System (HVAC).

Airline Respirator — *See* Airline Respirator System and Supplied Air Respirator (SAR).

Airline Respirator System — System in which breathing air is continuously supplied to the respirator user from a remote source of air. *Also known as* Supplied Air Respirator System. *See* Supplied Air Respirator.

Airport — Land used for aircraft takeoffs and landings. *Also known as* Aerodrome or Airfield.

Airport Control Tower — Building or unit built to house traffic control service for the movement of aircraft and vehicles in an airport operations area.

Airport Emergency Plan — Plan formulated by airport authorities to ensure prompt response to all emergencies and other unusual conditions, in order to minimize personal and property damage.

Airport Familiarization — Process of teaching personnel to become familiar with airport buildings, runways and taxiways, access roads, and surface features that may enhance or obstruct the prompt and safe response to accidents/incidents on the airport.

Airport Fire Protection — Specialized branch of the fire service dealing with airports and aircraft.

Airport Firefighter — Firefighter trained to prevent, control, or extinguish fires that are in or adjacent to aircraft. *Also known as* ARFF Firefighter.

Airport Flight Information Service — Air traffic services units that provide airport flight information service, search and rescue service, alerting service to aircraft at non-controlled airports, and assistance to aircraft in emergency situations.

Airport Ground Control — Control of aircraft and vehicular traffic on the ground operating in the airport movement area by the airport control tower.

Airport Operation Area (AOA) — Area of an airport where aircraft are expected to operate, such as taxiways, runways, and ramps.

Airport/Community Emergency Plan (A/CEP) — Plan formulated by airport and local community authorities to ensure prompt response to all emergencies and other unusual conditions in order to minimize the extent of personal and property damage.

Air-Pressure Sprinkler System — Sprinkler system in which air pressure is used to force water from a storage tank into the system.

Air-Purifying Respirator (APR) — Respirator that removes contaminants by passing ambient air through a filter, cartridge, or canister; may have a full or partial facepiece.

Air-Reactive Material — Substance that reacts or ignites when exposed to air at normal temperatures. *Also known as* Pyrophoric. *See* Reactive Material, Reactivity, and Water-Reactive Material.

Airspeed — *See* Aircraft Velocity.

Air-Supply Unit — Apparatus designed to refill exhausted SCBA air cylinders at the scene of an ongoing emergency.

Air-Supported Structure — Membrane structure that is fully or partially held up by interior air pressure.

Airway — (1) Metal or plastic framework designed to fit the curvature of the mouth and throat to prevent air passageways from closing. (2) Passage for carrying air from the nose or mouth to the lungs. (3) Channel of a designated radio frequency for broadcasting or other radio communications. (4) Designated route along which airplanes fly from airport to airport.

Aisle — Passageway between sections of seats in rows.

AISO — *see* Assistant Incident Safety Officer.

Alarm — Any signal or message from a person or device indicating the existence of a fire, medical emergency, or other situation requiring the need for emergency fire services response.

Alarm Assignment — Predetermined number and type of fire units assigned to respond to an emergency.

Alarm Center — *See* Telecommunications Center.

Alarm Check Valve — Type of check valve installed in the riser of an automatic sprinkler system that transmits a water flow alarm when the water flow in the system lifts the valve clapper.

Alarm Circuit — (1) Electrical circuit connecting two points in a fire alarm system; for example, from the signal device to the fire station, from the central alarm center to all fire stations, or from the sending device to the audible alarm services. (2) The circuit on a fire alarm system that connects the alarm initiating devices, such as the smoke detectors to the fire alarm control panel.

Alarm Signal — Signal given by a fire detection and alarm system when there is a fire condition detected.

Alarm System — System by which occupants and/or emergency personnel can be alerted to the existence of a hostile fire.

Alarm-Indicating Device — Bell, horn, chime, loudspeaker, or similar device that is actuated by a signal from an alarm-initiating device.

Alarm-Initiating Device — Alarm system component that transmits a signal when a change occurs; change may be the result of an action such as the activation of a manual fire alarm box, the presence of products of combustion in the atmosphere, or the automatic activation of a supervisory switch.

Alcohol-Resistant Aqueous Film Forming Foam Concentrate (AR-AFFF) — Aqueous film forming foam that is designed for use with polar solvent fuels. *See* Aqueous Film Forming Foam and Foam Concentrate.

ALI — *See* Automatic Location Identification.

Alkali — Strong base. *See* Acid, Base, Caustic, and pH.

All Clear — (1) Signal that a danger has passed. (2) Signal given to the incident commander that a specific area has been checked for victims and none have been found, or that all found victims have been extricated from an entrapment.

All Hands — Fire service jargon for an emergency incident engaging all companies on the first-alarm assignment; may be followed by multiple alarms.

Allergen — Material that can cause an allergic reaction of the skin or respiratory system. *Also known as* Sensitizer.

Allergic Reaction — Local or general (systemic) reaction to an allergen; usually characterized by hives, tissue swelling, or difficulty breathing.

All-Hazard Concept — Provides a coordinated approach to a wide variety of incidents; all responders use a similar, coordinated approach with a common set of authorities, protections, and resources.

Allied Professional — Individual with the training and expertise to provide competent assistance and direction at haz mat and WMD incidents.

Allocated Resources — Resources dispatched to an incident that have not been checked in with the incident commander.

Alloy — Substance or mixture composed of two or more metals (or a metal and nonmetallic elements) fused together and dissolved into each other to enhance the properties or usefulness of the base metal.

ALPA — See Air Line Pilots Association, International.

Alpha Particle — Energetic, positively charged particles (helium nuclei) emitted from the nucleus during radioactive decay that rapidly lose energy when passing through matter. *See* Alpha Radiation, Beta Particle, and Gamma Rays.

Alpha Radiation — Consists of particles having a large mass and a positive electrical charge; least penetrating of the three common forms of radiation. It is normally not considered dangerous to plants, or to animals or people unless it gets into the body. *See* Beta Radiation, Gamma Radiation, and Radiation (2).

ALS — *See* Advanced Life Support.

ALS Base — An acronym for Attack, Lobby, Staging, and Base. It serves as a guide to covering the initial phase of high-rise fire attack and support functions with personnel who may be quickly assigned to begin suppression and basic support functions.

Alternate Airport — Airport to which an aircraft may proceed if a landing at the intended airport becomes inadvisable.

Alternating Current (AC) Circuit — Electrical circuit in which the current can move through the circuit in both directions and the flow can be constantly reversing.

Alternatives — Possible answers in a multiple-choice test item.

Altitude — Geographic position of a location or object in relation to sea level. The location may be either above, below, or at sea level.

Aluminize — To coat with aluminum.

Aluminum Alloy Ladder — Ladder made of aluminum and other materials, such as magnesium, to make the ladder lightweight but strong.

Alveoli — Air sacs of the lungs; place where oxygen is passed to the blood and carbon dioxide is passed from the blood.

Ambient Conditions — Common, prevailing, and uncontrolled atmospheric weather conditions. The term may refer to the conditions inside or outside of the structure.

Ambient Temperature — Temperature of the surrounding environment.

Ambu-Bag — Trade name for a device that is used to provide the manual ventilation of a patient during cardiopulmonary resuscitation.

Ambulance — Ground vehicle that provides patient transport capability and is equipped with basic or advanced life support equipment and personnel.

American Association of Airport Executives (AAAE) — Professional organization for airport management that offers or co-sponsors emergency response and ARFF related training conferences.

American Board of Criminalistics (ABC) — National peer-review group that certifies forensic scientists in specific disciplines, including fire debris analysis.

American Conference of Governmental Industrial Hygienists® (ACGIH®) — Organization that promotes the free exchange of ideas and experiences and the development of standards and techniques in industrial health. *See* Biological Exposure Indices (BEI®).

American Fire Sprinkler Association (AFSA) — Nonprofit, international association representing open shop fire sprinkler contractors, dedicated to the educational advancement of its members and the promotion of the use of automatic fire sprinkler systems.

American National Standards Institute (ANSI) — Voluntary standards-setting organization that examines and certifies existing standards and creates new standards.

American Paradigm of Fire — A belief held by many people that fire cannot or will not happen to them.

American Society for Testing and Materials (ASTM) — Voluntary standards-setting organization that sets guidelines on characteristics and performance of materials, products, systems and services; for example the quality of concrete or the flammability of interior finishes.

American Society of Mechanical Engineers (ASME) — Voluntary standards-setting organization concerned with the development of technical standards, such as those for respiratory protection cylinders.

American Wire Gauge (AWG) — Measurement unit for the diameter of wire. Larger AWG numbers indicate smaller diameters than smaller AWG numbers; for example, No. 14 AWG wire is smaller than No. 8 AWG wire.

Americans with Disabilities Act (ADA) of 1990 - Public Law 101-336 — Federal statute intended to remove barriers, physical and otherwise, that limit access by individuals with disabilities.

Amidships — Center of a vessel's length, halfway between the bow and the stern.

Ammeter — (1) Instrument for measuring electric current in amperes. (2) Gauge that indicates both the amount of electrical current being drawn from and provided to the vehicle's battery.

Ammonium Nitrate and Fuel Oil (ANFO) — High explosive blasting agent made of common fertilizer mixed with diesel fuel or oil; requires a booster to initiate detonation. *See* Detonation, Explosive (1), and High Explosive.

Ampacity — Current-carrying capacity of conductors or equipment; expressed in amperes.

Amperage — Strength of an electrical current, expressed in amperes.

Ampere — Basic unit of electrical current; amount of current sent by one volt through one ohm of resistance. May be abbreviated either by A or I.

Amphitheater Room Setup — Room arrangement in which the chairs are positioned in a slight semicircle to provide for better eye contact between the educator and the audience and to improve the audience's line of sight to a screen or video monitor. *Also known as* Auditorium-Style Setup.

Amputation — Complete removal of an appendage.

Analysis — Ability to divide information into its most basic components. *See* Cost-Benefit Analysis and Impact Analysis.

Anaphylactic Shock — Shock caused by a severe allergic reaction. *Also known as* Anaphylaxis

Anarchism — Political belief that society should be organized without a coercive, compulsory government.

Anatomy — Structure of the body or the study of body structure.

Anchor — (1) Metal device used to hold down the ends of trusses or heavy timber members at the walls. (2) Reliable or principal support. (3) Something that serves to hold an object firmly. (4) A single object used to secure a rope rescue system. *Also known as* Anchor Points. (5) Heavy device used to hold a ship or boat in position.

Anchor Light — Light a vessel carries when at anchor; must be visible for 2 miles (3.22 km) at night in every direction. Vessels over 150 feet (45.7 m) must carry two lights visible for 3 miles (4.83 km).

Anchor Point — (1) Solid base or point from which pulling or pushing operations can be initiated. (2) Point from which a fire line is begun; usually a natural or man-made barrier that prevents fire spread and the possibility of the crew being "flanked" while constructing the fire line. Examples include lakes, ponds, streams, roads, earlier burns, rockslides, and cliffs.

Anchor System — Total combination or anchor points, slings, and carabiners used to create attachment points for a rope rescue system.

Anchorage — Designated areas, identified on navigational charts, where ships may safely anchor.

Ancillary Components — Supplemental written materials that help students meet the learning objectives; may include information sheets, study guides, skills sheets, work or activity sheets, and assignment sheets.

Ancillary Ladder — Small ladder attached to an elevating platform to be used as an escape route for platform passengers in the event of a mechanical failure of the aerial device.

Andragogy — Educational term referring to the study of adult education and its methods of teaching and learning.

ANFO — *See* Ammonium Nitrate and Fuel Oil.

Angina Pectoris — Spasmodic pain in the chest caused by insufficient blood supply to the heart; aggravated by exercise or tension and relieved by rest or medication.

Angle of Approach — (1) On a vehicle, the smallest angle made between the road surface and a line drawn from the front point of ground contact of the front tire to any projection of the apparatus ahead of the front axle — the front overhang. (2) Angle formed by level ground and a line from the point where the front tires of a vehicle touch the ground to the lowest projection at the front of the apparatus. The angle of approach should be at least 16 degrees.

Angle of Approach/Departure — Relationship described in degrees that is created by an incline from or to a road surface.

Angle of Departure — (1) On a vehicle, the smallest angle made between the road surface and a line drawn from the rear point of ground contact of the rear tire to any projection of the apparatus behind the rear axle — the rear overhang. (2) Angle formed by level ground and a line from the point where the rear tires of a vehicle touch the ground to the lowest projection at the rear of the apparatus. The angle of departure should be at least 8 degrees.

Angle of Inclination — Pitch for portable non-self-supporting ground ladders. The preferred angle of inclination is 75 degrees.

Angle of Loll — Angle at which an imbalanced vessel is leaning and to which the vessel will stabilize. *See* List and Loll.

Angle of Repose — Greatest angle above the horizontal plane at which loose material, such as soil, will lie without sliding.

ANI — *See* Automatic Number Identification.

Annealed — Soft state in metal caused by controlled application of heat and cold.

Annealed Glass — Glass that has slowly cooled during the forming process to relieve internal stresses of the quenching process; commonly found glass that breaks into large pieces and shards when broken.

Annual Leave — Vacation time allowed emergency services personnel per year.

Annunciator Panel — Electrical device used to indicate the source or location of an activated fire alarm initiating device or the status of the system. The panel may include individual lights located on a schematic map and an audible alarm signal.

ANSI — *See* American National Standards Institute.

Antenna — Device connected to a receiver, transmitter, or transceiver that is intended to radiate the transmitted signal and/or to receive a signal.

Anterior — Situated in front of, or in the forward part of. In anatomy, used in reference to the belly surface of the body.

Anthrax — Non-contagious, potentially fatal disease caused by breathing, eating or absorbing through cuts in the skin the bacteria known as *Bacillus anthracis*.

Antibiotic — Antimicrobial agent made from a mold or a bacterium that kills or slows the growth of bacteria; examples include penicillin and streptomycin. Antibiotics are ineffective against viruses.

Antidote — Substance that counteracts the effects of a poison or toxin.

Anti-Electrocution Platform — Slide-out platform mounted beneath the side running board or rear step of an apparatus equipped with an aerial device. This platform is designed to minimize the chance of the driver/operator being electrocuted should the aerial device come in contact with energized electrical wires or equipment.

Antifogging Chemical — Chemical used to prohibit fogging inside the facepiece.

Anti-Lock Braking Systems — An electronic system that monitors wheel spin. When braking and a wheel are sensed to begin locking up, the brake on that wheel is temporarily released to prevent skidding.

Anti-Shim Device — *See* Dead Latch.

Antisubmarine Device — Any device designed to prevent a driver from sliding forward and becoming wedged or trapped beneath the dashboard of a vehicle.

Anxiety — Feeling of apprehension, uncertainty, or fear.

Aorta — Largest artery in the body; originates at the left ventricle of the heart.

Apartment — Subdivision of residential property classification consisting of structures that contain three or more living units equipped with independent bathroom and cooking facilities. *Also known as* Apartment House, Garden Apartment, or Tenement.

Apnea — Cessation of breathing; the absence of respiration.

A-Post — Front post area of a vehicle where the door is connected to the body.

Apparatus — Motor-driven vehicle or vehicles designed for fighting fires; different types include engines, water tenders, and ladder trucks.

Apparatus Bay — Area of the fire station where apparatus are parked. *Also known as* Apparatus Room.

Apparatus Engine — Diesel or gasoline engine that powers the apparatus drive train and associated fire equipment. *Also known as* Power Plant.

Appliance — Generic term applied to any nozzle, wye, siamese, deluge monitor, or other piece of hardware used in conjunction with fire hose for the purpose of delivering water.

Application — (1) Lesson plan component in which the instructor provides opportunities for participants to practice, or to apply cognitive information to skills learned in a lesson. *See* Lesson Plan. (2) The third of the four teaching steps in which students use or apply what the educator has taught; the step in which students practice using new ideas, information, techniques, and skills. *See* Application Step.

Application Rate — Minimum amount of foam solution that must be applied to an unignited fire, spill, or spill fire to either control vapor emission or extinguish the fire; measured per minute per square foot (or square meter) of area to be covered.

Application Step — Third step, in the four-step teaching method of conducting a lesson, in which the learner is given the opportunity to apply what has been learned and to perform under supervision and assistance.

Applicator Pipe — Curved pipe attached to a nozzle for precisely applying water over a burning object.

Applied Learning — Making information relevant so that the proposed learner understands why she or he should receive and process the material.

Apprenticeship — Labor organization professional development program requiring at least three years of fire service experience supplemented with related technical instruction. Apprentices are subject to probationary periods, the length of which is stipulated by local programs. The fire service apprenticeship training program was developed by the International Association of Fire Chiefs and the International Association of Firefighters, and was accepted by the Department of Labor's Bureau of Apprenticeship and Training on July 11, 1975.

Approach Clothing — Special personal protective clothing designed to protect the firefighter from radiant heat while approaching the fire. It typically consists of standard turnout gear with an aluminized outer coating.

Approach Lights — System of lights arranged to assist an airplane pilot in aligning his or her aircraft with the runway for landing.

Approach Sequence — Order in which aircraft are positioned while on approach or while awaiting approach clearance.

Approach-Avoidance — Decision-making problem; refers to an inner conflict within the person in charge that results in an inability to make a decision.

Approved — Acceptable to the authority having jurisdiction.

Apron — Airport area intended to accommodate aircraft for purposes of loading or unloading passengers or cargo, refueling, parking, or maintenance. *Also known as* Ramp.

Aqueous Film Forming Foam (AFFF) — Synthetic foam concentrate that, when combined with water, can form a complete vapor barrier over fuel spills and fires and is a highly effective extinguishing and blanketing agent on hydrocarbon fuels. *See* Alcohol-Resistant AR-AFFF Concentrate, Foam Concentrate, and Foam System.

AR-AFFF — *See* Alcohol-Resistant Aqueous Film Forming Foam.

ARB — *see* Accident Review Board.

Arc — High-temperature luminous electric discharge across a gap or through a medium such as charred insulation.

Arc Mapping — Visual documentation of the path of electrical arcs at a scene.

Arc-Fault Circuit Interrupter (AFCI) — Electronic device, generally part of a circuit breaker, that detects arcing conditions caused by an energized conductor contacting either a neutral conductor or a grounded object.

Arch — Curved structural member in which the interior stresses are primarily compressive. Arches develop inclined reactions at their supports.

Arched Roof — Curved or arch shaped roof resembling the top half of a horizontal cylinder; typically found on supermarkets, auditoriums, bowling centers, sports arenas, and aircraft hangars.

Area Ignition — Simultaneous or nearly simultaneous ignition of several individual wildland fires that are spaced in such a way as to add to and influence the main body of the fire and each other in a way that produces a hot, fast-moving fire or blowup throughout the area. Area ignition is a cause of blowup and great fire spread.

Area of Origin — The general location (room or area) where the ignition source and the material first ignited actually came together for the first time.

Area of Refuge — (1) Space where a victim may find safety from hazards at an incident. *Also known as* Sheltered Areas. (2) Space protected from fire in the normal means of egress either by an approved sprinkler system, separation from other spaces within the same building by smokeproof walls, or location in an adjacent building. (3) Two-hour-rated building compartment containing one elevator to the ground floor and at least one enclosed exit stairway. (4) Area where persons who are unable to use stairs can temporarily wait for instructions or assistance during an emergency building evacuation. (5) In wildland fire fighting, a safe area.

Area Separation Wall — Wall that completely separates a building's compartments, dividing the building into distinct areas.

ARFF — *See* Aircraft Rescue and Fire Fighting.

ARFF Firefighter — *See* Airport Firefighter.

ARFFWG — *See* Aircraft Rescue and Fire Fighting Working Group.

Armormax® — A combination of numerous synthetic fibers used to form an opaque composite ballistic armor.

Around-the-Pump Proportioner — Apparatus-mounted foam proportioner in which a small quantity of water is diverted from the apparatus pump through an inline proportioner; there it picks up the foam concentrate and carries it to the intake side of the pump. It is the most common apparatus-mounted foam proportioner in service. *See* Foam Proportioner and Proportioning.

Arrest — (1) Sudden cessation or stoppage. (2) Restricting a person's movement or freedom, usually in a legal or law enforcement action.

Arresting System — Device used to engage an aircraft and absorb forward momentum in case of an aborted takeoff or landing.

Arrhythmia — Any disturbance in the rhythm of the heart.

Arrow Pattern — *See* Pointer Pattern.

Arson — Crime of willfully, maliciously, and intentionally starting an incendiary fire or causing an explosion to destroy one's property or the property of another. Precise legal definitions vary among jurisdictions, wherein it is defined by statutes and judicial decisions. *See* Firesetting.

Arson Hotline — Telephone line and operation set up for the purpose of receiving information, often given anonymously, on arson crimes.

Arson Immunity Law — Law stating that insurance companies must release all information and documentation when requested to a public entity when there is reason to suspect a fire under investigation was intentionally set. The released information is considered confidential until it is used in court proceedings, and the insurance company that released the information is granted immunity under the law for breaking any confidentiality requirements it may have to its clients.

Arson Investigator — Public sector fire investigator who primarily investigates intentionally set fires; may be tasked with locating and arresting arsonists and interviewing suspects. May also have limited law enforcement powers, such as the power of arrest and authorization to carry a weapon.

Arson Kit — Kit containing equipment used to detect, collect, protect, and preserve evidence of arson and to aid in determining the cause of a fire.

Arson Module — The Arson Module consists of two parts: a local investigation module that permits a fire department or arson investigation unit to document certain details concerning the incident; and a juvenile firesetter section that identifies key items of information that could be used for local, state, and national intervention programs.

Arson Strike Force — Special purpose, short-term mobilization of a team or teams of investigators, together with allied resources, that applies high intensity investigative efforts to a major arson incident or series of incidents.

Arson Task Force — Legal or quasi-legal bodies or private advisory committees established to set policy and implement new programs based upon information gathered about local arson activity. Arson task forces are not investigative units.

Arsonist — Person who commits an act of arson.

Arteriosclerosis — Generic name for several conditions that cause the walls of the arteries to become thickened, hard, and inelastic.

Artery — Blood vessel that carries oxygen-rich blood away from the heart.

Articulated Transit Bus — Passenger-carrying bus constructed with two sections, a tractor and a trailer, which are connected by a pivoting joint.

Articulating Aerial Platform — Aerial device that consists of two or more booms that are attached with hinges and operate in a folding manner. A passenger-carrying platform is attached to the working end of the device.

Articulating Boom — Arm portion or structural support member of an aerial device consisting of two or more sections that are hinged and rotate in a vertical plane in a folding manner.

Articulation — (1) The action or manner of jointing or interrelating. (2) A joint or juncture between bones or cartilages.

Artifacts — Remains of materials involved in the fire that are in some way related to ignition, development, or spread of the fire or explosion.

Artificial Respiration — Movement of air into and out of the lungs by artificial means. *Also known as* Artificial Resuscitation, Pulmonary Resuscitation, and Rescue Breathing.

Artificial Resuscitation — *See* Artificial Respiration.

Asbestos — Fibrous carcinogenic substance (noncombustible magnesium silicate minerals) used for fireproofing, brake linings, roofing compositions, and other purposes such as insulation and ceiling materials in older buildings. Inhaled asbestos fibers travel to the lungs, causing scarring, reduced lung capacity, and cancer.

Ascender — Mechanical contrivance, used when climbing rope, that allows upward but not downward movement.

Ascending Device — Mechanical contrivance that allows upward but not downward movement when climbing rope.

ASCLD — *See* Association of Crime Laboratory Directors.

Ash — Powdery residue left when organic material is burned completely or is oxidized by chemical means.

Ashore — Leaving a vessel and stepping on land; opposite of aboard.

ASME — *See* American Society of Mechanical Engineers.

Aspect — (1) Position facing a particular direction; exposure. (2) Compass direction toward which a slope faces.

Asphyxia — Suffocation.

Asphyxiant — Any substance that prevents oxygen from combining in sufficient quantities with the blood or from being used by body tissues. *See* Chemical Asphyxiant and Simple Asphyxiant.

Asphyxiation — Fatal condition caused by severe oxygen deficiency and an excess of carbon monoxide and/or other gases in the blood.

Aspirate — To inhale foreign material into the lungs.

Aspiration — Adding air to a foam solution as the solution is discharged from a nozzle. *Also known as* Aeration.

Aspirator — Suction device for removing undesirable material from a patient's throat.

Assay — To analyze or estimate.

Assembly — (1) All component or manufactured parts necessary for and fitted together to form a complete machine, structure, unit or system. (2) Occupancy classification of buildings, structures, or compartments (rooms) that are used for the gathering of 50 or more persons. *See* Occupancy Classification.

Assembly Area — Area designated in the employee emergency action plan in which employees displaced by an evacuation are to assemble.

Assessment — Process used to find out the knowledge, skills, and abilities possessed by a learner; can be accomplished by observation or by special assessment activities such as quizzes and tests.

Assessment Instrument — A tool to help the frontline JFIS identify why a juvenile is setting fires and his or her risk of repeat behaviors. An assessment instrument should be validated by mental health clinicians and use a scoring process to determine levels of risk for repeat fire-setting behaviors.

Assessment Stop — Distant location at which first responders can safely stop and evaluate the situation, complete donning their protective clothing and SCBA, and report conditions to the telecommunications center.

Assigned Resources — Resources on an incident that have been checked in and assigned an objective.

Assignment — Work that must be performed by learners outside class in order to reach a skill level, meet an objective, and/or prepare for the next lesson.

Assistant Incident Safety Officer (AISO) — Individual(s) who reports to the Incident Safety Officer and assist with monitoring hazards and safe operations for designated portions of the operation at large or complex incidents.

Assisting Agency — Agency directly contributing suppression, rescue, support, or service resources to another agency.

Association of Crime Laboratory Directors (ASCLD) — Association that provides criteria used to judge whether a facility and its practices provide an atmosphere conducive to the quality of work necessary in the forensic field.

Association of Flight Attendants-CWA (AFA-CWA) — The world's largest labor union for flight attendants, representing over 50,000 flight attendants at 22 airlines.

Asthma — Respiratory condition marked by attacks of labored breathing, wheezing, a sense of constriction in the chest, and coughing or gasping.

ASTM — *See* American Society for Testing and Materials.

Astragal — Molding that covers the narrow opening between adjacent double doors in the closed position.

ATF — *See* Bureau of Alcohol, Tobacco, Firearms and Explosives.

Atherosclerosis — Common form of arteriosclerosis characterized by fat deposits in the walls of the arteries.

Athwartship — Direction from side to side; to move across a vessel is to move athwartships.

Atmosphere — Area within the confined space where dust, vapors, mists, or other hazardous materials may exist.

Atmospheric Ceiling — Level in the atmosphere at which a heated column ceases to rise.

Atmospheric Displacement — System or method of applying water fog in a superheated area, causing the water to be converted into steam that expands and displaces the atmosphere in a burning room or building.

Atmospheric Pressure — Force exerted by the atmosphere at the surface of the earth due to the weight of air. Atmospheric pressure at sea level is about 14.7 psi (101 kPa) and is measured as 760 mm of mercury on a barometer. Atmospheric pressure increases as elevation is decreased below sea level and decreases as elevation increases above sea level.

Atmospheric Stability — Degree to which vertical motion in the atmosphere is enhanced or suppressed. Vertical motion and smoke dispersion are enhanced in an unstable atmosphere. Stability suppresses vertical motion and limits smoke dispersion. *See* Inversion (1).

Atmospheric Storage Tank — Class of fixed facility storage tanks. Pressures range from 0 to 0.5 psi (0 to 3.4 kPa) {0 to 0.03 bar}. *Also known as* Nonpressure Storage Tank. *See* Cone Roof Storage Tank, External Floating Roof Tank, Floating Roof Storage Tank, Horizontal Storage Tank, Internal Floating Roof Tank, Lifter Roof Storage Tank, Low-Pressure Storage Tank, and Pressure Storage Tank.

Atmospheric Temperature — Measure of the warmth or coldness of the air.

Atomic Number — Number of protons in an atom.

Atomic Weight — Physical characteristic relating to the mass of molecules and atoms. A relative scale for atomic weights has been adopted, in which the atomic weight of carbon has been set at 12, although its true atomic weight is 12.01115.

Atrium — (1) Upper chamber of the left or right side of the heart. (2) Open area in the center of a building, extending through two or more stories, similar to a courtyard but usually covered by a skylight, to allow natural light and ventilation to interior rooms.

ATSDR — *See* Agency for Toxic Substances and Disease Registry.

Attack — (1) To set upon forcefully. (2) Any action to control fire. (3) In ICS/IMS, used to describe the units attacking the fire.

Attack Hose — (1) Hose between the attack pumper and the nozzle(s); also, any hose used in a handline to control and extinguish fire. Minimum size is 1½ inch (38 mm). (2) Hose that is used by trained firefighters to combat fires.

Attack Line — (1) Hoseline connected to a pump discharge of a fire apparatus ready for use in attacking a fire; may or may not be preconnected. In contrast, supply lines are connected to a water supply with a pump. (2) Fire streams used to attack, contain, or prevent the spread of a fire.

Attack Methods — (1) Tactics for interior fire-suppression operations, including direct, indirect, and combination attacks. (2) Tactics for wildland fire-suppression operations, including aerial, direct, flank, frontal, indirect, mobile, pincer, and tandem attacks.

Attack Pumper — (1) Pumper that is positioned at the fire scene and is directly supplying attack lines. (2) Light truck equipped with a small pump and water tank. *Also known as* Midi-pumper or Mini-pumper.

Attention Seeking Firesetter — Starts a fire in an attempt to bring attention to a stressful life situation such as depression, anger, or abuse.

Attic — Concealed and often unfinished space between the ceiling of the top floor and the roof of a building. *Also known as* Cockloft or Interstitial Space.

Attic Fold — Method of folding a salvage cover that aids in spreading within the tight confines of an attic, where lateral movement is difficult.

Attic Ladder — Term commonly used for a folding ladder or combination ladder that is especially useful for inside work, and is used to access an attic through a scuttle or similar restricted opening. Attic ladders generally come in 6- to 14-foot (1.8-4.2 m) lengths.

Attractive Nuisance — A location or feature that is likely to require technical rescue if appropriate caution is not exercised by children or adults who are drawn to the area.

Atypical Stressful Event — term used in National Fire Protection Association® (NFPA®) standards to describe incidents that have a likelihood of causing critical incident stress.

Audible Alarm — Bell, whistle, or other sound-producing alerting device attached to a self-contained breathing apparatus, personal alert safety system, or fixed fire protection system.

Audience — Person or persons receiving a message. *Also known as* Receiver.

Audiometric Test — Examinations used to determine the extent of temporary or permanent shifts in thresholds of hearing acuity. These tests make it possible to grade occupational noise exposures and, when necessary, recommend appropriate hearing conservation procedures.

Audiovisual Materials — Instructional materials that can be heard as well as seen.

Auditorium Raise — Method of extending a ladder perpendicularly and holding it in place from four opposite points of the compass by four guy ropes attached to the top of the ladder. *Also known as* Church Raise or Steeple Raise.

Auditorium-Style Room Setup — *See* Amphitheater Room Setup.

Auger — (1) Screwlike shaft that is turned to move grain or other commodities through a farm implement. (2) A drilling device that uses screw threads to dig deep into soil or other dense surfaces. Can be used as anchors in a rope rescue system.

Auger Wagon — Large wagon containing a power take-off-driven auger for unloading purposes. Widely used in agriculture to transport and unload grain, silage, loose forage, stover, and other loose materials.

Authority — Relates to the empowered duties of an official to perform certain tasks. In the case of a fire inspector, the level of an inspector's authority is commensurate with the enforcement obligations of the governing body.

Authority Having Jurisdiction (AHJ) — An organization, office, or individual responsible for enforcing the requirements of a code or standard, or approving equipment, materials, an installation, or a procedure.

Autoclave — A device that uses high-pressure steam to sterilize objects.

Autocratic Leadership — Leadership style in which the leader makes decisions independently of others, informing them only after the decision has been made.

Autoexposure — *See* Lapping.

Autoignition — Initiation of combustion by heat but without a spark or flame. (NFPA® 921)

Autoignition Temperature — The lowest temperature at which a combustible material ignites in air without a spark or flame. (NFPA® 921)

Autoinjector — Spring-loaded syringe filled with a single dose of a life-saving drug.

Automated External Defibrillator (AED) — Cardiac defibrillator designed for layperson use that analyzes the cardiac rhythm and determines if defibrillation is warranted.

Automatic Aid — Written agreement between two or more agencies to automatically dispatch predetermined resources to any fire or other emergency reported in the geographic area covered by the agreement. These areas are generally located near jurisdictional boundaries or in jurisdictional "islands."

Automatic Alarm — (1) Alarm actuated by heat, gas, smoke, flame-sensing devices, or waterflow in a sprinkler system; the alarm is then conveyed to local alarm bells or the fire station. (2) Alarm box that automatically transmits a coded signal to the fire station, telecommunications center, or alarm company to give the location of the alarm box.

Automatic Closing Door — Self-closing door normally held in the open position by an automatic releasing device such as a magnetic hold-open device. When the door is released by the hold-open device, it closes.

Automatic Fire Detection Systems — Heat, gas, and flame detectors used in nonresidential buildings.

Automatic Hydrant Valve — Valve that opens automatically when connected to a hydrant, to allow water to flow into the supply line.

Automatic Location Identification (ALI) — Enhanced 9-1-1 feature that displays the address of the party calling 9-1-1 onscreen for use by the public safety telecommunicator; usually used in tandem with automatic number identification (ANI) services. This feature is also used to route calls to the appropriate public safety answering point (PSAP) and can store information in its database regarding all emergency services (police, fire, and medical) that respond to that address.

Automatic Nozzle — Fog stream nozzle that automatically corrects itself to provide a good stream at the proper nozzle pressure.

Automatic Number Identification (ANI) — Enhanced 9-1-1 feature that displays the phone number of the party calling 9-1-1 onscreen for use by the public safety telecommunicator; usually used in tandem with automatic location identification (ALI) services.

Automatic Oscillating Foam Monitor — Large-capacity foam system that is designed to operate automatically when a fire-detection system activates; may be found in aircraft hangars, tank farms, and loading racks. *See* Foam Monitor, Manual Foam Monitor, and Remote-Controlled Foam Monitor.

Automatic Sprinkler Kit — Kit containing the tools and equipment required to close an open sprinkler.

Automatic Sprinkler System — System of water pipes, discharge nozzles, and control valves designed to activate during fires by automatically discharging enough water to control or extinguish a fire. *Also known as* Sprinkler System. *See* Riser and Sprinkler.

Automatic Suppression Systems — Fire suppression systems that sense heat, smoke, or gas, and activate automatically. These include sprinkler, standpipe, carbon dioxide, and halogenated systems, as well as fire pumps, dry chemical agents and their systems, foam extinguishers, and combustible metal agents.

Automatic Vehicle Locator (AVL) — Computer system onboard an apparatus that uses GPS coordinates to deliver and log real-time location and driving information (speed, direction of travel) about the apparatus over wireless networks.

Autonomic Nervous System — Part of the nervous system concerned with the regulation of body functions not controlled by conscious thought.

Autorotation — Flight condition in which the lifting rotor of a rotary wing aircraft is driven entirely by action of the air when in flight, or, as in the case of a helicopter, after an engine failure.

Auxiliary — (1) Additional fire fighting equipment or staffing that are not part of the regular complement assigned to the fire service. (2) A group organized to assist the fire department.

Auxiliary Alarm System — System that connects the protected property with the fire department alarm communications center by a municipal master fire alarm box or over a dedicated telephone line.

Auxiliary Deadbolt — Deadbolt bored lock. *Also known as* Tubular Deadbolt.

Auxiliary Hydraulic Pump — Electrically operated, positive displacement pump used to supply hydraulic oil through the hydraulic system of an aerial device in the event that the main hydraulic pump fails.

Auxiliary Lock — Lock added to a door to increase security.

Auxiliary Power Unit (APU) — (1) Power unit installed in most large aircraft to provide electrical power and pneumatics for ground power, air conditioning, engine start, and backup power in flight. (2) Mobile units that are moved from one aircraft to another to provide a power boost during engine startup.

Available Fire Flow — Actual amount of water available from a given hydrant; determined by testing.

Available Resources — Resources not assigned to an incident and available for an assignment.

Average Daily Consumption — Average of the total amount of water used each day during a one-year period.

Avoidance — (1) Socio-psychological concept that shows that people feel they can protect themselves psychologically by denying unpleasant situations; thus, during the first moments of a fire, people tend to search for other, safer explanations for the cues they see, smell, and hear. (2) Effect of a decision-making problem. The person in charge might try to avoid being in charge, especially when faced with an approach-avoidance dilemma.

Avulsion — Forcible separation or detachment; the tearing away of a body part.

Awareness Campaign — A series of marketing strategies designed to inform a community about its local juvenile firesetting problem and potential solutions.

Awareness Level — Lowest level of training established by the National Fire Protection Association® for personnel at hazardous materials incidents. *See* Operations Level.

Awareness Materials — Fire and life safety teaching materials that attempt to make the audience more aware of a problem or situation. *Also known as* Promotional Materials.

AWG — *See* American Wire Gauge.

Awning Window — Type of swinging window that is hinged at the top and swings outward, often having two or more sections.

Axe — Forcible entry tool that has a pick or flat head and a blade attached to a wood or fiberglass handle. *Also known as* Firefighter's Axe.

Axial Load — Load applied to the center of the cross section of a member and perpendicular to that cross section. It can be either tensile or compressive and creates uniform stresses across the cross section of the material.

Heavy (heavy) aircraft are capable of takeoff weight of 300,000 pounds (136 078 kg) or more, whether or not the aircraft is operating at this weight during a particular phase of flight.

Large (medium) aircraft are capable of takeoff weight of more than 12,500 pounds (5 670 kg), maximum certified takeoff weight up to 300,000 pounds (136 078 kg).

Note: This category includes flipcharts, mark-and-wipe boards, and chalkboards, even though there is no audio element involved.

Small (light) aircraft are capable of takeoff weight of 12,500 pounds (5 670 kg) or less.

B

BA — Short for breathing apparatus. *See* Self-Contained Breathing Apparatus (SCBA).

Baby Bangor — Short-length tapered-truss wood ladder. *Also known as* Attic Ladder.

Back Burn — Process of burning vegetation in advance of an oncoming wildland fire in order to establish a firebreak that will stop the spread of the fire.

Back Flushing — Cleaning a fire pump or piping by flowing water through it in the opposite direction of normal flow.

Back Pressure — Pressure loss or gain created by changes in elevation between the nozzle and pump.

Backdraft — Instantaneous explosion or rapid burning of superheated gases that occurs when oxygen is introduced into an oxygen-depleted confined space. The stalled combustion resumes with explosive force; may occur because of inadequate or improper ventilation procedures. *See* Flashover and Rollover.

Backfill — Coarse dirt or other material used to build up the ground level around foundation walls, in order to provide a slope for drainage away from the foundation.

Backfire — Fire set along the inner edge of a control line to consume the fuel in the path of a wildland fire and/or change the direction of force of the fire's convection column.

Backfiring — (1) Technique used in the indirect attack method for wildland fires; involves intentionally setting a fire between the control line and the advancing fire to deprive the fire of fuel. The intent is for the backfire to meet the advancing fire some distance from the control line. Backfiring can be dangerous and is illegal in some places. (2) Leakage of the fuel-air explosion past the valves in a gasoline engine due to worn valves or bad valve timing.

Backflow Preventer — A check valve that prevents water from flowing back into a system and contaminating it.

Backing Fire — Fire spreading (or ignited to spread) into (against) the wind or downslope. A fire spreading on level ground in the absence of wind is a backing fire.

Backing Material — Material used to take up space or fill gaps behind shoring system parts.

Backlash — (1) Sudden violent backward movement or reaction. (2) Reverse bouncing motion that occurs when the motion of an aerial device is abruptly halted.

Backpack — (1) Tank-type extinguisher carried on the firefighter's back by straps. The unit has a pump built into the nozzle, and is used extensively to fight wildland fires. (2) Pack used to carry hose on firefighters' backs. (3) Assembly that holds the air cylinder and regulator of the self-contained breathing apparatus to the wearer. *Also known as* Backplate.

Backplate — *See* Backpack (3).

Backsplash — Vertical surface at the back of a countertop.

Backstay — Line made of rope or wire supporting a mast (vertical pole); extends from the top of the mast to the stern.

Backup Knot — A second knot tied in the short tail of a knot or bend to ensure that the primary knot or bend cannot come undone by itself during use.

Backwards — Slang for making a reverse lay; as in "Lay a backwards."

Bacteria — Microscopic, single-celled organisms. *See* Rickettsia and Virus.

Badge — Indicator of rank worn on a firefighter's or an officer's uniform.

Baffle — (1) Intermediate partial bulkhead that reduces the surge effect in a partially loaded liquid tank. (2) Divider used to separate beds of hose into two or more compartments. (3) Device to deflect, check, or regulate flow. (4) Partition placed in vehicular or aircraft water tanks to reduce shifting of the water load when starting, stopping, or turning.

Bagging — For salvage and overhaul purposes it is the spreading of a salvage cover to catch water from above, particularly in an attic.

Bag-Valve Mask — Portable artificial ventilation unit consisting of a face mask, a one-way valve, and an inflatable bag; can be used on a nonbreathing or breathing patient.

Balanced Pressure Proportioner — Foam concentrate proportioner that operates in tandem with a fire water pump to ensure a proper foam concentrate-to-water mixture.

Bale Hook — Tool used for moving bales or boxed goods, and for moving and overhauling stuffed furniture or mattresses. *Also known as* Baling Hook or Hay Hook.

Ball Valve — Valve having a ball-shaped internal component with a hole through its center that permits water to flow through when aligned with the waterway.

Ballast — Additional weight placed low in the vessel's hull to improve its stability; may be steel, concrete, or water. *See* Ballasting, Ballast Tank, and Trim.

Ballast Tank — Watertight compartment that holds liquid ballast. *See* Trimming Tank.

Ballasting — Process of filling empty tanks with seawater to increase a vessel's stability. *See* Ballast, Ballast Tank, and Trim.

Ballistics — Science of projectiles, their motion, and their effects.

Balloon Throw — Method of spreading a salvage cover; uses air trapped under the cover to float it into place over the materials to be protected.

Balloon-Frame Construction — Type of structural framing used in some single-story and multistory wood frame buildings; studs are continuous from the foundation to the roof, and there may be no fire stops between the studs.

Baluster — Vertical member supporting a handrail.

Balustrade — Entire assembly of a handrail including its supporting members (newel posts and balusters).

Band — Range of frequencies defined between two definite limits.

Bandage — Material used to hold a dressing in place.

Banding Method — Means of attaching a coupling to a fire hose using tightly wound strands of narrow-gauge wire or steel bands.

Bang Out — Slang for "to dispatch" or "to be dispatched."

Bangor Ladder — *See* Pole Ladder.

Bank-Down Application Method — Method of foam application that may be employed on an ignited or un-ignited Class B fuel spill. The foam stream is directed at a vertical surface or object that is next to or within the spill area; foam deflects off the surface or object and flows down onto the surface of the spill to form a foam blanket. *Also known as* Deflection. *See* Rain-Down Application Method and Roll-On Application Method.

Bar Joist — Open web truss constructed entirely of steel, with steel bars used as the web members.

Barge — Long, large vessel used for transporting goods on inland waterways; usually flat-bottomed, self-propelled, or towed or pushed by another vessel. *See* Cargo Vessel and Lighter.

Barometer — Instrument used to measure atmospheric pressure.

Barrel — Measure of liquid volume used in the marine industry; for petroleum, 1 barrel = 42 U.S. gallons (159 liters).

Barrel Strainer — Cylindrical strainer that is attached to a hard suction hose to prevent the induction of foreign debris into the pump during drafting operations.

Barrier — Any obstruction of the spread of fire; typically an area devoid of combustible fuel.

Bar-Screw Jack — Jack used to hold loads under compression; commonly used in shoring work or other similar evolutions.

Basal Skull Fracture — Fracture involving the base of the cranium.

Base — (1) Bottom of something; a foundation or support. (2) Location at which the primary Incident Management Logistics functions are coordinated and administered; the incident command post may be co-located with the base. There is only one base per incident. (3) At a high-rise fire, the location where reserve companies are staged until needed. *Also known as* Level II Staging. (4) Lowest or widest section of a non-self-supporting extension ladder. Also, the bottom end of any non-self-supporting ground ladder. (5) Any alkaline or caustic substance; corrosive water-soluble compound or substance containing group-forming hydroxide ions in water solution that reacts with an acid to form a salt. *See* Acid, Alkali, Caustic, Corrosive, and pH.

Base Area — At a high-rise fire, the location where reserve companies are staged until needed.

Base Leg — Flight path at right angles to the landing runway off the approach end.

Base Rails — Lower chords of the aerial ladder to which the rungs, trusses, and other portions of the ladder are attached. *Also known as* Beams.

Base Section — *See* Bed Section.

Base Station Radio — Fixed, nonmobile radio at a central location.

Baseline Data — Data and statistics gathered before an education program starts; educators compare baseline data with data collected after the program has concluded in order to determine educational gain.

Baseline Data — Data collected before a lesson or program is implemented.

Baseline Knowledge — What a person knows about a topic before a presentation is conducted.

Basement Plans — Drawings showing the below-ground view of a building. The thickness and external dimensions of the basement walls are given, as are floor joist locations, strip footings, and other attached foundations.

Basic Life Support (BLS) — Emergency medical treatment administered without the use of adjunctive equipment; includes maintenance of airway, breathing, and circulation, as well as basic bandaging and splinting.

Basket Stabilizer — Device used to support the basket portion of an elevating platform device in the stowed position during road travel.

Batch Mixing — Production of foam solution by adding an appropriate amount of foam concentrate to a water tank before application; the resulting solution must be used or discarded following the incident. *See* Premixing.

Batt Insulation — Blanket insulation cut in widths to fit between studs, and in short lengths to facilitate handling.

Battalion — Fire department organizational subdivision consisting of several fire service companies in a designated geographic area. A battalion is usually the first organizational level above individual companies or stations. *Also known as* District.

Battalion Chief — Chief officer assigned to command a fire department battalion. *Also known as* District Chief.

Batten — (1) Thin iron bar used to hold down the coverings of hatches on merchant vessels. (2) Strip of wood used to keep cargo away from the hull of a vessel or to prevent it from shifting.

Batten Door — *See* Ledge Door.

Battering — Act of creating an opening in a building component by striking and breaking it with a tool, such as a sledge or ram.

Battering Ram — Solid steel bar with handles and guards, a fork on one end, and a blunt end on the other, used to break down doors or create holes in walls. The tool weighs 30 to 40 pounds (13.6 to 18.1 kg) and can be operated by one or more firefighters.

Battery — (1) Unlawful, intentional and unauthorized touching of or application of force to a person without his or her consent. (2) Number of similar articles, items, or devices arranged, connected, or used together; for example, a grouping of artillery pieces for tactical purposes. (3) The electrical power supply for a vehicle, handlights, or other electrically powered device.

Battery Bank — Group of vehicle batteries clustered in one location.

Battle's Sign — Purplish discoloration above the bone behind the ear indicating a skull fracture.

Bay — Compartment or section in a fire station where the fire apparatus is parked.

BDA — *See* Bidirectional Amplifier.

Beam — (1) Structural member subjected to loads, usually vertical loads, perpendicular to its length. (2) Main structural member of a ladder supporting the rungs or rung blocks. *Also known as* Rail or Side Rail. (3) Width of a vessel measured at the widest point.

Beam Block — *See* Truss Block.

Beam Bolts — Bolts that pass through both rails at the truss block of a wooden ladder to tie the two truss rails together.

Beam Raise — Raising a ladder to the vertical position with only one beam in contact with the ground, instead of with both beams on the ground as with a flat raise.

Bearing Wall — *See* Load-Bearing Wall.

Bearing Wall Structures — Common type of structure that uses the walls of a building to support spanning elements, such as beams, trusses, and pre-cast concrete slabs.

Becket Bend — Knot used for joining two ropes; particularly well suited for joining ropes of unequal diameters or joining a rope and a chain. *Also known as* Sheet Bend.

Becquerel (Bq) — International System unit of measurement for radioactivity, indicating the number of nuclear decays/disintegrations a radioactive material undergoes in a certain period of time. *See* Curie (Ci), Radiation (2), and Radioactive Material (RAM).

Bed Ladder — Lowest section of a multi-section ladder.

Bed Ladder Pipe — Non-telescoping section of pipe, usually 3 or 3½ inches (76 mm or 89 mm) in diameter, attached to the underside of the bed section of the aerial ladder for the purpose of deploying an elevated master stream.

Bed Section — Bottom section of an extension ladder. *Also known as* Base Section.

Bedded Position — Extension ladder with the fly section(s) fully retracted.

Behavior — In psychology, any response or reaction to a stimulus, such as instruction.

Behavior Change — Change in a person's actions because of an increase in knowledge.

Behavioral Objective — Measurable and precise statement of intent that specifically describes behavior that the learner is expected to exhibit as a result of instruction. Behavioral objectives also indicate the conditions under which the behavior is to be performed (given certain equipment, a specific time period for completion, etc.) and the required standard of performance — a percentage (75 percent), a number (9 out of 10), a time constraint (within 1 minute), or an NFPA® (or OSHA, SOP, etc.) requirement. The term is often interchangeably used with the terms educational objective, outcome objective, and enabling objective. *See* Enabling Objective, Objective (1), and Performance Objective.

BEI® — *See* Biological Exposure Indices.

Belay — Climber's term for a safety line.

Below Minimum — Weather conditions below the minimum prescribed by regulation for a particular operation, such as the takeoff or landing of aircraft.

Below-Grade Hazards — Includes open pits creating flooding, or contaminated atmospheres.

Below-Grade Operation — (1) Rescue activity that occurs at the bottom or slope of an open pit or trench, for example, freeing an equipment operator from an overturned backhoe that lies on the slope of an excavation. (2) Per OSHA 1926.650, any open operation such as that in an open trench for footings and foundations.

Belowground Operation — (1) Activity that occurs below the surface of the earth, for example, rescue and search operations in a mine or mine shaft. (2) Per OSHA 1926.800, operation with earth cover such as mining and tunneling. *Also known as* Underground Operation.

Belt System — *See* Loop System.

Belt Weather Kit — Belt-mounted case with pockets fitted for anemometer, compass, sling psychrometer, slide rule, water bottle, pencils, and book of weather report forms.

Bench Trial — Court proceeding in which the judge alone acts as the trier of fact.

Benchmark — (1) Permanently affixed mark such as a stake driven into the ground that establishes the exact elevation of a place; used by surveyors in measuring site elevations or as a starting point for surveys. (2) Anything that serves as a standard against which others may be measured.

Bend — Two rope ends tied together.

Bent — Supporting legs of a bridge in a plane perpendicular to its length.

Benzene (C_6H_6) — Highly toxic carcinogen produced by the burning of PVC plastics or gasoline. Inhalation of high levels can cause unconsciousness and death from respiratory paralysis.

Berm — Outside or downhill side of a ditch or trench; a mound or wall of earth.

Bernoulli's Equation — Mathematical expression of the principle of conservation of energy applied to hydraulics.

Berth — Mooring or docking a vessel alongside a pier, wharf, or bulkhead. *See* Berthing Area and Mooring.

Berthing Area — Space at a wharf or pier for docking a vessel; place where a vessel comes to rest.

Berthing Space — Bed or bunk space on a vessel.

Beta Particle — Particle that is about 1/7,000th the size of an alpha particle but has more penetrating power. The beta particle has a negative electrical charge. *See* Alpha Particle, Beta Radiation, and Gamma Rays.

Beta Radiation — Type of radiation that can cause skin burns. *See* Alpha Radiation, Gamma Radiation, and Radiation (2).

Bias — Highly personal or unreasoned distortion of judgment; prejudice.

Bid Bond — Deposit provided to the apparatus purchaser from the manufacturer in order to ensure that the manufacturer will take the bid made if offered. This prevents damages to the purchaser should the bidder default on any part of the deal. The bond is returned to the manufacturer when the contract is executed.

Bidirectional Amplifier (BDA) — device similar to a cellular broadcast tower that uses a reception antenna, a signal amplifier, and an internal rebroadcast antenna to boost cellular phone reception within one building.

Bifold Doors — Doors designed to fold in half vertically.

Big Line — Slang for a hoseline of at least 2-inch (51 mm) diameter, especially when used as a handline.

Big Stick — Slang for a mechanically raised main ladder or an aerial ladder truck. Originally, aerial ladders were made of wood, hence the term "big stick."

Bight — Element of a knot formed by simply bending the rope back on itself (creating a u-shaped loop) while keeping the sides parallel.

Bile — Fluid secreted by the liver that is concentrated and stored in the gall bladder.

Bilge — Lowest inner part of a vessel's hull; flat part of the bottom of a vessel.

Bilge Pump — Small pump, located in the bilge, used to remove internal water.

Bill of Lading — Shipping paper used by the trucking industry (and others) indicating origin, destination, route, and product; placed in the cab of every truck tractor. This document establishes the terms of a contract between a shipper and a carrier. It serves as a document of title, contract of carriage, and receipt for goods. *See* Lading and Shipping Papers.

Billy Pugh Net — Rope net or basket designed to be suspended beneath a helicopter for transporting personnel and/or equipment.

Bimetallic — Strip or disk composed of two different metals that are bonded together; used in heat-detection equipment.

Biochemical — Involving chemical reactions in living organisms.

Biodegradable — Capable of being broken down into innocuous products by the actions of living things, such as microorganisms.

Biological Agent — Viruses, bacteria, or their toxins which are harmful to people, animals, or crops. When used deliberately to cause harm, may be referred to as a Biological Weapon.

Biological Attack — Intentional release of viruses, bacteria, or their toxins for the purpose of harming or killing citizens. *See* Terrorism and Weapons of Mass Destruction.

Biological Death — Condition present when irreversible brain damage has occurred, usually 4 to 10 minutes after cardiac arrest.

Biological Exposure Indices (BEI®) — Guidance value recommended for assessing biological monitoring results that is established by the American Conference of Governmental Industrial Hygienists® (ACGIH®).

Biological Toxin — Poison produced by living organisms. *See* Poison and Toxin.

Biological Weapon — *See* Biological Agent.

Bird Box — *See* Connection Box.

Bird Cage Construction — *See* Unibody Construction.

Bitter End — *See* Working End.

Bitts — Single or twin set of upright wood or steel posts located on deck along the sides of a vessel; used for securing mooring lines. *See* Bollard.

Bituminous Material — Refers to materials that contain tar or asphalt-like materials. Bituminous materials used in roofing include tar and tar-impregnated papers.

Black — Area already burned by a wildland fire. *Also known as* Burn.

Blacken — To "knock down" a fire; to reduce a fire by extinguishing all visible flame. As the flame is extinguished, the fire is said to be blackened.

Blacklining — Ensuring that there are no unburned fuels adjacent to the control line by burning out such areas; burning out adjacent to a control line in order to widen and strengthen it.

Bladder-Tank Balanced-Pressure Proportioner — Type of mobile or fixed foam system that uses water to displace foam concentrate in a storage tank and force it into a proportioning system. *See* Foam Proportioner and Proportioning.

Blanch — To become white or pale.

Blank — Thin piece of metal inserted between flanges in a pipe system to isolate part of the system.

Blank flange — Device that attaches to the end of a pipe in order to cap the end.

Blanket — (1) Thick layer of insulating material between two layers of heavy waterproof paper. (2) Layer of foam over a fuel.

Blast Area — Area affected by the blast wave from an explosion.

Blast Pressure Wave — Shock wave created by rapidly expanding gases in an explosion.

Blasting Cap — *See* Detonator.

Blast-Pressure Front — Expanding edge of the pressure in a detonation or deflagration that causes the majority of damage in an explosion.

Bleed — (1) Process of releasing a liquid or gas under pressure, such as releasing air from the regulator or cylinder of a self-contained breathing apparatus. (2) Internal or external loss of blood.

Bleeder Valve — Valve on a gate intake that allows air from an incoming supply line to be bled off before allowing the water into the pump.

Blended Gasoline — Gasoline that has oxygen added to it to increase the efficiency of the combustion of the fuel. *Also known as* Reformulated Gasoline.

BLEVE — *See* Boiling Liquid Expanding Vapor Explosion.

Blind Hoistway — Used for express elevators that serve only upper floors of tall buildings. There are no entrances to the shaft on floors between the main entrance and the lowest floor served.

Blister Agent — Chemical warfare agent that burns and blisters the skin or any other part of the body it contacts. *Also known as* Vesicant. *See* Chemical Warfare Agent.

Blitz Attack — To aggressively attack a fire from the exterior with a large diameter (2½-inch [65 mm] or larger) fire stream.

Block — In personnel management, a division of an occupational analysis consisting of a group of related tasks with one factor in common. *See* Unit.

Block and Tackle — Series of pulleys (sheaves) contained within a wood or metal frame; used with rope to provide a mechanical advantage for pulling operations.

Block Creel Construction — Method of manufacturing rope without any knots or splices; a continuous strand of fiber runs the entire length of the rope's core.

Block Grant — Annual government grant to help local authorities provide general services for the public good, with few restrictions.

Blog — Abbreviation for *web log*; refers to a list of journal entries or articles posted by a single author or group of authors. Includes comment sections where readers can engage in conversation about entries.

Blood Agent — *See* Chemical Asphyxiant.

Blood Poisons — *See* Chemical Asphyxiant.

Blood Pressure (BP) — Pressure exerted by the flow of blood against the arterial walls.

Blood Volume — Total amount of blood in the heart and the blood vessels.

Bloodborne Pathogens — Pathogenic microorganisms that are present in the human blood and can cause disease in humans. These pathogens include (but are not limited to) hepatitis B virus (HBV) and human immunodeficiency virus (HIV).

Blow-Down Valve — Manually operated valve that has the function of quickly reducing tank pressure to atmospheric pressure.

Blower — Large-volume fan used to blow fresh air into a building or other confined space; often used in positive-pressure ventilation (PPV). Blowers are most often powered by gasoline engines, but some have electric motors.

Blowup — Sudden, dangerously rapid increase in fireline intensity at a wildland fire; caused by any one or more of several factors, such as strong or erratic wind, steep uphill slopes, large open areas, and easily ignited fuels. Blowup is sufficient to preclude direct attack or to change the incident action plan; often accompanied by violent convection and may have other characteristics of a firestorm. *See* Flare-Up.

BLS — *See* Basic Life Support.

Blunt Start — *See* Higbee Cut.

Board of Appeals — Group of people, usually five to seven, with experience in fire prevention, building construction, and/or code enforcement, who are legally constituted to arbitrate differences of opinion between fire inspectors and building officials, property owners, occupants or builders.

Boat — (1) Small craft capable of being carried on board a vessel. (2) Naval slang for a submarine.

Boat Deck — Uppermost deck on which lifeboats and other lifesaving appliances are stowed; used as a promenade space on passenger vessels. *See* Deck.

Boat Hook — Long pole with distinctive hook at the end used for fending off other boats and retrieving or picking up mooring lines.

Boatswain — Petty officer on a merchant vessel who has charge of the deck crew, hull maintenance, and related work. *Also known as* Bosun.

BOCA — *See* Building Officials and Code Administrators International, Inc.

Body Bag — Rubber or plastic bag used to remove the bodies of deceased victims; widely used in disasters.

Body Language — Nonverbal communication including but not limited to body posture and gestures; represents a large portion of communication in human interactions.

Body Substance Isolation (BSI) — (1) The practice of taking proactive, protective measures to isolate body substances in order to prevent the spread of infectious disease. (2) Comprehensive method of infection control in which every patient is assumed to be infected; personal protective equipment is worn to prevent exposure to bodily fluids and bloodborne and airborne pathogens.

Body-on-Chassis Construction — Method of school bus or recreational vehicle construction where the manufacturer installs the body unit onto a commercially available chassis constructed by another manufacturer.

Bogie — Tandem arrangement of aircraft landing gear wheels with a central strut. The bogie swivels up and down so that all wheels stay on the ground as the attitude of the aircraft changes, or the slope of the ground surface changes.

Boiler Room — Compartment containing boilers but not containing a station for operating or firing the boilers.

Boilerplate — Standardized or formulaic language.

Boiling Liquid Expanding Vapor Explosion (BLEVE) — Rapid vaporization of a liquid stored under pressure upon release to the atmosphere following major failure of its containing vessel. Failure is the result of over-pressurization caused by an external heat source, which causes the vessel to explode into two or more pieces when the temperature of the liquid is well above its boiling point at normal atmospheric pressure. *See* Boiling Point.

Boiling Point — Temperature of a substance when the vapor pressure exceeds atmospheric pressure. At this temperature, the rate of evaporation exceeds the rate of condensation. At this point, more liquid is turning into gas than gas is turning back into a liquid. *See* Condensation, Physical Properties, and Vapor Pressure.

Boilover — Overflow of burning crude oil from an open top container when the hot oil reaches the water level in the tank. The water flashes to steam causing a violent expulsion of the material as a froth.

Bollard — Stout vertical post (single or double) on a pier or wharf used for securing a vessel's mooring lines; common along piers where large vessels are moored. *See* Bitts.

Bolster — *See* Chair.

Bolt Cutters — Cutting tool designed to make a precise, controlled cut; used for cutting wire, fencing, bolts, and small steel bars.

Bolted Fault — Condition occurring when two conductors in a circuit come into firm, direct contact with each other. *Also known as* Dead Short.

Bomb Line — Slang for a portable master stream device that is preconnected to a short length (less than 200 feet [61 m]) of hose for rapid deployment.

Bomb Squad — Crew of emergency responders specially trained and equipped to deal with explosive devices.

Bombproof Anchor Point — (1) Slang reference to an anchor that is absolutely immovable, such as a huge boulder, a large tree, or a fire engine. (2) Any anchor point capable of withstanding forces in excess of those that might be generated by the rescue operation or even catastrophic failure (and resultant shock load) of a raising or lowering system.

Bonding — (1) Connection of two objects with a metal chain or strap in order to neutralize the static electrical charge between the two; similar to *Grounding*. (2) Gluing two objects together.

Boneyarding — During mop-up, spreading materials that are no longer burning in an area within the black that has been cleared of all burning or hot fuels.

Boom — (1) Pole rigged for use as a crane on board a vessel. (2) Floating object used to confine materials on the surface of the water.

Booms — Telescoping or articulating arm portions of an elevating platform aerial device.

Booster Apparatus — *See* Brush Apparatus.

Booster Hose — Non-collapsible rubber-covered, rubber-lined hose usually wound on a reel and mounted somewhere on an engine or water tender; used for the initial attack and extinguishment of incipient and smoldering fires. This hose is most commonly found in ½-, ¾-, and 1-inch (13 mm, 19 mm, and 25 mm) diameters and is used for extinguishing low-intensity fires and mop-up. *Also known as* Booster Hoseline, Hard Line, and Red Line.

Booster Pump — Fire pump used to boost the pressure of the existing water supply within a fixed fire protection system.

Booster Reel — Mounted reel on which booster hose is carried.

Booster Tank — *See* Water Tank.

Bored Lock — *See* Cylindrical Lock.

Botts' Dots — Round, nonreflective raised pavement markers.

Bounce Flash — Lighting technique used to reduce glare or reflection by pointing the flash at a nearby surface rather than at the subject.

Bourdon Gauge — Most common type of gauge used to measure water pressures.

Bourdon Tube — Part of a bourdon gauge that has a curved, flat tube that changes its curvature as pressure changes. This movement is then transferred mechanically to a pointer on the dial.

Bow — Front end or forward part of a vessel; opposite of the stern.

Bow Thruster — Large propeller mounted in a tunnel located in the forward part of the vessel used to assist the vessel in docking and undocking; reduces the need for assistance from tugs.

Bowline Knot — Knot used to form a loop; it is easy to tie and untie, and does not constrict.

Bowstring Truss — Lightweight truss design noted by the bow shape, or curve, of the top chord.

Box — Shortened term for a public or private fire alarm box.

Box Alarm — (1) Signal transmitted from a fire alarm box. (2) Predetermined response assignment to an emergency call.

Box Crib — Stabilization platform constructed by creating opposing layers of pieces of cribbing.

Box Lock — Lock mortised into a door. *Also known as* Mortise Lock.

Box Stabilizer — Two-piece aerial apparatus stabilization device consisting of an extension arm that extends directly out from the vehicle, and a lifting jack that extends from the end of the extension arm to the ground. *Also known as* H-Jack.

Box-Beam Construction — Method of construction for aerial device booms consisting of four sides welded together to form a box shape with a hollow center. Hydraulic lines, air lines, electrical cords, and waterways may be encased within the center or on the outside of the box beam.

Boyle's Law — Law stating that the volume of a gas varies inversely with the applied pressure. The formula is $P_1V_1 = P_2V_2$, where:

BP — *See* Blood Pressure.

B-Post — Post between the front and rear doors on a four-door vehicle, or the door-handle-end post on a two-door car.

Bq — *See* Becquerel.

Brace — Pieces of wood attached between components of shoring systems for stability. *Also known as* Lacing.

Brace Lock — Rim lock equipped with a metal rod that serves as a brace against the door.

Bracketing — Taking a photograph at the setting recommended by the camera meter and then manually adjusting the exposure setting one or two f-stops above and below the recommended exposure.

Braided Hose — Non-woven rubber hose manufactured by braiding one or more layers of yarn, each separated by a rubber layer, over a rubber tube and encased in a rubber cover.

Braided Rope — Rope constructed by uniformly intertwining strands of rope together (similar to braiding hair).

Braid-on-Braid Rope — Rope that consists of a braided core enclosed in a braided, herringbone patterned sheath.

Brainstorm — Process of identifying as many ideas as possible without any initial evaluation, debate, agreement, or consensus.

Brake Limiting Valve — Valve that allows the vehicle's brakes to be adjusted for the current road conditions.

Brakes — Long wooden handles on early hand pump fire engines that firefighters moved up and down to pump water from the reservoir to the pump's discharge.

Braking Distance — Distance the vehicle travels from the time the brakes are applied until it comes to a complete stop.

Braking Prusik — A Prusik attached onto a main line using a three-wrap Prusik hitch to grab the line and prevent it from moving.

Branch — Organizational level of an incident management system having functional/geographic responsibility for major segments of incident operations. The branch level is organizationally between section and division/sector/group.

Branch Circuit — Wiring between the point of application (outlets) and the final overcurrent device protecting the circuit.

Branch Line — Pipes in an automatic sprinkler system to which the sprinklers are directly attached.

Brands — Large, burning embers that are lifted by a fire's thermal column and carried away with the wind.

Brass — Brasswork or brass appliances carried on fire apparatus; may now be chrome-plated or made of lightweight alloys.

Breach — To make an opening in a structural obstacle (such as a masonry wall) without compromising the overall integrity of the wall to allow access into or out of a structure for rescue, hoseline operations, ventilation, or to perform other functions.

Breach of Duty — Any violation or omission of a legal or moral duty; neglect or failure to fulfill the duties of an office or employment in a just and proper manner.

Breaching — The act of creating a hole in a wall or floor to gain access to a structure or portion of a structure.

Break a Line — To disconnect hoselines for any purpose, especially to break and roll up hose after a fire operation; to disconnect a hose coupling.

Break Bulk Cargo — Loose, non-containerized cargo commonly packaged in bags, drums, cartons, crates, etc.

Break Bulk Carrier — Ship designed with large holds to accommodate a wide range of products such as vehicles, pallets of metal bars, liquids in drums, or items in bags, boxes, and crates. *See* Cargo Vessel.

Break Bulk Terminal — Shore facility handling cargo shipped in bags, steel drums, cartons, crates, or pallets. Typical cargoes are rolls of paper, bags of fertilizer, coils of wire, or packages of steel.

Breakaway/Frangible Fences and Gates — Fences and gates designed and constructed to collapse when impacted by large vehicles, in order to allow rapid access to accident sites.

Breaking — To destroy a structural obstacle in order to gain access to victims or perform other functions.

Breakover — *See* Slopover.

Breakover Angle — Angle formed by level ground and a line from the point where the rear tires of a vehicle touch the ground to the bottom of the frame at the wheelbase midpoint. This angle should be at least 10 degrees.

Breakthrough Time — Time required for a chemical to permeate the material of a protective suit.

Breast Timber — Strut that holds a horizontal compression load, keeping sheeting in place for shoring.

Breathing Air — Compressed air that is filtered and contains no more contaminants than are allowed by standards.

Breathing Apparatus Support Unit — Mobile unit designed and constructed for the purpose of providing specified level of breathing air support capacity and personnel capable of refilling self-contained breathing apparatus (SCBA) at remote incident locations. These units may be equipped with either compressor- or cascade-type reservicing systems.

Breathing Tube — Low-pressure hose that extends from the regulator to the SCBA facepiece.

Breathing-Air Compressor — Compressor specifically designed to compress air for breathing-air cylinders.

Bresnan Distributor Nozzle — Cellar nozzle in which the head rotates when water flows through it.

Brick Veneer — Single layer of bricks applied to the inside or outside surface of a wall for esthetic and/or insulation purposes.

Brick-Joisted — Brick or masonry wall structure with wooden floors and roof. Commonly known as ordinary construction.

Bridge — (1) To span a gap by placing a ladder, usually between two structures. (2) Control center on modern mechanized vessels; forward part of a vessel's superstructure. (3) Persons in charge of a vessel.

Bridge Truss — Heavy-duty truss, usually made of heavy wooden members with steel tie rods, that has horizontal top and bottom chords and steeply sloped ends.

Bridging — Construction technique of adding diagonal cross braces between joists.

British Thermal Unit (Btu) — Amount of heat energy required to raise the temperature of 1 lb of water 1°F. 1 Btu = 1.055 kilo joules (kJ).

Brix Scale — Measurement of the mass ratio of a foam concentrate to water in a finished foam solution.

Broadside Collision — *See* Side-Impact Collision.

Broken Stream — Stream of water that has been broken into coarsely divided drops.

Bronchial Asthma — Constriction of the bronchial tubes in response to irritation, allergies, or other stimulus.

Brush — Collective term that refers to stands of vegetation dominated by bushes, shrubby, woody plants, or small low-growing trees; usually of a type undesirable for livestock or timber management and of little or no commercial value.

Brush Apparatus — *See* Wildland Fire Apparatus.

Brush Burn — Scrape or adhesion created when the skin is rubbed across a rough surface; not actually a thermal or chemical burn. *Also known as* Abrasion.

Brush Hook — Heavy cutting tool designed primarily to cut brush at the base of the stem; used in much the same way as an axe. Has a wide blade generally curved to protect the blade from being dulled by rocks.

Brush Patrol — *See* Wildland Fire Apparatus.

Brush Pumper — *See* Wildland Fire Apparatus.

BSI — *See* Body Substance Isolation.

Btu — *See* British Thermal Unit.

Bucket Brigade — Early fire fighting technique in which people formed two lines between a fire and a water source. One line (usually composed of men) would pass buckets of water toward the fire to be applied onto the fire, and the other line (composed of women and children) would return the empty buckets to be refilled.

Buddy Breathing System — *See* Team Emergency Conditions Breathing.

Buddy System — Safety procedure used in rescue work; when rescuers work in a hazardous area, at least two rescuers must remain in contact with each other at all times.

Budget — Plan for action, with associated costs, for the coming year.

Budget Cycle — Specific timelines used by organizations and communities to receive and act on requests for resources to support general operations.

Buff — Person other than a firefighter who is interested in fires, fire departments, and firefighters as a hobby. *Also known as* Fire Fan or Spark.

Buggie — Slang for a chief's vehicle.

Bugles — Insignia depicting early speaking trumpets used to designate the rank of fire department personnel.

Building Code — Body of local law, adopted by states/provinces, counties, cities, or other governmental bodies to regulate the construction, renovation, and maintenance of buildings. *See* Code and Occupancy Classification.

Building Department — Local governmental agency responsible for enforcing various codes and regulations.

Building Engineer — Person who is familiar with and responsible for the operation of a building's heating, ventilating, and air-conditioning (HVAC) system and other essential equipment.

Building Hardening — The process in which a building is fortified against the threat of damage caused by fire or attack.

Building Marking System — Standardized system used to identify and document (on the actual structure) the location of victims and hazards within that structure.

Building Officials and Code Administrators International, Inc. (BOCA) — Organization that provides model codes for city and state adoption; the model codes are for building, mechanical, plumbing, and fire prevention. BOCA joined with the Southern Building Code Congress International (SBCCI) and the International Conference of Building Officials (ICBO) to form the International Code Council (ICC).

Building Packaging — Process of protecting the building from outside threats, such as inclement weather.

Building Permit — Authorization issued from the appropriate authority having jurisdiction (AHJ) before any new construction, addition, renovation, alteration, or demolition of buildings or structures occurs. *See* Authority Having Jurisdiction (AHJ).

Building Survey — Portion of the preincident planning process during which the company travels to a building and gathers the necessary information to develop a pre-incident plan for the building.

Built-Up Membrane Roof — Use of several overlapping layers of roofing felt applied to a roof deck with intervening layers of roofing cement. The layers are then saturated with a bituminous material that may be either tar or asphalt. *See* Built-Up Roof.

Built-Up Roof — Roof covering made of several alternate layers of roofing paper and tar, with the final layer of tar being covered with pea gravel or crushed slag.

Bulk Cargo — Homogeneous cargo (oil, grain, coal, bricks, lumber, or ore) stowed loose in a hold and not enclosed in any container such as boxes, bales, or bags.

Bulk Cargo Carrier — Ship carrying either liquid or dry goods stowed loose in a hold and not enclosed in any container. *See* Cargo Vessel.

Bulk Container — Cargo tank container attached to a flatbed truck or rail flat car used to transport materials in bulk. This container may carry liquids or gases. *See* Container (1).

Bulk Packaging — Packaging, other than a vessel or barge, including transport vehicle or freight container, in which hazardous materials are loaded with no intermediate form of containment; has (a) a maximum capacity greater than 119 gallons (450 L) as a receptacle for a liquid, (b) maximum net mass greater than 882 pounds (400 kg) and a maximum capacity greater than 119 gallons (450 L) as a receptacle for a solid, or (c) water capacity greater than 1,000 pounds (454 kg) as a receptacle for a gas. Reference: *Title 49 CFR 171.8*. *See* Nonbulk Packaging and Packaging (1).

Bulk Terminal — Handling area for cargoes (unpackaged commodities carried in holds and tanks of cargo vessels and tankers) that are loaded and unloaded by conveyors, pipelines, or cranes. A liquid bulk terminal handles cargoes such as fuel and lubricating oils and chemicals. A dry bulk terminal handles cargoes such as coal or grain. *See* Dry Bulk Terminal and Liquid Bulk Terminal.

Bulkhead — (1) Upright partition that separates one aircraft compartment from another. Bulkheads may strengthen or help give shape to the structure and may be used for the mounting of equipment and accessories. *See* Main Transverse Bulkheads and Main Watertight Subdivision. (2) Structure on the roof of a building through which the interior stairway opens onto the roof. *Also known as* Penthouse. (3) A vertical row of wood or metal pilings, or stone blocks along the shore line, that has been back-filled to protect the shore from erosion, or form a berth for shipping.

Bulldozer — Any tracked vehicle with a blade for exposing mineral soil. *Also known as* Dozer. *See* Dozer Tender and Dozer Transport.

Bulwark — Wall built around the edge of a vessel's upper deck.

Bumper — Structure designed to provide front- and rear-end protection of a vehicle.

Bumper Line — Preconnected hoseline located on the apparatus bumper.

Bumper Struts — Bumpers that incorporate energy absorbing struts to make them less vulnerable to damage in low-speed collisions.

Bung — Cork or other type of stopper used in a barrel, cask, or keg.

Bunk — Firefighter's bed.

Bunk Room — Dormitory area where firefighters sleep.

Bunker Clothes — *See* Personal Protective Equipment. *Also known as* Bunker Gear.

Buoyancy (B) — Tendency or capacity to remain afloat in a liquid or rise in air or gas as a result of the upward force of a fluid upon a floating object. *See* Center of Buoyancy. *Also known as* Buoyant.

Bureau of Alcohol, Tobacco, Firearms and Explosives (ATF) — Division of U.S. Department of Justice that enforces federal laws and regulations relating to alcohol, tobacco, firearms, explosives, and arson.

Bureau of Mines — Former name for the Mine Safety and Health Administration. *See* Mine Safety and Health Administration.

Bureaucratic Leadership — Style of leadership in which the leader has a low degree of concern for workers and production.

Burn — (1) To be on fire; to consume fuel during rapid combustion. (2) Geographical area over which a fire has passed. *Also known as* Black. (3) Tissue injury caused by heat, electrical, current, or chemicals.

Burn Building — Structure designed to contain live fires for the purpose of fire suppression training.

Burn Center — Medical facility especially designed, equipped, and staffed to treat severely burned patients.

Burn Pattern — See Fire Patterns.

Burnback Resistance — Ability of a foam blanket to resist direct flame impingement such as would be evident in a partially extinguished petroleum fire.

Burning Out — Intentionally setting a fire to natural cover fuels inside the control line to widen the line; used as a direct attack technique, usually within 10 feet (3 m) of the line. Burning out is done on a small scale in order to consume unburned fuel and aid control-line construction. Burning out should not be confused with "backfiring," which is a larger-scale tactic to eliminate large areas of unburned fuels in the path of a fire, or to change the direction of force of a convection column.

Burning Point — See Fire Point.

Burning Velocity — Velocity of the flame front in an explosion relative to the unburned gases ahead of it.

Burnout — (1) Building that has been denuded of almost all combustible material. Also refers to a burned wildland area. (2) A work-related psychological disorder resulting from stress.

Burns, Degree of — First degree: reddened skin; second degree: blisters; third degree: deep skin destruction. Major types of burns: heat, chemical, electrical, and radiation.

Burst Test — Destructive test on a 3-foot (.9 m) length of hose to determine its maximum strength.

Butt — (1) One coupling of a fire hose. (2) Hydrant outlet. (3) Heel (lower end) of a ladder. (4) Act of steadying a ladder that is being climbed.

Butt Spurs — Metal safety plates or spikes attached to the butt end of ground ladder beams.

Butterfly Roof — V-shaped roof style resembling two opposing shed roofs joined along their lower edges.

Butterfly Valve — Control valve that uses a flat circular plate in a pipe that rotates 90 degrees across the cross section of the pipe to control the flow of water.

Buttress — Structure projecting from a wall designed to receive lateral pressure action at a particular point.

Byline — Line at the beginning of a news story, magazine article, or book giving the author's name.

Bypass Breathing — Emergency procedure in which the self-contained breathing apparatus (SCBA) wearer closes the mainline valve and opens the bypass valve for air when a regulator malfunctions.

Bypass Valve — Valve on a self-contained breathing apparatus (SCBA) that when opened allows air to bypass its normal route through the regulator; used when a regulator malfunctions.

Bypass-Type Balanced-Pressure Proportioner — Foam proportioning system that discharges foam through a pump separate from the water supply pump; most commonly found in airport crash vehicles and in fixed-site facilities. It is one of the most accurate types of foam proportioning systems in use. See Foam Proportioner and Proportioning.

P_1 = original pressure

P_2 = final pressure

V_1 = original volume

V_2 = final volume

C

C.A.F.I. — See Canadian Association of Fire Investigators.

C.A.N. Report — Situational report given periodically as needed throughout the incident to update Incident Command. This acronym stands for Conditions, Actions, and Needs.

CAA — See Civil Aviation Authority.

Cabin — (1) Aircraft passenger compartment that may be separated and may contain a cargo area. (2) See Accommodation Spaces.

Cable Hanger — Device used to test the structural strength of aerial ladders.

Cables — Flexible structural members used to support roofs, brace tents, and restrain pneumatic structures.

CABO — See Council of American Building Officials.

CAD — See Computer-Aided Design.

CAFS — See Compressed Air Foam System.

Caisson — (1) Watertight structure within which construction work is carried out underwater. (2) Protective sleeve used to keep water out of an excavation for a pier. (3) Protective hardened steel sleeves located inside automobile B- and C- posts that are designed to protect seatbelt pretensioners from being cut.

Calcination — Process of driving free and chemically bound water out of gypsum; also describes chemical and physical changes to the gypsum component itself.

Calcined — Process that heats a substance to a high temperature but below the melting or fusing point, causing loss of moisture, reduction or oxidation, and decomposition of carbonates and other compounds.

Calendering — Fire hose inner tube manufacturing process in which rubber is pressed between opposing rollers to produce a flat sheet. A tube is then formed by lapping and bonding together the edges of the sized sheet.

Calibrate — Operations to standardize or adjust a measuring instrument.

Calibration — Set of operations used to standardize or adjust the values of quantities indicated by a measuring instrument.

Calibration Curve — Generic method for identifying the concentration of a substance, such as foam, in an unknown sample by comparing the unknown to a set of standard samples of known concentration.

California FIRESCOPE Incident Command System — *See* Incident Command System (ICS).

Call Back — Process of notifying off-duty firefighters to return to their stations for service.

Call Box — *See* Telephone Alarm Box.

Call Firefighter — *See* Paid-On-Call Firefighter.

Calorie — Amount of heat needed to raise the temperature of one gram of water one degree Celsius. *See* Joule.

Cam — Part of a mortise lock cylinder that moves the bolt or latch as the key is turned.

Camber — Low vertical arch placed in a beam or girder to counteract deflection caused by loading.

Camlock Fastener — Trade name given to a quick-disconnect screw-type fastener, designed to open with a quarter or half turn (similar to Dzus fasteners).

Can Man — Slang for a firefighter, usually from a truck or ladder company, whose role is to carry a pump can and some other tool, often a pike pole.

Canadian Association of Fire Investigators (C.A.F.I.) — Professional organization for fire investigators in Canada offering training and professional development opportunities.

Canadian Centre for Occupational Health and Safety (CCOHS) — Canadian federal government agency that provides information and policy development regarding work-related injury, illness prevention initiatives, and occupational health and safety information.

Canadian Charter of Rights and Freedoms — Portion of the Canadian Constitution containing due process clauses. Section 7 states that "Everyone has the right to life, liberty and security of the person and the right not to be deprived thereof except in accordance with the principles of fundamental justice."

Canadian Coast Guard (CCG) — Marine law enforcement and rescue agency in Canada; responsible for the safety, order, and operation of maritime traffic.

Canadian Electrical Code® (CEC®) — Manual providing safety standards for installation of electrical wiring and electrical systems in Canada.

Canadian Nuclear Safety Commission — Agency responsible for regulating almost all uses of nuclear energy and nuclear materials in Canada.

Canadian Standards Association (CSA) — Canadian standards-writing organization.

Canadian Transportation Emergency Centre (CANUTEC) — Canadian center that provides fire and emergency responders with 24-hour information for incidents involving hazardous materials; operated by Transport Canada, a department of the Canadian government. *See* Chemical Transportation Emergency Center (CHEMTREC®) and Emergency Transportation System for the Chemical Industry (SETIQ).

Candidate Physical Ability Test (CPAT) — Optional, nationally recognized physical fitness examination for firefighter candidates which is oriented toward firefighter job tasks with established benchmarks.

Canine Search — Use of disaster-trained search dogs and handlers for the location of victims.

Canister Apparatus — Type of breathing apparatus that uses filtration, adsorption, or absorption to remove toxic substances from the air; generally referred to as a gas mask. Canister apparatus are not acceptable for use in fire fighting operations or IDLH atmospheres.

Cannula — Tube used to enter a duct or cavity. A nasal cannula is often used to administer supplemental oxygen.

Canopy — (1) Transparent enclosure over the cockpit of some aircraft. (2) Level or area containing the crowns of the tallest vegetation containing the leaves of trees and brush present (living or dead), usually above 20 feet (6.1 m).

Cant Strip — Angular board installed at the intersection of a roof deck and a wall to avoid a sharp right angle when the roofing is installed.

Canteen Unit — Emergency vehicle that provides food, drinks, and other rehabilitative services to emergency workers at extended incidents.

Cantilever — (1) Projecting beam or slab supported at one end. (2) Type of collapse void in which one end of a floor or roof section that has collapsed remains suspended and unsupported.

Cantilever Fire Wall — Free standing fire wall that is commonly found in large churches and shopping malls.

Cantilever Operation — *See* Unsupported Tip.

Cantilever Roof — Roof structure extending from the edge of the building that is anchored at only one end.

CANUTEC — *See* Canadian Transport Emergency Centre.

Capability Assessment for Readiness (CAR) — Self-assessment survey-type instrument conducted by states to assess their operational readiness and emergency management capabilities; created by the U.S. Federal Emergency Management Agency (FEMA), in partnership with National Emergency Management Association (NEMA) and state emergency managers in 1997.

Capacity — Maximum ability of a pump or water distribution system to deliver water. Also the maximum quantity of water that can be contained in a tank.

Capacity Indicator — Device installed on a tank to indicate capacity at a specific level.

Capacity Stencil — Number stenciled on the exterior of a tank car to indicate the volume of the tank.

Capillaries — Tiny blood vessels in the body's tissues in which the exchange of oxygen and carbon dioxide take place.

Capital Budget — Budget intended to fund large, one-time expenditures, such as those for fire stations, fire apparatus, or major pieces of equipment.

Capital Grant — Money gained through a large-scale fund-raising activity in which the funds raised will support a building or other so-called "capital expense," such as a large computer system.

Capital Improvement Program (CIP) — 5 to 10 year program intended to identify large equipment purchases or project expenses, schedule their purchase, and identify the needed funding sources.

Captain — (1) Rank used in some departments for a company officer. (2) Commander of a vessel. *See* Master.

Captain of the Port (COTP) — U.S. Coast Guard officer who has broad powers over all vessels in a port area in the U.S.; equivalent to Harbormaster in the United Kingdom.

CAR — *See* Capability Assessment for Readiness.

CAR — *See* Customer Approval Rating.

Car Terminal — Facility for loading and unloading vessels specially designed to transport automobiles.

Carabiner — Steel or aluminum D-shaped snap link device for attaching components of rope rescue systems together. In rescue work, carabiners should be of a positive locking type, with a 5,000-pound (2 300 kg) minimum breaking strength. *Also known as* Biner, Crab, or Snap Links.

Carbon-Based Fuels — Fuels in which the energy of combustion derives principally from carbon; includes materials such as wood, cotton, coal, or petroleum.

Carbon Dioxide (CO_2) — Colorless, odorless, heavier than air gas that neither supports combustion nor burns; used in portable fire extinguishers as an extinguishing agent to extinguish Class B or C fires by smothering or displacing the oxygen. CO_2 is a waste product of aerobic metabolism. *See* Combustion and Fire Exinguisher.

Carbon Dioxide System — Extinguishing system that uses carbon dioxide as the primary extinguishing agent; designed primarily to protect confined spaces because the gaseous agent is easily dispersed by wind.

Carbon Monoxide (CO) — Colorless, odorless, dangerous gas (both toxic and flammable) formed by the incomplete combustion of carbon. It combines with hemoglobin more than 200 times faster than oxygen does, decreasing the blood's ability to carry oxygen.

Carbon Monoxide Detector — Device designed to detect the presence of carbon monoxide (CO) and sound an alarm to prevent personnel within the structure from being poisoned by the colorless and odorless gas.

Carbon Monoxide Poisoning — Sometimes lethal condition in which carbon monoxide molecules attach to hemoglobin, decreasing the blood's ability to carry oxygen.

Carbonaceous — Made of or containing carbon.

Carboxyhemoglobin (COHB) — Hemoglobin saturated with carbon monoxide and therefore unable to absorb needed oxygen.

Carboy — Cylindrical container of about 5 to 15 gallons (19 L to 57 L) capacity used for pure or corrosive liquids. Made of glass, plastic, or metal, with a neck and sometimes a pouring tip; cushioned in a wooden box, wicker basket, or special drum. *See* Container.

Carcinogen — Cancer-producing substance.

Cardiac Arrest — Sudden cessation of heartbeat.

Cardiac Monitoring — Monitoring the status of the electrical activity of the heart.

Cardiogenic Shock — Shock caused when the heart fails to pump effectively.

Cardiopulmonary Resuscitation (CPR) — Application of rescue breathing and external cardiac compression used on patients in cardiac arrest to provide an adequate circulation and oxygen to support life.

Cardiopulmonary System — Heart and lungs.

Cardiovascular System — Body's system of blood vessels and associated organs that support the flow of blood through the body.

Career Fire — Jargon used to describe a large fire.

Career Fire Department — Fire department composed of full-time, paid personnel.

Career Firefighter — Person whose primary employment is as a firefighter within a fire department. *Also spelled* Career Fire Fighter.

Cargo Container — *See* Container (1).

Cargo Manifest — Document or shipping paper listing all contents carried by a vehicle or vessel on a specific trip.

Cargo Plan — View of a vessel showing all the storage space available for cargo; shows the amount and type of cargo carried, its destination, and how it will be stowed.

Cargo Tank — *See* Cargo Tank Truck.

Cargo Tank Truck — Motor vehicle commonly used to transport hazardous materials via roadway. *Also known as* Cargo Tank, Tank Motor Vehicle, and Tank Truck. *See* Compressed-Gas Tube Trailer, Corrosive Liquid Tank, Cryogenic Liquid Tank, Dry Bulk Cargo Tank, Elevated Temperature Materials Carrier, High-Pressure Tank, Low-Pressure Chemical Tank, and Nonpressure Liquid Tank.

Cargo Vessel — Ship used to transport cargo (dry bulk, break bulk, roll-on/roll off, and container) via waterways. *See* Barge, Break Bulk Carrier, Bulk Cargo Carrier, Container Vessel, and Roll-on/Roll-off Vessel.

Carline Supports — Structural members used in the construction of buses. They are designed to strengthen the sidewall of the bus where it might come into contact with a car during a collision.

Carotid Artery — Principal artery of the neck, easily felt on either side of the trachea.

Carriage — Main support for the stair treads and risers. *Also known as* Stringer.

Carryall — Waterproof carrier used to carry and catch debris or used as a water sump basin for immersing small burning objects.

Cartridge Filter Respirators — Type of respiratory protection that utilizes special filter cartridges to filter out specific airborne contaminants.

CAS® Number — Number assigned by the American Chemical Society's Chemical Abstract Service that uniquely identifies a specific compound.

Cascade Air Cylinders — Large air cylinders that are used to refill smaller SCBA cylinders.

Cascade System — Three or more large, interconnected air cylinders, from which smaller SCBA cylinders are recharged; the larger cylinders typically have a capacity of 300 cubic feet (8,490 L).

Case — Housing for any locking mechanism.

Case File — (1) The collection of documents comprising information concerning a particular investigation. (2) A file that includes all documentation from an agency pertinent to a specific juvenile firesetting case.

Case Identifiers — Alphabetic and/or numeric characters used to identify a case.

Case Law — Laws based on judicial interpretations and decisions rather than created by legislation.

Case Management Information — Data that is specific to an individual firesetter and his or her family. This might include names, addresses, specific incident numbers, etc. This information is critical in tracking the individual case through the program.

Case Study — Description of a real or hypothetical problem that an organization or an individual has dealt with and may face in the future.

Case-Hardened Steel — Steel with a surface that is much harder than the interior metal because the surface layer was hardened via special heat treating methods. Used in the construction of vehicles.

Casement Window — Window hinged along one side, usually designed to swing outward, with the screen on the inside.

Cast Coupling — Coupling manufactured by pouring molten metal into a mold, allowing it to cool and harden, then removing the hardened coupling from the mold.

Caster — Roller on the bottom of a chair or piece of furniture. *Also spelled* Castor.

Cast-in-Place Concrete — Common type of concrete construction. Refers to concrete that is poured into forms as a liquid and assumes the shape of the form in the position and location it will be used.

Catalyst — Substance that modifies (usually increases) the rate of a chemical reaction without being consumed in the process.

Catch a Hydrant — Process in which a firefighter dismounts the fire apparatus at the hydrant, connects the fire hose to the hydrant, and turns on the water.

Catch Basin — *See* Portable Tank.

Catchall — Retaining basin, usually made from salvage covers, to impound water dripping from above.

Caternary Wire System — Series of overhead wires used to transmit electrical power to buses, locomotives, and trams at a distance from the energy supply point. *See* Caternary System.

Caulk — Non-hardening paste used to fill cracks and crevices. *Also spelled* Calk.

Caustic — (1) Corrosive material that burns or destroys tissue by chemical action, as opposed to heat. (2) Substance having the destructive properties of a base. See Acid, Alkali, and Base.

Cave-In — Collapse of unsupported trench walls.

Cavitation — Condition in which vacuum pockets form due to localized regions of low pressure at the vanes in the impeller of a centrifugal pump, causing vibrations, loss of efficiency, and possibly damage to the impeller. *See* Centrifugal Pump and Impeller.

Cavity — Hollow or space, especially within the body or one of its organs. For example, the abdominal cavity is bounded by the abdominal walls, the diaphragm, and the pelvis.

CBL — *See* Competency-Based Learning.

CBRNE — *See* Chemical, Biological, Radiological, Nuclear, and Explosive.

CCG — *See* Canadian Coast Guard.

CCOHS — *See* Canadian Centre for Occupational Health and Safety.

CDC — *See* Centers for Disease Control and Prevention.

CD-R — Abbreviation for Compact Disc-Recordable.

CD-ROM — Abbreviation for Compact Disc-Read-Only Memory.

CD-RW — Abbreviation for Compact Disc-Rewritable.

Cease-and-Desist Order — Court order prohibiting a person or business from continuing a particular course of conduct.

CEC® — *See Canadian Electrical Code®.*

Ceiling — (1) Height of the base of the lowest layer of clouds when over half of the sky is obscured; reported as "broken," "overcast," "obscuration," or "partial obscuration." (2) Non-load-bearing structural component separating a living/working space from the underside of the floor or roof immediately above.

Ceiling Concentration — Maximum allowable concentration of dust, vapors, mists, or other hazardous materials that may exist in a confined space.

Ceiling Jet — Horizontal movement of a layer of hot gases and combustion by-products from the center point of the plume, when the vertical development of the rising plume is redirected by a horizontal surface such as a ceiling.

Cell — (1) Small cavity or compartment. (2) Smallest structural unit of living matter capable of functioning independently.

Cell Electrolyte Level — In apparatus terms, the level of water that is within the vehicle's batteries.

Cellar Pipe — Special nozzle for attacking fires in basements, cellars, and other spaces below the attack level.

Cellulosic Materials — Organic materials, such as cotton or wood, composed of cells.

Celsius — International temperature scale on which the freezing point is 0°C (32°F) and the boiling point is 100°C (212°F) at normal atmospheric pressure at sea level. *Also known as* Centigrade. *See* Fahrenheit Scale and Temperature.

Celsius Scale — International temperature scale on which the freezing point is 0°C (32°F) and the boiling point is 100°C (212°F) at normal atmospheric pressure at sea level. *Also known as* Centigrade Scale. *See* Fahrenheit Scale and Temperature.

Cement — Any adhesive material or variety of materials which can be made into a paste with adhesive and cohesive properties to bond inert aggregate materials into a solid mass by chemical hardening. For example, portland cement is combined with sand and/or other aggregates and water to produce mortar or concrete.

Cementitious — Containing or composed of cement; having cementlike characteristics.

Census Data — Demographical information about people and communities that is collected by the US Census Bureau every ten years.

Census Tracts — Defined geographical areas within a city, town, county, or village. Each tract carries a numerical identification.

Center of Buoyancy — Geometrical center of the underwater volume of a body; considered to be the point through which all forces of buoyancy are acting vertically upwards with a force equal to the weight of a body. *See* Buoyancy.

Center of Gravity — Point through which the weight of a load can be considered concentrated, so that if supported at that point, the load will maintain balance.

Center Rafter Cut — *See* Louver Cut.

Centerline — Imaginary line running the length of a vessel, from the point of the bow to the center of the stern; equidistant from the port and starboard sides of a vessel.

Centers for Disease Control and Prevention (CDC) — U.S. government agency for the collection and analysis of data regarding disease and health trends.

Centigrade — *See* Celsius.

Centigrade Scale — *See* Celsius Scale.

Central Fire Station — Headquarters station that contains administrative offices, special equipment, fire apparatus, and personnel.

Central Neurogenic Hyperventilation — Abnormal pattern of breathing seen in severe illness or injury involving the brain and characterized by very heavy, rapid breathing.

Central Processing Unit (CPU) — Part of a computer that actually possesses information.

Central Station Alarm System — System that functions through a constantly attended location (central station) operated by an alarm company. Alarm signals from the protected property are received in the central station and are then retransmitted by trained personnel to the fire department alarm telecommunications center.

Central Station Monitoring — Alarm systems that are monitored by a third party (usually a private alarm company) at a constantly attended location (central station) instead of a direct connection. Alarm signals from the protected property are received in the central station and are then retransmitted by trained personnel to the fire department alarm telecommunications center.

Central Station System — Alarm system that functions through a constantly attended location (central station) operated by an alarm company. Alarm signals from the protected property are received in the central station and are then retransmitted by trained personnel to the fire department alarm communications center.

Centrifugal Pump — Pump with one or more impellers that rotate and utilize centrifugal force to move the water. Most modern fire pumps are of this type. *See* Impeller, Multistage Centrifugal Pump, Self-Priming Centrifugal Pump, and Single-Stage Centrifugal Pump.

CERCLA — *See* Comprehensive Environmental Response, Compensation and Liability Act.

CERT — *See* Community Emergency Response Team.

Certificate of Occupancy — Issued by a building official after all required electrical, gas, mechanical, plumbing, and fire protection systems have been inspected for compliance with the technical codes and other applicable laws and ordinances.

Certification — (1) A certified statement. (2) Refers to a manufacturer's certification; for example, that a ladder has been constructed to meet requirements of NFPA® 1931. (3) Issuance of a document that states one has demonstrated the knowledge and skills necessary to function in a field.

Certification Tests — Pre-service tests for aerial device, ladder, pump, and other equipment conducted by an independent testing laboratory prior to delivery of an apparatus. These tests ensure that the apparatus or equipment will perform as expected after being placed into service.

Certified Fire and Explosion Investigator Program — Fire investigator certification program offered by the National Association of Fire Investigators (NAFI) through its National Certification Board in the U.S.

Certified Fire Investigator Program (CFI) — Fire investigator certification program offered by the International Association of Arson Investigators (IAAI).

Certified Shop Test Curves — Results, which are plotted on a graph, of the test performed by the manufacturer on its pump before shipping.

Cervical Collar — Device used to immobilize and support the neck.

Cervical Spine — First seven bones of the vertebral column, located in the neck.

C-Factor — Factor used in hydraulic formulas (usually the Hazen-Williams formula) to account for the roughness of the inner surface of piping or fire hose. The C-factor decreases as the sediment, incrustation, and tuberculation within the pipe increases.

CFEI — *See* Certified Fire and Explosion Investigator Program.

CFI — *See* Certified Fire Investigator Program.

CFM — *See* Cubic Feet Per Minute.

CFR — (1) *See* Code of Federal Regulations. (2) *See* Crash Fire Rescue.

CGA — *See* Compressed Gas Association.

Chafing Block — Blocks placed under hoselines to protect the hose covering from damage due to rubbing against the ground or concrete.

Chain Hose Tool — Tool used to carry, secure, and otherwise aid in handling hose.

Chain of Command — (1) Order of rank and authority in the fire and emergency services. (2) The proper sequence of information and command flow as described in the National Incident Management System - Incident Command System (NIMS-ICS).

Chain of Custody — Continuous changes of possession of physical evidence that must be established in court to admit such material into evidence. In order for physical evidence to be admissible in court, there must be an evidence log of accountability that documents each change of possession from the evidence's discovery until it is presented in court.

Chain Reaction — Series of self-sustaining changes, each of which causes or influences a similar reaction.

Chain Saw — Gas- or electric-powered saw that operates by rotating a chain of small cutting blades around an oblong bar.

Chair — In construction, device of bent wire used to hold reinforcing bars in position in reinforced concrete. *Also known as* Bolster.

Chamois — Soft pliant leather used for drying furniture and contents or for removing small amounts of water.

Change Order — Client's written order to a contractor, issued after execution of a construction contract, which authorizes a change in the construction work, project completion time, and/or cost of the project. *See* Contractor.

Channeling Devices — Items such as signs, road flares, and cones intended to guide traffic away from the active zones at an incident or accident.

Char — Carbonaceous material formed by incomplete combustion of an organic material, commonly wood; the remains of burned materials.

Char Gauge — Blunt-ended, thin probe similar to dial calipers or tire-tread gauges that is inserted into blistered char to measure the depth of char.

Charge — (1) To pressurize a fire hose or fire extinguisher. (2) The law that the law enforcement agency believes the defendant has broken.

Charged Building — Building heavily laden with heat, smoke, and gases, and possibly in danger of having a backdraft.

Charged Line — Hose loaded with water under pressure and prepared for use.

Charging Station — Group of equipment assembled in a location to refill self-contained breathing apparatus cylinders.

Charles' Law — Scientific law that says the increase or decrease of pressure in a constant volume of gas is directly proportional to corresponding increase or decrease of temperature. If a gas is confined so it cannot expand, its pressure will increase or decrease in direct proportion to temperature. The formula is stated as $P_1T_1 = P_2T_2$, where:

Charter Warning — Name for the advisement of rights read to a suspect in Canada. *See* Canadian Charter of Rights and Freedoms and Miranda Warning.

Chase — Vertical or horizontal space in a building used to route pipes, wires, ducts, or other utility or mechanical systems. Usually surrounded with a fire-rated enclosure. *See* Pipe Chase.

Chassis — Basic operating system of a motor vehicle consisting of the frame, suspension system, wheels, and steering mechanism, but not the body.

Chauffeur — *See* Fire Apparatus Driver/Operator.

Cheater Bar — Piece of pipe added to a prying tool to lengthen the handle and provide additional leverage.

Check Valve — Automatic valve that permits liquid flow in only one direction. For example, the inline valve that prevents water from flowing into a foam concentrate container when the nozzle is turned off or there is a kink in the hoseline.

Check-In — Process or location used by assigned resources to report in at an incident.

Checklists — Detailed lists generally prepared for the maintenance of equipment or apparatus or for installed fire protection equipment to ensure that the inspector does not overlook an item that needs to be checked regularly. They may also be used during pre-incident planning and fire prevention inspections.

Checkrail Window — Type of window usually consisting of two sashes, known as the upper and lower sashes, that meet in the center of the window. Checkrail or double-hung windows may be made of either wood or metal, but the construction design is quite similar.

Checks — Cracks or breaks in wood.

Chemical Agent — Chemical substance that is intended for use in warfare or terrorist activities to kill, seriously injure, or incapacitate people through its physiological effects.

Chemical Asphyxiant — Substance that reacts to prevent the body from being able to use oxygen. *Also known as* Blood Poison, Blood Agent, or Cyanogens Agent. *See* Asphyxiant.

Chemical Attack — Deliberate release of a toxic gas, liquid, or solid that can poison people and the environment. *See* Chemical Warfare Agent, Terrorism, and Weapon of Mass Destruction.

Chemical Burns — Burns caused by contact with acids, lye, and vesicants such as tear gas, mustard gas, and phosphorus.

Chemical Carrier — Tank vessel that transports multiple specialty and chemical commodities. *See* Tanker.

Chemical Chain Reaction — One of the four sides of the fire tetrahedron representing a process occurring during a fire. Vapor or gases are distilled from flammable materials during initial burning; atoms and molecules are then released from these vapors and combine with other radicals to form new compounds; these compounds are again disturbed by the heat, releasing more atoms and radicals that again form new compounds and so on. Interrupting the chain reaction will stop the overall reaction; this is the extinguishing mechanism utilized by several extinguishing agents. *See* Self-Sustained Chemical Reaction.

Chemical Change — When a substance changes from one type of matter to another.

Chemical Compound — Homogeneous substance consisting of two or more elements and having properties different from the constituent elements.

Chemical Degradation — Process that occurs when the characteristics of a material are altered through contact with chemical substances.

Chemical Entry Suit — Protective apparel designed to protect the firefighter's body from certain liquid or gaseous chemicals. May be used to describe both Level A and Level B protection.

Chemical Explosion — Rapid, exothermic reactions in which an ignition source initiates the explosion (or an increase in temperature self-initiates it), and combustion propagates along the blast-pressure front of the reaction in all directions.

Chemical Flame Inhibition — Extinguishment of a fire by interruption of the chemical chain reaction.

Chemical Foam — Foam produced as a result of a reaction between two chemicals, an alkaline solution and an acid solution, which unite to form a gas (carbon dioxide) in the presence of a foaming agent that traps the gas in fire-resistive bubbles. Chemical foam is not commonly used today. *See* Mechanical Foam.

Chemical Heat Energy — Heat produced from a chemical reaction including combustion, spontaneous heating, heat of decomposition, and heat of solution; sometimes occurs as a result of a material being improperly used or stored. Some materials may simply come in contact with each other and react, or they may decompose and generate heat.

Chemical Pellet — A pellet of solder, under compression, within a small cylinder that melts at a predetermined temperature, allowing a plunger to move down and release the valve cap parts.

Chemical Properties — Relating to the way a substance is able to change into other substances. Chemical properties reflect the ability to burn, react, explode, or produce toxic substances hazardous to people or the environment. *See* Physical Properties.

Chemical Protective Clothing (CPC) — Clothing designed to shield or isolate individuals from the chemical, physical, and biological hazards that may be encountered during operations involving hazardous materials. *See* Level A Protection, Personal Protective Equipment (PPE), and Special Protective Clothing (1).

Chemical Reaction — Change in the composition of matter that involves a conversion of one substance into another.

Chemical Transportation Emergency Center (CHEMTREC®) — Center established by the American Chemistry Council that supplies 24-hour information for incidents involving hazardous materials. *See* Canadian Transport Emergency Centre (CANUTEC) and Emergency Transportation System for the Chemical Industry (SETIQ).

Chemical Warfare Agent — Chemical substance intended for use in warfare or terrorist activity; designed to kill, seriously injury, or seriously incapacitate people through its physiological effects. *See* Blister Agent, Chemical Agent, Chemical Attack, Choking Agent, Nerve Agent, and Vomiting Agent.

Chemical, Biological, Radiological, Nuclear, or Explosive (CBRNE) Type Weapon — *See* Weapon of Mass Destruction.

CHEMTREC® — *See* the Chemical Transportation Emergency Center®.

Chest Compression — The act of forcefully compressing the heart in a rhythmic manner in order to circulate blood throughout the body.

Chevron Room Setup — Room arrangement in which the chairs are positioned in a fan or V-formation, thus placing more members of the audience closer to the educator and allowing participants to see each other. *Also known as* Fan-Style Room Setup or Herringbone Room Setup.

Cheyne-Stokes Respiration — Abnormal breathing pattern characterized by rhythmic increase and decrease in depth of ventilations, with regularly recurring periods during which breathing stops.

Chief — (1) Incident Management System title for individuals responsible for command of the functional sections: operations, planning, logistics, and finance/administrative. (2) Short for chief of department or fire chief. (3) Term used to verbally address any chief officer.

Chief Complaint — Problem for which a patient seeks help; usually stated in a word or short phrase.

Chief of Department — Highest ranking member of the fire department; in some instances, designated as the director or administrator.

Chief Engineer — Senior engineering officer responsible for the satisfactory working and upkeep of the main and auxiliary machinery on board a vessel.

Chief Officer — (1) Any of the higher officer grades, from district or battalion chief to the chief of the fire department. (2) Deck officer immediately responsible to a vessel's master on board a merchant vessel; officer next in rank to the master. *Also known as* Chief Mate, First Mate, or Mate.

Chief Steward — Person in charge of the steward's department, responsible for the comfort and service of passengers on passenger vessels; obtains and regulates the issue of provisions and stores, and is in charge of the inspection and proper storage of provisions.

Chief's Aide — *See* Adjutant.

Child Protective Services — A branch of the Department of Social Services. Its mission is to safeguard a juvenile when abuse, neglect, or abandonment is suspected, or when there is no family to take care of the juvenile.

Chimney — Steep, narrow draws or canyons in which heated air rises rapidly as it would in a flue pipe.

Chimney Effect — Created when a ventilation opening is made in the upper portion of a building, and air currents throughout the building are drawn in the direction of the opening. Also occurs in wildland fires when the fire advances up a V-shaped drainage swale. *See* Stack Effect.

Chimney Rods — Poles connected to a chimney brush.

Chlorinated Polyethylene (CPE) — Widely used synthetic roofing material used in single-ply membrane roofs.

Chock — (1) Cast metal ring mounted to the deck edge to control a mooring line or prevent chafing of the line; closed chock requires one end of the mooring line to pass through the center of the chock, open chock allows the line to be dropped in from the top. (2) Piece of wood or other material placed at the side of cargo to prevent rolling or moving sideways. *See* Fairlead. (3) *See* Wheel Chock.

Choking Agent — Chemical warfare agent that attacks the lungs, causing tissue damage. *See* Chemical Warfare Agent.

Chord — Top or bottom longitudinal member of a truss; main members of trusses, as distinguished from diagonals.

Chronic — Marked by long duration; recurring over a period of time. *See* Acute.

Chronic Health Effects — Long-term health effects resulting from exposure to a hazardous substance. *See* Acute Health Effects.

Chronic Health Hazards — Hazards that may cause long-term health effects from either a one-time or repeated exposure to a substance.

Chronic Obstructive Pulmonary Disease (COPD) — Term for several diseases that result in obstructive problems in the airways.

Chuck — Portable fire hydrant carried on the apparatus, with one or more gated connections for the hose. The device screws into a special flush hydrant connection on the water main or a special main. *Also known as* Air Line Connection.

Chuck Key — Key used to tighten or loosen the bit in a power drill.

Church Raise — *See* Auditorium Raise.

Churning — (1) Movement of smoke being blown out of a ventilation opening, only to be drawn back inside by the negative pressure created by the ejector because the open area around the ejector has not been sealed. *Also known as* Recirculation. (2) Rotation of a centrifugal pump impeller when no discharge ports are open so that no water flows through the pump.

Chute — Salvage cover arrangement that channels excess water from a building. A modified version can be made with larger sizes of fire hose.

Ci — *See* Curie.

Cilia — Tiny hairlike projections that help move mucus from the lungs.

CIP — *See* Capital Improvement Program

Circle or Walk-Around Method — An inspection method in which the driver or inspector starts at one point of the apparatus and continues in either a clockwise or counterclockwise direction inspecting the entire apparatus.

Circle System — *See* Loop System.

Circuit — Complete path of an electrical current.

Circuit Breaker — Device (basically an on/off switch) designed to allow a circuit to be opened or closed manually, and to automatically interrupt the flow of electricity in a circuit when it becomes overloaded.

Circular Saw — Gas- or electric-powered saw whose circular blade rotates at a high speed to produce a cutting action; a variety of blades may be used, depending on the material being cut. *Also known as* Rotary Rescue Saw.

Circular-Shaped Pattern — Fire pattern that appears on the undersides of horizontal surfaces such as ceilings or tables; formed when the plume generated by a fire spreads out across the horizontal surface.

Circulating Feed — Fire hydrant that receives water from two or more directions.

Circulating Hydrant — Fire hydrant that is located on a secondary feeder or distributor main that receives water from two directions.

Circulating System — *See* Loop System.

Circulation Relief Valve — Small relief valve that opens and provides enough water flow into and out of the pump to prevent the pump from overheating when it is operating at churn against a closed system.

Circulator Valve — Device in a pump that routes water from the pump to the supply in order to keep the pump cool when hoselines are shut down.

Circulatory System — Bodily system consisting of the heart and blood vessels.

Circumstantial Evidence — Evidence presented in a trial that tends to prove a factual matter through inference by proving other events or circumstances.

CIS — *See* Critical Incident Stress.

CISD — *See* Critical Incident Stress Debriefing.

Cistern — Water storage receptacle that is usually underground and may be supplied by a well or rainwater runoff.

Citation — Legal reprimand for failure to comply with existing laws or regulations; notice of a violation of law. *See* Violation.

Citizens Band (CB) Radio — Low-power radio transceiver that operates on frequencies authorized by the Federal Communications Commission (FCC) for public use with no license requirement.

Civil Aviation Authority (CAA) — United Kingdom's regulatory aviation authority; responsible for regulating all UK civil aviation functions, including airspace policy, safety regulations, economic regulation, and consumer protection.

Civil Liability — Legal responsibility for fulfilling a specified duty or behaving with due regard for the rights and safety of others.

Civil Support Team (CST) — Provides military (usually National Guard) support to civil authorities such as emergency managers.

Civil Wrong — Wrongdoing for which an action for damages may be brought. *Also known as* Tort.

Cladding — Exterior finish or skin.

Clapper Valve — Hinged valve that permits the flow of water in one direction only.

Clappered Siamese — Hose appliance that has one discharge and two or more intakes equipped with hinged gates that prevent water from being discharged through an open intake.

Class A Fire — Fires involving ordinary combustibles such as wood, paper, cloth, and similar materials.

Class A Foam — *See* Class A Foam Concentrate.

Class A Foam Concentrate — Foam specially designed for use on Class A combustibles. Class A foams, hydrocarbon-based surfactants are essentially wetting agents that reduce the surface tension of water and allow it to soak into combustible materials more easily than plain water. Class A foams are becoming increasingly popular for use in wildland and structural fire fighting. *Also known as* Class A Foam. *See* Finished Foam and Foam Concentrate.

Class A Fuels — Ordinary combustible solids such as wood, grass, rubber, cloth, paper, and plastics.

Class A Poison — Poisonous gases or liquids, of which a very small amount of the gas or vapor of the liquid is dangerous to life.

Class B Fire — Fires of flammable and combustible liquids and gases such as gasoline, kerosene, and propane.

Class B Foam — *See* Class B Foam Concentrate.

Class B Foam Concentrate — Foam fire-suppression agent designed for use on ignited or un-ignited Class B flammable or combustible liquids. *Also known as* Class B Foam. *See* Finished Foam and Foam Concentrate.

Class B Poison — Toxic substance that presents a severe health hazard if released during transportation.

Class C Fire — Fires involving energized electrical equipment.

Class Continuity — Principle of instruction which states that all information throughout a course should be in presented in a logical, understandable pattern.

Class D Fire — Fires of combustible metals such as magnesium, sodium, and titanium.

Class I Harness — Ladder-belt-type harness that is worn around the wearer's waist and used only to secure firefighters to a ladder or other object. *Also known as* Pompier Belt. *See* Life Safety Harness.

Class II Harness — Sit-type harness designed to support the weight of two people (victim and rescuer). *See* Life Safety Harness.

Class III Harness — Sit-type harness designed to support two people. However, this type has additional support over the shoulders that is designed to prevent the wearer from becoming inverted on the rope. *See* Life Safety Harness.

Class K Fire — Fires in cooking appliances that involve combustible cooking media, such as vegetable or animal oils and fats; commonly occurring in commercial cooking facilities such as restaurants and institutional kitchens.

Claustrophobia — Pathological fear of confined spaces.

Claw Tool — Forcible entry tool having a hook and a fulcrum at one end and a prying blade at the other.

Clean Agent — Fire suppression media that leaves little or no residue when used.

Clean Agent Fire Suppression System — System that uses special extinguishing agents that leave little or no residue.

Clean-Agent Fire-Extinguishing System — System that uses special extinguishing agents that leave little or no residue.

Clean-Burn Pattern — Fire pattern found on noncombustible surfaces where there has been direct contact with or intense radiant heat on the surface; the direct flame contact burns away any accumulated soot or smoke deposits on the surface, leaving demarcation lines.

Cleanout Fitting — Fitting installed in the top of a tank to facilitate washing the tank's interior.

Clear Dimensions — Interior compartment measurements made from the inside surface of one wall to the inside surface of the opposite wall.

Clear Text — Use of plain English in radio communications transmissions. No 10-codes or agency specific codes are used when using clear text.

Clear Vision — Central focus of human vision; provides sharp, in-focus pictures.

Clear Width — Actual unobstructed opening size of an exit. *See* Exit.

Cleat — (1) Fitting consisting of two arms fastened on deck, around which mooring lines may be secured. (2) Strip of wood or metal to give additional strength, prevent warping, or hold in place. (3) Small pieces of wood used to secure other parts of a shoring system.

Clerestory — Windowed space that rises above lower stories to admit air, light, or both.

Clevice — U-shaped shackle attached with a pin or bolt to the end of a chain.

Clevis Hook — Type of hook attached at the end of a rope, web sling, or chain engineered for rapid connection; commonly used during vehicle extrication and rescue incidents.

Clinical Death — Absence of life signs; no breathing and no pulse or blood pressure. Occurs immediately after the onset of cardiac arrest or during many other types of medical emergencies.

Clinical Social Worker — A frontline health professional who often serves as a case manager to assist clients with information, referral, and direct help in dealing with the mental health intervention process. A clinical social worker can interview and counsel clients, but cannot write prescriptions.

Clinician — A degreed health professional who is licensed to perform specific services.

Clinker — Stony matter fused together by heat.

Closed Fracture — Fracture in which there is no break in the overlying skin.

Closed Sprinkler — Sprinkler that is equipped with a heat-sensitive element, such as a fusible link or frangible bulb, that is rated at a fixed temperature; when heat rises past the preset temperature, the link melts or the bulb bursts, causing the head to open. May be used in wet-pipe, dry-pipe, or pre-action sprinkler systems. *See* Foam-Water Sprinkler and Open Sprinkler.

Closed-Circuit Self-Contained Breathing Apparatus — SCBA that recycles exhaled air; removes carbon dioxide and restores compressed, chemical, or liquid oxygen. Not approved for fire fighting operations. *Also known as* Oxygen-Breathing Apparatus (OBA) or Oxygen-Generating Apparatus.

Cloud — Ball-shaped pattern of an airborne hazardous material where the material has collectively risen above the ground or water at a hazardous materials incident. *See* Cone, Hemispheric Release, and Plume.

Clove Hitch — Knot that consists of two half hitches; its principal use is to attach a rope to an object such as a pole, post, or hose.

CNC — *See* Condensation Nuclei Counter.

CNG — *See* Compressed Natural Gas.

CO — *See* Carbon Monoxide.

CO$_2$ — *See* Carbon Dioxide.

Coach Space — Standard seating areas within a train car or airliner.

Coaching — (1) Process in which instructors direct the skills performance of individuals by observing, evaluating, and making suggestions for improvement. (2) Intensive tutoring given to subordinates regarding a particular activity.

Coalition — Formal, mutual relationship between or among organizations who agree to help each other in specific ways to reach a specific goal; relationship is often formalized with a written agreement that spells out what kind of help each organization will provide the other.

Coaming — Raised framework around deck or bulkhead openings; used to prevent entry of water.

Coarse Aggregates — Crushed stone, gravel, cinders, shale, lava, pumice, vermiculite, etc.

Cockloft — Concealed space between the top floor and the roof of a structure. *See* Attic.

Cockpit — Fuselage compartment occupied by pilots while flying the aircraft.

Cockpit Voice Recorder — Recording device installed in most large civilian aircraft to record crew conversation and communications; intended to assist in an accident investigation by helping to determine the probable cause of the accident.

Code — A collection of rules and regulations that has been enacted by law in a particular jurisdiction. Codes typically address a single subject area; examples include a mechanical, electrical, building, or fire code. *See* Building Code, Regulation, and Standard.

Code 1 — Operation of an emergency vehicle under non-emergency response conditions. Driver proceeds at his or her convenience, no warning devices are being used, and all traffic laws are followed.

Code 2 — Operation of an emergency vehicle in which the driver proceeds immediately, obeys all traffic laws, and uses visual warning devices but no audible warning devices. This is prohibited in most jurisdictions.

Code 3 — Operation of an emergency vehicle under emergency response conditions using visual and audible warning devices.

Code Enforcement — Process of enforcing a body of law aimed at reducing fire and life-safety hazards as well as mandating the proper installation and maintenance of building/structure fire and life-safety features to provide adequate community fire prevention.

Code Enforcement Officer — *See* Inspector.

Code for Safety to Life from Fire in Buildings and Structures — Old title of NFPA® 101, *Life Safety Code®*.

Code of Ethics — statement of behavior that is right and proper conduct for an individual functioning within an organization or society as a whole.

Code of Federal Regulations (CFR) — Rules and regulations published by executive agencies of the U.S. federal government. These administrative laws are just as enforceable as statutory laws (known collectively as federal law), which must be passed by Congress.

Coefficient of Discharge — Correction factor used in hydraulic calculations to account for irregularities in the shape of the hydrant discharge orifice. Frequently applied in computing the flow from a hydrant.

Coercion — Act of forcing or compelling someone (by use of threats, authority, or any other means) to comply, perform an act, or make a choice.

Coercive Power — Power to punish or impose sanctions on those who fail to behave in a prescribed manner.

COFC — *See* Container-on-Flatcar.

Coffer Dam — (1) Watertight enclosure, usually made of sheet piling, that can be pumped dry to permit construction inside. (2) Narrow, empty space (void) between compartments or tanks of a vessel that prevents leakage between them; used to isolate compartments or tanks.

COG — *See* Continuity of Government.

Cognition — Concept that refers to all forms of knowing, including perceiving, imagining, reasoning, and judging.

Cognitive Domain of Learning — Core domain of learning that involves knowledge recall and use of intellectual skill. Examples of cognitive process include how a person comprehends information, organizes subject matter, applies knowledge, and chooses alternatives.

Cognitive Evaluation — Assessment of knowledge that shows cognitive or knowing level by requiring that learners respond appropriately to questions on various types of tests.

Cognitive Learning Domain — Learning that relates to knowledge and intellectual skills (facts and information), emphasizing thought rather than feeling or movement and involving the learning of concepts and principles. *See* Learning Domain.

COHB — *See* Carboxyhemoglobin.

Cohesion — Act of binding together substances of like composition.

Cohesiveness — Property of sticking or binding together; often refers to the consistency of soil.

Coil Spring Suspension — Suspension system consisting of numerous spirally-bound, elastic steel bodies that recover their shape after being compressed, bent, or stretched.

Cold Smoke — Smoke from a fire that lacks any substantial heat.

Cold Trailing — Constructing a minimum fire line along the perimeter of a wildland fire after the perimeter is relatively cold, in order to ensure no further advance of the fire. Accomplished by carefully inspecting and feeling with the hand to detect any fire, digging out every live spot, and trenching any live edge.

Cold Zone — Safe area outside of the warm zone where equipment and personnel are not expected to become contaminated and special protective clothing is not required; the incident command post and other support functions are typically located in this zone. *Also known as* Support Zone.

Collaboration — Partnering with one or more person to accomplish a specific project.

Collaborative Partnership — Plans implemented by a partnership of agencies working to reduce risks of common interest.

Collapse Zone — Area beneath a wall in which the wall is likely to land if it loses structural integrity.

Collapsible Ladder — *See* Folding Ladder.

Collar Method — Means of attaching a coupling to a hose with a two- or three-piece collar, which is bolted into place.

Collision Beam — (1) Structural member within a vehicle door designed to prevent the door from collapsing inward if struck. (2) Heavy-gauge steel member strategically located in the sidewall of a bus. Collision beams limit penetration of an object into the passenger compartment of a vehicle.

Collision Bulkhead — Stronger-than-normal bulkhead located forward to control flooding in the event of a head-on collision.

Colorimetric Tube — Small tube that changes color when contaminated air is drawn through it. *Also known as* Detector Tube.

Column — Vertical supporting member.

Column Footing — Square pad of concrete that supports a column.

Combination Aircraft — Large aircraft with a passenger cabin in the front of the aircraft and a separate cargo compartment in the rear of the aircraft. *Also known as* Combies.

Combination Apparatus — Piece of fire apparatus designed to perform more than one function; usually called Triple Combinations, Quads, or Quints.

Combination Attack — Extinguishing a fire by using both a direct and an indirect attack; this method combines the steam-generating technique of a ceiling level attack with an attack on the burning materials near floor level. A water or foam stream is moved around a compartment in an O, T, or Z pattern; this movement allows the extinguishing agent to be applied to the fire and to the surrounding uninvolved fuel. *See* Attack Methods (1).

Combination Detection and Alarm Systems — Systems that have both fire and burglar alarms, with the fire alarm signal overriding the burglar alarm.

Combination Detector — Alarm-initiating device that is capable of detecting an abnormal condition by more than one means. The most common combination detector is the fixed-temperature/rate-of-rise heat detector.

Combination Fire Department — Organization in which some of the firefighters receive pay while other personnel serve on a voluntary basis. In Canada, a combination fire department is called a Composite Department.

Combination Ladder — Ladder that can be used as either a single, extension, or A-frame ladder.

Combination Lay — Hose lay in which two or more hoselines are laid in either direction - water source to fire or fire to water source.

Combination Nozzle — Nozzle designed to provide either a solid stream or a fixed spray pattern suitable only for mop-up. Not to be confused with an adjustable fog nozzle.

Combination Packaging — Shipping container consisting of one or more inner packagings secured in a nonbulk outer packaging. *See* Packaging (1).

Combination Smoke Detector — Smoke-sensing device consisting of both photoelectric and ionization smoke detectors.

Combination Spreader/Shears — Powered hydraulic tool consisting of two arms equipped with spreader tips that can be used for pulling or pushing. The insides of the arms contain cutting shears.

Combination System — Water supply system that is a combination of both gravity and direct pumping systems. It is the most common type of municipal water supply system.

Combine — Large, self-propelled machine that cuts, threshes, and cleans crops as it drives across a field.

Combined-Agent Vehicle — Type of rapid intervention vehicle that is designed to apply multiple types of fire-extinguishing agents on aircraft crash incidents.

Combplate — Grooved plate at the top of an escalator. The grooves in the plate mesh with matching ridges in the stair treads to prevent shoes from being caught in the crevice as the stair treads move under the plate.

Combustible Gas Detector — Indicates the explosive levels of combustible gases.

Combustible Liquid — Liquid having a flash point at or above 100°F (37.8°C) and below 200°F (93.3°C). *See* Flammable Liquid and Flash Point.

Combustion — A chemical process of oxidation that occurs at a rate fast enough to produce heat and usually light in the form of either a glow or flame. (Reproduced with permission from NFPA® 921-2011, Guide for Fire and Explosion Investigations, Copyright©2011, National Fire Protection Association®)

Come-Along — Manually operated pulling tool that uses a ratchet/pulley arrangement to provide mechanical advantage.

Comm Center — *See* Telecommunications Center.

Command — (1) Act of directing, ordering, and/or controlling resources by virtue of explicit legal, agency, or delegated authority. (2) Term used on the radio to designate the incident commander. (3) Function of NIMS-ICS that determines the overall strategy for the incident, with input from throughout the ICS structure.

Command Post (CP) — Designated physical location of the command and control point where the incident commander and command staff function during an incident, and where those in charge of emergency units report to be briefed on their respective assignments. The command post may be co-located with the base. *See* Incident Command Post.

Command Staff — In a fully developed fireground organization, the Information Officer (PIO), Safety (ISO), and Liaison Officer, who report directly to the IC.

Commercial Aviation Aircraft — Airline, commuter, cargo, and fire fighting aircraft.

Commercial Chassis — Truck chassis produced by a commercial truck manufacturer. These chassis are in turn outfitted with a rescue or fire fighting body.

Commercial Motor Coach — Custom-built bus designed to carry groups of people to a specific destination, usually a long distance away. These buses may run regularly scheduled routes, or they may be specially chartered. *Also known as* Charter Bus and Touring Bus.

Commissioner — Member of city or county government; the fire commissioner represents the fire department on the government ruling body. In some cases, there is no commissioner, and the fire chief is the ranking official directly responsible to the government.

Commodity Flowcharting — Portrays the movement of a tangible item, such as money or stolen property, through a system.

Common Brick — Fired clay brick with a plain, unfinished surface.

Common Conductor — *See* Grounded Conductor.

Common Fire Hazards — Those that are prevalent in almost all occupancies and encourage a fire to start.

Common Freight — Goods, other than passengers' baggage, transported to a specific destination by a regularly scheduled hauler, such as a bus or train.

Common Hazard — Condition likely to be found in almost all occupancies and generally not associated with a specific occupancy or activity.

Common Law — Law not created by legislative action but based on certain commonly held customs, traditions, and beliefs within a particular culture.

Common Path of Travel — Route of travel used to determine measured egress distances in code enforcement. The common path of travel is considered to be down the center of a straight corridor and a 1-foot (0.3 m) radius around each corner. *Also known as* Normal Path of Travel.

Communicable Disease — Disease that is capable of being transmitted from one person to another.

Communication — (1) Two-way process of transmitting and receiving some type of message. (2) Exchange of ideas and information that conveys an intended meaning in a form that is understood. (3) Ongoing process that educators and their audiences use to complete the exchange of information and attitudes about fire and life safety.

Communication Barriers — Poor listening habits or environments that get in the way of communication.

Communications Center — *See* Telecommunications Center.

Communications Unit — (1) Functional unit within the service branch of the logistics section of the incident management system; responsible for the incident communications plan, installation and repair of communications equipment, and operation of the incident communications center. (2) Vehicle used to provide the major part of an incident communications center.

Community Analysis — Process of creating a risk profile that identifies leading risks, who is affected, and where problems are occurring.

Community Emergency Response Team (CERT) — Groups of people within neighborhoods, community organizations, or the work place who are trained by the emergency services in basic response skills.

Community Master Plan — Medium- to long-range plan for the growth and development of a community; generally written through the interaction of all city agencies that may be affected by future developments.

Community Stakeholder — Person or group that is affected by or has a vested interest in a specific issue.

Community-Based Program — Activities that support an overall risk-reduction strategy and occur throughout the community.

Community-Risk Reduction — The process of addressing the larger issue of preventable injury that is occurring in a community. The process involves identifying leading risks and creating mitigation strategies through the use of integrated prevention interventions.

Companionway — Interior stair-ladder, usually enclosed, that is used to travel from deck to deck.

Company — (1) Basic fire fighting organizational unit consisting of firefighters and apparatus; headed by a company officer. (2) Term that encompasses the whole crew of a vessel.

Company Log — Record of the activities of a fire company; usually kept by a company officer.

Company Officer — Individual responsible for command of a company. This designation is not specific to any particular fire department rank (may be a firefighter, lieutenant, captain, or chief officer if responsible for command of a single company).

Comparison Sample — Evidence collected from undamaged areas or materials to offer a comparison to similar materials damaged by a fire.

Compartment — Interior space (room) of a vessel; numbered from forward to aft with odd numbers on starboard side and even numbers on port side.

Compartment Syndrome — Result of traumatic injury where the patient's muscle tissue becomes swollen and tightly encased. At four to six hours, crushed tissue begins to die and release toxins, which decreases the potential for saving the limb.

Compartmentalization — Systematic venting of a structure by controlling which windows and doors are opened at any given time.

Compartmentation — (1) Series of barriers designed to keep flames, smoke, and heat from spreading from one room or floor to another. (2) Subdividing of a vessel's hull by transverse watertight bulkheads; may allow a vessel to stay afloat under certain flooding conditions. *See* One-Compartment Subdivision.

Compartmentation Systems — Series of barriers designed to keep flames, smoke, and heat from spreading from one room or floor to another; barriers may be doors, extra walls or partitions, fire-stopping materials inside walls or other concealed spaces, or floors.

Compensation/Claims Unit — Functional unit within the finance/administrative section of an incident management system; responsible for financial concerns resulting from injuries or fatalities at an incident.

Compensatory Measure — Measure intended to compensate for a code deficiency; may be either administrative or physical. Provides approximately the same level of safety performance as the intent of the deficient element that it replaces.

Compensatory Time — Often referred to as "comp time." Work time earned by an individual that is space-banked by the employer. This time off may be used by the employee in place of vacation, holiday, or sick leave.

Competency-Based Learning (CBL) — Training that emphasizes knowledge and skills that are required on the job. Course objectives involve specific, criteria-based competence in performing tasks or understanding concepts that learners will use in their daily work. *Also known as* Criterion-Referenced and Performance-Based Learning.

Competent Ignition Source — A competent ignition source will have sufficient temperature and energy and be in contact with the fuel long enough to raise it to its ignition temperature.

Competent Person — A person with the ability and authority to identify and address hazards in an environment.

Complaint — (1) Objection to an existing condition that is brought to the attention of a fire prevention bureau or a building department, usually by a citizen. (2) In court, a formal charge or accusation.

Complement — (1) All firefighters assigned to a working unit, or the number of units assigned to a given alarm. (2) Equipment assigned to a piece of apparatus.

Complex Buildings — Structures such as manufacturing, health care, or multi-use buildings that are more structurally complex than single-family dwellings.

Complex Loop — Piping system that is characterized by one or more of the following: more than one inflow point, more than one outflow point, and/or more than two paths between inflow and outflow points. *Also known as* Grid System or Gridded Piping System.

Complex Mechanical Advantage System — A rope rigging system using a series of pulleys that move in different directions. The mechanical advantage is summative as multiple systems are joined.

Complex/Problematic Firesetting Case — A juvenile firesetting situation that is motivated or exacerbated by cognitive, social, or environmental dysfunction.

Compliance — Meeting the minimum standards set forth by applicable codes or regulations. *See* Code and Regulation.

Composite Cylinder — Lightweight air cylinder made of more than one material; often aluminum wrapped with fiberglass.

Composite Department — *See* Combination Fire Department.

Composite Materials — Plastics, metals, ceramics, or carbon-fiber materials with built-in strengthening agents. These materials are much lighter and stronger than the metals formerly used for such aircraft components as panels, skin, and flight controls.

Composite Packaging — Single container made of two different types of material. *See* Packaging (1).

Composite Panels — Produced with parallel external face veneers bonded to a core of reconstituted fibers.

Compound — Substance consisting of two or more elements that have been united chemically.

Compound Fracture — Open fracture; a fracture in which there is an open wound of the skin and soft tissues.

Compound Gauge — Pressure gauge capable of measuring above and below atmospheric pressure; commonly used to measure the intake pressure on a fire pump.

Compound Mechanical Advantage System — A rope rigging system using a series of travelling pulleys that move in the same direction. The mechanical advantage is multiplicative as multiple systems are joined.

Compound Tackle — Two or more blocks reeved with more than one rope.

Comprehensive Environmental Response, Compensation and Liability Act (CERCLA) — U.S. law that created a tax on the chemical and petroleum industries, and provided broad federal authority to respond directly to releases or threatened releases of hazardous substances that may endanger public health or the environment.

Comprehensive Test — Type of test typically given in the middle (midterm) or at the end (final) of instruction that measures terminal performance of program participants and whether they have achieved program objectives. *See* Test.

Compress — Folded cloth or pad used for applying pressure to stop hemorrhage, or as a wet dressing.

Compressed Air — Air under greater than atmospheric pressure; used as a portable supply of breathing air for SCBA or to operate pneumatic tools.

Compressed Air Foam System (CAFS) — Generic term used to describe a high-energy foam-generation system consisting of an air compressor (or other air source), a water pump, and foam solution that injects air into the foam solution before it enters a hoseline.

Compressed Gas — Gas that, at normal temperature, exists solely as a gas when pressurized in a container, as opposed to a gas that becomes a liquid when stored under pressure. *See* Gas, Liquefied Compressed Gas, and Nonflammable Gas.

Compressed Gas Association (CGA) — Trade association that writes standards pertaining to the use, storage, and transportation of compressed gases.

Compressed Natural Gas — Natural gas that is stored in a vessel at pressures of 2,400 to 3,600 psi (16 800 kPa to 25 200 kPa).

Compressed Gas (Tube) Trailer — Cargo tank truck that carries gases under pressure; may be a large single container, an intermodal shipping unit, or several horizontal tubes. *Also known as* Tube Trailer. *See* Cargo Tank Truck.

Compression — Vertical and/or horizontal forces that tend to push the mass of a material together; for example, the force exerted on the top chord of a truss.

Compressor — Machine designed to compress air or gas.

Computer-Aided Design (CAD) — Computer technology used to design, draw, or draft technical products, rooms, or entire buildings in two or three dimensions; it is also used to create animations from static drawings and to convert existing drawings and diagrams into computer models.

Computer-Based Training (CBT) — A variety of self-study in which the student completes work on a computer with minimal communication with an instructor. *Also known as* E-learning, Blended E-learning, or Online Instruction.

Computer-Generated Slide Presentations ☒ Computer presentations that are sequenced and displayed like traditional slideshows that use a slide projector with a carousel. Popular software for creating and viewing these presentations include Microsoft PowerPoint® and Apple Keynote®.

Concealed Space — Structural void that is not readily visible from a living/working space within a building, such as areas between walls or partitions, ceilings and roofs, and floors and basement ceilings through which fire may spread undetected; also includes soffits and other enclosed vertical or horizontal shafts through which fire may spread.

Concentrated Load — Load that is applied at one point or over a small area.

Concentration — (1) Quantity of a chemical material inhaled for purposes of measuring toxicity. (2) Percentage (mass or volume) of a material dissolved in water (or other solvent). *See* Dose and Lethal Concentration, 50 Percent Kill (LC_{50}).

Concrete — Strong, hard building material produced from a mixture of portland cement and an aggregate filler/binder to which water is added to form a wet, moldable slurry that sets into a rigid building material. Concrete is fireproof, watertight, and comparatively inexpensive to make. The aggregates used in concrete are inert mineral ingredients that reduce the amount of cement that otherwise would be needed. In structural concrete, the filler/binder is usually sand and/or gravel. Lightweight concrete, used as sound-proofing material, may use sand and/or vermiculite.

Concrete Admixture — Substance added to concrete to aid in imparting color, waterproofing, controlling workability, controlling the hardening process, and entraining air.

Concrete Block — Large rectangular brick used in construction; the most common type is the hollow concrete block. *Also known as* Concrete Masonry Units (CMU).

Concrete Block Brick Faced — Wall construction system that includes one wythe of concrete blocks with a brick wythe attached to the outside. *See* Course, Header Course, and Wythe.

Concrete Operational Stage of Intellectual Development — Third stage of intellectual development; spans ages 7 to 11 years. Leaner can perform operations logically with information as long as it can be applied to specific or concrete examples.

Condensation — Process of going from the gaseous to the liquid state.

Condensation Nuclei Counter (CNC) — Quantitative fit-test protocol using a counting instrument that quantitatively fit-tests respirators with the use of a probe. The probed respirator has a special sampling device installed on the respirator that allows the probe to sample the air from inside the mask.

Condition — Part of an educational objective that describes under what provisions and with what resources the learner should be able to act or fulfill the objective.

Conduction — Physical flow or transfer of heat energy from one body to another, through direct contact or an intervening medium, from the point where the heat is produced to another location, or from a region of high temperature to a region of low temperature. *See* Convection, Heat, Heat Transfer, Law of Heat Flow, and Radiation.

Conductivity — Ability of a substance to conduct an electrical current.

Conductivity Readings — Form of nondestructive testing used on aluminum aerial devices. Changes in the integrity of material in a certain area will be reflected by a divergence of conductivity readings.

Conductor — Substance or material that transmits electrical or thermal energy. *See* Dielectric, Semiconductor, and Thermocouple.

Cone — Triangular-shaped pattern of an airborne hazardous material release with a point source at the breach and a wide base downrange. *See* Cloud, Hemispheric Release, and Plume.

Cone of Learning — Visual representation that depicts in what percentage of information human beings retain using their senses alone and in combination.

Cone Roof Storage Tank — Fixed-site vertical atmospheric storage tank that has a cone-shaped pointed roof with weak roof-to-shell seams that are intended to break when excessive overpressure results inside. Used to store flammable, combustible, and corrosive liquids. *Also known as* Dome Roof Tank. *See* Atmospheric Storage Tank, External Floating Roof Tank, and Internal Floating Roof Tank.

Conference Discussion — Discussion in which a group attempts to solve a common problem.

Confidentiality — A principle of law and professional ethics that recognizes the privacy of individuals.

Confine a Fire — To restrict the fire within determined boundaries established either prior to the fire or during the fire. *Also known as* Confinement.

Confined Space — Space or enclosed area not intended for continuous occupation, having limited (restricted access) openings for entry or exit, providing unfavorable natural ventilation and the potential to have a toxic, explosive, or oxygen-deficient atmosphere.

Confinement — (1) The process of controlling the flow of a spill and capturing it at some specified location. (2) Fire fighting operations required to prevent fire from extending from the area of origin to uninvolved areas or structures. *See* Confine a Fire and Containment.

Conflagration — Large, uncontrollable fire covering a considerable area and crossing natural fire barriers such as streets; usually involves buildings in more than one block and causes a large fire loss. Forest fires can also be considered conflagrations.

Connection Box — Contains fittings for trailer emergency and service brake connections and an electrical connector to which the lines from the towing vehicle may be connected. *Formerly known as* Bird Box, Junction Box, or Light Box.

Consensus — General agreement; the judgment or agreement arrived at by most of those in a group.

Consensus Standard — Rules, principles, or measures that are established though agreement of the members of the standards-setting organization.

Consent — In terms of legal right of entry, refers to the granting of access to a scene by the lawful owner of the property where the incident occurred; consent is a courtesy on the part of the private property owner and as such may be withheld at any time.

Conservation of Energy — Law of physics that states that the total amount of energy in an isolated system remains constant. As a result, energy cannot be created or destroyed.

Consignee — Person who is to receive a shipment.

Consist — Rail shipping paper that contains a list of cars in the train by order; indicates the cars that contain hazardous materials. Some railroads include information on emergency operations for the hazardous materials on board with the consist. *Also known as* Train Consist. *See* Shipping Papers and Waybill.

Consistency — Quality of finished foam that has small bubbles of equal size - an important quality for all types of foam. *See* Finished Foam.

Constant Pressure Relay — Method of establishing a relay water supply utilizing two or more pumpers to supply the attack pumper. This method reduces the need for time-consuming and often confusing fireground calculations of friction loss.

Constitutional Law — Law based on the Constitution; all state/provincial laws must be consistent with the respective federal constitution.

Constrict — To be made smaller by drawing together or squeezing.

Construction Classification — Classification given to a particular building by a building code, based on its construction materials, methods, and ability to resist the effects of a fire situation. *See* Building Code.

Construction Plan — Visual depiction of how a building and all of its many components are to come together; serves as a medium for conveying information to those who need it for the construction of a building. *See* Floor Plan, Plot Plan, and Site Plan.

Consumable Materials — Instructional materials limited to one-time use because they are designed to be "consumed" or used up.

Consumer Product Safety Commission (CPSC) — U.S. government agency charged with protecting the public from unreasonable risks of serious injury or death from more than 15,000 types of consumer products under the agency's jurisdiction, including hazardous materials intended for consumer purchase and use. CPSC also operates the National Electronic Injury Surveillance System (NEISS) database, collecting data based on a sample of hospital emergency rooms and focusing on the role of consumer products in fire and burn injuries.

Contact Paper — Vinyl- or paper-like material that has a strong adhesive pre-applied to one side.

Contagious — Capable of transmission from one person to another through contact or close proximity.

Contained Fire — Fire whose progress has been stopped but for which the control line is not yet finished.

Container — (1) Article of transport equipment that is: (a) of a permanent character and strong enough for repeated use; (b) specifically designed to facilitate the carriage of goods by one or more modes of transport without intermediate reloading; and (c) fitted with devices permitting its ready handling, particularly its transfer from one mode to another. The term "container" does not include vehicles. *Also known as* Cargo Container or Freight Container. (2) Box of standardized size used to transport cargo by truck or railcar when transported overland or by cargo vessels at sea; sizes are usually 8 by 8 by 20 feet or 8 by 8 by 40 feet (2.4 m by 2.4 m by 6 m or 2.4 m by 2.4 m by 12.2 m). *See* Bulk Container, Carboy, Container Specification Number, Container Terminal, Container Vessel, Dewar, Intermodal Container, Reefer Container, Refrigerated Intermodal Container.

Container Chassis — Trailer chassis consisting of a frame with locking devices for securing and transporting a container as a wheeled vehicle.

Container Ship — Ship specially equipped to transport large freight containers, typically in vertical container cells; containers are usually loaded and unloaded by special cranes. *Also known as* Container Vessel.

Container Specification Number — Shipping container number preceded by letters "DOT" that indicates the container has been built to U.S. federal specifications.

Container Terminal — Facility for loading and unloading cargoes shipped in standard 20 foot or 40 foot long containers and their stowage; usually accessible by truck, railroad, and marine transportation. *See* Container and Reefer Container.

Container Vessel — *See* Container Ship.

Container-on-Flatcar (COFC) — Rail flatcar used to transport highway transport containers.

Containment — The act of stopping the further release of a material from its container. *See* Confinement.

Contaminant — Foreign substance that compromises the purity of a given substance. *See* Contamination.

Contamination — (1) Impurity resulting from mixture or contact with a foreign substance. (2) General term referring to anything that can taint physical evidence during a fire investigation. *See* Contaminant, Decontamination, and Surface Contamination.

Contents — Furnishings, merchandise, and any machinery or equipment not part of the building structure.

Continuity of Government (COG) — Process to ensure continuation of the government. *See* Continuity of Operations (COOP).

Continuity of Operations (COOP) — Defining plans for the continuity of operations in all state agencies and an overall plan for the continuity of government; includes emergency evacuation procedures, firm definitions of the leader's emergency powers, line of succession for all agency officials, and plans and programs for business continuity in the private sector. *See* Continuity of Government (COG).

Continuous Fuels — Fuels distributed uniformly over an area, thereby providing a continuous path for fire to spread. *See* Fuel Continuity.

Continuous Halyard — Halyard on which both ends are attached to the bottom rung of the fly section of an extension ladder. The rope is run from the bottom rung of the fly section, down around the bottom rung of the bed section, and back up to the bottom rung of the fly section.

Continuum of Care — The level and intensity of services needed to facilitate a positive outcome.

Contractor — Individual who performs construction on a property for a predetermined cost or fee.

Contractual Entry — Legal entry to a scene by those with a privately defined jurisdiction, such as private investigators representing an insurance company that insures the property where the incident occurred.

Contractual Sleeve Binding — Method of attaching couplings to fire hose with a tension ring that compresses a nylon sleeve to lock the hose onto the coupling shank.

Control — Point in time when progress of a fire has been halted, such as when the perimeter spread of a wildland fire has been halted and can reasonably be expected to hold under foreseeable conditions. When the fire is under control, the release of fire fighting resources can begin.

Control Agent — Material used to contain, confine, neutralize, or extinguish a hazardous material or its vapor.

Control Center — Telecommunications center used by the fire service for emergency communications. There are also mobile command posts that can be taken directly to the fire scene and function as the incident operational control center.

Control a Fire — To complete control line around a fire, any spot fire there from, and any interior island to be saved; to burn out any unburned area adjacent to the fire side of the control lines; and to cool down all hot spots that are immediate threats to the control line until the lines can reasonably be expected to hold under foreseeable conditions.

Control Line — Inclusive term for all constructed or natural barriers (or combinations thereof) and treated fire edges that ultimately contain and control the fire; not to be confused with fire line.

Control Pedestal — Central location for most or all of the aerial device controls. Depending on the type and manufacturer of the apparatus, the control pedestal may be located on the turntable, on the rear or side of the apparatus, or in the elevating platform. *Also known as* Pedestal.

Control Zones — *See* Hazard-Control Zones.

Control Zones — System of barriers surrounding designated areas at emergency scenes intended to limit the number of persons exposed to the hazard, and to facilitate its mitigation. At a major incident there will be three zones — restricted (hot), limited access (warm), and support (cold). *See* Hazard-Control Zones.

Controlled Airport — Airport with an operating control tower, typically staffed by FAA personnel.

Controlled Breathing — Technique for consciously reducing air consumption by forcing exhalation from the mouth and allowing natural inhalation through the nose.

Controlled Burning — (1) Fires intentionally set in vegetative fuels for the purpose of burning debris or accumulations of wildland fuels, such as grass and brush, to reduce available fuel and prevent the occurrence of uncontrolled wildland fires. May be done as part of a fuel-management program to prevent or reduce the rate of spread of wildland fires. See Prescribed Burning. (2) Any burn that is safely set and controlled for the purposes of fire and emergency services training.

Controller — Electric control panel used to switch a fire pump on and off and to control its operation.

Convection — Transfer of heat by the movement of heated fluids or gases, usually in an upward direction. *See* Conduction, Heat, Heat Transfer, Law of Heat Flow, and Radiation.

Convection Column — Rising column of heated air or gases above a continuing heat or fire source. *Also known as* Thermal Column.

Convenience Outlet — Electrical outlet that can be used for lamps and other appliances.

Convenience Stair — Stair that usually connects two floors in a multistory building.

Convergent Volunteers — Individuals who arrive at the scene of a disaster or emergency in order to help, but were not dispatched to the scene through official channels.

Convulsant — Poison that causes convulsions. *See* Poison.

Convulsion — Seizure or violent involuntary contraction (or series of contractions) of the voluntary muscles.

Cooling — (1) Act of lowering the temperature of the fuel and adjacent surfaces. (2) Reduction of heat by the quenching action or heat absorption of the extinguishing agent.

COOP — *See* Continuity of Operations.

Cooperating Agency — Agency supplying assistance to the incident control effort, other than direct suppression, rescue, support, or service functions; for example, the Red Cross, a law enforcement agency, or a telephone company.

Cooperative Agreement — Written agreement between fire protection agencies agreeing to cooperate in actions or share resources for a common good.

Cooperative Learning — Activity-based instruction that seeks to involve all students in the learning process.

Cope Steel — To cut a flange section in order to avoid interference with other structural members.

Copyright Law — Law designed to protect the competitive advantage developed by an individual or organization as a result of their creativity. *See* Law and Statute.

Corbel — Bracket or ledge made of stone, wood, brick, or other building material projecting from the face of a wall or column used to support a beam, cornice, or arch. *See* Corbelling.

Corbelling — Use of a corbel to provide additional support for an arch. *See* Corbel.

Core Temperature — Body temperature measured in deep structures such as the lungs or liver.

Corner Fittings — Strong metal devices located at the corners of a container having several apertures that normally provide the means for handling, stacking, and securing the freight container.

Corner Structures — Vertical frame components located at the corners of a container; integral with the corner fittings.

Cornice — Concealed space near the eave of a building; usually overhanging the area adjacent to exterior walls.

Coronary Arteries — Blood vessels that supply blood to the walls of the heart.

Coroner — Official chiefly responsible for investigating deaths, particularly under unusual circumstances, and determining the cause of death. *See* Medical Examiner.

Corpus Delicti — Evidence of substantial and fundamental facts necessary to prove the commission of a crime.

Corrective Maintenance — Reactive maintenance or repairs that are performed after a breakdown or mechanical failure.

Corrosive — Capable of causing corrosion by gradually eroding, rusting, or destroying a material. *See* Acid, Base, and Corrosive Material.

Corrosive Liquid Tank — Cargo tank truck that carries corrosive liquids, usually acids. *See* Cargo Tank Truck.

Corrosive Material — Gaseous, liquid, or solid material that can burn, irritate, or destroy human skin tissue and severely corrode steel. *Also known as* Corrosive. *See* Hazardous Material.

Corrugated — Formed into ridges or grooves; serrated.

Corrugated Hose — Hose shaped into folds or parallel and alternating ridges and grooves to improve flexibility.

Cost Unit — Functional unit within the finance/administrative section of an incident management system; responsible for tracking costs, analyzing cost data, making cost estimates, and recommending cost-saving measures.

Cost-Benefit Analysis — (1) Systematic methodology to compare costs and benefits to make cost-effective funding decisions on projects. (2) Examination of the proposed expense of an effort and deciding if the overall benefit is worth the investment of money and/or time. *See* Analysis and Impact Analysis.

COTP — *See* Captain of the Port.

Council of American Building Officials (CABO) — Former umbrella organization for BOCA (Building Officials and Code Administrators), ICBO (International Conference of Building Officials), and SBCCI (Southern Building Code Congress International). These agencies have merged to form the International Code Council (ICC).

Counseling — Advising learners or program participants on their educational progress, career opportunities, personal anxieties, or sudden crises in their lives.

Counterbalance Valve — Valves designed to prevent unintentional or undesirable motion of an aerial device from position.

Counterflow — Moving in opposite directions, for example, one group of people attempting to climb stairs while another group is trying to traverse down the stairs.

Countermeasures — Devices or systems designed to prevent sensor-guided weapons from locking onto and destroying a target.

County — Political subdivision of a state, province, or territory for administrative purposes and public safety. *Also known as* Parish.

Coupling — Fitting permanently attached to the end of a hose; used to connect two hoselines together, or to connect a hoseline to a device such as a nozzle, appliance, discharge valve, or hydrant.

Course — (1) Horizontal layer of individual masonry units. *See* Header Course and Wythe. (2) Series of lessons that lead to the completion of a discipline or certification.

Course Consistency — Principle of instruction that states that information throughout a course should have the same level of accuracy, be presented with the same equipment and training aids, and maintain a similar level of learning.

Course Description — Relates the basic goals and objectives of the course in a broad, general manner. It is designed to provide a framework and guide for further development of the course and also communicate the course content.

Course Objectives — Specific identification of the planned results of a course of instruction.

Course Outline — List of jobs and information to be taught to fulfill previously identified needs and objectives.

Courts of Queen's Bench — Canadian equivalent of the U.S. Federal Court System.

Cover — (1) Practice of moving unassigned fire companies into stations that have been emptied by another emergency. (2) To cover exposures by placing primary fire streams in advantageous positions to protect buildings or rooms exposed to heat and fire. (3) To protect with a salvage cover. (4) General term used to describe brush, grasses, and other natural ground covers.

Cover Letter — Letter explaining or containing additional information about an accompanying communication.

Covered Floating Roof Tank — *See* Internal Floating Roof Tank.

Covert — Not in the open; secret.

Cowl Flaps — Adjustable sections or hinged panels on the engine cowling of reciprocating engines; used to control the engine temperature.

Cowling — Removable covering around aircraft engines.

CP — *See* Command Post.

CPAT — *See* Candidate Physical Ability Test.

CPC — *See* Chemical Protective Clothing.

CPE — *See* Chlorinated Polyethylene.

C-Post — Post nearest the rear door handle on a four-door vehicle. On a two-door vehicle, the rear roof post is considered to be the C-post.

CPR — *See* Cardiopulmonary Resuscitation.

CPSC — *See* Consumer Product Safety Commission.

CPU — *See* Central Processing Unit.

Cradle — Rest designed to support the free end of the aerial device during road travel.

Crash Fire Rescue (CFR) — Old term used to describe the fire and rescue services provided at airport facilities. *Currently known as* Aircraft Rescue and Fire Fighting.

Crawl Space — Area between ground and floor, ceiling and floor above, or ceiling and roof, or any other structural void with a vertical dimension that does not allow a person to stand erect within the space. These spaces often contain ductwork, plumbing, and wiring.

Crazing — Formation of patterns of short cracks throughout a pane of glass, such as windows and mirrors, from the heat of fire. It is thought to be the result of heating of one side of a pane while the other side remains cool.

Creeping Fire — Fire burning with a low flame height and spreading slowly.

Creosote — Highly flammable byproduct of combustion composed of tars and other hydrocarbons that are distilled from carboniferous fuels as they burn; generally collects on cooler surfaces of flues and chimneys.

Crew — Organized group or specific number of emergency services personnel, under the leadership of a crew leader or other designated supervisor, that has been assembled for an assignment such as search, ventilation, or hoseline deployment and operations. The number of personnel in a crew should not exceed recommended span-of-control limits of three to seven people. Sometimes referred to as a "company" in municipal fire departments.

Crew List — Part of a vessel's papers listing the names and nationalities of every member of the crew, the capacity in which each member serves, and the amount of wages each member receives.

Crew Transport — Any vehicle capable of transporting a specified number of fire crew personnel in a specified manner.

Cribbing — (1) Lengths of solid wood or plastic, usually 4- X 4-inches (100 mm by 100 mm) or larger, used to stabilize vehicles and collapsed buildings during extrication incidents. (2) Process of arranging planks into a crate-like construction.

Crime Concealment Fire — Intentionally set fire intended to destroy evidence of another crime such as a homicide or burglary.

Criminal Law — Law intended to protect society by identifying certain conduct as criminal, and specifying the sanctions to be imposed on those who engage in criminal activity.

Criminal Search Warrant — Warrant issued with the intent of collecting evidence specifically to prove that a fire or explosion was intentionally set; issued when an administrative search leads to an investigator having probable cause that an arson or other crime was committed.

Crisis Communication — Preincident, incident, and postincident information or threat advisories provided to the public and news media.

Crisis Counseling — Programs to help relieve grieving, stress, or mental health problems caused or aggravated by a disaster or its aftermath.

Criterion — (1) Standard on which a decision or judgment is based. (2) One of the three requirements of evaluation. (3) The standard against which learning is compared after instruction. (4) The expected learning outcome; examples are Behavioral Objectives or NFPA® standards. Plural for the term is *criteria*.

Criterion-Referenced Assessment — Measurement of individual performance against a set standard or criteria, not against other students. Mastery learning is the key element to criterion-referenced testing. *Also known as* Criterion-Referenced Testing. *See* Mastery and Test.

Criterion-Referenced Learning — *See* Competency-Based Learning.

Critical Angle — Angle between legs of an anchor; it must be less than 90 degrees to avoid excessive loading of individual anchor points and components in an anchor system.

Critical Angle of List — Point at which critical events will occur. Not a point that remains constant in all cases; it is determined by stability calculations made by qualified personnel along with their professional judgment. *See* List and Heel.

Critical Criteria — Step or steps on a practical skills test that must be completed accurately in order for the student to pass the test.

Critical Flow Rate — Minimum flow rate at which extinguishment can be achieved.

Critical Incident Stress (CIS) — Physical, mental, or emotional tension caused when persons have been exposed to a traumatic event where they have experienced, witnessed, or been confronted with an event or events that involve actual death, threatened death, serious injury, or threat of physical integrity of self or others. *See* Post-Traumatic Stress Disorder (PTSD) and Stress (3).

Critical Incident Stress Debriefing (CISD) — Counseling designed to minimize the effects of psychological/emotional post-incident trauma on those at fire and rescue incidents who were directly involved with victims suffering from particularly gruesome or horrific injuries. *See* Post-Traumatic Incident Debriefing.

Critical Incident Stress Management (CISM) — Comprehensive crisis intervention system composed of 7 elements: pre-crisis preparation, a disaster or large scale incident, defusing, critical incident stress debriefing, one-on-one crisis intervention/counseling, family/organizational crisis intervention, and follow-up/referral mechanisms.

Critical Infrastructure — Systems, assets, and networks, whether physical or virtual, so vital that the incapacity or destruction of such systems and assets would have a debilitating impact on security, national economic security, national public health or safety, or any combination of those matters. *See* Infrastructure.

Critical Radiant Flux — Description of the amount of heat required to ignite a floor covering specimen in the critical radiant flux test; expressed as Btu per ft^2 (Watts per cm^2).

Critical Rescue and Fire Fighting Access Area (CRFFAA) — Rectangular area surrounding a runway. Its width extends 500 feet (150 m) outward from each side of the runway centerline, and its length extends 3,300 feet (1 100 m) beyond each runway end.

Cross Contamination — (1) Contamination of people, equipment, or the environment outside the hot zone without contacting the primary source of contamination. *Also known as* Secondary Contamination. *See* Contamination, Decontamination, and Hazard-Control Zones. (2) Evidence in one location at the scene that is moved to another location at the scene. (3) In terms of health safety, any spread of a harmful contaminant into an environment in which the contaminant should not normally be found.

Cross Main — Pipe connecting the feed main to the branch lines on which the sprinklers are located.

Crossover Line — Pipe that is installed in a bulk storage tank piping system that allows product unloading from either side of the tank.

Cross-training — Training emergency services personnel to function in more than one professional capacity; occurs most often when personnel are trained as firefighters and EMTs or paramedics.

Crosswind — Wind that is blowing in a direction from the side of an aircraft or foam stream; can affect a foam distribution pattern. *See* Downwind, Headwind, and Wind.

Croup — Common viral infection seen in small children; characterized by spasm of the larynx and resulting upper airway obstruction.

Crow Bar — Prying tool with a blade at either end; one end is significantly curved to provide additional mechanical advantage.

Crowd Control — Limiting the access of nonemergency personnel to the emergency scene.

Crown Fire — Fire that advances from top to top (canopy) of closely spaced trees or shrubs, more or less independent of a surface fire. Crown fires are sometimes classed as running or dependent to distinguish the degree of independence from the surface fire.

Crown Out — Fire that rises from ground level into the tree crowns and advances from treetop to treetop.

Cruising — Driving a vehicle in such a manner that an even speed and engine rpm is maintained. It is best to operate at 200 to 300 rpm below the maximum rpm recommended by the manufacturer.

Crush Points — Places within the frame of a vehicle that are designed to collapse, crush, deform, and otherwise absorb (not transmit) forces so as to minimize the impact on the passengers.

Crush Syndrome — Potentially fatal condition that occurs as a result of crushing pressure on a part of the body, typically the lower extremities. When blood flow to and from the injured area is absent for four to six hours, the injured tissue begins to die, giving off toxins; a sudden release of pressure may allow the toxins to flow into the bloodstream and to have an effect on other bodily organs.

Crushable Bumpers — Polystyrene foam or fluoroelastomer devices designed to absorb energy by flexing when struck.

Cryogenic Liquid — *See* Cryogens.

Cryogenic Liquid Storage Tank — Heavily insulated, vacuum-jacketed tanks used to store cryogenic liquids; equipped with safety-relief valves and rupture disks. *See* Cryogen.

Cryogenic Liquid Tank — Cargo tank truck that carries gases that have been liquefied by temperature reduction. *See* Cryogens and Cargo Tank Truck.

Cryogenics — The study of materials at very low temperatures.

Cryogens — Gases that are converted into liquids by being cooled below -130°F (-90°C). *Also known as* Refrigerated Liquids and Cryogenic Liquid. *See* Cryogenic Liquid Storage Tank and Cryogenic Liquid Tank.

CSA — *See* Canadian Standards Association.

CSR — Abbreviation for Customer Satisfaction Rating. *See* Customer Approval Rating (CAR).

CST — *See* Civil Support Team.

CT — Abbreviation used for a measure of toxicity determined by multiplying exposure concentration (C) in ppm by the time of exposure (T) in minutes and expressed as the CT product or ppm per minute.

Cubic Feet per Minute — Measure of a volume of material flowing past or through a specified measuring point.

Culture — The shared assumptions, beliefs, and values of a group or organization.

Cumulonimbus Clouds — Ultimate growth of a cumulus cloud into an anvil-shaped cloud with considerable vertical development, usually with fibrous ice crystal tops, and usually accompanied by lightning, thunder, hail, and strong winds.

Cumulus Clouds — Principal low-cloud type shaped into individual cauliflower-like cells of sharp nonfibrous outline, and having less vertical development than cumulonimbus clouds.

Curb Weight — Weight of an empty fire apparatus off the assembly line with no tools, water, equipment, or passengers.

Curbside — Side of a trailer nearest the curb when it is traveling in a normal forward direction (right-hand side); opposite to roadside.

Curie (Ci) — English System unit of measurement for radioactivity, indicating the number of nuclear decays/disintegrations a radioactive material undergoes in a certain period of time. *See* Becquerel, Radiation (2), and Radioactive Material (RAM).

Curing — (1) Maintaining conditions to achieve proper strength during the hardening of concrete. (2) Manufacturing step in making fire hose; the process of applying heat and pressure to "set" the shape of the tube and to increase its smoothness.

Curiosity-Motivated Firesetter — Child who is exploring his or her interest in fire through experimentation.

Curling — Method for raising a one-firefighter ladder from a flat rest position in preparation for carrying.

Current — (1) Rate of electrical flow in a conductor; measured in amperes. (2) The horizontal movement of water.

Curriculum — (1) Broad term that refers to the sequence of presentation, the content of what is taught, and the structure of ideas and activities developed to meet the learning needs of learners and achieve desired educational objectives; also the teaching and learning methods involved, how learner attainment of objectives is assessed, and the underlying theory and philosophy of education. (2) Series of courses in which students are introduced to skills and knowledge required for a specific discipline.

Curriculum Development — Using analysis, design, and evaluation to create a series of presentations that adhere to the four teaching steps (preparation, presentation, application, and evaluation) and address the learning needs of a particular audience or program. *Also known as* Instructional Design.

Curtain Boards — *See* Draft Curtains.

Curtain Door — Door used as a barrier to fire, consisting of interlocking steel plates or of a continuous formed spring steel "curtain." Curtain doors are often mounted in pairs, one door on the inside and the other on the outside of an opening.

Curtain Wall — Non-load-bearing exterior wall attached to the outside of a building with a rigid steel frame. Usually the front exterior wall of a building intended to provide a certain appearance.

Custom Chassis — Truck chassis designed solely for use as a fire or rescue apparatus.

Customer Approval Rating (CAR) — Organizational rating that adds a quantifiable raking to qualitative evaluations. *Also known as* Customer Satisfaction Rating (CSR).

Customer Service — Quality of an organization's relationship with individuals who have contact with the organization. There are internal customers such as the various levels of personnel and trainees, and external customers such as other organizations and the public. Customer service is the way these individuals, personnel, and organizations are treated, and their levels of satisfaction.

Cutters — *See* Powered Hydraulic Shears.

Cutting Tool — Hand or power tool used to cut a specific kind of material.

Cyanogen Agent — *See* Chemical Asphyxiant.

Cyanosis — Blueness of the skin due to insufficient oxygen in the blood.

Cyber Terrorism — Premeditated, politically motivated attack against information, computer systems, computer programs, and data which result in violence against noncombatant targets by sub-national groups or clandestine agents.

Cylinder — (1) Component of a locking mechanism that contains coded information for operating that lock, usually with a key. (2) Air tank portion of a self-contained breathing apparatus.

Cylinder Guard — Metal plate that covers a lock cylinder to prevent forceful removal.

Cylinder Plug — Part of a lock cylinder that receives the key. *Also known as* Key Plug.

Cylinder Pressure Gauge — Gauge attached to the cylinder outlet that indicates the pressure in the cylinder.

Cylinder Shell — External case of a lock cylinder.

Cylindrical Lock — Lock having the lock cylinder contained in the knob. *Also known as* Bored Lock.

P_1 = original pressure

P_2 = final pressure

T_1 = original temperature

T_2 = final temperature

D

Damage — Loss, injury, or deterioration caused by the negligence, design, or accident of one person to another in respect to another person's property.

Damage Control Locker — Compartment containing fire fighting/emergency equipment.

Damages — Compensation to a person for any loss, detriment, or injury whether to his person, property, or rights through the unlawful act, omission, or negligence of another.

Damping Mechanism — Structural element designed to control vibration.

Dangerous Cargo Manifest — Invoice of cargo used on ships, containing a list of all hazardous materials on board and their location on the ship.

Dangerous Goods — (1) Any product, substance, or organism included by its nature or by regulation in any of the nine United Nations classifications of hazardous materials. (2) Alternate term used in Canada and other countries for hazardous materials. (3) Term used in the U.S. and Canada for hazardous materials aboard aircraft. *See* Hazardous Material (1).

Dangerous Goods Guide to Initial Emergency Response (IERG) — Canada's equivalent of the DOT *Emergency Response Guidebook*.

Darcy-Weisbach Method — Technique used to determine pressure loss due to fluid friction in a piping system.

Data — Information used as a basis for reasoning or calculation. The singular form of the term is *datum*.

Database — Computer software program that serves as an electronic filing cabinet; used to create forms, record and sort information, develop mailing lists, organize libraries, customize telephone and fax lists, and track presentation and program outcomes.

dB — *See* Decibel.

DBH — *See* Diameter at Breast Height.

Dead Fuels — Fuels with no living tissue in which moisture content is governed almost entirely by atmospheric moisture (relative humidity and precipitation), dry-bulb temperature, and solar radiation.

Dead Latch — Sliding pin or plunger that operates as part of a dead-locking latch bolt. *Also known as* Anti-Shim Device.

Dead Load — Weight of the structure, structural members, building components, and any other features permanently attached to the building that are constant and immobile.

Dead Locking — *See* Latch Bolt.

Dead Shore — *See* Vertical Shore.

Deadbolt — Movable part of a deadbolt lock that extends from the lock mechanism into the door frame to secure the door in a locked position.

Dead-End Corridor — Corridor in which egress is possible in only one direction. *See* Egress.

Dead-End Hydrant — Fire hydrant located on a dead-end main that receives water from only one direction.

Dead-End Main — Water main that is not looped and in which water can flow in only one direction.

Deadman Switch — Foot pedal located below the aerial device control pedestal; must be depressed in order to operate the aerial device controls.

Dead-Man Valve — Spring-loaded valve that controls the flow of fuel from the loading rack to the tank vehicle or rail tank car. It is designed to shut off immediately when the operator releases the handle.

Debriefing — A gathering of information from all personnel that were involved in incident operations; in terms of incident-related stress, helps personnel process the scope of an event after a few days have passed.

Decay Stage — Stage at which the heat release rate is declining and the amount of fire diminishes as the fuel diminishes.

Decibel (dB) — (1) Unit used to express relative difference in power between acoustic and electrical signals; equal to 10 times the logarithm of the ratio of the two levels. (2) Degree of loudness; unit for expressing the relative intensity of sounds, on a scale from 0 for the least perceptible sound to about 130 for the average pain level.

Deck — Continuous, horizontal surface (floor) running the length of a vessel; some may not extend the whole length of a vessel, but they always reach from one side to the other. *See* Boat Deck, Main Deck, Poop Deck, Tank Top, Tween Deck, Upper Deck, and Weather Deck.

Deck Gun — *See* Turret Pipe.

Deck Pipe — *See* Turret Pipe.

Deckhead — *See* Overhead (1).

Decking — *See* Sheathing (2).

Declared Emergency — Aircraft emergency in which the crew is aware that there is a problem and notifies airport authorities before preparing to land.

Decoding — Translating a message to find its meaning.

Decomposition — Chemical change in which a substance breaks down into two or more simpler substances. Result of oxygen acting on a material that results in a change in the material's composition; oxidation occurs slowly, sometimes resulting in the rusting of metals. *See* Oxidation and Pyrolysis.

Decon — *See* Decontamination.

Decontaminate — To remove a foreign substance that could cause harm; frequently used to describe removal of a hazardous material from a person, clothing, or area.

Decontamination — Process of removing a hazardous foreign substance from a person, clothing, or area. *Also known as* Decon. *See* Contamination, Decontamination Corridor, Definitive Decontamination, Emergency Decontamination, Gross Decontamination, Mass Decontamination, Patient Decontamination, and Technical Decontamination.

Decontamination Corridor — Area where decontamination is conducted. *See* Decontamination.

Dedicated Railcar — Car set aside by the product manufacturer to transport a specific product. The name of the product is painted on the car.

Deep-Seated Fire — Fire that has moved deep into piled or bulk materials such as hay, baled cotton, or paper.

Defamation — Publication of anything that injures the good name or reputation of a person or brings disrepute to a person. *See* Libel and Slander.

Defendant — Party accused of alleged wrongdoing in a civil proceeding, or the party accused of a felony (indictable offense) in a criminal proceeding.

Defense Mechanisms — Systems, such as nasal hair, mucus, or cilia, that protect the body from invasion by foreign particles and injury.

Defensive Fire Attack — Exterior fire attack that is limited to controlling the spread of a fire, with an emphasis on exposure protection. *Also known as* Defensive Attack.

Defensive Mode — Commitment of a fire department's resources to protect exposures when the fire has progressed to a point where an offensive attack is not effective; deploying resources to limit the growth of an emergency incident rather than mitigating it. *See* Defensive Strategy.

Defensive Operations — Operations in which responders seek to confine the emergency to a given area without directly contacting the hazardous materials involved. *See* Nonintervention Operations and Offensive Operations.

Defensive Strategy — Overall plan for incident control established by the incident commander that involves protection of exposures, as opposed to aggressive, offensive intervention. *See* Nonintervention Strategy, Offensive Strategy, and Strategy.

Defibrillation — The delivery of a measured dose of electrical current to restore normal function of the heart; the machine that performs this task is called a *defibrillator*.

Definitive Decontamination — Decontaminating further after technical decontamination; may involve sampling and/or lab testing and is usually conducted by hospital staff or other experts. *See* Decontamination and Technical Decontamination.

Deflagration — (1) Chemical reaction producing vigorous heat and sparks or flame and moving through the material (as black or smokeless powder) at less than the speed of sound. A major difference among explosives is the speed of this reaction. (2) Intense burning, a characteristic of Class B explosives.

Deflector — Part of the sprinkler assembly that creates the discharge pattern of the water.

Deformation — (1) Alteration of form or shape. (2) Projection on the surface of reinforcing bars to prevent the bars from slipping through the concrete. *Also known as* Set.

Defusing — Informal discussion with incident responders conducted after the incident has been terminated, either at the scene or after the units have returned to quarters. During the discussion commanders address possible chemical and medical exposure information, identify damaged equipment and apparatus that require immediate attention, identify unsafe operating procedures, assign information gathering responsibilities to prepare for the post-incident analysis, and reinforce the positive aspects of the incident.

Degradation — *See* Chemical Degradation.

Dehydration — Process of removing water or other fluids.

Deionized Water — Water from which ionic salts, minerals, and impurities have been removed by ion exchange.

Delayed Treatment — Classification for patients with serious but not life threatening injuries; these patients may need additional care but may not need that care immediately.

Delegation — Providing subordinates with the authority, direction, and resources needed to complete an assignment.

Delinquent Firesetter — Individual who starts fires because of boredom or peer or media influences; fires may also be set to commit destruction of property, conceal a crime, or as a rite of passage into a group.

Deluge Sprinkler System — Fire-suppression system that consists of piping and open sprinklers. A fire detection system is used to activate the water or foam control valve. When the system activates, the extinguishing agent expels from all sprinkler heads in the designated area. *See* Dry-Pipe Sprinkler System, Preaction Sprinkler System, and Wet-Pipe Sprinkler System.

Deluge Valve — Automatic valve used to control water to a deluge sprinkler system.

Demand Valve — Valve within the self-contained breathing apparatus regulator that lets breathing air pass to the wearer when the wearer inhales.

Demand-Type Breathing Apparatus — Breathing apparatus with a regulator that supplies air to the facepiece only when the wearer inhales or when the bypass valve has been opened; no longer approved for fire fighting or IDLH situations. *Also known as* Negative-Pressure Breathing Apparatus.

Demobilization Unit — Functional unit within the planning section of an incident management system; responsible for assuring orderly, safe, and efficient demobilization of resources committed to the incident. *Also known as* Demob.

Democratic Leadership — Leadership style in which the leader is team-oriented and gives authority to the group; the group makes suggestions and decisions. *Also known as* Participative Leadership.

Demographic Information — Data that reports the general circumstances of an event and information about the participants. This information is pertinent in determining if there is a local community profile for a firesetter.

Demography — The statistical study of human population, especially the size, density, distribution, and other vital statistics of a group of people.

Demonstration — Instructional /teaching method in which the instructor/educator actually performs a task or skill, usually explaining the procedure step-by-step.

Density — Mass per unit of volume of a substance; obtained by dividing the mass by the volume.

Deodorization — Action taken following a fire to remove smoke odors from an atmosphere.

Department of Defense (DoD) — Administrative body of the executive branch of the U.S. Federal Government that encompasses all branches of the U.S. military.

Department of Energy (DOE) — Administrative body of the executive branch of the U.S. Federal Government that manages national nuclear research and defense programs, including the storage of high-level nuclear waste.

Department of Homeland Security (DHS) — U.S. agency that has the missions of preventing terrorist attacks, reducing vulnerability to terrorism, and minimizing damage from potential attacks and natural disasters; includes the Federal Emergency Management Agency (FEMA), U.S. Coast Guard (USCG), and Office for Domestic Preparedness (ODP).

Department of Housing and Urban Development (HUD) — U.S. government agency that oversees home ownership, low-income housing assistance, fair housing laws, homelessness, aid for distressed neighborhoods, and housing development programs.

Department of Justice (DOJ) — Administrative body of the executive branch of the U.S. Federal Government that assigns primary responsibility for operational response to threats or acts of terrorism within U.S. territory to the Federal Bureau of Investigation (FBI). *See* Terrorism.

Department of Juvenile Justice (DJJ) — A branch of state government that enforces laws violated by juveniles.

Department of Labor (DOL) — Administrative body of the executive branch of the U.S. Federal Government that is responsible for overseeing labor policy, regulation, and enforcement.

Department of Transportation (DOT) — U.S. federal agency that is responsible for transportation policy, regulation, and enforcement; regulates the transportation of hazardous materials. *Formerly known as* Interstate Commerce Commission.

Dependable Lift — Height a column of water may be lifted in sufficient quantity to provide a reliable fire flow. Lift may be raised through a hard suction hose to a pump, taking into consideration the atmospheric pressure and friction loss within the hard suction hose; dependable lift is usually considered to be 14.7 feet (4.48 m). *See* Lift.

Deposition — Process during which the witness answers questions under oath posed by the attorneys for each party; sworn testimony taken out of court. Not specifically delineated as a separate portion of discovery in Canada.

Depression — Emotional state in which there are extreme feelings of sadness, dejection, lack of worth, and emptiness. Mild symptoms include lack of motivation and inability to concentrate; more serious symptoms include sleeping and eating disorders and other severe changes of bodily functions, such as inactivity.

Depth of Calcination — Measurement of the depth of fire damage in gypsum board.

Depth of Char — Measurement of the depth of fire damage in wood.

Depth of Field — Range that is in focus both in front of and behind the subject of a photograph.

Dermis — True skin; a dense, elastic layer of fibrous tissue that lies beneath the epidermis (the outer skin). It is laced with blood vessels, nerve fibers, and receptor organs for sensations of touch, pain, heat, and cold. It also contains muscular elements, hair follicles, and oil and sweat glands.

Descent Device — Friction or mechanical device used to control a descent down a fixed line or to lower a load.

Desiccant — Substance that has a high affinity for water and is used as a drying agent.

Design Build — Concept involving the use of a single organization to both design and build a facility, rather than engaging separate firms to perform these activities; may be used to refer a design-build project or a design-build firm.

Designated Employee — Employee trained to use portable fire extinguishers or small hoselines to fight incipient fires in the employee's immediate work area.

Designated Length — Length marked on the ladder.

Desorption — Release of a substance from or through a surface. *Opposite of* Absorption *and* Absorption.

Det Cord — *See* Detonator Cord.

Detailed View — Additional, close-up information shown on a particular section of a larger drawing. *See* Elevation View, Plan View, and Sectional View.

Detector Tube — *See* Colorimetric Tube.

Detention Window — Window designed to prevent exit through the opening.

Determinants — Internal and external factors that influence how an organization develops and operates.

Deterrence — Discouraging or preventing someone from acting in a hazardous manner, or preventing a hazardous incident from happening; maintenance of legal powers for the purpose of inhibiting hazardous actions.

Detonation — (1) Supersonic thermal decomposition, which is accompanied by a shock wave in the decomposing material. (2) Explosion with an energy front that travels faster than the speed of sound. (3) High explosive that decomposes extremely rapidly, almost instantaneously. *See* Explosive (1) and (2) and High Explosive.

Detonator — Device used to trigger less sensitive explosives, usually composed of a primary explosive; for example, a blasting cap. Detonators may be initiated mechanically, electrically, or chemically. *Also known as* Initiator. *See* Detonation.

Detonator Cord — (1) Flexible explosive tape put around the outer edge of the inside of the canopy of some military aircraft to separate the Plexiglas® from the metal frame, in order to facilitate rescue or egress. (2) A flexible cord containing a center core of high explosive used to detonate other explosives. *Also known as* Det Cord.

Deutsches Institut fur Normung (DIN) — Nongovernmental organization established in Germany to develop consensus standards to ensure quality and conformity in materials, testing, and processes; similar to ANSI in the U.S.

Dew Point — Temperature at which the water vapor in air precipitates as droplets of liquid.

Dewar — All-metal container designed for the movement of small quantities of cryogenic liquids within a facility; not designed or intended to meet Department of Transportation (DOT) requirements for the transportation of cryogenic materials. *See* Container and Cryogen.

Dewatering — Process of removing water from a vessel or building.

DFDR — *See* Digital Flight Data Recorder.

DHS — *See* Department of Homeland Security.

Diabetes Mellitus — Complex disorder that is mainly caused by the failure of the pancreas to release enough insulin into the body; high levels of sugar in the blood and urine. Symptoms include the need to urinate often, increased thirst, weight loss, and increased appetite.

Diameter at Breast Height (DBH) — Means by which the relative size of trees is expressed. If a tree trunk measured at breast height is 20 inches (508 mm) or more in diameter, it is considered a large tree. A tree with a DBH of less than that is considered a small tree.

Diaphoresis — Profuse sweating that occurs with a fever, physical exertion, exposure to heat, or stress.

Diaphragm — Dome-shaped muscle that separates the chest cavity from the abdominal cavity. This muscle has holes through which pass the large artery (aorta), esophagus, and large vein (vena cava).

Diatomaceous Earth — Light siliceous material consisting chiefly of the skeletons of diatoms (minute unicellular algae); used especially as an absorbent or filter. *Also known as* Diatomite.

Diatomite — *See* Diatomaceous Earth.

Dicing — Ventilation exit opening created by making multiple cuts in the sheathing perpendicular to the ridge beam.

Dielectric — Material that is a poor conductor of electricity; usually applied to tools that are used to handle energized electrical wires or equipment. *See* Conductor.

Dielectric Heating — Heating that occurs as a result of the action of pulsating either direct current (DC) or alternating current (AC) at high frequency on a nonconductive material.

Diesel Exhaust Fluid (DEF) — A solution of 67.5 percent deionized water and 32.5 percent high-purity urea that is used in selective catalyst reductant (SCR) systems to reduce the level of NO_x in diesel exhaust.

Differential Dry-Pipe Valve — Valve in a dry-pipe sprinkler system in which relatively low air pressure is used to hold the valve closed and thus hold the water back.

Differential Manometer — Device whose primary application is to reflect the difference in pressures between two points in a system.

Diffuser — Equipment used for breaking up the stream of water from a fire hydrant. Also effective for diverting the flow of debris from the hydrant water stream.

Diffusion — (1) Process by which oxygen moves from alveoli to the blood cells in the thin-walled capillaries. (2) Process by which hazardous materials pass through protective clothing.

Digester — Large, circular container used at sewage treatment plants to cleanse raw sewage.

Digital Flight Data Recorder (DFDR) — Digital recording device on large civilian aircraft to record data such as aircraft airspeed, altitude, heading, and acceleration, in order to be used as an aid to accident investigation. *Also known as* Black Box.

Digital Memory Card — Storage device used on DSLR cameras and other digital devices to store captured images and data files; can be removed from the camera and easily uploaded to a computer. Also allows for immediate review, editing, and/or deletion of captured images.

Digital Single-Lens Reflex (DSLR) Camera — Digital camera that uses a mechanical mirror system and prism to direct light from the lens to an optical viewfinder on the back of the camera.

Dike — (1) Temporary dam constructed of readily available objects to obstruct the flow of a shallow stream of water to a depth that will facilitate drafting operations. (2) Temporary or permanent barrier that contains or directs the flow of liquids.

Diked Area — Area surrounding storage tanks or loading racks that is designed to retain spilled fuel and fire-extinguishing agents such as water and foam. *See* Nondiked Area.

Dilution — Application of water to a water-soluble material to reduce the hazard. *See* Dissolution and Water Solubility.

Dimensioning — Indicating or determining size and position in space relative to existing conditions.

Dimpled — Depressed or indented, as on a metal surface, to aid in gripping.

DIN — *See* Deutsches Institut fur Normung.

Dip Tube — Tube installed in a pressurized container from the top to the bottom to permit expelling of the contents, liquid or solid, out of the top of the container.

Direct Attack — (1) In structural fire fighting, an attack method that involves the discharge of water or a foam stream directly onto the burning fuel. *See* Attack Methods (1). (2) In wildland fire fighting, an operation where action is taken directly on burning fuels by applying an extinguishing agent to the edge of the fire or close to it. *See* Attack Methods (2).

Direct Current (DC) Circuit — Electrical circuit in which the current moves through the circuit in only one direction.

Direct Evidence — Type of evidence provided by a witness who obtained it through his or her senses.

Direct Injection — Application method where foam concentrate is injected directly into the water stream at the pump before it enters the hoseline. *See* Semisubsurface Injection and Subsurface Injection.

Direct Inspection — Used when an evaluator wants to physically examine an environment in search of proof that behavioral change has occurred.

Direct Lines — Phone lines leased or dedicated to a specific purpose used as point-to- point communication. These lines are not connected to the public telephone network and therefore do not have a dial tone. *Also known as* private lines.

Direct Loss — Loss caused directly by a fire; can be either primary damage or secondary damage.

Direct Order — Command or assignment to a subordinate that specifies the desired behavior or outcome.

Direct Pumping System — Water supply system supplied directly by a system of pumps rather than elevated storage tanks.

Direct Reading Conductivity Meter — Device designed to directly measure the specific conductivity of a solution and provide a reading on a display.

Direct-Connect Alarms — Alarm systems that are connected directly to a local police or fire department from the protected property.

Directional Anchor — Anchor that is capable of supporting a load in only a limited direction.

Directional Foam Spray Nozzle — Foam delivery device that consists of a small foam nozzle used to protect areas such as loading racks by applying the extinguishing agent onto the surface beneath tanker trucks.

Directive — Authoritative instrument or order issued by a superior officer.

Director — Title for individuals responsible for command of a branch in an incident management system.

Directory — Computer table of identifiers and references to the corresponding items or data; for example a computer listing of files stored on a hard drive.

Dirty Bomb — *See* Radiological Dispersal Device (RDD).

Disability — According to the *Americans with Disabilities Act (ADA)*, a person has a disability if he or she has a physical or mental impairment that substantially limits a major life activity.

Disaster Recovery Center (DRC) — Facility established in a centralized location within or near the disaster area at which disaster victims (individuals, families, or businesses) apply for disaster aid.

Discharge Outlet, Type I — Foam delivery device that conducts and delivers finished foam onto the burning surface of a liquid without submerging it or agitating the surface; no longer manufactured and considered obsolete, but may still be found in some fixed-site applications.

Discharge Outlet, Type II — Foam delivery device that delivers finished foam onto the surface of a burning liquid, partially submerges the foam into the surface, and produces limited agitation on the surface of the burning liquid.

Discharge Outlet, Type III — Foam delivery device that delivers finished foam in a manner that causes it to fall directly onto the surface of the burning liquid and does so in a way that causes general agitation; includes master streams and handlines.

Discharge Velocity — Linear velocity or rate at which water flows through and travels from an orifice.

Discipline — To maintain order through training and/or the threat or imposition of sanctions; setting and enforcing the limits or boundaries for expected performance.

Disclosure — Legal term referring to the act of giving out information either voluntarily or to meet legal requirements or agency policy requirements.

Discontinuity — Interruption of the typical structure of a weldment, such as inhomogeneity in the mechanical, metallurgical, or physical characteristics of the material or weldment.

Discovery — Means by which the plaintiff (one party) obtains information from the opposing party (defendant) to prove its allegation.

Discrimination — Measure of the extent to which any item in a test is answered more or less successfully by learners who do well or poorly on a test overall, thereby discriminating between them. Ideally, items that do not discriminate would be omitted from revised versions of the test. *Also known as* Discrimination Index.

Discussion — (1) Instructional method in which an instructor generates interaction with and among a group. There are several formats of discussion: guided, conference, case study, role-play, and brainstorming. In each type, it remains the responsibility of the instructor to steer the group discussion or activity to meet lesson objectives. (2) Teaching method by which students contribute to the class session by using their knowledge and experience to provide input. (3) The exchange of ideas between an educator and the audience. (4) Two-way communication between sender and receiver. See Case Study and Role-Playing.

Disentanglement — Aspect of vehicle extrication relating to the removal and/or manipulation of vehicle components to allow a properly packaged patient to be removed from the vehicle. Sometimes referred to as *removing the vehicle from the patient*.

Disinfect — Destroy, neutralize, or inhibit the growth of harmful microorganisms.

Dislocation — State of being misaligned; the condition that results when the surfaces of two bones are no longer in proper contact.

Dispatch — (1) To direct fire companies to respond to an alarm. (2) Radio designation for the dispatch center; for example, "Engine 65 to Dispatch, send me a second alarm."

Dispatcher — Person who works in the telecommunications center and processes information from the public and emergency responders.

Dispatching — Process by which an alarm is received at the telecommunications center, retransmitted to the emergency responders, and acknowledged when received.

Dispersion — Act or process of being spread widely. See Engulf (1) and Vapor Dispersion.

Displaced Runway Threshold — Temporary relocation of a runway threshold (beginning or end) because of maintenance or other activity on the runway.

Displacement — (1) Volume or weight of a fluid displaced by a floating body of equal weight. (2) Amount of water forced into the pump, thus displacing air.

Disseminate — To spread about or scatter widely.

Dissipate — To cause to spread out or spread thin to the point of vanishing.

Dissolution Act or process of dissolving one thing into another, such as dissolving a gas in water. See Concentration (2) and Dilution.

Distance Learning — Generic term for instruction that occurs when the student is remote from the instructor, and a medium such as the Internet, Interactive Television, or mail service is used to maintain communication between the two and to maintain assignment submission. See Open Learning.

Distention — State of being expanded or swollen, particularly of the abdomen.

Distortion — State of being twisted out of normal or natural shape or position.

Distracters — Possible answers in a multiple-choice test item that are incorrect but plausible.

Distress — More stress than a person can reasonably be expected to handle; in other words, when stress controls the person.

Distribution System — That part of an overall water supply system that receives the water from the pumping station and delivers it throughout the area to be served.

Distributor Nozzle — Nozzle used to create a broken stream that is usually used on basement fires.

District — See Response District.

District Survey — Evaluation of an entire response district to identify hazards on a broader scale than preincident planning; may also involve identifying standards of coverage needs for the district or jurisdiction.

Diuretic — Product that tends to increase the flow of urine.

Diurnal — (1) Daily; especially pertaining to cyclic actions of the atmosphere that are completed within 24 hours and that recur every 24 hours. (2) A tide pattern that has one high and one low in a 24-hour period.

Divert — Actions to control movement of a hazardous material to an area that will produce less harm.

Diverter Valve — See Selector Valve.

Diving Accident Network — Hotline that can advise physicians and rescue personnel about diving mishaps.

Division — NIMS-ICS organizational level having responsibility for operations within a defined geographic area. It is composed of a number of individual units that are assigned to operate within a defined geographical area.

Division of Labor — Subdividing an assignment into its constituent parts in order to equalize the workload and increase efficiency.

DJJ — See Department of Juvenile Justice.

Documentation — Written notes, audio/videotapes, printed forms, sketches and/or photographs that form a detailed record of the scene, evidence recovered, and actions taken during the search of an incident or crime scene.

Documentation Unit — Functional unit within the planning section of an incident management system; responsible for recording/protecting all documents relevant to the incident.

DoD — *See* Department of Defense.

DOE — *See* Department of Energy.

Dog — To lock levers or bolts and thumbscrews on watertight doors.

Dog the Hatches — Close the doors.

Dogs — *See* Pawls.

DOJ — *See* Department of Justice.

DOL — *See* Department of Labor.

Dollies — *See* Supports.

Domains of Learning — Areas of learning and classification of learning objectives; often referred to as cognitive (knowledge), affective (attitude), and psychomotor (skill performance) learning domains.

Dome Roof — Hemispherical roof assembly, usually supported only at the outer walls of a circular or many-sided structure.

Dome Roof Tank — *See* Cone Roof Storage Tank.

Domestic Consumption — Water consumed from the water supply system by residential and commercial occupancies.

Domineering Attitude — Effect of a decision-making problem in which the leader tries to control every facet of the operation.

Donning Mode — State of positive-pressure SCBA when the donning switch is activated.

Donning Switch — Device on a positive-pressure regulator that, when activated, stops airflow while the unit is being donned. Airflow is resumed with the user's first inhalation.

Donut Roll — Length of hose rolled up for storage and transport.

Door Closer — Mechanical device that closes a door. *Also known as* Self-Closing Door.

Door Control — Fire fighting tactic intended to reduce available oxygen to a fire and create a controlled flow path in a structure for tactical ventilation, firefighter survivability, and occupant survivability.

Door Hold-Open Device — Mechanical device that holds a door open and releases it upon a signal. Usually a magnetic device used to hold fire or smoke door assemblies in the open position.

Door/Roof Posts — Structural members that surround the doors and support the roofs of vehicles. *Also known as* Pillars.

Door/Window Schedule — Table that lists the door/window location on a plan, the dimensions of the opening, and any other construction information. *See* Schedule.

Doorjamb — Sides of the doorway opening.

Dormitory — Subdivision of residential property classification with structures, or spaces in structures, that provide group sleeping accommodations for more than 16 unrelated people; these occupants are housed in one room or a group of rooms under joint occupancy and single management. These rooms lack individual cooking facilities, and meals may or may not be provided.

Dose — Quantity of a chemical material ingested or absorbed through skin contact; employed in measuring toxicity. *See* Concentration (1) and Lethal Dose, 50 Percent Kill (LD_{50}).

Dosimeter — Detection device used to measure an individual's exposure to an environmental hazard such as radiation or sound.

DOT — *See* Department of Transportation.

DOT 3AA — DOT specification for type and material of steel self-contained breathing apparatus or cascade cylinder construction.

Double Bottom — Extra watertight floor within a vessel above the outer watertight hull; void or tank space between the outer hull of a vessel and the floor of a vessel. *Also known as* Inner Bottom or Tank Top.

Double Figure-Eight Knot — Knot used to tie ropes of equal diameters together.

Double-Acting Hydraulic Cylinder — Hydraulic cylinder capable of transmitting force in two directions.

Double-Edge Snap Throw — Method of spreading a salvage cover similar to the single- edge snap throw; intended to cover two groupings located on either side of a narrow aisle.

Double-Hung Window — Window having two vertically moving sashes.

Doubles — Truck combination consisting of a truck tractor and two semi-trailers coupled together. *Formerly known as* Double-Bottom or Double-Trailer.

Double-Trailer — See Doubles.

Doughnut-Shaped Pattern — Ring-shaped fire pattern formed on a floor when a pool of flammable liquid is ignited; the center of the ring is protected from fire damage while the edge of the ring will show a circular demarcation line.

Downstream — (1) Direction of airflow from a high-pressure source to a low-pressure source; for example, when the facepiece is downstream from the air cylinder. (2) Direction in which the current of a moving body of water is flowing.

Downwind — Wind that is blowing in a direction from behind a person or aircraft. See Crosswind, Headwind, and Wind.

Downwind Leg — Flight path parallel to the landing runway in the direction opposite to landing.

Dozer — See Bulldozer.

Dozer Tender — Ground vehicle (service unit) with personnel capable of maintenance, minor repairs, and limited fueling of bulldozers.

Dozer Transport — Heavy vehicle carrying a bulldozer to an incident.

Draft — (1) Process of acquiring water from a static source and transferring it into a pump that is above the source's level; atmospheric pressure on the water surface forces the water into the pump where a partial vacuum was created. (2) Vertical distance between the water surface and the lowest point of a vessel; depth of water a vessel needs in order to float. Draft varies with the amount of cargo, fuel, and other loads on board. See Draft Marks.

Draft Curtains — Noncombustible barriers or dividers hung from the ceiling in large open areas that are designed to minimize the mushrooming effect of heat and smoke and impede the flow of heat. Also known as Curtain Boards and Draft Stops.

Draft Marks — Numerals on the ends of a vessel indicating the depth of the vessel in the water. See Draft.

Draft Stops — See Draft Curtains.

Drafting Operation — See Draft.

Drafting Pit — Underground reservoir of water from which to draft for pumper testing; usually located at a training center.

Drag — (1) Procedure of dragging hooks through water to find drowning victims. (2) Rescue procedure for removing victims from a fire area.

Drag Chute — Parachute device installed on some aircraft that is deployed on landing roll to aid in slowing the aircraft to taxi speed.

Drain Valve — (1) Valve on a pump discharge that facilitates the removal of pressure from a hoseline after the discharge has been closed. (2) Valve on an elevated waterway system to facilitate the drainage of water from the system before stowing.

Drainage Dropout Rate — See Drainage Time.

Drainage Time — Amount of time it takes foam to break down or dissolve. Also known as Drainage, Drainage Dropout Rate, or Drainage Rate. See Quarter-Life.

Drawn to Scale — Dimensions that are reduced proportionately on construction plans to the actual size of the building or component. See Construction Plan.

DRC — See Disaster Recovery Center.

Dressing — Clean or sterile covering applied directly to a wound; used to stop bleeding and prevent contamination of the wound.

Drift Smoke — Smoke that has been transported from its point of origin and in which convective, columnar motion no longer dominates.

Drill — Exercise conducted to practice and/or evaluate training already received; the process of skill maintenance.

Drill Schedule — Calendar for training sessions in manipulative skills for firefighters or fire companies.

Drill Tower — A tall training structure, typically more than three stories high, used to create realistic training situations, especially ladder and rope evolutions. Also known as Tower.

Drive Train — See Power Train.

Driver — (1) Engine or motor used to turn a pump. (2) See Fire Apparatus Driver/Operator.

Driver Reaction Distance — Distance a vehicle travels while a driver is transferring the foot from the accelerator to the brake pedal after perceiving the need for stopping.

Driver/Operator — See Fire Apparatus Driver/Operator.

Driver's Side — Side of a vehicle that is on the same side as the steering wheel.

Drivewheel Horsepower — Power available at the wheels to move the vehicle.

Drop — To drop water or retardant from an aircraft.

Drop Bar — Metal or wooden bar that serves as a locking device when placed or dropped into brackets across a swinging door.

Drop Frame — Two-level frame section of a trailer that provides proper coupler height at the forward end and a lower height for the remainder of the length.

Drop Height — Most effective and safest altitude of an aircraft when fire-extinguishing agents are dropped on wildland fires.

Drop Panel — Type of concrete floor construction in which the portion of the floor above each column is dropped below the bottom level of the rest of the slab, increasing the floor thickness at the column.

Drop Zone — Target area for air tankers, helicopters, and cargo dropping.

Drop-Forged Coupling — Coupling made by raising and dropping a drop hammer onto a block of metal as it rests on a forging die, thus forming the metal into the desired shape.

Droplet Contact — Means of transmitting a communicable disease indirectly by spray droplets from an infected person's coughing or sneezing.

Drowning Machine — Colloquial term for the convection currents resulting when water floods over a low-head dam.

Dry Adiabatic Lapse Rate — Rate of decrease in temperature of a mass of dry air as it is lifted adiabatically through an atmosphere in hydrostatic equilibrium.

Dry Air Mass — Portion of the atmosphere that has a relatively low dew point temperature and where the formation of clouds, fog, or precipitation is unlikely.

Dry Bulk Cargo Tank — Cargo tank truck that carries small, granulated, solid materials; generally does not carry hazardous materials, but in some cases may carry fertilizers or plastic products that can burn and release toxic products of combustion. See Cargo Tank Truck.

Dry Bulk Carrier — Cargo tank that carries small, granulated, solid materials; generally does not carry hazardous materials, but in some cases may carry fertilizers or plastic products that can burn and release toxic products of combustion.

Dry Bulk Terminal — Facility equipped to handle dry goods, such as coal or grain, that are stored in tanks and holds on a vessel. See Bulk Terminal.

Dry Chemical — (1) Any one of a number of powdery extinguishing agents used to extinguish fires; the most common include sodium or potassium bicarbonate, monoammonium phosphate, and potassium chloride. (2) Extinguishing system that uses dry chemical powder as the primary extinguishing agent; often used to protect areas containing volatile flammable liquids.

Dry Chemical Fire Suppression System — System that uses dry chemical powder as the primary extinguishing agent; often used to protect areas containing volatile flammable liquids.

Dry Dock — Enclosed area into which a vessel floats but where water is then removed, leaving the vessel dry for repairs, cleaning, or construction.

Dry Foam — Foam that has a very high air-to-foam solution ratio. This foam will cling to horizontal surfaces.

Dry Hoseline — Hoseline without water in it; an uncharged hoseline.

Dry Hydrant — (1) Permanently installed pipe that has pumper suction connections installed at static water sources to speed drafting operations. (2) Hydrant that is permanently out of service; these may be found in old abandoned complexes or similar structures. It is important to make a distinction between these two uses of the term.

Dry Lightning Storm — Lightning storm during which little or no rain reaches the ground.

Dry Powder — Extinguishing agent suitable for use on combustible metal fires.

Dry Standpipe System — (1) Standpipe system that has closed water supply valves or that lacks a fixed water supply. (2) Any of several standpipe systems in which the piping contains water only when actually being used.

Dry Thunderstorm — Storm, including lightning, during which little or no rain reaches the ground.

Dry-Barrel Hydrant — Fire hydrant that has its operating valve at the water main rather than in the barrel of the hydrant. When operating properly, there is no water in the barrel of the hydrant when it is not in use. These hydrants are used in areas where freezing may occur.

Dry-Pipe Sprinkler System — Fire-suppression system that consists of closed sprinklers attached to a piping system that contains air under pressure. When a sprinkler activates, air is released, activating the water or foam control valve and filling the piping with extinguishing agent. Dry systems are often installed in areas subject to freezing. See Deluge Sprinkler System, Preaction Sprinkler System, and Wet-Pipe Sprinkler System.

Drysuit — Protective outwear worn during water-based rescue operations; provides an impermeable barrier between the wearer and the surrounding water. May be used in ice rescue or as protection from contaminants in water.

Drywall — System of interior wall finish using sheets of gypsum board and taped joints. See Wallboard.

DSLR — *See* Digital Single-Lens Reflex Camera.

Dual Pumping — Operation where a strong hydrant is used to supply two pumpers by connecting the pumpers intake-to-intake. The second pumper receives the excess water not being pumped by the first pumper, which is directly connected to the water supply source. Sometimes incorrectly referred to as *tandem pumping*.

Dual-Issue Leadership — Leadership style in which the leader has a high degree of concern for both workers and production.

Duct — (1) Tube or passage that confines and conducts airflow throughout the aircraft for pressurization, air conditioning, etc. (2) Channel or enclosure, usually of sheet metal, used to move heating and cooling air through a building. (3) Hollow pathways used to move air from one area to another in ventilation systems.

Ductile — Capable of being shaped, bent, or drawn out.

Due Course of Law — *See* Due Process.

Due Process — Conduct of legal proceedings according to established rules and principles for the protection and enforcement of private rights, including notice and the right to a fair hearing before a tribunal with the power to decide the case. *Also known as* Due Course of Law or Due Process of Law. *See* Due Process Clause.

Due Process Clause — Constitutional provision that prohibits the government from unfairly or arbitrarily depriving a person of life, liberty, or property. *See* Due Process.

Due Process of Law — *See* Due Process.

Duff — Matted, partly decomposed leaves, twigs, and bark beneath the litter of freshly fallen twigs, needles, and leaves lying under trees and brush.

Dummy Coupler — Fitting used to seal the opening in an air brake hose connection (gladhands) when the connection is not in use; a dust cap.

Dump Line — *See* Waste Line.

Dunnage — Loose packing material (usually wood boards and wedges) that is placed around cargo in a vessel's hold to support, protect, or prevent it from moving while the vessel is at sea.

Duplex Occupancy — Two-family dwelling in which the families live side by side; one family occupies the left half of the structure and the other occupies the right half.

Durable Agents — *See* Gelling Agents.

Dust — Solid particle that is formed or generated from solid organic or inorganic materials by reducing its size through mechanical processes such as crushing, grinding, drilling, abrading, or blasting.

Dust Devil — *See* Whirlwind.

Dust Explosion — Rapid burning (deflagration), with explosive force, of any combustible dust. Dust explosions generally consist of two explosions: a small explosion or shock wave creates additional dust in an atmosphere, causing the second and larger explosion.

Dutchman — Extra fold placed along the length of a section of hose as it is loaded, so that its coupling rests in proper position.

Duty — (1) Obligation that one has by law or contract. (2) Fire-related responsibility assigned to a member by the fire brigade organizational statement.

Dye-Penetrant Testing — Form of nondestructive testing in which the surface of the test material is saturated with a dye or fluorescent penetrant, and a developer is applied. Dyes bleed visibly to the surface indicating defects; fluorescents show the defective areas under ultraviolet light.

Dynamic — Amount of stretch built into a rope. Dynamic ropes have a large amount of stretch, in order to reduce the shock on the climber and anchor systems; they are used for recreational climbing, where long falls may occur.

Dynamic Load — Loads that involve motion. They include the forces arising from wind, moving vehicles, earthquakes, vibration, or falling objects, as well as the addition of a moving load force to an aerial device or structure. *Also known as* Shock Loading.

Dynamic Rope — Rope designed to stretch under load, reducing the shock of impact after a fall.

Dynamic Stress — Stress imposed on an aerial device while it is in motion, resulting from a dynamic load.

Dyspnea — Painful or difficult breathing; rapid, shallow respirations.

Dzus Fastener — Trade name given to a half-turn fastener with a slotted head. This type of fastener is used on engine cowlings, cover plates, and access panels throughout the aircraft.

E

EAP — *See* Employee Assistance Program.

Eave — The edge of a pitched roof that overhangs an outside wall. Attic vents in typical eaves provide an avenue for an exterior fire to enter the attic.

Ebb Tide — Falling tide.

EBS — Emergency building shores.

EBSS — *See* Emergency Breathing Support System.

Eccentric Load — Load perpendicular to the cross section of the structural member, but which does not pass through the center of the cross section. An eccentric load creates stresses that vary across the cross section, and may be both tensile and compressive.

ECO — *See* Entry Control Officer.

Economic Factor — Includes expenses caused by the loss of tools, apparatus, equipment, manpower, property, and systems, in addition to legal expenses.

Economic Incentives — Used to support the prevention interventions. Examples of positive incentives include reduced insurance premiums for buildings having fire-resistive construction, sprinklers, and automatic notification systems. Examples of negative incentives would be fines for violating fire codes.

Economizer — Assembly of coils in a vessel's stack (chimney), designed to transfer heat rising up the stack to water within the tubes. *See* Fiddley and Stack.

Eddy — A current of water moving in opposition to the main stream, caused by contact with an obstacle.

Edema — Condition in which fluid escapes into the body tissues and causes local or generalized swelling.

Education — (1) Process of teaching, instructing, or training individuals in new skills or additional knowledge, or preparing individuals for some kind of action or activity; what teachers do to bring about learning in their students. May or may not involve formal classroom instruction. (2) The acquisition of knowledge, usually through academic means such as college or university courses.

Education Interventions — Designed to raise awareness, provide information, impart knowledge, and ultimately produce a desired behavior. Education provides a foundation for the entire prevention intervention system.

Education Program — A series of lessons and activities designed to impact a specific risk-reduction outcome. Several programs are often used in tandem to support a broader-based risk-reduction strategy.

Educational Materials — Printed matter, audiovisual materials, and "props" that an educator uses to enhance delivery of a lesson.

Educational Objective — Teaching and learning goal that answers the question: "What will happen as a result of the education program?" *Also known as* Instructional Objective. *See* Behavioral Objective and Learning Objective.

Educator — Person charged with the responsibilities of conducting program presentations, directing the instructional process, teaching and demonstrating skills, imparting new information, leading discussions, and evaluating mastery to ensure that learning has taken place.

Eduction — Process used to mix foam concentrate with water in a nozzle or proportioner; concentrate is drawn into the water stream by the Venturi method. *Also known as* Induction. *See* Venturi Principle.

Eductor — (1) Portable proportioning device that injects a liquid, such as foam concentrate, into the water flowing through a hoseline or pipe. *See* Foam Proportioner and Proportioning. (2) Syphon used to remove water from flooded basements. (3) Venturi device that uses water pressure to draw foam concentrate into a water stream for mixing; also enables a pump to draw water from an auxiliary source. *Also known as* Inductor.

EEES — *See* Emergency Evacuation Elevator Systems.

EEO — *See* Equal Employment Opportunity.

Efflorescence — White soluble salt crystals consisting of calcium and magnesium sulfates that form as white powder on the surface of masonry walls; caused by water that penetrates the masonry.

Effluent — Fluid that flows from a pipe or similar outlet; most commonly used to describe waste products from industrial processes.

Egress — (1) Place or means of exiting a structure or vehicle. (2) Escape or evacuation. *See* Access, Exit, and Means of Egress.

EIFS — *See* Exterior Insulation and Finish Systems.

Ejection Seat — Aircraft seat capable of being ejected in an emergency to catapult the occupant clear of the aircraft.

EKG — *See* Electrocardiogram.

Elastomer — Generic term for the a group of synthetic rubber-like materials such as butyl rubber, neoprene, and silicone rubber used in facepiece seals, low-pressure hoses, and similar SCBA components.

Electric Arc — Visible discharge of electricity across a gap or between electrodes.

Electric Fence — Livestock-retaining fence that uses an electric current to deliver a shock, in order to discourage animals from trying to escape.

Electrical Burns — Burns caused by contact with electrical current or power, such as high-power wires or lightning.

Electrical Heat Energy — Heat energy that is electrical in origin; includes resistance heating, dielectric heating, heat from arcing, and heat from static electricity. Poorly maintained electrical appliances, exposed wiring, and lightning are sources of electrical heat energies.

Electrical Service — Conductor and equipment for delivering energy from the electrical supply system to the wiring system of the premises.

Electrical Shock — Injury caused by electricity passing through the body; severity of injury depends upon the path the current takes, the amount of current, and the resistance of the skin.

Electrical Systems — Wiring systems designed to distribute electricity throughout a building or vehicle.

Electrocardiogram (EKG) — Test used to observe the function of the heart. *Also known as* EKG.

Electrode — Conductor used to establish electrical contact in a circuit.

Electrolysis — Chemical change, especially decomposition of water and other inorganic compounds in a water solution (electrolyte), when an electric current is passed through the substance.

Electrolyte — (1) Substance that dissociates into ions in solution or when fused, thereby becoming electrically conducting. (2) Energy component within the human body that can be lost through sweating.

Electromagnetic Pulse (EMP) — Burst of electromagnetic energy produced by a nuclear explosion; EMPs damage electronic systems by causing voltage and current surges.

Electromotive Force (EMF) — *See* Voltage.

Electron — Subatomic particle that possesses a negative electric charge.

Electronic Bulletin Board — Computer application that allows network users to communicate in specific subject areas, such as fire safety, education, public safety, health, or travel. Users can post messages, and then other users of the bulletin board can respond.

Element — Most simple substance that cannot be separated into more simple parts by ordinary means.

Elevated Master Stream — Fire stream in excess of 350 gpm (1 400 L/min) that is deployed from the tip of an aerial device.

Elevated Storage — Water storage reservoir located above the level of the system it is being used to supply, in order to take advantage of head pressure.

Elevated Temperature Material — Material that when offered for transportation or transported in bulk packaging is (a) in a liquid phase and at temperatures at or above 212°F (100°C), (b) intentionally heated at or above its liquid phase flash points of 100°F (38°C), or (c) in a solid phase and at a temperature at or above 464°F (240°C). *See* Bulk Packaging, Elevated Temperature Materials Carrier, and Flash Point.

Elevated Temperature Materials Carrier — Cargo tank truck or cargo truck carrying large metal pots that transport elevated-temperature materials.

Elevating Master Stream Device — *See* Water Tower.

Elevating Platform — Work platform attached to the end of an articulating or telescoping aerial device.

Elevating Water Device — Articulating or telescoping aerial device added to a fire department pumper to enable the unit to deploy elevated master stream devices; these aerial devices range from 30 to 75 feet (9 m to 23 m) in height.

Elevation — (1) Height of a point above sea level or some other datum point. (2) Drawing or orthographic view of any of the vertical sides of a structure, or vertical views of interior walls.

Elevation Cylinder — Hydraulic cylinders used to lift the aerial device from its bed to a working position. *Also known as* Hoisting Cylinder.

Elevation Head — The height of a water supply above the discharge orifice.

Elevation Loss — *See* Elevation Pressure.

Elevation Pressure — Gain or loss of pressure in a hoseline due to a change in elevation. *Also known as* Elevation Loss.

Elevation View — Architectural drawing that shows the vertical view of a building, including floors, building height, and grade of surrounding ground. *See* Detailed View, Plan View, and Sectional View.

Elevator — (1) Hinged, movable control surface at the rear of the horizontal stabilizer of an aircraft. It is attached to the control wheel or stick and is used to control the pitch up or down or to hold the aircraft in level flight. (2) Passenger-carrying car in a multistory building. (3) Tall structure used to store grain or feed at an agricultural site.

Elliptical — (1) Large, cylindrical, oblong water tank that is used on tankers or tenders. (2) Having the shape of an ellipse; an elliptical cross section is frequently used for the tank-on-tank vehicles.

ELT — *See* Emergency Locator Transmitter.

Embezzlement Fire — Type of concealment fire designed to cover a "paper trail" of financial documents that incriminate the arsonist.

Embolism — Sudden blocking of an artery or vein by a clot or foreign material that has been carried by the blood.

Emergency — Sudden or unexpected event or group of events that require immediate action to mitigate.

Emergency Breathing Support System (EBSS) — Escape-only respirator that provides sufficient self-contained breathing air to permit the wearer to safely exit the hazardous area; usually integrated into an airline supplied-air respirator system.

Emergency Decontamination — The physical process of immediately reducing contamination of individuals in potentially life-threatening situations, with or without the formal establishment of a decontamination corridor. *See* Decontamination and Decontamination Corridor.

Emergency Evacuation Elevator Systems (EEES) — System of elevators within a high-rise building that is designed to report to the floor of alarm activation as well as two floors above and below to help evacuate occupants. After passengers are discharged at the floor of exit discharge, the elevator cars remain at that level for use by emergency response personnel.

Emergency Lighting System — (1) System of interior and exterior low-power incandescent and/or fluorescent lights that are designed to assist passengers in locating and using aircraft emergency exits, but that are not bright enough to assist aircraft rescue and fire fighting personnel in carrying out search and rescue operations. (2) Battery-operated floodlights in a building that are designed to activate when normal power supply is interrupted.

Emergency Locator Transmitter (ELT) — Radio transmitter carried by most aircraft. The radio is activated by impact forces; once activated, the ELT transmits a variable tone on emergency frequencies to aid in location of the accident site.

Emergency Management — Process of managing all types of emergencies and disasters by coordinating the actions of numerous agencies through all phases of disaster or emergency activity.

Emergency Medical Services (EMS) — (1) Medical treatment provided by emergency responders. (2) Initial medical evaluation/treatment provided to employees and others who become ill or are injured.

Emergency Medical Technician (EMT) — Professional-level provider of basic life support emergency medical care; requires certification by some authority.

Emergency Operations — Activities involved in responding to the scene of an incident and performing assigned duties in order to mitigate the emergency.

Emergency Operations Center (EOC) — Facility that houses communications equipment, plans, contact/notification list, and staff that are used to coordinate the response to an emergency.

Emergency Resource List — Directory held by a local jurisdiction that includes contact information for specific equipment not normally carried in the jurisdiction's inventory; should include agreements, contact numbers and equipment information. Jurisdictions should contract with more than one company for equipment, to provide redundancy in case the primary contractor is unable to supply the needed resource.

Emergency Responder — Qualified member of a fire and emergency services organization that provides search and rescue, fire suppression, medial, hazardous materials, or specialized protection services. The organization may be publicly or privately managed and funded.

Emergency Response — Because risk will never be completely eradicated, communities must have an adequately staffed, trained, and equipped emergency response system.

Emergency Response Guidebook (ERG) — Manual that aids emergency response and inspection personnel in identifying hazardous materials placards and labels; also gives guidelines for initial actions to be taken at hazardous materials incidents. Developed jointly by Transport Canada (TC), U.S. Department of Transportation (DOT), the Secretariat of Transport and Communications of Mexico (SCT), and with the collaboration of CIQUIME (Centro de Información Química para Emergencias).

Emergency Response Organization — Fire brigade or emergency medical response team.

Emergency Response Plan — Document that contains information on the actions that may be taken by a governmental jurisdiction to protect people and property before, during, and after an emergency.

Emergency Traffic — Urgent radio traffic; a request for other units to clear the radio waves for an urgent message. *Also known as* Priority Traffic.

Emergency Transportation System for the Chemical Industry (SETIQ) — Emergency response center for Mexico. *See* Canadian Transport Emergency Centre (CANUTEC) and Chemical Transportation Emergency Center (CHEMTREC®).

Emergency Truck — Van or similar-type vehicle used to carry portable equipment and personnel.

Emergency Valve — Self-closing tank outlet valve. *See* Emergency Valve Operator and Emergency Valve Remote Control.

Emergency Valve Operator — Device used to open and close emergency valves. *See* Emergency Valve and Emergency Valve Remote Control.

Emergency Valve Remote Control — Secondary means, remote from tank discharge openings, for operation in event of fire or other accident. *See* Emergency Valve Operator and Emergency Valve.

Emergency Vehicle Operator's Course (EVOC) — Training that originated with the National Highway Traffic and Safety Administration in the U.S. The training certifies drivers of emergency vehicles; courses may vary depending upon response discipline and jurisdiction.

Emergency Vehicle Technician (EVT) — Individual trained to perform emergency response vehicle inspections, diagnostic testing, maintenance, repair procedures, and operational testing. The NFPA® recognizes three levels of EVTs.

Emergency Voice Communications Systems (EVCS) — A secure, two-way communications system within high-rise buildings or large sites that assists firefighters in communicating during emergencies at locations where radio communications may not be effective.

Emergency-Relief Device — Device that is designed to relieve pressure on a vessel or container to prevent overpressurization; may be a rupture disk, relief valve, or similar device. *See* Rupture Disk.

Emetic — Agent that causes vomiting.

EMF — *See* Electromotive Force.

EMP — *See* Electromagnetic Pulse.

Empennage — Aircraft tail assembly, including the vertical and horizontal stabilizers, elevators, and rudders.

Emphysema — Lung disease in which there is destruction of the alveoli, resulting in labored breathing and increased susceptibility to infection; a chronic obstructive pulmonary disease.

Employee Assistance Program (EAP) — Program to help employees and their families with work or personal problems.

Employee Emergency Action Plan — OSHA-required plan that all employers must devise to inform all their employees of how they are to react to a fire or other emergency in the workplace. Employers with more than ten employees must put their plan in writing and keep it available to their employees.

EMS — *See* Emergency Medical Services.

EMT — *See* Emergency Medical Technician.

Emulsifier — Compound that supports one insoluble liquid in suspension in another liquid; for example, foam concentrate that is designed to mix with the fuel that it is covering, break the fuel into small droplets, and encapsulate it. The resulting emulsion is rendered nonflammable. *See* Foam Concentrate.

Emulsion — An insoluble liquid suspended in another liquid. *See* Insoluble.

Enabling Legislation — Legislation that gives appropriate officials the authority to implement or enforce the law.

Enabling Objective — *See* Behavioral Objective and Learning Objective.

Encapsulating — Completely enclosed or surrounded, as in a capsule.

Enclosed Structure — Any structure that may expose occupants to hazards such as trapped heat or accumulations of smoke or toxic gases.

Encoding — Putting a message into words.

Encrypting — Type of data security in which a signal is converted to bits of data, to which an algorithm is then applied.

End Impact — Ultimate goal; a long-term way of evaluating program effectiveness. For the fire and life safety educator, the end impact is the decrease of fires and injuries.

End-of Service-Life Indicator (ESLI) — Visual indicator that alerts the user when the APR canister or cartridge has reached its limit and is no longer providing breathable air.

End-of-Service-Time Indicator (ESTI) — Warning device that alerts the user that the respiratory protection equipment is about to reach its limit and that it is time to exit the contaminated atmosphere; its alarm may be audible, tactile, visual, or any combination thereof.

Endothermic Reaction — Chemical reaction in which a substance absorbs heat energy.

Energizers — Quick, attention-getting exercises that raise the energy level of participants (and perhaps the educator as well), promote readiness for learning, create excitement, overcome the effects of fatigue, and develop a sense of shared fun. *Also known as* Spice.

Energy — Capacity to perform work; occurs when a force is applied to an object over a distance, or when a chemical, biological, or physical transformation is made in a substance.

Energy-Absorbing Liner — Portion of helmet designed to cushion blows to the head.

Enforcement Interventions — All the ways in which people are required to act to mitigate risk. Examples related to fire safety include ordinances, laws, and building codes that require the installation of smoke detection and sprinkler systems.

Engine — Ground vehicle providing specified levels of pumping, water, and hose capacity, and staffed with a minimum number of personnel. *Also known as* Fire Department Pumper.

Engine Company — Group of firefighters assigned to a fire department pumper who are primarily responsible for providing water supply and attack lines for fire extinguishment.

Engine House — Firehouse, fire station, or fire hall.

Engine Numbers — For identification, engines of multiengine aircraft are numbered consecutively 1, 2, 3, 4, etc., as seen from the pilot's seat. They are numbered left to right across the aircraft even though some may be mounted on the wings or on the tail of the aircraft.

Engine Pressure — *See* Net Pump Discharge Pressure.

Engineer — (1) Fire protection or fire prevention personnel qualified by professional engineering credentials. (2) A member of the engineering profession. (3) *See* Fire Apparatus Driver/Operator.

Engineered Construction — structures comprised primarily of composite materials such as laminate beams or oriented strand board (OSB) and lightweight steel or wood components. Structural components are often installed prefabricated and may be secured with adhesives.

Engineering Controls — Barrier to a hazard that is built into the design of a building, apparatus or piece of equipment, for example fire doors, smoke evacuation systems, or sprinkler systems.

Engineering Interventions — Modifications made to vehicles, products, materials, and processes to make them less hazardous or to alter an environment to make it safer.

Engulf — (1) Dispersion of material as defined in the General Emergency Behavior Model (GEBMO); an engulfing event occurs when matter and/or energy disperses and forms a danger zone. (2) To flow over and enclose; in the fire service, this refers to being enclosed in flames. *See* Dispersion and Vapor Dispersion.

Enhanced 9-1-1 — Emergency telephone service that provides selective routing, automatic number identification (ANI), and automatic location identification (ALI).

Enhanced Performance Glass (EPG) — Laminated glass that provides higher levels of security as well as additional impact protection and sound proofing.

Enhanced Strike Team — Engine strike team to which a water tender is assigned to operate as part of the team.

Entombed — Condition of being trapped and/or pinned inside a collapsed structure by components of the structure itself.

Entrain — To draw in and transport solid particles or gasses by the flow of a fluid.

Entrainment — The drawing in and transporting of solid particles or gases by the flow of a fluid.

Entry Clothing — Personal protective clothing that is designed for entering into total flame and for specialized work inside industrial furnaces and ovens. *Also known as* Fire Entry Suit.

Entry Control Board — Record-keeping clipboard equipped with a clock, tables, and slots for tallies; used by entry control officers in the United Kingdom, Australia, and New Zealand to keep track of all firefighters wearing SCBA.

Entry Control Officer (ECO) — Command position at a confined space rescue operation responsible for keeping account of rescuers who enter the hazard zone.

Entry Point — Ventilation opening through which replacement air enters the structure; usually the same opening that rescue or attack crews use to enter the structure. *Also known as* Entry Opening.

Envelopment — Attacking key or critical segments around the entire fire perimeter at the same time.

Envenomization — Poisonous effects caused by the bites, stings, or deposits of insects, spiders, snakes, or other poison-carrying animals.

Environment — (1) Circumstances, objects, and conditions by which one is surrounded. (2) The physical area in which an evaluation is done.

Environment Canada — Agency responsible for preserving and enhancing the quality of the natural environment (including water, air, and soil quality), conserving Canada's renewable resources, and coordinating environmental policies and programs for the federal government of Canada.

Environmental Change — Change in a learner's surroundings, particularly the home or workplace, after a fire and life safety outreach activity. *See* Outreach Activity.

Environmental Lapse Rate — Rate of temperature change with elevation, determined by the vertical distribution of temperature at a given time and place.

Environmental Protection Agency (EPA) — U.S. federal regulatory agency designed to protect the air, water, and soil from contamination; responsible for researching and setting national standards for a variety of environmental programs.

EOC — *See* Emergency Operations Center.

EPA — *See* Environmental Protection Agency.

EPDM — *See* Ethylene Propylene Diene Monomer.

Epicenter — Center of a seated explosion, located at the center point of the seismic shock wave created by the explosion.

Epidemic — Occurrence of more cases of disease than expected in a given area, or among a specific group of people, over a particular period of time.

Epidemiologist — A professional who scientifically analyzes the occurrence of morbidity and mortality.

Epidemiology — Scientific study of the causes, distribution, and control of diseases among populations.

Epilepsy — Chronic brain disorder marked by seizures; usually associated with alteration of consciousness, abnormal motor behavior, and psychic or sensory disturbances.

Equal — In terms of specifying apparatus, it means the same level of quality, standard, performance, or design - but not necessarily identical.

Equal Employment Opportunity (EEO) — Personnel management responsibility to be sensitive to the social, economic, and political needs of a jurisdiction or labor market.

Equal Employment Opportunity Law — Law that applies to protected groups of individuals who have experienced past workplace discrimination.

Equilibrium — When the support provided by a structural system is equal to the applied loads.

Equipment — General term for portable tools or appliances carried on the fire apparatus that are not permanently attached to or part of the apparatus.

Equipment Strip — Removal of essential fire fighting tools and equipment at the fire scene before a pumper proceeds to the water source.

Equivalency — Alternative practices that are acceptable for meeting a minimum level of code mandated fire protection. In prescriptive code practice, equivalencies are difficult to produce; compensatory measures are more common.

ERG — *See Emergency Response Guidebook.*

Ergonomic Risk Factors — Aspects of a job task that might cause biomechanical stress to a worker; *also known as* Ergonimc Stressors *or* Ergonomic Factors.

Ergonomics — Applied science of equipment and workplace design intended to maximize productivity by reducing operator fatigue and discomfort.

Escape Route — Pathway to safety; can lead to an already burned area, a previously constructed safety area, a meadow that will not burn, or a natural rocky area that is large enough to allow evacuating personnel to take refuge without being burned. When escape routes deviate from a defined physical path, they must be clearly marked (flagged).

Escape Time — Time required for an individual to exit a hazardous atmosphere without incurring injury or death.

Escape Trunk — Vertical, enclosed shaft with a ladder, providing an escape path for crew stationed in low areas of a vessel.

ESLI — *See* End-of-Service-Life Indicator.

Esophagus — Portion of the digestive tract that lies between the throat and the stomach.

Established Burning — Fire stage in which fuel continues to burn without an external heat source.

ESTI — *See* End-of-Service-Time Indicator.

Ethylene Propylene Diene Monomer (EPDM) — Synthetic, M-class rubber used in single-ply membrane roofs.

Etiologic Agent — Living microorganism, such as a germ, that can cause human disease; a biologically hazardous material.

Etiological Hazards — Harmful viruses and bacteria; when used deliberately, they are *also known as* Biological Weapons.

Eustress — Just enough stress to allow one to perform well; in other words, the person controls the stress.

Eutectic Alloying — When two metals form an alloy that has a lower melting temperature than either of the two original metals.

Evacuation — Controlled process of leaving or being removed from a potentially hazardous location, typically involving relocating people from an area of danger or potential risk to a safer place. *See* Shelter in Place.

Evacuation Chute — Aircraft-door-connected escape slides that when deployed will inflate and extend to the ground; pneumatic in operation, most are deployed automatically by opening the door, though some require manual activation, normally a short pull on a lanyard. Many may be disconnected from the aircraft and used for a flotation device in water crashes. *Also known as* Evacuation Slide.

Evacuation System — System intended to allow people to escape to safety during a fire; includes egress systems (exit access, exit, and exit discharge) as well as doors, panic hardware, horizontal exits, stairs, smokeproof towers, fire-escape stairs, escalators, moving sidewalks, elevators, windows, and exit lighting and signs.

Evaluation — (1) Systematic and thoughtful collection of information for decision-making; consists of criteria, evidence, and judgment. (2) Last of the four teaching steps in which the fire and life safety educator finds out whether the educational objectives have been met. (3) Process that examines the results of a presentation or program to determine whether the participants have learned the information or behaviors taught; consists of criteria, evidence, and judgment. *See* Formative Evaluation or Test, Lesson Plan, and Summative Evaluation.

Evaluation Instrument — (1) Physical means used to evaluate or test a learner's mastery of the educational objectives taught; may be written, oral, or performance-based. (2) An instrument designed to collect data so changes in knowledge, behaviors, environments, and lifestyles can be evaluated.

Evaluation Plan — Component of an intervention strategy that includes the problem statement, goal, and a series of measurable objectives that support the goal.

Evaluation Step — Fourth step in conducting a lesson in which the student demonstrates that the required degree of proficiency has been achieved.

Evaluation Strategy — (1) Plan for conducting an evaluation; includes the type of evaluation instrument to be used, the information or behaviors to be evaluated, and the methods used to interpret the evaluation instrument. (2) Plan that details how a risk-reduction program will be examined for effectiveness.

Evaporation — Process of a solid or liquid turning into gas.

EVCS — *See* Emergency Voice Communications Systems.

Event Flowcharting — Method for chronologically displaying the movements of events or occurrences, either over time or through a system.

Evidence — (1) One of three requirements of evaluation; the information, data, or observation that allows the investigator to compare what was expected to what actually occurred. (2) In law, something legally presented in court that bears on the point in question. (3) Information collected and analyzed by an investigator.

Evolution — *See* Training Evolution.

EX — Rating symbol used on lift trucks that are safe for use in atmospheres containing flammable vapors or dusts.

EX Symbol — Rating symbol used on lift trucks that are safe for use in atmospheres containing flammable vapors or dusts. *See* Lift Truck.

Excavation — Opening in the ground that results from a digging effort.

Excelsior — Slender, curled wood shavings used for starting fires or packing fragile items.

Excepted Packaging — Container used for transportation of materials that have very limited radioactivity. *See* Industrial Packaging, Packaging (1), Strong, Tight Container, Type A Packaging, and Type B Packaging.

Exclusion Zones — Area usually within the hot zone which personnel should avoid regardless of their level of protective clothing and equipment.

Exclusionary Evidence — Evidence collected to show that a particular device or scenario can be ruled out, in relation to the ignition or fire spread scenario.

Exclusionary Rule — Judicially established evidentiary rule that excludes from admission at trial any evidence seized in a manner considered unreasonable by an interpretation of the Fourth Amendment of the U.S. Constitution.

Executive Summary — A condensed, single page synopsis of the written narrative.

Exhalation Valve — One-way valve that lets exhaled air out of the self-contained breathing apparatus facepiece.

Exhaust Area — Area behind a jet engine where hot exhaust gases present a danger to personnel.

Exhaust Opening — Intended and controlled exhaust locations that are created or improved at or near the fire to allow products of combustion to escape the building.

Exhaust System — Ventilation system designed to remove stale air, smoke, vapors, or other airborne contaminants from an area. *See* Heating, Ventilating, and Air-Conditioning (HVAC) System.

Exhauster — Device that speeds the discharge of air from a dry-pipe sprinkler system.

Exigent Circumstances — Right of entry stating that the fire department does not require a warrant to enter a property to suppress a fire, or to remain on the property for a reasonable amount of time afterward in order to determine the origin and cause of the fire.

Exit — Portion of a means of egress that is separated from all other spaces of the building structure by construction or equipment, and provides a protected way of travel to the exit discharge. *See* Clear Width, Egress, Exit Discharge, Means of Egress, and Travel Distance.

Exit Access — Portion of a means of egress that leads to the exit; for example, hallways, corridors, and aisles. *See* Exit and Means of Egress.

Exit Capacity — According to building code requirements, the maximum number of people who can discharge through a particular exit; determined by the useable width of the exit. *See* Exit.

Exit Device — *See* Panic Hardware.

Exit Discharge — That portion of a means of egress that is between the exit and a public way. *See* Exit and Means of Egress.

Exit Lighting — Lighting intended to help people see on their way out of a structure.

Exit Opening — In ventilation, the opening that is made or used to release heat, smoke, and other contaminants into the atmosphere.

Exit Sign — Lighted sign indicating the direction/location of an exit from a structure.

Exit Stairs — Stairs that are used as part of a means of egress. The stairs may be part of either the exit access or the exit discharge, when conforming to requirements in the NFPA®101, *Life Safety Code®*. *See* Exit, Exit Access, Exit Discharge, Means of Egress, and NFPA® 101, *Life Safety Code®*.

Exothermal — Characterized by or formed with the evolution of heat.

Exothermic Cutting Device — Cutting torch that uses a chemical reaction to produce heat and flame.

Exothermic Reaction — Chemical reaction between two or more materials that changes the materials and produces heat.

Expander — (1) Device that enlarges the expansion rings used for securing threaded couplings to fire hose. (2) Inner component of a screw-in expander coupling.

Expansion Joint — Flexible joint in concrete used to prevent cracking or breaking because of expansion and contraction due to temperature changes.

Expansion Ratio — Ratio of the finished foam volume to the volume of the original foam solution. *Also known as* Expansion. *See* Aeration and Foam Expansion.

Expansion Ring — Malleable metal band that binds fire hose to a threaded coupling by compressing the hose tightly against the inner surface of the coupling.

Expansion Ring Method — Means of attaching a threaded coupling to a fire hose; a metal expansion ring is placed inside the end of the hose and then expanded, in order to compress the hose tightly against the inner surface of the coupling.

Expellant Gas — Any of a number of inert gases that are compressed and used to force extinguishing agents from a portable fire extinguisher. Nitrogen is the most commonly used expellant gas. *See* Inert Gas.

Expert Power — Sufficiently strong perception that a leader's expertise, knowledge, and abilities will produce a desirable outcome, so that others will willingly follow that leader.

Expert Witness — Person with sufficient skill, knowledge, or experience in a given field so as to be capable of drawing inferences or reaching conclusions or opinions that an average person would not be competent to reach.

Explosion — physical or chemical process that results in the rapid release of high pressure gas into the environment. *See* Backdraft and Boiling Liquid Expanding Vapor Explosion.

Explosion-Dynamics Analysis — Process of following force vectors backwards toward the initial compartment where an explosion occurred.

Explosion-Proof Equipment — Equipment encased in a rigidly built container so that it can withstand an internal explosion and prevent ignition of a surrounding flammable or explosive atmosphere.

Explosive — (1) Any material or mixture that will undergo an extremely fast self-propagation reaction when subjected to some form of energy. *See* Detonation, High Explosive, and Magazine. (2) Materials capable of burning or bursting suddenly and violently. *See* Detonation, Low Explosive, and Magazine.

Explosive Atmosphere — Any atmosphere that contains a mixture of fuel to air that falls within the explosive limits for that particular material.

Explosive Breathing Technique — Individual emergency conditions breathing technique used by a firefighter wearing SCBA during accidental submersion; the firefighter holds his or her breath, rapidly inhales and exhales, and then holds breath again.

Explosive Device — Contrivance built to cause an explosion; commonly known as a bomb, though this is not always an accurate description. *See* Incendiary Device.

Explosive Limit — *See* Flammable Limit.

Explosive Range — Range between the upper and lower flammable limits of a substance. *See* Flammable Range.

Explosive Spalling — Spalling that occurs violently, throwing out bits of concrete-like projectiles.

Exposure — Contact with a hazardous material, causing biological damage, typically by swallowing, breathing, or touching (skin or eyes). Exposure may be short-term (acute exposure), of intermediate duration, or long-term (chronic exposure).

Exposure Bag — Special neoprene bag into which a person is placed for field treatment of hypothermia.

Exposure Concentration — Measure of toxicity; expressed in parts per million (ppm) and abbreviated C.

Exposure Control Program — Organizational program that provides resources, training, and equipment to firefighters in order to protect them from exposure to chemical and biological hazards in the workplace including hazardous materials, infectious diseases, and bloodborne pathogens.

Exposure Fire — A fire ignited in fuel packages or buildings that are remote from the initial fuel package or building of origin.

Exposure Limit — Maximum length of time an individual can be exposed to an airborne substance before injury, illness, or death occurs.

Exposure Protection — Covering any object in the immediate vicinity of the fire with water or foam.

Exposure Time — Length of exposure; expressed in minutes, and abbreviated T.

Exposures — (1) Structures or separate parts of the fireground to which a fire could spread. (2) People, property, systems, or natural features that are or may be exposed to the harmful effects of a hazardous materials emergency. (3) NFPA® defines exposure as the heat effect from an external fire that might cause ignition of or damage to an exposed building. (4) Direction in which a slope faces. (5) General surroundings of a site with special reference to its openness to winds and sunshine.

Extend — (1) To extend a hoseline by adding hose, straightening, or rerouting the hose already laid. (2) To increase the reach of an extension ladder or aerial device by raising the fly section.

Extended-Attack Fire — (1) Wildland fire that has not been contained or controlled by initial-attack forces and for which more fire fighting resources are arriving, en route, or being ordered by the incident commander. (2) Situation in which a fire cannot be controlled by initial-attack resources within a reasonable period of time.

Extension Cylinders — Hydraulic cylinders that control the extension and retraction of the fly sections of an aerial device.

Extension Fly Locks — Devices that prevent the fly sections of a ground or aerial ladder from retracting unexpectedly.

Extension Ladder — Variable-length ladder of two or more sections that can be extended to a desired height.

Extension Ram — Powered hydraulic tool designed for straight pushing operations; may extend as far as 63 inches (1 600 mm). *Also known as* Ram.

Exterior — (1) The outside of a building. (2) The outside of a vehicle; body panels, glass, bumpers, and other components.

Exterior Exposure — Building or other combustible object located close to the fire building that is in danger of becoming involved due to heat transfer from the fire building.

Exterior Fire Protection — Protection of structures from the exterior with no interior fire fighting.

Exterior Insulation and Finish Systems (EIFS) — Exterior cladding or covering systems composed of an adhesively or mechanically fastened foam insulation board, reinforcing mesh, a base coat, and an outer finish coat. *Also known as* Synthetic Stucco.

Exterior Stairs — Stairs separated from the interior of a building by walls.

External Customers — Citizens of the service area protected by the organization.

External Floating Roof Tank — Fixed-site vertical storage tank that has no fixed roof but relies on a floating roof to protect its contents and prevent evaporation. *Also known as* Open-Top Floating Roof Tank. *See* Cone Roof Tank, Floating Roof Storage Tank, and Internal Floating Roof Tank.

External Organizations — Groups from outside the organization with which the educator is associated.

External Respiration — Inhalation and exhalation of air into and from the lungs.

External Water Supply — Any water supply to a fire pump from a source other than the vehicle's own water tank, or any water supply to an aerial device from a source other than the vehicle's own fire pump.

Extinguish — To put out a fire completely.

Extinguisher — Portable fire fighting appliance designed for use on specific types of fuel and classes of fire.

Extinguisher Hose — Braided, rubber-covered hose used on extinguishers that is made to withstand pressures up to 1,250 psi (8 618 kPa).

Extinguishing Agent — Any substance used for the purpose of controlling or extinguishing a fire.

Extreme Rescue Load — Normally considered to be 600 pounds (272.15 kg).

Extremely Hazardous Substance — Chemicals determined by the Environmental Protection Agency (EPA) to be extremely hazardous to a community during an emergency spill or release, because of their toxicity and physical/chemical properties. There are 402 chemicals listed under this category. *See* Hazardous Substance.

Extrication — Incident in which a trapped victim must be removed from a vehicle or other type of machinery.

Extrication Collar — Device used for spinal immobilization.

Extrication Group — Group within the incident management system that is responsible for extricating victims.

Extrude — To shape heated plastics or metal by forcing them through dies.

Extruded Coupling — Coupling manufactured by the process of extrusion.

F

FAA — *See* Federal Aviation Administration.

Facade — Fascia added to some buildings with flat roofs to create the appearance of a mansard roof. *Also known as* False Roof or Fascia.

Face — *See* Wall.

Facepiece — Part of a self-contained breathing apparatus that fits over the face; includes the head harness, facepiece lens, exhalation valve, and connection for either a regulator or a low-pressure hose. *Also known as* Mask.

Faceshield — Protective shield attached to the front of a fire helmet. *Also known as* Helmet Faceshield.

Facilitator — (1) Instructor role in which he or she encourages productive interaction of group members; instructor does not display personal expertise, but channels and enhances the expertise of others. (2) Person in a group whose basic job is to stimulate others to participate in the group discussion.

Facilities Unit — Functional unit within the support branch of the logistics section of an incident management system; provides fixed facilities for an incident, including the incident base, feeding areas, sleeping areas, sanitary facilities, and a formal command post.

FACP — *See* Fire Alarm Control Panel.

Fact Sheet — List of facts and material that allow a reporter to write an in-depth article; may be several pages long, and includes historical perspective, anecdotal material, statistics, and local data.

Fact Testimony — When a witness is presenting factual information based on his or her personal observations. *Also known as* Lay Testimony.

Factory Certified — Qualification of fire department personnel who attend special manufacturers' repair schools to become formally qualified in certain testing and maintenance procedures.

Factory Raise — *See* Hotel Raise.

FACU — *See* Fire Alarm Control Unit.

Fahrenheit Scale — Temperature scale on which the freezing point is 32°F (0°C) and the boiling point at sea level is 212°F (100°C) at normal atmospheric pressure. *See* Celsius Scale and Temperature.

Failure Point — Point at which material ceases to perform satisfactorily; depending on the application, this can involve breaking, permanent deformation, excessive deflection, or vibration.

Fainting — Momentary loss of consciousness caused by insufficient blood supply to the brain. *Also known as* Syncope.

Fair Labor Standards Act — Federal law requiring employers to compensate employees at a pay/comp rate of time and half for services performed off-duty in the local community on behalf of the agency where the individual is employed.

Fair Use — Doctrine of the Copyright Act that grants the privilege of copying materials to persons other than the owner of the copyright, without consent, when the material is used in a reasonable manner.

Fairing — An auxiliary structure or the external surface either attached to, or a part of the roof of a large truck that serves to reduce drag.

Fairlead — Chock or opening, sometimes fitted with a roller device designed to lead a rope or line from one part of a vessel to another (change line direction); also controls lines and minimizes chafing. *See* Chock (1).

Fall Line — Portion of a rope, in a block and tackle system, that runs between the standing block and the leading block or the power source. *Also known as* Pull Line, Pulling Line, and Weft Yarn.

Fallout — Radioactive particles descending from a cloud after a nuclear detonation.

False Ceiling — Additional suspended ceiling below the true original ceiling, forming a concealed space.

False Front — Additional facade on the front of a building applied after the original construction for decoration; often creates a concealed space.

False Roof — *See* Facade.

Family Dynamics — Structure and characteristics of a person's living environment, including relatives, caregivers, or other relationships and their interactions with each other.

Family Educational Rights and Privacy Act (FERPA) — Legislation that provides that an individual's school records are confidential and that information contained in those records may not be released without the individual's prior written consent.

Family Medical Leave Act — Federal law requiring employers to allow employees specific use of personal sick leave to care for an ill spouse or other immediate family member.

Family-of-Eight Knots — Series of rescue knots based on a figure-eight knot.

Fan — Generic term used interchangeably for both blowers and smoke ejectors.

Fan Light — Semicircular window, usually over a doorway, with muntins radiating like the ribs of a fan.

Fan-Style Room Setup — *See* Chevron Room Setup.

Fantail — Back part of a vessel that hangs out over the water; a stern overhang.

Farm Implement — Any piece of farm machinery.

Fascia — (1) Flat horizontal or vertical board located at the outer face of a cornice. (2) Broad flat surface over a storefront or below a cornice.

Fast — To securely attach a vessel to a wharf or dock.

Fast Attack Mode — When the first-arriving unit at a fire makes a quick offensive attack on the fire. *Also known as* Fast Attack.

Fast Track — Method to reduce the overall time for completion of a project through phasing the design and construction process.

Fatality — Someone who has died as the result of the incident.

Fax — To send a hard copy facsimile (copy) of print or illustrative material via telephone lines.

FBARS — *See* Fire Breathing Air Replenishment Systems.

FBI — *See* Federal Bureau of Investigation.

FCC — *See* Federal Communications Commission.

FDC — *See* Fire Department Connection.

FDIC — *See* Fire Department Instructors Conference.

Febrile Convulsions — Convulsions brought on by fever, occurring most often in children.

Federal Aviation Administration (FAA) — Subdivision of the U.S. Department of Transportation that is involved with the regulation of civil aviation.

Federal Bureau of Investigation (FBI) — Agency within the U.S. Department of Justice that investigates the theft of hazardous materials, collects evidence for crimes, and prosecutes criminal violation of federal hazardous materials laws and regulations. The FBI is also the lead agency on terrorist incident scenes. *See* Terrorism.

Federal Communications Commission (FCC) — U.S. government agency charged with the control of all radio and television communications; acts as the main regulator of radio frequencies in both the public and private sectors.

Federal Emergency Management Agency (FEMA) — Agency within the U.S. Department of Homeland Security (DHS) that is responsible for emergency preparedness, mitigation, and response activities for events including natural, technological, and attack-related emergencies.

Federal Freedom of Information Act (FOIA) — Legislation used as a model for many state laws designed to make government information available to the public.

Feed Main — Pipe connecting the sprinkler system riser to the cross mains; these cross mains directly service a number of branch lines on which the sprinklers are installed.

Feedback — (1) Student responses generated by questions, discussions, or opportunities to perform that demonstrate learning or understanding. (2) In communications, responses that clarify and ensure that the message was received and understood.

Feeder Line — *See* Relay-Supply Hose.

Felony — Serious crime punishable by a fine, incarceration, and/or death, depending on the severity of the crime and the jurisdiction. See Indictable Offense.

FEMA — See Federal Emergency Management Agency.

Female Coupling — Threaded swivel device on a hose or appliance with internal threads designed to receive a male coupling with external threads of the same thread and diameter.

Femoral Artery — Principal artery of the thigh; allows pulse to be felt in the groin area.

Femur — Bone that extends from the pelvis to the knee; longest and largest bone in the body.

Fender — (1) Exterior body portion of a vehicle adjacent to the front or rear wheels. (2) Buffer between the side of a vessel and a dock, or between two vessels to lessen shock and prevent chafing. (3) Body material that surrounds the front tires; starts at the front of the vehicle, proceeds around the front tire and ends at the fire wall.

FERPA — See Family Educational Rights and Privacy Act.

FFAPR — See Full-Facepiece Air-Purifying Respirator.

FFFP — See Film Forming Fluoroprotein Foam.

FIBC — See Flexible Intermediate Bulk Container.

Fiber — Solid particle whose length is several times greater than its diameter.

Fiberboard — Lightweight insulation board made of compressed cellulose fibers; often used in suspended ceilings.

Fiberglass — Composite material consisting of glass fibers embedded in resin.

Fiber-Optic Search Device — Very small camera on a flexible arm that allows viewing in a confined space with limited access.

Fibrillation — Rapid, ineffective contraction of the heart.

Fibula — Smaller of the two bones of the lower leg.

Fiddley — Vertical space extending from the engine room to a vessel's stack (chimney). See Economizer and Stack.

FIDO — See Fire Incident Data Organization.

Field Mobile Mechanic — Motor-driven vehicle designed and constructed for the purpose of providing a specified level of equipment capacity and mechanically trained personnel.

Field Notes — Written record of an incident safety officer, or other officer's, observations during an incident; often used as the basis for a postincident analysis (PIA) report.

Field Sketch — Rough drawing of an occupancy that is made during an inspection. The field sketch is used to make a final inspection drawing.

Field Unit — See Brush Apparatus or Wildland Fire Apparatus.

Fifth Wheel — Device used to connect a truck tractor or converter dolly to a semitrailer in order to permit articulation between the units. It is generally composed of a lower part consisting of a trunnion, plate, and latching mechanism mounted on the truck tractor (or dolly), and a kingpin assembly mounted on the semitrailer. See Semitrailer.

Fifth-Wheel Pickup Ramp — Steel plate designed to lift the front end of a semitrailer to facilitate the engagement of the kingpin into the fifth wheel.

Figure-Eight — Forged metal device in the shape of an eight; used to help control the speed of a person descending a rope.

Figure-Eight Knot — Knot used to form a loop in the end of a rope; should be used in place of a bowline knot when working with synthetic fiber rope.

Figure-Eight Plate — Forged metal device in the shape of the number eight (8); used to help control the speed of a person descending a rope.

File Server — Computer that contains program files available to all workstations in a computer network.

File Sharing — Practice of making files or documents on one computer or server available to the general public, or to a selected group of individuals who are given access to the files. Allows users at remote locations to have access to the same materials without the need to put those materials on CD-ROM, memory drives, or other media.

Fill Hose — Short section of hose carried on apparatus, equipped with booster tanks to fill the tank from a hydrant or another truck.

Fill Opening — Opening on top of a tank used for filling the tank; usually incorporated into a manhole cover.

Fill Site — Location at which tankers/tenders will be loaded during a water shuttle operation.

Filler Yarn — Threads running crosswise in fabrics or woven hose.

Fillet Weld — Weld made in the interior angle of two pieces of metal placed at right angles to each other.

Fill-The-Box — Slang for a request by a unit responding on a reduced assignment to send the balance of the assignment.

Film Forming Fluoroprotein Foam (FFFP) — Foam concentrate that combines the qualities of fluoroprotein foam with those of aqueous film forming foam. *See* Foam Concentrate.

Filter Basket — Fine mesh screen that is attached to the end of the foam concentrate pickup hose to prevent sediment from entering the system and clogging the proportioner.

Filter Breathing — Individual emergency conditions breathing technique used by a firefighter with a depleted air supply. The firefighter inserts the regulator end of the low-pressure hose into a pocket or glove, or inside the turnout coat, to help filter smoke particles and to protect the firefighter from inhaling superheated air.

Filter Canister — Filtration device containing chemicals to filter out harmful substances through adsorption or absorption on negative-pressure respirators. This should not be used for fire fighting or Immediately Dangerous to Life or Health (IDLH) atmospheres.

Fin — Fixed or adjustable airfoil attached longitudinally to an aircraft to provide a stabilizing effect in flight.

Final Approach — Portion of the landing pattern in which the aircraft is lined up with the runway and is heading straight in to land.

Final Disposition — Point in the chain of custody where a piece of evidence is determined to no longer have value as evidence; options at this point may include permanent storage, return to the owner, or authorized destruction.

Finance/Administrative Unit — Component of an incident management system responsible for documenting the ongoing financial impact of the incident; oversees the documentation of time and costs associated with personnel, as well as the documentation of private resources used throughout the incident. Includes the time unit, procurement unit, compensation/claims unit, and the cost unit.

Financial Profiling — Investigative tool that allows the investigator to organize and display the financial data of an individual or organization onto a graph or chart.

Fine Aggregates — Usually sand.

Finger — Long, narrow extension of a fire projecting from the main body of a wildland fire.

Finish — (1) Arrangement of hose usually placed on top of a hose load and connected to the end of the load. *Also known as* Hose Load Finish. (2) Fine or decorative work required for a building or one of its parts. (3) Finishing material used in painting.

Finished Foam — (1) Extinguishing agent formed by mixing a foam concentrate with water and aerating the solution for expansion. *Also known as* Foam. *See* Foam Blanket, Foam Concentrate, and Foam Solution. (2) Completed product after air is introduced into the foam solution.

Finished Solution — Extinguishing agent formed by mixing foam concentrate with water and aerating the solution for expansion.

Fink Truss — Monoplane truss common in residential construction, in which all chords and diagonal members are in the same plane.

Fire — Rapid oxidation process, which is a chemical reaction resulting in the evolution of light and heat in varying intensities. (NFPA® 921)

Fireboat — Vessel or watercraft designed and constructed for the purpose of fighting fires; provides a specified level of pumping capacity and personnel for the extinguishment of fires in the marine environment.

Fire Alarm — (1) Call announcing a fire. (2) Bell or other device summoning a fire company to respond to a fire or other emergency.

Fire Alarm Control Panel (FACP) — System component that receives input from automatic and manual fire alarm devices and may provide power to detection devices or communication devices. The fire alarm control unit can be a local control unit or a master control unit. *Also known as* Fire Alarm Control Unit.

Fire Alarm Control Unit (FACU) — *See* Fire Alarm Control Panel (FACP). The main fire alarm system component that monitors equipment and circuits, receives input signals from initiating devices, activates notification appliances ,and transmits signals off-site

Fire Alarm Signal — Continuous rapid ringing of a vessel's bell for a period of not less than 10 seconds, supplemented by the continuous ringing of the general alarm bells for not less than 10 seconds.

Fire Alarm System — (1) System of alerting devices that takes a signal from fire detection or extinguishing equipment and alerts building occupants or proper authorities of a fire condition. (2) System of interconnected alarm-initiating and alarm-indicating devices designed to alert personnel to the existence of a fire in the protected premises. The alarm system may or may not be connected to a fire suppression system. (3) System used to dispatch fire department personnel and apparatus to emergency incidents. *Also known as* Fire Alarm Signaling System. *See* Fire Alarm Control Unit (FACU), Fire Detection System, Initiating Device, and Signaling Device.

Fire and Life Safety Educator I — Those who have demonstrated the ability to coordinate and deliver existing educational programs and information.

Fire and Life Safety Educator II — Those who have demonstrated the ability to prepare educational programs and information to meet identified needs.

Fire and Life Safety Educator III — Those who have demonstrated the ability to create, administer, and evaluate educational programs and information.

Fire and Life Safety Inspector — *See* Inspector.

Fire and Life Safety Strategy — A series of activities, programs, and initiatives designed to reduce identified risks within a community.

Fire Apparatus — Any fire department emergency vehicle used in fire suppression or other emergency situations.

Fire Apparatus Driver/Operator — Firefighter who is charged with the responsibility of operating fire apparatus to, during, and from the scene of a fire operation, or at any other time the apparatus is in use. The driver/operator is also responsible for routine maintenance of the apparatus and any equipment carried on the apparatus. This is typically the first step in the fire department promotional chain. *Also known as* Chauffeur, Driver/Operator, or Engineer.

Fire Behavior — (1) Manner in which fuel ignites, flames develop, and heat and fire spread; sometimes used to refer to the characteristics of a particular fire. (2) Manner in which a fire reacts to the variables of fuel, weather, and topography. *See* Fire Dynamics.

Fire Blanket — Blanket stored in a case in a kitchen or similar location that is intended to be wrapped around a victim whose clothing catches fire.

Fire Bomb — Incendiary device used to start an arson fire; usually hand-thrown so that it will break, spill its flammable contents, and ignite.

Fire Breathing Air Replenishment Systems (FBARS) — A standpipe for air permanently installed within a high-rise building or a large horizontal structure. Air is pumped into the system by a fire department's mobile air truck on the ground, providing an immediate and continuous supply of breathing air to the responders. Air bottles can then be refilled in a matter of seconds at fill stations located throughout a high-rise building.

Fire Brigade — (1) Organization of industrial plant personnel trained to use fire fighting equipment within the plant, and who are assigned to carry out fire prevention activities and basic fire fighting duties and responsibilities. The full-time occupation of brigade members may or may not involve fire suppression and related activities. (2) Term used in some countries, outside the U.S., in place of fire department.

Fire Brigade Apparatus — Emergency response vehicles used by fire brigade personnel for fire suppression, rescue, or other specialized functions.

Fire Brigade Management Official — Individual designated by the senior facility manager as the person responsible for the organization, management, and operation of the industrial fire brigade.

Fire Brigade Organizational Statement — OSHA-required statement that must be prepared by all employers who choose to have a fire brigade. The document must include a statement that establishes the existence of the brigade; the basic organizational structure; the type, amount, and frequency of training to be provided; the expected number of members; and the functions that the brigade is to perform at the workplace.

Fire Broom — Broom used in wildland fire fighting.

Fire Bucket — Bucket with a round bottom, usually painted red and marked with the word *fire* to discourage use for purposes other than fire fighting; frequently kept filled with water, sand, or other fire extinguishing material. *Also known as* Fire Pail.

Fire Buff — (1) A fire department enthusiast. (2) A person attracted to fires.

Fire Bug — Common slang term to describe an arsonist or pyromaniac; especially a repeating fire setter. Also describes a person who not only sets fires but also enjoys watching them.

Fire Building — (1) Building in which a fire originated and is in progress. (2) Training building in which fire fighting is practiced. *Also known as* Burn Building.

Fire Camp — Camp near a large wildland fire for coordinating agencies, communications, logistics, and support.

Fire Cause — (1) Agency or circumstance that started a fire or set the stage for one to start; source of a fire's ignition. (2) The sequence of events that allows the ignition source of ignition and the material first ignited to come together. (3) The combination of fuel supply, heat source, and a hazardous act that results in a fire.

Fire Cause Classification — One of four established classifications for the cause of a fire; accidental, natural, incendiary, or undetermined.

Fire Cause Determination — Process of establishing the cause of a fire incident through careful investigation and analysis of the available evidence.

Fire Command Center — Designated room or area in a structure where the status of the fire detection, alarm, and protection systems is displayed and the systems can be manually controlled; may be staffed or unstaffed and can be accessed by the fire department.

Fire Commissioner — In the U.S., a politically appointed or elected manager of the fire department; usually has authority over the fire chief, who is a civil servant. In Canada, the position is responsible for fire prevention activities. *See* Fire Marshal.

Fire Control Plan — Set of general arrangement plans for each deck that illustrate fire stations, fire-resisting bulkheads, and fire-retarding bulkheads, together with particulars of fire detecting systems, manual alarm systems, fire extinguishing systems, fire doors, means of access to different compartments, and ventilating systems (including locations of dampers and fan controls). Plans are stored in a prominently marked weather-tight enclosure outside the house, for the assistance of land-based fire fighting personnel.

Fire Control Zone — Uninvolved area immediately surrounding a fire, wide enough to protect fire brigade members and others from the adverse effects of the fire.

Fire Corps — National program in which citizen advocates assist fire departments with nonoperational roles.

Fire Curtain — Aluminized device on a rod, designed to be unrolled to reflect radiant heat from operators or crew members on some apparatus, bulldozers, or tractor-plows.

Fire Cut — Angled cut made at the end of a wood joist or wood beam that rests in a masonry wall to allow the beam to fall away freely from the wall in case of failure of the beam. This helps prevent the beam from acting as a lever to push against the masonry.

Fire Damper — (1) Device that automatically interrupts airflow through all or part of an air-handling system, thereby restricting the passage of heat and the spread of fire. (2) Device installed in air ducts that penetrate fire-resistant-rated vertical or horizontal assemblies; prohibits the transfer of heat or flames through the ducts at the point where the duct passes through the assembly. *See* Duct and Smoke Damper.

Fire Department Connection (FDC) — Point at which the fire department can connect into a sprinkler or standpipe system to boost the water pressure and flow in the system. This connection consists of a clappered siamese with two or more 2½-inch (64 mm) intakes or one large-diameter (4-inch [102 mm] or larger) intake. *Also known as* Fire Department Sprinkler Connection. *See* Standpipe System.

Fire Department Emergency Communications Systems — Detection and alarm systems that transfer information from whomever reports a fire to fire service personnel.

Fire Department Instructors Conference (FDIC) — Annual meeting of fire department training officials.

Fire Department Physician — Physician designated by a fire department to treat members of the department.

Fire Department Pumper — Fire apparatus having a permanently mounted fire pump with a rated discharge capacity of 750 gpm (3 000 L/min) or greater. This apparatus may also carry water, hose, and other portable equipment.

Fire Department Sprinkler Connection — *See* Fire Department Connection.

Fire Department Water Supply Officer — Officer in charge of all water supplies at the scene of a fire; duties include placing pumpers at the most advantageous hydrants or other water sources, and directing supplementary water supplies, including water shuttles and relay pumping operations. This may also be a permanent, full-time staff position with responsibility for coordinating, with other local agencies, water supply projects of concern to the fire department.

Fire Detection System — System of detection devices, wiring, and supervisory equipment used for detecting fire or products of combustion, and then signaling that these elements are present. *See* Fire Alarm Control Unit (FACU), Fire Alarm System, Initiating Device, and Signaling Device.

Fire Devil — Small cyclone or twister that forms when heated fire gases rise and cooler air rushes into the resulting low-pressure areas; most common in forest fires but can also be encountered in large structural fires.

Fire District — Designated geographic area where fire protection is provided, usually through a supporting tax, or an area where fire prevention codes are enforced.

Fire Door — Specially constructed, tested, and approved fire-rated door assembly designed and installed to prevent fire spread by automatically closing and covering a doorway in a fire wall to block the spread of fire through the door opening. *See* Fire Wall.

Fire Door Assembly — Listed and labeled assembly consisting of door, frame, and hardware. Rated in terms of hours of fire protection; for example, a one-hour fire door assembly.

Fire Drill — Training exercise to ensure that the occupants of a building can exit the building in a quick and orderly manner in case of fire.

Fire Dynamics — Applying the tools of chemistry and physics to gain a technical understanding of how fires ignite, grow, and spread. *Also known as* Fire Behavior.

Fire Edge — Boundary of a fire at a given moment.

Fire Effect — In fire investigations, the changes to materials or loss of materials, such as melting, calcinations, or char; sometimes confused with Fire Pattern.

Fire Entry Suit — *See* Entry Clothing.

Fire Escape — (1) Means of escaping from a building in case of fire; usually an interior or exterior stairway or slide, independently supported and made of fire-resistive material. (2) Traditional term for an exterior stair, frequently incorporating a movable section of noncombustible construction, that is intended as an emergency exit; usually supported by hangers installed in the exterior wall of the building. The traditional fire escape is narrower than required stairways and is no longer allowed by codes, except as an existing feature. They have been associated with frequent failures.

Fire Extinguisher — Portable fire fighting device designed to combat incipient fires. *See* Carbon Dioxide (CO_2) and Incipient Phase.

Fire Fighting Boots — Protective footwear meeting the design requirements of NFPA®, OSHA, and CAN/CSA Z195-02 (R2008).

Fire Fighting Ensemble — Protective gear used for fire fighting that consists of protective pants, coat, boots, gloves, hood, and helmet.

Fire Flank — Side of a wildland fire.

Fire Flow — (1) Quantity of water available for fire fighting in a given area. It is calculated in addition to the normal water consumption in the area. (2) The amount of water required to extinguish a fire in a timely manner.

Fire Flow Testing — Procedure used to determine the rate of water flow available for fire fighting at various points within the distribution system.

Fire Front — Part of a fire within which continuous flaming combustion is taking place; assumed to be the leading edge of the fire perimeter. In surface fires, the fire front may be mainly smoldering combustion.

Fire Gases — Gases produced as combustion occurs.

Fire Guard — Person trained and assigned to watch for fires and life safety hazards during specified periods or events.

Fire Hall — *See* Fire Station.

Fire Hazard — Any material, condition, or act that contributes to the start of a fire or that increases the extent or severity of fire. *See* Hazard.

Fire Hazard Severity Rating System — System of adjectives used to describe fire danger to the public; ranges from *low* to *extreme*.

Fire Hook — Hook, on the end of a rope or pole, used to pull combustible roofing materials such as thatch from burning buildings. These devices could also be used to hook into the gables of a building, or a ring installed on the gable, to pull that end of the building down. Because these hooks were carried on early ladder trucks, these trucks became known as "hook and ladder companies". The pike pole is a modern variation of the fire hook. *See* Pike Pole.

Fire Hose — Flexible, portable tube manufactured from water tight materials in 50 to 100 foot (15 to 30 m) lengths that is used to transport water from a source or pump to the point it is discharged to extinguish fire.

Fire Hose Float — Water-rescue flotation device made from inflated fire hose.

Fire House — *See* Fire Station.

Fire Hydrant — Upright metal casing that is connected to a water supply system and is equipped with one or more valved outlets to which a hoseline or pumper may be connected to supply water for fire fighting operations. *Also known as* Hydrant.

Fire Hydraulics — Science that deals with water in motion, as it applies to fire fighting operations.

Fire in the United States — USFA publication based largely on fire data submitted to the National Fire Incident Reporting System (NFIRS) by roughly 13,000 fire departments; presented in an easy-to-understand format that relies heavily on the extensive use of charts and graphs.

Fire Incident Data Organization (FIDO) — NFPA® organization that maintains a database that is one of the main sources of information (data, statistics) about fires in the U.S. Provides an in-depth look at reported fires involving three or more civilian deaths, one or more firefighter deaths, or large dollar loss, as well as some other particularly interesting types of fires, such as high-rise fires.

Fire Inspector — Fire personnel assigned to inspect property with the purpose of enforcing fire regulations.

Fire Investigator — Public or private sector individual tasked with discovering the origin and cause of a fire, as well as who may be responsible or liable for a fire.

Fire Line Tape — Plastic marking tape strung around an emergency scene to keep bystanders away from the action.

Fire Lines — (1) Boundaries established around a fire area to prevent access except for emergency vehicles and persons having a right and need to be present. (2) In wildland fire fighting, part of a control line that is scraped or dug to mineral soil; also, a general term for the area where fire fighting activities are taking place. The wildland equivalent of the term *fireground* as used in structural fire fighting.

Fire Load — (1) The amount of fuel within a compartment; expressed in pounds per square foot and obtained by dividing the amount of fuel present by the floor area. Fire load is used as a measure of the potential heat release of a fire within a compartment. (2) Maximum amount of heat that can be produced if all the combustible materials in a given area burn. *Also known as* Fuel Load.

Fire Loss In Canada — One of the main sources of information (data, statistics) about fires in Canada; compiled from data provided by the Association of Canadian Fire Marshals and Fire Commissioners and the government agency Statistics Canada.

Fire Main System — System that supplies water to all areas of a vessel; composed of fire pumps, piping (main and branch lines), control valves, hose, and nozzles.

Fire Mark — Distinctive metal or wooden marker once produced by insurance companies for identifying their policyholders' buildings.

Fire Marshal — Highest fire prevention officer of a state, province, county, or municipality. In Canada, this officer is *also known as* the Fire Commissioner.

Fire Medical Apparatus — Motor-driven vehicle, designed and constructed for the purpose of providing medical care with a specified level of equipment capacity and medically trained personnel; this unit is not used to transport patients to medical facilities. *Also known as* Non-Transport Fire Medical Apparatus.

Fire Pail — *See* Fire Bucket.

Fire Partition — Fire barrier that extends from one floor to the bottom of the floor above or to the underside of a fire-rated ceiling assembly; provides a lower level of protection than a fire wall. An example is a one-hour rated corridor wall. *See* Fire Wall.

Fire Patterns — (1) Visible or measurable physical effects that remain after a fire. (2) The apparent and obvious design of burned material and the burning path of travel from a point of fire origin. *Previously known as* Burn Patterns.

Fire Perimeter — Edge of a wildland fire.

Fire Plow — Heavy-duty plowshare or disc plow pulled by a tractor to construct a fireline.

Fire Plug — Wooden plug inserted into holes drilled into early wooden water main systems; when the plug was removed, water could be obtained for fire fighting purposes. Fire hydrants evolved from fire plugs. *See* Fire Hydrant.

Fire Point — Temperature at which a liquid fuel produces sufficient vapors to support combustion once the fuel is ignited. Fire point must exceed 5 seconds of burning duration during the test. The fire point is usually a few degrees above the flash point. *Also known as* Burning Point. *See* Flash Point.

Fire Police — Members, usually of a volunteer fire department, who respond with the fire department and assist the police with traffic control, crowd control, and scene preservation and security; common only in the mid-Atlantic states of the U.S. *Also known as* Special Police.

Fire Prevention — (1) Part of the science of fire protection that deals with preventing the outbreak of fire by eliminating fire hazards through such activities as inspection, code enforcement, education, and investigation programs. (2) Division of a fire department responsible for conducting fire prevention programs of inspection, code enforcement, education, and investigation. *Also known as* Fire Prevention Bureau.

Fire Prevention Bureau — Division of the fire department responsible for conducting fire prevention programs of inspection, code enforcement, education, and investigation. *Also known as* Fire Prevention.

Fire Prevention Code — Body of law enacted for the purpose of enforcing fire prevention and safety regulations, thus providing fire safety through the elimination of fire hazards and maintenance of fire protection equipment. *See* Code.

Fire Prevention Week — Week proclaimed each year by the President of the U.S. to commemorate the anniversary of the great Chicago conflagration on October 9, 1871; takes place the week in which October 9 falls.

Fire Protection — Actions taken to limit the adverse environmental, social, political, economic, and life-threatening effects of fire.

Fire Protection Engineer (FPE) — Graduate of an accredited institution of higher education who has specialized in engineering science related to fire protection.

Fire Pump — (1) Water pump used in private fire protection to provide water supply to installed fire protection systems. (2) Water pump on a piece of fire apparatus. (3) Centrifugal or reciprocating pump that supplies seawater to all fire hose connections aboard a ship.

Fire Report — Official report on a fire kept as a permanent record; generally prepared by the officer in charge of the fire operation.

Fire Resistance — The ability of a structural assembly to maintain its load-bearing ability under fire conditions.

Fire Resistant — Capacity of structural components to resist higher heat temperatures for certain periods of time. Lesser degree of resistance to fire than *fireproof*.

Fire Resistive — Ability of a structure or a material to provide a predetermined degree of fire resistance; usually according to building and fire prevention codes and given in hour ratings.

Fire Retardant — Any substance, except plain water, that when applied to another material or substance will reduce the flammability of fuels or slow their rate of combustion by chemical or physical action.

Fire Risk — Probability that a fire will occur and the potential for harm it will create in terms of the number of incidents, injuries, or deaths per capita. (**NOTE:** Depending on the source of information, the per capita unit may be one thousand people or one million people.)

Fire Science — Study of the behavior, effects, and control of fire.

Fire Season — Period of the year during which wildland fires are likely to occur, spread, and damage wildland areas sufficiently to warrant organized fire suppression.

Fire Separation — Horizontal or vertical assembly of fire-resistant materials that is designed to impede or slow the spread of fire.

Fire Service — Organized fire prevention, fire protection, fire training, and fire fighting services, including the members of such organizations individually and collectively who assist in preventing and combating fires.

Fire Service Hose — Specially constructed hose designed to withstand the hazards of the fire scene.

Fire Shelter — Aluminized tent carried by firefighters that offers personal protection by means of reflecting radiant heat and providing a volume of breathable air in a fire-entrapment situation.

Fire Standard Certified (FSC) — Indication on packs of cigarettes in the U.S. that indicates that the cigarettes are "fire-safe" and will self-extinguish if left lit and unattended.

Fire Station — (1) Building in which fire suppression forces are housed. *Also known as* Fire Hall or Fire House. (2) Location on a vessel with fire fighting water outlet (fire hydrant), valve, fire hose, nozzles, and associated equipment.

Fire Stop — Solid materials, such as wood blocks, used to prevent or limit the vertical and horizontal spread of fire and the products of combustion; installed in hollow walls or floors, above false ceilings, in penetrations for plumbing or electrical installations, in penetrations of a fire-rated assembly, or in cocklofts and crawl spaces.

Fire Storm — Violent convective atmospheric disturbance caused by a large continuous area of intense fire (conflagration), in which a rising column of heated air creates intense winds toward the fire center, encompassing the entire fire area. Often characterized by destructively violent surface indrafts, near and beyond the perimeter, and sometimes by tornado-like whirls.

Fire Stream — Stream of water or other water-based extinguishing agent after it leaves the fire hose and nozzle until it reaches the desired point.

Fire Suppressant — Agent used to extinguish the flaming and smoldering phases of combustion by direct application to the burning fuel.

Fire Suppression — All work and activities connected with fire-extinguishing operations, beginning with discovery and continuing until a fire is completely extinguished.

Fire Suppression System — System designed to act directly upon the hazard to mitigate or eliminate it, not simply to detect its presence and/or initiate an alarm.

Fire Swatter — Fire-suppression tool consisting of a flap of belting fabric fastened to a long handle; used in direct attack for beating out flames along a fire edge.

Fire Tetrahedron — Model of the four elements/conditions required to have a fire. The four sides of the tetrahedron represent fuel, heat, oxygen, and self-sustaining chemical chain reaction.

Fire Trap — Slang for an old structure in such a deteriorated state that it is highly susceptible to fire, has inadequate protective equipment and exits, and is considered likely to contribute to major loss of life in case of fire.

Fire Triangle — Plane geometric model of an equilateral triangle that is used to explain the conditions/elements necessary for combustion. The sides of the triangle represent heat, oxygen, and fuel. The fire triangle was used prior to the general adaptation of the fire tetrahedron, which includes a chemical chain reaction. *See* Fire Tetrahedron.

Fire Tube — *See* Heating Tube.

Fire Wall — (1) Fire rated wall with a specified degree of fire resistance, built of fire-resistive materials and usually extending from the foundation up to and through the roof of a building, that is designed to limit the spread of a fire within a structure or between adjacent structures. (2) Bulkhead separating an aircraft engine from the aircraft fuselage or wing. (3) The partition between the engine compartment and the passenger compartment of a vehicle. It is designed to protect vehicle occupants from the engine and its associated hazards.

Fire Wall Assembly — All of the components needed to provide a separating fire wall that meets the requirements of a specified fire-resistance rating.

Fire Wardens — Individuals appointed by the city of New Amsterdam and other early American cities in an attempt to prevent fires by inspecting chimneys and hearths. They were empowered to cite residents for failing to meet the city fire codes. Similar to the fire wards of early Boston.

Fire Wards — Individuals appointed by the city of Boston after 1631 in an attempt to prevent fires. They were provided badges and staffs of office and assigned to different parts of the city. Similar to the fire wardens of New Amsterdam and other early American cities.

Fire Watch — Usually refers to someone who has the responsibility to tour a building or facility on at least an hourly basis, look for actual or potential fire emergency conditions, and send an appropriate warning if such conditions are found.

Fire Weather — Weather conditions that influence fire ignition, behavior, and suppression.

Fire Weather Forecast — Weather prediction specially prepared for use in wildland fire control.

Fire Wire — Length of wire rope or chain hung from the bow and stern of a vessel in port to allow the vessel to be towed away from the pier in case of fire. *Also known as* Emergency Towing Wire or Fire Warp.

Fireball — Brief, roughly spherical fire suspended in the air, resulting from ignitable gases released during an explosion; may occur during or after an explosive reaction.

Fire-Behavior Forecast — Prediction of probable fire behavior, usually prepared by a fire-behavior officer in support of fire-suppression or prescribed-burning operations.

Fireboat — Vessel or watercraft designed and constructed for the purpose of fighting fires; provides a specified level of pumping capacity and personnel for the extinguishment of fires in the marine environment. *Also known as* Marine Unit.

Firebreak — Natural or man-made (constructed) barrier that is devoid of vegetation and stops or slows the advance of a wildland fire.

Firefighter — Active member of the fire department. *Also spelled* Fire Fighter.

Firefighter's Axe — *See* Axe.

Firefighter's Carry — One of several methods of lifting and carrying a disabled victim to safety.

Firefighter's Smoke Control Station (FSCS) — Interface between the smoke management system and the fire response forces.

Fire-Gas Detector — Device used to detect gases produced by a fire within a confined space.

Fireground — Area around a fire and occupied by fire fighting forces.

Fireground Commander — *See* Incident Commander.

Fireground Perimeter — Work area surrounding the fire building.

Fireline Intensity — Rate of heat energy released per unit time per unit length of fire front. Numerically, it is the product of the heat of combustion, quantity of fuel consumed in the fire front, and the rate of spread of a fire in Btu per second per foot (kilojoules per second per meter) of fire front.

Fireplay — Child's involvement with fire materials without the approval or supervision of parents; usually involves lighting matches or lighters.

Fireproof — Obsolete term for resistance to fire; inappropriate because all materials except water will burn. Other terms such as *fire resistive* or *fire resistant* should be used.

Fire-Protection Rating — Designation indicating the duration of a fire-test exposure to which a fire door assembly or fire window assembly was exposed, and for which it successfully met all acceptance criteria.

Fire-Protection System — System designed to protect structure and minimize loss due to fire.

Fire-Rated Glass — Type of glass that does not use interior wires. These fire-rated glass panels are made from a combination of glass and plastic.

Fire-Resistance Directory — Directory that lists building assemblies that have been tested and given fire-resistance ratings. Published by Underwriters Laboratories.

Fire-Resistance Rating — Rating assigned to a material or assembly after standardized testing by an independent testing organization; identifies the amount of time a material or assembly will resist a typical fire, as measured on a standard time-temperature curve.

Firesetter — Person who starts a fire, usually deliberately and maliciously.

Firesetter with Special Needs — Individual who suffers from some level of cognitive disability or challenge such as ADHD. Any level of firesetter may have a special need that should be considered during a screening process.

Firesetting — Any unsanctioned incendiary use of fire, including both intentional and unintentional involvement, whether or not an actual fire and/or explosion occurs. *See* Arson.

Firewhirl — Spinning vortex column of ascending hot air and gases rising from a fire and carrying smoke, debris, and flame aloft. *Also known as* Fire Devil or Fire Whirlwind.

Firing Out — Act of lighting backfires with a torch, fusee, or other device, to accomplish burning out or backfiring that will impede the growth of an uncontrolled wildland fire. *See* Backfire, Backfiring, and Burning Out.

First Aid — Immediate medical care given to a patient until he or she can be transported to a medical facility.

First Alarm — Initial fire department response to a report of an emergency.

First Responder (EMS) — (1) First person arriving at the scene of an accident or medical emergency who is trained to administer first aid and basic life support. (2) Level of emergency medical training, between first aider and emergency medical technician levels, that is recognized by the authority having jurisdiction (AHJ).

First-Degree Burn — Burn affecting only the outer skin layers; will cause redness and pain but not blisters or scars.

First-Due — Apparatus that should reach the scene of an emergency first based on the pre-fire attack plan. *Also known as* First-In.

Fiscal — Having to do with finances and money.

Fissionable — Capable of splitting the atomic nucleus and releasing large amounts of energy.

Fitness-for-duty Evaluation — Health evaluation administered by a fire department physician to determine an individual's ability to perform fire service tasks; evaluations may be physiological or psychological in nature depending on need.

Fitting — Device that facilitates the connection of hoselines of different sizes to provide an uninterrupted flow of extinguishing agent. *See* Appliance.

Five-Minute Escape Cylinder — Small air cylinder used as a backup air source with airline respirators.

Five-Step Planning Process — (1) Systematic planning and action process composed of five steps: 1) identification of major fire problems, 2) selection of the most cost-effective objectives for the education program, 3) design of the program itself, 4) implementation of the program plan, and 5) evaluation of the fire safety program to determine impact. (2) Nationally recognized process proven to be successful in guiding risk-reduction efforts.

Fixed Foam Extinguishing System — Complete installation of piping, foam concentrate storage, water supply, pumps, and delivery systems used to protect a specific hazard such as a petroleum storage facility. *See* Portable Foam Extinguishing System and Semifixed Foam Extinguishing System.

Fixed Monitor System — Fire suppression system employing stationary master stream devices (monitors) in areas where large quantities of water or foam will be needed in the event of a fire.

Fixed Stare — Condition while operating an apparatus in which the driver/operator becomes absorbed with particular details rather than maintaining the constant eye movement that ensures safe driving.

Fixed Window — Window that is set in a fixed or immovable position and cannot be opened for ventilation.

Fixed-Flow Direct-Injection Proportioner — Fixed-site foam system that provides foam at a preset rate. Foam concentrate and water are provided to the proportioner by two separate pumps that operate at preset rates. *See* Foam Proportioner and Proportioning.

Fixed-Foam Fire-Extinguishing System — Complete installation of piping, foam concentrate storage, water supply, pumps, and delivery systems used to protect a specific hazard such as a petroleum storage facility.

Fixed-Temperature Device — Fire alarm initiating device that activates at a predetermined temperature.

Fixed-Temperature Heat Detector — Temperature-sensitive device that senses temperature changes and sounds an alarm at a specific point, usually 135°F (57°C) or higher.

Flag State — Nation in which a vessel is registered.

Flagger — Individual assigned to direct the flow of traffic using a flag, flashlight, paddle, wand, sign, or other device.

Flail Chest — Condition in which several ribs are broken, each in at least two places.

Flame — Visible, luminous body of a convective flow of burning gas emitting radiant energy including light of various colors given off by burning gases or vapors during the combustion process.

Flame Depth — Depth of the fire front; horizontal distance between the leading and trailing edges of the fire front.

Flame Detector — Detection device used in some fire detection systems (generally in high-hazard areas) that detect light/flames in the ultraviolet wave spectrum (UV detectors) or detect light in the infrared wave spectrum (IR detectors).

Flame Front — (1) Outermost edge or surface of the flame. (2) Combustion area that lags behind the blast-pressure front in a deflagration; marks where expanding gases are igniting during an explosion. *See* Blast-Pressure Front.

Flame Height — Average maximum vertical extension of flames at the leading edge of the fire front; occasional flashes that rise above the general level of flames are not considered. This distance is less than the flame length if flames are tilted due to wind or slope.

Flame Impingement — Points at which flames contact the surface of a container or other structure.

Flame Interface — Area or surface between the gases or vapors and the visible flame.

Flame Length — Distance between the flame tip and the midpoint of the flame depth at the base of the flame (generally the ground surface); an indicator of fire intensity.

Flame Out — Unintended loss of combustion in turbojet engines resulting in the loss of engine power.

Flame Propagation Rate — Velocity at which combustion travels through a gas or over the surface of a liquid or solid.

Flame Resistant — Materials that are not susceptible to combustion to the point of propagating a flame after the ignition source is removed.

Flame Speed — Velocity of the flame front relative to the center of the explosion.

Flame Spread — Movement of a flame away from the ignition source.

Flame Spread Rating — (1) Numerical rating assigned to a material based on the speed and extent to which flame travels over its surface. (2) Measurement of the propagation of flame on the surface of materials or their assemblies as determined by recognized standard tests. *See* Steiner Tunnel Test.

Flame Test — Test designed to determine the flame spread characteristics of structural components or interior finishes.

Flame Zone — Area of any individual fire where flames are visible.

Flameover — Condition that occurs when a portion of the fire gases trapped at the upper level of a room ignite, spreading flame across the ceiling of the room. *See* Rollover.

Flammability — Fuel's susceptibility to ignition.

Flammable — Capable of burning and producing flames. *See* Flammable Gas, Flammable Liquid, Flammable Solid, and Nonflammable.

Flammable and Explosive Limits — Upper and lower concentrations of a vapor, expressed in percent mixture with an oxidizer, that will produce a flame at a given temperature and pressure.

Flammable Gas — Any material (except an aerosol) that is a gas at 68°F (20°C) or less and that (a) is ignitable and will burn at 14.7 psi (101.3 kPa) when in a mixture of 13 percent or less by volume with air, or (b) has a flammable range at 14.7 psi (101.3 kPa) by volume with air at least 12 percent regardless of the lower limit. *See* Flammable, Gas, Lower Flammable Limit (LFL), Nonflammable Gas, and Upper Flammable Limit (UFL).

Flammable Limit — Percentage of a substance in air that will burn once it is ignited. Most substances have an upper flammable limit (too rich) and a lower flammable limit (too lean). *Also known as* Explosive Limit. *See* Flammable Range, Lower Flammable Limit (LFL), and Upper Flammable Limit (UFL).

Flammable Liquid — Any liquid having a flash point below 100°F (37.8°C) and a vapor pressure not exceeding 40 psi absolute (276 kPa) {2.76 bar}. *See* Combustible Liquid, Flammable, Flash Point, Polar Solvent Fuel, and Vapor Pressure.

Flammable Material — Substance that ignites easily and burns rapidly.

Flammable Range — Range between the upper flammable limit and lower flammable limit in which a substance can be ignited. *Also known as* Explosive Range. *See* Flammable Limit.

Flammable Solid — Solid material (other than an explosive) that (a) is liable to cause fires through friction or retained heat from manufacturing or processing, or (b) ignites readily and then burns vigorously and persistently, creating a serious transportation hazard. *See* Flammable.

Flammable/Combustible Liquid Pit — Training prop designed to provide controlled burns of flammable or combustible liquids; used in training for the extinguishment of flammable/combustible liquid fires.

Flammable/Explosive Range — Percentage of a gas vapor concentration in the air that will burn if ignited.

Flank — One side of a wildland fire.

Flank Attack — Attacking a fire on the flanks or sides of a wildland fire and working along the fire edge toward the head of the fire. *Also known as* Flanking Attack. *See* Attack Methods (2).

Flanking — Attacking the sides of the fire from a less active area or from an anchor point; the intent being to have the two crews attacking the flanks meet at the head of the fire.

Flanks of Fire — Parts of a fire's perimeter that are roughly parallel to the main direction of spread.

Flaps — Adjustable airfoils attached to the leading or trailing edges of aircraft wings to improve aerodynamic performance during takeoff and landing. They are normally extended during takeoff, landing, and slow flight.

Flare-Up — Sudden acceleration in rate of spread or intensification of a fire. Unlike blowup, a flare-up is of relatively short duration and does not radically change existing control plans. *See* Blowup.

Flaring — Short-lived, high intensity fire in a small area at a wildland fire.

Flash Fire — Type of fire that spreads rapidly through a vapor environment.

Flash Fuels — Wildland ground cover fuels that are easily ignited and burn rapidly; examples include grass, leaves, pine needles, fern, tree moss, and some kinds of slash. *Also known as* Flashy Fuels.

Flash Point — Minimum temperature at which a liquid gives off enough vapors to form an ignitable mixture with air near the surface of the liquid. *See* Fire Point.

Flash Resistant — Aircraft materials that are not susceptible to burning violently when ignited.

Flashback — Spontaneous reignition of fuel when the blanket of extinguishing agent breaks down, or is compromised through physical disturbance.

Flashing — (1) Liquid-tight rail on top of a tank that contains water and spillage and directs it to suitable drains; may be combined with DOT overturn protection. (2) Sheet metal used in roof and wall construction to keep water out.

Flashing Drain — Metal or plastic tube that drains water and spillage from flashing to the ground.

Flashover — (1) Stage of a fire at which all surfaces and objects within a space have been heated to their ignition temperature, and flame breaks out almost at once over the surface of all objects in the space. (2) Rapid transition from the growth stage to the fully developed stage. *See* Backdraft, Ignition Temperature, and Rollover.

Flat — Two-family dwelling in which the families live one above the other; one family occupies the entire ground floor (and basement if there is one) while the second family occupies the entire second floor (and attic if there is one) of a building.

Flat Load — Arrangement of fire hose in a hose bed or compartment, in which the hose lies flat with successive layers one upon the other.

Flat Plate — Plain floor slab about 8 inches (203 mm) thick that rests on columns spaced up to 22 feet (6.7 m) apart and depends on diagonal and orthogonal patterns of reinforcing bars for structural support because the slab lacks beams; simplest and most economical floor system.

Flat Raise — Raising a ladder with the heel of both beams touching the ground.

Flat Roof — Roof that has a slope not exceeding 2 inches (51mm) vertical for each 12 inches (305 mm) horizontal; generally has a slight pitch to facilitate runoff.

Flat Slab — Type of concrete floor construction that provides a flat surface for the underside of the floor finish.

Flat-Head Axe — Axe with a cutting edge on one side of the head and a blunt or flat head on the opposite side.

Flat-Slab Concrete Frame — Construction technique consisting of concrete slabs supported by concrete columns.

Flesch Grade Level Index — *See* Flesch Reading Ease.

Flesch Reading Ease — Readability index. *Also known as* Flesch Grade Level Index.

Flexible Intermediate Bulk Container (FIBC) — *See* Intermediate Bulk Container (IBC).

Flight — (1) Series of steps between takeoff and landing of an aircraft. (2) Series of stairs or steps between two landings.

Flight Controls — General term applied to devices that enable the pilot to control the direction of flight and attitude of the aircraft.

Flight Data Recorder — Recording device on large civilian aircraft to record aircraft airspeed, altitude, heading, acceleration, and other flight data to be used as an aid to accident investigation. *Also known as* Black Box.

Flight Deck — Cockpit on a large aircraft that is separated from the rest of the cabin.

Flight Service Station — Facility from which aeronautical information and related aviation support services are provided to aircraft. This also includes airport and vehicle advisory services for designated uncontrolled airports.

Flipchart — Teaching aid consisting of a large easel and tablet.

Floating Dock Strainer — Strainer designed to float on top of the water; used for drafting operations. This eliminates the problem of drawing debris into the pump and reduces the required depth of water needed for drafting.

Floating Foundation — Foundation for which the volume of earth excavated will approximately equal the weight of the building supported. Thus, the total weight supported by the soil beneath the foundation remains about the same, and settlement is minimized because of the weight of the building.

Floating Pump — Small, portable pump that floats on the water source.

Floating Ribs — Two lowest pairs of ribs; so called because they are connected only to vertebrae in the back.

Floating Roof Storage Tank — Atmospheric storage tank that stands vertically, and is wider than it is tall. The roof floats on the surface of the liquid to eliminate the vapor space. *See* Atmospheric Storage Tank and Lifter Roof Storage Tank.

Flood Tide — Rising tide.

Floor Plan — Architectural drawing showing the layout of a floor within a building as seen from above. It outlines the location and function of each room. *See* Construction Plan, Plot Plan, and Site Plan.

Floor Runner — Heavy plastic or canvas placed on a floor to protect the floor's surface or covering from firefighter traffic; used during salvage operations.

Flow — Motion characteristic of water.

Flow Hydrant — Hydrant from which the water is discharged during a hydrant flow test. *See* Test Hydrant.

Flow Path — Composed of at least one inlet opening, one exhaust opening, and the connecting volume between the openings. The direction of the flow is determined by difference in pressure. Heat and smoke in a high pressure area will flow toward areas of lower pressure.

Flow Pressure — Pressure created by the rate of flow or velocity of water coming from a discharge opening. *Also known as* Plug Pressure.

Flow Test — Tests conducted to establish the capabilities of water supply systems. The objective of a flow test is to establish quantity (gallons or liters per minute) and pressures available at a specific location on a particular water supply system.

Flowmeter — Mechanical device installed in a discharge line that senses the amount of water flowing and provides a readout in units of gallons per minute (liters per minute).

Fluid — Any substance that can flow, has a definite mass and volume at constant temperature and pressure but no definite shape, and that is unable to sustain shear stresses. Fluids include both gases and liquids.

Fluid Mechanics — Branch of physics dealing with the behavior of fluids, particularly with respect to their reaction to forces applied to them.

Fluid-Applied Membrane — Roof coating material that is applied as a liquid and allowed to cure; includes neoprene, silicone, poly urethane, and butyl rubber. Typically applied to curved roof surfaces such as domes.

Fluorinated Surfactants — (1) Chemicals that lower the surface tension of a liquid, in this case fire fighting foams. (2) Surface-active substance where the hydrophobic (incapable of dissolving in water) part of the substance molecule contains fluorine; has the ability to lower aqueous surface tension, improve wetting, and remain chemically stable when exposed to heat, acids, and bases as well as reducing and oxidizing agents.

Fluoroprotein Foams — Protein foam concentrate with synthetic fluorinated surfactants added. These surfactants enable the foam to shed, or separate from, hydrocarbon fuels. *See* Foam Concentrate.

Flush Bolt — Locking bolt that is installed flush within a door.

Flush Hydrant — Hydrant installed in a pit below ground level such as near the runway area of airports or other locations where aboveground hydrants would be unsuitable.

Fly Rope — *See* Halyard.

Fly Section — Extendable section of ground extension or aerial ladder. *Also known as* Fly.

Flyer — Another term for a brochure.

Flying Shore — Shore for vertical surfaces, such as a wall, that is braced against another vertical surface.

FM — *See* FM Global.

FM Global (FM) — Fire research and testing laboratory that provides loss control information for the Factory Mutual System and anyone else who may find it useful.

FMD — *See* Foot and Mouth Disease.

Foam — Extinguishing agent formed by mixing a foam concentrate with water and aerating the solution for expansion; for use on Class A and Class B fires. Foam may be protein, fluoroprotein, film forming fluoroprotein, synthetic, aqueous film forming, high expansion, alcohol type, or alcohol-resistant type. *Also known as* Finished Foam. *See* Foam Concentrate and Foam System.

Foam Blanket — Covering of foam applied over a burning surface to produce a smothering effect; can be used on un-ignited surfaces to prevent ignition. *See* Foam Stability.

Foam Chamber — Foam delivery device that is mounted on storage tanks; applies foam onto the surface of the fuel in the tank. *See* Alcohol-Resistant AFFF Concentrate, Aqueous Film Forming Foam, Class A Foam Concentrate, Class B Foam Concentrate, Emulsifier, Film Forming Fluoroprotein Foam, Fluoroprotein Foam, Foam Solution, Protein Foam Concentrate, and Synthetic Foam Concentrate.

Foam Concentrate — (1) Raw chemical compound solution that is mixed with water and air to produce finished foam; may be protein, synthetic, aqueous film forming, high expansion, or alcohol types. (2) Raw foam liquid as it rests in its storage container before the introduction of water and air. *See* Aqueous Film Forming Foam, Class A Foam Concentrate, Class B Foam Concentrate, Foam, Foam Solution, and Foam System.

Foam Eductors — Type of foam proportioner used for mixing foam concentrate in proper proportions with a stream of water to produce foam solution. *See* Foam Proportioner and Foam Solution.

Foam Expansion — Result of adding air to a foam solution consisting of water and foam concentrate. Expansion creates the foam bubbles that result in finished foam or foam blanket. *See* Aeration and Expansion Ratio.

Foam Monitor — Master stream appliance used for the application of foam solution. *See* Automatic Oscillating Foam Monitor, Manual Foam Monitor, and Remote-Controlled Foam Monitor.

Foam Proportioner — Device that injects the correct amount of foam concentrate into the water stream to make the foam solution. *See* Foam Eductor and Foam Solution.

Foam Solution — Result of mixing the appropriate amount of foam concentrate with water; foam solution exists between the proportioner and the nozzle or aerating device that adds air to create finished foam. *See* Foam Concentrate, Foam Eductor, and Foam Proportioner. (2) Mixture of foam concentrate and water before the introduction of air.

Foam Stability — Relative ability of a foam to withstand spontaneous collapse or breakdown from external causes. *See* Foam Blanket.

Foam System — Extinguishing system that uses a foam such as aqueous film forming foam (AFFF) as the primary extinguishing agent; usually installed in areas where there is a risk of flammable liquid fires. *See* Aqueous Film Forming Foam, Flammable Liquid, Foam, and Foam Concentrate.

Foam Tanker — *See* Foam Tender.

Foam Tender — Apparatus that is specially designed to transport large quantities of foam concentrate to the scene of an incident; vehicle may be equipped with transfer pumps and have a tank ranging from 1,500 gallons (5 678 L) to 8,000 gallons (30 283 L). *Also known as* Foam Tanker or Mobile Water Supply Apparatus. *See* Foam Trailer and Mobile Foam Extinguishing System.

Foam Trailer — Foam concentrate tank mounted on a trailer; can be easily towed to the desired location; usually found at fixed-site facilities and used to supply master stream and subsurface injection systems. *See* Foam Tender and Mobile Foam Extinguishing System.

Foam-Water Sprinkler — Deluge-type open sprinkler that mixes water with the foam solution as it passes through the sprinkler. *See* Closed Sprinkler and Open Sprinkler.

Focal Length — Distance behind the lens where light from an object is sharply focused when the lens is set to infinity.

Foehn Wind — Wind that occurs when stable, high-pressure air is forced across and then down the lee slopes of a mountain range. The descending air is warmed and dried due to adiabatic compression; locally called by various names such as Santa Ana, Mono, or Chinook. *Also known as* Gravity Wind.

Fog Nozzle — Adjustable pattern nozzle equipped with a shutoff control device that can provide either a fixed or variable spray pattern; breaks the foam solution into small droplets that mix with air to form finished foam. *See* Multiagent Nozzle.

Fog Stream — Water stream of finely divided particles used for fire control.

FOIA — *See* Federal Freedom of Information Act.

Foil Back — Blanket or batt insulation with one surface faced with metal foil that serves as a vapor barrier and heat reflector.

Fold-A-Tank — *See* Portable Tank.

Folding Door — Door that opens and closes by folding.

Folding Jack — Common type of lifting jack. The frame of the folding jack is made of metal bars of equal lengths, fastened in the center to form Xs. This jack has limited use and is considered safe only for light loads.

Folding Ladder — Single-section, collapsible ladder that is easy to maneuver in restricted places such as access openings for attics and lofts.

Follow-Up — Act of maintaining contact with a juvenile firesetter and his or her family over a designated period of time.

Food Dispenser Unit — Vehicle capable of dispensing food to incident personnel.

Food Unit — Functional unit within the service branch of the logistics section of an incident management system; responsible for providing meals for personnel involved with incident.

Foot and Mouth Disease (FMD) — Acute viral disease of domestic and wild cloven-hoofed animals characterized by fever, lameness, and vesicular lesions on the feet, tongue, mouth and teats. FMD is considered to be one of the most contagious, infectious diseases known.

Foot Pads — Feet mounted on the butt of the ladder by a swivel to facilitate the placement of ladders on hard surfaces.

Footing — (1) Part of the building that rests on the bearing soil and is wider than the foundation wall. (2) Base for a column. (3) Method for securing the base of a ladder.

Footplate — A 4-inch (102 mm) metal plate that runs around the bottom edge of any railing on a balcony or elevated walkway to prevent someone's foot from slipping off such as on an elevating platform to prevent a firefighter's feet from slipping off the edge of the platform. *Also known as* Kickplate.

Force — (1) To break open, into, or through. (2) Simple measure of weight, usually expressed in pounds or kilograms.

Forced Ventilation — Any means other than natural ventilation; may involve the use of fans, blowers, smoke ejectors, and fire streams. *Also known as* Mechanical Ventilation.

Force-Vector Analysis — Graphical tool investigators can use to determine the direction of blast pressure in an explosion.

Forcible Entry — (1) Techniques used by fire personnel to gain entry into buildings, vehicles, aircraft, or other areas of confinement when normal means of entry are locked or blocked. (2) Entering a structure or vehicle by means of physical force, characterized by prying open doors and breaking windows, leaving visible indicators of illegal entry if pry marks and certain window breakage are present upon the first firefighters' arrival.

Forcing Foam Maker — *See* High Back-Pressure Foam Maker.

Fording — Ability of an apparatus to traverse a body of standing water. Apparatus specifications should list the specific water depths through which trucks must be able to drive.

Forecastle (Fo'c's'le Or Fok-Sul) — Section of the upper deck located at the bow of a vessel; forward section of the main deck; a superstructure at the bow of a ship where maintenance shops, rope lockers, and paint lockers are located.

Foreign Object Damage — Damage attributed to a foreign object that can be expressed in physical or economic terms that may or may not degrade the product's required safety and/or performance characteristics.

Foreign Object Debris — Any substance, debris, or article that is alien to the vehicle or system and would potentially cause damage.

Foreman — Rank used for a company officer in some departments.

Forensic Anthropology — Science of establishing the identities of deceased victims.

Forensic Pathology — Science of determining the cause and manner of death.

Forensic Science — (1) Application of scientific procedures to the interpretation of physical events such as those that occur at a crime scene or fire scene. (2) The art of reconstructing past events and then explaining that process and one's findings to investigators and triers of fact. *Also known as* Criminalistics or Forensics.

Forensic Scientist — One who applies scientific procedures to the interpretation of physical events such as those that occur at fire scenes; one who is adept at reconstructing past events and then explaining that process and findings to investigators and triers of facts. *Also known as* Criminologist.

Forensics — *See* Forensic Science.

Forepeak — Watertight compartment at the extreme forward end of a vessel; usually used for storage. *See* Afterpeak.

Foreseeability — (1) Concept that instruction should be based not only on dangerous conditions that may exist in training but also on anticipating what firefighters might face on the job. (2) Legal concept that states that reasonable people should be able to foresee the consequences of their actions and take reasonable precautions.

Forestry Hose — Lightweight, small diameter, unlined, single-jacket hose with lightweight couplings used to combat fires in forests and in other wildland settings.

Forged Couplings — Coupling formed by pounding a hot metal pellet into a forging die, which forms the metal into the desired shape.

Fork Pockets — Transverse structural apertures (openings) in the base of the container that permit entry of forklift devices.

Formal Decontamination — *See* Technical Decontamination.

Formal Operational Stage of Intellectual Development — Fourth and highest stage of intellect; usually occurs between ages 11 and 15. Learners can think systematically to develop hypotheses about why something is happening the way it is. They can then test the hypotheses with deductive reasoning.

Formal Proposal — Written request for funding that describes the educator's organization, the problem to be solved, the solution to the problem (project plan), and the benefits of the project to the prospective donor, the target audience, and the community.

Formaldehyde (HCHO) — Colorless gas with a characteristic pungent odor produced when wood, cotton, and newspaper burn; an eye, nose, and throat irritant, and a probable carcinogen.

Formative Evaluation or Test — (1) Ongoing, repeated assessment during a course to evaluate student progress; may also help determine any needed changes in instructional content, methods, training aids, and testing techniques. (2) Evaluation of a new or revised program in order to form opinions about its effects and effectiveness as it is in the process of being developed and tested (piloted). Its purpose is to gather information to help improve the program while in progress. *Also known as* Process Evaluation. *See* Evaluation.

Former — Frame of wood or metal that is attached to the truss of the fuselage or wing of an aircraft in order to provide the required aerodynamic shape.

Formula — A mathematical computation; used frequently in the fire service primarily for applications such as determining pressures, flows, and friction loss.

Formwork — Temporary structure used to provide shape to liquid concrete while it hardens; can be made of wood, metal, or plastic. Because the formwork must be strong enough to support the weight of the wet concrete, it becomes a major structural system in itself.

Forward — Front or nose section of an aircraft or toward that area. *Also known as* Fore.

Four-Step Method of Instruction — Teaching method based upon four steps: preparation, presentation, application, and evaluation. May be preceded by a pretest.

Four-Way Hydrant Valve — Device that permits a pumper to boost the pressure in a supply line connected to a hydrant without interrupting the water flow.

FPE — *See* Fire Protection Engineer.

Fragmentation — In terms of explosions, describes the process of a confining vessel losing its structural integrity and becoming shrapnel. Low-order damage involves large pieces of debris and shrapnel due to fragmentation, while high-order damage typically involves small, more widespread debris. *See* Shrapnel.

Fragmentation Effect — Process in which bomb parts and components are blown outwards by the explosion in the form of fragments, shards, or Shrapnel.

Frame — (1) Part of an opening that is constructed to support the component that closes and secures the opening such as a door or window. (2) The chassis of some automobiles. (3) Structural member of a vessel's framework that attaches perpendicularly to the keel to form the ribs of the vessel. *See* Keel.

Frangible — Breakable, fragile, or brittle.

Frangible Bulb — Small glass vial fitted into the discharge orifice of a fire sprinkler. The glass vial is partly filled with a liquid that expands as heat builds up. At a predetermined temperature, vapor pressure causes the glass bulb to break, which activates the device, causing water to flow.

Free Play — In the context of aerial apparatus inspection, refers to the distance an activation device such as a wheel or lever will move while the system it is part of is inactive.

Free Radical — (1) Atom or group of atoms that has at least one unpaired electron and is therefore unstable and highly reactive. (2) Electrically charged, highly reactive parts of molecules released during combustion reactions.

Free Surface Effect — Tendency of a liquid within a compartment to remain level as a vessel moves, which allows the liquid to move unimpeded from side to side. Loose water anywhere in a vessel impairs stability by raising the center of gravity. *See* Center of Buoyancy, Center of Gravity, and Stability.

Freeboard — Vertical distance between a vessel's lowest open deck and the water surface; measured near the center of the vessel's length where the deck is closest to the water.

Free-Burning Stage — Second stage of burning in a confined area in which the fire burns rapidly, using up oxygen and building up heat that accumulates in upper areas at temperatures that may exceed 1,300° F (700° C).

Freedom Fighters — Members of a militant organization fighting to establish a separate country or state for their nationality.

Freeflow — Continuous flow of air from the regulator, usually venting into the atmosphere.

Freelance — To operate independently of the incident commander's command and control.

Freewheeling Brainstorming — Type of brainstorming in which group members speak their ideas spontaneously, and ideas are recorded on flipcharts right away; most spontaneous form of brainstorming.

Freezing Point — Temperature at which a liquid becomes a solid at normal atmospheric pressure. *See* Melting Point.

Freight Container — *See* Container (1).

Frequency — The number of cycles (360 degree coil turn) that occur within an electrical generator; measured in Hertz (Hz).

Frequency-Sharing Agreement — Written agreement between agencies that are licensed to use a communications frequency that allows the other agency to use the frequency under specified conditions.

Friction — Resistance between two surfaces moving while in contact with each other. May generate heat and contribute to combustion.

Friction Burn — *See* Abrasion.

Friction Loss — That part of the total pressure lost as water moves through a hose or piping system; caused by water turbulence and the roughness of interior surfaces of hose or pipe.

Friendly Fire — Fully contained and controlled fire started for useful and nondestructive purposes.

Fringe Benefit — Employment benefit (pension, health insurance, paid holiday) granted by an employer to an employee without affecting the employee's basic wage rate; any additional benefit.

Fringe Vision — Vision surrounding central clear vision that alerts people to areas that may require attention; detects objects over a wide area, much like a wide angle lens in photography.

Front — (1) In meteorology, the boundary between two air masses of differing densities. (2) Section of the vehicle faced by the driver during normal travel or operation; generally indicated by the headlights.

Front Bumper Well — Hose or tool compartment built into the front bumper of a fire apparatus.

Front Member — Firefighter working at the front side of a ladder.

Front of Ladder — Climbing side; the side away from a building.

Front Stringer — Stringer that supports the side of the stairs with a balustrade.

Frontal Attack — Attack directed at the head of a wildland fire, from an anchor point at or near the head of the fire, which then proceeds to the flanks. *See* Attack Methods (2).

Frontal Winds — Winds generated by the movement of an air mass (front) across the earth's surface.

Front-Impact Air Bags — Supplemental restraint system designed to deploy air bags to absorb passenger impacts during a collision. These air bags are activated through a system of inertia switches located forward of the passenger compartment and by microelectronic controls that may be located under the front seats or in the console between the front seats.

Front-Impact Collision — Occurs when a vehicle collides head on with another vehicle or object.

Front-Mount Pump — Fire pump mounted in front of the radiator of a vehicle and powered off the crankshaft.

Frostbite — Local tissue damage caused by prolonged exposure to extreme cold. *See* Hypothermia.

FSC — *See* Fire Standard Certified.

F-Stop — Setting on an SLR or DSLR camera that increases or decreases the camera's aperture size to adjust the amount of light entering the camera.

Fuel — (1) Flammable and combustible substances available for a fire to consume. (2) Material that will maintain combustion under specified environmental conditions. (NFPA® 921)

Fuel Break — Wide strip or block of land on which the native vegetation has been modified (removed or widely separated) so that fires burning into them can be more readily extinguished. It may or may not have a fireline constructed in it prior to fire occurrence.

Fuel Characteristics — Factors that make up fuels such as compactness, loading, horizontal continuity, vertical arrangement, chemical content, size and shape, and moisture content.

Fuel Continuity — Degree or extent of continuous or uninterrupted distribution of fuel particles in a fuel bed, thus affecting a fire's ability to sustain combustion and spread. This applies to aerial fuels as well as surface fuels.

Fuel Crib — Uniform stacking of wood material where each layer is perpendicular to the layer directly beneath it. Wood is spaced uniformly throughout the crib with separation between the wood material equal to the thickness of the wood and the dimension of the wood is consistent throughout the crib. This is utilized to replicate fire sizes.

Fuel Load — The total quantity of combustible contents of a building, space, or fire area, including interior finish and trim, expressed in heat units of the equivalent weight in wood.

Fuel Loading — Amount of fuel present, expressed quantitatively in terms of weight of fuel per unit area. This may be available fuel (consumable fuel) or total fuel and is usually dry weight.

Fuel Management — Manipulation of fuel prior to an incident to prevent the occurrence or slow the spread of wildland fire. *Also known as* Vegetation Management or Weed Abatement.

Fuel Model — Simulated fuel complex for which all fuel descriptors required for the solution of a mathematical rate-of-spread model have been specified.

Fuel Moisture — Quantity of moisture in fuel expressed as a percentage of the weight when thoroughly dried at 212° F (100° C).

Fuel Orientation — Position of the fuel (rug as a vertical wall hanging as opposed to rug on the floor, for example).

Fuel Siphoning — Unintentional release of fuel from an aircraft caused by overflow, puncture, loose cap, etc. *Also known as* Fuel Venting.

Fuel Tender — Any vehicle capable of supplying fuel to ground or airborne equipment.

Fuel Venting — *See* Fuel Siphoning.

Fuel Volume — Quantity of fuel per unit area; usually expressed in tons per acre (tonnes per hectare).

Fuel-to-Air Ratio — Percentage of a fuel-gas suspended in air; ratios within the flammable range can sustain combustion when met with a competent ignition source.

Fuel-Controlled — A fire with adequate oxygen in which the heat release rate and growth rate are determined by the characteristics of the fuel, such as quantity and geometry. (NFPA® 921)

Fuel-Gas Migration — Tendency of leaking fuel gases to move through pipes, sewer systems, permeable soil, or HVAC ducts, spreading flammable gases to areas often great distances away from the source.

Fulcrum — Support or point of support on which a lever turns in raising or moving a load.

Fulcrum-Type Stabilizer — Stabilizing device that extends at an angle down and away from the chassis of an aerial fire apparatus. *Also known as* A-Frame Stabilizer.

Full Frames — Automobile construction (also known as body-on-frame construction) consisting of a steel ladder frame that is constructed using two parallel beams that run along the long axis of the vehicle to form a chassis. Cross members are bolted and welded between these beams to provide rigidity and support. This chassis then supports the powertrain, and the vehicle body is bolted to the frame. *Also known as* Rigid Frames.

Full Structural Protective Clothing — Protective clothing including helmets, self-contained breathing apparatus, coats and pants customarily worn by firefighters (turnout or bunker coats and pants), rubber boots, and gloves. It also includes covering for the neck, ears, and other parts of the head not protected by the helmet or breathing apparatus. When working with hazardous materials, bands or tape are added around the legs, arms, and waist. *See* Personal Protective Equipment (PPE).

Full Trailer — Truck trailer constructed so that all of its own weight and that of its load rests upon its own wheels; that is, it does not depend upon a truck tractor to support it. A semitrailer equipped with a truck tractor is considered a full trailer. *See* Semitrailer and Truck Tractor.

Full-Cycle Machine — Machine with a clutch that, when tripped, cannot be disengaged until the crankshaft has completed a full revolution and the press slide a full stroke.

Full-Duplex — Radio operating system in which two frequencies are used to communicate over one channel in a radio. This system allows operation in both directions at the same time.

Full-Facepiece Air-Purifying Respirator (FFAPR) — Filter, canister, or cartridge air-purifying respirator that covers the entire face from the forehead to the chin, providing protection to the eyes, nose, and mouth.

Full-Room-Involvement Pattern — Fire pattern that occurs after flashover or after a fire has burned for long periods of time, in which almost all vertical and horizontal surfaces in the compartment will show signs of damage.

Full-Trailer Tank — Any vehicle with or without auxiliary motive power, equipped with a cargo tank mounted thereon or built as an integral part thereof, used for the transportation of flammable and combustible liquids or asphalt; constructed so that practically all of its weight and load rests on its own wheels; a trailer with axles at the front and rear of the frame capable of supporting the entire weight of the tank.

Fully Developed Stage — All combustible materials in the compartment are burning and releasing the maximum amount of heat possible.

Fully Involved — When an entire area of a building is completely involved in heat and flame.

Fume — Suspension of particles that form when material from a volatilized (vapor state) solid condenses in cool air.

Fume Test — Qualitative test of a self-contained breathing apparatus (SCBA) facepiece fit in which a smoke tube is used to check for leakage around the facepiece.

Functional Fixity — Decision-making problem characterized by the tendency to use an object only for its designed purpose.

Functional Supervision — Organizational principle that allows workers to report to more than one supervisor without violating the unity of command principle; workers report to their primary supervisor for most of their activities but report to a second supervisor for activities that relate to an assigned function only, and both supervisors coordinate closely.

Funding — Fire department (or other organization) budget plus grants and in-kind contributions.

Furring — Wood strips fastened to a wall, floor, or ceiling for the purpose of attaching a finish material.

Fuse — Single-acting protective device designed to open a circuit on a predetermined overcurrent; types include Edison-based fuses and cartridge fuses.

Fused Head — Automatic sprinkler head that has operated due to exposure to heat.

Fused Sprinkler — Automatic sprinkler that has operated due to exposure to heat.

Fusee — (1) Colored flare designed as a railway warning device used to ignite backfires and other prescribed fires. (2) Friction match with a large head capable of burning in a wind.

Fuselage — Main body of an aircraft to which the wings and tail are attached. The fuselage houses the crew, passengers, and cargo.

Fusible Element — Dissimilar metals that fuse or melt when exposed to heat and allows the circuits to open or close and transmit a signal to a FACU; commonly found in fixed temperature heat detectors.

Fusing System — Electrical and/or mechanical mechanism by which an explosive device is detonated. Typically composed of a power source and some kind of switch.

G

Gabled Roof — Style of pitched roof with square ends in which the end walls of the building form triangular areas beneath the roof.

Gage — *See* Gauge.

Gage Lines — Term used in steel construction to describe lines parallel to the length of a member on which holes for fasteners are placed. The gage distance is the normal distance between the gage line and the edge or back of the member.

Galley — (1) On an aircraft, the food storage and preparation area. (2) On a ship, the kitchen.

Gallon — Unit of liquid measure. One U.S. gallon (3.785 L) has the volume of 231 cubic inches (3 785 cubic centimeters). One imperial gallon equals 1.201 U.S. gallons (4.546 L).

Gallons per Minute (GPM) — Unit of volume measurement used for water movement.

Gambrel Roof — Style of gabled roof on which each side slopes at two different angles; often used on barns and similar structures.

Gamma Radiation — (1) Electromagnetic wave with no electrical charge. This type of radiation is extremely penetrating. (2) Very high-energy ionizing radiation composed of gamma rays. *See* Alpha Radiation, Beta Radiation, Ionizing Radiation, and Radiation.

Gamma Rays — (1) One of three types of radiation emitted by radioactive materials. Gamma rays have extremely short wavelengths and very high energy; they are the most penetrating and potentially lethal of the three types of radiation. (2) High-energy photon (packet of electromagnetic energy) emitted from the nucleus of an unstable (radioactive) atom. *See* Alpha Particle, Beta Particle, Gamma Radiation, and Radioactive Material (RAM).

Gang Nail — Form of gusset plate. These thin steel plates are punched with acutely V-shaped holes that form sharp prongs on one side that penetrate wooden members to fasten them together.

Gangrene — Local tissue death as the result of an injury or inadequate blood supply; often caused by frostbite.

Gangway — Opening in the railings on the side of a vessel for a ladder or ramp providing access to a vessel from the shore.

Gantry — Overhead cross-girder structure on which a traveling crane is mounted or from which heavy tackle is suspended. Supporting towers at each end of the structure are on wheels.

Gap Analysis — Comparison between standards/regulations/best practices and actual behaviors within the organization to determine where real world performance differs from best practice.

Gas — Compressible substance, with no specific volume, that tends to assume the shape of a container. Molecules move about most rapidly in this state. *See* Compressed Gas, Flammable Gas, and Liquefied Gas.

Gas and Vapor Testing System — Detection and alarm system used mainly in industry and manufacturing where there is a potential for large collections of combustible or flammable gases.

Gas Chromatogram — Chart from a gas chromatograph tracing the results of analysis of volatile compounds by display in recorded peaks. *See* Gas Chromatograph.

Gas Chromatograph — Device to detect and separate small quantities of volatile liquids or gases through instrument analysis.

Gas Chromatography — Characterizing volatilities and chemical properties of compounds that evaporate enough at low temperatures (120° F or 49° C) to provide detectable quantities in the air, through the use of instrument analysis in a gas chromatograph. *See* Gas Chromatograph.

Gas Mask — *See* Canister Apparatus.

Gas/Vapor Explosion — Chemical, nonseated deflagration that occurs when a fuel-gas mixed with air in the proper ratio rapidly ignites; may reach the level of a detonation if the fuel-gas mixture ignites in a confined space, such as in a gas pipe.

Gas-Free Certificate — Document stating that an authorized and trained person has evaluated the atmosphere of a space, tank, or container, using approved equipment and methods, and determined that the atmosphere is safe for a specific purpose. *Also known as* Certified Gas-Free or Gas Certificate.

Gaskets — Rubber seals or packings used in joints and couplings to prevent the escape or inflow of fluids (liquids and gases). For example, gaskets are used at joints in self-contained breathing apparatus to prevent the escape or inflow of gases and in fire hose couplings and pump intakes to prevent the leakage of water at connections.

Gas-Sensing Detector — Detection and alarm device that uses either a semiconductor principle or a catalytic-element principle to detect fire gases.

Gate Valve — Control valve with a solid plate operated by a handle and screw mechanism. Rotating the handle moves the plate into or out of the waterway. *See* Butterfly Valve.

Gated Wye — Hose appliance with one female inlet and two or more male outlets with a gate valve on each outlet.

Gatekeeper — Primary care physician or local agency responsible for coordinating and managing the health care needs of members. Generally, in order for specialty services such as mental health and hospital care to be covered, the gatekeeper must first approve the referral.

Gauge — (1) Instrument used to indicate the magnitude of a variable quantity such as pressure. (2) Measurement for wire diameter. *Also known as* Gage.

GB — *See* Sarin.

GEBMO — *See* General Emergency Behavior Model.

Gelling Agents — Superabsorbent liquid polymers capable of absorbing hundreds of times their own weight in water. These gels can be used as fire suppressants and fire retardants. Gels function by entrapping water in their structure rather than air, as is the case with fire fighting foams. *Also known as* Durable Agents.

General Alarm — Larger than normal complement of firefighters and equipment assigned to an incident of large magnitude. In smaller departments this might mean that all available units are assigned to the incident.

General Aviation — All civil aviation operations other than scheduled air services and nonscheduled operations for remuneration or hire.

General Aviation Aircraft — Aircraft used for pleasure or training, business (also known as executive or corporate aircraft), or agricultural purposes.

General Emergency Behavior Model (GEBMO) — Model used to describe how hazardous materials are accidentally released from their containers and how they behave after the release. *See* Engulf (1).

General Municipal Fund — The largest and most important accounting activity managed by a community or organization. The fund provides revenue and expenditures of unrestricted municipal purposes and all financial transactions not accounted for in any other fund.

General Operating Guideline (GOG) — Written guideline describing the desired outcomes to be pursued under given circumstances. A GOG allows the responsible individual more decision-making latitude than is allowed by a Standard Operating Procedure (SOP).

General Order — Standing order, usually written, that is communicated through channels to all units and remains in effect until further notice.

General Staff — Group of incident management personnel: the incident commander, operations section chief, planning section chief, logistics section chief, and finance/administrative section chief.

General Support Grant — Money given with "no strings attached"; general support grant money may be applied to any legitimate operating expense, including salaries.

General Use — Rescue hardware and equipment that is designed, tested, and labeled for lifting and lowering specific loads.

General Winds — Large-scale winds caused by high- and low-pressure systems but generally influenced and modified in the lower atmosphere by terrain.

Generator — (1) Any coil of conductors that is rotated within a magnetic field. As the coil is turned in the field, a voltage is produced. (2) Portable device for generating auxiliary electrical power; generators are powered by gasoline or diesel engines and typically have 110- and/or 220-volt capacity outlets.

Genetic Effect — Mutations or other changes that are produced by irradiation of the germ plasma; changes produced in future generations.

Gentrification — Process of restoring rundown or deteriorated properties by more affluent people, often displacing poorer residents.

Geographic Information Systems (GIS) — Computer software application that relates physical features on the earth to a database to be used for mapping and analysis. The system captures, stores, analyzes, manages, and presents data that refers to or is linked to a location.

G-forces — The forces acting on a body as a result of acceleration or gravity, informally described in units of acceleration equal to one g.

GHS — *See* Globally Harmonized System of Classification and Labeling of Chemicals (GHS).

Gibbs Cam — Ascender for rope climbing.

Gin Pole — Vertical lifting device that may be attached to the front or the rear of the apparatus; consists of a single pole that is attached to the apparatus at one end and has a working pulley at the other. Guy wires may also be used to stabilize the pole.

Girders — (1) Large, horizontal structural members used to support joists and beams at isolated points along their length. (2) Steel frame members in a bus that run from front to rear to strengthen and shape the roof bows. *Also known as* Stringers.

Girth Hitch — Method of attaching a piece of rope or webbing to an anchor; if attached improperly, it can create a weak link in an anchor system.

GIS — *See* Geographic Information Systems.

Gladhands — Fittings for connection of air brake lines between vehicles. *Also known as* Hand Shakes, Hose Couplings, and Polarized Couplings.

Glare — Uncomfortably bright light, either direct or reflected.

Glass Door — Door consisting primarily of glass; usually set in a metal frame.

Glazier's Tool — Tool used for removing windows from their mountings.

Glazing — Glass or thermoplastic panel in a window that allows light to pass.

Global Positioning System (GPS) — System for determining position on the earth's surface by calculating the difference in time for the signal from a number of satellites to reach a receiver on the ground.

Globally Harmonized System of Classification and Labeling of Chemicals (GHS) — International classification and labeling system for chemicals and other hazard communication information, such as material safety data sheets.

Glovebox — Sealed container designed to allow a trained scientist to manipulate microorganisms while being in a different containment level than that of the agent he or she is manipulating; built into the sides of the glovebox are two glove ports arranged so that one can place one's hands into the ports and into gloves, to perform tasks inside the box without breaking the seal.

Gloves — Part of the firefighter's protective clothing ensemble necessary to protect the hands.

Glucose — Simple sugar.

Glue-Laminated Beam — (1) Wooden structural member composed of many relatively short pieces of lumber glued and laminated together under pressure to form a long, extremely strong beam. (2) Term used to describe wood members produced by joining small, flat strips of wood together with glue. *Also known as* Glued-Laminated Beam or Glulam Beam.

Go Rescue — Method of rescue where a rescuer enters the water and maneuvers toward a victim in a water environment; this is the most dangerous tactic in water rescue.

Goal — Broad, general, non-measurable statement of desired achievement or educational intent that usually expresses what a training organization or instructor intends to do for the learner; different from the term *objective*, which states specifically what the learner will do.

GOG — *See* General Operating Guideline.

Going Fire — Slang for a fire of such size and complexity as to be uncontrollable by initial attack units; commonly used term for large wildland fires.

Governor — Built-in pressure-regulating device to control pump discharge pressure by limiting engine rpm.

GPM — *See* Gallons Per Minute.

GPS — *See* Global Positioning System.

Gradability — Ability of a piece of apparatus to traverse various terrain configurations.

Grade — (1) Natural, unaltered ground level. (2) Surface level of pavement or stable earth.

Grade Beam — Concrete wall foundation in the form of a strong reinforced beam that rests on footings or caissons spaced at intervals.

Grade D Breathing Air — Classification of allowable contamination levels in breathing air. Compressed Gas Association (CGA) Grade D allows no more than 20 ppm carbon monoxide, 1,000 ppm carbon dioxide, and 5 mg/m^3 oil vapor.

Gradient Wind — Upper-level winds caused by air movement from a high- or low-pressure system; sometimes covering as much as 300 miles (480 km), with speeds of 5 to 30 miles per hour (8 km/h to 48 km/h), and gradual shifts in direction. Gradient winds flow clockwise around high-pressure cells and counterclockwise around low-pressure cells.

Grading Schedule — Schedule of deficiency points by which insurance engineers grade the fire defenses of a community.

Grading System — System used to convert achievements to grades or class standing.

Grain Bin — Large, cylindrical tank used to store harvested grain, corn, or other similar commodities.

Grant — Donated funding from a government or private source, typically secured through a competitive application process; funds do not have to be repaid, and are separate from an organization's operational or capital budget.

Grantsmanship — Art of raising funds by developing grant proposals and receiving grants.

Grass Roots — Society at the local level as distinguished from the centers of political leadership.

Gravity (G) — Force acting to draw an object toward the earth's center; force is equal to the object's weight. *See* Center of Gravity.

Gravity Circle — Theoretical safety zone that surrounds the center of gravity on an aerial apparatus.

Gravity System — Water supply system that relies entirely on the force of gravity to create pressure and cause water to flow through the system. The water supply, which is often an elevated tank, is at a higher level than the system.

Gravity Tank — Elevated water storage tank for fire protection and community water service. A water level of 100 feet (30 m) provides a static pressure head of 43.4 psi (300 kPa) minus friction losses in piping when water is flowing.

Gravity Wind — *See* Foehn Wind.

Green — Area of unburned fuels, not necessarily green in color, adjacent to but not involved in a wildland fire.

Green Design — Term used to describe the incorporation of such environmental principles as energy efficiency and environmentally friendly building materials into design and construction.

Green Fuels — Vegetation that has a high moisture content and will not easily burn.

Green Wood — Wood with high moisture content.

Greenbelt — Landscaped and perhaps irrigated fuel break that is regularly maintained; sometimes put to an additional use, such as a golf course, park, playground, or pasture. Greenbelts may also be dedicated but unmaintained open space within or between developments.

Grid — Column layout method used by designers to structure the parts of a page; typical grids are one-, two-, or three-column formats, with uniform margins and uniform spacing between columns.

Grid Map — Plan view of an area subdivided into a system of squares (numbered and lettered) to provide quick reference to any point.

Grid System — Water supply system that utilizes lateral feeders for improved distribution.

Grid System Water Mains — Interconnecting system of water mains in a crisscross or rectangular pattern. *Also known as* Grid System.

Gridded Piping System — *See* Complex Loop.

Grillage Footing — Footing consisting of layers of beams placed at right angles to each other and usually encased in concrete.

Grommets — Reinforced eyelets in salvage covers through which fasteners may be passed, allowing the hanging of covers with ropes or pike poles over wall shelving and in other difficult areas.

Gross Decontamination — Quickly removing the worst surface contamination, usually by rinsing with water from handheld hoselines, emergency showers, or other water sources. *See* Decontamination.

Gross Negligence — Willful and wanton disregard. *See* Negligence.

Gross Vehicle Weight Rating (GVWR) — Maximum weight at which a vehicle can be safely operated on roadways; includes the weight of the vehicle itself plus fuel, passengers, cargo, and trailer tongue weight.

Gross Weight — Weight of a vehicle or trailer together with the weight of its entire contents.

Ground Bus — Part of an electrical service panel where the neutral service wire is connected to the earth by a ground wire.

Ground Cover Fire — *See* Wildland Fire.

Ground Fault — (1) Accidental grounding of an electrical conductor. (2) Current that flows to ground outside of the normal current path, such as through a ground conductor, metal pipe, or person.

Ground Fault Circuit Interrupter (GFCI) — Device designed to protect against electrical shock; when grounding occurs, the device opens a circuit to shut off the flow of electricity. *Also known as* Ground Fault Indicator (GFI) Receptacle.

Ground Fire — *See* Wildland Fire.

Ground Fuel — *See* Surface Fuel.

Ground Gradient — Electrical field that radiates outward from where the current enters the ground; its intensity dissipates rapidly as distance increases from the point of entry.

Ground Jack — *See* Stabilizer (1).

Ground Ladder — Ladder specifically designed for fire service use that is not permanently attached (either mechanically or physically) to fire apparatus and does not require mechanical power from the apparatus for ladder use and operation.

Ground Support Unit — Functional unit within the support branch of the logistics section of an incident management system; responsible for fueling/maintaining/repairing vehicles and transporting personnel and supplies.

Grounded Conductor — Conductor in a branch circuit that carries the return current but that is not energized. The covering will be white or natural gray in color, and will be connected to the service neutral. *Also known as* Common Conductor or Neutral Conductor.

Grounding — Reducing the difference in electrical potential between an object and the ground by the use of conductors.

Grounding Conductor — Conductor in a branch circuit that connects the exposed metal parts of appliances to the ground system of the service, in order to minimize the chance of electric shock. This conductor carries no current unless a fault has occurred. The conductor will be bare or have a green, or green with yellow-stripe covering.

Group — NIMS-ICS organizational subunit responsible for a number of individual units that are assigned to perform a particular specified function (such as ventilation, salvage, water supply, extrication, transportation, or EMS) at an incident.

Group Commander — Person in charge of a group within the incident management system.

Group Supervisor — Person in charge of a group within the incident command system.

Grouping — Furniture, stock, and merchandise moved in a compact arrangement to facilitate protection by the least number of salvage covers.

Growth Ring — Layer of wood (as an annual ring) produced during a single period of growth.

Growth Stage — Stage in which a fire is developing within the compartment, drawing air into the plume above the fire, and spreading heat to other fuels in the compartment.

Guards — Protective coverings over dangerous pieces of machinery.

Guide — (1) Document that provides direction or guiding information; does not have the force of law but may provide the basis for what is reasonable in cases of negligence. (2) Device to hold sections of an extension ladder together while allowing free movement.

Guided Discussion — Type of discussion in which a group exchanges ideas directed toward reaching a common goal or conclusion.

Guideline — Statement that identifies a general philosophy; may be included as part of a policy.

Gunning Fog Index — Type of readability index.

Gunwale — Raised edge along the side of a vessel that prevents loose items on deck from falling overboard. *Also known as* Fishplate or Gunnel.

Gusset Plates — Metal or wooden plates used to connect and strengthen the joints of two or more separate components (such as metal or wooden truss components or roof or floor components) into a load-bearing unit.

Guy Ropes — Ropes attached between the tip of a raised aerial device and an object on the ground to stabilize the device during high wind conditions; should be used only if approved by the manufacturer of the aerial device.

GVWR Placard — Placard that commercial vehicles are required to display in plain view.

Gypsum — Hydrated calcium sulfate used for gypsum plaster and wallboard.

Gypsum Board — Widely used interior finish material; consists of a core of calcined gypsum, starch, water, and other additives that are sandwiched between two paper faces. *Also known as* Gypsum Drywall, Plasterboard, Sheetrock®, and Wallboard.

H

HAD — *See* Heat Actuated Devices.

Hailing — Calling out to victims and listening for responses during physical search.

Half Hitch — Knot typically used to stabilize long objects that are being hoisted; always used in conjunction with another knot.

Halfboard — Device used for spinal immobilization and patient removal; can be used as a lifting harness.

Half-Duplex — Radio operating system in which two frequencies are used to communicate over one channel; allows operation in both directions, but not simultaneously.

Half-Life — Time required for half the amount of a substance in or introduced into a living system or ecosystem to be eliminated or disintegrated by natural processes. For example, the period of time required for any radioactive substance to lose half of its strength or reduce by one-half its total present energy.

Halligan Tool — Prying tool with a claw at one end and a spike or point at a right angle to a wedge at the other end. *Also known as* Hooligan Tool.

Halogenated Agent System — Extinguishing system that uses a halogenated gas as the primary extinguishing agent; usually installed to protect highly sensitive electronic equipment.

Halogenated Agents — Chemical compounds (halogenated hydrocarbons) that contain carbon plus one or more elements from the halogen series. Halon 1301 and Halon 1211 are most commonly used as extinguishing agents for Class B and Class C fires. *Also known as* Halogenated Hydrocarbons.

Halogenated Extinguishing Agents — Chemical compounds (halogenated hydrocarbons) that contain carbon plus one or more elements from the halogen series. Halon 1301 and Halon 1211 are most commonly used as extinguishing agents for Class B and Class C fires. *Also known as* Halogenated Hydrocarbons.

Halogenated Hydrocarbons — *See* Halogenated Agents.

Halogens — Name given to the family of elements that includes fluorine, chlorine, bromine, and iodine.

Halon — Halogenated agent; extinguishes fire by inhibiting the chemical reaction between fuel and oxygen. *See* Halogenated Agents.

Halyard — Rope used on extension ladders to extend the fly sections. *Also known as* Fly Rope.

Hand Crew — Personnel who have been trained for operational assignments on an incident, primarily to clear vegetation with hand tools.

Hand Shakes — *See* Gladhands.

Hand Tool — Tool that is manipulated and powered by human force.

Handcuff Knot — Knot tied in a bight with two adjustable loops in opposing directions; used during rescues to secure hands or feet, so that a victim can be raised or dragged to safety.

Handi-Talki — *See* Portable Radio.

Handline — (1) Fireline at a wildland fire constructed with hand tools. (2) Small hoseline (2 inch [65 mm] or less) that can be handled and maneuvered without mechanical assistance.

Handline Nozzle — Any nozzle that can be safely handled by one to three firefighters and flows less than 350 gpm (1 400 L/min).

Handrail — Top piece of a balustrade that is grasped when ascending or descending a stairway. Handrails may be attached to the wall in closed stairways.

Handsaw — Saw that is operated by hand, with no other power source. Especially useful for cutting objects that require a controlled cut but are too big to fit in the jaws of a scissors-type cutter, or unsuitable for cutting with a power saw.

Harassment — Course of conduct directed at a specific person that causes substantial emotional distress in said person and serves no legitimate purpose.

Harbormaster — Person in charge of a port (anchorages, dock spaces, etc.) in the United Kingdom; equivalent to U.S. Coast Guard Captain of the Port.

Hard Hose — Noncollapsible hose; may be used to describe booster line (hard line) hose or hard intake/suction hose.

Hard Line — *See* Booster Hose.

Hard News — News that has a time value; must be delivered immediately or it will become stale and no longer newsworthy.

Hard Suction Hose — Rigid, noncollapsible hose that operates under vacuum conditions without collapsing, allowing a pumping apparatus or portable pump to "draft" water from static or nonpressurized sources (lakes, rivers, wells, etc.) that are below the level of the fire pump, usually available in 10-foot (3 m) sections.

Hard Target — Term used to define a facility or other target that is well defended or protected against a potential adversary attack; examples of hard targets include most military installations and secured government facilities.

Hardware — (1) General term for small pieces of equipment made with metal, such as hand tools. (2) Computer system components such as the electronic parts, keyboard, disk drive, and other physical items. (3) Ancillary equipment used in rope systems, such as carabiners, pulleys, and figure-eight plates.

Hardy Cross Method — Iterative technique used for solving the complicated problems involving gridded water supply systems.

Hasp — Fastening device consisting of a loop, eye, or staple and a slotted hinge or bar; commonly used with a padlock.

Hasty Search — A quick physical search conducted by a team to survey an incident site for visible victims.

Hatch — (1) Square or rectangular access opening in the ceiling or roof of a building, fitted with removable covers for the purpose of providing access and ventilation to the cockloft or roof. Hatches are usually locked on the inside. *Also known as* Scuttle or Scuttle Hatch. (2) Opening in the deck of a vessel that leads to a vertical space down through the various decks (hatchway); covered by a hinged or sliding hatch cover.

Hauling Prusik — Prusik attached onto a main line using a three-wrap Prusik hitch to grab the line and put it into motion. *See* Prusik.

Hauling System — Mechanical advantage system designed for lifting a load; constructed of rope and appropriate hardware.

Hay Hook — *See* Bale Hook.

Hazard — Condition, substance, or device that can directly cause injury or loss; the source of a risk. *See* Fire Hazard, Hazard Assessment, Hazard Class, Hazard or Risk Analysis, Hazardous Material, or Target Hazard.

Hazard and Risk Assessment — Formal review of the hazards and risks that may be encountered by firefighters or emergency responders; used to determine the appropriate level and type of personal and respiratory protection that must be worn. *Also known as* Hazard Assessment. *See* Hazard, Hazard or Risk Analysis, Risk Management Plan, and Target Hazard.

Hazard Area — Established area from which bystanders and unneeded rescue workers are prohibited. *See* Hazard-Control Zones.

Hazard Class — Group of materials designated by the Department of Transportation (DOT) that shares a major hazardous property such as radioactivity or flammability. *See* Hazard, Hazardous Chemical, Hazardous Material, and Hazardous Substance.

Hazard Identification — Process of defining and describing a hazard, including its physical characteristics, magnitude and severity, probability and frequency, causative factors, and locations or areas affected.

Hazard or Risk Analysis — Identification of hazards or risks and the determination of an appropriate response; combines the hazard assessment with risk management concepts. *See* Hazard, Hazard and Risk Assessment, and Risk Management Plan.

Hazard-Control Zones — System of barriers surrounding designated areas at emergency scenes, intended to limit the number of persons exposed to a hazard and to facilitate its mitigation. A major incident has three zones: Restricted (Hot) Zone, Limited Access (Warm) Zone, and Support (Cold) Zone. EPA/OSHA term: Site Work Zones. *Also known as* Control Zones and Scene Control Zones. *See* Hazard Area and Initial Isolation Zone.

Hazardous Atmosphere — Atmosphere that may or may not be immediately dangerous to life or health but that is oxygen deficient, that contains a toxic or disease-producing contaminant, or that contains a flammable or explosive vapor or gas. *See* Immediately Dangerous to Life or Health (IDLH).

Hazardous Chemical — Any chemical that is a physical hazard or health hazard to people, as defined by the Occupational Safety and Health Administration (OSHA). *See* Hazardous Material.

Hazardous Material — Any substance or material that poses an unreasonable risk to health, safety, property, and/or the environment if it is not properly controlled during handling, storage, manufacture, processing, packaging, use, disposal, or transportation. *See* Corrosive Material, Dangerous Goods (1), Hazardous Chemical, Hazardous Substance, Hazardous Waste, Material, and Product.

Hazardous Materials Company — Any piece of equipment having the capabilities, PPE, equipment, and complement of personnel as specified in the Hazardous Materials Company Types and Minimum Standards. The personnel complement shall include one member who is trained to a minimum level of Assistant Safety Officer - Hazardous Materials.

Hazardous Materials Incident — Emergency, with or without fire, that involves the release or potential release of a hazardous material. *See* Hazardous Material.

Hazardous Materials Regulations (HMR) — Regulations for the safe handling and transport of hazardous materials, developed and enforced by the U.S. Department of Transportation (DOT).

Hazardous Materials Task Force — Group of resources with common communications and a leader; may be pre-established and sent to an incident, or formed at the incident.

Hazardous Materials Technician — Individual trained to use specialized protective clothing and control equipment to control the release of a hazardous material. Hazardous materials technicians can specialize in four areas: Cargo Tank Specialty, Intermodal Tank Specialty, Marine Tank Vessel Specialty, and Tank Car Specialty.

Hazardous Materials Transportation Act (HMTA) — Law enacted in 1975 whose primary objective was to provide adequate protection against the risks to life and property inherent in the commercial transportation of hazardous material by improving the regulatory and enforcement authority of the Secretary of Transportation.

Hazardous Substance — Any substance designated under the U.S. Clean Water Act and the Comprehensive Environmental Response, Compensation and Liability Act (CERCLA) as posing a threat to waterways and the environment when released. *See* Extremely Hazardous Substance and Hazardous Material.

Hazardous Waste — Discarded material with no monetary value that can have the same hazardous properties it had before being used. Regulated by the U.S. Environmental Protection Agency (EPA) because of public health and safety concerns; regulatory authority is granted under the Resource Conservation and Recovery Act. *See* Hazardous Material.

Hazardous Waste Operations and Emergency Response (HAZWOPER) — U.S. regulations in Title 29 (Labor) *CFR* 1910.120 for cleanup operations involving hazardous substances and emergency response operations for releases of hazardous substances. *See Code of Federal Regulations (CFR)*.

Hazen-Williams Formula — Empirical formula for calculating friction loss in water systems; fire protection industry standard. To comply with most nationally recognized standards, the Hazen-Williams formula must be used.

HAZWOPER — *See* Hazardous Waste Operations and Emergency Response.

HBV — *See* Hepatitis B Virus.

HCN — *See* Hydrogen Cyanide.

Head — (1) Front and rear closure of a tank shell. (2) Alternate term for pressure, especially pressure due to elevation. For every 1-foot increase in elevation, 0.434 psi is gained (for every 1-meter increase in elevation, 9.82 kPa is gained). *Also known as* Head Pressure. (3) Top of a window or door frame. (4) Most active part of a wildland fire; the forward advancing part. *Also known as* Head of a Fire.

Head Harness — Straps that hold the self-contained breathing apparatus (SCBA) facepiece in place. *Also known as* Spider Strap.

Head of a Fire — *See* Head.

Head Pressure — Pressure exerted by a stationary column of water, directly proportional to the height of the column. *See* Head (2).

Head Protection Systems (HPS) — Air bags that deploy from a narrow opening between the headliner and the top of the door frame to protect passengers' heads.

Head-End Power — Power developed by generators in a train's locomotive to support the energy needs of the other cars in the train consist.

Header — (1) Gathering unit portion of a combine. (2) Term used to describe the looming up of smoke from a fire. (3) Surface contact piece that collects load from uppermost area of a vertical shoring system.

Header Course — (1) Course of bricks with the ends of the bricks facing outward. (2) Masonry unit laid flat on its bed across the width of a wall, with its face perpendicular to the face of the wall; used to bond two wythes. *See* Course and Wythe.

Headline — (1) Head of a newspaper story or article, usually printed in larger type, that introduces and gives the gist of the story or article that follows. (2) To publicize highly.

Head-on Collision — Collision in which the front of one vehicle strikes either the front of another vehicle or a stationary object.

Headwind — Wind that is blowing toward the face of a person or the front of an aircraft. *See* Crosswind, Downwind, and Wind.

Health and Fitness Coordinator (HFC) — Individual who, under the supervision of the fire department physician, is responsible for all physical fitness programs in the fire and emergency services organization.

Health and Safety Officer (HSO) — Officer authorized to manage the organization's health and safety program; performs the duties, functions, and responsibilities described in NFPA® 1521, *Standard for Fire Department Safety Officer*. Must meet the qualifications of this standard, or an approved equivalent.

Health Canada — Agency responsible for developing health policy, enforcing health regulations, promoting disease prevention, and enhancing healthy living in Canada.

Health Department — Agency that focuses on public health issues in a given region.

Health Hazard — Material that may directly affect human health once it enters or comes in contact with the body. *See* Physical Hazard.

Health Insurance Portability and Accountability Act (HIPAA) — Congressional law established to help ensure the portability of insurance coverage as employees move from job to job. In addition to improving efficiency of the health care payment process, it also helps protect a patient's/client's privacy. The law also applies to information pertinent to juvenile firesetting situations.

Hearing Protection — Device that limits noise-induced hearing loss when firefighters are exposed to extremely loud environments, such as apparatus engine noise, audible warning devices, and the use of power tools and equipment.

Heart — Hollow muscular organ that receives the blood from the veins, sends it through the lungs to be oxygenated, then pumps it to the arteries.

Heart Attack — *See* Acute Myocardial Infarction.

Heat — Form of energy associated with the motion of atoms or molecules in solids or liquids that is transferred from one body to another as a result of a temperature difference between the bodies, such as from the sun to the earth. To signify its intensity, it is measured in degrees of temperature. (2) Form of energy associated with the motion of atoms or molecules and capable of being transmitted through solid and fluid media by conduction, through fluid media by convection, and through empty space by radiation. *See* Conduction, Convection, Heat Transfer, Pyrolysis, and Radiation.

Heat Actuated Devices (HAD) — Thermostatically controlled detection devices used to activate fire equipment, alarms, or appliances.

Heat Cramps — Heat illness resulting from prolonged exposure to high temperatures; characterized by excessive sweating, muscle cramps in the abdomen and legs, faintness, dizziness, and exhaustion. *See* Heat Exhaustion, Heat Rash, Heat Stress, and Heat Stroke.

Heat Detector — Alarm-initiating device that is designed to be responsive to a predetermined rate of temperature increase or to a predetermined temperature level.

Heat Energy Applied — Sum of the temperature of the heat source and the time of exposure.

Heat Exhaustion — Heat illness caused by exposure to excessive heat; symptoms include weakness, cold and clammy skin, heavy perspiration, rapid and shallow breathing, weak pulse, dizziness, and sometimes unconsciousness. *See* Heat Cramps, Heat Rash, Heat Stress, and Heat Stroke.

Heat Flux — The measure of the rate of heat transfer to or from a surface, typically expressed in kilowatts/ m2.

Heat Flux History — Amount of heat flux exposed to materials over the duration of a fire.

Heat from Arcing — Type of electrical heating that occurs when the current flow is interrupted.

Heat of Combustion — Total amount of thermal energy (heat) that could be generated by the combustion (oxidation) reaction if a fuel were completely burned. The heat of combustion is measured in British Thermal Units (Btu) per pound, kilojoules per gram, or Megajoules per kilogram. *See* Combustion and Heat.

Heat of Decomposition — Release of heat from decomposing compounds, usually due to bacterial action.

Heat of Friction — Heat created by the movement of two surfaces against each other.

Heat of Hydration — During the hardening of concrete, heat is given off by the chemical process of hydration.

Heat of Ignition — Heat energy that brings about ignition; comes from various forms and usually from a specific object or source. Therefore, the heat of ignition is divided into two parts: (a) equipment involved in ignition and (b) form of heat of ignition.

Heat of Solution — Heat released by the solution of matter in a liquid.

Heat of Vaporization — Quantity of heat required to transform a liquid into a vapor.

Heat Protective Shield — Reflective shield around an elevating platform that protects firefighters from the effects of radiated heat.

Heat Rash — Condition that develops from continuous exposure to heat and humid air; aggravated by clothing that rubs the skin. Reduces the individual's tolerance to heat. *See* Heat Cramps, Heat Exhaustion, Heat Stress, and Heat Stroke.

Heat Release Rate (HRR) — Total amount of heat released per unit time. The heat release rate is typically measured in kilowatts (kW) or Megawatts(MW) of output.

Heat Resistance — Foam's ability to resist the actual heat of the liquid or surface on which it is applied.

Heat Sensor Label — Label affixed to the ladder beam near the tip to provide a warning that the ladder has been subjected to excessive heat.

Heat Shadowing — Fire pattern left when an object blocks a fire's radiant heat from reaching a combustible surface behind the object. *See* Protected Area.

Heat Stratification — *See* Thermal Layering (of Gases).

Heat Stress — Combination of environmental and physical work factors that compose the heat load imposed on the body; environmental factors include air, temperature, radiant heat exchange, air movement, and water vapor pressure. Physical work contributes because of the metabolic heat in the body; clothing also has an effect. *See* Heat Cramps, Heat Exhaustion, Heat Rash, and Heat Stroke.

Heat Stroke — Heat illness in which the body's heat regulating mechanism fails; symptoms include (a) high fever of 105° to 106° F (40.5° to 41.1° C), (b) dry, red, hot skin, (c) rapid, strong pulse, and (d) deep breaths or convulsions. May result in coma or even death. *Also known as* Sunstroke. *See* Heat Cramps, Heat Exhaustion, Heat Rash, and Heat Stress.

Heat Transfer — Flow of heat from a hot substance to a cold substance; may be accomplished by convection, conduction, or radiation. *See* Conduction, Convection, Heat, and Radiation.

Heat Treatment — Controlled cooling or quenching of heated metals, in order to harden the metal; usually accomplished by immersion in a liquid quenching medium.

Heat Wave — Movement of radiated heat through space until it reaches an opaque object.

Heating Tube — Tube installed inside a tank to heat the contents. *Also known as* Fire Tube.

Heating, Ventilating, and Air Conditioning (HVAC) System — Mechanical system used to provide environmental control within a structure, and the equipment necessary to make it function; usually a single, integrated unit with a complex system of ducts throughout the building. *Also known as* Air-Handling System. *See* Mechanical System.

Heavy Content Fire Loading — Storing of combustible materials in high piles that are placed close together.

Heavy Equipment — Ground vehicles used in the suppression of wildland fires. Includes bulldozers, tractors, plows, and transport vehicles, but not fire apparatus.

Heavy Equipment Transport — Any ground vehicle capable of transporting a dozer or tractor.

Heavy Fuels — Massive natural cover fuels such as logs, snags, and large limbs. Heavy fuels are not easy to ignite; once ignited, they burn slowly and hot.

Heavy Metal — Generic term for toxic elements such as lead, cadmium, or mercury; may also be applied to compounds containing these elements. *Also known as* Toxic Element.

Heavy Rescue Vehicle — Large rescue vehicle that may be constructed on a custom or commercial chassis. Carries additional equipment that includes A-frames or gin poles, cascade systems, larger power plants, trench and storing equipment, small pumps and foam equipment, large winches, hydraulic booms, large quantities of rope and rigging equipment, air compressors, and ladders.

Heavy Stream — *See* Master Stream.

Heavy Timber Construction — Type of construction in which the structural frame is composed of large wooden beams, columns, and trusses. *See* Type IV Construction.

Heavy-Duty Appliances — Master stream equipment.

Heel — (1) Base or butt end of a ground ladder. (2) To steady a ladder while it is being raised. (3) Rear portion of a wildland fire. *Also known as* Rear. (4) Angle at which a vessel leans to one side due to wind, waves, or turning of the vessel; measured in degrees. *See* Critical Angle of List, Heeling, and List.

Heel Plate — Metal reinforcement at the heel or butt of a ladder; generally shaped to give the ladder more stability.

Heeling — (1) Tipping or leaning to one side. (2) Causing a vessel to list (continuous lean to one side).

Heelman — Firefighter who carries the butt end of the ladder and/or who subsequently heels or secures it from slipping during operations.

Heimlich Maneuver — Technique to clear an obstruction from a patient's airway.

Helibase — Main location on an incident for parking, fueling, maintaining, and loading helicopters.

Helical Flow — Corkscrew-like flow of water against a solid boundary.

Helicopter — Rotary-wing aircraft ranging in size from small, single-seat aircraft to large transports that carry up to 50 passengers. It can deliver firefighters, water or chemical retardants (either a fixed tank or bucket system), or internal or external cargo.

Helicopter Tender — Ground service vehicle capable of supplying fuel and support equipment to helicopters.

Helispot — Temporary landing spot for helicopters.

Helitack Crew — Crew of individuals assigned to support helicopter operations.

Helitanker — Air Tanker Board certified helicopter equipped with a fixed tank and capable of delivering a minimum of 1,100 gallons (4 164 L) of water, retardant, or foam.

Helmet — Headgear worn by firefighters that provides protection from falling objects, side blows, elevated temperatures, and heated water.

Helmet Faceshield — *See* Faceshield.

Helmet Identification Shield — Insignia or plaque fastened to the front of the firefighter's helmet; typically displays the name of the city, and the firefighter's initials, unit, and rank.

Hematotoxic Agent — Chemical that damages the blood. *Also known as* Hemotoxin.

Hemispheric Release — Semicircular or dome-shaped pattern of airborne hazardous material that is still partially in contact with the ground or water. *See* Cloud, Cone, and Plume.

Hemispherical — Shaped like half of a sphere.

Hemispherical Head — End of a tank that is shaped like half of a sphere; usually found on pressure tanks such as MC-331 high-pressure tanks.

Hemoglobin — Oxygen-carrying component of red blood cells.

Hemorrhage — Profuse discharge of blood.

HEPA — *See* High-Efficiency Particulate Air Filter.

Hepatitis B Virus (HBV) — Bloodborne virus that causes liver infection and may lead to liver failure, liver cancer, cirrhosis of the liver, or permanent scarring of the liver. Most adults can make full recovery from HBV, but children and infants are much more likely to develop a chronic infection.

Hepatotoxic Agent — Chemical that damages the liver. *Also known as* Hepatotoxin.

Herbicides — Chemicals designed to control or eliminate all or certain kinds of plants.

Herringbone Room Setup — *See* Chevron Room Setup.

Hertz (Hz) — Measurement unit of frequency. *See* Frequency.

Hierarchy of Controls — Widely accepted system designed to mitigate and/or eliminate exposure to hazards (risks) in the workplace.

Higbee Cut — Special cut at the beginning of the thread on a hose coupling that provides positive identification of the first thread to eliminate cross threading. *Also known as* Blunt Start.

Higbee Indicators — Notches or grooves cut into coupling lugs to identify by touch or sight the exact location of the Higbee Cut.

High Angle — Vertical or near-vertical environment in which rescuers must be secured with rope for safety. The majority of the rescue load is supported by the rope system.

High Back-Pressure Foam Maker — In-line aspirator used to deliver foam under pressure. High back-pressure aspirators supply air directly to the foam solution through a venturi action. *Also known as* Forcing Foam Maker.

High Explosive — Explosive that decomposes extremely rapidly (almost instantaneously) and has a detonation velocity faster than the speed of sound. *See* Ammonium Nitrate and Fuel Oil (ANFO), Detonation, Explosive (1), and Low Explosive.

High Mobility Multipurpose Wheeled Vehicles (HMMWV or Humvee) — Four-wheel drive military vehicle that replaced earlier Jeep and MUTT vehicles.

High Speed Turnoff/Taxiway — Curved or angled taxiway designed to expedite aircraft turning off the runway after landing.

High-Efficiency Particulate Air (HEPA) Filter — Respiratory filter that is certified to remove at least 99.97 % of monodisperse particles of 0.3 micrometers in diameter.

High-Expansion Foam — Foam concentrate that is mixed with air in the range of 200 parts air to 1 part foam solution (200:1) to 1,000 parts air to 1 part foam solution (1,000:1). *See* Low-Expansion Foam, Mechanical Blower, and Medium-Expansion Foam.

High-Hazard Training — Training activities that involve both known and potentially unknown risks. Examples include evolutions or exercises in live fire suppression, hazardous materials mitigation, above- and below-grade rescue, and the use of power tools.

High-Impact Crashes — Aircraft crashes with severe damage to the fuselage and with a significantly reduced likelihood of occupant survival.

High-Order-Explosion Damage — Damage usually resulting from a detonation, including small shattered debris, widespread damage, and near or total destruction of the confining vessel. *See* Low-Order-Explosion Damage.

High-Pressure Air — Air pressurized to 3,000 to 5,000 psi (20 684 kPa to 34 473 kPa); used to differentiate from older air cylinders using a pressure range from 1,800 to 2,200 psi (12 411 kPa to 15 168 kPa).

High-Pressure Fog — Fog stream operated at high pressures and discharged through small diameter hose.

High-Pressure Hose — Hose leading from the air cylinder to the regulator; may be at cylinder pressure or reduced to some lower pressure.

High-Pressure Nozzle — Fire stream nozzle that is designed to be operated in excess of the 100 psi (689 kPa) to which ordinary fog nozzles are designed.

High-Pressure Tank ▢ Cargo tank truck that carries liquefied gases. *See* Cargo Tank Truck.

High-Rack Storage — Warehousing storage of materials on high, open racks that may be as high as 100 feet (30 m).

High-Rise Building — Building that requires fire fighting on levels above the reach of the department's equipment. The *Uniform Building Code (UBC)* defines a high-rise building as one greater than 75 feet (23 m) in height, but other fire and building codes may define the term differently. *Also known as* High-Rise.

High-Rise Pack — Special kit for high-rise operations containing hose, adapters, nozzle, and spanner wrenches.

High-Strength Low-Alloy (HSLA) Steel — Alloy steel developed to provide better mechanical properties or greater resistance to corrosion than carbon steel; different from other varieties of steels in that it is designed to possess specific mechanical properties.

High-Value District — Section of a city in which valuable property is concentrated and in which additional companies and apparatus are needed to combat a fire; usually the central business district of a community.

High-Voltage — Any voltage in excess of 600 volts.

Hinged Door — *See* Swinging Door.

Hip — Junction of two sloping roof surfaces forming an exterior angle.

Hip Roof — Pitched roof that has no gables. All facets of the roof slope down from the peak to an outside wall.

HIPAA — *See* Health Insurance Portability and Accountability Act.

Hitch — (1) Temporary knot that falls apart if the object held by the rope is removed. (2) Connecting device at the rear of a vehicle used to pull a full trailer with provision for easy coupling. (3) Loop that secures the rope but is not part of a standard rope knot.

HIV — *See* Human Immunodeficiency Virus.

H-Jack — *See* Box Stabilizer.

HMR — *See* Hazardous Materials Regulations.

HMTA — *See* Hazardous Materials Transportation Act.

Hog — (1) Vertical distance of a vessel's keel at amidships above a vessel's keel at the bow and stern. (2) To strain a vessel in a manner that tends to make the bow and stern lower than the middle portion, which has greater buoyancy. *Also known as* Hogging. See Sag and Sagging.

Hogging — *See* Hog.

Hoisting Cylinder — *See* Elevation Cylinder.

Hold-Down Locks — Locks that secure an aerial device, such as an aerial ladder, in its cradle during road travel.

Holdfast — Constructed anchor for a guy line.

Hollow Square Setup — Room arrangement in which the chairs are positioned as the outside sides of a square; similar to the U-shaped arrangement, but with this setup, the instructor/fire and life safety educator cannot walk into the center of the group.

Home Ignition Zones — Area surrounding a home during a wildfire which indicates that the home should be evaluated for protection from an encroaching wildfire.

Home Inspection — Process of educating residents about fire hazards by examining a residence for existing fire hazards and poor safety practices; also used as an evaluation instrument to determine the extent to which fire and life safety behaviors are being implemented in the community.

Home Safety Council (HSC) — National non-profit organization solely dedicated to preventing home-related injuries. HSC helps facilitate national programs and partnerships to educated people of all ages to be safer in and around their homes.

Homogeneous — Description of a substance that has uniform structure or composition throughout.

Hook and Ladder — Old term for an aerial ladder truck.

Hooking Up — Slang for connecting a fire department pumper to a hydrant or connecting a discharge hose to the pumper.

Hooks — Curved metal devices installed on the tip end of roof ladders to secure the ladder to the highest point on the roof of a building.

Hooligan Tool — *See* Halligan Tool.

Hopcalite® — Catalytic chemical that converts carbon monoxide to carbon dioxide.

Hopper — (1) Receptacle used for temporary storage. (2) Tank holding a liquid and having a device for releasing its contents through a pipe. (3) Freight car with a floor sloping to one or more hinged doors for discharging bulk contents. (4) Funnel shaped bin used to store dry solid materials, such as corn, which discharge from the bottom.

Hopper Window — Type of swinging window that is hinged along the bottom edge and usually designed to open inward.

Horizontal Motion — Side-to-side, swaying motion.

Horizontal Pressure Vessel — Pressurized storage tanks characterized by rounded ends; capacity may range from 500 to 40,000 gallons (1 893 L to 151 416 L). Propane, butane, ethane, and hydrogen chloride are examples of materials stored in these tanks. *See* Pressure Vessel and Spherical Pressure Vessel.

Horizontal Shore — Support for vertical surfaces, such as a wall, that is braced against another vertical surface. *Also known as* Flying Shore.

Horizontal Smoke Spread — Tendency of heat, smoke, and other products of combustion to rise until they encounter a horizontal obstruction. At this point they will spread laterally (ceiling jet) until they encounter vertical obstructions and begin to bank downward (hot gas layer development).

Horizontal Split-Case Pump — Centrifugal pump with the impeller shaft installed horizontally and often referred to as a split-case pump. This is because the case in which the shaft and impeller rotates is split in the middle and can be separated, exposing the shaft, bearings, and impeller.

Horizontal Storage Tank — Atmospheric storage tank that is laid horizontally and constructed of steel. *See* Atmospheric Storage Tank.

Horizontal Strut — Horizontal load-bearing timber placed between two wallplates in a horizontal shoring system.

Horizontal Ventilation — Any technique by which heat, smoke, and other products of combustion are channeled horizontally out of a structure by way of existing or created horizontal openings such as windows, doors, or other openings in walls. Typically portions of one or more of the horizontal openings will also serve as an air inlet.

Horseshoe Load — Arrangement of fire hose in a hose bed or compartment in which the hose lies on edge in the form of a horseshoe.

Hose Bed — Main hose-carrying area of a pumper or other piece of apparatus designed for carrying hose. *Also known as* Hose Body.

Hose Belt — Leather belt or nylon strap used for securing and handling charged hoselines, tools, or tying off a ladder. *See* Hose Strap or Rope Hose Tool.

Hose Bin — Tray or compartment, often located on the running board or over a hose bed, for carrying extra hose.

Hose Body — *See* Hose Bed.

Hose Bridge — Device placed alongside or astride hose that is laid across a street to permit traffic to drive over the hose without damaging it. *Also known as* Hose Ramp.

Hose Cabinet — Recessed wall cabinet that contains a wall hydrant and preconnected fire hose for incipient fire fighting. *Also known as* Hose Rack.

Hose Cap — Threaded female fitting used to cap a hoseline or a pump outlet.

Hose Clamp — Mechanical or hydraulic device used to compress fire hose to stop the flow of water.

Hose Control Device — Device used to hold a charged hoseline in a stationary position for an extended period of time.

Hose Couplings — Metal fasteners or devices attached to the ends of a length of fire hose, used to connect lengths of hose together.

Hose Dryer — Enclosed cabinet containing racks on which fire hose can be dried.

Hose Hoist — *See* Hose Roller.

Hose Jacket — (1) Outer covering of a hose. (2) Device clamped over a hose to contain water at a rupture point or to join hose with damaged or dissimilar couplings.

Hose Lay — (1) Arrangement of connected lengths of fire hose and accessories on the ground at a wildland fire, beginning at the first pumping unit and ending at the point of water delivery. (2) Connected lengths of hose from water source to pumping engine. (3) Layouts of hose from a fire pump to the place where the water needs to be.

Hose Layout, Complicated — Hose layout that includes the use of multiple lengths of unequal hoselines, unequal wyed or manifold lines, siamesed lines, or master stream devices. Such a layout requires the pump operator to perform complicated calculations in order to supply the lines properly.

Hose Layout, Simple — Hose layout that includes the use of single hoselines or multiple, wyed, siamesed, or manifold lines of equal length.

Hose Load Finish — *See* Finish.

Hose Pack — Compact bundle of hose, usually bound to facilitate moving.

Hose Plug — Threaded male fitting used to cap off a pump intake.

Hose Rack — Device used to hold a length of hose preconnected to a standpipe or other source of water for incipient fire fighting. *See* Hose Cabinet.

Hose Ramp — *See* Hose Bridge.

Hose Record — Individual history of a section of hose from the time it is purchased until it is taken out of service.

Hose Reel — Cylindrical device upon which fire hose is manually or mechanically rolled for later deployment.

Hose Roller — Metal device with a roller that can be placed over a windowsill or roof's edge to protect a hose and make it easier to hoist. *Also known as* Hose Hoist.

Hose Strap — Strap or chain with a handle suitable for placing over a ladder rung; used to carry and secure a hoseline. *See* Hose Belt or Rope Hose Tool.

Hose Test Gate Valve — Special valve designed to prevent injury caused by a burst hoseline during hose testing.

Hose Tool — *See* Hose Strap.

Hose Tower — Part of a fire station or building designed so that fire hose can be hung vertically to drain and dry.

Hose Trough — *See* Hose Tube.

Hose Tube — Housing used on tank and bulk commodity trailers for the storage of cargo handling hoses. *Also known as* Hose Trough.

Hose Wringer — Device used to remove water and air from large diameter hose.

Hoseline — Flexible conduit (fire hose) used to transport water from a source of supply to a point of application, usually onto a fire. May be used to deliver water from a hydrant or other source to a pumper, from a pumper to a nozzle or other appliance, or directly from the source to the application.

Hoseline Tee — Fitting that may be installed between lengths of hose to provide an independently controlled outlet for a branch line.

Hot Conductor — *See* Ungrounded Conductor.

Hot Refuel/Defuel — Refueling or defueling of an aircraft while the engines are operating. *Also known as* Rapid Refuel/Defuel.

Hot Smoldering Phase — Phase or stage of fire in which the level of oxygen in a confined space is below that needed for flaming combustion; characterized by glowing embers, high heat at all levels of the room, and heavy smoke and fire gas production.

Hot Spot — Particularly active area of a wildland fire.

Hot Work — (1) Any operation that requires the use of tools or machines that may produce a source of ignition. (2) In maritime terms, any construction, alteration, repair, or shipbreaking operation involving riveting, welding, burning, or similar fire-producing operations.

Hot Zone — Potentially hazardous area immediately surrounding the incident site; requires appropriate protective clothing and equipment and other safety precautions for entry. Typically limited to technician-level personnel. *Also known as* Exclusion Zone.

Hotel — Subdivision of residential property classification consisting of structures or groups of structures with more than 16 sleeping units, primarily used as lodging by transients; these structures must be under single management, and meals may or may not be provided. *Also known as* Apartment Hotel, Club, Inn, Lodging House, or Motel.

Hotel Raise — Method of raising a fire department extension ladder in line with several windows so that individuals can simultaneously escape from more than one floor. *Also known as* Factory Raise.

Hot-Gas-Layer Pattern — Fire pattern formed by radiant heat in the hot-gas layer during a fire before flashover; these patterns are found when fires are extinguished before the fire has reached flashover.

Hotline — Telephone line and operation set up for receiving information on one particular subject, often relating to criminal behavior and given anonymously.

Hotshot Crew — Highly trained fire fighting crew used primarily in handline construction.

Hotspotting — Checking the spread of fire at points of more rapid spread or special threat only.

Hourglass Pattern — Fire pattern that occurs when a smaller fire burns directly adjacent to a horizontal surface, leaving an inverted V-pattern on the wall with a traditional V-pattern above the inverted V.

House — Structure located above the main deck. *See* Superstructure.

House Lights — Lights throughout the fire station that are controlled from the alarm or watch desk, which makes it possible to illuminate the entire station when an alarm is received or in case of emergency.

House Line — Permanently fixed, private standpipe hoseline.

House Watch — Duty of maintaining the fire station alarm center for a prescribed period of time.

Household Fire Warning Systems — Detection and alarm systems that include single- and multiple-station smoke detectors as well as more complicated combination systems.

HPS — *See* Head Protection Systems.

HRR — *See* Heat Release Rate.

HSC — *See* Home Safety Council.

HSO — *See* Health and Safety Officer.

HUD — *See* Department of Housing and Urban Development.

Hull — Main structural frame or body of a vessel below the weather deck.

Human Factors — (1) Attributes or characteristics that cause an individual to be involved in more accidents than others. (2) Natural desire to conserve human resources as and prevent needless suffering caused by physical pain or emotional stress.

Human Immunodeficiency Virus (HIV) — Sexually transmitted and bloodborne virus that is the cause of AIDS.

Hunter Model of Instruction — Method of instruction that emphasizes practice or application to achieve mastery of skills; developed by Madeline Hunter of UCLA.

Hybrid Construction — Type of building construction that uses renewable, environmentally friendly or recycled materials. *Also known as* Natural or Green Construction.

Hybrid Modular Structure — Structure consisting of the elements of both modular design and panelized construction. Core modular units are assembled first and panels are added to complete the structure.

Hydrant Adapter — Adapter, fitting, or coupling to connect hose or pumper intake hose to a fire hydrant.

Hydrant Hose House — Small, fully enclosed structure enclosing a fire hydrant and containing some amount of fire hose and appropriate tools and appliances.

Hydrant Pressure — Amount of pressure being supplied by a hydrant without assistance.

Hydrant Wrench — Specially designed tool used to open or close a hydrant and to remove hydrant caps.

Hydration — (1) Act or process of combining with water. (2) Condition of having adequate fluid in body tissues through adequate fluid intake. (3) Chemical process in which concrete changes to a solid state and gains strength.

Hydraulic Braking Systems — A braking system that uses a fluid in a closed system to pressurize wheel cylinders when activated.

Hydraulic Calculations — Process of using mathematics to solve problems involving fire hydraulics.

Hydraulic Jack — Lifting jack that uses hydraulic fluid power supplied from a manually operated hand lever.

Hydraulic Pump — Positive displacement-type pump that imparts pressure on hydraulic oil within the hydraulic system.

Hydraulic Reservoir — Supplies the hydraulic fluid that is moved in and out of a hydraulic system; fluid displaced from the system flows back into the reservoir for storage before being recirculated through the system.

Hydraulic Shoring — Shores or jacks with movable parts that are operated by the action of hydraulic fluid.

Hydraulic System — (1) Aircraft system that transmits power by means of a fluid under pressure. (2) Aerial apparatus system that provides power to the stabilizers and aerial device.

Hydraulic Ventilation — Ventilation accomplished by using a spray stream to draw the smoke from a compartment through an exterior opening.

Hydraulics — Branch of fluid mechanics dealing with the mechanical properties of liquids and the application of these properties in engineering.

Hydrocarbon — Organic compound containing only hydrogen and carbon and found primarily in petroleum products and coal.

Hydrocarbon Fuel — Petroleum-based organic compound that contains only hydrogen and carbon; may also be used to describe those materials in a fuel load which were created using hydrocarbons such as plastics or synthetic fabrics. *See* Liquefied Compressed Gas, Liquefied Petroleum Gas, and Polar Solvent Fuel.

Hydrodynamics — The study of liquids in motion.

Hydrogen Chloride (HCL) — Gas produced by the combustion of polyvinyl chlorides; when inhaled, it mixes with the moisture in the respiratory tract and forms hydrochloric acid.

Hydrogen Cyanide (HCN) — Colorless, toxic, and flammable liquid until it reaches 79o F (26o C). Above that temperature, it becomes a gas with a faint odor similar to bitter almonds; produced by the combustion of nitrogen-bearing substances.

Hydrogen Sulfide (H_2S) — Colorless gas with a strong rotten-egg odor produced when rubber insulation, tires, and woolen materials burn, and by the decomposition of sulfur-bearing organic material; dangerous because it quickly deactivates the sense of smell. It is commonly called *silo gas*, although it is actually one of several components of silo gas.

Hydrokinetics — Branch of hydraulics having to do with liquids (water) in motion, particularly in relation to forces created by or applied to the liquid in motion.

Hydrologic Data — Information related to the movement of currents and tides in bodies of water.

Hydrolyze — To cause or undergo a chemical process of decomposition involving the splitting of a bond and the addition of the element of water.

Hydrophobic — Incapable of mixing with water.

Hydroplaning — Condition in which moving tires (automobile or aircraft) are separated from pavement surfaces by steam and/or water or liquid rubber film, resulting in loss of mechanical braking effectiveness.

Hydrostatic Test — Testing method that uses water under pressure to check the integrity of pressure vessels.

Hydrostatics — Branch of hydraulics dealing with the properties of liquids (water) at rest, particularly in relation to pressures resulting from or applied to the static liquid.

Hygroscopic — Ability of a substance to absorb moisture from the air.

Hyperglycemia — Excessive sugar in the blood due to lack of insulin to metabolize the sugar. *Also known as* Diabetic Ketoacidosis.

Hypergolic — Chemical reaction between a fuel and an oxidizer that causes immediate ignition on contact, without the presence of air. An example is the contact of fuming nitric acid and UDMH (unsymmetrical dimethyl hydrazine). *See* Hypergolic Materials.

Hypergolic Materials — Materials that ignite when they come in contact with each other. The chemical reactions of hypergolic substances vary from slow reactions that may barely be visible to reactions that occur with explosive force. *See* Hypergolic.

Hypertension — High blood pressure; blood higher pressure than normal.

Hyperthermia — Abnormally high body temperature.

Hyperventilation — Rapid breathing that overoxygenates the blood.

Hypoglycemia — Potentially life-threatening condition in which the level of sugar in the blood is abnormally low.

Hypotension — Low blood pressure; blood pressure lower than normal.

Hypothermia — Abnormally low body temperature. *See* Frostbite.

Hypothesis — Theory proposed about phenomena at a scene which requires further investigation and must be proved or disproved based upon evidence collected from the scene.

Hypovolemia — Decreased blood volume.

Hypovolemic Shock — Shock caused by loss of blood.

Hypoxia — Potentially fatal condition caused by lack of oxygen.

I

I Beam — Steel or wooden structural member consisting of top and bottom flanges joined by a center web section, so that the cross section resembles a capital I.

IAAI — *See* International Association of Arson Investigators.

IAFF — *See* International Association of Fire Fighters.

IAFPA — *See* International Aviation Fire Protection Association.

IAP — *See* Incident Action Plan.

IBC — (1) *See* International Building Code. (2) *See* Industrial Bulk Container. (3) *See* Intermediate Bulk Container.

IC — *See* Incident Commander.

ICAO — *See* International Civil Aviation Organization.

ICC — (1) *See* International Code Council. (2) *See* Interstate Commerce Commission.

Ice Shrugging — Method to remove ice from an aerial device.

ICS — *See* Incident Command System.

Identification Number — Serial number placed on each ground ladder by the manufacturer.

Identification Power — That which stems from the human tendency to follow or mimic those who are admired or respected.

Idle Thrust/RPM — Aircraft engine running at the lowest possible speed.

IDLH — *See* Immediately Dangerous to Life or Health.

IED — *See* Improvised Explosive Device.

IFR — *See* Instrument Flight Rules.

IFSAC — *See* International Fire Service Accreditation Congress.

IFSTA — *See* International Fire Service Training Association.

Ignition — The process of initiating self-sustained combustion. (NFPA® 921)

Ignition Sequence — History of the fire, beginning when the ignition source and the first fuel ignited meet at the area of origin, and proceeding through the entire duration of fire spread through the scene.

Ignition Source — Mechanism or initial energy source employed to initiate combustion, such as a spark that provides a means for the initiation of self-sustained combustion. *See* Combustion, Ignition, and Ignition Temperature.

Ignition Temperature — Minimum temperature to which a fuel (other than a liquid) in air must be heated in order to start self-sustained combustion independent of the heating source. *See* Autoignition, Autoignition Temperature, Flashover, and Ignition.

Ike-O-Hook — Steel hook with an eyelet on one end; used for hanging salvage covers and other devices from pike poles or ropes.

Illegal Clandestine Lab — Laboratory established to produce or manufacture illegal or controlled substance such as drugs, chemical warfare agents, explosives, or biological agents. *See* Chemical Warfare Agent, Explosive (2), and Meth Lab.

Illegal Dump — Site where chemicals are disposed of illegally.

Illicit — Illegal, Unlawful.

Illicit Clandestine Laboratory — Laboratory that produces illegal or controlled substances, such as drugs, explosives, biological agents, or chemical warfare agents. *Also known as* Illegal Clandestine Laboratory. *See* Meth Lab.

Illumination Unit — Portable light generating unit capable of providing 3 to 6 lights of 500 watts each with extension cords from 500 to 1000 feet (152 m to 305 m) for the purpose of providing a specified level of illumination capacity.

Illustrated Lecture — Instructional technique in which audiovisual training aids accompany a lecture, clarifying information and facilitating interaction with the students.

Illustration Method — Instructional method that uses the sense of sight. The instructor or educator provides information coupled with visuals such as drawings, pictures, slides, transparencies, film, models, and other visual aids to illustrate a lecture and help clarify details or processes.

ILS — *See* Instrument Landing System.

IM Portable Tank — *See* Nonpressure Intermodal Tank.

Image Stabilization (IS) — Vibration reduction feature available on some lenses and DSLR cameras. *Also known as* Optical Stabilization (OS) and Vibration Reduction (VR).

IMC — *See* International Mechanical Code.

Immediate Treatment — Classification for patients with the most serious injuries at an incident, who will require packaging and movement to a health care facility as soon as possible.

Immediately Dangerous to Life and Health (IDLH) — Description of any atmosphere that poses an immediate hazard to life or produces immediate irreversible, debilitating effects on health; represents concentrations above which respiratory protection should be required. Expressed in parts per million (ppm) or milligrams per cubic meter (mg/m^3); companion measurement to the permissible exposure limit (PEL). *See* Hazardous Atmosphere, Permissible Exposure Limit (PEL), Recommended Exposure Limit (REL), Short-Term Exposure Limit (STEL), and Threshold Limit Value (TLV®).

Immiscible — Incapable of being mixed or blended with another substance. *See* Insoluble, Miscibility, and Soluble.

Immobilization — To hold a part firmly in place, as with a splint.

Immunity — Freedom from legal liability for an act or physical condition; opposite of *liability*. See Liability.

Immunization — Process or procedure by which a subject (person, animal, or plant) is rendered immune or resistant to a specific disease. This term is often used interchangeably with vaccination or inoculation, although the act of inoculation does not always result in immunity.

IMO — See International Maritime Organization.

IMO Type 5 — See Pressure Intermodal Tank.

Impact Analysis — Process used to quantify the potential negative effects of a disaster. *Also known as* Business Impact Analysis in the private sector. See Analysis and Cost-Benefit Analysis.

Impact Evaluation — Measuring knowledge gain, behavioral change, and modifications to living conditions or lifestyles.

Impact Hammer — See Pneumatic Chisel.

Impact Load — Dynamic and sudden load placed on a rope, typically during a fall.

Impaled — (1) Condition resulting when a patient's head or other appendage pierces a stationary object such as a windshield. (2) Condition resulting from a foreign object becoming lodged in some portion of a patient's body.

Impaled Object — Object that has caused a puncture wound and remains embedded in the wound.

Impeachment — Process of showing the judge or jury that a witness has changed his or her opinion or prior testimony.

Impeller — Vaned, circulating member of the centrifugal pump that transmits motion to the water. See Centrifugal Pump, Multistage Centrifugal Pump, Self-Priming Centrifugal Pump, and Single-Stage Centrifugal Pump,.

Impeller Eye — Intake orifice at the center of a centrifugal pump impeller.

Impingement — See Flame Impingement.

Impinging Stream Nozzle — Nozzle that drives several jets of water together at a set angle in order to break water into finely divided particles.

Implosion — Rapid inward collapsing of the walls of a vessel or structure because the walls are unable to sustain a vacuum.

Impounded Water Supply — Generally used to describe an open, standing, man-made reservoir, but can be used to describe any type of standing, static water supply.

Improvised Explosive Device (IED) — Device that is categorized by its container and the way it is initiated; usually homemade, constructed for a specific target, and contained in almost anything. Not deployed in a conventional military fashion. See Explosive (1) and (2).

Improvised Nuclear Device (IND) — (1) Device that results in the formation of a nuclear-yield reaction (nuclear blast); low-yield device is called a *mininuke*. See Radiation (2) and Suitcase Bomb. (2) Illicit nuclear weapon bought, stolen, or otherwise originating from a nuclear state; or a weapon fabricated by a terrorist group from illegally obtained fissile nuclear weapons material that produces a nuclear explosion.

IMS — See Incident Management System.

In Service — Operational and available for an assignment.

Inappropriate Response — Reaction to a decision-making problem in which the person in charge and subordinates might try to hide their fear or revulsion by joking, getting angry, or rationalizing the problem away.

Inboard/Outboard — Location in relation to the centerline of the fuselage; for example, inboard engines are those closest to the fuselage, whereas outboard engines are those farthest away.

Incapacitant — Chemical agent that is temporarily disabling for several hours or days after exposure. See Riot Control Agent.

Incendiarism — Deliberate setting of a fire or fires by a human being.

Incendiary — (1) An incendiary agent, such as a bomb. (2) A fire deliberately set under circumstances in which the responsible party knows it should not be ignited. (3) Relating to or involving the deliberate burning of property.

Incendiary Device — (1) Contrivance designed and used to start a fire. (2) Any mechanical, electrical, or chemical device used intentionally to initiate combustion and start a fire. *Also known as* Explosive Device. See Incendiary Thermal Effect.

Incendiary Fire Cause — Classification referring to a fire deliberately set under circumstances in which the responsible party knows that the fire should not be ignited.

Incendiary Thermal Effect — (1) Thermal heat energy resulting from the fireball created by the burning of combustible gases or flammable vapors and ambient air at very high temperatures during an explosion. (2) Description of the brief but intense heat released during an explosion. In detonations, this heat is unlikely to ignite secondary fires; in deflagrations, this heat is more likely to ignite secondary fires.

Inches of Mercury — Scale used in measuring negative pressure; used to measure barometric pressure.

Incidence Reporting System — Database used to track reported incidents of juvenile firesetting that will provide valuable information about the entire problem of juvenile firesetting in a community. An incidence reporting system specifically for juvenile-set fires will provide information about the nature of the problem in the community.

Incident — Emergency or non-emergency situation or occurrence (either human-caused or natural phenomenon) that requires action by emergency services personnel to prevent or minimize loss of life or damage to property and/or natural resources.

Incident Action Plan (IAP) — Written or unwritten plan for the disposition of an incident; contains the overall strategic goals, tactical objectives, and support requirements for a given operational period during an incident. All incidents require an action plan. On relatively small incidents, the IAP is usually not in writing; on larger, more complex incidents, a written IAP is created for each operational period and disseminated to All units assigned to the incident. When written, the plan may have a number of forms as attachments. *Also known as* Building Emergency Action Plan.

Incident Base — Location at the incident where the primary logistics functions are coordinated and administered. Incident name or other designator is added to the term "base." The incident command post may be co-located with the base. There is only one base per incident. *Formerly known as* Fire Camp.

Incident Command Post (ICP) — Location at which the incident commander and command staff direct, order, and control resources at an incident; may be co-located with the incident base.

Incident Command System (ICS) — Standardized approach to incident management that facilitates interaction between cooperating agencies; adaptable to incidents of any size or type.

Incident Commander (IC) — Person in charge of the incident command system and responsible for the management of all incident operations during an emergency.

Incident Investigation — Act of investigating or gathering data to determine the factors that contributed to a fatality, injury, or property loss, or to determine fire cause and origin.

Incident Management System (IMS) — (1) System described in NFPA® 1561, *Standard on Emergency Services Incident Management System*, that defines the roles, responsibilities, and standard operating procedures used to manage emergency operations. Such systems may also be referred to as Incident Command Systems (ICS). (2) Management system developed by the National Fire Service Incident Management System Consortium, combining pre-existing command systems into one.

Incident Management Team (IMT) — The Incident Commander and appropriate Command and General staff personnel assigned to an incident. IMTs are generally grouped in five types: Types I and II are national teams, Type III are state and regional, Type IV are large-department specific, and Type V are for smaller jurisdictions.

Incident Safety Officer (ISO) — Member of the command staff responsible for monitoring and assessing safety hazards and unsafe conditions during an incident, and developing measures for ensuring personnel safety. The ISO is responsible for the enforcement of all mandated safety laws and regulations and departmental safety-related standard operating procedures. On very small incidents, the incident commander may act as the ISO.

Incident Safety Plan — Written document at complex incidents which outlines hazards identified during pre-incident surveys and size-up. It also defines planned mitigation strategies for those hazards.

Incidental Release — Spill or release of a hazardous material where the substance can be absorbed, neutralized, or otherwise controlled at the time of release by employees in the immediate release area, or by maintenance personnel who are not considered to be emergency responders.

Incident-Related Stress — Physical and psychological stress related to emergency response; could be associated with especially traumatic events involving fatality or serious injury; may also stem from the normal demands of emergency response over time.

Incipient Fire Fighting — Activities involved in fighting incipient phase fires inside or outside of buildings or other enclosed structures.

Incipient Phase — First phase of the burning process in a confined space, in which the substance being oxidized is producing some heat, but the heat has not spread to other substances nearby. During this phase, the oxygen content of the air has not been significantly reduced.

Incipient Phase Fire — Fire that is in the initial or beginning stage and that can be controlled or extinguished by portable fire extinguishers or small hoselines, without the need to wear protective clothing or breathing apparatus or to take evasive action such as crawling to avoid smoke.

Incipient Stage — First stage of the burning process in a compartment in which the substance being oxidized is producing some heat, but the heat has not spread to other substances nearby. During this phase, the oxygen content of the air has not been significantly reduced and the temperature within the compartment is not significantly higher than ambient temperature.

Inclined Plane — A simple machine that uses a slanted surface to raise objects. Augers are related machines.

Inclinometer — Instrument that measures the angle at which a vessel is leaning to one side or the other.

Incomplete Combustion — Result of inefficient combustion of a fuel; the less efficient the combustion, the more products of combustion are produced rather than burned during the combustion process.

Increaser — Adapter used to attach a larger hoseline to a smaller one; has female threads on the smaller side and male threads on the larger side.

Incrustation — Deposit on the inner wall of a water pipe creating additional friction and loss of pressure.

Incumbent Physical Ability Test (IPAT) — Physical fitness test developed within an individual jurisdiction to assess fitness-for-duty of potential firefighter candidates.

Incursion — Any occurrence in the airport runway environment involving an aircraft, vehicle, person, or object on the ground that creates a collision hazard or results in a loss of required separation with an aircraft taking off, intending to take off, landing, or intending to land.

IND — *See* Improvised Nuclear Device.

Indemnify — One party agreeing to compensate another party for losses or damages that are incurred if specific actions or events occur.

Independent Learning — *See* Self-Directed Learning.

Index Gas — Commonly encountered gas, such as carbon monoxide in fires, whose concentration can be measured. In the absence of devices capable of measuring the concentrations of other gases present, the CO measurement may be assumed to indicate their concentrations as well.

Indicating Valve — Water main valve that visually shows the open or closed status of the valve.

Indicator — Visual remains at a fire scene revealing the fire's progress and action.

Indicator Action — Part of a behavioral objective that tells how a student will show a desired behavior so that it can be observed and measured.

Indictable Offense — *See* Felony.

Indictment — Formal written accusation charging the defendant with a crime.

Indirect Attack — (1) In structural fire fighting, a form of fire attack that involves directing fire streams toward the ceiling of a compartment in order to generate a large amount of steam in order to cool the compartment. Converting the water to steam displaces oxygen, absorbs the heat of the fire, and cools the hot gas layer sufficiently for firefighters to safely enter and make a direct attack on the fire. *See* Attack Methods (1). (2) In wildland fire fighting, a method of controlling a fire in which a control line is constructed or located some distance from the edge of the main fire, and the fuel between the two points is burned. *See* Attack Methods (2).

Indirect Loss — Loss indirectly associated with a fire.

Individual Container — Product container used to transport materials in small quantities; includes bags, boxes, and drums. *See* Packaging (1).

Individual Emergency Conditions Breathing — Procedures or techniques performed by an individual during emergencies where SCBA malfunctions, remaining air supply is inadequate for escape, or air supply is depleted.

Individualized Instruction — Adapting teaching methods to suit individual students' specific learning styles, so that students will be better able to achieve lesson objectives.

Induction — *See* Eduction.

Industrial — Occupancy classification whose primary objective is the manufacturing or distribution of products. *See* Occupancy Classification.

Industrial Bulk Container (IBC) — Large-capacity bulk storage container used for foam concentrate, usually in quantities of 250 to 450 gallons (946 L to 1 703 L).

Industrial Consumption — Water consumed from the water supply system by industrial facilities.

Industrial Fire Brigade — Team of employees organized within a private company, industrial facility, or plant who are assigned to respond to fires and emergencies on that property.

Industrial Fire Department — Full-time emergency response organization providing fire suppression, rescue, and related activities at a commercial, institutional, or industrial facility or facilities under the same ownership and management. While the industrial fire department is generally trained and equipped for specialized operations based on site-specific hazards present at the facility, it may also respond off-site under a mutual aid agreement.

Industrial Firefighters — Full-time emergency response fire fighters that provide fire suppression, rescue, and related activities at a commercial, institutional, or industrial facility or facilities under the same ownership and management.

Industrial Hose — Fire hose, usually of lighter construction than fire service hose, used by industrial fire brigades.

Industrial Occupancy — Industrial, commercial, mercantile, warehouse, utility power station, institutional or similar facilities.

Industrial Packaging — Container used to ship radioactive materials that present limited hazard to the public and the environment, such as smoke detectors. *See* Excepted Packaging , Packaging (1), Strong, Tight Container, Type A Packaging, and Type B Packaging.

Industry Standard — Set of published procedures and criteria that peer, professional, or accrediting organizations recognize as acceptable practice.

Inert Gas — Gas that does *not* normally react chemically with another substance or material; any one of six gases: helium, neon, argon, krypton, xenon, and radon. *See* Expellant Gas and Simple Asphyxiant.

Inertia — Tendency of a body to remain in motion or at rest until it is acted upon by force.

Inertia Light — Light mounted in the aircraft structure so that a sharp deceleration, such as a crash situation, will activate the light. It can also be turned on manually and removed from the mounting to be used as a portable flashlight.

Inerting — Introducing a nonflammable gas (such as nitrogen or carbon dioxide) to a flammable atmosphere in order to remove the oxygen and prevent an explosion.

Infection (Exposure) Control Officer — Designated staff member who supervises and reviews the infection control plan within a fire service organization; may or may not be among the duties of the health safety officer (HSO).

Infection Control Plan — Policies and procedures managed as part of an exposure control program to protect members from contracting infections in the workplace; includes training on the plan and supervision of the plan.

Infectious — Transmittable; able to infect people.

Infectious Agent — Biological agent that causes disease or illness to its host. *See* Biological Attack.

Inference — Conclusion that is derived from a set of premises.

Inference Development — Development of a meaningful hypothesis, conclusion, prediction, or estimate based on the data available and the knowledge, experience, and expertise of the investigator.

Inferior — Near the feet; below.

In-Flight Emergency — Fire or other emergency that occurs when an aircraft is in-flight; includes hydraulic failure, engine failure, landing-gear malfunction, and other system malfunctions.

Inflow — Flow of grain toward the vortex of the discharge funnel in a grain bin.

Informal Proposal — In-person request for funding.

Information Officer (IO) — *See* Public Information Officer.

Information Presentation — Lesson plan format for a presentation or delivery that covers theory and technical knowledge, such as the facts of fire growth and spread, and provides the background information (cognitive domain) that is often essential to performance skills development (psychomotor domain). *Also known as* Technical Lesson.

Information Sheet — Instructional fact sheet or type of handout used to present ideas or information that is not in printed form or otherwise available to the student to the learner. An information sheet provides additional background information on a topic, which supplements what is provided in the text or other course resources.

Informational Materials — Fire and life safety teaching materials that suggest an action or provide facts and figures. *Also known as* Promotional Materials.

Infrared (IR) Wave Spectrum — Wavelengths longer than visible light, lying at the red end of the spectrum. *Also known as* Infrared.

Infrared Analyzer — Instrument used to monitor gas and vapor exposures by measuring the infrared energy absorbed by the contaminant.

Infrared Radiation — Radiation with a wavelength outside the visible spectrum at the red end of the spectrum. Thermal radiation from free-burning fires is an example of infrared radiation.

Infrared Scanner — Device that detects radiant heat emitted by concealed materials by converting infrared energy to an electrical signal; used primarily by the fire service to detect hidden fires.

Infrared Thermometer — Non-contact measuring device that detects the infrared energy emitted by materials and converts the energy factor into a temperature reading. *Also known as* Temperature Gun.

Infrared Victim Locating Device — Search device that detects heat and may be useful in locating victims.

Infrastructure — Public services of a community that have a direct effect on the quality of life; includes communication technologies such as phone lines, vital services such as water supplies, and transportation systems such as airports, highways, and waterways. *See* Critical Infrastructure.

Ingestion — Taking in food or other substances through the mouth. *See* Routes of Entry.

Inhalation — Taking in materials by breathing through the nose or mouth. *See* Routes of Entry.

Inhalation Injuries — Injuries as a result of the patient inhaling products other than the normal products of respiration; dust and hazardous atmospheres are the primary concerns in structural collapse situations.

Inhalation Tube — *See* Low-Pressure Hose.

Inhalator — Mechanical device for administering breathing oxygen to an individual who is breathing.

Inhibitor — Material that is added to products that easily polymerize in order to control or prevent an undesired reaction. *Also known as* Stabilizer (2). *See* Polymerization.

Initial Action — *See* Initial Attack.

Initial Attack — Control efforts taken by the resources that are the first to arrive at an incident. This includes hoselines employed to prevent further extension of fire and to safeguard life and property while additional lines are being laid and other forces put in motion. *Also known as* Initial Action.

Initial Attack Apparatus — Fire apparatus whose primary purpose is to initiate a fire attack on structural and wildland fires and support associated fire department actions. *Also known as* Midi-pumper or Mini-pumper.

Initial Attack Incident — Incident that can be handled by the resources in the first-alarm assignment.

Initial Isolation Distance — Distance within which all persons are considered for evacuation in all directions from a hazardous materials incident. *See* Initial Isolation Zone, Isolation Perimeter, Protective Action Distance, and Protective Action Zone.

Initial Isolation Zone — Circular zone, with a radius equivalent to the initial isolation distance, within which persons may be exposed to dangerous concentrations upwind of the source and may be exposed to life-threatening concentrations downwind of the source. *See* Hazard-Control Zones, Initial Isolation Distance, and Isolation Perimeter.

Initiating Device — Alarm system component that transmits a signal when a change occurs; change may be the result of an action such as the activation of a manual fire alarm box, the presence of products of combustion in the atmosphere, or the automatic activation of a supervisory switch. *Also known as* Alarm-Initiating Device. *See* Fire Alarm System, Fire Detection System, and Signaling Device.

Initiators — Cylinder-shaped explosive or gas pressure devices used to create gas or mechanical pressure in order to activate another device; usually found in the seat and canopy ejection mechanism of jet military fighter aircraft.

Injection — (1) Method of proportioning foam that uses an external pump or head pressure to force foam concentrate into the fire stream at the correct ratio for the flow desired. *See* Proportioning. (2) Process of taking in materials through a puncture or break in the skin. *See* Routes of Entry.

Injector Lines — Small tubes or hoselines that inject fuel into the combustion chamber of an engine at high pressures. These pressures may be in excess of 1,200 psi (8 400 kPa).

Injury in America — Resource book for fire and life safety educators that pinpoints the effects of injury and shows how fire and burns fit into the larger picture of injuries.

Injury/Loss Statistics — Facts and figures that reflect the effects of change, are used for long-term evaluation of overall programs, and are the most reliable indicator of the success of a program.

In-Kind Contribution — Donations of services, time, or products; do not involve money.

In-Line Eductor — Eductor that is placed along the length of a hoseline.

In-Line Proportioner — Type of foam delivery device that is located in the water supply line near the nozzle. The foam concentrate is drawn into the water line using the Venturi method. *See* Foam Proportioning, Foam Proportioner, and Venturi Principle.

In-Line Relay Valve — Valve placed along the length of a supply hose that permits a pumper to connect to the valve to boost pressure in the hose.

Inn — *See* Hotel.

Inpatient Mental Health Treatment — Hospitalization in a facility where 24-hour supervised care is present. This type of center offers treatment in cases where a child is in crisis and possibly a danger to him/herself or others.

Insecticides — Chemicals designed to control or eliminate certain kinds of insects.

In-Service Training — Formal or informal training received while on the job; in the fire service, this training generally occurs in-house at the station rather than at formal training facilities.

Inside Hand or Foot — Hand or foot closest to the ladder, or closest to the other member of a two-firefighter team.

Inside Ladder Width — Distance between the inside edge of one beam and the inside edge of the opposite beam.

Insoluble — Incapable of being dissolved in a liquid (usually water). *See* Emulsion, Immiscible, Miscibility, and Soluble.

Inspection — Formal examination of an occupancy and its associated uses or processes to determine its compliance with the fire and life safety codes and standards. *See* Building Code, Code, and NFPA® 101, *Life Safety Code®*.

Inspection Holes — Small openings created by the rescuers in walls, roofs, floors, or other collapse debris; used to check for the presence of victims that may be in close proximity to where the breaching operations are progressing.

Inspector — Person who is trained and certified to perform fire prevention and life safety inspections of all types of new construction and existing occupancies. *Also known as* Code Enforcement Officer or Fire and Life Safety Inspector.

Institutionalized Behaviors — Collective support shown for a project by an organization. This includes the investment of time, people, money, and equipment to support the project.

Instruction Order — Organization of jobs or ideas according to learning difficulty, so that learning proceeds from the simple to the complex.

Instructional Design — *See* Curriculum Development.

Instructional Development Process — Process of designing classroom instruction that consists of three major components: analysis, design, and evaluation.

Instructional Materials — Materials that an instructor may use to ensure and/or enhance a good learning experience for students; includes lesson plans, computer-generated slide presentations, lesson outlines, and student worksheets.

Instructional Model — Series of steps that guide development of a program of instruction. It includes the steps of performing a needs analysis, planning the program, developing objectives, completing a task analysis, designing a lesson plan, and creating evaluation instruments.

Instructional Objective — *See* Behavioral Objective and Educational Objective.

Instructor — Individual deemed qualified by the authority having jurisdiction to deliver instruction and training in fire and emergency services; charged with the responsibility to conduct the class, direct the instructional process, teach skills, impart new information, lead discussions, and cause learning to take place.

Instructor Information — Component of the lesson plan that lists lesson resources such as personnel, texts, references, sources, instructional methods, learning activities, training locations, etc. *See* Lesson Plan.

Instrument Flight Rules (IFR) — Regulations governing the operation of an aircraft in weather conditions with visibility below the minimum required for flight under visual flight rules.

Instrument Landing — Landing an aircraft by relying only on instrument data.

Instrument Landing System (ILS) — Electronic navigation system that allows aircraft to approach and land during inclement weather conditions.

Instrument Reaction Time — Elapsed time between the movement of an air sample into a monitoring/detection device and the reading provided to the user. Also known as Response Time.

Insulating Glass — Two panes of glass separated by an air space and sealed around the edge.

Insulators — Materials with atomic structures that do not allow the easy movement of electrons; opposite of *conductors*.

Insurance Institute for Highway Safety (IIHS) — Independent research organization funded by a host of well-known insurance companies; focuses on crash avoidance and the crashworthiness of vehicles.

Insurance Services Office (ISO) — Private national insurance organization that evaluates and rates fire defense for all communities through the fire-suppression rating schedule. Also serves as an advisory organization to other property-liability insurance companies. *Also known as* Rating Bureau.

In-Swinging Door — Door that swings away from someone who stands on the outside of the opening.

Intake — (1) Inlet for water into the fire pump. (2) Process of collecting the comprehensive background information from the juvenile's family or caregiver regarding the incident(s) that brought the juvenile to the program.

Intake Area — Area in front of and to the side of a jet engine that might be unsafe for personnel.

Intake Form — Form used to gather information about the firesetter and the firesetting incident.

Intake Hose — Hose used to connect a fire department pumper or a portable pump to a nearby water source; may be soft sleeve or hard suction hose.

Intake Pressure — Pressure coming into the fire pump.

Intake Relief Valve — Valve designed to prevent damage to a pump from water hammer or any sudden pressure surge.

Intake Screen — Screen used to prevent foreign objects from entering a pump.

Integral Construction — *See* Unibody Construction.

Integral Frame — *See* Unitized Body.

Integrated Prevention Interventions — Process of combining education, technology, codes, standards, supporting incentives, and emergency response to address community risk.

Intelligence — (1) Information concerning incident management, operational security, an enemy, or an area. (2) Information about criminal or terrorist activities gathered from multiple agencies and shared to aid with investigations.

Intelligence Section — Section within an incident management system that may be activated when an incident is heavily influenced by intelligence factors, or when there is a need to manage and/or analyze a large volume of classified or highly sensitive intelligence or information. Particularly relevant to a terrorism incident for which intelligence plays a crucial role. *See* Incident Management System (IMS).

Intensity — *See* Fireline Intensity.

Intercostal Muscles — Muscles between the ribs.

Interface — Area between the fuel-rich area and the air-rich area where the two are mixing.

Interior — Internal section of a vehicle; composed of the passenger compartment and possibly a storage compartment.

Interior Access Vehicle — Fire apparatus designed to provide a raised platform for aircraft fire fighting operations that will elevate firefighters to an even level with the aircraft compartment.

Interior Exposure — Areas of a fire building that are not involved in fire but that are connected to the fire area in such a manner that may facilitate fire spread through any available openings.

Interior Finish — Exposed interior surfaces of buildings, including fixed or movable walls and partitions, columns, and ceilings. Commonly refers to finish on walls and ceilings, but not floor coverings.

Interior Structural Fire Fighting — Fire suppression and/or rescue activities within buildings or enclosed structures involving a fire that is beyond the incipient phase.

Interlocking Deadbolt — *See* Jimmy-Resistant Lock.

Intermediate Bulk Container (IBC) — Rigid (RIBC) or flexible (FIBC) portable packaging, other than a cylinder or portable tank, that is designed for mechanical handling with a maximum capacity of not more than three 3 cubic meters (3,000 L, 793 gal, or 106 ft^3) and a minimum capacity of not less than 0.45 cubic meters (450 L, 119 gal, or 15.9 ft^3) or a maximum net mass of not less than 400 kilograms (882 lbs). *See* Packaging (1).

Intermodal Container — Freight containers designed and constructed to be used interchangeably in two or more modes of transport. *Also known as* Intermodal Tank Container. *See* Container (2), Container Vessel, Intermodal Reporting Marks, and Refrigerated Intermodal Container.

Intermodal Reporting Marks — Series of letters and numbers stenciled on the sides of intermodal tanks that may be used to identify and verify the contents of the tank or container. *See* Intermodal Container and Railcar Initials and Numbers.

Intermodal Tank Container — *See* Intermodal Container.

Internal Customers — Employees and membership of the organization.

Internal Floating Roof Tank — Fixed-site vertical storage tank that combines both the floating roof and the closed roof design. *Also known as* Covered Floating Roof Tank. *See* Atmospheric Storage Tank, Cone Roof Tank, External Floating Roof Tank, and Floating Roof Storage Tank.

Internal Resources — Those resources immediately available for fire department use without special arrangements; includes equipment, personnel, capabilities, and supplies.

Internal Respiration — Exchange of oxygen and carbon dioxide in the bloodstream at the cellular level.

International Air Transport Association (IATA) —

Global trade organization that represents approximately 230 airlines. Its mission is to represent, lead, and serve the airline industry around the world.

International Association of Arson Investigators (IAAI) — Professional organization for fire investigators that offers training, certification, and opportunities to serve on committees to set standards for the fire investigation profession.

International Association of Fire Chiefs (IAFC) — Professional organization that provides leadership to career and volunteer chiefs, chief fire officers, and managers of emergency service organizations throughout the international community through vision, information, education, representation, and services to enhance their professionalism and capabilities.

International Association of Fire Fighters (IAFF) — Professional organization that represents career fire fighters and paramedics in labor/management relations through local labor unions. The association also gathers data on on-duty firefighter deaths and injuries.

International Aviation Fire Protection Association (IAFPA) — Professional and fraternal association of international airport, municipal, and military fire and emergency services professionals. The IAFPA was formed in 2000 by a group of airport/municipal fire service professionals and industry specialists.

International Building Code® (IBC®) — Code that is dedicated to providing safety regulations for life safety, structural, and fire protection issues that occur throughout the life of a building. *See* Building Code, Code, International Fire Code® (IFC®), and National Electrical Code® (NEC®).

International Civil Aviation Organization (ICAO) — United Nations agency responsible for developing and adopting standards and approved practices for international civil aviation that relate to aircraft rescue and fire fighting (ARFF), air navigation, preventing unlawful interference of air traffic, and facilitating border-crossing procedures.

International Code Council (ICC) — Organization that develops the *International Building Code® (IBC®)* and the *International Fire Code® (IFC®)*, for city and state adoption. Was formed by the merger of the Building Officials and Code Administrators (BOCA) International, Inc., the International Conference of Building Officials (ICBO), and the Southern Building Code Congress International (SBCCI). *See* Building Officials and Code Administrators (BOCA), *International Building Code® (IBC®)*, International Conference of Building Officials (ICBO), *International Fire Code® (IFC®)*, and Southern Building Code Congress International (SBCCI).

International Conference of Building Officials (ICBO) — Organization that provides the *Uniform Building Code (UBC)* for city and state adoption, and produces fire codes in conjunction with the Western Fire Chiefs Association. The ICBO joined with the Building Officials and Code Administrators International, Inc. (BOCA) and the Southern Building Code Congress International (SBCCI) to form the International Code Council (ICC).

International Convention for The Safety of Life at Sea (SOLAS) — International convention dealing with maritime safety; covers a wide range of measures designed to improve the safety of shipping. The first version was adopted in 1914. Since then, four more versions have been adopted. The present version was adopted in 1974 and became effective in 1980. The Protocol of 1978 and Amendments of 1990 and 1991 have since been added.

International Fire Code® (IFC®) — Code that is dedicated to ensuring public safety through the implementation of a unified set of fire protection measures and restrictions. *See* Code, *International Building Code® (IBC®)*, and *National Electrical Code® (NEC®)*.

International Fire Service Accreditation Congress (IFSAC) — Peer driven, self governing system that accredits both public fire service certification programs and higher education fire-related degree programs.

International Fire Service Training Association (IFSTA) — Nonprofit educational alliance organized to develop training materials for the fire and emergency services.

International Maritime Organization (IMO) — Specialized agency of the United Nations devoted to maritime affairs. It first met in 1959. Over the years, IMO has developed and promoted the adoption of more than 30 conventions and protocols, as well as 700 codes and recommendations dealing with maritime safety. Its main purpose is to ensure safer shipping and cleaner oceans.

International Mechanical Code (IMC) — Code that establishes minimum safeguards for heating, ventilating, and air conditioning (HVAC) systems and is published by the International Code Council (ICC). *See* Code and International Code Council (ICC).

International Plumbing Code (IPC) — Code that establishes minimum requirements for plumbing systems and is published by the International Code Council (ICC). *See* Code and International Code Council (ICC).

International Shore Connection (ISC) — Pipe flange with a standard size and bolt pattern allowing land-based fire department personnel to charge and supply a vessel's fire main.

International Society of Fire Service Instructors (ISFSI) — Professional organization for fire service instructors that provides professional development opportunities for its membership.

International Standards Organization (ISO) — World's largest developer and publisher of international standards. In relation to photography, sets the standards for film speeds; these standards have carried over to digital cameras as the ISO number, which mimics the film speed settings in film cameras.

International System of Units — Modern metric system, based on units of ten.

Interoperability — Ability of two or more systems or components to exchange information and use the information that has been exchanged.

Interpersonal Skills — JFIS I interventionist or other program personnel should be trained to interact well with children and to communicate effectively with adults.

Interrogation — Formal line of questioning of an individual who is suspected of committing a crime or who may be reluctant to provide answers to the investigator's questions.

Interrupt Rating — Highest current at rated voltage that a device is intended to interrupt under standard test conditions.

Interstate Commerce Commission (ICC) — *See* Department of Transportation (DOT).

Interstitial Space — In building construction, refers to generally inaccessible spaces between layers of building materials. May be large enough to provide a potential space for fire to spread unseen to other parts of the building. *See* Attic.

Intervention — Formal response to firesetting behavior; may include education alone or be combined with referral to counseling or medical or social services. Juvenile justice sanctions are sometimes used (or required) as part of an intervention process.

Intervention Strategy — Action plan that describes how a risk-reduction initiative will be implemented and evaluated.

Interventionist — Person who is qualified to perform a specific task, such as a juvenile firesetter intervention specialist; works directly with juvenile firesetters and their families to prevent acts of recidivism. The term interventionist is used interchangeably with practitioner.

Interview — (1) Questioning an individual for the purpose of obtaining information related to an investigation. (2) Process of meeting with the juvenile firesetter and his or her family to determine the severity of the problem.

Intravenous Fluids — Fluids administered directly into a vein or veins to maintain a patient's medications or hydration.

Intravenous Line (IV) — Catheter placed in the patient's vein to allow fluid replacement and the administration of medications.

Intrinsically Safe ¾ Describes equipment that is approved for use in flammable atmospheres; must be incapable of releasing enough electrical energy to ignite the flammable atmosphere. *Formerly known as* Explosion Proof.

Intrinsically Safe Equipment — Equipment designed and approved for use in flammable atmospheres that is incapable of releasing sufficient electrical energy to cause the ignition of a flammable atmospheric mixture.

Intumescent Coating — Coating or paintlike product that expands when exposed to the heat of a fire; creates an insulating barrier that protects the material that is underneath.

Invasion of Privacy — Wrongful intrusion into a person's private activities by the government or other individuals.

Inventory — Detailed, written listing of equipment and materials on hand at a given time.

Inverse Square Law — Physical law that states that the amount of radiation present is inversely proportional to the square of the distance from the source of radiation.

Inversion — (1) Increase of temperature with height in the atmosphere. Vertical motion in the atmosphere is inhibited allowing for smoke buildup. A "normal" atmosphere has temperature decreasing with height. *See* Atmospheric Stability. (2) Atmospheric phenomenon that allows smoke to rise until its temperature equals the air temperature and then spreads laterally in a horizontal layer. *Also known as* Night Inversion.

Inverted-Cone Pattern — Fire pattern formed on vertical surfaces by small fires or low-heat-release-rate fires; the pattern appears as an inverted V on the vertical surface.

Inverter — Step-up transformer that converts a vehicle's 12- or 24-volt DC current into 110- or 220-volt AC current.

Investigation — Official inquiry into a fire's cause.

Investigative Fire Mode — Situation in which products of combustion or flames are not immediately visible to the first-arriving units, and firefighters must investigate to determine the cause of the alarm.

Involuntary Muscle — Muscle that acts without voluntary control.

Involved — Actual room, portion, or area of building involved in, or affected by fire.

Ion — Atom that has lost or gained an electron, thus giving it a positive or negative charge.

Ionization — (1) Process by which an object or substance gains or loses electrons, thus changing its electrical charge. (2) Process in which a charged portion of a molecule (usually an electron) is given enough energy to break away from the atom; results in the formation of two charged particles or ions: (a) a molecule with a net positive charge, and (b) a free electron with a negative charge. (3) Physical process of converting an atom or molecule into an ion by adding or removing charged particles, such as electrons or other ions.

Ionization Detector — Type of smoke detector that uses a small amount of radioactive material to make the air within a sensing chamber conduct electricity. *Also known as* Ionization Smoke Detector.

Ionization Potential — Energy required to free an electron from its atom or molecule.

Ionization Smoke Alarm — Type of smoke detector that uses a small amount of radioactive material to ionize air particles as they enter a sensing chamber; the ionized air particles then combine with products of combustion that enter the chamber, reducing the number of ionized particles. When the number of ionized particles falls below a given threshold, an alarm is initiated. *Also known as* Ionization Detector.

Ionization Smoke Detector — *See* Ionization Detector.

Ionizing Radiation — Radiation that causes a chemical change in atoms by removing their electrons. *See* Nonionizing Radiation and Radiation (2).

IPC — *See* International Plumbing Code.

Ipecac Syrup — Medication used to induce vomiting.

IR — *See* Infrared Wave Spectrum.

Iris — Colored portion of the eye that surrounds the pupil.

Ironing — Flattening deformation of aerial device base rails caused by the pressure exerted when the device is extended and retracted.

Irritant — Liquid or solid that, upon contact with fire or exposure to air, gives off dangerous or intensely irritating fumes. *Also known as* Irritating Material.

Irritating Agent — *See* Riot Control Agent.

ISC — *See* International Shore Connection.

ISFSI — *See* International Society of Fire Service Instructors.

Island — Unburned area within a fire perimeter.

ISO — (1) *See* Insurance Services Office. (2) *See* International Standards Organization.

ISO Setting — On a DSLR camera, changes the camera's sensitivity to light; mimics different film speeds established on film cameras.

Isoamyl Acetate — Banana oil; used for an odor test of facepiece fit.

Isolate — (1) To set apart. (2) Second of three steps (locate, isolate, mitigate) in one method of sizing up an emergency situation.

Isolated Flames — Flames in the hot gas layer that indicate the gas layer is within its flammable range and has begun to ignite; often observed immediately before a flashover. *See also* Rollover.

Isolation Area — Predetermined area designated for temporary parking of aircraft experiencing emergencies that may adversely affect the safety of people and property.

Isolation Perimeter — Outer boundary of an incident that is controlled to prevent entrance by the public or unauthorized persons. *See* Initial Isolation Distance and Initial Isolation Zone.

Isotope — Atoms of a chemical element with the usual number of protons in the nucleus, but an unusual number of neutrons; has the same atomic number but a different atomic mass from normal chemical elements. *See* Radionuclide.

J

J — *See* Joules.

JAA — *See* Joint Aviation Authority.

Jack — Portable device used to lift heavy objects with force applied by a lever, screw, or hydraulic press.

Jack Pads — *See* Stabilizer Pad.

Jack Plates — *See* Stabilizer Pad.

Jacket — Metal cover that is used to protect the insulation of a tank.

Jackknife — Condition of truck tractor/semitrailer combination when their relative positions to each other form an angle of 90 degrees or less about the trailer kingpin, such as turning the tractor portion of a tractor-tiller aerial apparatus at an angle from the trailer to increase stability when the aerial device is being used.

Jacob's Ladder — Flexible ladder made of rope or chain but having solid rungs (wood or iron); used for boarding a vessel or scaling the sides of a vessel.

Jacobs Engine Brake® — Device that mounts on, or within, the engine overhead.

Jalousie Window — Window consisting of narrow, frameless glass panes set in metal brackets at each end that allow a limited amount of axial rotation for ventilation.

Jamb — *See* Frame.

Jargon — The specialized or technical language of a trade, profession, or similar group.

JATO — *See* Jet-Assisted Takeoff.

Jaws — *See* Powered Hydraulic Spreaders.

Jet — Term used in England for a fire stream.

Jet Pump — Water-operated pump that creates a suction by using the Venturi Principle.

Jet Ratio Controller — Type of foam eductor that is used to supply self-educting master stream nozzles; may be located at distances up to 3,000 feet (914 m) from the nozzle.

Jet Siphon — Section of pipe or hard suction hose with a 1-inch (25 mm) discharge line inside that bolsters the flow of water through the tube. The jet siphon is used between portable tanks to maintain a maximum amount of water in the tank from which the pumper is drafting.

Jet-Assisted Takeoff (JATO) — Rocket or auxiliary jet used to augment normal aircraft thrust for takeoffs.

Jettison — To selectively discard or throw away objects or items in order to lighten a vessel's load during an emergency; for example, external fuel tanks or canopies from an aircraft or cargo from ship.

Jetway — Enclosed ramp between a terminal and an aircraft for loading and unloading passengers.

Jib — Lever used with a block and tackle to lift or lower.

JIC — *See* Joint Information Center.

Jihad — Jihad means to *strive* or *struggle* in Arabic. It encompasses a set of actions that is designed to make a person more pious, to seek perfection in the way of Allah, to expand Islam throughout the world by good example, and to defend Muslims against aggressors. Jihad may be directed against the temptations of evil, aspects of one's own self, or against a visible enemy. In the context of terrorism, radical Islamists use an extreme interpretation of jihad to justify violence against perceived enemies of Islam.

Jimmy — (1) To pry apart; usually to separate the door from its frame, allowing the latch or bolt to clear its strike. (2) A tool used to pry open locks.

Jimmy-Resistant Lock — Auxiliary lock having a bolt that interlocks with its strike, and therefore resists prying. *Also known as* Interlocking Deadbolt or Vertical Deadbolt.

JIS — *See* Joint Information System.

Job — Slang used in the Eastern U.S. fire service to describe a working fire.

Job Breakdown Sheet — Instructional sheet that lists a job and breaks it down into step-by-step procedures and required knowledge. It is designed to assist in teaching and learning a psychomotor objective. *See* Key Points.

Job Performance Requirement (JPR) — Statement that describes a specific job task, lists the items necessary to complete the task, and defines measurable or observable outcomes and evaluation areas for the specific task.

Job Safety Analysis — Method of analyzing occupational hazards and working toward their solution.

Job Task Analysis — Process of evaluating firefighter job tasks and determining the best medical, physical fitness, safety, and health requirements and programs to help ensure that firefighters can perform those tasks safely and without injury.

Jockey Pump — Small-capacity, high-pressure pump used to maintain constant pressures on the fire protection system. A jockey pump is often used to prevent the main pump from starting unnecessarily. *See* Pressure Maintenance Pump.

Joiner Construction — Bulkheads that subdivide the ship into compartments but do not contribute to the structural strength of the ship. *Also known as* Nonstructural Bulkheads.

Joint Aviation Authority (JAA) — Organization representing the civil aviation regulatory authorities of various European states; it is associated with the European Civil Aviation Conference (ECAC). The purpose of the JAA is to develop and implement common aviation standards and procedures.

Joint Information Center (JIC) — Facility established to coordinate all incident-related public information activities.

Joint Information System (JIS) — Integrates incident information and public affairs into a cohesive organization designed to provide consistent, coordinated, and timely information during incident operations.

Joint Photographic Experts Group (JPEG) — Group who created the photo compression standard commonly known as the JPEG (also known as .jpg). JPEG's are RAW digital images that have been compressed to reduce their size and make them more editable and useable in digital format.

Joint Service Lightweight Integrated Suit Technology (JSLIST) — Chemical-protective, universal, lightweight, two-piece, front-opening suit that can be worn as an overgarment or as a primary uniform over underwear. The JSLIST liner consists of a non-woven front laminated to activate carbon spheres and bonded to a knitted back that absorbs chemical agents.

Joists — Horizontal structural members used to support a ceiling or floor. Drywall materials are nailed or screwed to the ceiling joists, and the subfloor is nailed or screwed to the floor joists.

Joule (J) — Unit of work or energy in the International System of Units; the energy (or work) when unit force (1 newton) moves a body through a unit distance (1 meter). Takes the place of calorie for heat measurement (1 calorie = 4.19 J). *See* Calorie.   Joules are defined in terms of mechanical energy. It is equal to the energy expended in applying a force of one newton through a distance of one meter. However, it is more useful for firefighters to think about the energy required to increase temperature. 4.2 Joules are required to raise the temperature of one gram of water one degree Celsius.

Journal — Book in which all activities of a fire shift are recorded. *Also known as* Day Book, Log Book, or Record Book. *See* Log.

JPR — *See* Job Performance Requirement.

JSLIST — *See* Joint Service Lightweight Integrated Suit Technology.

Judge-Made Law — *See* Judiciary Law.

Judgment — One of three requirements of evaluation; the decision-making ability of the instructor to make comparisons, discernments, or conclusions about the instructional process and learner outcomes.

Judicial System — System of courts set up to interpret and administer laws and regulations.

Judiciary Law — Law established by judicial precedent and decisions. *Also known as* Judge-Made Law or Unwritten Law. *See* Law.

Jumar — Ascender for rope climbing.

Jump Seats — Seats on a fire apparatus that are behind the front seats; usually open and facing to the rear.

Junction Box — *See* Connection Box.

Jurisdiction — (1) Legal authority to operate or function. (2) Boundaries of a legally constituted entity. *See* Authority and Authority Having Jurisdiction (AHJ).

Jury Trial — Court proceeding in which a jury acts as the trier of fact.

Juvenile Firesetter — Person though the age of 18, or as defined by the authority having jurisdiction, who is involved in the act of firesetting.

Juvenile Firesetter Intervention Specialist I — Those who have demonstrated the ability to conduct an intake and interview with a firesetter and his or her family using prepared forms and guidelines and who, based on program policies and procedures, determine the need for referral for counseling, and implement educational intervention strategies to mitigate effects of firesetting behavior.

Juvenile Firesetter Intervention Specialist II — Those who have demonstrated the ability to manage juvenile firesetting intervention program activities and the activities of Juvenile Firesetter Intervention Specialist I.

K

Kalamein Door — *See* Metal-Clad Door.

Kasch Step Test — Medical test used to measure cardiovascular fitness.

Keel — Principal structural member of a vessel, running fore and aft extending from bow to stern. Forms the backbone of a vessel to which frames are attached; lowest member of a vessel framework. *See* Frames.

Kelly Day — Rotating off-duty shift in addition to the normal off-duty schedule of the firefighter.

Kelly Tool — Prying tool similar to a claw tool, but with an adze blade at one end and a forked blade at the other end.

Kendrick Extrication Device (KED)® — Device used to immobilize a patient's spine when the patient is in a sitting position and/or partially accessible.

Kerf Cut — Single cut the width of the saw blade made in a roof to check for fire extension.

Kernmantle Rope — Rope that consists of a protective shield (mantle) over the load-bearing core strands (kern).

Kevlar® — Trademarked name of a lightweight, very strong, para-aramid synthetic fiber; used in many products including bicycle tires, sails, body armor, and armor components for vehicles.

Key — Device that allows a person to lock and unlock a locking mechanism. When the key is inserted into the plug of a lock it causes internal pins or disks to align in a manner that allows the plug to turn within the lock cylinder.

Key Box — *See* Key Safe.

Key Plug — *See* Cylinder Plug.

Key Points — (1) Factors that condition or influence operations within an occupation; information that must be known to perform correctly the steps in a procedure. *See* Job Breakdown Sheet. (2) Important cognitive information on a skill sheet that students need to know in order to perform a task or operational step; generally appears on the right-hand side of a skills sheet.

Key Safe — Boxlike container that holds keys to the building, usually mounted on or in the front wall; requires a master key to open. *Also known as* Key Box or Knox Box (tradename).

Key Tool — Tool for manipulating an exposed lock mechanism so that the latch or deadbolt is retracted from its strike; used in conjunction with a K-Tool.

Key-In-Knob Lock — Lock in which the lock cylinder is within the knob.

Keystoning — Distortion of the projected transparency image that happens when the projector and screen are not perpendicular to each other.

Keyway — Opening in a cylinder plug that receives the key.

Kick Panels — Vertical panel walls of a vehicle that are enclosed by several structural members.

Kickback — A sudden forceful recoil or violent jerk caused by a chain saw binding in the material it is cutting.

Kickplate — *See* Footplate.

Kiln Dried — Term applied to lumber that has been dried by artificially controlled heat and humidity to a prescribed moisture content.

Kilopascal (kPa) — Metric unit of measure for pressure; 1 psi = 6.895 kPa, 1 kPa = 0.1450 psi.

Kilowatt (KW) — Measurement of rate of heat release measured in the number of Btu per second (equivalent to ten 100-watt light bulbs).

Kinematic Viscosity — Ratio of a fluid's absolute viscosity (lbf sec/ft^2) to its mass density (lbf sec^2/ft^4).

Kinematics — One branch of the study of dynamics that defines the motion of objects without addressing mass and force or the factors that lead to the motion. In terms of accidents, kinematics describe the effects of collisions on vehicles.

Kinematics of Injury — Types of injuries suffered by vehicle occupants that vary depending upon the type of collision incurred.

Kinetic Energy — Energy possessed by a moving object because of its motion.

Kingpin — Attaching pin on a semitrailer that connects with pivots within the lower coupler of a truck tractor or converter dolly while coupling the two units together.

Kink — Severe bend in a hoseline that increases friction loss and reduces the flow of water through the hose.

Kink Test — Test of hose under extreme conditions to ensure performance by folding the hose over on itself, securing it to maintain the kink, and pressurizing. Pressures used vary with the type of hose tested.

Kip — Unit of weight equal to 1,000 pounds; used to express deadweight load.

Kit — Collection of tools or equipment kept in one location for a specific purpose.

Knee Bolsters — Type of antisubmarine device. *See* Antisubmarine Device.

Knee-Foot Lock — Leg position with the knee against the front of a ladder beam and the instep of the foot hooked around the rear of the butt spur on the same beam; used to secure one beam of a ladder while operating the fly section.

Kneeling — Ability of some buses to lower the front end of the bus to curb level for ease of passenger boarding.

Knock Down — Reduction of most flame and heat generation on the more vigorously burning parts of a fire edge, using an extinguishing agent such as water, in order to bring the fire to an overhaul stage.

Knot — (1) Term used for tying a rope around itself. (2) International nautical unit of speed; 1 knot = 6,076 feet or 1 nautical mile per hour (1.15 miles or 1.85 kilometers per hour).

Knowledge Change — Increase in a learner's understanding of fire and life safety practices.

Known-To-Unknown — Method of sequencing instruction so that information begins with the familiar or known and progresses to the unfamiliar or unknown while making relationships that enable learners to become familiar with the unknown.

Knox Box — *See* Key Safe.

Knurled — Having a series of small ridges or beads, as on a metal surface, to aid in gripping.

kPa — *See* Kilopascal.

Kraft Paper — Strong brown paper made of sulfate pulp.

K-Tool — V-blade tool that is designed to pull lock cylinders from a door with only minimal damage to the door itself.

Kussmaul's Respiration — Deep, rapid respirations characteristic of hyperglycemia (or diabetic ketoacidosis) that result as the body tries to eliminate excess carbon dioxide.

KW — *See* Kilowatt.

L

Label — Four-inch-square diamond-shaped marker required by federal regulations on individual shipping containers that contain hazardous materials, and are smaller than 640 cubic feet (18 m³). *See* Marking, Package Markings, and Placard.

Labeled — Equipment, materials, or assemblies to which has been attached a label, symbol, or other identifying mark of a testing organization that indicates compliance with certain test performance standards.

Labeled Assembly — *See* Rated Assembly.

Laceration — Jagged tear or wound.

Lactic Acid — Hygroscopic organic acid normally present in tissue.

Ladder — (1) Any fire department ladder of varying length, type, or construction consisting of two rails or beams with steps or rungs spaced at intervals. Ladders are manufactured in a number of lengths and can be manually or power raised. (2) Any stairway or ladder (often nearly vertical) onboard a vessel.

Ladder Bed — Rack or racks in which ladders are carried on a ladder truck.

Ladder Belt — Belt with a hook that secures the firefighter to the ladder.

Ladder Carry — Any organized system for carrying ladders.

Ladder Company — Group of firefighters assigned to a fire department aerial apparatus equipped with a compliment of ladders; primarily responsible for search and rescue, ventilation, salvage and overhaul, forcible entry, and other fireground support functions. *Also known as* Truck Company.

Ladder Cribbing — Pieces of cribbing that have been attached to two lengths of webbing to form the appearance of a ladder.

Ladder Float — Inflated tire or inner tube fastened to a ladder and used to rescue persons from water or ice.

Ladder Fuels — Fuels that provide vertical continuity between strata, thereby allowing fire to carry from surface fuels into the crowns of trees or shrubs with relative ease. They help initiate and assure the continuation of crowning.

Ladder Gin Pole — Gin pole in which the load-supporting member is a straight ladder.

Ladder Locks — *See* Pawls.

Ladder Nesting — Positioning of different width ladders, one partially within another, for storage on apparatus.

Ladder Pipe — Master stream nozzle mounted on an aerial ladder.

Ladder Rig — Term for a simple 2:1 pulley system, used in a vertical configuration to raise and lower loads.

Ladder Spur — Spiked device that is attached to the foot of a ladder to provide good traction on soft ground.

Ladder Stop — Blocks that limit the travel of the fly sections on an extension ladder to prevent the sections from being separated.

Lading — Freight or cargo that makes up a shipment. *See* Bill of Lading and Shipping Papers.

Lag Screw — Large wood screw with a hexagonal or square head for turning with a wrench.

Lagging — Heavy sheathing used in underground work to withstand earth pressure.

Laid Rope — Rope constructed by twisting several groups of individual strands together.

Laissez-Faire Leadership — Style of leadership in which the leader shares responsibility with the group, often relying on other people for suggestions and delegating a limited amount of decision making.

Lake Test — Method of testing a salvage cover for leaks by forming a catchall to hold a small "lake" of water and observing for leakage on the underside.

Lamella Arch — Special type of arch constructed of short pieces of wood called lamellas.

Laminar Flow — Smooth, non-turbulent flow of a fluid that occurs at low velocities.

Laminate Beams — Structural members created from several layers of plywood or oriented strand board (OSB).

Laminated Glass — *See* Safety Glass.

Laminated Plastic — Sheet material made of lamination cloth or other fiber impregnated with plastic and brought to the desired thickness or shape with heat and/or pressure.

Lamination — (1) Bonding or impregnating superposed layers with resin and compressing under heat. (2) One of several layers of lumber making up a laminated beam.

Land Line — Term for a wire-connected telephone.

Landing — Horizontal platform where a flight of stairs begins or ends.

Landing Gears — *See* Supports.

Landing Roll — Distance from the point of touchdown to the point where the aircraft is brought to a stop or exits the runway.

Landing Site — Area in which a helicopter will land during air medical evacuations.

Lantern Roof — Roof style consisting of a high gabled roof with a vertical wall above a downward-pitched shed roof section on either side.

Lapping — Means by which fire spreads vertically from floor to floor in a multistory building. Fire issuing from a window laps up the outside of the building and enters the floor(s) above, usually through the windows. *Also known as* Autoexposure.

Lapse Rate — Change of an atmospheric variable (temperature unless specified otherwise) with height.

Large Diameter Hose (LDH) — Relay-supply hose of 3½ to 6 inches (90 mm to 150 mm) in diameter; used to move large volumes of water quickly with a minimum number of pumpers and personnel.

Large Handline — Fire hose/nozzle assembly capable of flowing up to 300 gpm (1 140 L/min).

Lash — To secure or tie anything down, or to something else, with rope or line.

Latch — Spring-loaded part of a locking mechanism that extends into a strike within the door frame.

Latch Bolt — Latch with a shim or plunger that causes the latch to operate in a manner similar to a deadbolt. The latch plunger prevents "loiding" of the latch.

Latent Heat of Vaporization — Quantity of heat absorbed by a substance at the point at which it changes from a liquid to a vapor.

Lateral — Toward the side of the human body.

Lath — (1) Closely spaced narrow strips of wood used to fasten covering material to a wall or ceiling. In older buildings, wood lath was used to support plaster finish on the walls and ceilings. (2) Used to hold salvage covers, sheeting, or tar paper in place when covering a building opening during overhaul.

Law — Rules of conduct that are adopted and enforced by an authority having jurisdiction that guide society's actions. There are three types of laws: legislative, administrative, and judiciary. *See* Administrative Law, Copyright Law, Judiciary Law, Legislative Law, Ordinance, and Statute.

Law of Association — Principle that learning comes easier when new information is related to similar things already known.

Law of Conservation of Mass — Theory that states that mass is neither created nor destroyed in any ordinary chemical reaction; mass that is lost is converted into energy in the form of heat and light.

Law of Effect — Notion that learning is more effective when a feeling of satisfaction, pleasantness, or reward accompanies or is a result of the learning process.

Law of Exercise — Idea that repetition is necessary for the proficient development of a mental or physical skill.

Law of Heat Flow — Natural law that specifies that heat tends to flow from hot substances to cold substances. This phenomenon is based on the supposition that one substance can absorb heat from another. *See* Conduction, Convection, and Radiation.

Law of Intensity — Premise that if the experience is real, there is more likely to be a change in behavior or learning.

Law of Readiness — Principle that a person learns when physically and mentally adjusted or ready to receive instruction.

Law of Recency — Principle that the more recently the reviews, warm-ups, and makeup exercises are practiced before using the skill, the more effective the performance will be.

Law of Specific Heat — (1) Measure of the heat-absorbing quality of a substance as measured in Btu's or kilojoules. (2) Relative quantity of heat required to raise the temperature of substances, or the quantity of heat that must be removed to cool a substance.

Lay — To lay out hose in a predetermined sequence for fire fighting.

Lay Testimony — *See* Fact Testimony.

Layering — Deposition of fire debris in identifiable layers, such as above or below a floor assembly, ceiling materials, or roof assembly

Layout — Distribution of hose at the scene of a fire.

LC_{50} — *See* Lethal Concentration, 50 Percent Kill.

LD_{50} — *See* Lethal Dose, 50 Percent Kill.

LDH — *See* Large Diameter Hose.

Leach — To pass out or through by percolation (gradual seepage).

Lead — Introductory section of a news story, typically the first few sentences.

Leader — Individual responsible for command of a crew, task force, strike team, or functional unit.

Leader's Guide — Publication of the National Safe Kids Campaign® that identifies and discusses seven fundamental steps in coalition building.

Leadership — Knack of getting other people to follow you and to do willingly the things that you want them to do.

Leading Block — Pulley or snatch block used to change the direction of the fall line in a block and tackle system. This does not affect the mechanical advantage of the system.

Leading/Trailing Edge Devices — Forward and rear edges of aircraft wings normally extended for takeoff and landings to provide additional lift at low speeds and to improve aircraft performance.

Leaf Spring Suspension — Type of suspension system consisting of several long, narrow, layers of elastic metal bracketed together.

Lean-to Collapse — Type of structural collapse where one end of a floor or roof section support fails while the other end remains secured to a wall. The floors and roof drop in large sections and form voids.

Learn Not To Burn® Curriculum — National Fire Protection Association® curriculum that establishes 22 key fire safety behaviors for school children, as well as three "local option" behaviors.

Learn Not To Burn®: The Pre-School Program — NFPA® curriculum that addresses fire and life safety issues for preschool children.

Learning — Relatively permanent change in behavior that results from learning new information, practicing skills, or developing attitudes following some form of instruction.

Learning Activity — Any component of a lesson designed to enhance knowledge, attitude, or skill development. Activities are used to facilitate achievement of learning objectives.

Learning Contract — Formal agreement between learner and instructor that establishes an amount of work that must be finished in order to successfully complete a course.

Learning Disability — Cognitive disorder that diminishes a person's capacity to interpret what he or she sees and hears, and/or to link information from different parts of the brain.

Learning Domain — Distinct sphere or area of knowledge, such as the affective, cognitive, and psychomotor domains. *See* Affective Learning Domain, Cognitive Learning Domain, and Psychomotor Learning Domain.

Learning Environment — Physical facilities where learning takes place.

Learning Objective — Specific statement that describes the knowledge or skills that students should have acquired at the conclusion of a lesson. *See* Behavioral Objective, Competency, or Performance Objective.

Learning Outcome — Statement that broadly specifies what students will know or be able to do once learning is complete.

Learning Plateau — A break or leveling of a student's progress in a training course or class.

Learning Style — Learner's habitual manner of problem-solving, thinking, or learning, though the learner may not be conscious of his or her style and may adopt different styles for different learning tasks or circumstances.

L-E-A-S-T Method — Progressive discipline method used in the classroom; stands for leave it alone, eye contact, action, stop the class, and terminate.

Lecture — Instructional method utilizing one-way communication in which an instructor or educator provides material verbally by telling, talking, and explaining but allows no exchange of ideas or verbal feedback.

Ledge Door — Door constructed of individual boards joined within a frame. *Also known as* Batten Door.

Ledger — (1) Horizontal framework member, especially one attached to a beam side that supports the joists. (2) Book in which financial records are kept.

Lee — *See* Leeward.

Leeward — Protected side; the direction opposite from which the wind is blowing. *Also known as* Lee.

Left-Hand Door — *See* Third Door (2).

Leg Lock — Method of entwining a leg around a ladder rung to ensure that the individual cannot fall from the ladder, thus freeing the climber's hands for working.

Legal Precedent — History of rulings made in courts of law that can be referenced and used to make court decisions in future cases or influence laws outside of the court system.

Legend — Explanatory list of symbols on a map or diagram. *See* Title Block.

Legislative Law — Law made by federal, state/province, county/parish, and city legislative bodies that have powers to make statutory laws. *See* Law and Statute.

Legislative Strategy — Compromise between what some parties want and what all parties can live with.

Legislator — An elected official who makes laws.

Legitimate Power — Power that stems from any or all of three sources: shared values, acceptance of social structure, or the sanctions of a legitimizing agent.

Legs — *See* Supports.

LEL — *See* Lower Explosive Limit.

Lens — Clear portion of the self-contained breathing apparatus (SCBA) mask.

Lens Fogging — Condensation on the inside of the facepiece lens caused by moisture in the wearer's exhalations.

LEPC — *See* Local Emergency Planning Committee.

LERP — *See* Local Emergency Response Plan.

Lesson Plan — Teaching outline or plan for teaching that is a step-by-step guide (or general guidelines) for presenting a lesson or presentation. It contains information and instructions on what will be taught and the teaching procedures to be followed. It covers lessons that may vary in length from a few minutes to several hours. *See* Application, Evaluation (1), Instructor Information, Level of Instruction, Preparation, Presentation, Summary, and Time Frame.

Lethal — Deadly; resulting in death.

Lethal Concentration, 50 Percent Kill (LC_{50}) — Concentration of an inhaled substance that results in the death of 50 percent of the test population. LC_{50} is an inhalation exposure expressed in parts per million (ppm), milligrams per liter (mg/liter), or milligrams per cubic meter (mg/m^3); the lower the value, the more toxic the substance. *See* Concentration (1).

Lethal Dose, 50 Percent Kill (LD_{50}) — Concentration of an ingested or injected substance that results in the death of 50 percent of the test population. LD_{50} is an oral or dermal exposure expressed in milligrams per kilogram (mg/kg); the lower the value, the more toxic the substance. *See* Dose.

Level A Protection — Highest level of skin, respiratory, and eye protection that can be given by personal protective equipment (PPE), as specified by the U.S. Environmental Protection Agency (EPA); consists of positive-pressure self-contained breathing apparatus, totally encapsulating chemical-protective suit, inner and outer gloves, and chemical-resistant boots. *See* Chemical Protective Clothing (CPC), Personal Protective Equipment (PPE), and Special Protective Clothing.

Level B Protection — Personal protective equipment that affords the highest level of respiratory protection, but a lesser level of skin protection; consists of positive-pressure self-contained breathing apparatus, hooded chemical-protective suit, inner and outer gloves, and chemical-resistant boots. *See* Chemical Protective Clothing (CPC), Personal Protective Equipment (PPE), and Special Protective Clothing.

Level C Protection — Personal protective equipment that affords a lesser level of respiratory and skin protection than levels A or B; consists of full-face or half-mask APR, hooded chemical-resistant suit, inner and outer gloves, and chemical-resistant boots.

Level D Protection — Personal protective equipment that affords the lowest level of respiratory and skin protection; consists of coveralls, gloves, and chemical-resistant boots or shoes.

Level I Staging — Used on all multiple-company emergency responses. The first-arriving vehicles of each type proceed directly to the scene, and the others stand by a block or two from the scene and await orders. Units usually stage at the last intersection on their route of travel before reaching the reported incident location.

Level II Fire and Life Safety Educator — Manager who exhibits proficiency in creating and leading community risk-reduction programs.

Level II Staging — Used on large-scale incidents where a larger number of fire and emergency services companies are responding; these companies are sent to a specified remote location to await assignment. *See* Base.

Level of Learning — Lesson plan component that states the learning level that participants will reach by the end of the lesson; may be based on the taxonomy of learning domains or on performance of job requirements. *See* Lesson Plan.

Lever — Device consisting of a bar pivoting on a fixed point (fulcrum), using power or force applied at a second point to lift or sustain an object at a third point.

Leverage — Action or mechanical power of a lever.

Lexan® — Polycarbonate plastic used for windows; has one-half the weight of an equivalent-sized piece of glass, yet is 30 times stronger than safety glass and 250 times stronger than ordinary glass. It cannot be broken using standard forcible entry techniques.

LFL — *See* Lower Flammable Limit.

Liability — (1) All types of debts and obligations one is bound in justice to perform; a condition of being responsible for a possible or actual loss, penalty, evil, expense, or burden; a condition that creates a duty to perform an act immediately or in the future. *See* Vicarious Liability. (2) To be legally obligated or responsible for an act or physical condition; opposite of *immunity*. *See* Immunity.

Liability of Injury — Legal responsibility and accountability for an act or process related to a program.

Liaison Officer — Point of contact for assisting or coordinating agencies; member of the command staff.

Libel — Written or oral defamatory statement; the act, tort, or crime of making or publishing a libel against someone. *See* Defamation, Slander, and Tort.

Lieutenant — Rank used in some fire departments for company officers.

Life Belt — Wide, adjustable belt with a snap hook that can be fastened to the rungs of a ladder to secure a firefighter to the ladder while leaving the firefighter's hands free for working. The formal term for *life belt* is Class I Life Safety Harness.

Life Net — Canvas device with a folding circular metal frame and spring action used to catch persons who jump from buildings; not considered safe or effective for jumps from above the fourth floor of a building.

Life of Foam — Period of time that the foam blanket remains in place until more foam must be applied.

Life Safety — Refers to the joint consideration of the life and physical well-being of individuals, both civilians and firefighters.

***Life Safety Code*®** — *See* NFPA® 101, *Life Safety Code*®.

Life Safety Harness — Harness that meets the requirements of NFPA® 1983, *Standard on Life Safety Rope and Equipment for Emergency Services*. *See* Class I Harness, Class II Harness, and Class III Harness.

Life Safety Rope — Rope designed exclusively for rescue and other emergency operations; used to raise, lower, and support people at an incident or during training. Must meet the requirements established in NFPA® 1983, *Standard on Life Safety Rope and Equipment for Emergency Services*. *See also* Lifeline.

Lifeline — Non-load-bearing rope attached to a firefighter during search operations to act as a safety line.

Lift — (1) Apparatus for raising an automobile. (2) Component of the total aerodynamic force acting on an airplane or airfoil that is perpendicular to the relative wind; for an airplane this constitutes the upward force that opposes the pull of gravity. (3) Dimension from the top of one pouring of concrete in a form to the top of the next pouring; for example, "pour concrete in 8-inch (203 mm) lifts." *See* Dependable Lift, Maximum Lift, and Theoretical Lift.

Lift-On/Lift-Off (LO/LO) — Refers to a vessel capable of loading and unloading its own cargo without shoreside crane assistance.

Lift Slabs — System of concrete construction in which the floor slabs are poured in place at the ground level and then lifted to their position by hydraulic jacks working simultaneously at each column.

Lift Truck — Small truck for lifting and transporting loads. *See* EX Symbol.

Lifter Roof Storage Tank — Atmospheric storage tank designed so that the liquid-sealed roof floats on a slight cushion of vapor pressure. When the vapor pressure exceeds a designated limit, the roof lifts to relieve the excess pressure. *See* Atmospheric Storage Tank and Floating Roof Storage Tank.

Light — Visible radiation produced at the atomic level, such as a flame produced during the combustion reaction.

Light Attack Vehicle — *See* Initial Attack Apparatus or Mini-pumper.

Light Box — *See* Connection Box.

Light Detector — *See* Flame Detector.

Light Duty Policy — Department policy that indicates certain assignments that may be given to firefighters who are recovering from injuries; the assignments should be evaluated to ensure that they will not interfere with injury rehabilitation.

Light Fuels — Fast-drying fuels, with a comparatively high surface-area-to-volume ratio, that are generally less than ½ inch (6.35 mm) in diameter and have a time lag of 1 hour or less. These fuels readily ignite and are rapidly consumed by fire when dry.

Light Meter — Device used to measure the amount of light; used to determine the exposure of a photograph. DSLR cameras and other cameras often have light meters integrated in the camera that determine the exposure automatically.

Light Rescue Vehicle — Small rescue vehicle usually built on a 1-ton or 1½-ton chassis; designed to handle only basic extrication and life-support functions and carries only basic hand tools and small equipment.

Light Shaft — *See* Light Well.

Light Well — Vertical shaft at or near the center of a building to provide natural light and/or ventilation to offices or apartments not located on an outside wall. *Also known as* Light Shaft.

Lighter — Large boat or barge (usually non-powered), typically used for conveying cargo to and from vessels in harbor, transporting coal or construction materials, or transporting garbage. *See* Barge.

Light-Frame Construction — Method for construction of wood-frame buildings; replaced the use of heavy timber wood framing.

Light-Gauge Steel Joist — Joist system for supporting metal decks or wood-panel flooring systems; produced from cold-rolled steel and available in several cross-sectional varieties. *See* Open-Web Joist.

Lightly Trapped — Victims who are trapped by furniture or debris within a structure that has remained standing.

Lightweight Steel Truss — Structural support made from a long steel bar that is bent at a 90-degree angle with flat or angular pieces welded to the top and bottom.

Lightweight Transparent Armor® (LTA) — Polycarbonate bulletproof glass sheets bonded together to form windows and windshields for armored vehicles and structures.

Lightweight Wood Truss — Structural supports constructed of 2 x 3-inch or 2 x 4-inch (51 mm by 76 mm or 51 mm by 102 mm) members that are connected by gusset plates.

Likelihood of Survival — Determination of the possibility of survival for victims in a collapsed structure.

Limited Access (Warm) Zone — Large geographical area between the support zone and the restricted zone, for personnel who are directly aiding rescuers in the restricted zone. This includes personnel who are handling hydraulic tool power plants, fire personnel handling standby hoselines, and so on. This zone should contain the decontamination area, the safe haven, and the Haz Mat control officer. Personnel in this zone should not get in the way of rescuers working in the restricted zone.

Line — (1) Hoseline. (2) Rope or lifeline. (3) Rope when in use, such as a main line or safety line. (4) Length of rope in use on a vessel.

Line Functions — Personnel who provide emergency services to external customers (the public).

Line of Lividity — Colored area of the corpse that is noticeably contrasted from the rest of the body caused by the pooling of blood.

Line of Sight — Unobstructed, imaginary line between an observer and the object being viewed.

Line Organization — Portion of the fire department directly involved in providing fire suppression and rescue services.

Lined Hose — Fire hose composed of one or two woven outside jackets and an inside rubber lining.

Line-Item Budget — Budget that details the department's proposed expenditures line by line; the most common type of fire department budget.

Lineman's Gloves — Special gloves insulated for protection against electrical current.

Linen Hose — Fire hose made of linen or flax fabric without a rubber lining; used for standpipe cabinets and forestry operations.

Line-of-Duty — During the performance of fire department duties.

Line-of-Duty Death (LODD) — Firefighter or emergency responder death resulting from the performance of fire department duties.

Link Analysis — Method of computing, organizing, and utilizing data relating to an investigation; allows the analysis and presentation of complex data in a clear and concise manner.

Lintel — Support for masonry over an opening; usually made of steel angles or other rolled shapes, singularly or in combination.

LIP Service — Emergency incident management priorities of **L**ife safety, **I**ncident stabilization and control, and **P**rotection of property and the environment.

Lipid Pneumonia — Pneumonia that may follow the aspiration of an oily substance such as mineral oil.

Liquefied Compressed Gas — Gas that under the charging pressure is partially liquid at 70°F (21°C). *Also known as* Liquefied Gas. *See* Compressed Gas, Gas, Liquefied Natural Gas (LNG), and Liquefied Petroleum Gas (LPG).

Liquefied Flammable Gas Carrier — Tanker used to transport liquefied natural gas (LNG) and liquefied petroleum gas (LPG) (such as propane and butane); generally uses large insulated spherical tanks for product storage. *See* Tanker.

Liquefied Gas — Confined gas that at normal temperatures exists in both liquid and gaseous states. *See* Compressed Gas, Gas, Liquefied Compressed Gas, and Liquefied Petroleum Gas (LPG).

Liquefied Natural Gas (LNG) — Natural gas stored under pressure as a liquid. *See* Gas, Hydrocarbon Fuel, and Liquid Compressed Gas.

Liquefied Petroleum Gas (LPG) — Any of several petroleum products, such as propane or butane, stored under pressure as a liquid. *See* Gas, Hydrocarbon Fuel, and Liquefied Compressed Gas.

Liquid — Incompressible substance with a constant volume that assumes the shape of its container; molecules flow freely, but substantial cohesion prevents them from expanding as a gas would.

Liquid Oxygen (LOX) — Oxygen that is stored under pressure as a liquid.

Liquid Propellants — Liquids used in rockets as fuels and oxidizers.

Liquid-Splash Protective Clothing — Chemical-protective clothing designed to protect against liquid splashes per the requirements of NFPA® 1992, *Standard on Liquid Splash-Protective Suits for Hazardous Chemical Emergencies*; part of an EPA Level B ensemble.

List — Continuous lean or tilt of a vessel to one side due to an imbalance of weight within the vessel. *See* Angle of Loll, Critical Angle of List, Heel, Heeling, and Loll.

Listed — Refers to a device that has been tested by the Underwriters' Laboratories Inc. Factory Mutual System and certified as having met minimum criteria.

Listening — Process of receiving, attending to, and assigning meaning to auditory stimuli; a process of steps that gives information that listeners try to understand.

Litter — (1) Top layer of forest floor composed of loose debris of dead sticks, branches, twigs, and recently fallen leaves or needles, on top of a duff layer; decomposition does little to alter its structure. (2) Inappropriately discarded rubbish. (3) *See* Stretcher.

Live Fuels — Living plants, such as trees, grasses, and shrubs, in which the seasonal moisture content cycle is controlled largely by internal physiological mechanisms rather than by external weather influences.

Live Load — (1) Items within a building that are movable but are not included as a permanent part of the structure; merchandise, stock, furnishings, occupants, firefighters, and the water used for fire suppression are examples of live loads. (2) Force placed upon a structure by the addition of people, objects, or weather. *See* Dead Load and Load.

Live-Fire Exercises — Training exercises that involve the use of an unconfined open flame or fire in a structure or other combustibles to provide a controlled burning environment. *Also known as* Live Burn Exercises.

Livestock — Cattle, horses, sheep, and other useful animals kept on a ranch or farm.

LNG — *See* Liquefied Natural Gas.

LO/LO — *See* Lift-on/Lift-off.

Load — (1) The sum of the wattages of the various devices being served by a circuit. (2) Any effect that a structure must be designed to resist, such as gravity, wind, earthquakes, or soil pressure. *See* Dead Load and Live Load.

Load Line — *See* Plimsoll Mark.

Load Monitor — Device that "watches" an electrical system for added loads that may threaten to overload the system.

Load Sequencer — Device in an electrical system that turns lights on at specified intervals, so that the start-up load for all of the devices does not occur at the same time.

Load Shedding — When an overload condition occurs, the load monitor will shut down less important electrical equipment to prevent the overload.

Load Testing — Aerial device test intended to determine whether or not the device is capable of safely carrying its rated weight capacity.

Load-Bearing Frame Members — Portions of the frame that provide direct support to attached members.

Load-Bearing Wall — Wall that supports itself, the weight of the roof, and/or other internal structural framing components, such as the floor beams and trusses above it; used for structural support. *Also known as* Bearing Wall. *See* Pony Wall.

Loading Rack — Fixed facility where either truck or railroad tank cars are bulk loaded with flammable and combustible liquids.

Loading Site — In a tanker/tender shuttle operation, the location where apparatus tanks are filled from the water supply. *See* Fill Site.

Lobby Control — In high-rise fire fighting, the individual responsible for, and the process of, taking and maintaining control of the lobby and elevators; includes establishing internal communications, coordinating the flow of personnel and equipment up interior stairways to upper levels, and coordinating with building engineering personnel.

Lobbying — Educating a person or an organization about your position on an issue, or urging them to adopt your position; more specifically, conducting activities aimed at influencing public officials, especially members of a legislative body.

Lobbyist — People compensated by an organization to influence decision-makers to favor the special interests of the sponsoring organization.

Local Alarm System — *See* Protected Premises Fire Alarm System.

Local Application System — Fixed-site fire-suppression system that is required to cover a protected area with 2 feet (0.6 m) of foam depth within 2 minutes of system activation; foam supply must support the continuous operation of the system for at least 12 minutes. *See* Total Flooding System.

Local Emergency Planning Committee (LEPC) — Community organization responsible for local emergency response planning. Required by SARA Title III, LEPC's are composed of local officials, citizens, and industry representatives with the task of designing, reviewing, and updating a comprehensive emergency plan for an emergency planning district; plans may address hazardous materials inventories, hazardous material response training, and assessment of local response capabilities. *See* Local Emergency Response Plan (LERP).

Local Emergency Response Plan (LERP) — Plan detailing how local emergency response agencies will respond to community emergencies; required by U.S. Environmental Protection Agency (EPA) and prepared by the Local Emergency Planning Committee (LEPC).

Local Energy System — Auxiliary fire alarm system that has its own power source.

Local Winds — Winds generated over a comparatively small area, and whose speed and direction are influenced by local conditions such as topography, fires, and weather fronts. They do not fit within the general pressure pattern and possess no peculiarities.

Locard Exchange Principle — Investigative principle that states that whenever a person comes into contact with a scene, he or she leaves something at the scene and also takes something from it.

Location Marker — Device, such as a reflective marker or flag, used to mark the location of a fire hydrant for quicker identification during a fire response.

Lock — (1) Device for fastening, joining, or engaging two or more objects together such as a door and frame. (2) *See* Pawls.

Lock Mechanism — Moving parts of a lock, which include the latch or bolt, lock cylinder, and articulating components.

Locking In — *See* Leg Lock.

Lockout/Tagout Device — Device used to secure a machine's power switches, in order to prevent accidental re-start of the machine.

LODD — *See* Line-of-Duty Death.

Lodging House — *See* Hotel.

Log — (1) Record book. (2) To record information in a log or record book.

Logistics — Process of managing the scheduling of limited materials and equipment to meet the multiple demands of training programs and instructors.

Logistics Section — Section responsible for providing facilities, services, and materials for the incident; includes the Communications unit, Medical unit, and Food unit within the service branch and the supply unit, facilities unit, and ground support unit within the support branch. *Also known as* Logistics.

Logroll — Method for placing a patient onto a backboard by turning the patient as a unit, first onto the side, then onto the back.

Loiding — Method of slipping or shimming a spring latch from its strike with a piece of celluloid such as a credit card.

Loll — Neutral equilibrium when vessel comes to rest, within a range of stability as opposed to a point of stability; that is, instead of being stable when upright, the vessel may be stable within 1 degree to port or starboard sides, and thus will lean either port or starboard. *See* Angle of Loll and List.

Loma Prieta — Name given to the October 1989 earthquake in central California that occurred during the World Series baseball game in San Francisco.

Long Backboard — Board used to package a patient with suspected spinal injury.

Long Spine Board (LSB) — A victim packaging device used in extrication. The board is slightly longer and wider than an average person to accommodate immobilization straps and cervical immobilization devices. May be used with a Stokes litter for vertical movement. *Also known as* a Long Backboard.

Long Ton — Unit of weight used in the marine industry; 1 long ton = 2,240 pounds or 1 tonne (1,016 kilograms). A short ton = 2,000 pounds or 0.9 tonne (907 kilograms).

Long-Duration Apparatus — Breathing apparatus that supplies the wearer with air for more than 30 minutes.

Longeron — Longitudinal members of the framing of an aircraft fuselage or nacelle; usually continuous across a number of bulkheads or other points of support.

Longitudinal Hose Bed — Hose bed located to the side of the main hose bed; designed to carry preconnected attack hose.

Longitudinal Stability — Ability of a vessel to return to an upright position when forced from its rest condition by pitching. *See* Stability and Static Stability.

Longshoreman — Worker who loads and unloads cargo from a vessel. *Also known as* Stevedore.

Long-Term Memory — Large cognitive storage area where information can be processed and applied.

Lookout — (1) Location from which fires can be detected and reported. (2) Fire crew member assigned to observe the fire from a vantage point and warn the crew when there is danger of becoming trapped.

Lookout Tower — Tower or station, usually on a high place, used for spotting wildland fires; also used for pinpointing lightning strikes and identifying approaching frontal systems.

Loop System — Water main arranged in a complete circuit so that water will be supplied to a given point from more than one direction. *Also known as* Belt System, Circle System, or Circulating System.

Loose End — *See* Working End.

Loss Control — Practice of minimizing damage and providing customer service through effective mitigation and recovery efforts before, during, and after an incident.

Loss Control Factors — Factors that are useful in risk evaluation and that are the conditions unique to a particular scenario or location.

Loss Control Risk Analysis — Process in which specific potential risks are identified and evaluated before an incident has occurred. The goal of this process is to develop strategies to minimize the impact of these risks.

Loss Control Strategies — Actions that reduce or eliminate all or part of a loss control risk. Loss control strategies are developed by using the information from the loss control risk analysis.

Louver Cut — Rectangular exhaust opening cut in a roof, allowing a section of roof deck (still nailed to a center rafter) to be tilted, thus creating an opening similar to a louver. *Also known as* Center Rafter Cut and Louver Vent.

Low Angle — Environment with a shallow grade that does not require the use of rope systems, but movement would be hazardous or difficult without the aid of a rope.

Low Explosive — (1) Explosive that decomposes or burns rapidly, but does not produce an explosive effect unless it is confined. *See* Detonation, Explosive (2), and High Explosive. (2) Explosive material that deflagrates, producing a reaction slower than the speed of sound.

Low-Density Combustible Fiberboard — Building material, often used for interior finishes, that is usually highly combustible.

Lower Airway — Portion of the respiratory system below the epiglottis.

Lower Explosive Limit (LEL) — *See* Lower Flammable Limit.

Lower Flammable (Explosive) Limit (LFL) — Lower limit at which a flammable gas or vapor will ignite and support combustion; below this limit the gas or vapor is too *lean* or *thin* to burn (too much oxygen and not enough gas, so lacks the proper quantity of fuel). *Also known as* Lower Explosive Limit (LEL). *See* Flammable Gas, Flammable Limit, and Upper Flammable Limit (UFL).

Lower In — Procedure for positioning the tip of a ladder against a building after raising.

Lowering — Procedure for removing a ladder from the raised position.

Low-Expansion Foam — Foam concentrate that is mixed with air, in the range of less than 20 parts air to 1 part foam solution (20:1). *See* High-Expansion Foam and Medium-Expansion Foam.

Low-Head Dam — Wall-like concrete structure across a river or stream that is designed to back up water; allows water to flow over the crest and drop into a lower level. *Also known as* Low-Water Dam.

Low-Impact Crashes — Aircraft crashes that do not severely damage or break up the fuselage and are likely to have a large percentage of survivors.

Low-Order-Explosion Damage — Damage typically associated with a deflagration that includes bulging of walls, walls fallen intact away from a structure, roofs lifted and dislodged, and large debris moved only short distance.

Low-Pressure Alarm — Alarm that sounds when SCBA air supply is low, typically 25 percent.

Low-Pressure Chemical Tank — Cargo tank truck designed to carry various chemicals such as flammables, corrosives, or poisons, with pressures not to exceed 40 psi (276 kPa) {2.76 bar} at 70°F (21°C). *See* Cargo Tank Truck.

Low-Pressure Hose — Hose containing pressure slightly above atmospheric pressure, leading from the regulator to the facepiece. *Also known as* Inhalation Tube.

Low-Pressure Storage Tank — Class of fixed-facility storage tanks that are designed to have an operating pressure ranging from 0.5 to 15 psi (3.45 kPa to 103 kPa) {0.03 bar to 1.03 bar}. *See* Atmospheric Storage Tank, Noded Spheroid Tank, Pressure Storage Tank, Pressure Vessel, and Spheroid Tank.

Low-Rise Elevator — Elevator that serves only the lower floors of a high-rise building.

Low-Voltage — Any voltage that is less than 600 volts and safe enough for domestic use, typically 120 volts or less.

LOX — *See* Liquid Oxygen.

LPG — *See* Liquefied Petroleum Gas.

Lugging — Condition that occurs when the throttle application is greater than necessary for a given set of conditions; may result in an excessive amount of carbon particles issuing from the exhaust, oil dilution, and additional fuel consumption. Lugging can be eliminated by using a lower gear and proper shifting techniques.

Lumber — Lengths of wood cut and prepared for use in construction.

Lungs — Paired organs of respiration that lie in the chest.

M

Machine-Guarding — Use of gates, covers, housings, deflectors, or other guards on power machinery to prevent the user from contacting moving parts or being struck by flying objects.

Machinery — All the equipment on a vessel; including but not limited to the main and auxiliary engines, pumps, deck winches, steering engine, and hoists.

Macro Lens — Camera lens that is designed for close-up photos.

Magazine — Storage facility approved by the Bureau of Alcohol, Tobacco, Firearms, and Explosives (ATF) for the storage of explosives. *See* Explosive.

Magnetic Particle Inspection — Form of nondestructive steel aerial device testing where the aerial device is magnetized and metal particles are applied. Deviances in the coating of the particles indicate flaws in the metal aerial device.

Magneto — Device used in gasoline engines that produces a periodic spark in order to maintain fuel combustion.

Main Deck — Uppermost continuous deck of a vessel that runs from bow to stern. *See* Deck.

Main Guideline — Special rope used in the United Kingdom, Australia, and New Zealand as a safety guideline to indicate a route between the entry control point and the scene of operations.

Main Line — Rope system built to support a rescuer and/or patient.

Main Transverse Bulkheads — Watertight bulkheads that subdivide a vessel into watertight compartments. *See* Bulkhead (1) and Main Watertight Subdivision.

Main Watertight Subdivision — Space between two main transverse watertight bulkheads. *See* Bulkhead (1) and Main Transverse Bulkheads.

Mainline Valve — Valve that when opened lets air from the cylinder travel its normal route through the regulator to the facepiece.

Maintenance — Keeping equipment or apparatus in a state of usefulness or readiness.

Major — Rank used by some fire departments for company officers.

Make the Fire — Order given to a specific unit to respond to a fire.

Makeup — All actions involved in connecting fire hose or apparatus to other equipment.

Making a Hydrant — Procedure for connecting to and laying hose forward from a fire hydrant.

Male Coupling — Hose coupling with external threads that fit into the threads of a female coupling of the same pitch and appropriate diameter and thread count.

Malicious — Describes a state of mind characterized by the intent to injure, vex, or annoy another person, to commit an unlawful act, or attempt to defraud; often an element of arson. Legal definitions of maliciousness vary from state to state.

Maltese Cross — Commonly used insignia of the fire service, worn on the uniform or the cap. The popular variety of the Maltese cross is actually a modification of the *cross patee* rather than the actual Maltese cross, which has eight points.

Mammalian Diving Reflex — Autonomous physiologic reaction to immersion in cold water, in which the blood and oxygen supply is shunted to the brain to keep the animal alive, although outward appearances may suggest death.

Manage — To provide direction and leadership in order to achieve organizational objectives through effective and efficient application of resources.

Management — Process of accomplishing organizational objectives through effective and efficient handling of resources; official, sanctioned leadership.

Management by Objectives (MBO) — Planning and control device used to organize resources and motivate personnel toward the fulfillment of specified objectives.

Management Information System (MIS) — Provides a means for tracking information about the program, for summarizing and analyzing the program's case load, and securing data for annual reports and evaluations.

Manager — Individual who accomplishes organizational objectives through effective and efficient handling of material and human resources.

Manhole — (1) Hole through which a person may go to gain access to an underground or enclosed structure. (2) Opening usually equipped with a removable, lockable cover, that is large enough to admit a person into a tank trailer or dry bulk trailer. *Also known as* Manway.

Manifest — *See* Cargo Manifest.

Manifold — (1) Hose appliance that divides one larger hoseline into three or more small hoselines. *Also known as* Portable Hydrant. (2) Top portion of the pump casing. (3) Device used to join a number of discharge pipelines to a common outlet.

Manila Rope — Rope made from manila fiber, which is grown in Manila in the Philippines. This type of rope is not suitable for life safety applications.

Manipulative Lesson — *See* Practical Demonstration.

Manipulative Skills — Skills that use the psychomotor domain of learning; refers to the ability to physically manipulate an object or move the body to accomplish a task.

Manipulative Training — Lesson or exercise in a training program, in which participants handle or learn to handle equipment or materials in a coordinated or skillful manner.

Manipulative-Performance Test — Practical competency-based test that measures mastery of the psychomotor objectives as they are performed in a job or evolution.

Mansard Roof — Roof style with characteristics similar to both gambrel and hip roofs. Mansard roofs have slopes of two different angles, and all sides slope down to an outside wall.

Manual Foam Monitor — Foam monitor that is operated by hand; may be found mounted on apparatus, in fixed locations to protect target hazards, or as a portable unit. *See* Automatic Oscillating Foam Monitor, Foam Monitor, and Remote-Controlled Foam Monitor.

Manual on Uniform Traffic Control Devices (MUTCD) — DOT Federal Highway Administration publication that identifies the types of traffic control devices that should be used to establish work areas and identify incident scenes, as well the methods for deploying these devices.

Manual Pull Station — Manual fire alarm activator.

Manual Stabilizer — Manually deployed stabilizing device for aerial apparatus that consists of an extension arm with a jack attached to the end of it.

Manufactured Home — Dwelling that is the assembly of four major components: the chassis and the floor, wall, and roof systems; although they are constructed of steel, wood, plywood, aluminum, gypsum wallboard, and other materials, they are basically frame construction. Characterized by small compartment sizes, low ceilings, and very lightweight construction throughout. *Also known as* Mobile Home.

Manufacturer's Tests — Fire pump or aerial device tests performed by the manufacturer prior to delivery of the apparatus.

Manway — *See* Manhole.

Marina — Special harbor with facilities constructed especially for yachts and other pleasure craft.

Marine Company — Personnel assigned to work on a fireboat.

Marine firefighter - Fire fighter personnel assigned to work on a fireboat for the purpose of extinguishment of fires in the marine environment.

Marine Unit — *See* Fire Boat.

Maritime Law — Laws relating to commerce and navigation on the high seas and other navigable waters; a court exercising jurisdiction over maritime cases. *Also known as* Admiralty Law.

Marking — Descriptive name, identification number, weight, or specification, along with instructions, cautions, or UN marks required on outer packagings of hazardous materials. *See* Label, Package Marking, and Placard.

Marrying Vehicles — Attaching two vehicles to one another in such a fashion that the two vehicles move as one stable object.

Mars Light — Single-beam, oscillating warning light; originally manufactured by the Mars Light Company.

Martial Law — System of rule that occurs when the military takes control of the administration of justice.

Mask — *See* Facepiece.

Maslow's Hierarchy of Needs — Theory put forth by psychologist Abraham Maslow stating that all human behavior is motivated by a drive to attain specific human needs in a progressive manner. The hierarchy begins with basic physiological needs and progresses through security, social, self-esteem, and self-actualization needs.

Masonry — Bricks, blocks, stones, and unreinforced and reinforced concrete products.

Mass Casualty Incident — Incident that results in a large number of casualties within a short time frame, as a result of an attack, natural disaster, aircraft crash, or other cause that is beyond the capabilities of local logistical support. *See* Multi-Casualty Incident.

Mass Communication — Rapid transmission of a warning directly to the general population.

Mass Decontamination — Process of decontaminating large numbers of people in the fastest possible time to reduce surface contamination to a safe level. It is typically a gross decon process utilizing water or soap and water solutions to reduce the level of contamination, with or without a formal decontamination corridor or line. *See* Decontamination, Decontamination Corridor, and Gross Decontamination.

Mass Media — News communications that are designed to reach a large number of people.

Mass Notification System (MNS) — System that notifies occupants of a dangerous situation and allows for information and instructions to be provided.

Mass Prophylaxis — Capability to protect the health of the population through administration of critical interventions (such as antibiotics, vaccinations, or antivirals), in order to prevent the development of disease among those who are exposed or potentially exposed to public health threats. This capability includes the provision of appropriate follow-up and monitoring of adverse events, as well as risk communication messages to address the concerns of the public.

Mass Transportation — Any mode of transportation designed to carry large numbers of people at the same time.

MAST — *See* Medical Antishock Trousers or Military Antishock Trousers.

Mast — Vertical pole, rising from the keel or deck of a vessel, that supports sailing rigging. Also used for radio antennas and signal flags.

Master — Commander of a merchant vessel. *See* Captain.

Master Stream — Large-caliber water stream usually supplied by combining two or more hoselines into a manifold device or by fixed piping that delivers 350 gpm (1 325 L/min) or more. *Also known as* Heavy Stream.

Master Stream Nozzle — Nozzle capable of flowing in excess of 350 gpm (1 325 L/min).

Mastery — High-level or nearly complete degree of proficiency in the performance of a skill, based on criteria stated in objectives; in training, the ability to perform at a designated skill level, which enables the learner to progress to the next designated skill level. A mastery test checks that learners have achieved the appropriate skill level (mastery). *See* Criterion-Referenced Testing.

Mastery Learning — Element of criterion-referenced or competency-based learning; outcomes of learning are expressed in minimum levels of performance for each competency.

Masthead — Printed, usually boxed section of a newspaper or periodical that gives the title and pertinent details of ownership, editorship, advertising rates, and subscription rates.

Mat Foundation — Thick slab beneath the entire area of a building; differs from a simple floor slab in its thickness and amount of reinforcement.

Matching Grant — Grant in which the funder agrees to give an amount that is equal to (or a specific ratio of) the amount that another funder gives.

Mate — *See* Chief Officer.

Material — Generic term used by first responders for a substance involved in an incident. *See* Hazardous Material and Product.

Material First Ignited — Fuel that is first set on fire by the heat of ignition. To be meaningful, both a type of material and a form of material should be identified.

Material Safety Data Sheet (MSDS) — *See* Safety Data Sheet.

Materials Needed — List of everything needed to teach a lesson, such as models, mock-ups, visual aids, equipment, handouts, and quizzes.

Matter — Anything that occupies space and has mass.

Mattress Chains — Light chains with hooks or locking devices used to bind a mattress in a roll for removal from a building.

Mattydale Hose Bed — *See* Transverse Hose Bed.

Maximum Allowable Quantity — Maximum amount of a hazardous material to be stored or used within a control area inside a building or an outdoor control area; maximum allowable quantity per control area is based on the material state (solid, liquid, or gas) and the material storage or use conditions (**Source:** *International Fire Code®*, 2006 edition).

Maximum Daily Consumption — Maximum total amount of water used during any 24-hour interval over a 3-year period.

Maximum Extended Length — Total length of an extension ladder with all sections fully extended and pawls engaged.

Maximum Lift — Maximum height to which any amount of water may be raised through a hard suction hose to a pump; determined by the ability of the pump to create a vacuum. *See* Lift.

May — Term used in NFPA® standards that denotes voluntary or optional compliance.

MAYDAY — Internationally recognized distress signal.

Maze — Training facility with or without smoke, lighted or unlighted, in which firefighters wearing SCBA must negotiate obstacles to perform certain tasks.

MBO — *See* Management By Objectives.

MDH — *See* Medium Diameter Hose.

Mean — Term that refers to the "average" of a set of scores; calculated by adding all of the set of scores (values) and dividing by the total number of scores. For example, if a set of scores is 98, 98, 95, 92, 92, 92, 89, 88, 87, 85, 85, 79, 75, 74, and 74, the mean or average score is 91.8.

Means of Egress — (1) Safe, continuous path of travel from any point in a structure to a public way. Composed of three parts: exit access, exit, and exit discharge. (2) Continuous and unobstructed way of exit travel from any point in a building or structure to a public way, consisting of three separate and distinct parts: exit access, exit, and exit discharge. (**Source:** NFPA® 101, *Life Safety Code®*). *See* Egress, Exit, Exit Access, Exit Discharge, Public Way, and Travel Distance.

Mechanical Advantage — (1) The gain in force obtained by moving the fulcrum closer to the object during levering. (2) Advantage created when levers, pulleys, and other tools are used to make work easier during rope rescue or while lifting heavy objects. (3) The ratio of the force applied by a simple machine, such as a lever or pulley, to the force applied to the machine by the user.

Mechanical Blower — High-expansion foam generator that uses a fan to inject the air into the foam solution as it passes through the unit. *See* High-Expansion Foam.

Mechanical Explosion — Explosion that is the result of an increase in pressure in a confined container; may or may not be a result of additional heat. *See* Boiling Liquid Expanding Vapor Explosion (BLEVE).

Mechanical Filter — Air-purification component that physically separates the greatest part of water, oil, and other contaminants from compressed air. May also refer to the filter on a negative-pressure respirator that performs the same task.

Mechanical Foam — Foam produced by a physical agitation of a mixture of foam concentrate, water, and air. *See* Chemical Foam.

Mechanical Heat Energy — Heat that is generated by friction or compression. Moving parts on machines, such as belts and bearings, are a source of mechanical heating.

Mechanical Shoe Seal — Fabric seal that is anchored to the top of the roof and rides on the inside of a large fuel storage tank wall. The actual mechanical shoe, also known as a pantograph, is attached below the fabric seal to keep the roof properly aligned within the tank. *Also known as* Pantograph Seal.

Mechanical System — Large equipment system within a building that may include climate-control systems; smoke, dust, and vapor removal systems; trash collection systems; and automated mail systems. Does not include general utility systems such as electric, gas, and water. *See* Heating, Ventilating, and Air-Conditioning (HVAC) System.

Mechanical Trauma — Injury, such as an abrasion, puncture, or laceration, resulting from direct contact with a fragment or a whole container.

Mechanical Ventilation — *See* Forced Ventilation.

Mechanism of injury — Forces placed on the victim's body by collapse or collision.

Media — *See* Medium.

Media Advisory — Advisory of specific event or program to be held in the future; in contrast, a *news release* discusses something that has already happened.

Media Kit — Packet containing information about the public fire and life safety educator's own organization.

Media Release — Prepared statement distributed to provide the media with information in a ready-to-use news story format. *Also known as* Press Release.

Medial — Toward the midline of the body.

Median — Middle score in a set of scores (values) that are arranged or ranked in size (order) from high to low. For example, if a set of scores is 98, 98, 95, 92, 92, 92, 89, 88, 87, 85, 85, 79, 75, 74, and 74, the median or middle score is 88.

Medic Unit — (1) Ambulance staffed by paramedics. (2) Nonpatient transport vehicle used by paramedics to respond to emergencies. *Also known as* Paramedic Unit.

Medical Antishock Trousers (MAST) — Inflatable trousers used to counteract the effects of heavy blood loss. *Also known as* Military Antishock Trousers.

Medical Evaluation — Annual evaluation performed by a physician or professional health-care provider to ensure that a fire and emergency services responder is physically fit to perform the duties assigned; required before a responder uses respiratory protection.

Medical Examination — Complete medical examination by a physician or professional health-care provider for entry-level emergency personnel and periodically for all personnel during their service careers; mandatory for personnel who will be using respiratory protection equipment.

Medical Examiner — Medically qualified government officer whose duty is to investigate deaths and injuries that occur under unusual or suspicious circumstances, to perform post-mortem examinations, and in some jurisdictions to initiate inquests. *Also known as* Coroner.

Medical Identification Bracelet (or Necklace) — Medical identification worn by individuals having an illness that requires certain care and treatment. *Also known as* Medic-Alert® Bracelet (or Necklace).

Medical Surveillance — Rehabilitation function during an incident intended to monitor responders' vital signs and incident-stress levels.

Medical Surveillance Program — Series of medical evaluations based upon medical fitness-for-duty requirements for firefighters which begin when the firefighter is hired and continue on a regular basis throughout the firefighter's career.

Medical Unit — Functional unit within the service branch of the logistics section of an incident command system; responsible for providing emergency medical treatment for emergency personnel. This unit does not provide treatment for civilians.

Medic-Alert® Bracelet (or Necklace) — *See* Medical Identification Bracelet (or Necklace).

Medium — Vehicle for sending a message; the plural for the term is *media*. *Also known as* Communications Vehicle.

Medium Diameter Hose (MDH) — 2½- or 3-inch (65 mm or 77 mm) hose that is used for both fire fighting attack and relay-supply purposes.

Medium Fuels — Material available to burn in a geographic area that is in the midrange of size such as various brush species; generally excludes short grasses and large trees.

Medium Rescue Vehicle — Rescue vehicle somewhat larger and better equipped than a light rescue vehicle; may carry powered hydraulic spreading tools and cutters, air bag lifting systems, power saws, oxyacetylene cutting equipment, ropes and rigging equipment, as well as basic hand equipment.

Medium-Duty-Chassis Ambulance — Ambulance built upon a medium truck chassis rated at over 15,000 pounds to under 32,000 pounds gross vehicle weight.

Medium-Expansion Foam — Foam concentrate that is mixed with air in the range of 20 parts air to 1 part foam solution (20:1) to 200 parts air to 1 part foam solution (200:1). *See* High-Expansion Foam and Low-Expansion Foam.

Medium-Pressure Air — Air pressurized from 2,000 to 3,000 psi (13 790 kPa to 20 684 kPa); used to distinguish specific types of breathing- air cylinders.

Megapixel — One million pixels; used as the reference for the number of pixels in a digital image. Also refers to the number of image sensor elements that a digital camera can display, which in turn describes the largest photograph that can be taken with that camera.

Megawatt (MW) — Measurement of rate of heat release equal to 1,000 kilowatts (equivalent to ten thousand 100-watt bulbs).

Melting Point — Temperature at which a solid substance changes to a liquid state at normal atmospheric pressure. *See* Freezing Point.

Member Assistance Program (MAP) — Program to help employees and their families with work or personal problems. *Previously known as* Employee Assistance Program (EAP).

Member Organization — Organization formed to represent the collective and individual rights and interests of the fire and emergency services organization, such as a labor union or fraternal organization.

Membrane — Thin sheath or layer of pliable material.

Membrane Ceiling — Usually refers to a suspended, insulating ceiling tile system.

Membrane Roof — Roof consisting of a single membrane laid in sheets on a roof deck, and attached using adhesives, gravel ballasts, mechanical fasteners, or heating the roof side of the membrane. Can be applied over existing roofs. *Also known as* Single-Ply Membrane Roof.

Membrane Structure — (1) Structure with an enclosing surface of a thin stretched flexible material. Examples include a simple tent or an air-supported structure. (2) Weather-resistant, flexible or semiflexible covering consisting of layers of materials over a supporting framework.

Memorandum of Understanding — Form of written agreement created by a coalition to make sure that each member is aware of the importance of his or her participation and cooperation.

Mentor — Trusted and friendly adviser or guide for someone who is new to a particular role.

Mentoring — Instructional method in which an individual, as trusted and friendly advisor or guide, sets tasks, coaches activities, and supervises progress of individuals in new learning experiences or job positions.

Mercantile — Occupancy Classification whose primary objective is the wholesale purchase and retail sale of goods for profit. *See* Occupancy Classification.

Message — Information, ideas, attitude, or opinion that is transmitted and received.

Metabolism — Conversion of food into energy and waste products.

Metacenter (m) — Point through which the force of buoyancy works; point of intersection of the vertical through the center of buoyancy of a floating body with the vertical through the new center of buoyancy when the body is displaced. *See* Center of Buoyancy and Metacentric Height.

Metacentric Height (gm) — Measure of a vessel's initial stability; distance of the metacenter above the center of gravity of a floating body. *See* Center of Buoyancy, Center of Gravity, and Metacenter.

Metal-Clad Door — Door with a metal exterior; may be flush type or panel type. *Also known as* Kalamein Door.

Meter Booting — The addition of insulators to an electrical meter by the utility company.

Meth — *See* Methamphetamine.

Meth Lab — Illicit clandestine laboratory that produces methamphetamine (meth). *See* Illegal Clandestine Lab.

Methamphetamine (Meth) — Central nervous system stimulant drug that is similar in structure to amphetamine.

Method of Instruction — Procedure, technique, or manner of instructing others that is determined by the type of learning to take place. Typical examples are lecture, demonstration, or group discussion.

Micro Siemen — One millionth of a siemen; a siemen is a Standard International (SI) unit of measurement of electrical conductance.

Microanalysis — Examination of items such as damaged electrical wiring, tool marks, and impressions from tires and shoes found at a fire scene; also involves the analysis of broken glass, smoking materials and matches, and hair or fibers found at the scene or on a suspect.

Micron — Unit of length equal to one-millionth of a meter.

Microwave — Term applied to radio waves in the frequency range of 1000 mhz and above.

Middle-of-the-Road Leader — Leadership style characterized by a leader who is moderately concerned with both production and relationships.

Midi-pumper — Apparatus sized between a mini-pumper and a full-sized fire department pumper, usually with a gross vehicle weight of 12,000 pounds (5 443 kg) or greater. The midi-pumper has a fire pump with a rated capacity generally not greater than 1,000 gpm (3 785 L/min). *See* Initial Attack Apparatus.

Midship Pump — Fire pumps mounted at the center of the fire apparatus.

Mil — One thousandth of an inch (0.001 inch [0.0254 mm]).

Milestone Budget Date — Date by which municipalities are required to initiate actions on budget proposals. Governances usually require municipalities to submit a balanced budget by a specific date.

Military Antishock Trousers (MAST) — *See* Medical Antishock Trousers.

Military Aviation Aircraft — Cargo, fighter, bomber, trainer, and special-mission aircraft.

Military Fire Department — Fire prevention/suppression unit operated by the U.S. Department of Defense (DoD); jurisdiction is usually limited to the confines of a military base or installation.

Military Specifications (MILSPECS) — Specifications developed by the U.S. Department of Defense (DoD) for the purchase of materials and equipment.

Mill — One thousandth of an inch (.001 inch [.0254 mm]).

Miller Board — Board used to package a patient with suspected spinal injury; may also be used with a harness for lifting the patient.

Millwork — Woodwork such as doors and trim.

MILSPECS — *See* Military Specifications.

Mine Rescue Drill — Special drill, operated by the mine emergency division of the U.S. Mine Safety and Health Administration, that can drill a 24-inch diameter (0.6 m) shaft through 50 feet (15 m) of solid limestone in one day.

Mine Resistant Ambush Protected (MRAP) Vehicles — Series of armored fighting vehicles designed to survive ambushes and improvised explosive device (IED) attacks.

Mine Safety and Health Administration (MSHA) — U.S. government organization that regulates mine safety.

Mineral Soil — Soil containing little or no combustible material.

Minimum Acceptable Standard — Lowest acceptable level of student performance.

Mininuke — *See* Improvised Nuclear Device (IND).

Mini-pumper — Small fire apparatus mounted on a pickup-truck-sized chassis, usually with a pump having a rated capacity less than 500 gpm (2 000 L/min). Its primary advantage is speed and mobility, which enables it to respond to fires more rapidly than larger apparatus. *Also known as* Light Attack Vehicle. *See* Initial Attack Apparatus.

Minor Treatment — Classification for patients with minor injuries; they may simply require first aid and may even be able to be transported to medical facilities in private vehicles without the care of EMS staff.

Miosis — Abnormal contraction of the pupils, resulting in a pinpoint appearance.

Miranda Warning — Rights read to suspects in the U.S.; based upon the Supreme Court decision *Miranda vs. Arizona*. Miranda language may vary among U.S. states.

MIS — *See* Management Information System

Miscibility — Two or more liquids' capability to mix together. *See* Immiscible, Insoluble, and Soluble.

Miscible — Materials that are capable of being mixed in all proportions.

Misdemeanor — Lesser crime usually punishable by a fine or a term of less than one year in jail or prison.

Mission Statement — Statement about the present state of the risk-reduction division; provides a meaning for the existence of risk-reduction work and allows the entire organization to understand the importance of the work being performed.

Mist — Finely divided liquid suspended in the atmosphere; generated by liquids condensing from a vapor back to a liquid, or by breaking up a liquid into a dispersed state by splashing, foaming, or atomizing. *See* Aerosol.

Mitigate — (1) To cause to become less harsh or hostile; to make less severe, intense or painful; to alleviate. (2) Third of three steps (locate, isolate, mitigate) in one method of sizing up an emergency situation.

Mixed Occupancy — Where two or more types or classes of occupancy exist in the same building or structure. Separate requirements are often impractical so the most restrictive fire and life safety requirements apply.

Mixture — Substance containing two or more materials not chemically united.

MNS — *See* Mass Notification System.

Mobile Attack — In wildland fire fighting, suppressing fire along a fire edge by driving mobile apparatus along the perimeter and simultaneously applying fire streams to knock down the fire. *Also known as* Pump and Roll.

Mobile Communications Unit — Unit designed and constructed for the purpose of providing specified level of incident radio communications capacity and personnel.

Mobile Data Computer (MDC) — Portable computer that, in addition to functioning as a Mobile Data Terminal, has programs that enhance the ability of responders to function at incident scenes.

Mobile Data Terminal (MDT) — Mobile computer that communicates with other computers on a radio system.

Mobile Foam Apparatus — *See* Mobile Foam Extinguishing System.

Mobile Foam Extinguishing System — Foam delivery system that is mounted on a fire apparatus or trailer. *See* Foam Tender and Foam Trailer.

Mobile Home — *See* Manufactured Home.

Mobile Kitchen Unit — Unit designed and constructed for the purpose of dispensing food for incident personnel, providing specified level of capacity.

Mobile Radio — (1) Radio service between a radio station at a fixed location and one or more mobile stations, or between mobile stations. (2) Radio mounted on an apparatus.

Mobile Water Supply Apparatus — Fire apparatus with a water tank of 1,000 gallons (3 785 L) or larger whose primary purpose is transporting water; may also carry a pump, some hose, and other equipment. *Also known as* Tanker or Tender.

Mock Incident — Simulated emergency that allows responders to test their skills under realistic conditions. *Also known as* Staged Incident.

Mock-Up — Working model for realistic training and drilling.

Mode — (1) Phase, step, or progression of applying fireground strategy. (2) Most frequent score (value) in a set of scores. For example, if a set of scores is 98, 98, 95, 92, 92, 92, 89, 88, 87, 85, 85, 79, 75, 74, and 74, the mode or most frequent score is 92.

Mode of Operation — Expressed speed and focus of a search operation. In a search and rescue incident, the two modes are Rescue or Recovery.

Model Code — Consensus-based standards or codes established to provide uniformity in regulations in regards to construction, design, and use. When adopted by the local jurisdiction, these codes become enforceable laws.

Model Release — Legal document that grants permission to use a person's image or voice in photos or recordings.

Modem — Device that converts digital data from a computer to an analog signal that can be transmitted on a telephone line.

Modular Building — Building assembled at the factory in two or more all inclusive sections. All utilities and millwork are also installed at the factory, and connected when the building is delivered to a site.

Moist Adiabatic Lapse Rate — Rate of decrease in temperature with increasing height of an air mass.

Moisture Barrier — (1) Liner within a piece of protective clothing that is designed to keep water out. (2) Backing found on building insulation that prevents moisture from entering the structure.

Moisture Content — Amount of moisture that is available in the environment.

Molecular Sieve — Air-purification component that chemically absorbs water from compressed air.

Molotov Cocktail — Crude bomb made of a breakable container, such as a bottle filled with a flammable liquid; usually fitted with a wick that is ignited just before the bottle is hurled, creating a fire bomb.

Monitor — (1) To measure radioactive emissions from a substance with monitoring device. (2) To closely follow radio communications. (3) To observe and record the activities of an individual performing a function. (4) Individual assigned to supervise an evacuation process for a specified area within a structure, such as a ward monitor or floor monitor.

Monitor Appliance — Master stream appliance whose stream direction can be changed while water is being discharged; can be fixed, portable, or a combination. *Also known as* Monitor.

Monitor Roof — Roof style similar to an exaggerated lantern roof, with a raised section along the ridge line, providing additional natural light and ventilation.

Monitor Valve — Multidirectional valve used to control the flow of hydraulic oil through a hydraulic system.

Monitor Vent — Structure, usually rectangular in shape, which penetrates the highest point of a roof to provide additional natural light and/or ventilation. May have metal, glass, wired glass, or louvered sides, which are counterweighted, hinged, and designed to stay in place when held shut with a fusible link; this type of monitor vent is designed to ventilate an area when heat fuses the link. *Also known as* Monitor.

Monocoque — Construction technique in which an object's external skin supports the structural load of the object.

Monopropellant — Chemical or mixture of chemicals that is stable under specific storage conditions, but reacts very rapidly under other conditions to produce large amounts of energetic (hot) gasses. Monopropellants, such as hydrazine, are commonly used in aircraft emergency power units.

Moody Diagram — Diagram used with the Darcy-Weisbach friction loss computation technique in fluid flow; relates the Reynolds number, pipe size, and roughness to a friction factor.

Mooring — (1) Permanent anchoring equipment (attached by a chain to a buoy) to which a vessel may connect a line, wire, or chain, eliminating the need to use the vessel's anchor. (2) Act of securing a vessel. (3) Location where a vessel is berthed. *See* Anchorage, Berth, and Berthing Area.

Mop-Up — (1) Overhaul of a fire or hazardous material scene. (2) In wildland fire fighting, the act of making a fire safe after it is controlled by extinguishing or removing burning material along or near the control line, felling dead trees (snags), and trenching logs to prevent rolling.

Mortar — Cement-like liquid material that hardens and bonds individual masonry units into a solid mass.

Mortise — (1) Notch, hole, or space cut into a door to receive a lock case, which contains the lock mechanism. (2) Hole, groove, or slot cut into a wooden ladder beam to receive a rung tenon. (3) Notch, hole, or space cut into a piece of timber to receive the projecting part (tenon) of another piece of timber.

Mortise Cylinder — Lock cylinder for a mortise lock.

Mortise Lock — Lock mortised into a door. *Also known as* Box Lock.

Motel — *See* Hotel.

Motivation — Internal process, arousal and maintenance of behavior, in which energy is produced by needs or expended in the direction of goals. Motivation usually occurs in someone who is interested in achieving a goal. *See* Preparation Step.

Motive Power Unit — Engine, locomotive, or other power unit that provides power to move a train.

Motor Nerves — Nerves that carry impulses from the brain to the muscles.

Motor Vehicle Accident (MVA) — Term used when one vehicle hits a stationary object or another vehicle.

Moulage Kit — Makeup kit containing appliqué wounds and stage makeup; used during casualty simulations to simulate wounds on a mannequin or actor.

Mouse — To tightly wrap or cover the open end of a hook with a material to prevent an object from slipping off the hook. Mousing the hook prevents the hook from accidentally slipping off its intended anchor point.

Mouth-to-Mouth Breathing — Form of resuscitation that involves placing one's mouth over the patient's mouth, then breathing into the patient. *Also known as* Mouth-to-Mouth Resuscitation.

Mouth-to-Mouth Resuscitation — *See* Mouth-to-Mouth Breathing.

Movement Area — Runways, taxiways, and other areas of an airport that are used for taxiing, hover taxiing, air taxiing, takeoff, and landing of aircraft; does not include loading ramps and aircraft parking areas.

Move-Up — Procedure where uncommitted apparatus are relocated to stations emptied by apparatus committed to a long-term incident.

Moving Pivot — Method for positioning a ladder parallel to the objective while raising it.

MSDs — *See* Musculoskeletal Disorders.

MSDS — *See* Safety Data Sheet.

MSHA — *See* Mine Safety and Health Administration.

MT — Prefix to the name of a tank vessel powered by diesel machinery.

Mucous Membrane — Membrane that lines many organs of the body and contains mucus-secreting glands.

Mullion — Vertical division between multiple windows or a double door opening.

Multiagent Nozzle — Device that is capable of simultaneously applying any two of the following: foam, halon substitute, or dry chemical extinguishing agents. *See* Fog Nozzle.

Multibolt Lock — High security lock that uses metal rods to secure the door on all sides.

Multi-Casualty Incident — Emergency incident involving 20 or more transportable patients; may be classified as Extended, Major, or Catastrophic. *Also known as* Multiple-Casualty Incident. *See* Mass Casualty Incident.

Multigas Detector — Personal device that checks air quality against a wide range of harmful gases.

Multi-Jurisdictional Incident — Incident that involves or threatens to involve property in more than one jurisdiction; for example, a fire or other emergency in an industrial complex that may threaten to spread into an adjacent municipality and/or a navigable waterway.

Multiloop — Preferred method of software to attach to an anchor point.

Multiple Alarm — Additional alarm, such as second or third, that is a call for additional assistance or response.

Multiple Jacket Hose — Type of hose construction consisting of a combination of two separately woven jackets (double jackets), or two or more interwoven jackets, and lined with an inner rubber tube.

Multiple Patient Incident (MPI) — Incident where the emergency response resources are not overtaxed by the number of patients involved.

Multiple Points of Origin — Two or more separate points of fire origin discovered at a fire scene; indicates a strong possibility of arson.

Multiple-Stage Pump — Any centrifugal fire pump having more than one impeller.

Multipurpose Fire Extinguisher — Portable fire extinguisher that is rated for Class A, Class B, and Class C fires. *Also known as* A:B:C Extinguisher.

Multistage Centrifugal Pump — Centrifugal fire pump having more than one impeller. *See* Centrifugal Pump, Impeller, Self-Priming Centrifugal Pump, and Single-Stage Centrifugal Pump.

Multiversal — Master stream appliance that may be removed from the pumper and anchored on the ground for use.

Municipal Fire Department — refers to a functional division of the local government, that the state or province authorizes to provide fire protection for functional divisions of the local government.

Munitions Loaders — Military vehicles used to transport and load/off-load bombs and other munitions from military aircraft.

Muntin — Small members dividing the glass panes in a window sash.

Muscle Cars — High performance American cars made from 1964 to 1974, or modeled on cars built during that time; usually 2-door, rear wheel drive, mid-sized vehicles with oversized, V8 engines.

Musculoskeletal Disorders (MSDs) — Injuries or disorders of the muscles, nerves, tendons, joints, cartilage, and supporting structures caused or increased by sudden exertion or prolonged exposure to repeated motions, force, vibration, or awkward posture.

Mushroom Capital — Flaring conical head on a concrete column.

Mushrooming — Tendency of heat, smoke, and other products of combustion to rise until they encounter a horizontal obstruction; at this point they will spread laterally until they encounter vertical obstructions and begin to bank downward.

Muster List — List of crew members/passengers and their duty/emergency stations on a vessel.

Mutagen — Material that causes changes in the genetic system of a cell, in ways that can be transmitted during cell division. The effects of a mutagen may be hereditary.

MUTCD — *See Manual on Uniform Traffic Control Devices.*

Mutual Aid — Reciprocal assistance from one fire and emergency services agency to another during an emergency, based upon a prearranged agreement; generally made upon the request of the receiving agency.

Mutual Aid Agreement — Written agreement between agencies and/or jurisdictions that they will assist one another on request by furnishing personnel, equipment, and/or expertise in a specified manner.

Mutual Company — Insurance company that is run to benefit the insured, and in which any revenue above operating expenses is returned to policyholders as dividends.

MV — Prefix to the name of a vessel powered by diesel machinery.

MVA — *See* Motor Vehicle Accident.

Mystery Nozzle — Older style, variable gallonage, adjustable fog stream nozzle.

N

9-1-1 — Emergency number used to summon police, fire, or medical assistance throughout the U.S. The *Omnibus Crime Control and Safe Streets Act of 1968* authorized and designated its use.

Nacelle — Housing of an externally mounted aircraft engine.

Nader Pin — Bolt on a vehicle's door frame that the door latches onto in order to close.

Nader Safety Lock — Vehicle door safety lock; required by law on all passenger vehicles built since 1973.

NAFI — *See* National Association of Fire Investigators.

Nailable — Construction term for the ability of a material to accept nails.

Nasal Cannula — Small tubular prong that fits into the patient's nostril to provide supplemental oxygen; usually there are two, one for each nostril.

National Association of Fire Investigators (NAFI) — Nonprofit association of fire investigators dedicated to the education of fire investigators worldwide; offers training and certification.

National Burn Information Exchange — A national burn registry that collects data to compare organizational treatment variables and survival.

National Cave Rescue Commission — Division of the National Speleological Society that specializes in cave rescue.

National Crime Information Center (NCIC) — Computerized index of criminal justice information available to federal, state, and local law enforcement and other criminal justice agencies. NCIC is operational 24 hours a day, 365 days a year.

National Defense Area (NDA) — Temporary establishment of "federal areas" for the protection or security of U.S. Department of Defense (DoD) resources. Normally, NDAs are established for emergency situations, such as accidents; NDAs may be established or discontinued, or have their boundaries changed as necessary to provide protection or security of DoD resources.

National Electrical Code® (NEC®) — NFPA® 70, *National Electrical Code®*, is the standard for electrical activity; contains basic minimum provisions considered necessary to safeguard persons and buildings. It was prepared by the NFPA® National Electrical Code Committee.

National Fire Academy (NFA) — Division of the U.S. Fire Administration that provides training and certification to members of the fire and emergency services, public and private, across the U.S. Through its courses and programs, this federal agency works to enhance the ability of emergency service providers and allied professionals to deal more effectively with fire and related emergencies.

National Fire Codes® (NFC®) — Series of codes and standards pertaining to fire protection adopted and published by the National Fire Protection Association® and periodically revised by various committees.

National Fire Danger Rating System (NFDRS) — Multiple index matrix designed to provide fire-control and land-management personnel with a systematic means of assessing various aspects of fire danger on a day-to-day basis. The system is used to classify wildland fuels based on similar burning characteristics.

National Fire Incident Reporting System (NFIRS) — National fire incident data collection system managed by the United States Fire Administration. Local fire departments forward incident data to a state coordinator. The coordinator collects statewide fire incident data and reports information to the USFA.

National Fire Protection Association® (NFPA®) — U.S. nonprofit educational and technical association devoted to protecting life and property from fire by developing fire protection standards and educating the public. Located in Quincy, Massachusetts. *See* NFPA® 704 Labeling System and NFPA® 704 Placard.

National Fire Sprinkler Association (NFSA) — U.S.-based non-profit organization that champions the cause of wide-spread acceptance of fire sprinklers for fire protection.

National Highway Traffic Safety Administration (NHTSA) — Agency within the U.S. Department of Transportation (DOT) that publishes annual summary reports of fatal highway accidents.

National Incident Management System - Incident Command System (NIMS-ICS) — The U.S. mandated incident management system that defines the roles, responsibilities, and standard operating procedures used to manage emergency operations; creates a unified incident response structure for federal, state, and local governments.

National Institute For Occupational Safety And Health (NIOSH) — U.S. government agency that helps ensure workplace safety; investigates workplaces, recommends safety measures, and produces reports about on-the-job fire injuries. Operates as part of the Centers for Disease Control and Prevention, within the U.S. Department of Health and Human Services.

National Professional Qualifications Board (Pro Board) — Nonprofit organization that provides accreditation to organizations that **certify** uniform members of public fire departments, both career and volunteer.

National Research Council (NRC) — Canada's publicly funded premier organization for scientific research and development.

National Response Center — U.S. federal organization that coordinates the response of numerous agencies to hazardous materials incidents.

National Response Framework (NRF) — Document that provides guidance on how communities, states, the U.S. federal government, and private-sector and nongovernmental partners conduct all-hazards emergency response.

National Safe Kids® Campaign — Nationwide coalition with the goal of reducing preventable injuries to children.

National Standard Thread (NST) — Screw thread of specific dimensions for fire service use as specified in NFPA® 1963, *Standard for Screw Threads and Gaskets for Fire Hose Connections*.

National Transportation Safety Board (NTSB) — Agency within the U.S. Department of Transportation (DOT) that maintains a fire-related database on aircraft and railway accidents, as well as highway accidents involving hazardous materials injuries.

National Wildland Fire Coordinating Group (NWCG) —Multiagency group that coordinates wildfire management programs, in order to increase effectiveness and promote cooperation. NWCG provides a forum to discuss wildfire-related issues, and acts as the certifying body for all courses in the National Fire Curriculum. Participating agencies are the Department of Agriculture Forest Service (FS); the U.S. Fire Administration (USFA); state forestry agencies through the National Association of State Foresters (NASF); and four Department of the Interior agencies - Bureau of Land Management (BLM), National Park Service (NPS), Bureau of Indian Affairs (BIA), and the Fish and Wildlife Service (FWS).

Natural Barrier — Area where the lack of flammable material obstructs the spread of wildland fires.

Natural Cover Fire — *See* Wildland Fire.

Natural Fiber Rope — Utility rope made of manila, sisal, or cotton; not accepted for life-safety applications.

Natural Fire Cause — Classification referring to fires where human intervention has not been involved in the ignition process; examples include fires caused by lightning, storms, or floods.

Natural Ventilation — Techniques that use the wind, convection currents, and other natural phenomena to ventilate a structure without the use of fans, blowers, smoke ejectors, or other mechanical devices.

Naval Architecture — Branch of knowledge concerned with the design and construction of things that float, such as vessels, submarines, docks, or yachts.

Navigable — Term for any body of water suitable for navigation by any particular vessel, although not necessarily all vessels.

NBIE — *See* National Burn Information Exchange.

NCIC — *See* National Crime Information Center.

Neat Cement — Pure cement uncut by a sand mixture.

Nebulizer — Electrically powered machine that turns liquid medication into a mist.

Needs Analysis — (1) Assessment of the gap between what exists and what should exist, with regard to an organization's staffing, services, equipment, or training. (2) Assessment of the gap between the training an organization provides and the training it should provide, either currently or in the future.

Needs Assessment — Analysis identifying life-support and critical infrastructure requirements.

Negative — Clear text radio response for "no."

Negative Buoyancy — Tendency to sink.

Negative Heat Balance — Condition that occurs in a fire when heat is dissipated faster than it is generated and, therefore, will not sustain combustion. *See* Positive Heat Balance.

Negative Pressure — Air pressure less than that of the surrounding atmosphere; a partial vacuum.

Negative-Pressure Phase — Portion of an explosion in which air rushes back toward the center; caused by the low pressure created from the positive-pressure phase.

Negative-Pressure Ventilation (NPV) — Technique using smoke ejectors to develop artificial air flow and to pull smoke out of a structure. Smoke ejectors are placed in windows, doors, or roof vent holes to pull the smoke, heat, and gases from inside the building and eject them to the exterior.

Neglect — Failure to act on behalf of, or in protection of, an individual in one's care.

Negligence — Breach of duty in which a person or organization fails to perform at the standard required by law, or that would be expected by a reasonable person under similar circumstances. *See* Gross Negligence, Proximate Cause, Standard of Care, and Tort.

Nephrotoxic Agent — Chemical that damages the kidneys. *Also known as* Nephrotoxin.

Nerve Agent — Toxic agent that attacks the nervous system by affecting the transmission of impulses. *See* Chemical Warfare Agent.

Nesting — *See* Ladder Nesting.

Net Pressure — *See* Net Pump Discharge Pressure.

Net Pump Discharge Pressure (NPDP) — Actual amount of pressure being produced by the pump; difference between the intake pressure and the discharge pressure. *Also known as* Engine Pressure or Net Pressure.

Network — Informal group of persons with a mutual interest who share ideas, information, and resources.

Networking — (1) Communicating and creating links between people and clusters of people. (2) Process of meeting others and determining resources that others have that can assist with the accomplishment of the risk-reduction initiatives.

Neurogenic Shock — Shock caused by the overexpansion of blood vessels due to damage to the brain, spinal cord, or other nerves.

Neurotoxic Agent — Chemical that damages the central nervous system. *Also known as* Neurotoxin.

Neutral Conductor — *See* Grounded Conductor.

Neutral Plane — Level at a compartment opening where the difference in pressure exerted by expansion and buoyancy of hot smoke flowing out of the opening and the inward pressure of cooler, ambient temperature air flowing in through the opening is equal.

Neutral Pressure Plane — Point within a building, especially a high-rise, where the interior pressure equals the atmospheric pressure outside. This plane will move up or down, depending on temperature and wind.

Neutron — Part of the nucleus of an atom that has a neutral electrical charge yet produces highly penetrating radiation; ultrahigh energy particle that has a physical mass like alpha or beta radiation but has no electrical charge. *See* Radiation (2).

Newel — Outer posts of balustrades and the stiffening posts at the angle and platform of stairways.

NFA — *See* National Fire Academy.

NFC — *See* National Fire Codes.

NFDRS — *See* National Fire Danger Rating System.

NFIRS — *See* National Fire Incident Reporting System.

NFPA® — *See* National Fire Protection Association®.

NFPA® 101, *Life Safety Code* — Widely used building standard that addresses the life safety aspects of building design in order to protect lives in the event of a fire. Formerly known as *Code for Safety to Life from Fire in Buildings and Structures.*

NFPA® 1035 — Document outlining the job performance requirements that can be used to determine whether an individual possesses the skills and knowledge to perform as a public or private fire and life safety educator.

NFPA® 704 Labeling System — Labeling system derived from NFPA® 704, *Standard System for the Identification of the Hazards of Materials for Emergency Response.* This labeling system is intended to aid in identifying hazardous materials in fixed facilities; its color-coded, symbol-specific placard is divided into sections that identify the degree of hazard with respect to health, flammability, reactivity, and special hazards. *See* NFPA® 704 Placard.

NFPA® 704 Placard — Color-coded, symbol-specific placard affixed to a structure to inform of fire hazards, life hazards, special hazards, and reactivity potential. The placard is divided into sections that identify the degree of hazard according to health, flammability, reactivity, and special hazards. *See* NFPA® 704 Labeling System.

NFPA® Fire Department Survey — Annual survey in which the NFPA® gathers data and statistics about fires in the U.S.

NFSA — See National Fire Sprinkler Association.

NHTSA — See National Highway Traffic Safety Administration.

Niells-Robertson Stretcher — Stretcher that immobilizes a patient, prevents further spinal damage, and protects the head; particularly effective for cave and confined-space rescues.

Night Inversion — See Inversion.

Night Latch — Button on a rim lock that prevents retracting the latch from the outside.

Night Order Book — Written instructions, special orders, or reminders from the captain or master for each officer taking night watch; placed in the chart room before the captain or master retires for the night.

NIMS-ICS — See National Incident Management System - Incident Command System.

NIOSH — See National Institute for Occupational Safety and Health.

Nitrogen — Inert gas that is commonly used as a propellant in portable fire extinguishers.

Nitrogen Oxides — Gases that consist of nitrogen and oxygen; commonly given off as a by-product of the combustion process.

Nitrogen-Bearing Substances — Substances that produce hydrogen cyanide; found in synthetic fibers such as nylon and polyurethane foam, some plastics (particularly in aircraft), and natural fibers such as wool, rubber, and paper.

Nitroglycerin — Viscous liquid used in the production of dynamite and other explosives; toxic and highly sensitive to heat and shock. It is also used medically as a drug to treat angina pectoris, usually taken under the tongue.

Noded Spheroid Tank Low-pressure fixed facility storage tank held together by a series of internal ties and supports that reduce stress on the external shell. See Low-Pressure Storage Tank, Pressure Storage Tank, and Spheroid Tank.

Noise Reduction Rating (NRR) — The measurement of how well a hearing protection device will block noise. The rating is stated in decibels (dB) and established by the U.S. Environmental Protection Agency (EPA).

NOMEX® Fire-Resistant Material — Flame-resistant fabric used to construct firefighter's personal protective equipment.

NOMEX® Hood — See Protective Hood.

Nominal Group Process — Classroom discussion format that requires students to follow a decision-making process similar to the processes that they will encounter in their professional duties.

Nomograph — Chart in which a straight line is drawn between two scales intersecting a third scale which satisfies an equation. Often used based on the Hazen-Williams formula, to assist in the determination of fire flows.

Nonaspirating Foam Nozzle — Nozzle that does not draw air into the foam solution stream. The foam solution is agitated by the nozzle design, causing air to mix with the solution after it has exited the nozzle. See Air-Aspirating Foam Nozzle.

Nonbearing Wall — See Nonload-Bearing Wall.

Nonbulk Packaging — Package that has the following characteristics: (a) maximum capacity of 119 gallons (450 L) or less as a receptacle for a liquid, (b) maximum net mass of 882 pounds (400 kg) or less and a maximum capacity of 119 gallons (450 L) or less as a receptacle for a solid, and (c) water capacity of 1,000 pounds (454 kg) or less as a receptacle for a gas. See Bulk Packaging and Packaging (1).

Noncombustible — Incapable of supporting combustion under normal circumstances. See Combustion and Nonflammable.

Nonconforming Apparatus — Apparatus that does not conform to NFPA® standards.

Nondestructive Testing — Method of testing metal objects that does not subject them to stress-related damage.

Nondiked Area — Any location where flammable or combustible liquids might be spilled but not contained within a system of predesigned barriers. See Diked Area.

Nondirectional Anchor — Anchor that is capable of supporting a load in any direction.

Nonflammable — Incapable of combustion under normal circumstances; normally used when referring to liquids or gases. See Flammable and Noncombustible.

Nonflammable Gas — Compressed gas not classified as flammable. See Compressed Gas, Flammable Gas, and Gas.

Nonintervention Mode See Nonintervention Strategy.

Nonintervention Operations — Operations in which responders take no direct actions on the actual problem. See Defensive Operations and Offensive Operations.

Nonintervention Strategy — Overall plan for incident control established by the incident commander in which responders take no direct actions on the actual problem. *Also known as* Nonintervention Mode. See Defensive Strategy, Offensive Strategy, and Strategy.

Nonionizing Radiation — Series of energy waves composed of oscillating electric and magnetic fields traveling at the speed of light; examples include ultraviolet radiation, visible light, infrared radiation, microwaves, radio waves, and extremely low frequency radiation. *See* Ionizing Radiation and Radiation (2).

Non-Lifeline Rope — Rope that does not meet the requirements set forth in NFPA® 1983, *Standard on Life Safety Rope and Equipment for Emergency Services.*

Nonliquefied Gas — Gas, other than a gas in a solution, that under the charging pressure is entirely gaseous at 70°F (21°C). *See* Gas and Liquefied Compressed Gas.

Nonload-Bearing Wall — Wall, usually interior, that supports only its own weight. These walls can be breached or removed without compromising the structural integrity of the building. *Also known as* Nonbearing Wall.

Nonmetallic (NM) Shielded Cable — Factory assemblies with two or more wires that have a nonmetallic outer sheath which is moisture resistant and fire retardant.

Nonpersistent Agent — Chemical agent that generally vaporizes and disperses quickly (in less than 10 minutes). *See* Persistent Agent and Vapor Pressure.

Nonpressure Intermodal Tank — Portable tank that transports liquids or solids at a maximum pressure of 100 psi (689 kPa) {6.9 bar}. *Also known as* IM Portable Tank. *See* Intermodal Container and Pressure Intermodal Tank.

Nonpressure Liquid Tank — Cargo tank truck used to carry flammable liquids (such as gasoline and alcohol), combustible liquids (such as fuel oil), Division 6.1 poisons, and liquid food products. *See* Cargo Tank Truck.

Nonpressure Storage Tank — *See* Atmospheric Storage Tank.

Nonprofit — Legal status that the U.S. Internal Revenue Service (IRS) may give to an organization that is not run on a for-profit basis; nonprofit, tax-exempt organizations are sometimes called 501(c) organizations, after the section of the Internal Revenue Code that describes them. Designation as a nonprofit requires the submission of a written application to the IRS for review.

Nonrated Concentrate — Foam concentrate that has not been tested or certified by Underwriters Laboratories Inc. *See* Foam Concentrate and Rated Concentrate.

Non-Rebreather Mask — Device used to deliver a high concentration of oxygen through the mouth and nose to injured or ill patients.

Nonseated Explosion — Explosion having no identifiable epicenter or seat; typically deflagrations in fuel-air mixtures and dust clouds. *See* Seated Explosion.

Nonstructured Question — Question that does not offer a list of answer choices. Respondents are simply asked to write their response to the question.

Nonthreaded Coupling — Coupling with no distinct male or female components. *Also known as* Sexless Coupling or Storz Coupling.

Nonverbal Cues — Messages without words; often transmitted in gestures, posture or body language, eye contact, facial expression, tone of voice, or appearance.

Normal Operating Pressure — Amount of pressure that is expected to be available from a hydrant, prior to pumping. *See* Static Pressure.

Normal Path of Travel — *See* Common Path of Travel.

Normalization of Deviation — State of a safety culture in which acting against SOP/Gs becomes normal behavior rather than an exception.

Norm-Referenced Assessment — Form of assessment in which a student's performance is compared to that of other students. Grades are determined by comparing scores to the class average, and assigning grades based on where students scored compared to that average. *Also known as* Norm-Referenced Testing. *See* Test.

Nosecup — Device inside a facepiece that directs the wearer's exhalations away from the facepiece lens, thus preventing internal fogging of the lens.

Nosing — Usually rounded edge of a stair tread that projects over a riser.

Noxious — Physically harmful or destructive to living beings; unwanted or troublesome.

Nozzle — Appliance on the discharge end of a hoseline that forms a fire stream of definite shape, volume, and direction.

Nozzle Pressure — Velocity pressure at which water is discharged from the nozzle.

Nozzle Reaction — Counterforce directed against a person holding a nozzle or a device holding a nozzle by the velocity of water being discharged.

Nozzleperson — Individual assigned to operate a fire department nozzle. *Also known as* Nozzleman.

NRC — (1) *See* National Response Center. (2) *See* Nuclear Regulatory Commission. (3) *See* National Research Council.

NRF — *See* National Response Framework.

NST — *See* National Standard Thread.

NTSB — *See* National Transportation Safety Board.

Nuclear Heat Energy — Creation of heat through the splitting apart or combining of atoms.

Nuclear Radiation — Emission of alpha, beta, and gamma radiation resulting from the decay of an atomic nucleus.

Nuclear Regulatory Commission (NRC) — U.S. agency that regulates commercial nuclear power plants and the civilian use of nuclear materials, as well as the possession, use, storage, and transfer of radioactive materials.

Nurse Tanker — Very large water tanker (generally 4,000 gallons [15 140 L] or larger) that is stationed at the fire scene and serves as a portable reservoir rather than as a shuttle tanker. *Also known as* Nurse Tender.

NWCG — *See* National Wildland Fire Coordinating Group.

O

Objective — (1) Purpose to be achieved by tactical units at an emergency. (2) Specific, measurable, achievable statement of intended accomplishment. (3) Purpose or educational goal of a presentation or program; a step necessary to achieve a stated goal. (4) Unbiased; dealing with facts of interpreting results without the distortion of personal feelings, prejudices, or interpretations. *See* Behavioral Objective, Enabling Objective, and Performance Objective.

Objective Test — Test or test items designed so that all qualified test developers agree on the correct answer. Test items and their answers are based on course objectives that were developed from some selected criterion or standard. Types of objective tests include multiple-choice, matching, true-false, and short answer/completion. *See* Test.

Observation — Actually seeing or watching a person's behavior in a natural setting. As a means of evaluation, direct observation provides very reliable information on the effects of programs.

Observational Survey — Method used when evaluators want to observe if behavioral, environmental, or lifestyle changes are occurring among a target population.

Occlusive Dressing — Watertight dressing for a wound.

Occupancy — (1) General fire and emergency services term for a building, structure, or residency. (2) Building code classification based on the use to which owners or tenants put buildings or portions of buildings. Regulated by the various building and fire codes. *Also known as* Occupancy Classification.

Occupancy Classification — Classification given to a structure by the model code used in that jurisdiction, based on the intended use for the structure. *See* Assembly, Building Code, Industrial, Mercantile, and Occupancy.

Occupant — Person who lives in, uses, occupies, or has other possession of an apartment, house, or other premise. *See* Occupancy (1) and Occupant Load.

Occupant Load — Total number of people who may occupy a building or portion of a building at any given time. *See* Occupant.

Occupant Services Unit — Sector/Group designated to provide information and support services to the victims of a fire. Among the services provided may be assistance in contacting relatives, public agencies, and/or charitable institutions for transportation, temporary shelter, and other basic needs.

Occupational Analysis — Method of gathering information about an occupation and developing a description of qualifications, conditions for performance, and an orderly list of duties.

Occupational Safety and Health Administration (OSHA) — U.S. federal agency that develops and enforces standards and regulations for occupational safety in the workplace.

Odor Test — Qualitative test of facepiece fit.

ODP — *See* Office for Domestic Preparedness.

Offensive Fire Attack — Aggressive, usually interior fire attack that is intended to stop the fire at its current location. *Also known as* Offensive Mode Attack.

Offensive Fire Fighting — Fire control activities intended to reduce the size of a fire and extinguish it.

Offensive Mode — *See* Offensive Strategy.

Offensive Mode Attack — *See* Offensive Fire Attack.

Offensive Operations — Operations in which responders take aggressive, direct action on the material, container, or process equipment involved in an incident. *See* Defensive Operations and Nonintervention Operations.

Offensive Strategy — (1) In wildland fire fighting, a direct attack on the fire perimeter by crews, engines, or aircraft, or an aggressive indirect attack such as backfiring. (2) Overall plan for incident control established by the incident commander (IC) in which responders take aggressive, direct action on the material, container, or process equipment involved in an incident. *See* Defensive Strategy, Nonintervention Strategy, and Strategy.

Off-Gassing — Emission of toxic gases, caused by the release of chemicals from non-metallic substances under ambient or greater pressure.

Office for Domestic Preparedness (ODP) — Former U.S. agency under the Department of Homeland Security that issued federal emergency responder guidelines for events involving weapons of mass destruction.

Officer — Any member of the fire and emergency services with supervisory responsibilities; company officer level and above.

Ohm — Basic unit of measurement of electrical resistance, symbolized either by Ω or R. One ohm is the resistance between two points in a conductor when one volt produces one ampere of current.

Ohm's Law — Mathematical relationship between a circuit's voltage (V), current (I), and resistance (R): V = IR.

Ohmmeter — Device designed to measure electrical resistance in a circuit.

OI — *See* Standard Operating Procedure (SOP).

Oil Tanker — Tank vessel specially designed for the bulk transport of petroleum products by sea. *Also known as* Tanker.

OJT — *See* On-the-Job Training.

Olfactory Fatigue — Gradual inability of a person to detect odors after initial exposure; can be extremely rapid with some toxins, such as hydrogen sulfide.

One- and Two-Family Dwellings — Subdivision of residential property classification consisting of structures that have one or two dwelling units; each is occupied by members of a single family, with as many as three non-family-members living in rented rooms.

One-Compartment Subdivision — Subdivision of a vessel by bulkheads that will result in a vessel remaining afloat with any one compartment flooded under certain conditions. *See* Compartmentation.

On-the-Job Training (OJT) — System of training firefighters that makes full use of personal contact between firefighters and their immediate supervisor; trains firefighters both physically and psychologically for the position they will perform.

Opacity — Capacity to obstruct the transmission of radiant energy-like heat.

Open Burning — Description of a fire burning completely in the open with no restrictions to its oxygen supply.

Open Butt — End of a charged hoseline that is flowing water without a nozzle or valve to control the flow.

Open Learning — Form of distance learning designed so that participants attend a minimum number of classes and complete reading and writing assignments that are turned in or mailed to the instructor. *See* Distance Learning.

Open Sprinkler — Sprinkler that lacks a heat-sensitive element and is open at all times; used on a deluge-type sprinkler system. *See* Closed Sprinkler and Foam-Water Sprinkler.

Open Stringer — Stringer that is notched to follow the lines of the treads and risers of a stairway.

Open Up — To ventilate a building or other confined space.

Open Web Joist — Joist with a web composed of materials that do not fill the entire web space. Examples include steel bars or tubes.

Open Web Truss — Structural assembly consisting of a top chord and a bottom chord connected by a triangulated series of web components such as bars or tubes.

Open-Circuit Airline Equipment — Airline breathing equipment that allows exhaled air to be discharged to the open atmosphere.

Open-Circuit Self-Contained Breathing Apparatus — SCBA that allows exhaled air to be discharged or vented into the atmosphere.

Open-Ended Question — Question that requires more than a yes or no answer.

Open-Head System — *See* Deluge Sprinkler System.

Opening — Beginning or motivational part of a fire and life safety educator's presentation.

Open-Top Floating Roof Tank — *See* External Floating Roof Tank.

Operating Budget — Budget intended to fund the day-to-day operations of the department or agency; usually includes the costs of salaries and benefits, utility bills, fuel, and preventive maintenance.

Operating Instruction (OI) — *See* Standard Operating Procedure (SOP).

Operation — One step in performing a job skill within an occupation. Operations are listed in the order in which they are performed.

Operation School Burning — Series of fire tests in Los Angeles that analyzed the physiological effects of fire in schools and determined tenability.

Operational Budget — Document that outlines operating expenses for any course, curriculum, or training program.

Operational Period — Period of time scheduled for execution of a specified set of operational goals and objectives as identified in the incident action plan (IAP). An operational period may be 12 hours, 24 hours, or any other arbitrary amount of time. A new IAP is created for each operational period.

Operational Readiness — Ready for or in condition to undertake a predetermined function.

Operational Readiness Inspection — Inspecting an apparatus and equipment on the apparatus to ensure that all equipment is in place, clean, and ready for service.

Operational Step — Smallest aspect of performing a task; to complete the task, students perform a series of operational steps in sequential order.

Operational Strategy — Overall plan for incident attack and control.

Operational Tactics — Methods of employing equipment and personnel to obtain optimum results when carrying out operational strategies.

Operational Tests — Tests to ensure that aerial device controls are working properly.

Operations Level — Level of training established by the National Fire Protection Association® allowing first responders to take defensive actions at hazardous materials incidents. *See* Awareness Level and Operations Plus.

Operations Plus — Level of training allowing first responders to take defensive actions at all hazardous materials incidents, plus offensive actions when dealing with gasoline, diesel fuel, natural gas, and liquefied petroleum gas (LPG).

Operations Section — Incident command system section responsible for all tactical operations at the incident. The Operations Section includes branches, divisions and/or groups, task forces, strike teams, single resources, and staging areas. *Also known as* Ops Section.

Operations Section Chief — Person responsible to the incident commander for managing all tactical operations directly applicable to accomplishing the incident objectives. *Also known as* Ops Chief or Ops Section Chief.

Operations Security (OPSEC) — Process of identifying critical information that hostile intelligence systems might obtain and use, and eliminating or reducing vulnerabilities that could be exploited to gain such information.

Opinion Testimony — When a credentialed expert takes the stand to present not just factual information, but interpretation of those facts based on his or her area of expertise.

OPSEC — *See* Operations Security.

Optical Stabilization (OS) — *See* Image Stabilization (IS).

Optimum Ratio — *See* Stoichiometric Ratio.

Oral Airway — Device inserted in the patient's upper airway to keep the tongue from blocking the airway.

Oral Test — Form of assessment in which candidates are tested in face-to-face discussion with an examiner or group of examiners. Candidates are usually assessed individually, often in conjunction with written and/or skill testing. *See* Test.

Order — Specific rule, regulation, or authoritative direction.

Ordinance — Local or municipal law that applies to persons and things of the local jurisdiction; a local agency act that has the force of a statute; different from law that is enacted by federal or state/provincial legislatures. *See* Law.

Ordnance — Bombs, rockets, ammunition, and other explosive devices carried on most military aircraft, ships, and combat vehicles

Organic Peroxide — Any of several organic derivatives of the inorganic compound hydrogen peroxide.

Organic Vapor/Acid Gas Cartridge — Device in respiratory protection equipment that absorbs harmful vapors and gases from breathable air, preventing inhalation. Each cartridge is a one-time use item that lasts 8 to 10 hours.

Organizational Protocol — Rules set by the JFIS organization or JFS coalition that guide how a juvenile firesetting intervention program will be administered.

Organophosphate Pesticides — Chemicals that kill insects by disrupting their central nervous systems; these chemicals inactivate acetylcholinesterase, an enzyme which is essential to nerve function in insects, humans, and many other animals. Because they have the same effect on humans, they are sometimes used in terrorist attacks.

Orientation — (1) Direction in which a building faces. (2) Relating blueprints to the actual structure with respect to direction. (3) Location or position relative to the points of the compass.

Oriented Strand Board (OSB) — Wooden structural panel formed by gluing and compressing wood strands together under pressure. This material has replaced plywood and planking in the majority of construction applications. Roof decks, walls, and subfloors are all commonly made of OSB.

Orifice — Opening, usually circular, through which water is discharged.

Orifice Plate Meter — Device used for measuring water flow that is similar in principle to a Venturi meter. The change of water velocity is accomplished by using a plate with an orifice that is smaller than the diameter of the pipe in which it is placed.

Origin — *See* Point of Origin.

O-Ring — Circular gasket with rounded edges used for sealing between two machined surfaces; usually made of rubber or silicone.

ORM-D — Other Regulated Material — Domestic. *See* Other Regulated Material.

Orthopedic Injuries — Injuries to the bones, mainly the extremities.

OS&Y Valve — Outside stem and yoke valve; a type of control valve for a sprinkler system in which the position of the center screw indicates whether the valve is open or closed. *Also known as* Outside Screw and Yoke Valve.

OSB — *See* Oriented Strand Board.

OSHA — *See* Occupational Safety and Health Administration.

Other Regulated Material — Material, such as a consumer commodity, that does not meet the definition of a hazardous material and is not included in any other hazard class but possesses enough hazardous characteristics that it requires some regulation; presents limited hazard during transportation because of its form, quantity, and packaging.

Outage — Difference between the full or rated capacity of a tank or tank car and its actual content.

Outboard — Anything that is on the seaside of a vessel; anything mounted outside the hull.

Outcome Evaluation — Measures changes in the occurrence of incidents over time; also involves documenting anecdotal success stories of how prevention efforts impacted community risk.

Outcome Objective — Desired student performance resulting from a lesson or presentation.

Outer Shell — Outer fabric of protective clothing.

Outlet Valve — Valve farthest downstream to which a discharge hose is attached in a tank piping system.

Out-of-Service — Unit that is not available for assignment to a response.

Out-of-Service Resources — Resources, such as companies or crews, assigned to an incident but unable to respond for mechanical, rest, or personnel reasons.

Outreach Activity — Method the public fire and life safety educator uses to reach an audience, generally a direct presentation.

Outrigger — *See* Stabilizer.

Outside Aid — Assistance from agencies, industries, or fire departments that are not part of the agency having jurisdiction over the incident.

Outside Hand or Foot — Hand or foot furthest from the ladder and furthest from the other member of a two-firefighter team.

Outside Sprinkler — System with open sprinklers, automatically or manually operated, to protect a structure or window openings against a severe exposure hazard.

Outside Standpipe — Exterior standpipe riser that is equipped with a fire department siamese connection.

Outside Width — Dimension from the outside surface of one ladder beam to the outside surface of the opposite ladder beam, or the widest point of a ladder including staypoles when provided, whichever is greater.

Overburden — Loose earth covering a building site.

Overcurrent — Current in excess of the rated current of equipment or the ampacity of a conductor; may be caused by an overload, short circuit, or ground fault.

Overfiring — Overheating of a solid-fuel room heater by heating the unit and its connections until they glow a dull red, thus subjecting the unit and surrounding materials to higher-than-expected heat.

Overhand Safety Knot — Supplemental knot tied to prevent the primary knot from failing; prevents the running end of the rope from slipping back through the primary knot.

Overhaul — Operations conducted once the main body of fire has been extinguished; consists of searching for and extinguishing hidden or remaining fire, placing the building and its contents in a safe condition, determining the cause of the fire, and recognizing and preserving evidence of arson.

Overhead — (1) Underside of a deck; ceiling of a vessel's compartment. *Also known as* Deckhead. (2) *See* Transparency.

Overhead Door — Door that opens and closes above a large opening, such as in a warehouse or garage, and is usually of the rolling, hinged-panel, or slab type.

Overhead Expenses — Expenses necessary for routine administration of a project, program, or department.

Overhead Hazards — Identified during recon, these include problems with overhead utilities as well as loose debris on the structure, damaged chimneys, and the weight or movement of unorganized rescuers above the recon team.

Overhead Question — Type of question or questioning method in which an instructor asks a question of the whole group rather than just one person, and to which anyone is free to respond.

Overlapping Rotation — Tactic in which personnel are rotated in and out of an assignment in overlapping shifts; used during prolonged rescue operations to maintain effectiveness.

Overload — Operation of equipment or a conductor in excess of its rated ampacity; continuous overload may result in overheating that damages the equipment.

Overpressure — Air pressure above normal or atmospheric pressure.

Override Collision — Occurs when a striking vehicle collides with another vehicle and comes to rest on top of the vehicle being struck.

Overrun — Area beyond the end of the runway that has been cleared of nonfrangible obstacles and strengthened to allow overruns without serious damage to the aircraft. *Also known as* Clearway.

Overrun Area — *See* Stopway Area.

Overt — Not secret; in the open.

Overthrottling — Injecting or supplying the diesel engine with more fuel than can be burned.

Overturn Protection — Structural component that provides protection for fittings on top of a tank in case of rollover. May be combined with flashing rail or flashing box.

Oxidation — Chemical process that occurs when a substance combines with an oxidizer such as oxygen in the air; a common example is the formation of rust on metal. *See* Decomposition and Pyrolysis.

Oxides of Nitrogen — Nitrogen oxide (NO2) and nitric oxide (NO); can mix with moisture in the air and respiratory tract and form nitric and nitrous acids that can burn the lungs.

Oxidizer — Any material that readily yields oxygen or other oxidizing gas, or that readily reacts to promote or initiate combustion of combustible materials. (Reproduced with permission from NFPA® 400-2010, Hazardous Materials Code, Copyright©2010, National Fire Protection Association®)

Oxidizing Agent — Substance that oxidizes another substance; can cause other materials to combust more readily or make fires burn more strongly. *Also known as* Oxidizer.

Oxyacetylene Cutting Torch — Commonly used torch that burns oxygen and acetylene to produce a very hot flame. Used as a forcible entry cutting tool for penetrating metal enclosures that are resistant to more conventional forcible entry equipment.

Oxygen (O_2) — A chemical element. Colorless, odorless, tasteless gas constituting 21 percent of the atmosphere.

Oxygen Mask — Device that fits over a patient's nose and mouth and is used to administer supplemental oxygen.

Oxygen Therapy — Administration of oxygen through a mask or tube in the nose to increase the amount of oxygen in the patient's blood.

Oxygenator — Simplified, convenient oxygen administration system for prolonged home use.

Oxygen-Deficient Atmosphere — Atmosphere containing less than the normal 19.5 percent oxygen. At least 16 percent oxygen is needed to produce flames or sustain human life.

Oxygen-Enriched Atmosphere — Area in which the concentration of oxygen is in excess of 21 percent by volume or 21.3 kPa; typically 23.5 percent for confined spaces, as defined by the Occupational Safety and Health Administration (OSHA).

Oxygen-Generating Apparatus — SCBA that chemically generates breathing oxygen. This apparatus is no longer acceptable within the fire service.

Oxygen-Generating Canister — Container of chemicals that generate oxygen when mixed with the moisture of an individual's breath.

Oxyhemoglobin — Combination of oxygen and hemoglobin.

P

Package Markings — Descriptive name, instructions, cautions, weight, and specification marks required on the outside of hazardous materials containers. *See* Label, Marking, and Placard.

Packaging — (1) Shipping containers and their markings, labels, and/or placards. *See* Bulk Packaging, Combination Packaging, Composite Packaging, Excepted Packaging, Individual Container, Industrial Packaging, Intermediate Bulk Container (IBC), Nonbulk Packaging, Strong, Tight Container, Type A Packaging, and Type B Packaging. (2) Readying a victim for transport.

Padlock — Detachable, portable lock with a hinged or sliding shackle.

Pager — Compact radio receiver used for providing one-way communications

Paid-on-Call Firefighter — Firefighter who receives reimbursement for each call that he or she attends. *Also known as* Call Firefighter.

Pancake Collapse — Situation where the weakening or destruction of bearing walls cause the floors or the roof to collapse, allowing debris to fall as far as the lower floor or basement. Typically, there will be little void space with these types of collapse. *Also known as* Pancake.

Pandemic — Epidemic occurring over a very wide area (several countries or continents), usually affecting a large proportion of the population.

Panel Cutter — Chisel-like tool used with a mallet or hammer to cut through sheet metal.

Panel Door — Door inset with panels, which are usually made of wood, metal, glass, or plastic.

Panel Points — Points where the load of roof panels is transferred to trusses.

Panelboard — Single or multiple panels that contain conductive bus bars and automatic overcurrent protection devices, such as circuit breakers or fuses. These panels may also contain manually operated switches.

Panelized Home — Home assembled on site consisting of constructed panels made of foam insulation sandwiched between sheets of plywood. The panels are assembled on-site and require no framing members.

Panic — Sudden, excessive feeling of alarm or fear, usually affecting a group of people; originates in some real or supposed danger that is vaguely apprehended. May lead to extravagant and injudicious efforts to secure safety.

Panic Hardware — Hardware mounted on exit doors in public buildings that unlocks from the inside and enables doors to be opened when pressure is applied to the release mechanism. *Also known as* Exit Device.

Pantograph — Mechanical linkage device which maintains electrical contact with a contact wire and transfers power from the wire to the traction unit of electric buses, locomotives, and trams.

Pantograph Seal — *See* Mechanical Shoe Seal.

PAPR — *See* Powered Air-Purifying Respirator.

PAR — *See* Personnel Accountability Report.

Paracargo — Anything intentionally dropped or intended for dropping from an aircraft by parachute, other retarding devices, or free fall.

Paradoxical Movement — Motion of an injured section of a flail chest; opposite to the normal movement of the chest wall.

Parallel Attack — Constructing a fireline parallel to a wildland fire's edge. After the line is constructed, the fuel inside the line is burned out.

Parallel Chord Truss — Truss constructed with the top and bottom chords parallel. These trusses are used as floor joists in multistory buildings and as ceiling joists in buildings with flat roofs.

Parallel Circuit — Circuit configuration in which different components in the circuit receive current from different pathways, which allows individual components in the circuit to continue to function even if another component fails.

Parallel Method — Constructing, with hand tools, a fire line parallel to a wildland fire's edge. After the line is constructed, the fuel inside the line is burned out.

Parallel Operation — Operation of a multistage pump in which each of its impellers receives water from a common source. Water flowing through the pump is divided among the stages or impellers and contributes volume directly to the discharge. *Also known as* Volume Operation.

Paramedic — (1) Certified emergency medical professional who provides advanced life support. (2) Professional level of certification for emergency medical personnel who are trained in advanced life support procedures.

Paramedic Engine — Fire engine company that carries firefighter/paramedics and paramedic equipment.

Paramedic Unit — *See* Medic Unit (1).

Parapet — (1) Portion of the exterior walls of a building that extends above the roof. A low wall at the edge of a roof. (2) Any required fire walls surrounding or dividing a roof or surrounding roof openings such as light/ventilation shafts.

Parapet Wall — Vertical extension of an exterior wall, and sometimes an interior fire wall, above the roofline of a building.

Parbuckling — The use of nets or cables to roll an item. In water rescue, this method is used roll victims into a watercraft.

Park — Rest position of a ladder with one beam resting on the ground and the rungs vertical and perpendicular to the ground.

Parquet Flooring — Flooring, usually of wood, laid in an alternating or inlaid pattern to form various designs. Flooring strips may be glued together to make square units.

Partial Thickness Burns (Second-Degree) — Burns involving several layers of skin.

Partially-Structured Question — Used in cases where the researcher has some idea of potential responses that may be generated by respondents.

Participative Leadership — *See* Democratic Leadership.

Particulate — Very small particle of solid material, such as dust, that is suspended in the atmosphere.

Particulate Air Filter — Portion of respiratory protective equipment that traps particulates from the air before they can be inhaled.

Partition Wall — Interior non-load-bearing wall that separates a space into rooms.

Partner — Person, group, or organization willing to join forces and address a community risk.

Partnerships — Joining forces with other groups to address common interests.

Part-Paid Firefighter — Firefighters paid on the basis of time that they are used.

Parts Per Million (ppm) — Method of expressing the concentration of very dilute solutions of one substance in another, normally a liquid or gas, based on volume; expressed as a ratio of the volume of contaminants (parts) compared to the volume of air (million parts).

Party Wall — Dividing wall that stands between two adjoining buildings or units, often on the property line, and is common to both buildings. A party wall is almost always a load-bearing wall and usually serves as a fire wall.

Pascal's Law — Law of physical science that states that pressure acts in all directions and not simply downward.

PASS — *See* Personal Alert Safety System.

Passageway — Any interior walkway, corridor, or hallway in a vessel.

Passenger Side — Side of a vehicle that is opposite of the steering wheel.

Passive Agent — Material that absorbs heat but does not participate actively in the combustion process.

Passive Smoke Control — Smoke control strategies that incorporate fixed components that provide protection against the spread of smoke and fire. Passive smoke control components include fire doors, fire walls, fire stopping of barrier penetrations, and stair and elevator vestibules.

Passive-Sentence Index — Index that measures the readability of a passage of text by determining the percentage of passive sentences it contains.

Pathogens — Organisms that cause infection, such as viruses and bacteria.

Pathological Firesetting — Firesetters who have transcended through the firesetting profiles whereby they are now setting fires as a way to release stress.

Pathology — Manifestation of a problem into a deviating condition.

Patient — Person who is receiving medical care.

Patient Assessment — Process of examining a patient to determine injuries or illness.

Patient Decontamination — Removing contamination from injured patients or victims. *See* Decontamination.

Patio Door — Sliding glass door that is commonly placed in an opening that accesses the patio or rear of a residence.

Patrol — (1) To travel over a given route to prevent, detect, and suppress fires. (2) In wildland fire fighting, to go back and forth vigilantly over a length of control line during and/or after construction, in order to prevent slopovers, suppress spot fires, and extinguish overlooked hot spots.

Pattern — (1) Shape of the water stream as it is discharged from a fog nozzle. (2) Distinctive markings left on a structure or contents after a fire.

Pawls — Devices attached to the inside of the beams on fly sections; used to hold the fly section in place after it has been extended. *Also known as* Dogs or Ladder Locks.

PC — Personal Computer.

PCB — *See* Polychlorinated Biphenyl.

PDP — *See* Pump Discharge Pressure.

Peak Hourly Consumption — Maximum amount of water used in a water distribution system during any hour of a day.

Pedestal — *See* Control Pedestal.

Peer Assistance — (1) Situation in which learners assist their peers in the learning process. (2) Refers to the process of having employees of equal status assist each other in the training process.

Peer Fitness Trainer — Firefighter-certified fitness trainers who oversee fitness programs for firefighter recruits as directed by the Health and Fitness Coordinator (HFC).

Peer Pressure — Decision-making problem; the tendency of a decision maker to bow to the will of the group rather than to lead the group.

PEL — *See* Permissible Exposure Limit.

PEL-C — *See* Permissible Exposure Limit/Ceiling Limit.

Pendant Sprinkler — Automatic sprinkler designed for placement and operation with the head pointing downward from the piping.

Penetrant — Water with added chemicals called wetting agents that increase water's spreading and penetrating properties due to a reduction in surface tension. *Also known as* Wet Water.

Penetration — Process in which a hazardous material enters an opening or puncture in a protective material. *See* Routes of Entry.

Penthouse — (1) Structure on the roof of a building that may be used as a living space, to enclose mechanical equipment, or to provide roof access from an interior stairway. (2) Room or building built on the roof, which usually covers stairways or houses elevator machinery, and contains water tanks and/or heating and cooling equipment. *Also known as* Bulkhead.

Per Capita — Per unit of population.

Percentage — Part of a whole expressed in hundredths.

Percentage Score — Way of interpreting evaluation results by expressing a part of a whole in hundredths.

Performance — Part of an educational objective that tells what learners must do to show what they have learned. *Also known as* Behavior.

Performance Bond — Binding financial agreement that ensures that the manufacturer will build the apparatus to the desired specifications. Usually, the amount of the bond is equal to the difference that would be required to have another manufacturer build the apparatus specified.

Performance Budget — Form of program budgeting in which the cost of each unit of performance (fire call, EMS call, code enforcement inspection, plan review, etc.) is identified, and total funding is based on projected performance levels.

Performance Evaluation — Evaluation of an individual's job performance as measured against one or more objective performance criteria.

Performance Levels — Desired level of ability required to perform a particular job as specified in a behavioral objective.

Performance Objective — Explicitly worded statement that specifies learners' behaviors (actions or performances), the conditions by which they will perform (what is given to them to perform), and the criteria they will meet (standards or similar measurable requirements). The objective states that the activity will be performed in some observable and measurable form. *See* Behavioral Objective, Learning Objective, and Objective (1).

Performance Requirements — Written list of expected capabilities for new apparatus. The list is produced by the purchaser and presented to the manufacturer as a guide for what is expected.

Performance Standards — Minimum level of knowledge and/or skill that a student must demonstrate to successfully complete a training or education session.

Performance Test — Test that measures a learner's ability and proficiency in performing a job or evolution by requiring that he or she handle equipment or materials in a coordinated, step-by-step process; measures learner or employee achievement of a psychomotor objective and holds the test-taker to either a speed standard (timed performance), a quality standard (minimum acceptable product or process standard), or both. Typically given at the end of instruction to measure final performance. *Also known as* Skills Test. *See* Test.

Performance-Based Learning — *See* Competency-Based Learning.

Perimeter — (1) Outer boundary of a fire or other incident scene established to secure the area, ensure safety, examine a scene, protect civilians, and/or collect evidence. (2) The perimeter of a wildland fire is the boundary of the fire; the total length of the outer edge of the burning or burned area.

Perimeter Control — Establishing and maintaining control of the outer edge or boundary of an incident scene.

Periphery-Deflected Fire Streams — Fire streams produced by deflecting water from the periphery of an inside circular stem in a fog nozzle against the exterior barrel of the nozzle.

Perjury — Lying under oath in a court or legal proceeding.

Permanent Deformation — Deformation remaining in any part of a ladder or its components after all test loads have been removed.

Permeation — Process in which a chemical passes through a protective material on a molecular level.

Permissible Exposure Limit (PEL) — Maximum time-weighted concentration at which 95 percent of exposed, healthy adults suffer no adverse effects over a 40-hour work week; an 8-hour time-weighted average unless otherwise noted. PELs are expressed in either parts per million (ppm) or milligrams per cubic meter (mg/m^3). They are commonly used by OSHA and are found in the NIOSH *Pocket Guide to Chemical Hazards*. *See* Immediately Dangerous to Life or Health (IDLH), Permissible Exposure Limit/Ceiling Limit (PEL-C), Recommended Exposure Limit (REL), Short-Term Exposure Limit (STEL), and Threshold Limit Value (TLV®).

Permissible Exposure Limit/Ceiling Limit (PEL-C) — Legal term for the maximum amount of a chemical substance or other hazard that an employee can be exposed to; typically expressed in parts per million (ppm) or milligrams per cubic meter (mg/m3). If exposed to this concentration for an entire 40-hour work week, 95% of healthy adults would not suffer health consequences. *See* Permissible Exposure Limit (PEL).

Permit-Required Confined Space — A category of confined space that includes an extra safety hazard in addition to the baseline definition of confined space. The parameters of the hazard must be indicated on the entry permit.

Peroxidizable Compound — Material apt to undergo spontaneous reaction with oxygen at room temperature and form peroxides and other products.

Persistence — Length of time a chemical agent remains effective without dispersing. *See* Dispersion, Nonpersistent Agent, and Persistent Agent.

Persistent Agent — Chemical agent that remains effective in the open (at the point of dispersion) for a considerable period of time (more than 10 minutes). *See* Dispersion, Nonpersistent Agent, and Persistence.

Personal Alert Device (PAD) — *See* Personal Alert Safety System.

Personal Alert Safety System (PASS) — Electronic lack-of-motion sensor that sounds a loud alarm when a firefighter becomes motionless. It can also be manually activated. *Also known as* Personal Alert Device. *See* Personal Protective Equipment.

Personal Fire Hazards — Common hazards that are caused by unsafe acts of individuals.

Personal Flotation Devices (PFD) — Life jackets, vests, or other devices that provide buoyancy for the wearer. Devices must be United States Coast Guard approved Type III or V when used for rescue operations.

Personal Line — Short, 20-foot (6 m) rope used in the United Kingdom, Australia, and New Zealand by an SCBA team member to maintain contact with another team member or the main guideline.

Personal Protective Clothing — Garments emergency responders must wear to protect themselves while fighting fires, mitigating hazardous materials incidents, performing rescues, and delivering emergency medical services.

Personal Protective Equipment (PPE) — General term for the equipment worn by fire and emergency services responders; includes helmets, coats, trousers, boots, eye protection, hearing protection, protective gloves, protective hoods, self-contained breathing apparatus (SCBA), personal alert safety system (PASS) devices, and chemical protective clothing. When working with hazardous materials, bands or tape are added around the legs, arms, and waist. *Also known as* Bunker Clothes, Chemical Protective Clothing, Full Structural Protective Clothing, Protective Clothing, Turnout Clothing, or Turnout Gear. *See* Chemical Protective Clothing (CPC) and Special Protective Clothing (1).

Person-Borne Improvised Explosives Device (PBIED) — Improvised explosive device carried by a person; employed by suicide bombers as well as individuals coerced into carrying the bomb against their will. *See* Explosion, High Explosive, and Improvised Explosive Device.

Personnel Accountability Report (PAR) — Roll call of all units (crews, teams, groups, companies, sectors) assigned to an incident. The supervisor of each unit reports the status of the personnel within the unit at that time, usually by radio. A PAR may be required by standard operating procedures at specific intervals during an incident, or may be requested at any time by the incident commander or the incident safety officer.

Personnel Accountability System — Method for identifying which emergency responders are working on an incident scene.

Personnel Carriers — Armored fighting vehicles designed to transport infantry to a battlefield.

Personnel Management — Decision-making concerning the effective use of human resources within an organization so that organizational and individual goals are met.

Personnel Record — Account of an individual employee's work history; includes personal data (such as name, address, date of employment, or job classification), citations, commendations, promotions, performance evaluations, letters of reprimand or other disciplinary documentation, and medical history.

Pertinent Medical Information (PMI) — A patient's personal data, condition, and medical history.

Pesticides — Chemicals designed and used to control or eliminate undesirable forms of life, such as plants and animal pests.

Petroleum Carrier — Tank vessel that transports crude or finished petroleum products. *See* Tanker.

pH — Measure of the acidity or alkalinity of a solution. *See* Acid, Alalki, and Base.

Phantom Box — Predetermined fire department response assignment to a given location that is not equipped with a fire alarm box.

Phase — Distinguishable part in a course, development, or cycle; aspect or part under consideration.

Phase I Operation — Emergency operating mode for elevators. Phase I operation recalls the car to a certain floor and opens the doors.

Phase II Operation — Emergency elevator operating mode that allows emergency use of the elevator with certain safeguards and special functions.

Phobia — Abnormal and persistent fear of a specific object or situation.

Phonetic Alphabet — Alphabet devised by the International Civil Aviation Organization for use in radiotelephone conversations. Words are used phonetically in place of letters; for example, A is alpha, B is bravo, etc.

Phosgene ($COCl_2$) — Toxic gas produced when refrigerants, such as freon, plastics containing polyvinyl chloride (PVC), or electrical wiring insulation, contact flames; may be absorbed through the skin as well as through the lungs.

Phosphine — Colorless, flammable, and toxic gas with an odor of garlic or decaying fish; ignites spontaneously on contact with air. Phosphine is a respiratory tract irritant that attacks the cardiovascular and respiratory systems, causing pulmonary edema, peripheral vascular collapse, and cardiac arrest and failure.

Photo Log — Assembled group of photographs with numerical references and descriptions of the photos.

Photoelectric Cell — Light-sensitive device used in some fire detectors. Cell initiates an alarm signal when light strikes it or is kept from striking it, depending upon the particular design.

Photoelectric Smoke Detector — Type of smoke detector that uses a small light source, either an incandescent bulb or a light-emitting diode (LED), to detect smoke by shining light through the detector's chamber. Smoke particles reflect the light into a light-sensitive device called a photocell. *Also known as* Photoelectric Smoke Alarm.

Photon — Weightless packet of electromagnetic energy, such as X-rays or visible light.

Physical Change — When a substance remains chemically the same but changes in size, shape, or appearance.

Physical Evidence — Tangible or real objects that are related to the incident.

Physical Fitness Plan — Individualized or department-wide plan that firefighters can follow to maintain fitness-for-duty and improve their overall health and well-being.

Physical Hazard — Material that presents a threat to health because of its physical properties. *See* Health Hazard and Physical Properties.

Physical Performance Assessment — Series of exercises that are performed and evaluated before beginning a physical fitness plan in order to individualize the plan and establish a baseline for evaluating progress.

Physical Performance Requirements — Fitness level benchmarks based upon recommended industry standards which firefighters must meet to be considered fit-for-duty.

Physical Properties — Properties that do not involve a change in the chemical identity of the substance, but affect the physical behavior of the material inside and outside the container, which involves the change of the state of the material. Examples include boiling point, specific gravity, vapor density, and water solubility. *See* Boiling Point, Chemical Properties, Specific Gravity, Vapor Density, and Water Solubility.

Physical Rehabilitation Program — Physical fitness training program designed for firefighters who do not meet or no longer meet the physical performance requirements associated with their job functions.

Physical Science — Study of the physical world around us; includes the sciences of chemistry and physics.

Physical Search — Performed by rescuers without outside search-specific resources; involves an organized approach to checking all areas of the structure. This is the most easily implemented type of search, as it can be done with available resources.

Physically Fit — As determined by a qualified physician, an individual who has no known physical or medical limitations that would interfere with the performance of sustained heavy work or the use of self-contained breathing apparatus (SCBA) that may be required during emergency operations.

Physiological — (1) Of or relating to an organism's healthy and normal functioning. (2) First and most basic group of needs in Maslow's Hierarchy of Needs; needs such as food, water, and shelter, which are related to personal survival and are essential to sustain life. Maslow's physiological needs also include escaping from a situation, such as a fire, that is immediately life-threatening.

Physiological Stress — Stress caused by physical exertion.

Picket — Steel rod or wooden stake driven into the ground to create an anchor, such as for a guy line.

Pick-Head Axe — Forcible entry tool that has a chopping blade on one side of the head and a sharp pick on the other side.

Pick-Up Plate — Sloped plate and structure of a trailer, which is located forward of the kingpin and designed to facilitate engagement of fifth wheel to kingpin.

Pickup Tube — Solid or flexible tube used to transfer foam concentrate from a storage container to the in-line eductor or proportioner.

Pictogram — Drawing or symbol that indicates information.

Pier — (1) Supporting section of a wall between two openings. (2) Short masonry column. (3) Load-supporting member constructed by drilling or digging a shaft, then filling the shaft with concrete. (4) Elevated working platform, usually made of wood or masonry, that extends outward from the shore into a standing body of water for use as a landing place for vessels; supported on pilings and open underneath, allowing the berthing of vessels alongside. *See* Wharf.

Piercing Nozzle — Nozzle with an angled, case-hardened steel tip that can be driven through a wall, roof, or ceiling to extinguish hidden fire. *Also known as* Piercing Applicator Nozzle or Puncture Nozzle.

Piezometer Tube — Device that uses the heights of liquid columns to illustrate the pressures existing in hydraulic systems.

Pig Rig — Term used for a simple 3:1 mechanical advantage pulley system. Used in a horizontal configuration to attach to a main line or directly to a load; more versatile than a simple 3:1 Z-pulley system.

Piggyback Transport — *See* Trailer-On-Flatcar (TOFC).

Pike Pole — Sharp prong and hook of steel, on a wood, metal, fiberglass, or plastic handle of varying length, used for pulling, dragging, and probing.

Pilaster — Rectangular masonry pillar that extends from the face of a wall to provide additional support for the wall; may also be for decorative use only, in which case it does not provide any support.

Piles — Wooden or steel beams used to support loads; piles are driven into the ground and develop their load-carrying ability either through friction with the surrounding soil or by being driven into contact with rock or a load-bearing soil layer.

Pilot — Person knowledgeable of the local waters who meets vessels and steers them safely into and out of port.

Pilot Course — First implementation of a newly developed course; intended to allow instructors to evaluate a new course and make changes for improvement.

Pilot Testing — Testing a specific intervention for effectiveness.

Piloted Ignition — Moment when a mixture of fuel and oxygen encounters an external heat (ignition) source with sufficient heat or thermal energy to start the combustion reaction.

PIN — *See* Product Identification Number.

Pin Lug Couplings — Hose couplings with round lugs in the shape of a pin.

Pincer Attack — Simultaneous attack on two or more sides of a wildland fireline. Similar to a wildland flank attack. *See* Attack Methods (2).

Pinch Point — Any point, other than the point of operation, at which it is possible for a part of the body to be caught between the moving parts of the machine or between a moving part and a stationary part where the user could receive a crushing injury to a portion of the body.

PIO — *See* Public Information Officer.

Pipe Chase — Concealed vertical channel in which pipes and other utility conduits are housed. Pipe chases that are not properly protected can be major contributors to the vertical spread of smoke and fire in a building. *Also known as* Chase.

Pipe Plugs and Caps — Devices used to stop broken water lines in order to minimize water damage.

Pipe Schedule — Thickness of the wall of a pipe.

Piston Pump — Positive-displacement pump using one or more reciprocating pistons to force water from the pump chambers.

Piston Valve — Valve with an internal piston that moves within a cylinder to control the flow of water through the valve.

Pitch — (1) In steel construction, the spacing between rivet centers. (2) Slope of a roof expressed as a ratio of rise to span. (3) Resin present in certain woods. (4) Asphaltic, tarlike liquid used to repair blacktopped streets. (5) Angle between horizontal and a ladder positioned for use.

Pitched Roof — Roof, other than a flat or arched roof, that has one or more pitched or sloping surfaces.

Pitot Gauge — Instrument that is inserted into a flowing fluid (such as a stream of water) to measure the velocity pressure of the stream; commonly used to measure flow; functions by converting the velocity energy to pressure energy that can then be measured by a pressure gauge. The gauge reads in units of pounds per square inch (psi) or kilopascals (kPa). *Also known as* Pitot Tube.

PIV — *See* Post Indicator Valve.

PIVA — *See* Post Indicator Valve Assembly.

Pivot — Method for turning a ladder on one beam when the ladder has been raised to a near vertical position.

Pivoting Deadbolt — Lock having a deadbolt that pivots 90 degrees, designed to fit a narrow-entry, stiled door.

Pivoting Window — Window that opens and closes either horizontally or vertically on pivoting hardware.

Pixel — Smallest item of information in a digital image; can be used to define the size of a digital image, the printing capabilities of a digital printer, or the screen resolution of high definition televisions and computer monitors.

Placard — Diamond-shaped sign that is affixed to each side of a structure or a vehicle transporting hazardous materials to inform responders of fire hazards, life hazards, special hazards, and reactivity potential. The placard indicates the primary class of the material and, in some cases, the exact material being transported; required on containers that are 640 cubic feet (18 m³) or larger. *See* Label, Marking, NFPA® 704 Placard, Package Marking, and Packaging.

Place — In identifying the place in a social marketing campaign, may refer to barriers that prevent the target audience from adopting the behavior. Place refers to accessibility of the supports that provide the target audience with the ability to make the change.

Plagiarism — To present as an original idea without crediting the source.

Plain Language — Communication that can be understood by the intended audience and meets the purpose of the communicator. For the purpose of the *National Incident Management System*, plain language is designed to limit the use of codes and acronyms during an incident response involving more than a single agency. *Also known as* Plain Text.

Plan of Operations — Clearly identified strategic goal and the tactical objectives necessary to achieve that goal; includes assignments, authority, responsibility, and safety considerations.

Plan View — Drawing containing the two-dimensional view of a building as seen from directly above the area. *See* Detailed View, Elevation View, and Sectional View.

Plancier — Board that forms the underside of an eave or cornice.

Plane of Weakness — Area created by the rescuers in a wall or floor that is weaker than the surrounding structural material.

Planning Meeting — Meeting held as needed at any point during an incident, in order to select specific strategies and tactics for incident control operations and for service and support planning.

Planning Model — Organized procedure that includes the steps of analyzing, designing, developing, implementing, and evaluating instruction; a systematic approach to the design, production, evaluation, and use of a system of instruction.

Planning Section — Incident command system section responsible for collection, evaluation, dissemination, and use of information about the development of the incident and the status of resources; includes the situation status, resource status, documentation, and demobilization units, as well as technical specialists.

Plans Review — Process of reviewing building plans and specifications to determine the safety characteristics of a proposed building; generally done before permission is granted to begin construction.

Plaster Hook — Barbed collapsible hook on a pole used to puncture and pull down materials.

Plasterboard — *See* Wallboard.

Plat — Drawing of a parcel of land that gives its legal description.

Plat Plan — *See* Plat.

Plate — Top (top plate) or bottom (soleplate) horizontal structural member of a frame wall or partition.

Plate Glass — Sheet glass that is ground, polished, and clear.

Platform — (1) Intermediate landing between floors to change the direction of a stairway or to break up excessively long flights. (2) Main deck of an offshore drilling rig. (3) Horizontal surface extending partway through a vessel, usually in the cargo space. (4) Any flat-topped vessel capable of providing a working area for personnel or vehicles.

Platform Frame Construction — (1) Type of framing in which each floor is built as a separate platform, and the studs are not continuous beyond each floor. *Also known as* Western Frame Construction. (2) A construction method in which a floor assembly creates an individual platform that rest on the foundation. Wall assemblies the height of one story are placed on this platform and a second platform rests on top of the wall unit. Each platform creates fire stops at each floor level restricting the spread of fire within the wall cavity.

Platoon — Entire shift of a fire department; may indicate those who are on or off duty.

Play a Stream — To direct a stream of water at the fire.

Playpipe — Base part of a three-part nozzle that extends from the hose coupling to the shutoff.

Plenum — Open space or air duct above a drop ceiling that is part of the air distribution system.

Plimsoll Mark — Symbol placed on the sides of a vessel's hull at amidships, indicating the maximum allowable draft of the vessel. *Also known as* Load Line.

Plot Plan — Architectural drawing showing the overall project layout of building areas, driveways, fences, fire hydrants, and landscape features for a given plot of land; view is from directly above. *See* Construction Plan, Floor Plan, and Site Plan.

Plug — (1) Fire hydrant. (2) Wooden peg used to stop a hole in a container. (3) Safety device that grounds a tank vehicle or rail tank car during the loading and unloading process and prevents static electricity buildup. (4) Patch to seal a small leak in a container.

Plug Pressure — *See* Flow Pressure.

Plume — (1) Irregularly shaped pattern of an airborne hazardous material where wind and/or topography influence the downrange course from the point of release. *See* Cloud, Cone, and Hemispheric Release. (2) The column of hot gases, flames, and smoke rising above a fire; also called convection column, thermal updraft, or thermal column. (NFPA® 921)

Plume-Generated Pattern — Any of a number of fire patterns created as a result of the plume of hot gases rising above an individual fire. *See* Circular-Shaped Pattern, Hourglass Pattern, Inverted-Cone Pattern, U-Pattern, and V-Pattern.

Plymetal Panels — Railcar floor panels constructed of plywood sheets covered by sheets of metal, usually aluminum.

Plywood — Wood sheet product made from several thin veneer layers that are sliced from logs and glued together.

PMI — *See* Protected Medical Information.

PMP — Prusik minding pulley. A specially designed pulley to work with tandem Prusiks in rope rescue systems.

Pneumatic — Operated by air or compressed air.

Pneumatic Chisel — Tool designed to operate at air pressures between 100 and 150 psi (700 kPa and 1 050 kPa); during periods of normal consumption, it will use about 4 to 5 cubic feet (113 L to 142 L) of air per minute. It is useful for extrication work. *Also known as* Air Chisel, Impact Hammer, or Pneumatic Hammer.

Pneumatic Lifting Bag — Inflatable, envelope-type device that can be placed between the ground and an object and then inflated to lift the object; it can also be used to separate objects. Depending on the size of the bag, it may have lifting capabilities in excess of 75 tons (68 040 kg).

Pneumatic Power — Power derived by using the properties of compressed air at rest or in motion; generally used with a pressure regulator.

Pneumatic Shoring — Shores or jacks with movable parts that are operated by the action of a compressed gas.

Pneumatic Tools — Tools that receive their operating energy from compressed air.

Pneumothorax — Accumulation of air in the pleural cavity, usually after a wound or injury that penetrates the chest wall or lacerates the lungs.

Point and Cut Off — Attacking several heads or fingers of a wildfire at the same time and then connecting the short line segments.

Point of No Return — Point at which air in the SCBA will last only long enough to exit a hazardous atmosphere.

Point of Operation — Location at which the intended work is being done.

Point of Origin — Exact physical location where the heat source and fuel come in contact with each other and a fire begins.

Pointer Pattern — Fire pattern created when structural components such as wood studs or trim are exposed to flame; sharp edges of the component are often burned away on the side of the component that faces the heat source. Also refers to a series of burned components that indicate a longer duration on one end of the series to shorter duration on the other. *Also known as* Arrow Pattern.

Poison — Any material that is injurious to health when taken into the body. *See* Convulsant and Toxin.

Poke-Through — Opening in a floor, ceiling, or wall through which ducting, plumbing, or electrical conduits pass. If these openings are not properly caulked or sealed, they can contribute significantly to the spread of smoke and fire in a building.

Polar Solvents — (1) Flammable liquids that have an attraction for water, much like a positive magnetic pole attracts a negative pole; examples include alcohols, esters, ketones, amines, and lacquers. (2) A liquid having a molecule in which the positive and negative charges are permanently separated, resulting in their ability to ionize in solution and create electrical conductivity. Water, alcohol, and sulfuric acid are examples of polar solvents.

Polarized Couplings — *See* Gladhands.

Pole — (1) Sliding pole from upper stories to the apparatus area of a fire station. (2) Ladder poles to assist in raising large ground ladders. (3) Pike pole.

Pole Ladder — Large extension ladder that requires tormentor poles to steady the ladder as it is raised and lowered. *Also known as* Bangor Ladder.

Police Power — (1) Authority that may be given to an inspector to arrest, issue summons, or issue citations for fire code violations. (2) Constitutional right of the government to impose laws, statutes, and ordinances, including zoning ordinances and building and fire codes, in order to protect the health, safety, morals, and general welfare of the public.

Policy — Organizational principle that is developed and adopted as a basis for decision-making.

Polychlorinated Biphenyl (PCB) — Toxic compound found in some older oil-filled electric transformers.

Polyethylene Membrane — Type of plastic sheet used for waterproofing.

Polymerization — Chemical reactions in which two or more molecules chemically combine to form larger molecules; this reaction can often be violent. *See* Inhibitor.

Polyvinyl Chloride (PVC) — Synthetic chemical used in the manufacture of plastics and single-ply membrane roofs.

Pompier Belt — *See* Class I Harness.

Pompier Ladder — Scaling ladder with a single beam and a large curved metal hook that can be put over windowsills for climbing.

Pony Wall — Non-load-bearing wall that is less than 36 inches (914 mm) high. *See* Load-Bearing Wall.

Poop Deck — Partial deck above main deck at stern. *See* Deck.

Porcelainize — To coat with a ceramic material.

Port — (1) General area of a shore establishment having facilities for the landing, loading/unloading, and maintenance of vessels; a harbor with piers. (2) Left-hand side of a vessel as a person faces forward. *Also known as* Port Side.

Port Authority — Agency entrusted with the duty or power of constructing, improving, managing, or maintaining a harbor or port. *Also known as* Harbor Authority, Harbor Board, Port Commission, or Port Trust.

Port of Registry — Port in which a vessel is registered. *Also known as* Home Port.

Port State — Nation in which a port is located.

Port State Authority — Government agency having authority over port operations.

Portable Basin — *See* Portable Tank.

Portable Equipment — Items carried on the fire or rescue apparatus that are not permanently attached to or a part of the apparatus.

Portable Fire Extinguisher — *See* Fire Extinguisher.

Portable Foam Extinguishing System — Foam extinguishing system that can be carried by hand, such as a handheld foam fire extinguisher. *See* Fixed Foam Extinguishing System and Semifixed Foam Extinguishing System.

Portable Hydrant — *See* Manifold.

Portable Ladder Pipe — Portable, elevated master stream device clamped to the top two rungs of the aerial ladder when needed and supplied by a 3- or 3½-inch (77 mm or 90 mm) fire hose.

Portable Pump — (1) Small fire pump, available in several volume and pressure ratings, that can be removed from the apparatus and taken to a water supply inaccessible to the main pumper. (2) In marine fire fighting, a small gasoline-driven pump used in emergencies to deliver water to a fire independent of a vessel's fire main system.

Portable Radio — Hand-held, self-contained transceiver radio used by personnel to communicate with each other when away from the vehicle radio. A portable radio draws upon power from its own battery and uses its case along with an antenna to make up the antenna assembly. Portable radios do not have much power; they generally transmit 1 to 5 watts and have limited coverage areas. Duration of battery and duty cycles depends on their power source. *Also known as* Handi-Talki.

Portable Source — Water that is mobile and may be taken directly to the location where it is needed; may be a fire department tanker or some other vehicle that is capable of hauling a large quantity of water.

Portable Tank — Collapsible storage tank used during a relay or shuttle operation to hold water from water tanks or hydrants; this water can then be used to supply attack apparatus. *Also known as* Catch Basin, Fold-a-Tank, Porta-Tank, or Portable Basin.

Porta-Power — Manually operated hydraulic tool that has been adapted from the auto body business to the rescue service. This device has a variety of tool accessories that allows it to be used in numerous applications.

Porthole — Circular window in the side of a vessel.

Portland Cement — Most commonly used cement, consisting chiefly of calcium and aluminum silicate. It is mixed with water to form a paste that hardens, and is therefore known as a hydraulic cement.

Position — (1) Specific assignment during a fire operation. (2) To spot an apparatus for maximum effective use.

Position Description — Written description of a specific employee position that spells out the expectations of the job, the activities that are needed to meet those expectations, and the qualifications needed to fill the job.

Positive Buoyancy — Tendency to float.

Positive Displacement Pumps — Self-priming pump that utilizes a piston or interlocking rotors to move a given amount of fluid through the pump chamber with each stroke of the piston or rotation of the rotors. Used for hydraulic pumps on aerial devices' hydraulic systems and for priming pumps on centrifugal fire pumps.

Positive Heat Balance — Situation that occurs when heat is fed back to the fuel; a positive heat balance is required to maintain combustion. *See* Negative Heat Balance.

Positive Pressure — Air pressure greater than that of the surrounding atmosphere.

Positive-Pressure Phase — Portion of an explosion in which gases are expanding outward from the center.

Positive-Pressure SCBA — Protective breathing apparatus that maintains a slight positive pressure inside the mask.

Positive-Pressure Test — Test to verify that there is positive pressure within a facepiece; after donning the facepiece, the wearer pulls the sealing surface of the facepiece away from the skin, allowing air to escape.

Positive-Pressure Ventilation (PPV) — Method of ventilating a room or structure by mechanically blowing fresh air through an inlet opening into the space in sufficient volume to create a slight positive pressure within and thereby forcing the contaminated atmosphere out the exit opening. *See* Ventilation.

Possible — Legal classification for an occurrence that is less likely than 50 percent.

Post Indicator Valve (PIV) — Type of valve used to control underground water mains that provides a visual means for indicating "open" or "shut" positions; found on the supply main of installed fire protection systems. The operating stem of the valve extends above ground through a "post," and a visual means is provided at the top of the post for indicating "open" or "shut." *See* Wall Post Indicator Valve.

Post Indicator Valve Assembly (PIVA) — Similar to a post indicator valve (PIV), except that the valve used is of the butterfly type; in contrast, the PIV and the wall post indicator valve (WPIV) use a gate valve.

Posterior — At or toward the back.

Post-Fire Operations — Overhaul after knockdown. Includes searching for and extinguishing hidden fire, determining the fire cause, identifying and preserving evidence of arson, and making the building and area safe for occupation; may also include returning to quarters, preparing equipment for future response, and writing incident reports.

Post-Incident Analysis — Overview and critique of an incident by members of all responding agencies, including dispatchers. Typically takes place within two weeks of the incident. In the training environment it may be used to evaluate student and instructor performance during a training evolution.

Postincident Critique — Meeting to discuss strategy and tactics, problems, SOP/G changes, or training changes derived from a postincident analysis report; usually led by a chief officer some time after a major incident.

Post-Incident Loss Control Activities — Those preparations needed in order to turn the property back over to the owner or occupant.

Post-Incident Stress — Psychological stress that affects emergency responders after returning from a stressful emergency incident.

Postmedical Surveillance — Evaluation of responders' vital signs, body temperature, and symptoms of dehydration, heat/cold stress, cardiovascular illness, stroke, or other ailment before leaving a scene to return to duty or cycling off a work shift.

Post-Tensioned Reinforcing — Technique used in post-tensioned concrete; reinforcing steel in the concrete is tensioned after the concrete has hardened.

Post-Traumatic Incident Debriefing — Counseling designed to minimize the effects of post-incident trauma. *Also known as* Critical Incident Stress Debriefing.

Post-Traumatic Stress Disorder (PTSD) — Disorder caused when persons have been exposed to a traumatic event in which they have experienced, witnessed, or been confronted with an event or events that involve actual death, threatened death, serious injury, or the threat of physical injury to self or others. *Also known as* Critical Incident Stress (CIS) and Post-Traumatic Stress Syndrome.

Pot Metals — Slang term for alloys consisting of inexpensive, low-melting point metals that are normally used for inexpensive castings; examples include zinc, lead, copper, and aluminum.

Potassium Iodide — Drug used as a blocking agent to prevent the human thyroid gland from absorbing radioactive iodine.

Potential Energy — Stored energy possessed by an object that can be released in the future to perform work once released.

Potentially Traumatic Event (PTE) — term developed by the National Fallen Firefighters Foundation (NFFF) in their programs to describe incidents that have the potential for critical incident stress.

Pounds Per Square Inch — Gauge (psig) — Pressure indicated on a gauge that does not include atmospheric pressure; at sea level, 0 psig is equal to 14.7 psia.

Pounds Per Square Inch (psi) — Unit for measuring pressure in the English or Customary System. Its International System equivalents are kilopascals (kPa) and bar. **Pounds Per Square Inch - Absolute (psia)** — Unit for measuring pressure in the English or Customary System; its International System equivalents are kilopascals (kPa) and bar. Absolute pressure equals atmospheric pressure (14.7 psi) plus the gauge pressure; at 100 psig, absolute pressure equals 114.7 psia.

Power — Amount of energy delivered over a given period of time.

Power Plant — *See* Apparatus Engine.

Power Take-Off (PTO) System — Mechanism that allows a vehicle engine to power equipment such as a pump, winch, or portable tool; it is typically attached to the transmission. Farm tractors are designed to operate the PTO shaft at either 540 or 1,000 revolutions per minute.

Power Tool — Tool that acquires its power from a mechanical device, such as a motor or pump.

Power Train — (1) Includes all of the parts that create and transfer power to the surface being traversed. *Also known as* Drive Train. (2) Means of transferring power from an engine to a pump; includes all power-transmitting components.

Powered Air-Purifying Respirator (PAPR) — Motorized respirator that uses a filter to clean surrounding air, then delivers it to the wearer to breathe; typically includes a headpiece, breathing tube, and a blower/battery box that is worn on the belt.

Powered Hydraulic Shears — Large rescue tool whose two blades open and close by the use of hydraulic power supplied through hydraulic hoses from a power unit.

Powered Hydraulic Spreaders — Large rescue tool whose two arms open and close by the use of hydraulic power supplied through hydraulic hoses from a power unit; capable of exerting in excess of 20,000 pounds (9 072 kg) of force at its tips. *Also known as* Jaws.

PPE — *See* Personal Protective Equipment.

ppm — *See* Parts Per Million.

PPV — *See* Positive-Pressure Ventilation.

Practical Demonstration — Method of teaching psychomotor skills, such as installing a smoke detector or using a portable fire extinguisher safely. *Also known as* Manipulative Lesson.

Practical Training Evolution — *See* Training Evolution.

Preaction Sprinkler System — Fire-suppression system that consists of closed sprinkler heads attached to a piping system that contains air under pressure and a secondary detection system; both must operate before the extinguishing agent is released into the system. Similar to a dry-pipe sprinkler system. *See* Deluge Sprinkler System, Dry-Pipe Sprinkler System, and Wet-Pipe Sprinkler System.

Preassembled Lock — Lock designed to be installed as a complete unit, requiring no assembly, within a door. *Also known as* Unit Lock.

Preattack Planning — *See* Pre-Incident Planning.

Precast Concrete — Method of building construction where the concrete building member is poured and set according to specification in a controlled environment and is then shipped to the construction site for use.

Precipitation — Any or all forms of water particles, liquid or solid, that fall from the atmosphere.

Preconnect — (1) Attack hose connected to a discharge when the hose is loaded; this shortens the time it takes to deploy the hose for fire fighting. (2) Soft-sleeve intake hose that is carried connected to the pump intake. (3) Hard suction hose or discharge hose carried connected to a pump, eliminating delay when hose and nozzles must be connected and attached at a fire.

Predetermined Procedures — *See* Standard Operating Procedure (SOP).

Predischarge Alarm — Alarm that sounds before a total flooding fire extinguishing system is about to discharge; this gives occupants the opportunity to leave the area.

Predischarge Warning Device — Alarm that sounds before a total flooding fire extinguishing system is about to discharge. This gives occupants the opportunity to leave the area.

Prefabricated Construction — Method of building construction in which the walls, floors, and ceilings are manufactured complete with plumbing, electrical wiring, and millwork (woodwork such as doors and trim). Once delivered to the site, the entire assembly is erected. *Also known as* Panelized Construction.

Prefire Inspection — *See* Pre-Incident Planning.

Prefire Planning — *See* Pre-Incident Planning.

Preincident Inspection — Thorough and systematic inspection of a building for the purpose of identifying significant structural and/or occupancy characteristics to assist in the development of a pre-incident plan for that building. *See* Pre-Incident Planning.

Preincident Plan — Document, developed during pre-incident planning that contains the operational plan or set procedures for the safe and efficient handling of emergency situations at a given location, such as a specific building or occupancy. *Also known as* Preplan.

Preincident Planning — Act of preparing to manage an incident at a particular location or a particular type of incident before an incident occurs. *Also known as* Prefire Inspection, Prefire Planning, Pre-Incident Inspection, Pre-Incident Survey, or Preplanning.

Preincident Survey — Assessment of a facility or location made before an emergency occurs, in order to prepare for an appropriate emergency response. *Also known as* Preplan.

Preliminary Hearing — Court proceeding in which the prosecution must establish that a crime has been committed and that there is probable cause to believe the defendant committed the crime.

Premedical Surveillance — Evaluation of responders' baseline vital signs before clearance to enter the hot zone; also includes hydration before beginning work at the scene.

Premise — Proposition from which inferences may be drawn to reach conclusions about a question or problem.

Premixing — Mixing pre-measured portions of water and foam concentrate in a container. Typically, the pre-mix method is used with portable extinguishers, wheeled extinguishers, skid-mounted twin-agent units, and vehicle-mounted tank systems. *See* Batch Mixing.

Preoperational Stage of Intellectual Development — Second stage of intellectual development; occurs from age two to seven. Learners represent the world with words, images, and drawings.

Preparation Step — First step in conducting a lesson, in which the job or topic to be taught is identified, a teaching base is developed, and students are motivated to learn.

Preparedness — Activities to ensure that people are ready for a disaster and respond to it effectively; includes determining what will be done if essential services are lost, developing a plan for contingencies, and practicing the plan.

Preplan — *See* Pre-Incident Plan and Pre-Incident Survey.

Preponderance of Evidence — In civil proceedings, a collection of evidence that proves that an alleged civil wrong is probable to have occurred.

Prerequisite — (1) Something that is necessary to an end or to carrying out a function; knowledge or skill required before the learner can acquire additional or more complex knowledge or skill. (2) Entry-level knowledge or abilities that a learner must have prior to starting a particular course or qualifying for a certain job or promotion. *See* Requisite.

Prescribed Burning — (1) Controlled application of fire to wildland fuels in either their natural or modified state, under specified environmental conditions, that allows the fire to be confined to a predetermined area and produces the fire behavior and fire characteristics required to attain planned fire treatment and resource-management objectives. (2) A written plan that describes specifically planned results and specific conditions as part of a vegetation-management program.

Prescriptive Test — Test given at the beginning of instruction to determine what students already know; alternatively, a test that is given remedially.

Prescriptive Training — Instructional approach that uses the four-step teaching method in a different order: evaluation, preparation, presentation, application, and reevaluation.

Presentation — (1) Second of the four teaching steps in which the educator teaches a class or individual and transfers facts and ideas. (2) Lesson plan component in which an instructor provides to, shares with, demonstrates to, and involves the participants in the lesson information. *See* Lesson Plan. (3) Single delivery of fire and life safety information. *Also known as* Lesson or Delivery.

Presentation Step — Second step in conducting a lesson, at which point new information and skills are presented to the learners.

Preservice Tests — Tests performed on fire pumps or aerial devices before they are placed into service; these tests consist of manufacturers' tests, certification tests, and acceptance tests.

Press Conference — Scheduled event intended for presenting prepared information to the media; may or may not include a question and answer section.

Press Release — *See* Media Release.

Pressure — Force per unit area exerted by a liquid or gas measured in pounds per square inch (psi) or kilopascals (kPa).

Pressure Differential — Effect of altering the atmospheric pressure within a confined space by mechanical means. When air is exhausted from within the space, a low-pressure environment is created and replacement air will be drawn in; when air is blown into the space, a high-pressure environment is created and air within will move to the outside.

Pressure Gauge — Device for indicating pressure; the most common pressure gauges use a dial face to indicate pressures, such as the pump discharge pressure.

Pressure Governor — Pressure control device that controls engine speed, eliminating hazardous conditions that result from excessive pressures.

Pressure Intermodal Tank — Liquefied gas container designed for working pressures of 100 to 500 psig (689 kPa to 3 447 kPa) {6.9 bar to 34.5 bar}. *Also known as* Spec 51, or internationally as IMO Type 5. *See* Intermodal Container and Nonpressure Intermodal Tank.

Pressure Maintenance Pump — Pump used to maintain pressure on a fire protection system in order to prevent false starts at the fire pump.

Pressure Operation — Operation of a two- (or more) stage centrifugal pump in which water passes consecutively through each stage (or impeller) to provide high pressures at a reduced volume. *Also known as* Series Operation.

Pressure Point — Point over an artery where the pulse may be felt; pressure on the point often helps to stop the flow of blood from a wound beyond that point.

Pressure Regulator — Device used to maintain a constant pressure within a pump while operating.

Pressure Relief Valve — Pressure control device designed to eliminate hazardous conditions resulting from excessive pressures by allowing this pressure to release in manageable quantities.

Pressure Storage Tank — Class of fixed facility storage tanks divided into two categories: low-pressure storage tanks and pressure vessels. *See* Atmospheric Storage Tank, Low-Pressure Storage Tank, and Pressure Vessel.

Pressure Tank — Water storage receptacle that uses compressed-air pressure to propel the water into the distribution system. Pressure tanks are generally small and provide only a limited amount of water for fire protection.

Pressure Tank Railcar — Tank railcars that carry flammable and nonflammable liquefied gases, poisons, and other hazardous materials. They are recognizable by the valve enclosure at the top of the car and the lack of bottom unloading piping.

Pressure Vessel — Fixed-facility storage tanks with operating pressures above 15 psi (103 kPa) {1.03 bar}. *See* Low-Pressure Storage Tank, Pressure Storage Tank, and Spherical Pressure Vessel.

Pressure-Demand Device — Self-contained breathing apparatus (SCBA) that may be operated in either positive-pressure or demand mode. This type of SCBA is presently being phased out and replaced with positive-pressure-only apparatus.

Pressure-Reducing Valve — Valve installed at standpipe connection that is designed to reduce the amount of water pressure at that discharge to a specific pressure, usually 100 psi (700 kPa). *See* Standpipe System.

Pressure-Relief Device — Automatic device designed to release excess pressure from a container.

Prestressing — Compressive force induced in the concrete before the load is applied by placing the reinforcing bars in concrete beams in tension before the concrete is poured, so that the member will develop greater strength after the concrete has set. This "pre" stress is applied by tightening or "pre" loading the reinforcing steel; the preloading of the steel creates compressive stresses in the concrete that counteract the tensile stresses, which result when loads are applied.

Pre-Suppression — Activities in advance of fire occurrence to ensure effective suppression action.

Pretensioned Reinforcing — Reinforcing method used with pretensioned concrete. Steel strands are stretched between anchors producing a tensile force in the steel; concrete is then placed around the steel strands and allowed to harden.

Pretensioner Device that takes up slack in a seatbelt; prevents the passenger from being thrown forward in the event of a crash.

Pretest/Posttest — Prescriptive evaluation instrument administered to students and used to compare knowledge or skills either before (pretest) or after (posttest) a presentation or program. Pretests check entry-level knowledge or abilities; pretest scores are compared with posttest scores to determine learners' progress. *See* Test.

Pretreating — Exposure protection tactic in which water, foam, or retardant is applied to unburned materials near the fire; this soaks the material, making it less likely to ignite.

Pretrip Road Worthiness Inspection — A visual inspection of an apparatus to ensure the major components of the chassis are present and in proper working condition.

Prevention — Actions to avoid, prevent, or intervene to stop an incident from occurring; predetermined action or process employed to deter, obstruct, negate, forestall, hinder, impede, or preclude the occurrence of a potential loss.

Prevention Intervention — Reducing risk through education, technology, fire codes, standards, supporting incentives, or emergency response.

Preventive Maintenance — Scheduled, ongoing, routine inspection and maintenance that is intended to prolong the life and to prevent the breakdown of apparatus, equipment, and facilities; does not involve repairing or replacing damaged or worn-out components.

Price — One of the four basic principles of any marketing campaign, which is commonly referred to as the "Four Ps." For social marketing, there may not necessarily be a dollars-and-cents price for a product. It may be a question of time or the effort that the target audience will have to put forth to make the change or adopt the desired behavior. Price in a social marketing campaign is not always easy to quantify, but it is still a determining factor for the target audience.

Prill — Spherical pellets.

Prilled Oxidizer — Solid material such as ammonium nitrate that is formed into spherical pellets through a process that involves spraying the material in liquid form and allowing the drops to solidify. *See* Oxidizer.

Primary (Or Universal) Prevention — Strategy which promotes the well-being of an already healthy population through activities designed to prevent events that might result in injuries or property loss. It also seeks to enhance well-being by reinforcing healthy behaviors and discouraging lifestyles that may eventually lead to injury or illness.

Primary Damage — Damage caused by a fire itself and not by actions taken to fight the fire.

Primary Explosive — High explosive that is easily initiated and highly sensitive to heat; often used as a detonator.

Primary Feeder — Large pipes (mains), with relatively widespread spacing, that convey large quantities of water to various points of the system for local distribution to the smaller mains.

Primary Label — Label placed on the container of a hazardous material to indicate the primary hazard. *See* Subsidiary Label.

Primary Search Rapid but thorough search to determine the location of victims; performed either before or during fire suppression operations. May be conducted with or without a charged hoseline, depending on local policy.

Primary Stakeholder — People, groups, or organizations that have a vested interest in a specific issue.

Prime — To create a vacuum in a pump by removing air from the pump housing and intake hose, which permits the drafting of water.

Primer — *See* Priming Device.

Primer Fluid Tank — Tank of fluid used to seal and lubricate the priming pump.

Primer Oil Tank — Tank of oil used to seal and lubricate the priming pump.

Priming Device — Any device, usually a positive-displacement pump, used to exhaust the air from inside a centrifugal pump and the attached hard suction; this creates a partial vacuum, allowing atmospheric pressure to force water from a static source through the suction hose into the centrifugal pump. *Also known as* Primer.

Priority Traffic — *See* Emergency Traffic.

Privacy Law — Federal and state/provincial statue that prohibits an invasion of a person's right to be left alone and also restricts access to personal information.

Private Branch Exchange (PBX) — Telephone switching system found in large businesses or organizations that allows many users to be reached by dialing a seven-digit number. Internal users of a PBX system can use special features such as hold, conference calling, or transfer; many times these systems are marketed under trade names such as Centrex.

Private Connection — Connections to water supplies other than the standard municipal water supply system; may include connection within a large industrial facility, a farm, or a private housing development.

Private Hydrant — Hydrant provided on private property or on private water systems to protect private property. *Also known as* Yard Hydrant.

Private Law — Portion of the law that defines, regulates, enforces, and administers relationships among individuals, associations, and corporations. *See* Public Law.

Probable — Legal classification that means *more likely than not* or at least 50.1 percent accurate.

Probable Cause — Sufficient information or facts to believe that it is probable (more likely than not) that a certain party is responsible for committing a felony (indictable offense).

Problem Firesetter — Includes the profiles of intentional firesetting and firesetters with special needs.

Problem-Related Data — Statistics that can be used to analyze incident occurrences, develop a risk profile, prioritize problems, and identify at-risk populations.

Procedure —Outline of the steps that must be performed in order to properly follow an organizational policy. Procedures help an organization to ensure that it consistently approaches a task in the correct way, in order to accomplish a specific objective.

Proceed with Caution — Order for incoming units to discontinue responding at an emergency rate; after the order has been given, units should turn off warning devices and follow routine traffic regulations. *Also known as* Reduce Speed.

Process Alarm — Device which monitors the activity of industrial equipment to alert workers to critical failures in machinery; may also deactivate or de-energize equipment in order to to prevent a fire.

Process Evaluation — *See* Formative Evaluation.

Procurement Unit — Functional unit within the finance section of an incident command system; responsible for financial matters involving vendors.

Product — (1) Generic term used in industry to describe a substance that is used or produced in an industrial process. *See* Hazardous Material and Material. (2) The product is what is being marketed. In commercial campaigns, this is usually an item or service that the target audience will purchase or use. For social marketing, the product is the behavior that members of the community should modify or adopt.

Product Identification Number (PIN) — Number assigned by the United Nations and used in the *Emergency Response Guidebook (ERG)* to identify specific product names.

Production Order — Order in which jobs must be done; more difficult jobs may have to be done first.

Products of Combustion — Materials produced and released during burning.

Program — (1) Comprehensive strategy that addresses fire and life safety issues via educational means. (2) Collection of curricula and the resources necessary to deliver the instruction for those curricula.

Program Budget — Budgetary system in which each major program (such as administration, suppression, prevention, EMS, or training) is funded independently of other departmental programs. The overall department budget is a composite of the individual program budgets.

Program Revision — Enhancement of a program so it is accurate, current, and effective in helping to achieve intended outcome.

Progress Chart — Chart designed to record the progress of an individual or group during a course of study.

Progress Test — Diagnostic device to measure the progress or improvement of learners throughout a course; helps guide instructors and participants in deciding which areas need more emphasis or learning time. Typically called a quiz and may be written or oral. *See* Test.

Progressive Hose Lay — Method used when fire apparatus cannot drive along a wildland fire's edge, making a mobile attack impossible. Consists of laying hose from a fire pump to the fire's edge, extinguishing fire in that area, connecting another section, advancing, extinguishing fire as far as the hose will reach, and repeating this process until the fire is extinguished.

Progressive Line Construction — System of organizing workers to build a fireline in which they advance without changing relative positions in line.

Progressive Method — Progressive method of constructing a fire line. Each member of a hand crew takes a few strokes to clear fuel or widen the break, advances a specified distance, takes a few strokes, and advances; this process is repeated until the fire is extinguished.

Project Grant — Money given for a very specific activity or for very specific expenses, such as the purchase of fire safety educational videotapes for use in community schools.

Projected Window — Type of swinging window that is hinged at the top and swings either outward or inward.

Promotion — Advertising that is done to create awareness of the issue and provide the solution.

Promotional Materials — *See* Awareness Materials and Informational Materials.

Prone — Position of lying face downward.

Prop — Object used during a fire and life safety presentation that the audience can see, touch, smell, or hear; for example, a burned remnant from a home fire, a piece of melted glass, a manual fire alarm pull station, or a smoke detector.

Prop Wash — Current of air created by the rotation of a propeller.

Propagation — Spread of combustion through a solid, gas, or vapor, or the spread of fire from one combustible to another.

Proper Seal — Result of the facepiece fitting snugly against the bare skin, preventing the entry of smoke, fumes, or gases.

Proportional Directional Control Valve — Valve that controls the flow of hydraulic fluid through a hydraulic system.

Proportioner — Device used to introduce the correct amount of agent, especially foam and wetting agents, into streams of water. *See* Foam Proportioner.

Proportioning — Mixing of water with an appropriate amount of foam concentrate in order to form a foam solution.

Proportioning Valve — Foam system valve that is used to balance or divide the air supply between the aeration system and the discharge manifold.

Proposal — Document or request that describes the accomplishments an applicant promises to achieve in return for the investment of the sponsor's funds.

Proprietary Alarm System — Fire protection system owned and operated by the property owner.

Props — *See* Supports.

Prosecute — To charge someone with a crime; a prosecutor tries a criminal case on behalf of the government.

Prospect — Local business or other resource that may be a possible source of funding.

Protected Area — Undamaged surface within an otherwise fire damaged area, possibly resulting from objects shielding the surface from the effects of the fire; generally used to refer the fire investigator to where large objects such as furniture were positioned before the fire. *See* Heat Shadowing.

Protected Corridor — Corridor with code required fire-rated walls; intended to protect occupants as they make their way to an exit.

Protected Medical Information (PMI) — Information of a patient that includes personal data (name, birth date, social security number, address), medical history, and condition.

Protected Premises Fire Alarm System — (1) Alarm system that alerts and notifies only occupants on the premises of the existence of a fire so that they can safely exit the building and call the fire department. If a response by a public safety agency (police or fire department) is required, an occupant hearing the alarm must notify the agency. (2) Combination of alarm components designed to detect a fire and transmit an alarm on the immediate premises. *Also known as* Local Alarm System and Protected Premises System.

Protected Stair Enclosure — Stair with code required fire-rated enclosure construction; intended to protect occupants as they make their way through the stair enclosure.

Protected Steel — Steel structural members that are covered with either spray-on fire proofing (an insulating barrier) or fully encased in an Underwriters Laboratories Inc. (UL) tested and approved system.

Protection Factor — Ratio of contaminants in the atmosphere outside the facepiece to the contaminants inside the facepiece; determined by quantitative fit testing from the manufacturer.

Protection Plates — Plates fastened to a ladder to prevent wear at points where it comes in contact with mounting brackets.

Protective Action Distance — Downwind distance from a hazardous materials incident within which protective actions should be implemented. *See* Initial Isolation Distance, Protective Action Zone, and Protective Actions.

Protective Action Zone — Area immediately adjacent to and downwind from the initial isolation zone, which is in imminent danger of being contaminated by airborne vapors within 30 minutes of material release. *See* Initial Isolation Zone, Protective Action Distance, and Protective Actions.

Protective Actions — Steps taken to preserve health and safety of emergency responders and the public. *See* Protective Action Distance and Protective Action Zone.

Protective Clothing — Includes the helmet, protective coat, protective trousers, protective hood, boots, gloves, self-contained breathing apparatus, and eye protection where applicable. *See* Personal Protective Equipment.

Protective Coat — Coat worn during fire fighting, rescue, and extrication operations.

Protective Gloves — Protective clothing designed to protect the hands.

Protective Hood — Hood designed to protect the firefighter's ears, neck, and face from heat and debris; typically made of Nomex®, Kevlar®, or PBI®, and available in long or short styles.

Protective Signaling Systems — Detection and alarm systems that transfer information from a specific place in a fire-involved building to some remote monitoring station.

Protective Trousers — Trousers worn to protect the lower torso and legs during emergency operations. *Also known as* Bunker Pants or Turnout Pants.

Protein Foam Concentrate — Foam concentrate that consists of a protein hydrolysate plus additives to prevent the concentrate from freezing, prevent corrosions on equipment and containers, prevent bacterial decomposition of the concentrate during storage, and control viscosity. *See* Foam Concentrate.

Proton — Subatomic particle that possesses a positive electric charge.

Proximal — Located near the trunk of the body.

Proximate Cause — Act that causes injury or damage. *See* Negligence.

Proximity Clothing — Special personal protective equipment with a reflective exterior that is designed to protect the firefighter from conductive, convective, and radiant heat while working in close proximity to the fire. *Also known as* Proximity Suit.

Proximity Fire Fighting — Activities required for rescue, fire suppression, and property conservation at fires that produce high radiant, conductive, or convective heat; includes aircraft, hazardous materials transport, and storage tank fires.

Prusik — Length of low stretch Kernmantle rope, six to eight millimeters in diameter, tied with a double overhand bend into a loop.

Prusik Hitch — Prusik loop wrapped three times around a line, forming a hitch that can slip along the rope or seize the rope and hold it.

Prusik Knot — Special knot used to assist a person climbing a rope.

Pry — To raise, move, or force with a prying tool.

Prying Tools — Hand tools that use the principle of leverage to allow the rescuer to exert more force than would be possible without the tool; typically long, slender, and constructed of hardened steel.

PSAP — *See* Public Safety Answering Point.

psi — *See* Pounds Per Square Inch.

psig — *See* Pounds Per Square Inch Gauge.

Psychiatrist — Medical doctor who specializes in mental health and can prescribe medications.

Psychological Stress — Mental stress.

Psychologist — Professional with a doctoral degree in psychology who specializes in therapy. He or she can help direct several levels of clinical intervention.

Psychomotor Learning Domain — Learning that involves physical, hands-on activities, or actions that a students must be able to do or perform. *See* Learning Domain.

PTE — *See* Potentially Traumatic Event.

PTO — *See* Power Take-Off.

PTSD — *See* Post-Traumatic Stress Disorder.

Public Domain — Works of artists, photographers, and authors that were published before 1923 or are no longer covered by any copyright ownership.

Public Fire Education — A systematic approach to designing, implementing, and evaluating community safety education programs.

Public Information Officer (PIO) — (1) Member of the command staff responsible for interfacing with the media, public, or other agencies requiring information direct from the incident scene. *Also known as* Information Officer (IO). *See* Incident Management System (IMS). (2) Those who have demonstrated the ability to conduct media interviews, prepare news releases, and advisories.

Public Law — Classification of law consisting of constitutional, administrative, criminal, and international law. *See* Private Law.

Public Policy — System of laws, regulatory measures, courses of action, and funding priorities by a government entity or its representatives. Public policy is often created through a legislative process at either the federal, state, or local level.

Public Safety Answering Point (PSAP) — Any facility or location at which 9-1-1 calls are answered either by direct calling, rerouting, or diversion.

Public Safety Department — Organization that combines the administrative, financial, and technical service and support for functions such as fire and rescue services, police, ambulance and emergency medical services, and emergency communications.

Public Service — (1) Slang used over the radio to have someone make a call by telephone. (2) Service rendered in the public interest.

Public Way — Parcel of land such as a street or sidewalk that is essentially open to the outside and is used by the public to move from one location to another. *See* Means of Egress.

Public-Duty Doctrine — States that a government entity (such as a state or municipality) cannot be held liable for an individual plaintiff's injury that results from a governmental officer or employee's breach of a duty that is owed to the general public, rather than to the individual plaintiff.

Pull Box — Manual fire alarm activator.

Pull Line — *See* Fall Line.

Pulley — (1) Small, grooved wheel through which the halyard is drawn on an extension ladder. (2) Wheel used to transmit power by means of a band, belt, cord, rope, or chain passing over its rim. (3) Steel or aluminum rollers used to change direction and reduce friction in rope rescue systems.

Pull-In Point — Rotating portion of a machine that is capable of entangling a victim if a body part or piece of clothing comes into contact with it.

Pulling Line — *See* Fall Line.

Pulmonary Artery — Major artery leading from the right ventricle to the lungs.

Pulmonary Edema — Accumulation of fluids in the lungs.

Pulmonary Resuscitation — *See* Artificial Respiration.

Pulmonary Veins — Veins that carry oxygenated blood from the lungs to the left atrium.

Pulse — Rhythmic throbbing caused by expansion and contraction of arterial walls as blood passes through them.

Pump — (1) A device that imparts pressure to water. (2) To supply water to hoselines. (3) Apparatus in a two-piece engine company that positions as the attack apparatus at the fire scene.

Pump and Roll — Ability of an apparatus to pump water while the vehicle is in motion. *See* Mobile Attack.

Pump Apparatus — Fire department apparatus that has the primary responsibility to pump water.

Pump Can — Water-filled pump-type extinguisher. *Also known as* Pump Tank.

Pump Capacity Rating — Maximum amount of water a pump will deliver at the indicated pressure.

Pump Charts — Charts carried on a fire apparatus to aid the pump operator in determining the proper pump discharge pressure when supplying hoselines.

Pump Controller — Electric control panel used to switch a fire pump on and off and to control its operation.

Pump Discharge Pressure (PDP) — Actual pressure of the water as it leaves the pump and enters the hoseline; total amount of pressure being discharged by a pump. In mathematical terms, it is the pump intake pressure plus the net pump discharge pressure. Measured in pounds per square inch.

Pump Drain — Drain located at the lowest part of the pump to help remove all water from the pump; this eliminates the danger of damage due to freezing.

Pump Operator — Firefighter charged with operating the pump and determining the pressures required to operate it efficiently.

Pump Panel — Instrument and control panel located on the pumper.

Pump Room — Compartment in tank vessels where the pumping plant for handling cargo is installed; pumps are placed as low as possible in order to facilitate draining. In oil tankers over 400 feet (122 m) long, two pump rooms are provided, along with a ballast pump in some cases.

Pump Tank — *See* Pump Can.

Pumper Outlet Nozzle — Fire hydrant outlet that is 4 inches (102 mm) in diameter or larger.

Pumper, Class A — Pumper that delivers its rated capacity of at least 750 gpm (3 000 L/min) at 150 psi (1 000 kPa) net pump pressure at a lift of not more than 10 feet (3 m) with a motor speed of not more than 80 percent of the certified peak of the brake horsepower curve. Will deliver 70 percent of rated capacity at 200 psi (1 350 kPa) and 50 percent of rated capacity at 250 psi (1 700 kPa).

Pumper, Class B — Pumper that delivers its rated capacity at 120 psi (800 kPa) net pump pressure at a lift of not more than 10 feet (3 m) with a motor speed of not more than 80 percent of the certified peak of the brake horsepower curve. Will deliver 50 percent of its rated capacity at 200 psi (1 350 kPa) and 33 1/3 percent of its rated capacity at 250 psi (1 700 kPa). Class B pumps have not been manufactured since the mid-1950s.

Pumper/Tender — Mobile water supply apparatus equipped with a fire pump. In some jurisdictions, this term is used to differentiate a fire pump equipped mobile water supply apparatus whose main purpose is to attack the fire. *Formerly known as* Pumper/Tanker.

Pumping Apparatus — Fire department apparatus that has the primary responsibility to pump water.

Pump-Off Line — Pipeline that usually runs from the tank discharge openings to the front of the trailer; most pumps are mounted on the tractor.

Punch Press — Large, heavy piece of industrial machinery used to cold-form sheet metal.

Puncture Nozzle — *See* Piercing Nozzle.

Purchase Order — Written document generated by the purchaser that allows vendors to invoice for monies owed. This action serves as an official record that the organization and vendor/contractor have conducted financial business.

Purging — Freeing from impurities by introducing fresh air; for example, ventilating a contaminated space.

Purification System — Series of mechanical and chemical filters through which compressed breathing air is passed to remove moisture, oil, carbon monoxide, and other contaminants.

Purlin — Horizontal member between trusses that support the roof.

Purpose Classification — Means of classifying tests based on when the test occurs during a course.

Purpose-Built Structure — Building specially designed for live-fire training; fires can be ignited inside the building multiple times without major structural damage.

PVC — *See* Polyvinyl Chloride.

Pyro — *See* Pyromaniac.

Pyrolize — Description of the process of a solid beginning to emit gases due to heat exposure.

Pyrolysis — The chemical decomposition of a solid material by heating. Pyrolysis precedes combustion of a solid fuel.

Pyromania — Psychological disorder in which the sufferer has an uncontrollable impulse to set fires, either to relieve tension or produce a feeling of euphoria.

Pyromaniac — Person with an uncontrollable impulse to set fires, or who enjoys setting fires. *See* Pyromania.

Pyrometer — Device used to measure temperatures by wavelength or electrical generation; pyrometers connected to thermocouples record the heat at various points.

Pyrophoric — Material that ignites spontaneously when exposed to air. *Also known as* Air-Reactive Material.

Pyrophoric Liquids — Liquids that ignite spontaneously in dry or moist air at or below 130°F (54°C).

Pyrophoric Materials — Elements that react and ignite on contact with air.

Pyrotechnics — Fireworks.

Pyroxylin Plastic — Nitrocellulose plastic that is extremely combustible and susceptible to deterioration and self-ignition; produces toxic fumes when burned.

Q

Q — Slang for a Federal Q2B Mechanical Coaster Siren.

Qassam Rocket — Improvised or homemade steel rocket filled with explosives that is produced by Hamas. Qassam rockets have no guidance system, and their warheads are typically filled with TNT and urea nitrate.

Quad — Four-way combination fire apparatus; combines the water tank, pump, and hose of a pumper with the ground ladder complement of a truck company. *Also known as* Quadruple Combination.

Quadrant — One of the four regions into which the abdomen may be divided for purposes of physical diagnosis.

Qualitative Analysis — Nonstatistical subjective analysis of educational materials that relies on a developer's experience, judgment, and interpretation of the material.

Qualitative Evaluation — Means of assessment that does not use numbers and that relies on the fire and life safety educator's experience, judgment, and interpretation; evaluation based on non-numerical analysis and intended to assess the quality of something.

Qualitative Fit Test (QLFT) — Respirator fit test that measures the wearer's response to a test agent, such as irritant smoke or odorous vapor. If the wearer detects the test agent, such as through smell or taste, the respirator fit is inadequate.

Quantitative Analysis — Statistical analysis of educational materials that includes formal testing of the products for ease of readability.

Quantitative Evaluation — Means of assessment that uses numbers and statistical data to compare different materials and methods, is likely to be more sophisticated than qualitative methods, and involves formal testing; intended to discover quantifiable data.

Quantitative Fit Test (QNFT) — Fit test in which instruments measure the amount of a test agent that has leaked into the respirator from the ambient atmosphere. If the leakage measures above a pre-set amount, the respirator fit is inadequate.

Quarantine — State of enforced isolation; restraint upon the activities or communication of persons or the transport of goods, designed to prevent the spread of disease or pests.

Quarter Drain Time — *See* Quarter-Life.

Quarter Panels — Rear sections of the vehicle's body shell, including the rear fender and the C-post.

Quarter-Life — Time required, in minutes, for one-fourth of the total liquid solution to drain from a foam blanket. *Also known as* 25 Percent Drain Time and Quarter Drain Time. *See* Drainage Time.

Quarters — Fire station or office.

Quarter-Turn Coupling — Nonthreaded (sexless) coupling with two hooklike lugs that slip over a ring on the opposite coupling and then rotate 90 degrees clockwise to lock.

Quench — To extinguish a fire by cooling.

Quick-Fill® System — Mine Safety Appliances Company (MSA) system that can be used as an emergency breathing system connection or that can be used to refill self-contained breathing apparatus (SCBA) cylinders during nonemergency conditions.

Quintuple Combination Pumper (Quint) — Apparatus that serves as an engine and as a ladder truck; equipped with a fire pump, water tank, ground ladders, hose bed, and aerial device.

R

R — *See* Roentgen.

Rabbet — Groove cut in the surface or on the edge of a board to receive another member.

Rabbeted Jamb — Jamb into which a shoulder has been milled to permit the door to close against the provided shoulder.

Rabbit Tool — Hydraulic spreading tool that is specially designed to open doors that swing inward.

Races — (1) Sliding channels between two sections of the aerial device. (2) Interior of a box-beam construction aerial device.

Rack — (1) Framework used to support ladders while being carried on fire apparatus. (2) Act of placing a ladder on an apparatus.

rad — *See* Radiation Absorbed Dose.

Radial Artery — One of the major arteries of the forearm, located where the pulse can be felt at the base of the thumb.

Radial Engines — Internal-combustion, piston-driven aircraft engines with cylinders arranged in a circle.

Radiant Heat — *See* Radiation.

Radiant Heat Flux — Measure of the rate of heat transfer to a surface, expressed in kilowatts/m^2, sec, or Btu/ft2 • sec.

Radiated Heat — *See* Radiation.

Radiation — (1) Transmission or transfer of heat energy from one body to another body at a lower temperature through intervening space by electromagnetic waves, such as infrared thermal waves, radio waves, or X-rays. *Also known as* Radiated Heat. (2) Energy from a radioactive source emitted in the form of waves or particles, as a result of the decay of an atomic nucleus; process known as *radioactivity. Also known as* Nuclear Radiation. *See* Alpha Radiation, Beta Radiation, Gamma Radiation, Ionizing Radiation, Nonionizing Radiation, Radiation Absorbed Dose (rad), Radioactive Material (RAM), and Radioactive Particles. (3) Heat transfer by way of electromagnetic energy. (NFPA® 921)

Radiation Absorbed Dose (rad) — English System unit used to measure the amount of radiation energy absorbed by a material; its International System equivalent is gray (Gy). *See* Radiation (2) and Radioactive Material (RAM).

Radiation detector — Device for detecting the presence and sometimes the amount of radiation.

Radiation Dose — Quantity of radiation energy absorbed into the body.

Radiation-Emitting Device (RED) — Powerful gamma-emitting radiation source used as a weapon.

Radiative Feedback — Radiant heat from the combustion process that provides energy for the continued vaporization of the fuel. *See* Vaporization.

Radiator Fill Line — Small waterline leading from the fire pump to the radiator of the apparatus; used to refill the radiator during pumping at a fire scene.

Radio Channel — Band of frequencies of a width sufficient to permit its use for radio communication.

Radio Repeater — Combination of a radio receiver and a radio transmitter that receives a weak or low-level signal and retransmits it at a higher level or higher power, so that the signal can cover longer distances without degradation.

Radio Systems Regulations — Federal Communications Commission (FCC) rules that govern the operation of radio systems.

Radio Transmitter — Device for producing radio frequency power for purposes of radio transmission.

Radioactive Material (RAM) — Material whose atomic nucleus spontaneously decays or disintegrates, emitting radiation. *See* Becquerel (Bq), Curie (Ci), and Radiation (2).

Radioactive Particles — Particles emitted during the process of radioactive decay. There are three types of radioactive particles: alpha, beta, and gamma. *See* Alpha Particle, Beta Particle, Gamma Rays, and Radiation (2).

Radioactivity — *See* Radiation (2).

Radiography — Process of making a picture on a sensitive surface by a form of radiation other than light.

Radioisotope Unstable or radioactive isotope (form) of an element that can change into another element by giving off radiation. *See* Radionuclide.

Radiological Dispersal Device (RDD) — Conventional high explosives wrapped with radioactive materials; designed to spread radioactive contamination over a wide area. *Also known as* Dirty Bomb.

Radiological Dispersal Weapons (RDW) — Devices that spread radioactive contamination without using explosives; instead, radioactive contamination is spread using pressurized containers, building ventilation systems, fans, and mechanical devices.

Radionuclide Any radioactive isotope (form) of any element. *See* Radioisotope.

Radiopharmaceutical — Radioactive drug used for diagnostic or therapeutic purposes.

Radius — Half the diameter of a circle; line or distance from the center of a circle to any point on its circumference.

Rafter — Inclined beam that supports a roof, runs parallel to the slope of the roof, and to which the roof decking is attached.

Rafter Cut — *See* Louver Cut.

Ragged Left — Format in a written text in which every line ends the same distance from the right-hand edge of the page or column.

Rail — (1) Horizontal member of a window sash. (2) Metal portion of a railroad track upon which the car wheels ride. *See* Beam.

Rail Tank Car — Railroad car that is designed to carry liquids in pressurized or unpressurized cylinders; may be constructed of steel, stainless steel, or aluminum.

Railcar Initials and Numbers — Combination of letters and numbers stenciled on rail tank cars that may be used to get information about the car's contents from the railroad's computer or the shipper. *Also known as* Reporting Marks. *See* Intermodal Reporting Marks.

Rain Roof — Second roof constructed over an existing roof.

Rain-Down Application Method — Foam application method that directs the stream into the air above the unignited or ignited spill or fire, allowing the foam to float gently down onto the surface of the fuel. *See* Bank-Down Application Method and Roll-On Application Method.

Raise — Any of several accepted methods of raising and placing ground ladders into service; for example the two-firefighter raise.

Raising System — Mechanical advantage system designed for lifting a load; constructed of rope and appropriate hardware. *Also known as* Hauling System.

Raker — Diagonal strut which collects load delivered by the wallplate in a raker shoring system, such as an angled timber bearing the load exerted on a raking shore.

Raker Shore — Shore footed on a horizontal surface used to brace a vertical surface.

Ram — *See* Extension Ram.

RAM — *See* Radioactive Material.

Ramp — (1) Area at airports intended to accommodate aircraft for purposes of loading or unloading passengers or cargo, refueling, parking, or maintenance. (2) Parking space between the fire station garage doors and the street. (3) Movable stairway or enclosure for loading and unloading passengers.

Ranking of Scores — Process of arranging a number of scores (values) in order from high to low, making it easy to determine median and mode.

Rapid Ascent Team (RAT) — Team of firefighters equipped with lightweight PPE and communications equipment whose job is to quickly ascent stairways in high-rise buildings to report conditions and ensure that civilians are evacuating safely.

Rapid Intervention Crew (RIC) — Two or more firefighters designated to perform firefighter rescue; they are stationed outside the hazard and must be standing by throughout the incident. *Previously known as* Rapid Intervention Team (RIT).

Rapid Intervention Vehicle (RIV) — Type of mobile foam apparatus that is used to provide a rapid attack on an aircraft crash incident for quick extinguishment and rescue.

Rapid Relief — Fast release of a pressurized hazardous material through properly operating safety devices caused by damaged valves, piping, or attachments or holes in the container.

Rappel — A system for controlled movement down a high angle using a friction device on lifeline looped at a secure anchor at the top of the angle and with one end attached to a harness around the body. The rappeller controls movement by loosening and tightening his or her hold on the unanchored rope.

Rappel Rack — A descent device that allows the user to apply or disengage friction bars to change the amount of weight that can be controlled by the device. This device does not twist the rope or generate much friction. *Also known as* Brake Bar Racks.

Rappelling — Technique of sliding or descending down a rope in a controlled manner using a friction device, such as descending down the side of a building or cliff, or landing firefighters from helicopters in hover during wildland fire fighting operations.

RAT — *See* Rapid Ascent Team.

Ratchet-Lever Jack — Type of lifting jack that uses the principles of leverage to operate. It is capable of lifting moderately heavy loads but tends to be unstable; generally recognized as the most dangerous of all jacks.

Rate Compensated Heat Detector — Temperature-sensitive device that sounds an alarm at a preset temperature, regardless of how fast temperatures change.

Rate Meter — Nuclear radiation detection device.

Rate of Spread (ROS) — Relative activity of a fire in extending its horizontal dimensions. Expressed as rate of increase of the total perimeter of a fire, as rate of forward spread of the fire front, or as rate of increase in area, depending on the intended use of the information; usually expressed in chains or acres (hectares) per hour for a specific period in the fire's history.

Rate of Vaporization — Speed at which a liquid evaporates or vaporizes. *See* Vaporization.

Rated Assembly — Assemblies of building components such as doors, walls, roofs, and other structural features that may be, because of the occupancy, required by code to have a minimum fire-resistance rating from an independent testing agency. *Also known as* Labeled Assembly.

Rated Concentrate — Foam concentrate that has been tested and certified by Underwriters Laboratories Inc. *See* Foam Concentrate and Nonrated Concentrate.

Rated Fire Barrier Assembly — Continuous membrane, such as a wall or floor assembly, that is designed and constructed to limit the spread of fire and restrict the spread of smoke; must achieve a specified fire-resistance rating and meet the test performance requirements of an independent testing organization.

Rated Fire Door Assembly — Door, frame, and hardware assembly that has a fire-resistive rating from an independent testing agency.

Rated Smoke Barrier Assembly — Continuous membrane, such as a wall, floor, or ceiling assembly, that is designed and constructed to restrict the spread of smoke, and meets certain performance testing requirements, typically from an independent testing organization.

Rate-of-Raise Alarm System — System that detects fire by an abnormal rate of heat increase; operates when a normal amount of air in a pneumatic tube or chamber expands rapidly when heated, exerting pressure on a diaphragm.

Rate-of-Rise Heat Detector — Temperature-sensitive device that sounds an alarm when the temperature changes at a preset value, such as 12°F to 15°F (7°C to 8°C) per minute.

Rating Bureau — *See* Insurance Services Office.

Raw Image — Digital image that has undergone no compression.

Raw Score — Score on a test that has not yet been statistically processed to make it comparable with other scores. For example, an individual who received 38 points on a test of 40 questions has a raw score of 38; to compare the score to 100 percent, divide the raw score (38) by the total number of possible points (40) to get the percentage score of 95. Raw scores and the percent scores can be compared to other learner scores (norm-referenced), or scores can be assigned to a selected mastery level based on a standard or job performance requirement (criterion-referenced); for example, 85 percent is considered mastery.

Razor Ribbon — Coil of lightweight, flexible metallic ribbon with extremely sharp edges; often installed on parapet walls and on fence tops to discourage trespassers.

RDD — *See* Radiological Dispersal Device.

Reach Rescue — Method of rescue that uses the lowest level of risk and resources; a rescuer establishes a secure position and extends an arm with or without a tool to a victim in a water environment.

Reach, Throw, Row, Go — The order of operations in water rescue incidents. "Helo" may be added at the end of this list if helicopter or other air support resources are available.

Reaction Distance — *See* Driver Reaction Distance.

Reactive Material — Substance capable of chemically reacting with other substances; for example, material that reacts violently when combined with air or water. *See* Air-Reactive Material, Reactivity, and Water-Reactive Material.

Reactivity — Ability of a substance to chemically react with other materials, and the speed with which that reaction takes place. *See* Chemical Reaction and Reactive Material.

Readability Index — System for measuring how easy or difficult a passage is to read.

Reading a Roof — Process of observing the surface and other important features of a roof from a point of safety, in order to assess the roof's condition before stepping onto it.

Reagent — Chemical that is known to react to another chemical or compound in a specific way, often used to detect or synthesize another chemical.

Rear — (1) Part of a wildland fire opposite the head; the slowest burning part of the fire. *See* Heel. (2) Opposite end of the vehicle from the front; usually indicated by the taillights.

Rear of Ladder — Side closest to the objective; the nonclimbing side.

Rear-Impact Collision — Collision in which one vehicle is struck in the rear by another vehicle, or in which a vehicle backs into an object. *Also known as* Rear-End Collision

Reasonable Accommodation — (1) Legal requirement (under Title VII of the *Civil Rights Act of 1964*) that employers make reasonable adjustments to an employee's work schedule or other job requirements to accommodate employee differences, such as religion, gender, and/or physical or mental disability. (2) Changes or adjustments in a work or school site, program, or job that makes it possible for an otherwise qualified employee or student with a disability to perform the duties or tasks required.

Rebar — Short for reinforcing bar. These steel bars are placed in concrete forms before the cement is poured. When the concrete sets (hardens), the rebar within it adds considerable strength and reinforcement.

Rebreather — Closed-circuit breathing apparatus.

Reburn — Burning of an area that has been previously burned but that contains flammable fuel that ignites when burning conditions are more favorable; area that has reburned.

Recall — (1) To call off-duty firefighters back to duty. (2) To order units responding to or on the scene of an emergency to return to their quarters.

Receiver — (1) Part of an automatic switching system that receives signals from a calling device or other source for interpretation and action. (2) Person or persons to whom a message is directed in the communications process. *Also known as* Audience.

RECEO Model — One of many models for prioritizing activities at an emergency incident: Rescue, Exposures, Confine, Extinguish, and Overhaul.

RECEO V/S Model — One of many models for prioritizing activities at an emergency incident: Rescue, Exposures, Confine, Extinguish, Overhaul, Ventilation, and Salvage.

Recidivism — Relapse into criminal behavior, such as firesetting.

Reciprocating Engine — Internal-combustion engine with cylinders arranged in opposition, and in in which the back and forth movement of the pistons causes the rotation of a crank shaft.

Reciprocating Saw — Electric saw that uses a short, straight blade that moves back and forth.

Recirculation — Movement of smoke being blown out of a ventilation opening only to be drawn back inside by the negative pressure created by the ejector because the open area around the ejector has not been sealed.

Recommended Exposure Limit (REL) — Recommended value expressing the maximum time-weighted dose or concentration to which workers should be exposed over a 10-hour period, as established by National Institute for Occupational Safety and Health (NIOSH). *See* Permissible Exposure Limit (PEL), Short-Term Exposure Limit (STEL), and Threshold Limit Value (TLV®).

Reconnaissance — Process of examining an area to obtain information about the current specific situation, probable fire behavior, and other information related to fire-suppression.

Reconstruction — Portion of an investigation in which the investigator attempts to determine the original position of the contents at the scene.

Record Book — *See* Journal.

Records — Permanent accounts of past events or of actions taken by an individual, unit, or organization.

Recovery — (1) Situation where the victim is determined or presumed to be dead, and the goal of the operation is to recover the body. (2) Process of restoring normal public or utility services following a disaster; activities necessary to rebuild after a disaster, such as rebuilding homes and businesses, clearing debris, repairing roads and bridges, and restoring water and other essential services.

RED — *See* Radiation-Emitting Device.

Red Blood Cells — Cellular components of the blood that transport oxygen from the lungs to body tissues and carbon dioxide from the tissues to the lungs.

Red Line — *See* Booster Hose.

Reduce Speed — *See* Proceed With Caution.

Reducer — Adapter used to attach a smaller hose to a larger hose; the female end has the larger threads, while the male end has the smaller threads. *See* Adapter, Fitting, and Increaser.

Reducing Agent — Fuel that is being oxidized or burned during combustion. *See* Fuel.

Reducing Wye — Wye that has two outlets smaller in diameter than the inlet valve. *Also known as* Leader Line Wye.

Redundancy — Secondary or backup systems that allow for uninterrupted use in the event of failure or damage to the primary system.

Reefer — *See* Refrigerated Intermodal Container.

Reefer Container — Cargo container having its own refrigeration unit. *See* Container and Container Terminal.

Reel Load — Arrangement of fire hose, especially large diameter hose, on a reel.

Reeves Stretcher — Device used to package a patient for removal from a confined environment.

Reeving — Threading rope or cable through a block.

Reference — Method by which an authority having jurisdiction (AHJ) refers to a code in a regulation and states that the code is legally enforceable.

References — Citations, bibliographies, and resources used in developing, planning, and researching course or lesson information.

Referral — Act or process by which an individual and/or family gains access to a program or community resources.

Reflex Time — Amount of time responding units need to reach the emergency scene and set up for operations.

Reformulated Gasoline — *See* Blended Gasoline.

Refractive Index — Measurement of the amount by which the speed of light (or other waves such as sound waves) is reduced inside a medium such as a finished foam solution.

Refractometer — Device used to measure the amount of foam concentrate in the solution; operates on the principle of measuring the velocity of light that travels through the foam solution.

Refrigerant — Substance used within a refrigeration system to provide the cooling action.

Refrigerated Intermodal Container — Cargo container having its own refrigeration unit. *Also known as* Reefer. *See* Container Vessel and Intermodal Container.

Refrigerated Liquids — *See* Cryogens.

Refrigerated Vessel — Vessel specially designed and equipped for the transportation of food products (such as meat, fruit, fish, butter, or eggs) under cold storage; cargo space is insulated for this purpose.

Refrigerating Plant — Installation of machinery for the purposes of cooling designated spaces aboard a vessel and manufacturing ice.

Refrigeration Unit — Cooling equipment used to maintain a constant temperature within a given space.

Refuse Chute — Vertical shaft with a self-closing access door on every floor; usually extending from the basement or ground floor to the top floor of multistory buildings.

Regenerative Braking — Mechanical system that reduces the speed of a vehicle by converting part of the vehicle's kinetic energy into another type of energy that can be fed back into a power system or stored for future use.

Reglet — Flat, narrow molding that forms a water seal for roofing in a parapet wall.

Regulations — Rules or directives of administrative agencies that have authorization to issue and enforce them. *See* Code and Standard.

Regulator — Device between the facepiece and air cylinder of the SCBA that reduces the pressure of the air coming from the cylinder.

Regulator Breathing — Emergency procedure in which the firefighter breathes directly from the regulator outlet if the low-pressure hose or facepiece is damaged.

Regulator Gauge — Gauge connected to the regulator of an SCBA that indicates the pressure of the air reaching the regulator; used as an indication of the air pressure in the air cylinder.

Rehab — *See* Rehabilitation.

Rehabilitation — (1) Activities necessary to repair environmental damage or disturbance caused by wildland fire or the fire-suppression activity. (2) Allowing firefighters or rescuers to rest, rehydrate, and recover during an incident; also refers to a station at an incident where personnel can rest, rehydrate, and recover. *Also known as* Rehab.

Reinforced Concrete — Concrete that is internally fortified with steel reinforcement bars or mesh placed within the concrete before it hardens. Reinforcement allows the concrete to resist tensile forces.

Rekindle — To reignite because of latent heat, sparks, or smoldering embers; rekindling can be prevented by proper overhaul.

REL — *See* Recommended Exposure Limit.

Relative Humidity — Measure of the moisture content (water quantity expressed in a percentage) in both the air and solid fuels.

Relay — (1) Use of two or more pumpers to move water distances that would require excessive pressures if only one pumper was employed. (2) To shuttle water between a source and an emergency scene using mobile water supply apparatus.

Relay Emergency Valve — Combination valve in an air brake system that controls brake application and provides for automatic emergency brake application should a trailer become disconnected from the towing vehicle.

Relay Operation — Using two or more pumpers to move water over a long distance by operating them in series; water discharged from one pumper flows through hoses to the inlet of the next pumper, and is then pumped to the next pumper in line. *Also known as* Relay Pumping.

Relay Pumping — *See* Relay Operation.

Relay Question — Type of question in which an instructor or educator returns or redirects a question from the audience back to the individual who asked the question, or to the group, to answer.

Relay Valve — Pressure-relief device on the supply side of the pump designed to protect the hose and pump from damaging pressure surges common in relay pumping operations.

Relay Valve (In-Line) — Special valve that is inserted in the middle of a long relay hose; allows an additional pumper to connect to the line to boost pressure without having to interrupt the current flow of water.

Relay-Supply Hose — Hose between the water source and the attack pumper, laid to provide large volumes of water at low pressure. *Also known as* Feeder Line or Supply Hose.

Reliability — A condition of validity; the extent to which a test or test item consistently and accurately produces the same results or scores when given to a set of learners on different occasions, marked by different assessors, or marked by the same assessors on different occasions.

Reliable Data — Statistics that come from an organization with proven expertise in collecting and disseminating data.

Relief Cut — Cut made to reduce resistance and to facilitate the bending of a portion of a car or other object.

Relief Valve — Pressure control device designed to eliminate hazardous conditions resulting from excessive pressures by allowing this pressure to bypass to the intake side of the pump.

rem — *See* Roentgen Equivalent in Man.

Remediation — Fixing or correcting a fault, error, or deficiency.

Remote Power Outlet (RPO) — AC power receptacle mounted on or in a vehicle and powered by an inverter from the vehicle's DC electrical system.

Remote Pressure Gauge — Pressure gauge that is not mounted on the regulator but can be seen by the SCBA wearer; commonly found on SCBA that have facepiece-mounted regulators.

Remote Receiving System — System in which alarm signals from the protected premises are transmitted over a leased telephone line to a remote receiving station with a 24-hour staff; usually the municipal fire department's alarm communications center sending an alarm signal to the FACU. *See* Remote Station Alarm System.

Remote Station Alarm System — System in which alarm signals from the protected premises are transmitted over a leased telephone line or by radio signal to a remote receiving station with a 24-hour staff; usually the municipal fire department's alarm telecommunications center. *Also known as* Remote Receiving System.

Remote-Controlled Foam Monitor — Large-capacity foam system that is operated by a remote control located away from the monitor; usually found on aircraft rescue apparatus, fire fighting apparatus, and fireboats. *See* Automatic Oscillating Foam Monitor, Foam Monitor, and Manual Foam Monitor.

Repair — To restore or put together something that has become inoperable or out of place.

Repeater System — System in which a radio message is received on one frequency (input) and then rebroadcast at a high energy level on another frequency (output); this enables two lower power field units to speak to each other without having to be nearby. Repeaters can be mounted in cars or be part of a complex network of other repeaters to provide adequate coverage.

Reporting Locations — Any one of six facilities/locations where incident-assigned resources may check in. The locations are: incident command post - resources status unit (RESTAT), base, camp, staging area, helibase, and division supervisor for direct line assignments.

Reporting Marks *See* Railcar Initials and Numbers.

Reports — Official accounts of an incident, response, or training event, either verbally or in writing.

Request for Proposal (RFP) — Public document that advertises an organizational need to manufacturers or individuals who may be able to meet that need and defines the specific requirements for an item that an organization intends to purchase through the bid process.

Requisite — Fundamental knowledge or essential skill one must have in order to perform a specific task. *See* Prerequisite. *Also known as* Requisite Knowledge.

Requisite Knowledge — *See* Requisite.

Rescue — Saving a life from fire or accident; removing a victim from an untenable or unhealthy atmosphere.

Rescue Breathing — *See* Artificial Respiration.

Rescue Company — Specialized unit of people and equipment dedicated to performing rescue and extrication operations at the scene of an emergency. *Also known as* Rescue Squad or Rescue Truck.

Rescue Eight — Device used to create friction for rappelling and as a connection point. This style of descent device has extra protrusions to prevent the rope from slipping out of place while the device is in use. *Also known as* Figure-Eight Plates, Rescue Eight with Ears, *or* Figure-Eight Descenders.

Rescue Knot — Knot that is easy to tie, easy to inspect visually, easy to untie after loading, and that remains tied and will not untie itself or loosen during use. Causes minimal loss in rope strength when tied.

Rescue Load — Normally considered to be 450 pounds (204 kg), or one fully equipped rescuer and patient.

Rescue Officer — Officer in charge of the rescue company.

Rescue Pumper — Specially designed apparatus that combines the functions of both a rescue vehicle and a fire department pumper.

Rescue Rope — Term used interchangeably with life safety rope.

Research and Special Programs Administration (RSPA) — U.S. Department of Transportation agency that carries out and enforces the hazardous materials regulations (HMR) through a program of regulation, enforcement, emergency response education and training, and data collection and analysis.

Reserve Apparatus — Apparatus not scheduled to respond to fires in normal or first-line duty but available for emergencies or replacing first-line equipment.

Reset — (1) To restore fire protection or detection equipment to original standby condition after operation. (2) To reactivate an inoperable fire alarm box or sprinkler system.

Residential Board and Care Facility — Subdivision of residential property classification consisting of a structure or part of a structure used for boarding or lodging four or more residents who are not related to the operators or owners, in order to provide personal care services.

Residential Occupancy — Occupancy that provides sleeping accommodations for routine residential purposes; includes all structures designed for providing sleeping accommodations.

Residential Sprinkler System — Wet- or dry-pipe fire suppression system that is built into a residential structure; activation of a sprinkler causes the extinguishing agent to flow from the open sprinkler.

Residential Treatment Center — Center which provides 24-hour care and can usually serve several young people at a time. Children with serious emotional disturbances receive constant supervision and care at an RTC. Treatment may include individual, group, and family therapy; cognitive / behavior therapy; special education; recreation therapy; and medical services. Residential treatment is usually more long-term than inpatient hospitalization.

Residual Pressure — Pressure measured at the hydrant to which a pressure gauge is attached while water is flowing from one or more other hydrants during a hydrant flow test. It represents the pressure remaining in the water supply system while the test water is flowing and is that part of the total pressure that is not used to overcome friction or gravity while forcing water through fire hose, pipe, fittings, and adapters.

Resistance — Opposition to the flow of an electric current in a conductor or component; measured in ohms (Ω).

Resistance Heating — Heat generated by passing an electrical current through a conductor, such as a wire or an appliance.

Resistance to Freezing — Foam concentrate's usefulness after it has frozen and thawed; most can be freeze protected, but some concentrates freeze at lower temperatures than others.

Resource Status Unit (RESTAT) — Functional unit within the planning section of an incident command system; responsible for recording and evaluating the status of resources committed to the incident, the impact that additional responding resources will have on the incident, and the anticipated resource needs.

Resources — (1) All of the immediate or supportive assistance available, or potentially available, for assignment to help control an incident; includes personnel, equipment, control agents, agencies, and printed emergency guides. (2) Locations that provide instructional training materials and support.

Respiration — Act of breathing; the exchange of oxygen and carbon dioxide in the body tissues and lungs.

Respirator — Device designed to protect the wearer from inhaling harmful air contaminants. There are two main categories of respirators are air-purifying respirators, which use cartridges or filters to remove contaminants, and air-supplied respirators, such as SCBAs and airline respirators, which provide an alternate supply of fresh air.

Respiratory Arrest — Cessation of breathing.

Respiratory Hazards — Exposure to conditions that create a hazard to the respiratory system, including products of combustion, toxic gases, and superheated or oxygen-deficient atmospheres.

Respiratory Protection Program — Systematic and comprehensive program of training in the use and maintenance of respiratory protection devices and related equipment.

Respiratory Protection Specialist School — School that provides 25 to 60 hours of additional or advanced training in the use of respiratory protection equipment. *Also known as* Smoke Divers School.

Respiratory System — System of organs that serve the function of respiration; consists of lungs, their nervous and circulatory supply, and the channels by which these are continuous with the outer air.

Responder Unit — Emergency medical unit that carries first aid and/or advanced life support equipment but is not equipped for patient transport.

Responding — Clear text radio term given when a unit is en route to an assignment.

Response — Call to respond.

Response District — Geographical area to which a particular apparatus is assigned to be first due on a fire or other emergency incident. *Also known as* District.

Response Objectives — Statements based on realistic expectations of what can be accomplished when all allocated resources have been effectively deployed that provide guidance and direction for selecting appropriate strategies and the tactical direction of resources.

Response Time — Time between when a fire company is dispatched and when it arrives at the scene of an emergency.

Response Time Index (RTI) — Numerical value representing the speed and sensitivity with which a heat responsive fire protection device, such as a fusible link, responds.

Responsibility — Act or duty for which someone is clearly accountable.

Responsible Party (RP) — An individual or group of people who are legally responsible or liable for a decision or action and therefore liable for the outcome.

Rest — Position of a ladder when both beams are resting on and parallel to the ground.

Restricted (Hot) Zone — (1) In a hazmat incident, the potentially hazardous area immediately surrounding the incident site requiring appropriate protective clothing and equipment and other safety precautions for entry; typically limited to technician-level personnel. *Also known as* Exclusion Zone. (2) In a rescue or extrication operation, the area where the extrication is taking place. Only personnel who are attending directly to the victims should be in this zone; this avoids crowding and confusion among rescuers.

Restricted Grant — *See* Project Grant.

Restrictive Bid — Bid that includes many specifications that only one manufacturer can meet; also known as Sole-Sourced Bid.

Resuscitation — Act of reviving an unconscious patient. *See* Artificial Respiration.

Retard Chamber — Chamber that catches and slows the excess water that may be sent through the alarm valve of an automatic sprinkler system during momentary water pressure surges; this reduces the chance of false alarm activation. The retard chamber is installed between the alarm check valve and alarm signaling equipment.

Retardant — *See* Fire Retardant.

Retardant Drop — Fire retardant cascaded from an air tanker or helicopter.

Retention — Characteristic of Class A foam and foam solution; its ability to remain on and in the fuel, reduce the fuel temperature, and increase the fuel moisture content.

Retroreflective Trim — Surfaces such as those used on road signs, emergency vehicle markings, or safety vests which are designed to reflect light along multiple planes at once, giving the surface the appearance of illumination.

Return-Air Plenum — Unoccupied space within a building through which air flows back to the heating, ventilating, and air-conditioning (HVAC) system; normally immediately above a ceiling and below an insulated roof or the floor above. *See* Heating, Ventilating, and Air-Conditioning (HVAC) System.

Returning — Clear text radio term used when a company is leaving the scene of an incident.

Reverse Curl — Method of returning a one-firefighter ladder to a flat rest position on the ground.

Reverse Hose Lay — *See* Reverse Lay.

Reverse Lay — Method of laying hose from the fire scene to the water supply.

Revolutions per Minute (rpm) — The number of revolutions that the blade of a cutting tool completes in one minute; a measurement of speed in both English and metric systems.

Revolving Door — Door made of three or four sections, or wings, arranged on a central pivot that operates by rotating within a cylindrical housing.

Reward Power — Power based on the subordinate's perception of the leader's ability to grant rewards, such as salary increases, promotions, and bigger budgets.

Reynolds Number (Re) — Mathematically calculated factor that determines the state of flow (laminar or turbulent) of a fluid.

RFP — *See* Request for Proposal.

Rhetorical Question — Question designed to stimulate thinking or motivate participants rather than to seek a correct answer; an answer is not necessarily required or expected.

Rhythm — Handling and climbing ladders with smooth motion.

Rib Cage — Skeletal framework of the chest; composed of the sternum, ribs, and thoracic vertebrae.

Ribbon — Narrow strip of board cut to fit into the edge of studding to help support joists.

RIBC — *See* Rigid Intermediate Bulk Container.

RIC — *See* Rapid Intervention Crew.

Rickettsia — Specialized bacteria that live and multiply in the gastrointestinal tract of arthropod carriers, such as ticks and fleas. *See* Bacteria and Virus.

Ridge — (1) Peak or sharp edge along the very top of a pitched roof of a building. *Also known as* Ridge Beam, Ridge Board, or Ridge Pole. (2) Horizontal line at the junction of the top edges of two sloping roof surfaces.

Ridge Beam — Highest horizontal member in a pitched roof to which the upper ends of the rafters attach. *Also known as* Ridge Board or Ridgepole. *See* Ridge.

Rig — Any piece of fire apparatus.

Rigging — (1) Ropes or cables used with lifting or pulling devices such as block and tackle. (2) The process of arranging ropes into a mechanical advantage system.

Right Justified — *See* Ragged Left.

Right of Entry — Legal access to private property obtained in one of five ways: exigent circumstances, consent, administrative search warrant, criminal search warrant, or contractual entry agreement.

Right of Privacy — Concept that means that an individual's records are confidential. *See* Family Education Rights and Privacy Act of 1974.

Righting Arm — Moment that tends to return a vessel to the upright position after any small rotational displacement. *Also known as* Righting Moment or Restoring Moment.

Rigid Conduit — Nonflexible steel tubing used for the passage of electrical conductors.

Rigid Intermediate Bulk Container (RIBC) — *See* Intermediate Bulk Container (IBC).

Rigor Mortis — Sign of death in a deceased individual in which the muscles cause the body to be stiff and difficult to move; begins within a few hours of death and recedes within a few days.

Rim Cylinder — Lock cylinder for a rim lock.

Rim Lock — Type of auxiliary lock mounted on the surface of a door.

Ring Stiffener — Circumferential tank shell stiffener that helps to maintain the tank cross section.

Riot Control Agent — Chemical compound that temporarily makes people unable to function, by causing immediate irritation to the eyes, mouth, throat, lungs, and skin. *Also known as* Irritating Agent or Tear Gas. *See* Incapacitant.

Rise — Vertical distance between the treads of a stairway, or the height of the entire stairway.

Riser — (1) Vertical part of a stair step. (2) Vertical water pipe used to carry water for fire protection systems above ground, such as a standpipe riser or sprinkler riser. (3) Pipe leading from the fire main to the fire station (hydrants) on upper deck levels of a vessel. *See* Automatic Sprinkler System and Standpipe System.

Risk — (1) Likelihood of suffering harm from a hazard; exposure to a hazard. The potential for failure or loss. (2) Estimated effect that a hazard would have on people, services, facilities, and structures in a community; likelihood of a hazard event resulting in an adverse condition that causes injury or damage. Often expressed as *high, moderate,* or *low* or in terms of potential monetary losses associated with the intensity of the hazard.

Risk Analysis — *See* Hazard or Risk Analysis.

Risk Assessment — (1) Determining the risk level or seriousness of a risk. (2) Process for evaluating risk associated with a specific hazard defined in terms of probability and frequency of occurrence, magnitude and severity, exposure, and consequences. *Also known as* Risk Evaluation.

Risk Identification — Stage of loss control risk analysis where planners identify risks that are specific to a particular facility, occupancy, or area.

Risk Level — Seriousness of a risk; determined by frequency and severity.

Risk Management — Analyzing exposure to hazards, implementing appropriate risk management techniques, and monitoring their results.

Risk Reduction Strategy — Series of integrated programs designed to impact a common goal.

Risk-Based Response — Method using hazard and risk assessment to determine an appropriate mitigation effort based on the circumstances of the incident.

Risk-Benefit Analysis — Comparison between the known hazards and potential benefits of any operation; used to determine the feasibility and parameters of the operation.

Risk-Management Plan — Written plan that identifies and analyzes the exposure to hazards, selects appropriate risk management techniques to handle exposures, implements those techniques, and monitors the results. *See* Hazard Assessment and Hazard Risk Analysis.

Risk-Reduction Initiative — Fire of life safety program that targets a specific issue and audience(s) and is terminated when program goals are achieved.

Risk-Reduction Program — Comprehensive strategy that addresses fire and life safety issues via educational means.

RIT — *See* Rapid Intervention Crew.

Rivet Construction — Multi-piece metal structure (aerial device) fastened together by rivets.

RO/RO — *See* Roll On/Roll Off Cargo and Roll On/Roll Off Vessel.

Road Performance Test — Series of tests required to determine the performance ability of fire apparatus.

Road Tests — Pre-service apparatus maneuverability tests designed to determine the road-worthiness of a new vehicle.

Roadside — Side of the trailer farthest from the curb when trailer is traveling in a normal forward direction; in North America, this is the left-hand side). Opposite to "curbside."

Rocker Panels — Narrow body panels on each side of an automobile, below the doors and between the kick panel and the quarter panel; usually rounded in shape.

Rocket-Propelled Grenade (RPG) — Common anti-tank projectile launched from a handheld tube.

Roentgen (R) — English System unit used to measure radiation exposure, applied only to gamma and X-ray radiation; the unit used on most U.S. dosimeters. *See* Gamma Radiation, Radiation (2), Radiation Absorbed Dose (rad), Radioactive Material (RAM), and Roentgen Equivalent in Man (rem).

Roentgen Equivalent in Man (rem) — English System unit used to express the radiation absorbed dose (rad) equivalence as pertaining to a human body; used to set radiation dose limits for emergency responders. Applied to all types of radiation. *See* Radiation (2), Radiation Absorbed Dose (rad), Radioactive Material (RAM), and Roentgen.

Role — Socio-psychological concept that says the role or status someone has in a building determines that person's response to a fire or other emergency.

Role Model — Individual to whom others look to as an example while learning or adopting a new role or job; the part an instructor plays, the image an instructor portrays, and the actions an instructor demonstrates to learners or program participants who look to their instructor as an example.

Role-Playing — Discussion in which a group acts out various scenarios. *See* Discussion.

Roll On/Roll Off Cargo (RO/RO) — Form of cargo handling using a vessel designed to carry vehicles that are loaded and unloaded by driving them onto/off the vessel by means of ramps.

Rolled Shape — Structural steel member made by passing a hot steel billet between shaped rollers until it reaches the required shape and dimensions.

Roll-On Application Method — Method of foam application in which the foam stream is directed at the ground at the front edge of the unignited or ignited liquid fuel spill; foam then spreads across the surface of the liquid. *Also known as* Bounce. *See* Bank-Down Application Method and Rain-Down Application Method.

Roll-On/Roll-Off Vessel (RO/RO Vessel) — Ship with large stern and side ramp structures that are lowered to allow vehicles to be driven on and off the vessel; for example, a vehicle ferry. *See* Cargo Vessel.

Rollover — (1) Condition in which the unburned fire gases that have accumulated at the top of a compartment ignite and flames propagate through the hot-gas layer or across the ceiling. These superheated gases are pushed, under pressure, away from the fire area and into uninvolved areas where they mix with oxygen. When their flammable range is reached and additional oxygen is supplied by opening doors and/or applying fog streams, they ignite and a fire front develops, expanding very rapidly in a rolling action across the ceiling. *See* Backdraft, Flashover, and Incipient Phase. (2) Involves a vehicle rolling sideways onto its side and possibly continuing onto its top, then the opposite side.

Rollover Protection — Roll bars and roll cages within automobiles that protect passengers in the event of a rollover.

Romax™ — Trade name for nonmetallic-shielded cable.

Roof — (1) Outside top covering of a building. (2) Vehicle body component above the passengers' heads that encloses the passenger compartment.

Roof Bows — Steel frame members that run horizontally from side to side.

Roof Covering — Final outside cover that is placed on top of a roof deck assembly; common roof coverings include composition or wood shake shingles, tile, slate, tin, or asphaltic tar paper. *See* Roof Deck.

Roof Deck — Bottom components of the roof assembly that support the roof covering; the roof deck may be constructed of such components as plywood, wood studs (2 inches by 4 inches [50mm by 100 mm] or larger), lath strips, and other materials.

Roof Decking — *See* Sheathing.

Roof Ladder — Straight ladder with folding hooks at the top end; the hooks anchor the ladder over the roof ridge.

Rookie Academy — Special school to indoctrinate newly appointed firefighters in the rudiments of all fire service subjects.

Rooming House — Subdivision of residential property classification consisting of structures with sleeping accommodations for up to 16 persons on either a temporary or permanent basis, where meals may or may not be provided, but lacking separate cooking facilities for each occupant.

Rope Hose Tool — Piece of rope spliced to form a loop through the eye of a metal hook; used to secure hose to ladders or other objects. *See* Hose Belt and Hose Strap.

Rope Log — Record of all use, maintenance, and inspection throughout a rope's working life; also includes the product label and manufacturer's recommendations.

Rope Rescue — Use of rope and related equipment to perform rescue.

ROS — *See* Rate of Spread.

Rotary Gauge — Gauge for determining the liquid level in a pressurized tank.

Rotary Gear Positive Displacement Pump — Type of positive displacement pump commonly used in hydraulic systems. The pump imparts pressure on the hydraulic fluid by having two intermeshing rotary gears that force the supply of hydraulic oil into the pump casing chamber.

Rotary Rescue Saw — *See* Circular Saw.

Rotary Vane Pump — Type of positive displacement pump commonly used in hydraulic systems. A rotor with attached vanes is mounted off-center inside the pump housing; pressure is imparted on the water as the space between the rotor and the pump housing wall decreases.

Rotational Collisions — Collisions caused by off center front or side impacts that forcefully turn the impacted vehicle horizontally, causing one or both of the vehicles to spin.

Rotational Locks — Locking mechanisms that prevent the aerial device turntable from rotating unexpectedly.

Rotor — Rotating airfoil assemblies that provide lift for helicopters and other rotary-wing aircraft.

Rotor Blast — Air turbulence occurring under and around the rotors of an operating helicopter. *Also known as* Rotor Downwash from the main rotor.

Rotor Downwash — *See* Rotor Blast.

Round Turn — Element of a knot that consists of further bending one side of a loop.

Round-Robin Brainstorming — Method of brainstorming in which each member of a group offers an idea in turn, with participants electing to pass on any round, until everyone has passed a turn; ideas are recorded as soon as they are stated.

Routes of Entry Means by which hazardous materials enter (or affect) the body; common routes are inhalation, ingestion, skin contact, injection, absorption, and penetration (for radiation). *See* Permeation and Radiation (2).

Row Rescue — Method of rescue that uses watercraft to approach victims in a water environment.

RPG — *See* Rocket-Propelled Grenade.

RPM — *See* Revolutions Per Minute.

RSPA — *See* Research and Special Programs Administration.

Rub Rail — Sixteen-gauge steel W-shaped rails placed the full length of the sidewalls on a bus; they are intended to minimize penetration during collision.

Rubric — Scoring tool that outlines criteria that must be present on exams that are more subjective such as short-answer tests, essay tests, or oral tests; the criteria should be tied to learning objectives.

Rudder — Upright movable part of the aircraft tail assembly that assists in the directional control of the aircraft. *Also known as* Vertical Stabilizer.

Run — (1) Response to a fire or alarm. (2) The horizontal measurement of a stair tread or the distance of the entire stair length.

Run Block — *See* Truss Block.

Rung — Step portion of a ladder running from beam to beam.

Rung Block — *See* Truss Block.

Rung Side — Front or climbing side of a ladder; the side away from the building or objective.

Rungs Away — Position of a raised truss ladder when the rungs are on the side furthest from the objective.

Rungs Down — Position of a truss ladder at rest when the rungs are on the side closest to the ground.

Rungs Up — Position of a truss ladder at rest when the rungs are on the side furthest from the ground.

Running Block — In a block and tackle system, the block attached to the load that is to be moved.

Running Fire — (1) Wildland fire that spreads rapidly. (2) Behavior of a fire spreading rapidly with a well-defined head.

Running Part — Free end of the rope used for hoisting, pulling, or belaying.

Runway — Defined rectangular area on airports prepared for the takeoff or landing of aircraft along its length.

Runway Threshold — Beginning or end of a runway that is usable for landing or takeoff.

Rupture Disk — Formed, thin metal membrane or diaphragm designed to burst at a predetermined pressure and temperature in order to prevent overpressurization of the attached vessel or container; a pressure-relief device. *See* Emergency-Relief Device.

S

S/S — Prefix to the name of a vessel with a steam propulsion plant.

S/V — Prefix to the name of a vessel propelled by sail.

Saddle — Depression or pass in a ridgeline; low area on a ridgeline between two higher points.

Saddle Burn — Saddle-shaped fire pattern that is the result of fire burning downward through the floor surface above the joist.

SAE — *See* the Society of Automotive Engineers.

Safe Refuge — *See* Safety Zone.

Safe Zone — The space within 2 ft (0.6 m) in all directions of an installed support component of an existing approved shoring system.

Safety — (1) Extra hitch tied in the end of a knot to prevent the end from being pulled through the knot. (2) Device designed to prevent an inadvertent or hazardous operation. (3) To insert locking pins into appropriate openings on ejection seats in military aircraft to render the seats safe to work on or around.

Safety Audit — Comprehensive compliance review of all organizational components that could contribute to firefighter safety including policies, procedures, practices, inspection reports, and firefighter behaviors.

Safety Bar — Hinged bar designed to protect firefighters from falling out of the open jump seat area of a fire apparatus.

Safety Belt — Life safety harness.

Safety Can — Flammable liquid container, usually five gallons (19 L) or less, that has a self-closing spout and has been approved by a suitable testing agency.

Safety Chain — Chain connecting two vehicles to prevent separation in the event the primary towing connection breaks.

Safety Data Sheet (SDS) — Reference material that provides information on chemical that are used, produced, or stored at a facility. Form is provided by chemical manufacturers and blenders; contains information about chemical composition, physical and chemical properties, health and safety hazards, emergency response procedures, and waste disposal procedures. *Also known as* Material Safety Data Sheet (MSDS) *or* Product Safety Data Sheet (PSDS).

Safety Gates — Protective guards that are placed over the apparatus jump seat opening to prevent firefighters from falling off the apparatus.

Safety Glass — Two sheets of glass laminated to a sheet of plastic sandwiched between them; the plastic layer makes the glass stronger and more shatter resistant. Most commonly used in windshields and rear windows. *Also known as* Laminated Glass.

Safety Glasses — *See* Safety Goggles.

Safety Goggles — Enclosed, but adequately ventilated goggles that have impact- and shatter-resistant lenses to protect the eyes from dusts, chips, and other small particles; should be OSHA approved. *Also known as* Safety Glasses.

Safety Guidelines — Rules, regulations, or policies created and/or adopted by an organization that list steps or procedures to follow that will aid in reducing, if not eliminating, accident or injury. *Also known as* Safety Plan.

Safety Island — *See* Safety Zone.

Safety Knot — Extra hitch or overhand knot tied in the loose end of a knot to prevent the working end from being pulled through the knot. *Also known as* Safety.

Safety Line — (1) Extra rope tied to the main hauling rope in a rope rescue operation. (2) System built to protect a rescuer and/or patient.

Safety Net — Net used to protect firefighters in the event of a fall during aboveground rope training evolutions.

Safety Officer — (1) Fire officer whose primary function is to administrate safety within the entire scope of fire department operations. *Also known as* Health and Safety Officer. (2) Member of the IMS command staff responsible to the incident commander for monitoring and assessing hazardous and unsafe conditions and developing measures for assessing personnel safety on an incident. *Also known as* Incident Safety Officer.

Safety Policy — Written policy that is designed to promote safety to departmental members.

Safety Program — Program that sets standards, policies, procedures, and precautions regarding the safe purchase, operation, and maintenance of the department's equipment, and educates employees on how to protect themselves from personal injury.

Safety Relief Valve — Device on cargo tanks with an operating part held in place by a spring; the valve opens at preset pressures to relieve excess pressure and prevent failure of the vessel.

Safety Shoes — (1) Rubber or neoprene foot plates, usually of the swivel type, attached to the butt end of the beams of a ground ladder. (2) Protective footwear meeting OSHA requirements.

Safety Zone — Recently burned area or one cleared of vegetation, used for escape in the event a line is outflanked or a spot fire outside a control line renders the line unsafe. In firing operations, crews progress so as to maintain a safety zone close at hand, allowing the fuels inside the control line to be consumed before going ahead. *Also known as* Safe Refuge *or* Safety Island.

Sag — (1) The vertical distance of the ship's keel at amidships below the ship's keel at the bow and stern. (2) To curve downward in the middle as a result of improper loading. *Also known as* Sagging. *See* Hog.

Sagging — *See* Sag.

Sail Area — Area of a vessel, when viewed from the side, that is above the waterline and is subject to the force of the wind.

Salamander — Portable heating device; generally found on construction sites.

Salvage — Methods and operating procedures by which firefighters attempt to save property and reduce further damage from water, smoke, heat, and exposure during or immediately after a fire; may be accomplished by removing property from a fire area, by covering it, or by other means.

Salvage Cover — Waterproof cover made of cotton duck, plastic, or other material used by fire departments to protect unaffected furniture and building areas from heat, smoke, and water damage; a tarpaulin. *Also known as* Tarp.

Salvage Kit — Assortment of tools and appliances used for a specific purpose during salvage.

Salvo Drop — Air tanker dropping its entire load of fire retardant at one time.

Sampling Error — Measure of the error created because estimates are based on a sampling of fire losses rather than on a complete census of the fire problem. *Also known as* Standard Error.

Sanction — Notice or punishment attached to a violation for the purpose of enforcing a law or regulation. *See* Citation and Violation.

Sandshoe — Flat, steel plate that serves as ground contact on the supports of a trailer; used instead of wheels, particularly where the ground surface is expected to be soft.

Sanitary Sewer — Underground pipe used to carry waste from toilets (water closets) and from other drains.

Sanitary Tee — Soil pipe fitting with a side outlet to form a tee shape; the side outlet has a smooth radius to permit unhampered flow in the fitting.

Sanitize — To make free from dirt or microorganisms that endanger health.

Saponification — Phenomenon that occurs when mixtures of alkaline based chemicals and certain cooking oils come into contact, resulting in the formation of a soapy film.

SAR — *See* Supplied Air Respirator.

SARA — *See* Superfund Amendments and Reauthorization Act.

Sarin (GB) — Fluorinated phosphinate that attacks the central nervous system; classified as a chemical warfare agent.

Sash — Framework in which panes of glass are set in a window or door.

Sash Cord — Cotton cord, usually ¼ inch (6.35 mm) in diameter, that is used for securing tiebacks, guy lines, hose bundles, or salvage covers by lacing through grommets.

Saturation Tactics — Practice of having a person (usually a child) who has played with fire repeat an act (such as lighting a match) until the person's curiosity is satisfied and the act becomes boring; no longer a recommended practice.

Save — Life that has been saved as a direct result of a public fire and life safety education program.

Sawtooth Roof — Roof style characterized by a series of alternating vertical walls and sloping roofs that resembles the teeth of a saw; this type of roof is most often found on older industrial buildings to provide light and ventilation.

Scald — Burn to the human body caused by contact with hot fluids or steam; or, to cause such a burn.

Scantlings — Dimensions of the various parts of a vessel (frames, girders, plating, etc.).

Scare Tactics — Practice of frightening a child who has played with fire by showing the child pictures of injured or dead children, pets, or burned toys, or by threatening the child with punishment, hospitalization, or painful treatment. Not a recommended practice.

Scarf Joint — Connection between two parts made by the cutting of overlapping mating parts and securing them by glue or fasteners so that the joint is not enlarged and the patterns are complementary.

SCBA — *See* Self-Contained Breathing Apparatus.

Scene Assessment — Initial observation and evaluation of an emergency scene; related more to incident stabilization than to problem mitigation.

Scene Control Zones — *See* Hazard-Control Zones.

Scene Control Zones — *See* Hazard-Control Zones.

Scene Management — Those elements of incident management that include keeping those not involved in the incident from entering unsafe areas and protecting those in potentially unsafe areas through evacuation or sheltering in place.

SCF — *See* Standard Cubic Foot.

SCFM — *See* Standard Cubic Feet Per Minute.

Schedule — Table or chart on plans that contains information relating to doors, windows, hardware, and room finishes. *See* Door/Window Schedule.

Schema — Refers to conceptual or knowledge structure (a mental map) in our memory system that we use to interpret information that is presented to our senses by the external environment. When our senses are presented with a new object or situation, we match it against our existing knowledge or schema and act upon it based on our experience. If we don't have the appropriate knowledge or experience, we change the schema or develop additional ones. We develop more sophisticated and differentiated schemata as we gain new knowledge and encounter more experiences. The plural form is *schemata*.

School Bus — As defined by U.S. Federal Motor Vehicle Safety Standards, a passenger motor vehicle designed to carry more than 10 passengers, in addition to the driver, and which the Secretary of Transportation determines is to be used for the purpose of transporting preschool, primary, and secondary school students to or from such schools or school-related events.

School-Based Program — Activities that support specific outcome objectives of a curriculum presented within a school system.

Scientific Method — Widely accepted, systematic approach to examining evidence in order to create hypotheses and draw conclusions about phenomena at a fire or explosion scene.

Scissor Stairs — Two sets of crisscrossing stairs in a common shaft. Each set serves every floor but on alternately opposite sides of the stair shaft; for example, one set would serve the west wing on even-numbered floors and the east wing odd-numbered floors, while the other set would serve floors opposite to the first set.

Sclera — White outer coat enclosing the eyeball, except the part covered by the cornea.

Scoop Stretcher — Device used to package patient for removal from a confined environment; may be placed around the patient without lifting the patient.

Scratch Line — Unfinished preliminary control line, hastily established or constructed as an emergency measure to check the spread of fire.

Screed — Two or more strips set at a desired elevation so that concrete may be leveled by drawing a leveling device over their surface; also the straightedge.

Screw Jacks — Long, nonhydraulic jacks that can be extended or retracted by turning a collar on a threaded shaft.

Screw-In Expander Method — Method of attaching threaded couplings to rubber-jacket booster hose with expanders that are screwed into place.

Scrub Area — Area within the span of reach of an aerial device.

SCUBA — *See* Self-Contained Underwater Breathing Apparatus.

Scupper — (1) Form of drain opening provided in outer walls at floor or roof level to remove water to the exterior of a building in order to reduce water damage. (2) Opening in the side of a vessel to allow water falling on deck to drain overboard.

Scuttle — Opening in the roof or ceiling that provides access to the roof or attic; fitted with removable covers that may be used for access or ventilation. *See* Hatch.

SDH — *See* Small Diameter Hose.

SDS — *See* Safety Data Sheet.

Sea Chest — (1) Enclosure attached to the inside of a vessel's underwater shell open to the sea and fitted with a portable strainer plate; passes seawater into the vessel for cooling, fire fighting, or sanitary purposes. (2) Storage chest for mariner's personal property.

Sealed, Pneumatic, Line-Type Heat Detector — Heat-sensitive device that depends on the presence of normal pressure at normal temperatures.

Search — Techniques that allow the rescuer to identify the location of victims and to determine access to those victims in order to remove them to a safe area.

Search and Rescue Boat — Watercraft designed and equipped to carry personnel during search and rescue operations such as boating accidents, flood evacuations, and dive rescues.

Search and Rescue Operation — Emergency incident operation consisting of an organized search for the occupants of a structure or for those lost in the outdoors, and the rescue of those in need.

Search Assessment — Search assessment and findings performed by search personnel.

Search Line — Nonload-bearing rope that is anchored to a safe, exterior location and attached to a firefighter during search operations to act as a safety line.

Search Warrant — Written order, in the name of the People, State, Province, Territory, or Commonwealth, signed by a magistrate, that commands a peace officer to search for personal property or other evidence and return it to the magistrate.

Seat Belt Pretensioners — Protective devices designed to tighten the belts as the front-impact air bags deploy.

Seat Catapult — Device for catapulting the seat from an aircraft in case of emergency.

Seat of Fire — Area in which the main body of fire is located.

Seated Explosion — Explosion with a clearly defined epicenter or seat, often a crater; typically associated with boiling liquid expanding vapor explosions (BLEVEs) and high explosives. *See* Nonseated Explosion.

Seaworthy — In fit condition to go safely to sea.

Secondary Collapse — Collapse that occurs after the initial collapse of a structure; common causes include aftershock (earthquake), weather conditions, and the movement of structural members.

Secondary Contamination — Contamination of people, equipment, or the environment outside the hot zone without contacting the primary source of contamination. *Also known as* Cross Contamination. *See* Contamination, Decontamination, and Hazard-Control Zones.

Secondary Damage — Damage caused by or resulting from those actions taken to fight a fire and leaving the property unprotected.

Secondary Decontamination — Taking a shower after having completed a technical decontamination process. *See* Decontamination and Technical Decontamination.

Secondary Device — Bomb or other weapon placed at the scene of an ongoing emergency response that is intended to cause casualties among responders; secondary explosive devices are designed to explode after a primary explosion or other major emergency response event has attracted large numbers of responders to the scene.

Secondary Duties — Actions required to restore the emergency scene to a safe condition.

Secondary Explosion — Explosion occurring as a direct result of an initial explosion or previous explosion.

Secondary Explosive — High explosive that is designed to detonate only under specific circumstances.

Secondary Feeder — Network of intermediate-sized pipes that reinforce the grid within the loops of the primary feeder system and aid in providing the required fire flow at any given point in a sprinkler system.

Secondary Line — Any fireline that is constructed at a distance from the fire perimeter, concurrently with or after a line has already been constructed on or near the perimeter of the fire; generally constructed as an insurance measure in case a fire escapes control by the primary line.

Secondary Prevention — Seeks to mitigate or modify events to reduce their severity. It targets high-risk conditions and populations.

Secondary Search — Slow, thorough search to ensure that no occupants were overlooked during the primary search; conducted after the fire is under control by personnel who did not conduct the primary search.

Second-Degree Burn — Burn that penetrates beneath the superficial skin layers to the dermis; produces edema, skin blisters, and possible scarring.

Section — Organizational level of an incident command system; has functional responsibility for primary segments of incident operations such as Operations, Planning, Logistics, and Finance/Administrative. This section level is organizationally between branch and incident commander.

Sectional View — Vertical view of a building as if it were cut into two parts; the purpose of a sectional view is to show the internal construction of each assembly. *See* Detailed View, Elevation View, and Plan View.

Sector — Either a geographic or function-based subdivision or assignment within the Fireground Command System or National Fire Service Incident Management System; may take the place of either a division, a group, or both.

Secure — (1) To make fast; for example, to secure a line to a cleat or other stationary object. (2) To close in a manner that prevents accidental opening or operation.

Security — Second need in Maslow's Hierarchy of Needs, concerned with personal safety and future physical comfort; similar to physiological needs, but encompassing long-term dangers or dangers that are not immediately threatening.

Security Window — Window designed to prevent illegal entrance to a building.

Sediment — Dirt and other foreign debris that may fall out of a fluid and collect in fluid-moving equipment.

Seismic Effect — Movement of a shock wave through the ground or structure after a large detonation; may cause additional damage to surrounding structures.

Seismic Forces — Forces produced by earthquakes; they are the most complex forces that can be exerted on a building.

Seismic Victim Locating Device — Device that detects minute vibrations and movement within a collapsed structure in order to help identify a victim's location.

Selective Catalyst Reductant (SCR) — A system that uses diesel exhaust fluid (DEF) to help further reduce emissions.

Selective Routing (SR) — Enhanced 9-1-1 feature that allows 9-1-1 calls to be routed to the appropriate public safety answering point (PSAP); this system allows for more than one PSAP to be located in an area where jurisdictions share the same central telephone office.

Selector Valve — Three-way valve on a fire department aerial apparatus that directs oil to either stabilizer control valves or the aerial device control valves. *Also known as* Diverter Valve.

Self-Actualization — Highest need in Maslow's Hierarchy of Needs; the need to realize one's full potential.

Self-Closing Door — (1) Door equipped with a door closer. (2) On a ship, an installation in which watertight doors are remotely operated by a hydraulic pressure system, allowing them to be closed simultaneously from the bridge or separately at the doors from either side of the bulkhead. *See* Door Closer.

Self-Contained Breathing Apparatus (SCBA) — Respirator worn by the user that supplies a breathable atmosphere that is either carried in or generated by the apparatus and is independent of the ambient atmosphere. Respiratory protection is worn in all atmospheres that are considered to be Immediately Dangerous to Life and Health (IDLH). *Also known as* Air Mask *or* Air Pack. *See* Supplied Air Respirator (SAR).

Self-Contained Underwater Breathing Apparatus (SCUBA) — Protective breathing apparatus designed to be used underwater by divers to allow the exploration of underwater environments. *Also known as* SCUBA Gear *or* Underwater Breathing Gear.

Self-Directed Learning — Method of instruction in which individual students work at their own pace to accomplish course objectives in any way they choose. Course objectives may be determined by the instructor or chosen by the student, but course content is always determined by the instructor. *Also known as* Independent Learning.

Self-Educting Master Stream Foam Nozzle — Large-capacity nozzle with built-in foam eductor.

Self-Educting Nozzle — Handline nozzle that has the foam eductor built into it.

Self-Esteem — Second highest need in Maslow's Hierarchy of Needs; the need to have self-respect and respect for others, to have social status within the group, and to be recognized for one's worth or value.

Self-Heating — The result of exothermic reactions, occurring spontaneously in some materials under certain conditions, whereby heat is generated at a rate sufficient to raise the temperature of the material (NFPA® 921).

Self-Presenters — People seeking medical attention who were not treated or decontaminated at the incident scene.

Self-Priming Centrifugal Pump — Centrifugal pump that uses an air-water mixture to reach a fully primed pumping condition. *See* Centrifugal Pump, Multistage Centrifugal Pump, and Single-Stage Centrifugal Pump.

Self-Sustained Chemical Reaction — One of the four sides of the fire tetrahedron representing a process occurring during a fire: Vapors or gases are distilled from flammable materials during initial burning; atoms and molecules are released from these vapors and combine with other radicals to form new compounds. These compounds are again disturbed by the heat, releasing more atoms and radicals that again form new compounds, restarting the chain reaction. Interrupting the chain will stop the overall reaction; this is the extinguishing mechanism utilized by several extinguishing agents. *Also known as* Chemical Chain Reaction.

Semantics — Study of meaning in words and symbols; refers to language, word meanings, and meaning changes due to context, all of which may be affected by an individual's background, knowledge, and experience.

Semiconductor — Material that is neither a good electrical conductor nor a good insulator, and therefore may be used as either in some applications. *See* Conductor and Thermistor.

Semifixed Foam Extinguishing System — Foam system that is designed to provide fire-extinguishing capabilities to an area but is not automatic in operation and depends on human intervention to place it into operation. *See* Fixed Foam Extinguishing System and Portable Foam Extinguishing System. *Also known as* Semifixed System.

Semifixed System — *See* Semifixed Foam Extinguishing System.

Semisubsurface Injection — Application method that discharges foam through a flexible hose that rises from the bottom of a storage tank, up through the fuel, and to the surface of the fuel; foam then blankets the surface of the fuel. *See* Direct Injection and Subsurface Injection.

Semitrailer — Freight trailer that when attached is supported at its forward end by the fifth wheel device of the truck tractor; occasionally used to refer to a trucking rig made up of a tractor and a semitrailer. *See* Fifth Wheel.

Semitrailer Tank — Any vehicle with or without auxiliary motive power, equipped with a cargo tank mounted thereon or built as an integral part thereof, that is used for the transportation of flammable and combustible liquids or asphalt; constructed so that when drawn by a tractor by means of a fifth-wheel connection, some part of its load and weight rests upon the towing vehicle.

Sender — Person who sends or transmits a message in the communications process. *Also known as* Source.

Senior Facility Manager — Individual with overall responsibility for a particular commercial, institutional, or industrial facility.

Sensitizer — *See* Allergen.

Sensorimotor Stage of Intellectual Development — First stage of intellectual development; occurs from birth to age two. Learner constructs understanding of the world by coordinating sensory experiences (seeing/hearing) with motor actions (doing).

Sensory Memory — Mental storage system for attention-getting sensory stimuli or input.

Separating — Act of creating a barrier between the fuel and the fire.

Septic Shock — Shock caused by a severe infection in the body.

Sequential Training — Preferred training method, in which the student is taken step by step from simple to complex exercises when learning to use equipment such as a self-contained breathing apparatus (SCBA).

Sergeant — Rank used by some fire departments for company officers or fire apparatus driver/operators.

Series Circuit — Circuit configuration in which the current flows through all the components.

Series Operation — See Pressure Operation.

Serology — Science of analyzing bodily fluids such as saliva, blood, urine, or semen.

Serrated — Notched or toothed edge.

Service Branch — Branch within the logistics section of an incident command system; responsible for service activities at an incident. Components include the communications unit, medical unit, and foods unit.

Service Learning — Educational trend that tries to connect young people to the community in which they live through community service projects.

Service Records — Detailed description of maintenance and repair work for a particular apparatus or piece of equipment.

Service Test — (1) Series of tests performed on apparatus and equipment in order to ensure operational readiness of the unit; should be performed at least yearly, or whenever a piece of apparatus or equipment has undergone extensive repair. (2) Series of tests performed on fire detection and/or suppression systems in order to ensure operational readiness. These tests should be performed at least yearly or whenever the system has undergone extensive repair or modification.

Session Guide — Plan for using a group of lesson plans or instructional materials during a predetermined period of instruction.

Set — (1) Individual incendiary fire. (2) Point or points of origin of an incendiary fire. (3) Material left to ignite an incendiary fire at a later time. (4) Individual lightning or railroad fires, especially when several are started within a short time. (5) Burning material at the points deliberately ignited for backfiring, slash burning, prescribed burning, and other purposes. See Permanent Deformation (1).

Setback — Distance from the street line to the front of a building.

SETIQ — See the Emergency Transportation System for the Chemical Industry (Mexico).

Sewer Drain Guard — Strainer to prevent debris from getting into the sewer system; used when utilizing soil pipes to remove water during salvage operations.

Sexless Coupling — See Nonthreaded Coupling.

Sexual Harassment — (1) Unwanted and unwelcome sexual behavior toward a worker by someone who has the power to reward or punish the worker. (2) Superior offering advancement or special treatment in return for sexual favors from a subordinate; also may refer to any situation in which an employee, regardless of gender, believes that the workplace is a hostile environment because of sexually offensive or sexist behavior.

Shackle — (1) Hinged part of a padlock. (2) A U-shaped metal device that is secured with a pin or bolt across the device's opening. (3) A hinged metal loop that is secured with a quick-release locking pin mechanism.

Shaft — Any vertical enclosure within a building; for example, a stairwell or elevator hoistway.

Shaft Alley — Narrow, watertight compartment between the engine room and the stern of a vessel that houses the propeller shaft. *Also known as* Shaft Tunnel.

Shaftway — Tunnel or alleyway through which the drive shaft or rudder shaft passes.

Shall — Common verb used in NFPA® standards denoting compulsory compliance. When used in codes, "shall" denotes a mandatory provision.

Shank — Portion of a coupling that serves as a point of attachment to the hose.

Shaped Charge — (1) Concave metal hemisphere (known as a liner) backed by a high explosive, seated in a steel or aluminum casing; when the high explosive is detonated, the metal liner is compressed and squeezed forward, forming a jet of superheated liquid metal whose tip may travel many times faster than the speed of sound. (2) An explosive that has been designed (or shaped) to direct the energy of the explosion in a single direction.

Shear Line — Space between the shell and the plug of a lock cylinder that is obstructed by tumblers in the locked position.

Shear Point — Hazard created by a reciprocal (sliding) movement of a mechanical component past a stationary point on a machine.

Shear Strength — Ability of a building component or assembly to resist lateral or shear forces.

Shear Stress — Stress resulting when two forces act on a body in opposite directions in parallel adjacent planes.

Shears — Powered hydraulic cutting tool that will cut most metals, other than case-hardened steel.

Sheathing — (1) Covering applied to the framing of a building to which siding is applied. (2) First layer of roof covering laid directly over the rafters or other roof supports; may be plywood, chipboard sheets, or planks that are butted together or spaced about 1 inch (25 mm) apart. *Also known as* Decking or Roof Decking.

Shed Roof — Pitched roof with a single sloping aspect, resembling half of a gabled roof.

Sheet Bend — *See* Becket Bend.

Sheeting — Wood planks and wood panels that support trench walls when held in place with shoring.

Sheetrock® — Brand name often used to describe any gypsum wallboard. *See* Wallboard.

Shelf Angles — Brackets fastened to the face of a building at or near floor levels to support masonry or wall facing materials.

Shell — (1) Outer component of a screw-in expander coupling. (2) Outer layer of fabric on personal protective clothing.

Shell Structure — Rigid, three-dimensional structure having an outer "skin" thickness that is small compared to other dimensions.

Shelter and Thermal Control — Process of protecting patients and rescuers from inclement weather and extreme temperatures.

Shelter in Place — Having occupants remain in a structure or vehicle in order to provide protection from a rapidly approaching hazard, such as a fire or hazardous gas cloud. *Opposite of* evacuation. *Also known as* Protection-in-Place, Sheltering, *and* Taking Refuge. *See* Evacuation.

Shielded Fire — Fire that is located in a remote part of the structure or hidden from view by objects in the compartment.

Shims — Angle-cut piece of timber used to ensure close contact between shoring and loads, fill in voids, or change the angle of thrust.

Shiplap Siding — Horizontal siding boards lapped over each other to provide a water shedding effect.

Shipping Papers — Shipping orders, bills of lading, manifests, waybills, or other shipping documents issued by the carrier. *See* Air Bill, Bill of Lading, Consist, Lading, and Waybill.

Shock — Failure of the circulatory system to produce sufficient blood to all parts of the body; results in depression of bodily functions, and eventually death if not controlled.

Shock Front — Boundary between the pressure disturbance created by an explosion (in air, water, or earth) and the ambient atmosphere, water, or earth.

Shock Loading — Loads that involve motion; includes the forces from wind, moving vehicles, earthquakes, vibration, falling objects, or the addition of a moving load force to an aerial device or structure. *Also known as* Dynamic Load.

Shock Wave — Blast pressure front moving faster than the speed of sound; becomes an amplitude wave that travels through solid objects and the ground.

Shoe — Metal plate used at the bottom of heavy timber columns.

S-Hook — S-shaped steel hook frequently used in salvage operations by placing through grommet holes when hanging salvage covers; also used to hang covers for drying after cleanup.

Shop — Fire department maintenance or repair area.

Shoring — General term used for lengths of timber, screw jacks, hydraulic and pneumatic jacks, and other devices that can be used as temporary support for formwork or structural components or used to hold sheeting against trench walls. Individual supports are called shores, cross braces, and struts. Commonly used in conjunction with Cribbing.

Shoring Block — Shim for a jack.

Shoring Timbers — Heavy timbers used to support bulkheads damaged by collision or to secure cargo; also any props or supports placed against or beneath anything to prevent sinking or sagging.

Short Circuit — Abnormal, low-resistance path between conductors that allows a high current flow that normally leads to an overcurrent condition.

Short-Jacking — Setting the stabilizers on one side of an apparatus shorter than the stabilizers on the other side; usually done when access for full stabilization is restricted.

Short-Term Exposure Limit (STEL) — Fifteen-minute time-weighted average that should not be exceeded at any time during a workday; exposures should not last longer than 15 minutes and should not be repeated more than four times per day with at least 60 minutes between exposures. *See* Immediately Dangerous to Life or Health (IDLH), Permissible Exposure Limit (PEL), Recommended Exposure Limit (REL), and Threshold Limit Value (TLV®).

Short-Term Memory — Information storage area with a finite amount of space. Information not acted upon or used by the learner within a relatively brief time period will be replaced.

Shoulder Carry — Procedure of carrying fire hose or a ground ladder on the shoulder.

Shove Knife — Tool for opening a latch on a lock.

Shrapnel — Descriptor of debris, large and small, carried by the blast-pressure front of an explosion.

Shrapnel Fragmentation — Small pieces of debris thrown from a container or structure that ruptures from containment or restricted blast pressure.

Shunt System — Auxiliary fire alarm system that connects a public fire alarm reporting system to initiating devices within a protected premises. When an initiating device in the protected property operates, it activates the public fire alarm sending an alarm to the public communication center.

Shutoff Nozzle — Type of nozzle that has a valve or other device for controlling the water supply; firefighters use it to control water supply at the nozzle rather than at the source of supply.

Siamese — Hose appliance used to combine two or more hoselines into one. The siamese has multiple female inlets and a single male outlet. An example of a siamese is a fire department connection. This hose is most commonly found in ½-, ¾-, and 1-inch (13 mm, 19 mm, and 25 mm) diameters and is used for extinguishing low-intensity fires and overhaul operations.

Side Rails — Upper and lower side rails, which are the main longitudinal frame members of a tank used to connect the upper and lower corner fittings, respectively. *See* Beam.

Side-Impact Collision — (1) Collision in which a vehicle is struck along its side by another vehicle. *Also known as* Broadside Collision or T-Bone Collision. (2) Collision in which a vehicle slides sideways into another object.

Side-Impact Protection Systems (SIPS) — Air bag systems designed to protect passengers during side-impact collisions; may be operated mechanically or powered by the vehicle's electrical system.

Sidewall Sprinkler — Sprinkler designed to be positioned at the wall of a room rather than in the center of a room; has a special deflector that creates a fan-shaped pattern of water that is projected into the room, away from the wall. *Also known as* Wall Sprinkler.

Signal Word — Government-mandated warnings provided on product labels that indicate the level of toxicity; for example CAUTION, WARNING, or DANGER.

Signaling Device — System component that generates an audible, visual, or motion signal intended to alert humans to the activation of an alarm-initiating device. *See* Fire Alarm System, Fire Detection System, and Initiating Device.

Significant New Alternatives Policy (SNAP) — EPA-mandated program for identifying and evaluating new alternative agents to replace ozone-depleting substances such as halon.

Silage — Contents of a silo.

Silent Alarm — *See* Still Alarm.

Siliceous Aggregate — Coarse material such as gravel, broken stone, or sand, with which silica, cement, and water are mixed to form concrete.

Sill — (1) Bottom rough structural member that rests on the foundation. (2) Bottom exterior member of a window or door or the masonry below.

Silo — Tall, round structure found on farms, at grain elevators, and at mills; used to store feed for livestock, and grain during shipment to and upon arrival at the mills.

Silo Gas — Collection of gases produced during the spoilage of crops in a silo; includes methane, carbon dioxide, nitrogen dioxide, and hydrogen sulfide.

Simple Asphyxiant Any inert gas that displaces or dilutes oxygen below the level needed by the human body. *See* Asphyxiant and Inert Gas.

Simple Firesetting Case — Juvenile firesetting situation motivated by curiosity and can be addressed effectively though educational intervention.

Simple Fracture — Fracture in which the skin overlaying the broken bone is intact.

Simple Loop — Loop in which there is exactly one inflow point and one outflow point, and exactly two paths between the inflow and outflow points.

Simple Machine — Any device that changes the direction or magnitude of a force. Simple machines include: Levers, Wedges, Pulleys, Inclined Planes, Screws

Simple Mechanical Advantage System — A rigging system using one or more pulleys that move a load in line with the anchor point. This type of system is often used with relatively light loads.

Simple Tackle — One or more blocks reeved with a single rope.

Simple Triage and Rapid Treatment (START) — Triage evaluation method for checking respiratory, circulatory, and neurological function, with the intention of categorizing patients in one of the four care categories: Minor, Delayed, Immediate, and Expectant. The START method is recommended for use by first-arriving responders for initial and secondary field triage.

Simple-To-Complex — Method of sequencing instruction so that information begins with the basic or simple information, or beginning steps, and progresses to the more difficult or complex information and processes.

Simplex — Radio operating system in which one frequency is used to communicate between radio units or stations; communication can only take place in one direction at a time, otherwise transmissions become unreadable and garbled.

Single Ladder — One-section nonadjustable ladder. *Also known as* Wall Ladder or Straight Ladder. *See* Straight Ladder.

Single Loop — Method of attaching software to an anchor.

Single Resource — Individual company or crew.

Single-Acting Hydraulic Cylinder — Hydraulic cylinder capable of transmitting force in only one direction.

Single-Edge Snap Throw — Method of spreading a salvage cover with a snap action; intended for spreading covers in narrow spaces.

Single-Issue Leader — Leader who is very concerned about either production needs or worker needs.

Single-Issue Leadership — Leadership style that is characterized by an overriding concern for either production or people.

Single-Jacket Hose — Type of hose construction consisting of one woven jacket; usually lined with an inner rubber tube.

Single-Lens Reflex (SLR) Camera — Film camera that uses a semi-automatic moving mirror system that allows the photographer to see exactly what will be captured on the film.

Single-Ply Membrane Roof — *See* Membrane Roof.

Single-Stage Centrifugal Pump — Centrifugal pump with only one impeller. *Also known as* Single-Stage Pump. *See* Centrifugal Pump, Multistage Centrifugal Pump, Self-Priming Centrifugal Pump.

Single-Stage Pump — *See* Single-Stage Centrifugal Pump.

Sinkhole — Natural depression in a land surface formed by the collapse of a cavern roof; generally occurs in limestone regions.

Siphon — (1) Section of hard suction hose or piece of pipe used to maintain an equal level of water in two or more portable tanks. (2) A method of utilizing atmospheric pressure to transfer a liquid over a small elevation from an upper level to a lower level.

Siphon Eductor — Water removal device that utilizes venturi action to evacuate water from basements, sumps, or low areas.

SIPS — *See* Side-Impact Protection Systems.

Siren — Audible warning device that makes a high-pitched or an alternating high- and low-pitched wailing sound when used by emergency vehicles.

Sisal Rope — Rope made from sisal fiber; most common substitute for manila rope. Sisal is a hard fiber with about three-fourths the tensile strength of manila; its most common use is in binder's twine, but it is sometimes used in larger ropes.

Site — Term used to indicate the location of a building, construction site, or an incident.

Site Characterization — Size-up and evaluation of hazards, problems, and potential solutions of a site.

Site Plan — Drawing that provides a view of the proposed construction in relation to existing conditions. Includes survey information and information on contours and grades; generally the first sheet on a set of drawings. *See* Construction Plan, Floor Plan, and Plot Plan.

Site Safety Plan — Facility plan that identifies potential hazards and risks to employees and the public at businesses that meet certain hazardous criteria such as hazardous waste storage facilities.

Site Work Zones — *See* Hazard-Control Zones.

Site-Specific Hazard — Hazard that sometimes or always exists at the facility for which the fire brigade is responsible, but does not exist in most other occupancies.

Situation Status Unit (SITSTAT) — Functional unit within the planning section of an incident command system; responsible for analysis of situation as it progresses. Reports to the planning section chief.

Situational Awareness — Perception of the surrounding environment and the ability to anticipate future events.

Size-Up — Ongoing evaluation of influential factors at the scene of an incident.

SKED® — Lightweight, compact device for patient packaging; shaped to accommodate a long backboard; may be used with a rope mechanical advantage system.

Skeleton Key — Key for a warded lock.

Skewed Data — Data that is inaccurate because of faults inherent to the reporting mediums, bias from evaluators, or false information.

Skid Load — System of loading fire hose so that the top layer can be pulled off at the fire.

Skid Unit — Fire fighting system or systems built on a frame that can be mounted in the bed of a pickup truck or larger vehicle.

Skills Test — Evaluation instrument used to assess an individual's ability to perform a specific physical behavior. *Also known as* Performance Test.

Skin — (1) Outer covering of the body, and the largest organ of the body. Consists of the dermis and the epidermis, and contains various sensory and regulatory mechanisms. (2) Outer covering of an aircraft, which includes the covering of wings, fuselage, and control surfaces.

Skin Contact — Occurrence in which a chemical or hazardous material (in any state — solid, liquid, or gas) contacts the skin or exposed surface of the body, such as the mucous membranes of the eyes, nose, or mouth. *See* Routes of Entry.

Skin Penetrating Agent Applicator Tool® (SPAAT) — Penetrating nozzle used on aircraft fires.

Skip Breathing — Emergency procedure in which the firefighter inhales normally, holds the inhalation for as long as it would take to exhale, takes another breath, and then exhales; used only when the firefighter is stationary and must wait for help.

Skip Shoring — Procedure for supporting trench walls with uprights and shores at spaced intervals.

Skull — Bony structure surrounding the brain; consists of the cranial bones, the facial bones, and the teeth.

Skylights — Roof structures or devices intended to increase natural illumination within buildings, either in rooms or over stairways and other vertical shafts that extend to the roof.

Slab — (1) Heavy steel plate used under a steel column. (2) Reinforced concrete floor. (3) Reinforced wall section in tilt-slab construction.

Slab and Beam Frame — Construction technique using concrete slabs supported by concrete beams.

Slab Door — Door that appears to be made of a single piece (slab) of wood; there are two types, hollow core and solid core.

Slack Adjusters — Devices used in an air brake system that connect between the activation pads and the brake pads that compensate for brake pad wear.

Slag — Hot molten metal that is a byproduct of welding or torch cutting operations.

Slander — False and defamatory oral statement about a person that damages his or her reputation. Definitions vary among states/provinces; may require proof or the presence of other factors. *See* Defamation, Libel, and Tort.

Slash — Debris left after logging, pruning, thinning, or brush cutting; includes logs, chunks, bark, branches, stumps, and broken understory trees or brush.

Sleeper — Compartment built into or behind the cab of a large truck, to be used by the driver for rest and relaxation.

Sleeve — Tube or pipe extending through a floor slab to provide openings for the passage of plumbing and heating pipes to be installed later.

Slide Pole — Brass or stainless steel pole that allows firefighters to quickly slide down to the apparatus bay from the floor above.

Sliding Door — Door that opens and closes by sliding across its opening, usually on rollers.

Sliding Fifth Wheel — Fifth-wheel assembly capable of being moved forward or backward on the truck tractor, in order to vary load distribution and adjust the overall length of combination.

Sliding Rope — *See* Rappel.

Sling — Assembly that connects the load to the material handling equipment. There are four common types of slings: chain, wire rope, synthetic round, and synthetic web.

Sling Psychrometer — Meteorological instrument used to determine relative humidity.

Slip Brainstorming — Type of brainstorming in which each person in the group independently and anonymously writes ideas on a slip of paper, and the slips are then collected and organized.

Slip Hook — Hook used on a chain that is designed to be fastened by slipping it over a link in the chain.

Slope — Natural or artificial topographic incline; degree of deviation from horizontal.

Slope Winds — Small-scale convective winds that occur due to local heating and cooling of a natural incline of the ground.

Slopover — (1) Fire edge that crosses a control line at a wildland fire. *Also known as* Breakover. (2) Situation that occurs when burning oil that is stored in a tank is forced over the edge of the tank by water that is heated to the boiling point and has accumulated under the surface of the oil.

SLR — *See* Single-Lens Reflex Camera.

Slump Test — Method of evaluating the moisture content of wet concrete by measuring the amount that a small, cone-shaped sample of the concrete settles or "slumps" after it is removed from a standard-sized test mold.

Slurry — (1) Watery mixture of insoluble matter such as mud, lime, or Plaster of Paris. (2) Thick mixture formed when a fire-retardant chemical is mixed with water and a viscosity agent. (3) Suspension formed by a quantity of powder mixed into a liquid in which the solid is only slightly soluble.

Small Diameter Hose (SDH) — Hose of ¾-inch to 2 inches (20 mm to 50 mm) in diameter; used for fire fighting purposes. *Also known as* Small Line.

Small Line — *See* Small Diameter Hose.

Smallpox — Serious, contagious, and sometimes fatal infectious disease; there is no specific treatment, and the only prevention is vaccination.

Smoke — Visible products of combustion resulting from the incomplete combustion of carbonaceous materials; composed of small particles of carbon, tarry particles, and condensed water vapor suspended in the atmosphere, which vary in color and density depending on the types of material burning and the amount of oxygen.

Smoke Alarm — Device designed to sound an alarm when the products of combustion are present in the room where the device is installed. The alarm is built into the device rather than being a separate system.

Smoke Building — *See* Smokehouse.

Smoke Curtains — Salvage covers placed in stairways, halls, or doors to prevent movement of smoke into clear areas.

Smoke Damper — Device that restricts the flow of smoke through an air-handling system; usually activated by the building's fire alarm signaling system. *See* Duct and Fire Damper.

Smoke Detector — Alarm-initiating device designed to actuate when visible or invisible products of combustion (other than fire gases) are present in the room or space where the unit is installed.

Smoke Diver — Highly trained user of a self-contained breathing apparatus (SCBA).

Smoke Diver School — School that provides the firefighter with 25 to 30 hours of sequential training in self-contained breathing apparatus use.

Smoke Ejectors — Electrically powered fans that have intrinsically safe motors that are placed in the smoke-filled atmosphere to push the smoke out. They can also be used to push fresh air into the structure. They require the use of electrical power cords and generators to operate.

Smoke Explosion — Form of fire gas ignition; the ignition of accumulated flammable products of combustion and air that are within their flammable range.

Smoke Jumpers — Wildland firefighters who are deployed into remote wildland fires and other emergencies by parachuting from aircraft.

Smoke Management System — System that limits the exposure of building occupants to smoke; may include elements such as compartmentation, control of smoke migration from the affected area, and a means of removing smoke to the exterior of the building.

Smoke Obscuration — Act of being hidden by smoke.

Smoke Room — Enclosed area into which mechanically generated smoke is introduced and in which firefighters perform training exercises while wearing SCBA.

Smoke Shaft — Fire resistive shaft or tower, with or without an exhaust fan at the top, for the purpose of removing smoke directly to the outside from any of the floors served that may become involved in a fire. Smoke shafts are often used in conjunction with a smokeproof enclosure.

Smoke Tower — Fully enclosed escape stairway that exits directly onto a public way; these enclosures are either mechanically pressurized or they require the user to exit the building onto an outside balcony before entering the stairway. *Also known as* Smokeproof Enclosure or Smokeproof Stairway.

Smoke Tube — Device containing stannic or titanium tetrachloride used to produce nontoxic smoke for testing a facepiece seal.

Smoke-Control System — Engineered system designed to control smoke by using mechanical fans to produce airflows and pressure differences across smoke barriers, in order to limit and direct smoke movement.

Smokehouse — (1) Specially designed fire training building that is filled with smoke to simulate working under live fire conditions; used for SCBA and search and rescue training. *Also known as* Smoke Building. (2) A building used for the smoking of meat.

Smokeproof Enclosures — Stairways that are designed to limit the penetration of smoke, heat, and toxic gases from a fire on a floor of a building into the stairway, and that serve as part of a means of egress. *See* Means of Egress.

Smoldering — Fire burning without flame and barely spreading.

Smoldering Phase — *See* Hot Smoldering Phase.

Smooth Bore Nozzle — Nozzle with a straight, smooth tip, designed to produce a solid fire stream.

Smothering — Act of excluding oxygen from a fuel.

Snag — Standing dead tree or part of a dead tree from which at least the leaves and smaller branches have fallen.

Snap Coupling — Coupling set with nonthreaded male and female components; when a connection is made, two spring-loaded hooks on the female coupling engage a raised ring around the shank of the male coupling.

Snap Link — *See* Carabiner.

Social Marketing — Application of commercial marketing strategies and techniques to health issues and social problems. While the focus of commercial marketing is to change people's behavior for the benefit of a producer or merchant, the focus of social marketing is to change people's behavior to benefit themselves, the community, or society as a whole.

Social Need — Middle need in Maslow's Hierarchy of Needs; the need to belong to a group or to have some means of identification.

Social Networking — Websites that allow users to be part of a virtual community. Users can communicate through private messages or real-time chat, and share photos, video, and audio.

Society of Automotive Engineers (SAE) — Organization of engineers in the automotive industry; the initials SAE coupled with a number (e.g. SAE 30) are used to indicate the viscosity of motor oil.

Sodium Saccharin — Chemical substance used in qualitative facepiece fit taste tests.

Soffit — Lower horizontal surface such as the undersurface of eaves or cornices.

Soft Sleeve Hose — Large diameter, collapsible piece of hose used to connect a fire pump to a pressurized water supply source; sometimes incorrectly referred to as *soft suction hose*.

Soft Target — Term used to define a facility or other target that is undefended or unprotected against attack from a potential adversary; examples include most public assembly areas such as churches, schools, bus stations, and shopping districts.

Soft Tissue Injury — Damage to human tissue that encloses bones or joints, such as muscles, tendons, or ligaments.

Software — (1) Computer program that performs a specific function or set of functions. (2) In rope rescue, refers to nylon webbing, rope, and harnesses.

SOG — *See* Standard Operating Guideline and Standard Operating Procedure.

Soil Stack — Vertical pipe which runs from the horizontal soil pipe to the house drain to carry waste, including that from water closets.

Solar Heat Energy — Energy transmitted from the sun in the form of electromagnetic radiation.

SOLAS — *See* International Convention for the Safety of Life at Sea.

Sole — Horizontal wooden member that rests on the top of a foundation wall of a building; the vertical framing of the exterior walls and the first-floor wooden floor joists are supported by these members. *Also known as* Sill.

Sole Plate — (1) Member against which the vertical load of a shore is ultimately exerted. (2) Surface contact piece that distributes loads delivered by struts.

Solicit — To approach with a request.

Solid — Substance that has a definite shape and size; the molecules of a solid generally have very little mobility.

Solid Core Door — Door whose entire core is filled with solid material.

Solid Stream — Hose stream that stays together as a solid mass, as opposed to a fog or spray stream; a solid stream is produced by a smooth bore nozzle and should not be confused with a straight stream.

Solubility — Degree to which a solid, liquid, or gas dissolves in a solvent (usually water).

Soluble — Capable of being dissolved in a liquid (usually water). *See* Immiscible, Insoluble, Miscibility, and Water Solubility.

Somatic — Pertaining to all tissues other than reproductive cells.

SOP — *See* Standard Operating Procedure.

Sorbent — (1) Material, compound, or system that holds contaminants by adsorption or absorption. In *adsorption*, the contaminant molecule is retained on the surface of the sorbent granule by physical attraction; in *absorption*, a solid or liquid is taken up or absorbed into the sorbent material. (2) Granular, porous filtering material used in vapor- or gas-removing respirators.

Sorption — Method of removing contaminants; used in vapor- and gas-removing respirators.

Sounding — (1) Striking the surface of a roof or floor to determine its structural integrity or locate underlying support members; the blunt end of a hand tool is used for this purpose. (2) Name of the measurement of the depth of water in which a vessel is floating.

Source — *See* Sender.

Southern Building Code Congress International (SBCCI) — Organization that provided the *Standard Building Code* along with mechanical, plumbing, and fire prevention codes for city and state adoption. SBCCI was found mostly in the southern states. It has joined with the Building Officials and Code Administrators International, Inc. (BOCA) and the International Conference of Building Officials (ICBO) to form the International Code Council (ICC).

SPAAT — *See* Skin Penetrating Agent Applicator Tool®.

Space frames — Aluminum skeletons that are similar to aircraft frames, upon which the aluminum, plastic, or composite skin of the vehicle's body is attached; the internal structure of these space frames provides the structural support for the vehicle, while the skin provides aerodynamics, styling, and protection from the elements.

Spacer — Length of timber that keeps a breast timber from shifting vertically.

Spalling — Expansion of excess moisture within masonry materials due to exposure to the heat of a fire, resulting in tensile forces within the material, and causing it to break apart. The expansion causes sections of the material's surface to violently disintegrate, resulting in explosive pitting or chipping of the material's surface.

Span of Control — Maximum number of subordinates that that one individual can effectively supervise; ranges from three to seven individuals or functions, with five generally established as optimum.

Spandrel — Part of a wall between the head of a window and the sill of the window above it.

Spanish Windlass — Tool, such as a stick or dowel, used for post-tensioning lines by twisting.

Spanner Wrench — Small tool primarily used to tighten or loosen hose couplings; can also be used as a prying tool or a gas key.

Spar — Principal, span-wide aircraft structural member of an airfoil or control surface.

Spark — (1) *See* Buff. (2) Small bit of solid material heated to incandescence.

Speaking Diaphragm — Device on some SCBA facepieces that aids oral communication.

Spec 51 — *See* Pressure Intermodal Tank.

Spec Building — Building built without a tenant or occupant. *Spec* is short for *speculation*.

Special Duty — Type of obligation that an inspector assumes by providing expert advice or assistance to a person; this obligation may make the inspector liable if it creates a situation in which a person moves from a position of safety to a position of danger by relying upon the expertise of the inspector.

Special Fire Hazards — Hazards that arise from or are related to the particular processes or operation in an occupancy.

Special Police — *See* Fire Police.

Special Protective Clothing — (1) Chemical protective clothing specially designed to protect against a specific hazard or corrosive substance. *See* Chemical Protective Clothing (CPC) and Personal Protective Equipment (PPE). (2) High-temperature protective clothing, including approach, proximity, and fire entry suits.

Special Rescue Technician — *See* Technical Rescuer.

Special Service — Fire company's assignment to a special detail, such as removing water from a basement or directing traffic.

Special Service Unit — *See* Emergency Truck.

Specific Gravity — Mass (weight) of a substance compared to the weight of an equal volume of water at a given temperature. A specific gravity less than 1 indicates a substance lighter than water; a specific gravity greater than 1 indicates a substance heavier than water. *See* Physical Properties and Vapor Density.

Specific Heat — Amount of heat required to raise the temperature of a specified quantity of a material, and the amount of heat necessary to raise the temperature of an identical amount of water by the same number of degrees.

Specification Marking — Stencil on the exterior of a tank car indicating the standards to which the tank car was built; may also be found on intermodal containers and cargo tank trucks.

Specification Marking — Stencil on the exterior of tank cars indicating the standards to which the tank car was built; may also be found on intermodal containers and cargo tank trucks.

Specifications — (1) Detailed information provided by a manufacturer on the function, care, and maintenance of equipment or apparatus. (2) Detailed list of requirements prepared by a purchaser and presented to a manufacturer or distributor when purchasing equipment or apparatus.

Spectrochemical Analysis — Test method by which contaminants suspended in oil can be detected; typically expresses contaminant level in parts per million (ppm).

Speed Brakes — *See* Spoilers.

Speedometer — Dashboard gauge that measures the speed at which the vehicle is traveling.

Spherical Pressure Vessel — Round-shaped fixed facility pressure vessel. *See* Pressure Vessel.

Spheroid Tank — Round- or oval-shaped fixed facility low-pressure storage tank. *See* Low-Pressure Storage Tank, Noded Spheroid Tank, and Pressure Storage Tank.

Sphygmomanometer — Device for measuring blood pressure; blood pressure cuff.

Spice — Attention-getter or energizer that gets the audience's attention during a fire and life safety presentation.

Spider Strap — *See* Head Harness.

Spinal Column — Flexible bony structure that supports the central part of the body and encloses the spinal cord.

Spinal Cord — Part of the central nervous system contained within the spinal column, extending from the base of the brain to the coccyx.

Spinal Immobilization — Stabilizing the patient's cervical spine (neck and back) to prevent severing of the spinal cord under the assumption, based on mechanism of injury or actual findings, that the patient has an injury. *Also known as* Spinal Precautions.

Spinal Precautions — *See* Spinal Immobilization.

Spirometer Test — Medical test used to measure pulmonary capacity.

Splash Guard — Deflecting shield sometimes installed on tank trailers to protect meters, valves, and other components.

Splash Pattern — Characteristic pattern left on a wall by an accelerant splashed there; usually in the shape of an inverted V.

Splice — (1) To join two ropes or cables by weaving the strands together. (2) Process of joining two covers into a larger or longer one with a leakproof seal.

Splint — Support used to immobilize a fracture or restrict movement of a body part.

Split Drop — Two retardant drops made from one compartment at a time, from an air tanker with a multi-compartment tank.

Split Lay — Hose lay deployed by two pumpers, one making a forward lay and the other making a reverse lay from the same point.

Split-Sash — Two-piece window split horizontally in the middle.

Splitter Valve — Valve installed to divide the pipeline manifold.

Spoil Pile — Excavated materials consisting of topsoil or subsoils that have been removed and temporarily stored during the digging of a trench.

Spoilage — Decomposition of grain or other perishable items.

Spoilers — Movable aerodynamic devices or plates on aircraft, which extend into the airstream to break up the smoothness of flow and thus increase drag and decrease lift. This process results in reduced airspeed during descent and assists in slowing the aircraft after landing. *Also known as* Speed Brakes.

Spoliation — Term which refers to evidence that is destroyed, damaged, altered, or otherwise not preserved by someone who has responsibility for the evidence.

Spontaneous Combustion — *See* Spontaneous Ignition.

Spontaneous Heating — Heating resulting from chemical or bacterial action in combustible materials that may lead to spontaneous ignition. *See* Spontaneous Ignition.

Spontaneous Ignition — Initiation of combustion of a material by an internal chemical or biological reaction that has produced sufficient heat to ignite the material (NFPA® 921). *Also known as* Spontaneous Combustion.

Spores — Airborne reproductive particles produced by plants that may or may not be airborne.

Spot Fire — Wildland fire started outside the perimeter of a main fire; typically caused by flying sparks or embers landing outside the main fire area. *See* Spotting.

Spotter — (1) Experienced firefighter who guides and directs bulldozer operations for wildland fire fighting operations. (2) Firefighter who walks behind a backing apparatus to provide guidance for the driver/operator. *Also known as* Swamper.

Spotting — (1) Positioning the apparatus in a location that provides the utmost efficiency for operating on the fireground. (2) Positioning a ladder to reach an object or person. (3) Behavior of a wildland fire that produces sparks or embers that are carried by the wind to start new fires beyond the main fire. *See* Spot Fire.

Spray — Application of water through specially designed nozzles in the form of finely divided particles.

Spray Curtain Nozzle — Fog nozzle mounted to the underside of an elevating platform to provide a protective shield against convected heat for firefighters operating in the platform.

Sprinkler — Water flow discharge device in a sprinkler system; consists of a threaded intake nipple, a discharge orifice, a heat-actuated plug, and a deflector that creates an effective fire stream pattern that is suitable for fire control. *See* Automatic Sprinkler System.

Sprinkler Block — *See* Sprinkler Wedge.

Sprinkler Connection — *See* Fire Department Connection.

Sprinkler Kit — Collection of sprinklers, wedges, tongs, and wrenches used to close or replace open sprinklers.

Sprinkler Riser — Vertical pipe used to carry water to the sprinkler system.

Sprinkler System — Fixed piping system that is designed to discharge water in the event of a fire. *See* Automatic Sprinkler System.

Sprinkler System Supervision Systems — *See* Waterflow Alarm.

Sprinkler Tongs — Tool used to stop the flow of water from a sprinkler.

Sprinkler Wedge — Wedge-shaped piece of wood used to stop the flow of water from individual sprinklers. *Also known as* Sprinkler Block.

Sprinkler Wrench — Special wrench designed for tightening or loosening sprinklers.

Spurs — Metal points at the end of a ladder or staypole.

Squad — *See* Rescue Company.

Squeegee — Rubber-edged broomlike device used in salvage to assist in the removal of water from floors; used by pushing the water to a drain, disposal area, or collection place.

Stability — Tendency of a floating vessel to return to an upright position when inclined from the vertical by an external force, such as winds or waves. When a vessel returns to or remains at rest after being acted upon, it is either in stable or neutral equilibrium; if it continues to move unchecked in reaction to the external force, it is in unstable equilibrium. If an unstable vessel does not find a point of stable or neutral stability, it continues to incline until it capsizes. *See* Free Surface Effect, Longitudinal Stability, and Static Stability.

Stabilization — (1) Process of providing additional support to key places between an object of entrapment and the ground or other solid anchor points, in order to prevent unwanted movement. (2) Stage of an incident when the immediate problem or emergency has been controlled, contained, or extinguished.

Stabilizer — (1) Device that transfers the center of gravity of an apparatus and prevents it from tipping as the aerial device, hydraulic lifting boom, gin pole, or A-frame is extended away from the centerline of the chassis. *Also known as* Outrigger or Stabilizing Jack. (2) Chemical added to an unstable substance to prevent a violent reaction. *Also known as* Inhibitor. (3) Airfoil on an airplane, used to provide stability; for example, the aft horizontal surface to which the elevators are hinged (horizontal stabilizer), and the fixed vertical surface to which the rudder is hinged (vertical stabilizer).

Stabilizer Foot — Flat metal plate attached to the bottom of the aerial apparatus stabilizer to provide firm footing on the stabilizing surface. *Also known as* Stabilizer Boot.

Stabilizer Pad — Unattached flat metal plate that is larger in area than the stabilizer foot; placed on the ground beneath the intended resting point of the stabilizer foot, in order to provide better weight distribution. *Also known as* Jack Pad or Jack Plate.

Stable Atmosphere — Condition of the atmosphere in which the temperature decrease with increasing altitude is less than the dry adiabatic lapse rate; in this condition, the atmosphere tends to suppress large-scale vertical motion.

Stack — Ducting through which exhaust gases and often supply gas are routed; a chimney. *See* Economizer and Fiddley.

Stack Action — See Stack Effect.

Stack Effect — (1) Tendency of any vertical shaft within a tall building to act as a chimney or "smokestack", by channeling heat, smoke, and other products of combustion upward due to convection. *Also known as* Stack Action. (2) Phenomenon of a strong air draft moving from ground level to the roof level of a building; affected by building height, configuration, and temperature differences between inside and outside air. *Also known as* Chimney Effect.

Stack Valve — Type of multidirectional valve used in an aerial device hydraulic system.

Staff/Support Functions — Personnel who provide administrative and logistical support to line units (internal customers).

Staged Incident — *See* Mock Incident.

Stages of Intellectual Development — Stages of intellect that build upon one another in sequential order.

Staging — (1) Standardized process or procedure by which available resources responding to a fire or other emergency incident are held in reserve at a location away from the incident while awaiting assignment. (2) In high-rise fire fighting, the incident management system (IMS) term for the area within the building where relief crews are assembled and spare equipment is stockpiled, usually established two floors below the fire floor. Staging may also include a first-aid station and rehab.

Staging Area — Prearranged, temporary strategic location, away from the emergency scene, where units assemble and wait until they are assigned a position on the emergency scene; these resources (personnel, apparatus, tools, and equipment) must then be able to respond within three minutes of being assigned. Staging area managers report to the incident commander or operations section chief, if one has been established.

Staging Area Manager — Company officer of the first-arriving company at the staging who takes command of the area and is responsible for communicating available resources and resource needs to the operations section chief.

Stagnant Organizational Culture — Group of people satisfied with the status quo of an organization and not interested in engaging in progressive change.

Stair Nosing — Plate that wraps around the front edge of the stair.

Stair Pressurization System — System that enables a stairwell to have separate air-handling controls that can be adjusted to increase the air pressure in the stairway so that the smoke, heat, and other products of combustion on the fire floor will not enter the stairwell.

Stakeholder — People, groups, or organizations that have a vested interest in a specific issue.

Stand By — (1) To remain immediately available. (2) To relocate to another fire station and cover that district for additional emergencies in the area. (3) To clear the airwaves for a broadcast.

Standard — (1) A set of principles, protocols, or procedures that explain how to do something or provide a set of minimum standards to be followed. Adhering to a standard is not required by law, although standards may be incorporated in codes, which are legally enforceable. *See* Code and Regulations. (2) Part of an educational objective that describes the minimum level of performance that a learner must meet in accomplishing the performance/behavior.

Standard Apparatus — Apparatus that conforms to the standards set forth by the National Fire Protection Association® standards on fire apparatus design.

Standard Cubic Feet Per Minute (SCFM) — Volume of material based on the standard cubic foot flowing past or through a specified measuring point.

Standard Cubic Foot (scf) — Amount of air in a cubic foot at 14.7 psia and 60 degrees Fahrenheit.

Standard Deviation — Average of the degree to which the scores in a test deviate from the mean.

Standard Error — *See* Sampling Error.

Standard of Care — Level of care that all persons should receive; care that does not meet this standard is considered inadequate. *See* Negligence.

Standard Operating Guideline (SOG) — *See* Standard Operating Procedure (SOP).

Standard Operating Procedure (SOP) — Formal methods or rules to guide the performance of routine functions or emergency operations. Procedures are typically written in a handbook, so that all firefighters can consult and become familiar with them. *Also known as* Operating Instruction (OI), Predetermined Procedures, *or* Standard Operating Guideline (SOG).

Standard Response — Predetermined amount of resources that will be dispatched to the report of an emergency.

Standard Thread — National Standard hose threads.

Standard Time-Temperature Curve — Plot on a graph of temperature versus time used in structural fire tests. *Also known as* Time-Temperature Curve.

Standard Transportation Commodity Code — Numerical code on the waybill used by the rail industry to identify the commodity. *Also known as* STCC Number.

Standardization — Process of making or creating things that meet an established criteria.

Standing Block — Block, in a block and tackle system, that is attached to a solid support and from which the fall line leads.

Standing Part — Middle of the rope, between the working end and the running part.

Standpipe Hose — Single-jacket lined hose that is preconnected to a standpipe; used primarily by building occupants to mount a quick attack on an incipient fire.

Standpipe System — Wet or dry system of pipes in a large single-story or multistory building, with fire hose outlets installed in different areas or on different levels of a building to be used by firefighters and/or building occupants. This system is used to provide for the quick deployment of hoselines during fire fighting operations. *See* Dry Standpipe System, Riser, and Wet Standpipe System.

Starboard — Right-hand side of a vessel as a person faces forward. *Also known as* Starboard Side.

Static — (1) Stationary; without movement. (2) Refers to the amount of stretch built into a rope. Static ropes have relatively little stretch; they are ideal for rescue work where stretch becomes a hazard and a nuisance, but should not be used where long falls could be anticipated before being stopped by the rope.

Static Electricity — Accumulation of electrical charges on opposing surfaces, created by the separation of unlike materials or by the movement of surfaces.

Static Load — Load that is steady, motionless, constant, or applied gradually.

Static Pressure — (1) Potential energy that is available to force water through pipes and fittings, fire hose, and adapters. (2) Pressure at a given point in a water system when no testing or fire protection water is flowing.

Static Rope — Rope designed not to stretch under load.

Static Source — Body of water that is not under pressure (other than atmospheric) or in a supply piping system, and must be drafted from in order to be used. Static sources include ponds, lakes, rivers, wells, etc.

Static Stability — Ability of a vessel to initially resist heeling from the upright position. Initial stability characteristics hold true only for relatively small angles of inclination; at larger angles (over 10 degrees), the ability of a vessel to resist inclining moments is determined by its overall stability characteristics. *Also known as* Initial Stability. *See* Longitudinal Stability and Stability.

Static Stress — Stress imposed on the aerial device when it is at rest.

Static Water Supply — Supply of water at rest that does not provide a pressure head for fire suppression but may be employed as a suction source for fire pumps; for example, water in a reservoir, pond, or cistern.

Station Bill — List of all crew members showing where they should be for the various operations involved in operating a vessel; shows the duty stations and duties of the crew, by rank.

Statistics — Numerical data.

Statute — Federal or state/provincial legislative act that becomes law; prescribes conduct, defines crimes, and promotes public good and welfare. *Also known as* Statutory Law. *See* Copyright Law, Law, and Legislative Law.

Staypole — Pole attached to long extension ladders to assist in raising and steadying the ladder; some poles are permanently attached, and some are removable. *Also known as* Tormentor Pole.

STCC Number — *See* Standard Transportation Commodity Code.

Steady-State Burning Phase — (1) Phase of the fire in which sufficient oxygen and fuel are available for fire growth and open burning, to a point where total involvement is possible. (2) Phase of the fire in which the rate of heat release is constant with respect to time.

Steam Conversion — Physical changing of water from a liquid to a gaseous form; water expands in size 1,700 times when it converts to steam.

Steamer Connection — Large-diameter outlet, usually 4½ inches (115 mm), at a hydrant or at the base of an elevated water storage container.

Steeple Raise — *See* Auditorium Raise.

Steering Gear — All the apparatus by which a vessel is steered; includes the wheel, rudder, and any ropes or chains connected to them.

Steiner Tunnel — Test apparatus used in the determination of flame spread ratings; consists of a horizontal test furnace 25 feet (7.6 m) long, 17½ inches (445 mm) wide, and 12 inches (305 mm) high that is used to observe flame travel. A 5,000 Btu (5 270 kJ) flame is produced in the tunnel, and the extent of flame travel across the surface of the test material is observed through ports in the side of the furnace.

Steiner Tunnel Test — Test to determine the flame-spread ratings of various materials. The test apparatus consists of a horizontal furnace 25 feet (7.6 m) long, 17½ inches (445 mm) wide, and 12 inches (305 mm) high; a 5,000 Btu (5 270 kj) flame is produced in the tunnel, and the extent of flame travel across the surface of the test material is observed through ports in the side of the furnace. *See* Flame Spread Rating.

STEL — *See* Short-Term Exposure Limit.

Stem — (1) Part of a lock cylinder that activates the bolt or latch as the key is turned. *Also known as* Tailpiece. (2) Rod-type portion of a hydrant, between the operating nut and the valve. (3) Introductory statement in a multiple-choice test item.

Stem Light — Elevating floodlighting tower.

Stem Wall — In platform frame construction, an exterior wall between the foundation and the first floor of a building.

Step Block — Piece of cribbing with a tapered end, specially designed for stabilization of automobiles. *Also known as* Step Chock.

Step Chock — *See* Step Block.

Stern — Back end or rear of a vessel.

Stevedore — *See* Longshoreman.

Stile — Vertical member of a window sash.

Still Alarm — Response to an emergency in which no audible alarm is sounded at dispatch; usually a one- or two-company response. *Also known as* Silent Alarm.

Stimsonite® Markers — Raised reflective marker typically used as a roadway safety device; blue-colored markers are used by the fire service to mark hydrant locations.

Stinger — Bright, one-directional, moving warning light.

Stoichiometric Ratio — Ideal fuel-to-air ratio at which complete combustion of fuels occurs without byproducts of combustion; does not naturally occur. *Also known as* Optimum Ratio.

Stokes Basket — Wire or plastic basket-type litter suitable for transporting patients from locations where a standard litter would not be easily secured, such as a pile of rubble, a structural collapse, or the upper floor of a building; may be used with a harness for lifting.

Stoma — Small opening, especially an artificially created opening.

Stop, Drop, and Roll — Fire and life safety behavior to be performed when one's clothing or hair catches fire.

Stops — Wood or metal pieces that prevent the fly section of a ladder from being extended too far.

Stopway Area — Area beyond the runway end capable of supporting aircraft that overshoot the runway on aborted takeoff or landing without causing structural damage to the aircraft. *Also known as* Overrun Area.

Storage Tank — Storage vessel that is larger than 60 gallons (227 L) and is located in a fixed location.

Story — Space in a building between two adjacent floor levels, or between a floor and the roof.

Storz Coupling — Nonthreaded (sexless) coupling commonly found on large-diameter hose. Nonthreaded fire hose couplings have been used in the North American fire and emergency services since the early 1900s. With this type of coupling, the mating of two couplings is achieved with locks or cams without the use of screw threads.

Straight Jack — Stabilizing device that extends straight down from the chassis.

Straight Ladder — One-section ladder. *Also known as* Single Ladder.

Straight Lay — Hose laid from the hydrant or water source to the fire.

Straight Stream — Most compact discharge pattern that a fog nozzle can produce; similar to but not as compact as a solid stream.

Strainers — Wire or other metal guards used to prevent debris from clogging the intake hose of fire pumps.

Straining Piece — Length of timber that keeps pressure on the breast timbers of a flying shore.

Strap — Metal piece used to hold together joints in heavy timber construction.

Straps — Strips of webbing with buckles for securing ladders, improvising step ladders, and other tying purposes.

Strategic Goals — Broad statements of desired achievement to control an incident; achieved by the completion of tactical objectives. *See* Strategy, Tactical Objectives, and Tactics.

Strategic Planning — Process for identifying long-term goals and objectives for a program or department, usually for a period of five years.

Strategy — Overall plan for incident attack and control established by the incident commander (IC). *See* Defensive Strategy, Nonintervention Strategy, Offensive Strategy, and Tactics.

Stratification — Formation of smoke into layers as a result of differences in density with respect to height, with low density layers on the top and high density layers on the bottom.

Stratum — Sheet-like layer of rock or earth; numerous other layers, each with different characteristics, are typically found above and below.

Street Clothes — Clothing that is anything other than chemical protective clothing or structural firefighters' protective clothing, including work uniforms and ordinary civilian clothing.

Stress — (1) Factors that work against the strength of any piece of apparatus or equipment. (2) State of tension put on a shipping container by internal or external chemical, mechanical, or thermal change. (3) Any condition causing bodily or mental tension. *See* Critical Incident Stress (CIS), Physiological Stress, Post-Traumatic Stress Disorder (PTSD), and Psychological Stress.

Stress Test — Test in which a person's vital functions are monitored while the person labors.

Stressed Skin — Outer surface of a structure when it provides lateral support.

Stressor — Any agent, condition, or experience that causes stress.

Stretch Hose — To lay out hose as a line or advance it into a building.

Stretcher — Portable device that allows two or more persons to move the sick or injured, by carrying or rolling, while keeping the patient immobile.

Strike — Metal plate mounted in the door frame that receives the latch or deadbolt.

Strike Team — Specified combinations of the same kind and type of resources with common communications and a leader; usually composed of either engines, hand crews, or bulldozers, but may be composed of any resource of the same kind and type. Exception — *see* Enhanced Strike Team.

Striking Tools — Tools characterized by large, weighted heads on handles; includes axes, battering rams, ram bars, punches, mallets, hammers, sledgehammers or mauls, chisels, automatic center punches, and picks.

Stringer — (1) Horizontal structural member supporting joists and resting on vertical supports. (2) General construction term referring to the member on each side of a stair that supports the treads and risers. *See* Girders and Longeron.

Strip Ventilation — *See* Trench Ventilation or Trenching (2).

Stripped Territory — Area that has been completely depleted of fire protection apparatus and staffing.

Strobe — High-intensity flashing light.

Strong Oxidizer — Substance that readily gives off large quantities of oxygen, thereby stimulating combustion; produces a strong reaction by readily accepting electrons from a reducing agent (fuel). *See* Oxidizer.

Strong, Tight Container — Packaging used to ship materials of low radioactivity. *See* Excepted Packaging, Industrial Packaging, Packaging (1), Type A Packaging, and Type B Packaging.

Structural Abuse — Using or changing a building beyond its originally designed capabilities.

Structural Collapse — Structural failure of a building or any portion of it resulting from a fire, snow, wind, water, or damage from other forces.

Structural Engineer — Licensed professional engineer trained in structural stability.

Structural Fire Fighting — Activities required for rescue, fire suppression, and property conservation in structures, vehicles, vessels, and similar types of properties.

Structural Firefighters' Protective Clothing — General term for the equipment worn by fire and emergency services responders; includes helmets, coats, pants, boots, eye protection, gloves, protective hoods, self-contained breathing apparatus (SCBA), and personal alert safety system (PASS) devices.

Structural Instability — Degree to which a specific structure has lost its integrity.

Structural Triage — Process of inspecting and classifying structures according to their defensibility/indefensibility, based on their situation, their construction, and the immediately adjacent fuels.

Structure — Constructed object; usually a building standing free and aboveground.

Structure Fire — Fire that involves a building, enclosed structure, vehicle, vessel, aircraft, or like property.

Structure Hazard Assessment — Assessment performed by a structural engineer and hazardous materials specialist to determine the current condition of the structure.

Structure Protection Group/Sector Supervisor — Individual responsible for supervising assigned strike teams, firefighters, or single resources in the defense of structures from wildland fire.

Structured Question — Question that offers the respondent a closed set of responses from which to choose.

Strut — (1) Aircraft structural components designed to absorb or distribute abrupt compression or tension, such as the landing gear forces. (2) In a shore, any member that holds either a vertical or horizontal compression load.

Stud — Vertical structural member within a wall in frame buildings; most are made of wood, but some are made of light-gauge metal.

Student — Any person involved in a formal learning process; the most important member of any class, in that all the activities and efforts are directed toward enabling him or her to learn.

Study Session — Formal, open meeting of a legislative body to study the merits of proposed legislation, and to ask questions of the fire and life safety code official and the public regarding the provisions of the proposed code. *See* Code.

Study Sheet — Instructional document designed to generate student interest in a topic and explain to students the specific areas to study.

Subbasement — Basement below the level of a first basement.

Subcutaneous Layer — Bottom layer of skin, consisting of fatty tissues that insulate the body and store excess calories.

Subhead — Headline that is subordinate to or of lesser importance than the main headline; generally printed in smaller type than the main headline.

Subjective Test — Type of test in which a learner is free to organize, analyze, revise, redesign, or evaluate a problem based on his or her perceptions and understanding; measures higher cognitive levels than other types. Common types of subjective tests are essays and term papers. Different but equally qualified assessors will judge the quality and characteristics of a learner's work differently (subjectivity) and may therefore award different scores. The results may be influenced by the subject being tested, the test itself, the tester, or other outside factors. *See* Test.

Sublimation — Vaporization of a material from the solid to vapor state without passing through the liquid state.

Submersible Pump — Pump capable of operating when placed underwater.

Subsidence — Sinking or settling of land due to various natural and human-caused factors, such as the removal of underground water or oil.

Subsidiary Label — Label indicating a secondary hazard associated with a material. *See* Primary Label.

Subsonic — Slower than the speed of sound.

Substance Abuse — Uncontrolled or excessive use of a drug by an individual.

Substrate — Layer of material between a roof deck and the roof covering that may or may not be bonded to the roof covering; the most common substrate is roofing felt or tar paper.

Subsurface Fire — Fire that consumes the organic material beneath the ground, such as a peat fire or burning roots.

Subsurface Fuel — All combustible materials below the surface litter that normally support smoldering combustion without flame; examples include tree or shrub roots, peat, and sawdust.

Subsurface Injection — Application method where foam is pumped into the bottom of a burning fuel storage tank and allowed to float to the top to form a foam blanket on the surface of the fuel. *See* Direct Injection and Semisubsurface Injection.

Sucking Chest Wound — Wound in which the chest wall is penetrated, causing air to accumulate in the pleural cavity.

Suction — (1) Misnomer used to describe the drafting process. (2) Inlet side of the pump that is better referred to as the *Intake*. (3) *See* Hard Suction Hose.

Suction Hose — Intake hose that connects pumping apparatus or portable pump to a water source.

Suffocate — To die from being unable to breathe; to be deprived of air or to stop respiration, as by strangulation or asphyxiation.

Suitcase Bomb — Small, suitcase-or backpack-sized nuclear weapon. *See* Improvised Nuclear Device.

Sulfur Dioxide (SO_2) — Colorless gas produced when sulfur-containing materials burn; its pervasive, highly irritating rotten-egg odor makes it detectable below its IDLH level of 100 ppm.

Summary — Lesson plan component in which an instructor restates or reemphasizes key points with the learners; this is accomplished by methods such as asking questions, guiding review, and having learners recall relationships, make comparisons, or draw conclusions. *See* Lesson Plan.

Summary Offense — Lowest form of offense in most legal systems; minor infractions of laws or local ordinances. Offender is typically served a citation on the spot where the violation occurred.

Summative Test — Evaluation that measures students' learning at the conclusion of a training session or course; the test results can also be used to measure the effects and effectiveness of a course or program. *Also known as* Posttest. *See* Evaluation (2).

Sump — (1) Low point of a tank at which the emergency valve or outlet valve is attached. (2) Area in the air-purification system that receives drainage.

Sump Basin — Pit or reservoir, often connected to a drain, that serves as a receptacle for water; can be improvised from salvage covers, ladders, and pike poles.

Sunset Provision — Clause in a law or ordinance that stipulates the periodic review of government agencies and programs in order to continue their existence. *See* Ordinance.

Sunshine Laws — Local, state, or provincial laws that require public notification and open attendance of government meetings.

Sunstroke — *See* Heat Stroke.

Super Bus — School bus with an extra-large carrying capacity; may carry up to 84 people seated and over 100 people if standing is permissible.

Superfund Amendments and Reauthorization Act (SARA) — U.S. law that reauthorized the Comprehensive Environmental Response, Compensation and Liability Act (CERCLA) to continue cleanup activities around the country; included several site-specific amendments, definition clarifications, and technical requirements. Enacted in 1986.

Superior — Near the head; above.

Superplasticizer — Admixture used with concrete or mortar mix to make it workable, pliable, and soft while using relatively little water.

Supersonic — Faster than the speed of sound.

Superstructure — Enclosed structure built above the main deck that extends from one side of a vessel to the other. *See* House.

Supervise — To oversee the work of others or another.

Supervised Circuit — Alarm circuit on which a minute electrical current is constantly flowing; when this current is shorted or interrupted, an alarm or trouble signal is initiated.

Supervisor — (1) A person who is responsible for directing the performance of other people or employees. (2) IMS term for the individual responsible for command of a division, group, or sector.

Supervisory Circuit — Electronic circuit within a fire protection system that monitors the system's readiness and transmits a signal when there is a problem with the system.

Supervisory Signal — Signal given by a fixed fire protection system when there is a problem with the system.

Supine — Lying horizontal in a face upward position.

Supplemental Pumping — Pumping water from a stronger point in the water system to the units at the fire, or pumping it back into the water system where it is weak. Used when a large fire overwhelms the water supply system.

Supplied Air Respirator (SAR) — Atmosphere-supplying respirator for which the source of breathing air is not designed to be carried by the user; not certified for fire fighting operations. *Also known as* Airline Respirator System. *See* Self-Contained Breathing Apparatus (SCBA).

Supply Hose — Hose that is designed for the purpose of moving water between a water source and a pump that is supplying attack hoselines or fire suppression systems. *Also known as* Feeder Line or Relay-Supply Hose.

Supply Unit — Functional unit within the support branch of the logistics section of an incident command system; responsible for ordering equipment/supplies required for incident operations.

Supply/Exhaust Ventilation — Combined supply and exhaust system of mechanical ventilation that is generally used in the ventilation of passenger quarters. *See* Ventilation.

Support (Cold) Zone — Area that surrounds the limited access (warm) zone and is restricted to emergency response personnel who are not working in either the restricted (hot) zone or the limited access (warm) zone. This zone may include the portable equipment and personnel staging areas and the command post; the outer boundary of this area should be cordoned off to the public.

Support Branch — Branch within the logistics section of an incident command system; responsible for providing the personnel, equipment, and supplies to support incident operations. Components include the supply unit, facilities unit, and ground support units.

Supported Tip — Operation of an aerial device with the tip of the device or the platform resting on another object, such as a window ledge or roofline.

Supports — Devices, generally adjustable in height, that are used to support the front end of a semitrailer in an approximately level position when disconnected from the towing vehicle. *Formerly known as* Dollies, Landing Fears, Legs, and Props.

Suppressant — *See* Fire Suppressant.

Suppressing — Preventing the release of flammable vapors, to reduce the possibility of ignition or reignition.

Suppression — *See* Fire Suppression.

Suppression System — System designed to act directly upon the hazard to mitigate or eliminate it, not simply to detect its presence and/or initiate an alarm.

Suppression-Generated Pattern — Fire pattern left as a result of the way the fire was extinguished during fire-suppression efforts.

Surcharge Load — Any load imposed on the surface of the soil that will affect the stability of a trench's walls/lip.

Surface Application — Application method where finished foam is applied directly onto the surface of the burning fuel or unignited fuel spill.

Surface Bolt — Sliding bolt installed on the surface of a door.

Surface Contamination — Contamination that is limited to the surface of a material. *See* Contaminant, Contamination, and Decontamination.

Surface Fire — Wildland fire that burns loose and includes dead branches, fallen leaves, needles, duff, stubble, grass, and low vegetation.

Surface Fuel — Fuel that contacts the surface of the ground; consists of duff, leaf and needle litter, dead branch material, downed logs, bark, tree cones, and low-stature living plants. These are the materials normally scraped away to construct a fireline. *Also known as* Ground Fuel.

Surface Hazards — Class of hazards that includes debris on the ground, trip hazards, and unstable footing.

Surface Systems — System of construction in which the building consists primarily of an enclosing surface, and in which the stresses resulting from the applied loads occur within the surface-bearing wall structures.

Surface Tension — (1) Force minimizing a liquid surface's area. (2) The effect of a surfactant on the water/concentrate solution; allows the water to spread more rapidly over the surface of Class A fuels and to penetrate organic fuels.

Surface Victims — Victims not trapped by the structure, usually found on top of structural debris; may have been injured by falling debris or by falling down.

Surface-Burning Characteristic — Speed at which flame will spread over the surface of a material.

Surface-To-Mass Ratio — Ratio of the surface area of the fuel to the mass of the fuel.

Surfactant — Chemical that lowers the surface tension of a liquid; allows water to spread more rapidly over the surface of Class A fuels and penetrate organic fuels. *See* Surface Tension.

Surveillance — Close watch kept over someone or something.

Surveillance Systems — Ongoing, systematic collection and analysis of data; may lead to actions taken to prevent and control an infectious disease. *See* Surveillance.

Survey — Evaluation instrument used to identify the behavior and/or attitude of an individual or audience, both before and after a presentation.

Survey Meter — Nuclear-radiation detection instrument.

Survey Questionnaire — Survey that uses a series of questions to obtain data from respondents.

Suspended Ceiling — Very common ceiling system composed of a metal framework suspended, by wires, from the underside of the roof or the floor above; the framework supports panels that constitute the finish of the ceiling. Typically found in office buildings and in the common areas of apartment buildings and hotels.

Suspension Harness — Web suspension network that supports the helmet on the firefighter's head and prevents the shell from striking the head when hit.

Suspicious Fires — Fires that may be incendiary or caused by arson.

Sustained Attack — Continuing fire-suppression action until the fire is under control.

Swamper — *See* Spotter (2).

Swash Plates — Metal plates in the lower part of tanks that prevent the surging of liquids with the motion of a vessel.

Sweep Pattern — Effective lateral range of an elevated master stream nozzle.

Sweetener — Component (generally charcoal) in an air-purification system that removes odors and tastes from the compressed breathing air.

Swinging Door — Door that opens and closes by swinging from one side of its opening, usually on hinges. *Also known as* Hinged Door.

Swiss Seat — Harness that keeps a person's center of gravity near normal while rappelling.

Switch List — List of railroad cars on a track and instructions as to where those cars go within the yard.

Switchable Regulator — Positive-pressure breathing apparatus regulator that has a switch to accommodate donning.

Swivel — Free-turning ring on all Storz fire hose couplings, and on the female coupling of threaded couplings.

Symptom — Sensation or awareness of a disturbance of a bodily function, as reported by the patient.

Syndromic Surveillance — Surveillance using health-related data that precede diagnosis and signal a sufficient probability of a case or an outbreak to warrant further public health response. *See* Surveillance.

Synergistic Effect — (1) Phenomenon in which the combined properties of substances have an effect greater than their simple arithmetical sum of effects. (2) Working together.

Synthesis — Process of combining elements to make a compound.

Synthetic Fiber Rope — Rope made from continuous, synthetic fibers running the entire length of the rope; it is strong, easy to maintain, and resists mildew and rotting.

Synthetic Foam Concentrate — Foam concentrate that is composed of a synthetically produced material, such as a fluorochemical or hydrocarbon surfactant, that forms a foam blanket across a liquid; performs similarly to a protein-based foam concentrate. Examples include aqueous film-forming foam and alcohol-resistant aqueous film-forming foam. *See* Foam Concentrate.

Synthetic Nylon — Artificial material that has replaced natural fiber in rope construction, due to its superior strength and durability.

Synthetic Stucco — *See* Exterior Insulation and Finish Systems.

System — Total combination of hardware, software, and anchors to create a main line or a safety line.

Systemic Effect — Damage spread through an entire system; opposite of a local effect, which is limited to a single location.

Systemic Hypothermia — *See* Hypothermia.

Systole — Rhythmic contraction of the heart by which blood is pumped throughout the body.

T

25 Percent Drain Time — *See* Quarter-Life.

Tachometer — Instrument that indicates the rotational speed of a shaft in revolutions per minute (rpm); usually used to indicate engine speed.

Tactical Box — Reduced assignment to a fire alarm.

Tactical Objectives — Specific operations that must be accomplished to achieve strategic goals; objectives must be both specific and measurable. *See* Strategic Goals and Tactics.

Tactical Ventilation — Planned, systematic, and co-ordinated removal of heated air, smoke, gases or other airborne contaminants from a structure, replacing them with cooler and/or fresher air to meet the incident priorities of life safety, incident stabilization, and property conservation.

Tactical Worksheet — Document that the IC may use on the fireground to track units and record field notes during an incident; could evolve into a written IAP if an incident escalates in size or complexity.

Tactics — Methods of employing equipment and personnel on an incident to accomplish specific tactical objectives, in order to achieve established strategic goals. *See* Strategic Goals, Strategy, and Tactical Objectives.

Tag Line — (1) Non-load-bearing rope attached to a hoisted object to help steer it in a desired direction, prevent it from spinning or snagging on obstructions, or act as a safety line. (2) As used in anchors, a length of rope or webbing used to extend anchor points closer to the actual rescue work site.

Tailboard — Back step of fire apparatus.

Tailpiece — Part of a lock cylinder that activates the bolt or latch as the key is turned. *Also known as* Stem.

Talk Group — Group of mobile radio units that are addressed as a single entity by a radio system; functionally equivalent to a conventional repeater channel.

Tally — Rectangular plastic identification tag used for entry control in the United Kingdom, Australia, and New Zealand.

Tandem — (1) Two-axle suspension or support. (2) Two or more units of any kind working one in front of the other to accomplish a specific fire-suppression job; this term can be applied to combinations of hand crews, engines, bulldozers, or aircraft.

Tandem Attack — Attack on a wildland fire using multiple resources, including apparatus, hand crews, or aircraft. *See* Attack Methods (2).

Tandem Prusik — Two Prusiks, one made from 65 inches (1.65 meters) of line and one made from 53 inches (1.35 meters) of line, to form a long and a short matched pair of Prusiks. Used together, attached to a line with a Prusik hitch, they form a braking device on a main line or safety line.

Tandem Pumping — Short relay operation in which the pumper taking water from the supply source pumps into the intake of the second pumper; the second pumper then boosts the pressure of the water even higher. This method is used when pressures higher than the capability of a single pump are required.

Tank Motor Vehicle — *See* Cargo Tank Truck.

Tank Top — Lowest deck; top plate of the bottom tanks. *See* Deck.

Tank Truck — Single self-propelled motor vehicle equipped with a cargo tank mounted thereon, and used for the transportation of flammable and combustible liquids or asphalt; a tank truck with two or three axles on which the tank is permanently affixed. *See* Cargo Tank Truck.

Tank Vehicle — Any tank truck, full-trailer tank, or tractor and semitrailer tank combination.

Tank Vessel — *See* Tanker.

Tanker — (1) Mobile water supply fire apparatus that carries at least 1,500 gallons (6 000 L) of water and is used to supply water to fire scenes that lack fire hydrants. *Also known as* Tender in ICS terms. (2) In the ICS, tanker refers to a water-transporting fixed-wing aircraft. (3) Vessel (ship) that exclusively carries liquid products in bulk. *Also known as* Tank Vessel. *See* Chemical Carrier, Liquefied Flammable Gas Carrier, Oil Tanker, and Petroleum Carrier.

Tanker Shuttle Operation — Method of supplying water to a fire scene; tankers carry water between a water source and the pumping engines, generally in a rotating order. *See* Water Shuttle Operation.

Tanker/Pumper — Mobile water supply apparatus equipped with a fire pump. In some jurisdictions, this term is used to differentiate a fire pump equipped mobile water supply apparatus whose main purpose is to shuttle water.

Tankerman — Person qualified and certified to perform all duties included in the handling of bulk liquid cargoes (petroleum products). *See* Oil Tanker.

Tape — (1) Tape recording of calls received and dispatched during telecommunications center operations. (2) The paper printout of code signals on some types of alarm systems.

Tapped Out — *See* Under Control.

Tar and Gravel Roof — *See* Built-Up Roof.

Tare — Weight of an empty vehicle or container; subtracted from gross weight to ascertain net weight.

Target Audience — Group of people who will receive the fire and life safety education presentation. This may not always be the group of people identified as a high-risk group; rather, it may be those who influence or control the high-risk group.

Target Hazard — Any facility in which a fire, accident, or natural disaster could cause substantial casualties or significant economic harm, through either property or infrastructure damage. *See* Hazard and Hazard Assessment.

Target Hazard — Facility in which there is a great potential likelihood of life or property loss in the event of a fire, terrorist attack, or natural disaster. *See* Hazard and Hazard Assessment.

Tarp — *See* Salvage Cover.

Task — Duty or job in an occupation that is performed regularly and requires psychomotor skills and technical information to meet occupational requirements.

Task Analysis — Systematic analysis of duties for a specific job or jobs, which identifies and describes all component tasks of that job; enables program developers to design appropriate training for personnel and trainees who must learn certain tasks to perform a job.

Task Force — (1) Group of individuals convened to analyze, investigate, or solve a particular problem. (2) Group of resources, with common communications and a leader, temporarily assembled for a specific mission. An example would be a group of firefighters and equipment assigned to a special task, such as backfiring; consists of three engines, a dozer, a hand crew, and a task force leader. (3) Any combination of single resources, within a reasonable span of control, assembled for a particular tactical need with common communications and a leader.

Tax-Exempt — *See* Nonprofit.

Taxiway — Specially designated and prepared surface on an airport for aircraft to taxi to and from runways, hangars, and parking areas.

Taxonomy — Classification system in which each separate class of items is given a name, and items within a class are more like one another than like items in other classes. Examples are the Dewey Decimal System and Bloom's Taxonomy of Objectives for the Cognitive Domain.

T-Bone Collision — *See* Side-Impact Collision.

TC — *See* Transport Canada.

Teaching — Method of giving instruction through various forms of communicating knowledge and demonstrating skill; successful teaching causes an observable change in behavior through activities that provide opportunities for learners to demonstrate knowledge and skill, and to receive feedback on progress toward expected behavior change.

Teaching Aids — *See* Instructional Materials.

Team Emergency Conditions Breathing — Procedures or techniques performed by a team of two individuals during emergencies when one person's SCBA malfunctions or does not have adequate air supply. *Also known as* Buddy Breathing.

Team Teaching — Instructional method in which a group of two or more instructors work together, combining their individual content, techniques, and materials in order to present information, demonstrate skills, and supervise practice of a class or several classes. The instructor with the expertise in a particular topic teaches that particular topic; remaining instructors share the responsibilities of assisting with instructional details and supervising practice of skills. A lead instructor organizes and coordinates the activities of all instructors.

Tear Gas — *See* Riot Control Agent.

Technical Assistance — Personnel, agencies, or printed materials that provide technical information on handling hazardous materials or other special problems.

Technical Decontamination — Using chemical or physical methods to thoroughly remove contaminants from responders (primarily entry team personnel) and their equipment; usually conducted within a formal decontamination line or corridor following gross decontamination. *Also known as* Formal Decontamination. *See* Decontamination, Decontamination Corridor, and Gross Decontamination.

Technical Lesson — *See* Information Presentation.

Technical Rescue — Application of special knowledge, skills and equipment to safely resolve unique and/or complex rescue situations. This term has often been used interchangeably with rope rescue; however, it can refer to other disciplines, such as trench or confined space rescue, which require advanced knowledge to perform.

Technical Rescue Specialist — *See* Technical Rescuer.

Technical Rescuer — Individual who has been trained to perform or direct a variety of unique and/or complex rescue situations, such as rope rescues (low- and high-angle), confined space, trench and excavation, structural collapse, mine and tunnel, and other rescue types. *Also known as* Special Rescue Technician or Technical Rescue Specialist.

Technical Safety Officer (TSO) — Individual assigned to function as the safety officer at technical rescue or hazardous materials incidents; assigned at the request of an incident safety officer to ensure the proper level of experience based upon incident type and conditions.

Technical Search — Use of specialized equipment to locate victims; examples of this equipment include seismic, acoustic, fiber-optic, and infrared devices.

Technical Skills — Skills involving manipulative aptitude.

Technical Specialists — Personnel with special skills that are activated only when needed; may be needed in the areas of fire behavior, water resources, environmental concerns, resource use, and training. Technical specialists report initially to the planning section of an incident management system but may be assigned anywhere within the organizational structure as needed.

Telecommunications Center — Facility (a building or portion of a building) that is specifically configured for the primary purpose of providing emergency communication services or public safety answering point (PSAP) services to one or more public safety agencies under the authority or authorities having jurisdiction. Serves as the point through which nearly all information flows, is processed, and is acted upon. *Also known as* Alarm Center, Comm Center, Communications Center, or Dispatch Center.

Telecommunications Device for the Deaf (TDD) — Special type of phone used by hearing-impaired individuals that sends text messages. Letters are transmitted in a specific code, to be decoded by another similar device at the telecommunications center. The *Americans with Disabilities Act of 1990* requires that all public safety telecommunications centers have one available for use.

Telecommunicator — Person who works in the telecommunications center and processes information from the public and from emergency responders. *Formerly known as* Dispatcher. *Also known as* Emergency Communications Technician.

Teleconferencing — Telephone service that allows multiple individuals at remote locations to have an audio-only meeting.

Telephone Alarm Box — Public fire alarm station with a telephone that provides a direct line to the telecommunicator. *Also known as* Call Box.

Telephone Tree — Passing information by means of people calling other people, who in turn call others; this process is repeated until all persons on the tree have been contacted.

Telephoto Lens — Fixed focal length lens with a long focal length and a shallow depth of field.

Telescoping Aerial Platform Apparatus — Aerial apparatus equipped with an elevating platform, as well as piping systems and nozzles for elevated master stream operations. These apparatus are not meant to be climbed and are equipped with a small ladder that is to be used only for escape from the platform in emergency situations.

Telescoping Boom — Aerial device raised and extended via sections that slide within each other.

Teletype (TTY) — Similar to a TDD device in which messages are sent over phone lines; however, teletype machines usually use dedicated phone lines and have other features, such as the ability to send messages to many users at once, query certain databases, or provide a secure network.

Temperature — Measure of a material's ability to transfer heat energy to other objects; the greater the energy, the higher the temperature. Measure of the average kinetic energy of the particles in a sample of matter, expressed in terms of units or degrees designated on a standard scale. *See* Celsius Scale and Fahrenheit Scale.

Temperature Bar — Reinforcing bar within concrete used to counteract stresses caused by temperature changes.

Temperature Gradient — Measure of the change in temperature as a function of distance from a particular location, typically expressed in units of temperature per a measure of distance.

Temperature Inversion — Meteorological condition in which the temperature of the air some distance above the earth's surface is higher than the temperature of the air at the surface; normally, air temperatures decrease as altitude increases. An air inversion traps air, gases, and vapors near the surface, and impedes their dispersion.

Temperature-Compensated Conductivity Meter — Device designed to measure the conductivity of a solution and adjust for conductivity variances at different temperatures.

Tempered Glass — Treated glass that is stronger than plate glass or a single sheet of laminated glass; safer than regular glass because it crumbles into chunks when broken, instead of splintering into jagged shards. Most commonly used in a vehicle's side and rear windows.

Tempered Plate Glass — Type of glass specially treated to become harder and more break-resistant than plate glass or a single sheet of laminated glass.

Temporal Artery — Any one of three arteries located on each side of the head, above and in front of the upper portion of the ear; supplies blood to the scalp.

Temporary Deformation — Alteration of form or shape that disappears entirely after a load has been removed.

Temporary Traffic Control Devices — Cones, flags, lighting, and other devices set up at a vehicle incident to temporarily divert traffic and create a safe work zone.

Tenability — Determination of whether or not people can remain unhurt or escape a fire area without serious injury.

Tenable Atmosphere — Capable of maintaining human life.

Tender — Term used within the incident command system for a mobile piece of apparatus that has the primary function of supporting another operation; examples include a water tender that supplies water to pumpers, or a fuel tender that supplies fuel to other vehicles. *See* Mobile Water Supply Apparatus and Tanker (1).

Tenement — *See* Apartment.

Tenon — Projecting member in a piece of wood or other material for insertion into a mortise to make a joint.

Tensile — Force of pulling apart or stretching.

Tensile Stress — Stress in a structural member that tends to stretch the member or pull it apart; often used to denote the greatest amount of tensile force a component can withstand without failure.

Tension — Vertical or horizontal forces that tend to pull things apart; for example, the force exerted on the bottom chord of a truss.

Tension Ring Method — Method used to attach a coupling to large diameter hose using a tension ring and contractual sleeve.

Tepee Cut — *See* Triangular Cut.

Teratogen — Chemicals that interfere with the normal growth of an embryo, causing malformations in the developing fetus.

TERC — See Transportation Emergency Rescue Committee.

Terminal — *See* Break Bulk Terminal, Bulk Terminal, Car Terminal, Container Terminal, and Dry Bulk Terminal.

Termination — The phase of an incident in which emergency operations are completed and the scene is turned over to the property owner or other party for recovery operations.

Terne Coated Steel — Cold rolled sheet steel that was hot-dip coated with a lead-tin alloy; this dull gray coating provides corrosion protection from contact with petroleum fuels.

Terrazzo — Mixture of a liquid and chipped marble that is poured as a floor; hardens to form a durable, long-lasting floor.

Territory — Specific geographical area to be covered by a responding company.

Terrorism — Unlawful force or violence against people or property to coerce or intimidate a government or its citizens, for social or political purposes (as defined by the U.S. Code of Federal Regulations).

Test — Any means of measuring some learner quality or ability. *See* Comprehensive Test, Criterion-Referenced Testing, Norm-Referenced Test, Objective Test, Oral Test, Performance Test, Pretest/Posttest, Progress Test, Subjective Test, and Written Test.

Test Hydrant — Fire hydrant used during a fire-flow test to read the static and residual pressures. *See* Flow Hydrant.

Test Item — Single question on a testing instrument that elicits a student response and can be scored for accuracy.

Test Item Analysis — Process that shows how difficult a test is, how much it discriminates between high and low scorers, and whether the alternatives used for distracters truly work.

Test Planning — Steps to determine the purpose and type of test, identify and define the learning objectives, prepare the test specifications, and construct the test items.

Test Planning Sheet — Planning form that lists and specifies the number of test items to be written and at what levels of learning in each content area. This planning sheet aids the test developer in ensuring that necessary numbers of questions are included for each objective or learning level in order to appropriately measure learning.

Testing Instrument — Series of test items that are based on learning objectives and collectively measure student learning on a specific topic. *Also known as* Evaluation Instrument or Test.

Tetrahedron — (1) Four-sided solid geometric figure that resembles a pyramid. (2) In fire science, a tetrahedron is used to represent the flaming mode of combustion consisting of fuel, heat, oxygen, and the uninhibited chain reaction. *See* Fire Tetrahedron. (3) A hollow four-sided object mounted on a central pivot used to indicate wind direction at some airports.

Theoretical Lift — Theoretical, scientific height that a column of water may be lifted by atmospheric pressure in a true vacuum; at sea level, this height is 33.8 feet (10 m). The height will decrease as elevation increases. *See* Lift.

Theoretical Mechanical Advantage — Advantage gained if all friction could be removed from a system; can never be attained in actuality.

Theory X — Style of leadership in which the leader believes that the average worker prefers to be directed and will avoid responsibility due to a general lack of ambition.

Theory Y — Style of leadership in which the leader believes that the average worker enjoys work, performs well with minimal supervision, will both seek and accept responsibility if given the opportunity, and will subscribe to organizational objectives if he associates those objectives with direct rewards.

Theory Z — Management style based on the belief that involved workers are the key to increased productivity and that there is a mutual loyalty between the company and the workers that often translates into lifetime employment and a close relationship between work and social life.

Thermal Balance — *See* Thermal Layering (of Gases).

Thermal Barrier — Heat protective barrier within protective clothing.

Thermal Belt — Elevation on a mountainous slope that typically experiences the least variation in diurnal temperatures and has the highest average temperatures and, thus, the lowest relative humidity. Its presence is most evident during clear weather with light wind.

Thermal Blocking — Phenomenon that occurs when a concealed hot spot contains enough heat to turn small amounts of penetrating water into steam, thereby preventing the water from cooling the material thoroughly.

Thermal Burns — Burns caused by contact with flames, hot objects, and hot fluids; examples include scalds and steam burns.

Thermal Column — Updraft of heated air, fire gases, and smoke directly above the involved fire area. *Also known as* Convection Column.

Thermal Conductivity — The propensity of a material to conduct heat within its volume. Measured in energy transfer over distance per degree of temperature.

Thermal Element — Device used in sprinklers and some fire detection equipment that is designed to activate when temperatures reach a predetermined level.

Thermal Energy — Kinetic energy associated with the random motions of the molecules of a material or object; often used interchangeably with the terms heat and heat energy. Measured in joules or Btu.

Thermal Equilibrium — The point at which two regions that are in thermal contact no longer transfer heat between them because they have reached the same temperature.

Thermal Expansion — Elongation or expansion of materials, such as steel, when exposed to heat.

Thermal Imager — Electronic device that forms images using infrared radiation. *Also known as* Thermal Imaging Camera.

Thermal Layering (of Gases) — Outcome of combustion in a confined space in which gases tend to form into layers, according to temperature, with the hottest gases found at the ceiling and the coolest gases at floor level. *Also known as* Heat Stratification or Thermal Balance.

Thermal Protective Performance (TPP) — Rating given to protective clothing to indicate the level of heat protection it affords the wearer.

Thermal Radiation — Transmission or transfer of heat energy, from one body to another body at a lower temperature, through intervening space by electromagnetic waves similar to radio waves or X-rays.

Thermal Updraft — Convection column of hot gases, smoke, and flames rising above a high-intensity fire; can create unsafe operating conditions for aircraft and reduce the effectiveness of finished foam.

Thermistor — Semiconductor made of substances whose resistance varies rapidly and predictably with temperature. *See* Semiconductor.

Thermocouple — Device for measuring temperature in which two electrical conductors of dissimilar metals, such as copper and iron, are joined at the point where the heat is to be applied. *See* Conductor.

Thermoelectric-Effect Heat Detector — Heat-sensitive device that measures electrical resistance changes that correspond with temperature changes.

Thermoplastic — Plastic that softens with an increase of temperature and hardens with a decrease of temperature but does not undergo any chemical change. Synthetic material made from the polymerization of organic compounds that become soft when heated and hard when cooled.

Thermoplastic Glazing — Plastic glazing made of acrylic, butyrate, or polycarbonate plastic; known for its resistance to breakage.

Thermosetting Plastics — Plastics that are hardened into a permanent shape in the manufacturing process and are not subject to softening when heated again.

THIRA — Acronym for Threat and Hazard Identification and Risk Assessment.

Third Door — (1) Automobile extrication technique used to free people who are trapped in the rear seat of a two-door vehicle. (2) An additional emergency exit door that may be found on the left side of some school buses, depending on local requirements. *Also known as* Left-Hand Door.

Third Rail — Electrically charged rail used to convey power to electrically powered trains; usually a contact shoe rides along the third rail to provide electrical pickup.

Third-Degree Burn — Full-thickness burn that penetrates the epidermis, dermis, and underlying tissue, leaving skin charred or white and leathery, and accompanied by an initial loss of pain or sensation to the area because of destroyed nerve cells.

Third-Party Testing Agency — Independent agency hired to perform nonbiased testing on a specific piece of apparatus.

Thread Gauge Device — Device that is screwed onto the discharge of a hydrant to check the condition of the threads and ensure that they are not damaged.

Threaded Coupling — Male or female coupling with a spiral thread.

Thready Pulse — Pulse that is very weak; characteristic of a person in shock.

Threat Zone — Operational zone designation that indicates an active shooter or explosion hazard.

Three Bight — Method of attaching a software loop to an anchor.

Three-Ply Process — Process of producing rubber-covered hose in which a nitrile rubber is vulcanized to the interior surface of a woven polyester tube.

Threshold Limit Value (TLV®) — Maximum concentration of a given material in parts per million (ppm) that may be tolerated for an 8-hour exposure during a regular workweek without ill effects. *See* Permissible Exposure Limit (PEL), Recommended Exposure Limit (REL), Short-Term Exposure Limit (STEL), Threshold Limit Value/Ceiling (TLV®/C), Threshold Limit Value/Short-Term Exposure Limit (TLV®/STEL), and Threshold Limit Value/Time Weighted Average (TLV®/TWA).

Threshold Limit Value/Ceiling (TLV®/C) — Maximum concentration of a given material in parts per million (ppm) that should not be exceeded, even instantaneously. *See* Permissible Exposure Limit (PEL), Recommended Exposure Limit (REL), Short-Term Exposure Limit (STEL), Threshold Limit Value (TLV®), Threshold Limit Value/Short-Term Exposure Limit (TLV®/STEL), and Threshold Limit Value/Time Weighted Average (TLV®/TWA).

Threshold Limit Value/Short-Term Exposure Limit (TLV®/STEL) — Fifteen-minute time-weighted average exposure. It should not be exceeded at any time nor repeated more than four times daily, with a 60-minute rest period required between each STEL exposure. These short-term exposures can be tolerated without suffering from irritation, chronic or irreversible tissue damage, or narcosis of a sufficient degree to increase the likelihood of accidental injury, impair self-rescue, or materially reduce worker efficiency. TLV/STELs are expressed in parts per million (ppm) and milligrams per cubic meter (mg/m³). *See* Permissible Exposure Limit (PEL), Recommended Exposure Limit (REL), Short-Term Exposure Limit (STEL), Threshold Limit Value (TLV®), Threshold Limit Value/Ceiling (TLV®/C), and Threshold Limit Value/Time Weighted Average (TLV®/TWA).

Threshold Limit Value/Time-Weighted Average (TLV®/TWA) — Maximum airborne concentration of a material to which an average, healthy person may be exposed repeatedly for 8 hours each day, 40 hours per week without suffering adverse effects. Based upon current available data and adjusted on an annual basis. *See* Permissible Exposure Limit (PEL), Recommended Exposure Limit (REL), Short-Term Exposure Limit (STEL), Threshold Limit Value (TLV®), Threshold Limit Value/Ceiling (TLV®/C), and Threshold Limit Value/Short-Term Exposure Limit (TLV®/STEL).

Throttle Control — Device that controls the engine speed.

Through the Roof — Description of a fire that has gained sufficient headway and vented itself by burning a hole through the roof.

Throw a Ladder — Raise a ladder quickly.

Throw Rescue — Method of rescue that uses minimal commitment and resources; a floating rope or flotation device is thrown to the victim and then pulled back to a place of safety.

Throwbag — Rope bag designed to deploy floating rope without tangling, from a protected position to a conscious receiving party in a water rescue environment.

Throwing Salvage Covers — To spread salvage covers by throwing them.

Thrust — Pushing or pulling force developed by an aircraft engine.

Thrust Reverser — Device or apparatus for diverting jet engine thrust in order to slow or stop the aircraft.

Thumbturn — Part of the lock, other than the key or knob, used to lock and unlock the door.

TIC — *See* Toxic Industrial Chemical.

Tie — (1) Metal strip used to tie masonry wall to the wood sheathing. (2) Device used to tie the two sides of a form together.

Tie In — (1) Securing oneself to a ladder; accomplished by using a rope hose tool or belt or by inserting one leg between the rungs. (2) Securing a ladder to a building or object.

Tie Rods — Metal rods running from one beam to the other.

Tied-Back Anchor — Method of building a strong anchor system from a weak or inadequate anchor point.

Tier — (1) Horizontal division of a multistory building, usually the stories in a steel-frame building. (2) Layer of hose loaded in the hose bed of a fire apparatus.

TIH — *See* Toxic Inhalation Hazard.

Tiller — Rear steering mechanism on a tractor-trailer aerial ladder truck.

Tiller Operator — Driver/operator of the trailer section of a tractor-tiller aerial ladder apparatus. *Also known as* Tillerman.

Tillered Truck — *See* Tractor-Drawn Apparatus.

Tillerman — *See* Tiller Operator.

Tilting Joint — Joint that allows movement in a tilt steering wheel.

Tilt-Up Construction — Type of construction in which concrete wall sections (slabs) are cast on the concrete floor of the building, then tilted up into the vertical position. *Also known as* Tilt-Slab Construction.

Tilt-Up Wall — Precast concrete wall that is raised or tipped up into position with a crane.

TIM — *See* Toxic Industrial Material.

Time Frame — Lesson plan component that lists the estimated time it will take to teach a lesson. *See* Lesson Plan.

Time Unit — Functional unit within the finance/administrative section; responsible for record keeping of time for personnel working at an incident.

Time-Temperature Curve — *See* Standard Time-Temperature Curve.

Tin-Clad Door — Similar to a metal-clad door, except covered with a lighter-gauge metal.

Tip — (1) Extreme top of a ladder. *Also known as* Top. (2) Slang for a nozzle.

Title Block — Small information section on the face of every plan drawing; contains such information as name of project, title of the particular drawing, the scale used, and date of drawing and/or revisions. *See* Legend.

Title VII — Part of the *Civil Rights Act of 1964* that prohibits discrimination based on race, color, religion, national origin, or gender.

TLV® — *See* Threshold Limit Value.

TLV®/C — *See* Threshold Limit Value/Ceiling.

TLV®/STEL — *See* Threshold Limit Value/Short-Term Exposure Limit.

TLV®/TWA — *See* Threshold Limit Value/Time-Weighted Average.

TOFC — *See* Trailer-on-Flatcar.

Toggle — (1) Hinge device by which a staypole is attached to a ladder. (2) Piece or device for holding or securing rope or chain by twisting. (3) Type of joint consisting of two levers joined at the end that exert outward pressure when a force is applied to straighten them.

Ton Container Pressurized tank with a capacity of 1 short ton or approximately 2,000 pounds (907 kg or 0.9 tonne).

Tone Out — To dispatch by activating pagers or station-alerting equipment using radio tones.

Tones — Radio signals that activate pagers or station-alerting systems.

Tongue — Rib on the edge of a ladder beam that fits into a corresponding groove or channel attached to the edge of another ladder beam; its purpose is to hold the two sections together while allowing the sections to move up and down.

Tongue and Groove — Projection on the edge of a board that fits into a recess in an adjacent board.

Tonnage — Amount of internal volume of the vessel, where 100 cubic feet (2.8 m^3) = one ton, used for determining port and canal charges.

Top Ventilation — *See* Vertical Ventilation.

TOPOFF Exercise — National-level, multiagency, multijurisdictional, real-time, and limited-notice weapons of mass destruction (WMD) response exercise designed to better prepare senior government officials to effectively respond to an actual terrorist attack.

Topography — Physical configuration of the land or terrain; often depicted using contour lines.

Topside — General term referring to the weather decks as opposed to below deck.

Torch — Professional firesetter, often for hire, who deliberately and maliciously sets fire to property.

Torching — (1) Burning of fuel at the end of the exhaust pipe or stacks of an aircraft engine due to excessive richness of the fuel/air mixture. (2) Vertical phenomenon in which a surface fire ignites the foliage of a tree or bush that becomes entirely involved in fire very quickly. A torching fire may or may not initiate a crown fire.

Tormentor Poles — *See* Staypole.

Torque — (1) Force that tends to create a rotational or twisting motion. (2) Measurement of engine shaft output. (3) Force that produces or tends to produce a twisting or rotational action.

Torque Box — Structural housing that contains the rotational system for the aerial device between the apparatus chassis frame rails and the turntable.

Torque Wrench — Specially designed wrench that may be set to produce a particular amount of torque on a bolt.

Torsional Load — Load offset from the center of the cross section of the member and at an angle to or in the same plane as the cross section; produces a twisting effect that creates shear stresses in a material.

Tort — Private or civil wrong or injury, including action for bad faith breach of contract, resulting from breach of duty that is based on society's expectations regarding interpersonal conduct; a violation of a duty imposed by general law upon all persons in a relationship that involves a given transaction. *See* Civil Wrong, Libel, Negligence, and Slander.

Tort Liability — Liability for a civil wrong or injury; noncriminal acts or failures to act that result in physical and/or monetary damages.

Total Energy — Total energy at any point in a system; the sum of the potential energy and kinetic energy at that point.

Total Flooding System — Fire-suppression system designed to protect hazards within enclosed structures; foam is released into a compartment or area and fills it completely, extinguishing the fire. *See* Local Application System.

Total Pressure — Total amount of pressure at any point in a system, including static pressure and velocity pressure.

Total Station — Surveying equipment attached to a computer-imaging system, used to create computer models of incident scenes.

Total Stopping Distance — Sum of the driver reaction distance and the vehicle braking distance.

Touch Off — Term for setting a fire or for describing a fire that firefighters believe has been purposely set.

Touring Bus — *See* Commercial Motor Coach.

Tourniquet — Any wide, flat material wrapped tightly around a limb to stop bleeding; used only for severe, life-threatening hemorrhage that cannot be controlled by other means.

Tow Bar — Device used to maintain the distance between a towed vehicle and the towing vehicle.

Towboat — *See* Tugboat.

Tower — *See* Drill Tower.

Tower Ladder — Telescoping aerial platform fire apparatus.

Toxic — Poisonous.

Toxic Atmosphere — Any area, inside or outside a structure, where the air is contaminated by a poisonous substance that may be harmful to human life or health if it is inhaled, swallowed, or absorbed through the skin. *See* Toxic Gas.

Toxic Element — *See* Heavy Metal.

Toxic Gas — Gas that contains poisons or toxins that are hazardous to life. Many gaseous products of combustion are poisonous; toxic materials generally emit poisonous vapors when exposed to an intensely heated environment. *See* Gas, Toxic Material, and Toxin.

Toxic Industrial Chemical (TIC) — *See* Toxic Industrial Material (TIM).

Toxic Industrial Material (TIM) — Industrial chemical that is toxic at a certain concentration and is produced in quantities exceeding 30 tons per year at any one production facility; readily available and could be used by terrorists to deliberately kill, injure, or incapacitate people. *Also known as* Toxic Industrial Chemical (TIC). *See* Chemical Attack and Terrorism.

Toxic Inhalation Hazard (TIH) — Volatile liquid or gas known to be a severe hazard to human health during transportation.

Toxic Material — Substance classified as a poison, asphyxiant, irritant, or anesthetic that can damage the environment or cause severe illness, poisoning, birth defects, disease, or death when ingested, inhaled, or absorbed by living organisms. *See* Asphyxiant, Toxic Gas, Toxic Inhalation Hazard (TIH), Toxicity, and Toxin.

Toxic Substances Control Act (TSCA) — Law giving the Environmental Protection Agency (EPA) the ability to track the 75,000 industrial chemicals currently produced or imported into the U.S.; enacted in 1976.

Toxicity — Ability of a substance to do harm within the body. *See* Toxic Inhalation Hazard (TIH), Toxic Material, and Toxin.

Toxin — Substance that has the property of being poisonous. *See* Biological Toxin, Poison, Toxic Gas, Toxicity, and Toxic Material.

TPP — *See* Thermal Protective Performance.

Trace Evidence — Physical evidence that results from the transfer of small quantities of materials; for example, hair, textile fibers, paint chips, glass fragments, or gunshot residue particles.

Track Traps — Areas which, from their nature or location, are likely to provide evidence of victim movement. For example, a muddy riverbank may allow shore-based rescuers to see footprints left by a victim who self-rescued and left the water.

Traction — (1) Act of exerting a pulling force. (2) The friction between a wheel and the surface on which it is rolling that permits the wheel to move forward and exert a pulling force.

Tractor-Drawn Apparatus — Truck equipped with steerable rear wheels on its trailer. *Also known as* Tillered Truck.

Tractor-Plow — Tractor with a plow for exposing mineral soil.

Tractor-Tiller Aerial Ladder — Aerial ladder apparatus that consists of a tractor power unit and trailer (tiller) section that contains the aerial ladder, ground ladders, and equipment storage areas. The trailer section is steered independently of the tractor by the tiller operator.

Trade Secret — Formula, practice, process, design, instrument, pattern, or compilation of information used by a business to obtain an advantage over competitors or customers.

Traffic Control — Important function of scene management that helps to control scene access and vehicular traffic in and out of the area. This function is generally handled by law enforcement personnel.

Traffic Control Device — Mechanical device that automatically changes traffic signal lights to favor the path of responding emergency apparatus.

Traffic Control Zone — Operational zone established on or near a roadway for the rerouting of traffic and protection of civilians and responders; may include a hot, warm, and cold zone depending on the incident.

Traffic Pattern — Traffic flow that is prescribed for aircraft landing or taking off from an airport.

Traffic Preemption Device — Wireless system (coded emitter or GPS activated) on apparatus that can communicate with traffic signals to request right-of-way at intersections for emergency vehicles.

Trail Drop — Dropping fire suppressant sequentially from tanks in aircraft; generally used with light fuels.

Trailer — (1) Combustible material, such as rolled rags, blankets, newspapers, or flammable liquid, often used in intentionally set fires to connect remote fuel packages (such as pools of ignitable liquid and other combustible materials) in order to spread fire from one point or area to other points or areas; frequently used in conjunction with an incendiary device. (2) Fire pattern left behind after the combustible material has burned. (3) Highway or industrial-plant vehicle designed to be hauled/pulled by a tractor.

Trailer-on-Flatcar (TOFC) — Rail flatcar used to transport highway trailers. *Also known as* Piggyback Transport.

Trailing Edge Devices — Rear edges of aircraft wings; normally extended for takeoff and landing to provide additional lift at low speeds and to improve aircraft performance.

Train Consist — *See* Consist.

Training — (1) Supervised activity or process for achieving and maintaining proficiency through instruction and hands-on practice in the operation of equipment and systems that are expected to be used in the performance of assigned duties. *Also known as* Drilling. (2) The transfer of knowledge regarding vocational or technical skills.

Training Aids — Broad term referring to any audiovisual aids, reprinted materials, training props, or equipment used to supplement instruction.

Training Concentrates — Foam concentrates that are specially designed for hydrocarbon fuel fire training; generally reproduce the white color, appearance, expansion ratio, and drain time of AFFF.

Training Evolution — Operation of fire and emergency services training covering one or several aspects of fire fighting. *Also known as* Practical Training Evolution.

Training Officer — Individual responsible for running a fire department's training program, to include the administration of all training activities; typically reports directly to the fire chief. *Also known as* Chief of Training or Drillmaster.

Training Session — For the intervention program, a training session provides the program participants with all the procedures and processes of the program.

Transcription — Method by which an authority having jurisdiction (AHJ) adopts a code in whole to become a new regulation.

Transfer — (1) To move a firefighter to a different unit. (2) The movement of companies or apparatus.

Transfer of Learning — Process of applying what has been learned in one situation to a new situation.

Transfer Valve — Valve used for placing multistage centrifugal pumps in either volume or pressure mode operation.

Transfilling System — Self-contained breathing apparatus designed so that two SCBAs can be connected by a hose, allowing the air pressure of the two SCBA cylinders to equalize; used as an EBSS to equalize air pressure of one cylinder with an adequate air supply and another cylinder with depleted or inadequate air supply.

Transformer — Device that uses coils and magnetic fields to increase (step-up) or decrease (step-down) incoming voltages.

Transient Evidence — Material that will lose its evidentiary value if it is unpreserved or unprotected; for example, blood in the rain.

Transit Bus — Vehicle designed to move a large number of people over relatively short distances; most commonly found in urban or metropolitan areas that operate a mass transit system.

Transit Time — Time that it takes a foam solution to pass from the proportioner to the nozzle.

Transition — (1) Passage from one state, stage, subject, or place to another. (2) Section of a tank that joins two unequal cross-sections.

Transmission of Heat — Flow of heat by conduction, convection, or radiation.

Transmit — To send out an alarm by vocal, visual, or audible means.

Transmit Site — Assembly of elements capable of functioning together to transmit signal waves.

Transmitter — Device for sending or transmitting codes or signals over alarm circuits when operated by any one of a group of actuating devices or for sending voice communications over the air.

Transparency — Educational visual aid printed on acetate or mylar and projected onto a screen for common viewing. *Also known as* Overhead.

Transparent Armor — Ballistic protection materials used as windows for vehicles.

Glossary

Transport Canada (TC) — Canadian agency responsible for developing and administering policies, regulations, and programs for a safe, efficient, and environmentally friendly transportation system; contributing to Canada's economic growth and social development; and protecting the physical environment.

Transport Index — Number placed on the label of a package expressing the maximum allowable radiation level in millirem per hour at 1 meter (3.3 feet) from the external surface of the package.

Transportation Area — Location where accident casualties are held after receiving medical care or triage before being transported to medical facilities.

Transportation Emergency Rescue Committee (TERC) — Organization founded in 1986 that serves as a competent source of guidance and information on transportation emergencies for those involved in providing emergency services; members share their vehicle extrication expertise by conducting schools, seminars, and competitive exercises.

Transportation Group — Group within the incident command system responsible for seeing that all patients are transported to the appropriate medical facility.

Transportation Security Administration (TSA) — U.S. Department of Homeland Security (DHS) agency that is responsible for the security of the national transportation systems (highways, buses, railroads, mass transit systems, ports, and airports); established following the events of September 11, 2001.

Transshipment — Transfer of cargo from one vessel to another before the place of destination has been reached.

Transverse — Athwartship (side to side) dimensions of a vessel.

Transverse Hose Bed — Hose bed that lies across the pumper body at a right angle to the main hose bed; designed to deploy preconnected attack hose to the sides of the pumper. *Also known as* Mattydale Hose Bed.

Trash Line — Small diameter, preconnected hoseline intended to be used for trash or other small, exterior fires.

Trauma Kit — Well-stocked medical first-aid kit.

Travel Corridors — Areas inside a building that are designed and used for travel by occupants; includes hallways, stairwells, access ramps, and elevator shafts.

Travel Distance — Distance from any given area in a structure to the nearest exit or to a fire extinguisher. *See* Exit and Means of Egress.

Tread — Horizontal face of a step.

Treatment Group — Group within the incident command system responsible for triage and the initial treatment of patients.

Tree System — Water supply piping system that uses a single, central feeder main to supply branches on either side of the main.

Tremie — Chute used to deliver concrete to the bottom of a caisson.

Trench — Temporary excavation in which the length of the bottom exceeds the width of the bottom; generally limited to excavations that are less than 15 feet (4.6 m) wide at the bottom and less than 20 feet (6 m) deep.

Trench Foot — Foot condition resulting from prolonged exposure to damp conditions or immersion in water; symptoms include tingling and/or itching, pain, swelling, cold and blotchy skin, numbness, and a prickly or heavy feeling in the foot. In severe cases, blisters can form, after which skin and tissue die and fall off.

Trench Jack — Jack used to keep sheeting or wales apart for the insertion of a breast timber in a trenching operation; also a jack that is used as a breast timber.

Trench Lip — Top edge of a trench.

Trench Ventilation — Defensive tactic that involves cutting an exit opening in the roof of a burning building, extending from one outside wall to the other, to create an opening at which a spreading fire may be cut off. *Also known as* Strip Vent. *See* Trenching (2).

Trenching — (1) In wildland fire fighting, digging a trench in a slope to catch any burning, rolling material that could cross the control line. (2) In strip or trench ventilation, the process of opening a roof area the width of the building, with a 2 foot (0.6 m) wide opening, to channel out fire and heat.

Triage — (1) System used for sorting and classifying accident casualties to determine the priority for medical treatment and transportation. (2) *See* Structural Triage.

Triage Tagging — Method used to identify accident patients as to extent of injury.

Triangular Cut — Triangular opening cut in a roll-up or tilt-slab door to provide access into the building or a means of egress for those inside. *Also known as* Tepee Cut.

Trier of Fact — Party that determines the facts in a legal case. In a jury trial, the jury is the trier of fact; in a bench (non-jury) trial, the judge is the trier of fact.

Trim — (1) Relation of a vessel's floating attitude to the water or to the longitudinal angle of a vessel; the difference between forward and aft draft readings. (2) To cause a vessel to assume a desirable position in the water by arrangement of ballast, cargo, or passengers. *See* Ballast and Trimming Tank.

Trimming Tank — Tank located near the ends of a vessel used to change the trim of a vessel by admitting or discharging water ballast. *See* Ballast Tank and Trim.

Trip Curve — Relationship between current and time that determines when circuit protective devices operate.

Triple Hydrant — Fire hydrant having three outlets, usually two 2½-inch (65 mm) outlets and one 4½-inch (115 mm) outlet.

Triple-Combination Pumper — Fire department pumper that carries a fire pump, hose, and a water tank.

Trouble Signal — Signal given by a fixed fire protection alerting system when a power failure or other system malfunction occurs.

Truck — (1) Self-propelled vehicle carrying its load on its wheels; primarily designed for transportation of property rather than passengers. (2) Slang term for an aerial apparatus such as a ladder truck.

Truck Company — Group of firefighters assigned to a fire department aerial apparatus equipped with a compliment of ladders; primarily responsible for search and rescue, ventilation, salvage and overhaul, forcible entry, and other fireground support functions. *Also known as* Ladder Company.

Truck Tractor — Powered motor vehicle designed to pull a truck trailer. *See* Truck Trailer.

Truck Trailer — Vehicle without motor power; primarily designed for transportation of property rather than passengers and drawn by a truck or truck tractor. *See* Full Trailer.

Trumpets — Symbolic insignia of rank used throughout the fire service; originated during the time when fire officers gave commands through speaking trumpets.

Trunk Pipeline — A main pipeline used for carrying large quantities of liquids, often with a diameter between 8 to 48 inches (200 to 1,220 mm).

Trunked Radio System — Radio system designed to use available frequencies more efficiently by use of a computerized radio controller. The radio controller assigns a group of radios to a certain frequency to allow communications when one of the radios transmits; the system then simultaneously switches all radios on that channel or talk group to the assigned frequency to listen to the transmission. These systems are typically very complex and can cover entire states and regions.

Trunnion — In a hydraulic cylinder, the pivoting end of the piston rod that is connected to the anchor ear by the heel pin.

Truss — (1) Structural member used to form a roof or floor framework; trusses form triangles or combinations of triangles to provide maximum load-bearing capacity with a minimum amount of material. Often rendered dangerous by exposure to intense heat, which weakens gusset plate attachment. (2) Beams consisting of one tensile chord, one compression chord, and truss blocks or spaces between the two.

Truss Block — Block used to separate the beams of a truss beam ladder. *Also known as* Beam Block and Run Block.

Truss Construction Ladder — Aerial device boom or ladder sections that are constructed by trussed metal pieces.

Trussed Rafter — Roof truss that serves to support the roof and ceiling construction.

TSA — *See* Transportation Security Administration.

TSCA — *See* Toxic Substances Control Act.

TSO — *see* Technical Safety Officer.

Tube Seal — Type of seal used on a floating roof fuel storage tank; the seal is constructed of urethane foam that is contained within an envelope, and is connected to the edge of the roof around the entire circumference of the tank. A secondary weather shield is usually installed above the main seal.

Tube Trailer — *See* Compressed Gas Tube Trailer.

Tubular Deadbolt — Deadbolting bored lock. *Also known as* Auxiliary Deadbolt.

Tubular Truss-Beam Construction — Similar in design to the truss construction of aerial ladders; tubular steel is welded to form a box shape, using cantilever or triangular truss design.

Tug — See Aircraft Tug.

Tugboat — Strongly built, powerful boat used for towing and pushing in harbors and inland waterways. *Also known as* Towboat.

Tumbler — Pin in the tumbler type of lock cylinder.

Turbidity — Muddy, cloudy, or murky condition of water caused by the stirring up of sediment.

Turbojet — Jet engine employing a turbine-driven compressor to compress the intake air, or an aircraft with this type of engine. Also known as a gas turbine.

Turbulence — (1) Irregular motion of the atmosphere usually produced when air flows over a comparatively uneven surface, such as the surface of the earth; when two currents of air flow past or over each other in different directions or at different speeds. (2) In terms of explosions, the changes in shape of the blast-pressure front as the expanding pressure is forced around objects and/or toward areas of ventilation; occurs in deflagrations but not in detonations.

Turbulent State — Fluid flow is in the turbulent state at higher velocities where there is no definite pattern to the direction of the water particles. Turbulent flow is reflected by a calculated Reynolds number in excess of 2,100.

Turn Out — Alerting of a fire company for a response.

Turnaround Maintenance Tag — Tag attached to the valve on the oxygen tank of closed-circuit self-contained breathing apparatus that tells when the unit was last serviced, lists what services were performed, and indicates that the unit is ready to perform.

Turnout Gear — Term used to describe personal protective clothing made of fire resistant materials that includes coats, pants, and boots. *Also known as* Turnout Boots, Turnout Clothing, Turnout Coat, Turnout Pants, Turnouts, or Bunker Gear. *See* Personal Protective Equipment.

Turntable — Rotational structural component of the aerial device. Its primary function is to provide continuous rotation on a horizontal plane.

Turret Pipe — *See* Turret/Turret Nozzle.

Turret/Turret Nozzle — Large, pre-plumbed master stream appliance connected directly to a pump that is mounted on a pumper, a trailer, and some airport rescue and fire fighting apparatus, and is capable of sweeping from side to side and designed to deliver large volumes of foam or water. *Also known as* Deck Gun, Deck Pipe, or Turret Pipe.

Tween Deck — Intermediate deck between the main deck and the bottom of a cargo hold; designed to support cargo so that the cargo at the bottom of the hold is not crushed by the weight of cargo above it. *See* Deck.

Twist Lock — (1) Mechanically operated device located on the corners of a container chassis and on automatic lifting spreaders; used for restraining a container during transport or transfer. (2) Type of positive connector used on most fire service extension cords; may be a two- or three-prong connector.

Two-Stage Centrifugal Pump — Centrifugal pump with two impellers.

Two-Way Radio — Voice network that provides an always-on connection that enables the user to just "push the button and talk;" allows either transmitting or receiving, but not both at once unless it is a full-duplex system. *Also known as* Dispatch Radio or Transceiver.

Tying In — (1) Securing oneself to a ladder by using a rope hose tool or belt, or by inserting one leg between the rungs. (2) Securing a ladder to a building or object. *See* Tie In (1) and (2).

Type A Packaging — Container used to ship radioactive materials with relatively high radiation levels. *See* Excepted Packaging, Industrial Packaging, Packaging (1), Strong, Tight Container, and Type B Packaging.

Type A School Bus — Van conversion-type school bus with a gross vehicle-weight rating of less than 10,000 pounds (4 536 kg).

Type B Packaging — Container used to ship radioactive materials that exceed the limits allowed by Type A packaging, such as materials that would present a radiation hazard to the public or the environment if there were a major release. *See* Excepted Packaging, Industrial Packaging, Packaging (1), Strong, Tight Container, and Type A Packaging.

Type B School Bus — Minibus-type vehicle with a gross vehicle-weight rating in excess of 10,000 pounds (4 536 kg).

Type C School Bus — Conventional school bus vehicle with a gross vehicle-weight rating well in excess of 10,000 pounds (4 536 kg); the engine is found ahead of the cab.

Type D School Bus — Cab-forward style school bus with a gross vehicle-weight rating well in excess of 10,000 pounds (4 536 kg); the engine is found in the front, rear, or midship of the vehicle.

Type I Construction — Construction type in which structural members, including walls, columns, beam, floors, and roofs, are made of noncombustible or limited combustible materials and have a specified degree of fire resistance. *Formerly known as* Fire Resistive Construction.

Type II Construction — Construction type that is similar to Type I except that the degree of fire resistance is lower. *Formerly known as* Noncombustible or Noncombustible/Limited Combustible Construction.

Type III Construction — Construction type in which exterior walls and structural members are made of noncombustible or limited combustible materials, but interior structural members, including walls, columns, beams, floors, and roofs, are completely or partially constructed of wood. *Formerly known as* Ordinary Construction.

Type IV Construction — Heavy timber construction in which interior and exterior walls and their associated structural members are made of noncombustible or limited combustible material; interior structural framing consists of heavy timber with minimum dimensions larger than those used in Type III construction. *Formerly known as* Heavy Timber Construction.

Type of Collapse — Manner in which a structure failed; this information is useful in determining possible victim locations and potential for additional collapse.

Type of Occupancy — Nature of business taking place in a given structure or area.

Type V Construction — Construction type in which exterior walls, bearing walls, floors, roofs, and supports are made completely or partially of wood or other approved materials of smaller dimensions than those used in Type IV construction. *Formerly known as* Wood Frame Construction.

Typical Tool Order — Order in which hand-crew members are assigned tools for varying types of wildland fuels; types of tools that are necessary varies depending on fuel type.

U

U.S. Bureau of Alcohol, Tobacco, Firearms and Explosives (ATF) — *See* Bureau of Alcohol, Tobacco, Firearms and Explosives (ATF).

U.S. Coast Guard (USCG) — U.S. military, multimission, and maritime service, whose mission is to protect the public, the environment, and U.S. economic interests. Operates in U.S. ports and waterways, along the coasts, in international waters, or in any maritime region as required to support national security.

U.S. Department of Defense (DoD) — *See* Department of Defense (DoD).

U.S. Department of Energy (DOE) — *See* Department of Energy (DOE).

U.S. Department of Homeland Security (DHS) — *See* Department of Homeland Security (DHS).

U.S. Department of Housing and Urban Development (HUD) — *See* Department of Housing and Urban Development (HUD).

U.S. Department of Justice (DOJ) — *See* Department of Justice (DOJ).

U.S. Department of Labor (DOL) — *See* Department of Labor (DOL).

U.S. Department of Transportation (DOT) — *See* Department of Transportation (DOT).

U.S. Environmental Protection Agency — *See* Environmental Protection Agency.

U.S. Fire Administration — U.S. agency whose aim is to reduce the nation's fire deaths. Promotes better fire prevention and control, supports existing programs of research, training, and education, and encourages new programs sponsored by state and local governments. Administers an extensive fire data and analysis program and co-administers a program concerned with firefighter health and safety. USFA is a division of the Federal Emergency Management Agency (FEMA), which itself is a division of the Department of Homeland Security (DHS). *Formerly known as* National Fire Prevention and Control Administration (NFPCA).

U.S. Occupational Safety and Health Administration (OSHA) — *See* Occupational Safety and Health Administration (OSHA).

UC — *See* Unified Command.

UEL — *See* Upper Explosive Limit.

UFL — *See* Upper Flammable Limit.

UL — *See* Underwriters Laboratories, Inc.

Ullage — Measure of the empty part of a partially-filled tank. *See* Ullage Hole.

Ullage Hole — Opening that leads to a liquid cargo tank, allowing the measurement of liquid cargo; usually located in the hatch cover. *See* Ullage.

Ultimate Capacity — Total capacity of a water supply system, including residential and industrial consumption, available fire flow, and all other taxes on the system.

Ultrahigh Frequency (UHF) — Radio band containing ultrahigh frequencies ranging from 400 MHz to 512 MHz; typically used in metropolitan areas. Includes portions that are licensed to broadcast television.

Ultrasonic Inspection — Nondestructive testing method in which ultrasonic vibrations are injected into an aerial device; deviance in the return of the waves is an indication of existing flaws.

Ultraviolet (UV) Wave Spectrum — Wavelengths shorter than visible light, lying at the violet end of the spectrum.

Uncontrolled Airport — Airport with no operating control tower.

Undeclared Emergency — Aircraft emergency that occurs without prior warning.

Under Control — Point in a fire incident when the fire's progress has been stopped; final extinguishment and overhaul can begin at this time. *Also known as* Tapped Out.

Undercarriage — (1) Portion of a vehicle's chassis (frame) that is located beneath the vehicle; consists of the chassis, drive train, and floor pan. (2) Landing gear of an aircraft.

Undercut Line — Fireline below a fire burning on a slope; should be trenched to catch rolling material. *Also known as* Underslung Line.

Underlayment — Floor covering of plywood or fiberboard installed to provide a level surface for carpet or other resilient flooring.

Underpinning — Strengthening an existing foundation.

Underride Collision — Collision in which the striking vehicle comes to rest under the vehicle that it struck.

Underslung Line — *See* Undercut Line.

Underwriters Laboratories, Inc. (UL) — Independent fire research and testing laboratory that certifies equipment and materials, which can be approved only for the specific use for which it is tested. Headquartered in Northbrook, Illinois.

Undetermined Fire Cause — Fire cause classification used when the specific cause has not been determined to a reasonable degree of probability.

Unfriendly Fire — Uncontained and uncontrolled fire of intentional or accidental origin that may cause injury or damage.

Ungrounded Conductor — Energized conductor in a branch circuit that supplies current to load devices; its covering will be some color other than white, gray, or green. *Also known as* Hot Conductor.

Unibody — *See* Unitized Body.

Unibody Construction — Method of automobile construction in which the frame and body form one integral unit; used on most modern cars. *Also known as* Bird Cage Construction, Integral Frame Construction, or Unitized Construction.

Unified Command (UC) — In the Incident Command System, a shared command role in which all agencies with geographical or functional responsibility establish a common set of incident objectives and strategies. In unified command there is a single incident command post and a single operations chief at any given time.

Uniform — Official dress uniform or work uniform; different from protective clothing.

Uniformly Distributed Load — Load in a building that is evenly distributed over a particular area.

Unincorporated — Portion of a state/province that is outside the jurisdiction of any municipality and does not provide an inspection program.

Unit — Division of a block within an occupation consisting of an organized grouping of tasks within that block. *See* Block.

Unit Loading Device Pallet or container used to facilitate the rapid loading and unloading of aircraft cargo. *Also known as* Unit Load Device.

Unit Lock — Lock designed to be installed in a cutout within the door, without requiring disassembly and reassembly of the lock.

Unit Vents — Vents normally constructed of metal frames and walls and operated by a hinged damper that is controlled either manually or automatically.

Unitized Body — Automobile construction in which a vehicle's stress bearing elements and sheet metal body parts are built together as one unit, instead of attaching the vehicle's body to a frame as in body-on-frame construction. *Also known as* Integral Frame or Unibody.

Unitized Construction — *See* Unibody Construction.

Unity of Command — Organizational principle in which workers report to only one supervisor in order to eliminate conflicting orders.

Universal Coupling — Coupling device that permits unlike couplings to be connected.

Universal Emergency Telephone Number — Typically an easy to remember and dial 3-digit telephone number that can be used to access emergency services personnel in the event of an emergency. 9-1-1 is the common universal emergency telephone number in the U.S. and Canada. In the United Kingdom, this number is 9-9-9; in other nations in the European Union, it is 1-1-2. *Also known as* Emergency Services Number, Emergency Telephone Number, or Universal Number.

Universal Joint — Multidirectional hinged joint used in the steering system and drivetrain of an automobile.

Universal Precautions — Set of precautions designed to prevent transmission of biological pathogens, especially bloodborne pathogens, when providing first aid or health care.

Universal Priorities — Operational priorities that apply to all emergency incidents. In order of importance (priority) they are: life safety, incident stabilization, and property conservation.

Unlined Hose — Fire hose without a rubber lining; most frequently used in interior standpipe systems and in wildland fire fighting.

Unloading Site — Site where tankers unload their water into portable tanks during a water shuttle operation. *Also known as* Dump Site.

Unplanned Ventilation — Any ventilation that occurs outside of planned tactics such as ventilation from failed structural components, wind conditions, or firefighters acting outside of assigned tactics.

Unprotected Openings — Openings in floors, walls, or partitions that are not protected against the passage of smoke, flame, and heat; generally used to refer to such openings in fire walls.

Unprotected Steel — Steel structural members that are not protected against exposure to heat.

Unresponsive — Unable to respond to verbal commands or stimuli, such as pain.

Unrestricted Grant — *See* General Support Grant.

Unstable Material — Materials that are capable of undergoing chemical changes or that can violently decompose with little or no outside stimulus.

Unsupported Tip — Operation of an aerial device with the tip of the device, or the platform if so equipped, in the air and not resting on another object. *Also known as* Cantilever Operation.

Unwritten Law — *See* Judiciary Law.

U-Pattern — Fire pattern left on a vertical surface from a fire plume some distance away; similar to a V-pattern but formed higher on the vertical surface.

Upper Airway — Portion of the respiratory system above the epiglottis.

Upper Coupler Assembly — Assembly consisting of an upper coupler plate, reinforcement framing, and a fifth-wheel kingpin mounted on a semitrailer. *Formerly known as* Upper Fifth Wheel Assembly.

Upper Deck — Topmost continuous deck running the entire length and width of a vessel. *See* Deck.

Upper Explosive Limit (UEL) — *See* Upper Flammable Limit (UFL).

Upper Fifth-Wheel Assembly — *See* Upper Coupler Assembly.

Upper Flammable Limit (UFL) — Upper limit at which a flammable gas or vapor will ignite; above this limit the gas or vapor is too *rich* to burn (lacks the proper quantity of oxygen). *Also known as* Upper Explosive Limit (UEL). *See* Flammable Gas, Flammable Limit, Flammable Range, and Lower Flammable Limit (LFL).

Upper Layer — Buoyant layer of hot gases and smoke produced by a fire in a compartment.

Upright Sprinkler — Sprinkler that sits on top of the piping and sprays water against a solid deflector; breaks up the spray into a hemispherical pattern that is redirected toward the floor.

Uprights — Planks that are held in place against sections of sheeting with shores. Uprights add strength to the shoring system; they distribute forces exerted by trench walls and counterforces exerted by shores over wider areas of the sheeting.

Upstream — Direction opposite the flow of a stream or of the airflow; in a self-contained breathing apparatus (SCBA), the regulator is upstream from the face piece.

Urban Search and Rescue (US&R) — Search and rescue efforts involving structural collapse and other urban environments.

Urban/Wildland Interface — *See* Wildland/Urban Interface.

URM — Unreinforced masonry.

US&R — *See* Urban Search and Rescue.

USCG — *See* U.S. Coast Guard.

USFA — *See* U.S. Fire Administration.

Utilidor — Insulated, heated conduit built below ground or supported above ground to protect enclosed water, steam, sewage, and fire lines from freezing.

Utilities — Services such as gas, electricity, and water that are provided to the public.

Utility Rope — Rope designed for any use except rescue; can be used to hoist equipment, secure unstable objects, or cordon off an area. *See* Non-Lifeline Rope.

UV — *See* Ultraviolet Wave Spectrum.

V

Vaccine — Biological agent that immunizes a person against a specific disease; prepared from the disease-causing biological agent itself or from a synthetic substitute.

Vacuum — (1) Space completely devoid of matter or pressure. (2) In the fire and emergency services, a pressure that is somewhat less than atmospheric pressure; a vacuum is needed to facilitate drafting of water from a static source.

Validity — Extent to which a test or other assessment technique measures the learner qualities (knowledge or skills) that it is meant to measure.

Value at Risk — One of several factors considered when planning loss control strategies; a subjective determination of the value of a structure, its contents, the operations associated with it, or its existence.

Valve — Mechanical device with a passageway that controls the flow of a liquid or gas.

Vapor — Gaseous form of a substance that is normally in a solid or liquid state at room temperature and pressure; formed by evaporation from a liquid or sublimation from a solid. *See* Vapor Density, Vapor Dispersion, Vaporization, Vapor Pressure, and Vapor Suppression.

Vapor Barrier — (1) Watertight material used to prevent the passage of moisture or water vapor through walls or roofs. (2) For personal protective equipment, material that prevents water from penetrating the clothing.

Vapor Density — Weight of pure vapor or gas compared to the weight of an equal volume of dry air at the same temperature and pressure. A vapor density less than 1 indicates a vapor lighter than air; a vapor density greater than 1 indicates a vapor heavier than air. *See* Physical Properties, Specific Gravity, and Vapor.

Vapor Dispersion Action taken to direct or influence the course of airborne hazardous materials. *See* Dispersion and Vapor.

Vapor Mitigating Foam — Foam concentrate designed solely for use on unignited spills of hazardous liquids; it is not effective for fire-suppression operations. *See* Foam Concentrate.

Vapor Pressure — (1) Measure of the tendency of a substance to evaporate. (2) The pressure at which a vapor is in equilibrium with its liquid phase for a given temperature; liquids that have a greater tendency to evaporate have higher vapor pressures for a given temperature. *See* Boiling Point and Vapor.

Vapor Recovery System (VRS) — System that recovers gasoline vapors emitted from a vehicle's gasoline tank during product dispensing.

Vapor Suppression — Action taken to reduce the emission of vapors at a hazardous materials spill. *See* Hazardous Material and Vapor.

Vaporization — Physical process that changes a liquid into a gaseous state; the rate of vaporization depends on the substance involved, heat, pressure, and exposed surface area. *See* Radiative Feedback, Vapor, and Vapor Density.

Vaporizing Liquid Agent — (1) Any liquid that evaporates at elevated temperatures. (2) One of several extinguishing agents used on Class B or Class C fires that produces vapors that are heavier than air and acts as a smothering vapor agent.

Vapor-Protective Clothing — Gas-tight chemical-protective clothing designed to meet NFPA® 1991, *Standard on Vapor-Protective Suits for Hazardous Chemical Emergencies*; part of an EPA Level A ensemble.

Vaportight Fixture — Fixture sealed to prevent an explosive atmosphere from entering the device's electrical contacts, where an ignition spark could be generated.

Variable-Flow Demand-Type Balanced-Pressure Proportioner — Foam proportioning system that is used in both fixed and mobile applications; a variable speed mechanism drives the foam pump and automatically monitors the flow of foam to produce an effective foam solution. *See* Foam Proportioner and Proportioning.

Variable-Flow Variable-Rate Direct-Injection System — Apparatus-mounted foam system that injects the correct amount of foam into the pump piping, thereby supplying all discharges with foam. The system automatically monitors the operation of the hoselines and maintains a consistent quality of foam solution. *See* Foam Proportioner and Proportioning.

Vaulted Surfaces — Ground above underground vaults such as underground parking structures, utility chases, drainage culverts, basements that extend under sidewalks, or underground transportation systems.

Vector — (1) Quantity in mathematics that has magnitude and direction; commonly represented by a directed line segment or arrow whose length represents the magnitude and whose orientation in space represents the direction. Common quantities with vector representations include force, pressure, and velocity. (2) An animate intermediary in the indirect transmission of an agent that carries the agent from a reservoir to a susceptible host. (3) Compass heading or course followed by or to be followed by an aircraft.

Vector Diagram — Visual representation of fire movement at a scene, generally from areas of greater damage to areas of less damage.

Vegetation Fire — *See* Wildland Fire.

Vegetation Management — *See* Fuel Management.

Vehicle Driving Course — Permanent or temporary training course used for apparatus and vehicle driving and operation.

Vehicle Identification Number (VIN) — Unique serial number used by automobile manufacturers to identify each individual vehicle.

Vehicle Rescue Technician (VRT) — Firefighter who is specially trained and certified to perform automobile extrications.

Vehicle Stabilization — Providing additional support to key places between a vehicle and the ground or other solid anchor points to prevent unwanted movement.

Vehicle-Borne Improvised Explosives Device (VBIED) — An improvised explosive device placed in a car, truck, or other vehicle, typically creating a large explosion. *Also known as* Car Bomb or Vehicle Bomb. *See* Explosion, High Explosive, and Improvised Explosive Device.

Vein — Blood vessel that carries blood from the tissues to the heart.

Velocity — Rate of motion in a given direction; measured in units of length per unit time, such as feet per second (meters per second) and miles per hour (kilometers per hour). *Also known as* Speed.

Veneered Product — Structural product that has an external surface material over an inner core.

Veneered Walls — Walls with a surface layer of attractive material laid over a base of a common material.

Vent — To release enclosed smoke and heat through an opening in the structure; the opening is made by chopping a hole in the roof, allowing for freer passage of air.

Vent Sector — Incident command system term for firefighters assigned to ventilate a structure.

Ventilation — (1) Systematic removal of heated air, smoke, gases or other airborne contaminants from a structure and replacing them with cooler and/or fresher air to reduce damage and facilitate fire fighting operations. (2) Process of replacing foul air in any of a vessel's compartments with pure air. *See* Positive-Pressure Ventilation and Supply/Exhaust Ventilation.

Ventilation-Controlled — Fire with limited ventilation in which the heat release rate or growth is limited by the amount of oxygen available to the fire. (NFPA® 921

Ventilation-Generated Pattern — Fire pattern that can vary widely in appearance and was created by ventilation introduced to a fire.

Venturi Meter — Device used to measure water velocity, when coupled with a differential manometer; consists essentially of a piece of pipe in which the cross-sectional area has been constricted.

Venturi Principle — Physical law stating that when a fluid, such as water or air, is forced under pressure through a restricted orifice, there is an increase in the velocity of the fluid passing through the orifice and a corresponding decrease in the pressure exerted against the sides of the constriction. Because the surrounding fluid is under greater (atmospheric) pressure, it is forced into the area of lower pressure. *Also known as* Venturi Effect.

Verdict — Decision made by a jury on matters submitted to them by a judge. *Also known as* Discovery of Truth.

Vermiculite — Expanded mica used for loose fill insulation and as aggregate in concrete.

Vertical Deadbolt — *See* Jimmy-Resistant Lock.

Vertical Shore — Shore that is applied vertically to support a horizontal load; for example, an unstable floor. *Also known as* Dead Shore.

Vertical Strut — Vertical load-bearing member that receives the load from the header in a vertical shoring system.

Vertical Ventilation — Ventilating at a point above the fire through existing or created openings and channeling the contaminated atmosphere vertically within the structure and out the top; done with openings in the roof, skylights, roof vents, or roof doors. *Also known as* Top Ventilation.

Vertical Zone — Area of a vessel between adjacent bulkheads.

Vertically Mounted Split-Case Pump — Centrifugal pump similar to the horizontal split-case, except that the shaft is oriented vertically and the driver is mounted on top of the pump.

Vertical-Shaft Turbine Pump — Fire pump originally designed to pump water from wells. Presently, it still has application when the water supply is from a nonpressurized source. Vertical-shaft pumps ordinarily have more than one impeller and are therefore multistage pumps.

Very High Frequency, High Band (VHF-HI) — Radio band containing very high frequencies (high range), ranging from 150 MHz to 176 MHz; characterized by relatively short radio waves and generally the most widely used band for public safety agencies.

Very High Frequency, Low Band (VHF-LO) — Radio band containing very low frequencies (low range), ranging from 30 MHz to 76 MHz; characterized by long radio waves and generally used by public safety agencies serving large geographic areas.

Vesicant — Agent that causes blistering. *See* Blister Agent.

Vessel — (1) General term for all craft capable of floating on water and larger than a rowboat. (2) Tank or container used to store a commodity that may or may not be pressurized.

VFR — *See* Visual Flight Rules.

Vibration Reduction (VR) — *See* Image Stabilization (IS).

Vibrator — Mechanical device used in placing concrete to make certain that it fills all voids.

Vicarious Liability — Liability imposed on one person for the conduct of another, based solely on the relationship between the two persons; indirect legal responsibility for the acts of another, such as the liability of an employer for acts of an employee. *See* Liability.

Victim — Person who suffers death, injury, or loss as a result of an act, circumstance, agency, or condition.

VIN — *See* Vehicle Identification Number.

Violation — Infringement of existing rules, codes, or laws. *See* Citation and Sanction.

Violent Rupture — Immediate release of chemical or mechanical energy caused by rapid cracking of the container.

Virga — Precipitation that evaporates before reaching the ground.

Virus — Simplest type of microorganism that can only replicate itself in the living cells of its hosts. Viruses are unaffected by antibiotics. *See* Bacteria and Rickettsia.

Viscosity — Liquid's thickness or ability to flow.

Viscous — Having a thick, sticky, adhesive consistency.

Visible Damage — Damage that is clearly evident by visual inspection, without the use of optical measuring devices.

Vision Statement — Description of the ultimate results of the work of the fire and life safety educators; the first step in developing goals, procedures, or policies for fire and life safety activities.

Visual Aid — *See* Instructional Materials.

Visual Approach — Approach to landing made by visual reference to the surface.

Visual Flight Rules (VFR) — Rules that govern the operation of an aircraft during visual flight.

Visual Inspection — Observation without the use of optical devices, except prescription lenses; may include physical and mechanical examination.

Visual Lead Time — Refers to driver/operators scanning far enough ahead of the apparatus to ensure that evasive action can be taken if it becomes necessary.

Vital Signs — Indicators of a patient's condition, as determined by temperature, pulse, respirations, and blood pressure.

Vitrified Clay Tile — Ceramic tile baked to become very hard and waterproof.

Void Space Nonstructural Entrapment — Situation in which victims are either trapped inside a collapsed structure, pinned by furniture or debris, or have no means of escape.

Volatile — Capable of changing into vapor quite readily at a fairly low temperature. *See* Vapor and Vaporization.

Volatile Organic Compounds (VOCs) — Organic chemicals that have high vapor pressures under normal conditions.

Volatility — Ability of a substance to vaporize easily at a relatively low temperature. *See* Vapor, Vaporization, and Volatile.

Volt — Basic unit of electrical potential; difference in potential (electromotive force) needed to create a current of one ampere through the resistance of one ohm. Most commonly abbreviated V, but may also be abbreviated E.

Voltage — Electrical force that causes a charge (electrons) to move through a conductor. Measured in volts (V). *Also known as* Electromotive Force (EMF). *See* Conductor.

Voltmeter — Device used for measuring existing voltage in an electrical system.

Volume Operation — *See* Parallel Operation.

Volunteer — Anyone inside or outside the fire department who helps with fire and life safety education programs.

Volunteer Fire Department — Organization of part-time firefighters who may receive monetary compensation for on-call time and/or fire fighting duty time.

Volunteer Firefighter — Active member of a fire department who may receive monetary compensation for on-call time and/or fire fighting duty time. *Also spelled* Volunteer Fire Fighter.

Volute — Spiral, divergent chamber of a centrifugal pump, in which the velocity energy given to water by the impeller blades is converted into pressure.

Vomiting Agent — Chemical warfare agent that causes violent, uncontrollable sneezing, cough, nausea, vomiting, and a general feeling of bodily discomfort. *See* Chemical Warfare Agent.

V-Pattern — Characteristic cone-shaped fire pattern left on a wall at or near its point of origin.

VRT — *See* Vehicle Rescue Technician.

V-Type Collapse — (1) Situation where heavy loads concentrated near the center of the floor, such as furniture and equipment, cause the floor to give way. A V-type collapse will result in void spaces near the walls. (2) Type of collapse void in which a floor section fails at the center and falls to the floor below; this results in two void spaces along the supporting outer wall.

Vulnerability — Degree to which an asset is exposed to damage; depends on its construction, contents, and economic value.

Vulnerability Assessment — (1) Act of predicting what could happen in the future within a given jurisdiction. (2) Extent of injury and damage that may result from a hazard emergency of a given intensity in a given area; should address effects of hazard emergencies on both existing and future-built environments.

W

Wagon — Special piece of fire apparatus that carries a large quantity of hose.

Wake Turbulence — Phenomena that result from the passage of an aircraft through the atmosphere; includes vortices, thrust stream turbulence, jet blast, jet wash, propeller wash, and rotor wash or downdraft.

Wale — Timber placed against sheeting planks in an excavation to keep the sheeting planks in place. *Also known as* Breast Timber.

Walk-Through — Initial assessment conducted by carefully walking through the scene to evaluate the situation, recognize potential evidence, and determine resources required; also, a final survey conducted to ensure the scene has been effectively and completely processed.

Wall — (1) Vertical component of a structure or compartment that is intended to enclose, divide, or protect a space. *See* Load-Bearing Wall and Non-Load-Bearing Wall. (2) Side of a trench from the lip to the floor. *Also known as* Face.

Wall Footing — Continuous strip of concrete that supports a wall.

Wall Hydrant — Water discharge protruding through and mounted on the wall of a building or pump house; is supplied by an interior fire pump and has one or more 2½-inch (6.3 cm) hose valves. Typically used for testing the fire pump.

Wall Ladder — Straight, single-section ladder.

Wall Post Indicator Valve (WPIV) — Similar to a post indicator valve (PIV) but mounted on the wall of the protected structure.

Wall Sprinkler — *See* Sidewall Sprinkler.

Wallboard — Fire-resistive building material that consists of highly compacted gypsum sandwiched between two layers of paper. *Also known as* Drywall, Plasterboard, or Sheetrock®.

Wallplate — In a raker shore or a flying shore, the continuous sheeting member placed immediately against the vertical surface that is being shored and that collects and distributes the load from the weakened wall.

Warded Lock — Simple type of mortise lock that requires the use of a skeleton key to open.

Warm Front — Leading edge of a warm air mass that displaces colder air. Its atmosphere is more stable than that of a cold front; winds associated with warm fronts are usually light, and mixing is limited.

Warm Gas Inhalator — Device that warms air for a hypothermia patient to breathe.

Warm Zone — Area between the hot and cold zones that usually contains the decontamination corridor; typically requires a lesser degree of personal protective equipment than the Hot Zone. *Also known as* Contamination Reduction Zone *or* Contamination Reduction Corridor.

Warning — Action which signifies that the event has been sighted or will likely occur soon.

Warning Devices — Audible or visual devices, such as flashing lights, sirens, horns, or bells, added to an emergency vehicle to gain the attention of drivers of other vehicles.

Warning Lights — Lights on the apparatus designed to attract the attention of other motorists.

Warp Yarn — Threads that run lengthwise in a woven fabric or hose.

Warrant — Written order from a judge or magistrate that authorizes an officer to make an arrest, search, or seizure, or to perform other acts in the interest of justice. Some states specify what evidence is needed for a warrant, as well as the type of magistrate who can issue certain types of warrants.

Wash Down — To flush spilled liquids from the roadway.

Waste Line — Secured hoseline that is used to handle excess water during a relay operation. *Also known as* Dump Line.

Watch — (1) Time during which a firefighter is assigned to the telecommunications center desk. (2) Period of duty for a crew member on a vessel. *See* Watch Officer. (3) Conditions have developed that make the event possible; watches are often issued for weather conditions such as tornados and floods.

Watch Desk — Communications desk of a fire station.

Watch Line — Charged hoseline that remains at the scene after the fire has been extinguished, with a detail of firefighters to stand guard against possible rekindling.

Watch Officer — Officer in charge of a watch; is responsible for the safe and proper navigation of the vessel during this time period. *Also known as* Officer of the Watch. *See* Watch.

Watchman — Employee assigned to patrol and guard a property against fire or theft.

Water Authority — Municipal authority responsible for the water supply system.

Water Curtain — Fan-shaped stream of water applied between a fire and an exposed surface to prevent it from igniting due to radiated heat; for example, the stream discharged from beneath an elevating platform to absorb radiant heat and protect the platform's occupants.

Water Department — Municipal authority responsible for the water supply system.

Water Distribution System — System designed to supply water for residential, commercial, industrial, and/or fire protection purposes; water is delivered through a network of piping and pressure-developing equipment.

Water Gel — Chemical solution that is gelled or partially solidified to make it easier to use or handle; for example, gelatin dynamite (gelignite).

Water Hammer — Force created by the rapid deceleration of water, causing a violent increase in pressure that can be powerful enough to rupture piping or damage fixtures. Generally results from closing a valve or nozzle too quickly.

Water Main — Principal pipe in a system of pipes for conveying water, especially one installed underground.

Water Mist Extinguisher — Fire extinguisher capable of discharging atomized water through a special applicator; pressurized water mist extinguishers use distilled water, whereas back-pump water mist extinguishers use ordinary water.

Water Motor Gong — Audible local alarm on an automatic sprinkler system; powered by a small water wheel that operates when water begins to flow in the system.

Water Retention — Foam's ability to retain its water content.

Water Shuttle Operation — Method of water supply by which mobile water supply apparatus continuously transport water between a fill site and the dump site located near the emergency scene.

Water Solubility — Ability of a liquid or solid to mix with or dissolve in water. *See* Dilution, Insoluble, Physical Properties, and Soluble.

Water Superintendent — Manager of the water department.

Water Supply — Any source of water available for use in fire fighting operations.

Water Supply Pumper — Pumper that takes water from a source and sends it to attack pumpers operating at the fire scene.

Water Tank — Water storage receptacle carried directly on the apparatus. NFPA® 1901 specifies that Class A pumpers must carry at least 500 gallons (2 000 L). *Also known as* Booster Tank.

Water Tender — Any ground vehicle capable of transporting large quantities of water. *Also known as* Tanker in some regions.

Water Thief — Any of a variety of hose appliances with one female inlet for 2½-inch (64 mm) or larger hose and with three gated outlets, usually two 1½-inch (38 mm) outlets and one 2½-inch (64 mm) outlet.

Water Tower — Aerial device primarily intended for deploying an elevated master stream; not generally intended for climbing operations. *Also known as* Elevating Master Stream Device.

Water Vacuum — Appliance designed to pick up water that is similar to a household vacuum cleaner.

Watercraft — Powered vehicles that travel over water.

Waterflow Alarm — Alarm-initiating device actuated by the movement (flow) of water within a pipe or chamber; most common installation is in the main water supply pipe of a sprinkler system. *Also known as* Sprinkler System Supervision System and Waterflow Detector.

Waterflow Detector — Detector that recognizes movement of water within the sprinkler or standpipe system; once movement is noted, the waterflow detector gives a local alarm and/or may transmit the alarm. *Also known as* Waterflow Device and Waterflow Indicator.

Waterflow Device —Initiating device that recognizes movement of water within the sprinkler or standpipe system. Once movement is noted, the waterflow device activates a local alarm and/or may transmit a signal to the FACU. *See* Waterflow Detector.

Waterflow Indicator — *See* Waterflow Detector.

Waterline — Level at which a vessel floats; line to which water raises on hull.

Water-Reactive Material — Substance, generally a flammable solid, that reacts when mixed with water or exposed to humid air. *See* Air-Reactive Material, Reactive Material, and Reactivity.

Watertight Bulkhead — Bulkhead (wall) strengthened and sealed to form a barrier against flooding. *See* Bulkhead (1).

Watertight Door — Door designed to keep out water; fitted to ensure integrity of the bulkheads (walls).

Watertight Transverse Bulkhead — Bulkhead (wall) that has no openings through it and extends from tank top to the main deck; designed to control flooding. *See* Bulkhead (1) and Watertight Bulkhead.

Waterway — Path through which water flows within a hose or pipe.

Watt (W) — The SI unit of power or rate of work equal to one joule per second (J/s).

Waybill — Shipping paper used by a railroad to indicate origin, destination, route, and product; a waybill for each car is carried by the conductor. *See* Consist and Shipping Papers.

Weapon of Mass Destruction (WMD) — Any weapon or device that is intended or has the capability to cause death or serious bodily injury to a significant number of people through the release, dissemination, or impact of toxic or poisonous chemicals or their precursors, a disease organism, or radiation or radioactivity; may include chemical, biological, radiological, nuclear, or explosive (CBRNE) type weapons.

Weaponize — To improve the ability of an agent to be delivered as an effective weapon; for example, reducing its particulate size in order to increase the likelihood of inhalation.

Wear Course — External covering on a roof that protects it from mechanical abrasion; the typical tar and gravel roof uses gravel as the wear course.

Weather Deck — All parts of the main deck and decks above that are exposed to the weather. *See* Deck.

Web — Wide vertical part of a beam between the flanges.

Web Conferencing — Meeting service that combines teleconferencing with an Internet-based sharing service, enabling users to communicate in real time while viewing and interacting with a computer-based presentation.

Web Member — Secondary members of a truss that are contained between the chords.

Webbing — Device used for creating anchors and lashings, or for packaging patients and rescuers; typically constructed from the same material as synthetic rope.

Wedge — Angle-cut piece of timber used to snug up loads, fill in voids, or change the angle of thrust.

Wedge — Shims used in pairs to provide close contact between a shoring system and the supported load. REVISED

Weed Abatement — *See* Fuel Management.

Weep Hole — Small holes in a masonry veneer wall that release accumulated water to the exterior.

Weeping — Giving off or leaking fluid slowly; for example, couplings leaking at the point of attachment.

Weft Yarn — *See* Filler Yarn.

Weld — Joint created between two metal surfaces when they are heated and the two metals run together.

Weldment — Structure formed by welding together two or more pieces.

Wellness Program — Ongoing program that provides information, education, and counselling to fire service members on topics such as good nutrition, tobacco cessation, injury prevention, and substance abuse.

Western Frame Construction — *See* Platform Frame Construction.

Wet Chemical Fire-Extinguishing System — *See* Wet Chemical System.

Wet Chemical System — Extinguishing system that uses a wet chemical solution as the primary extinguishing agent; usually installed in range hoods and associated ducting where grease may accumulate. *Also known as* Wet Chemical Fire-Extinguishing System.

Wet Down — (1) To wet down or dampen debris after fire has been controlled but not completely extinguished. (2) Ceremony used in some departments to celebrate the acquisition of a new piece of apparatus.

Wet Line — Line of water or water and chemical retardant sprayed along the ground that serves as a temporary fire-stop or containment line from which to ignite or stop a low-intensity fire.

Wet Standpipe System — Standpipe system that has water supply valves open and maintains water in the system at all times. *See* Dry Standpipe System and Standpipe System.

Wet Water — Wetting agent that is introduced to water to reduce its surface tension and improve its penetration qualities. *See* Penetrant.

Wet-Barrel Hydrant — Fire hydrant that has water all the way up to the discharge outlets; may have separate valves for each discharge or one valve for all the discharges. This type of hydrant is only used in areas where there is no danger of freezing weather conditions.

Wet-Pipe Sprinkler System — Fire-suppression system that is built into a structure or site; piping contains either water or foam solution continuously; activation of a sprinkler causes the extinguishing agent to flow from the open sprinkler. *See* Deluge Sprinkler System, Dry-Pipe Sprinkler System, and Preaction Sprinkler System.

Wetsuit — Protective outwear worn during water-based rescue operations; water permeable: allows water between the garment and the rescuer's skin to provide thermal insulation.

Wetting Agent — Chemical solution or additive that reduces the surface tension of water (producing wet water), causing it to spread and penetrate more effectively; may also produce foam through mechanical means. Detergent is a mild form of wetting agent. *See* Penetrant.

Wharf — Place for berthing ships along or at an angle from a shore; constructed by extending bulkheads out from the shore and back-filling the enclosed area, creating a flat surface for loading and unloading vessels. *See* Pier.

Wheel Block — *See* Wheel Chock.

Wheel Chock — Block placed against the outer curve of a tire to prevent the apparatus from rolling; can be wooden, plastic, or metal. *Also known as* Wheel Block.

Wheelbase — Distance between a vehicle's front and rear axles.

Whirlwind — Small rotating windstorm containing sand or dust. *Also known as* Dust Devil.

WHMIS — *See* Workplace Hazardous Materials Information System.

Whole-Part-Whole — Method of sequencing instruction; information begins with the complete picture or demonstration of a skill or with an overview of all information, progresses to provide a breakdown of all steps or details on segments of information, and returns to a complete demonstration or overview in a review or summary.

Wicking — Pattern that occurs when quantities of the ignitable liquid are absorbed by the material onto which the liquid is poured.

Wiki — Website that allows users to update, edit, or comment on the original content using their own Internet browser; allows for the rapid creation and deployment of websites and collaborative work on documents.

Wild Line — Uncontrolled hoseline and nozzle or butt that thrash about from the reaction of highly pressurized flowing water.

Wildfire — *See* Wildland Fire.

Wildfire Mop-Up — *See* Wildfire Overhaul.

Wildfire Overhaul — Operations involving extinguishing hot spots and hidden fires after the main body of fire has been knocked down. *Also known as* Wildfire Mop-Up.

Wildland Fire — Unplanned, unwanted, and uncontrolled fire in vegetative fuels such as grass, brush, or timberland involving uncultivated lands; requires suppression action and may threaten structures or other improvements. *Also known as* Ground Cover Fire, Ground Fire, Natural Cover Fire, Vegetation Fire, or Wildfire. *See* Structure Fire and Wildland/Urban Interface.

Wildland Fire Apparatus — Fire department apparatus designed specifically for fighting wildland fires; generally a light, mobile vehicle with limited pumping and water capability for off-road operations. *Also known as* Booster Apparatus, Brush Apparatus, Brush Patrol, Brush Pumper, and Field Unit.

Wildland Firefighter — Person trained to function safely as a member of a wildland fire-suppression crew. NFPA® identifies four levels of progression, Wildland Firefighter I through Wildland Firefighter IV; each level denotes a higher level of capability, supervision, or management.

Wildland/Urban Interface — Line, area, or zone where an undeveloped wildland area meets a human development area. *Also known as* Urban/Wildland Interface.

Winch — (1) Pulling tool that consists of a length of steel chain or cable wrapped around a motor-driven drum; most commonly attached to the front or rear of a vehicle. (2) On a ship, a stationary motor-driven hoisting machine with a vertical drum around which a rope or chain winds as a load is lifted; a special form of this type of winch using a horizontal drum is called a *windlass*.

Wind — Horizontal movement of air relative to the surface of the earth. *See* Crosswind, Downwind, and Headwind.

Wind Shakes — Damage done to timber by repeated flexing in the wind.

Wind Sock — Cone-shaped cloth sock at airports that is used to indicate wind direction and, to some extent, wind velocity.

Wind Tee — T-shaped indicator mounted horizontally on a pivot pole to swing freely in the wind; used as a wind direction indicator or landing direction indicator. May also be in the shape of a tetrahedron.

Winders — Radiating or wedge-shaped treads at the turn of a stairway.

Windpipe — Trachea.

Windshield Survey — Process of taking a moment to get a 360-degree view of the incident before exiting the vehicle.

Windward Side — (1) Unprotected side of the building the wind is striking. (2) Side or direction from which the wind is blowing.

Wing Tank — Tank located well outboard, next to the side shell plating of a vessel; often a continuation of the double bottom up the sides to a deck.

Wipe Sample — Sample collected by wiping an area's surface.

Wire Cutters — Tool with approved, insulated handles to cut wire.

Wire Nut — Approved, solderless, nonconductive connector used to connect one wire to another.

Wired Glass — Flat sheet glass containing an embedded wire mesh that increases its resistance to breakage and penetration; installed in exterior doors, windows, and skylights to increase interior illumination without compromising fire resistance and security. When the glass is exposed to the heat of a fire, it will typically break or crack due to thermal stresses; the wires give the glass dimensional stability, permitting it to act as a barrier to the fire even when the glass has failed. Can be either transparent or translucent.

Wired Telegraph Circuit Box — Alarm system operated by pressing a lever in the alarm box that starts a wound-spring mechanism; the rotating mechanism transmits a code by opening and closing the circuit.

Witness — Person called upon to provide factual testimony before a judge or jury.

WMD — *See* Weapon of Mass Destruction.

Wood Grain — Stratification of wood fibers in a piece of wood.

Wood Planer — Device used to smooth the surface of wooden boards; commonly found in woodworking shops.

Wooden Short Board — Device used as part of spinal immobilization.

Worker — *See* Working Fire.

Working End — End of the rope used to tie a knot. *Also known as* Bitter End, Loose End *or* Tail End.

Working Fire — Term used to describe a fire at which considerable fire fighting activity will be required. *Also known as* Worker.

Working Length — Length of a non-self-supporting ladder; measured along the beams from the butt to the point of bearing at the top.

Working Load — The manufacturer's recommended maximum load for a rope or other system component. The working load for components of a rope system supporting victims or responders is determined by dividing the minimum breaking strength by a safety factor of 15. For a lifeline with a breaking strength of around 9,000 pounds (4 090 kg), the working load is 600 pounds (272 kg).

Workplace Hazardous Materials Information System (WHMIS) — Canadian law requiring that hazardous products be appropriately labeled and marked.

Worksheet — Activity sheet that lists tasks to accomplish, guides activities, and enables learners to apply cognitive information in order to practice and develop skills.

Workshop — Beneficial to the intervention program. Workshops are generally small, participatory, informal, and generally short. Topics for each workshop may vary according to need.

Woven-Jacket Hose — Fire hose constructed with one or two outer jackets woven on looms from cotton or synthetic fibers.

WPIV — *See* Wall Post Indicator Valve.

Wrapped Hose — Nonwoven rubber hose manufactured by wrapping rubber-impregnated woven fabric around a rubber tube and encasing it in a rubber cover.

Wrecker — Vehicle that is usually equipped with a small crane, winch, or tilting bed assembly, and is used to transport damaged vehicles; the size and type of the wrecker depends on the type and size of the vehicle to be moved. *Also known as* Tow Truck.

Wristlet — Portion of the coat that prevents fire or water from entering the sleeve.

Written Evaluation Form — Preferred method for evaluation. At the end of the training session, participants are asked to provide input in writing on the quality of the training, the trainer, the location, the materials presented, the training methodologies, and suggestions for improvements for future training sessions.

Written Narrative — Chronicles a timeline of important events (past to present) relevant to the juvenile's firesetting problem. The narrative includes events that brought the juvenile into the program, a profile of the firesetting problem, intervention(s) performed, and referrals made.

Wye — Hose appliance with one female inlet and multiple male outlets; the outlets are usually smaller than the inlet. Outlets are also usually gated.

Wythe — Single vertical row of multiple rows of masonry units in a wall, usually brick. *See* Course and Header Course.

X

Xiphoid Process — Flexible cartilage at the lower tip of the sternum.

X-Ray — High-energy photon produced by the interaction of charged particles with matter.

X-Type — Type of gypsum wallboard formulated by adding noncombustible fibers to the gypsum, which is required to achieve fire resistance as designated in ASTM C 36.

Y

Yard Hydrant — *See* Private Hydrant.

Y-Branch — Plumbing drainage fitting with a branch or branches that extend at a 45-degree angle.

Z

Zero-Base Budget — Budgetary system in which each department or program theoretically terminates at the end of the budget cycle, and must justify its existence in order to receive funding in the next budget.

Zero Mechanical State — State a machine is in when all its power sources are neutralized and all its parts are stabilized.

Zoning Commission — Local agency responsible for managing land use by dividing the jurisdiction into zones, in which only certain uses, such as residential, commercial, or manufacturing, are allowed.

Zoom Lens — Lens designed to extend focal length; used for taking close-up pictures from greater distances.

Index

Courtesy of Ron Jeffers.

Index

A

Accountability Officer, 342
Ackerman, Peter, 100
ADA (Americans with Disabilities Act), 160
Administration, 59-64
    assistant/deputy chief, 62-63
    Fire Chief/Chief of Department, 63-64
    grants administrator, 59
    human resources/personnel services, 59
    ICS Finance/Administrative Unit, 337
    offices and buildings, 288
    Public Information Officer, 61-62
    Safety Officer, 60-61
Aerial apparatus, 265-269
    aerial ladder apparatus, 266
    aerial ladder platform apparatus, 266-267
    articulating aerial platform apparatus, 268
    defined, 265
    telescoping aerial platform apparatus, 268
    water towers, 268-269
Aerial ladder apparatus, 266, 268
Aerial ladder platform apparatus, 266-267
AFG (Assistance to Firefighters Grants), 59, 330
AFL-CIO (American Federation of Labor-Congress of Industrial Organizations), 134
AHJ (authority having jurisdiction), 160
Air pack, 100
Air purifying respirator (APR), 280, 297
Aircraft, fire fighting, 278
Aircraft hanger fire hazards, 150
Aircraft rescue and fire fighting (ARFF)
    airport fire departments, 21
    apparatus, 273-274
    as federal fire department, 21
Airport fire departments, 21
Airport firefighter, 64-65
Alarm and communications systems, 103-105
    mobile data communications systems, 105
    pagers/personal alerting systems, 105
    radio communications, 104
    telegraph alarm systems, 103
    telephone systems, 104
    town crier, 103
    watchmen and watchtowers, 103
Alarm boxes, 103
Alarm systems. *See* Fire detection, alarm, and suppression systems
Ambient temperature, 239
Ambulances, 271-272
American Association of Railroads Bureau of Explosives (BOE), 139
American Council on Education, 125
American Federation of Labor-Congress of Industrial Organizations (AFL-CIO), 134
American National Standards Institute (ANSI), 131
American Red Cross, 118
American Trucking Association (ATA), 139
Americans with Disabilities Act (ADA), 160
Animals in the fire service, 96
ANSI (American National Standards Institute), 131
APCO International Minimum Training Standards for Public Safety Telecommunicators (APCO ANS 3.103.1-2010), 73
Apparatus
    aerial apparatus, 265-269
        aerial ladder apparatus, 266
        aerial ladder platform apparatus, 266-267
        articulating aerial platform apparatus, 268
        defined, 265
        telescoping aerial platform apparatus, 268
        water towers, 268-269
    aircraft rescue and fire fighting apparatus, 273-274
    commercial fire apparatus, 262
    custom fire apparatus, 262
    defined, 260
    driver/operator, 53-54
    engine (pumper), 260-262
    fire boats, 276-277
    fire engine development, 85-86
    fire fighting aircraft, 278
    fire service ambulances, 271-272
    gasoline and diesel powered, 98-99
    hazardous materials response unit, 274-275
    maintenance of, 53
    maintenance personnel, 74-75
    mobile air supply unit, 275
    Mobile Command Post, 276
    mobile fire investigation unit, 277
    mobile water supply apparatus, 264
    power and light unit, 277
    quintuple apparatus, 269
    rescue apparatus, 269-271
    search and rescue boats, 276-277
    smaller fire apparatus, 263
    special units, 278-279
    wildland fire apparatus, 264-265
Applications, filing, 35
APR (air purifying), 280
APRs (air purifying respirator), 297
Arched roof, 226, 227
Arcing, 190
ARFF. *See* Aircraft rescue and fire fighting (ARFF)
Army Corps of Engineers, 123
Arson
    defined, 169
    International Association of Arson Investigators, 133, 169
    investigations, 169-170
    search warrant, 177
Articulating aerial platform apparatus, 268
Assistance to Firefighters Grants (AFG), 59, 330
Assistant Chief, 62-63
ATA (American Trucking Association), 139
Atmospheric pressure, 184
Augustus, Caesar, 81
Authority having jurisdiction (AHJ), 160
Autoignition, 189
Automatic aid, 326, 327-328

Automatic alarm systems, 243–244
Automatic sprinkler systems, 244–253
    applications, 250–251
    defined, 244
    failure or incomplete operation, 245
    Fire Sprinkler Initiative, 251
    history of, 245
    life safety, 246
    reliability of, 238
    sprinkler position, 246–249
    standpipe systems, 251–253
Auxiliary services, 244

B

Backdraft, 202, 203–204
Balloon-frame construction, 219–220
Barometer, 184
Barrel arched roof, 227
BATFE (Bureau of Alcohol, Tobacco, Firearms, and Explosives), 123
Battalion Chief, 55–57
Bed key, 90
Behavior of firefighters, 22–23
Beverly Hills Supper Club Fire, Southgate, Kentucky (1977), 107
BFRL (Building and Fire Research Laboratory), 126
Black Chief Officers Committee, 133
BLM (Bureau of Land Management), 129
BOCA (Building Officials and Code Administrators), 134–135
BOE (American Association of Railroads Bureau of Explosives), 139
Bomb squad, 120
Bond sales, 329
Booster hose, 261
Booster reel, 261
Booster tank, 261
Boots, 26
Borough fire departments, 15–16
Boston, Mutual Fire Society, 90
Bowstring roof, 227
Branch, defined, 339
Branch Director, 339
Breathing air system on aerial devices, 268
Breathing apparatus, 280–281. *See also* Self-contained breathing apparatus (SCBA)
Britain's first fire brigade, 84–85
British thermal unit (BTU), 188
Brush trucks, 265
BTU (British thermal unit), 188
Bucket brigade, 86, 87
Budget constraints of fire departments, 330
Building and Fire Research Laboratory (BFRL), 126
Building codes
    Building Officials and Code Administrators, 134–135
    common codes, 159
    consistent, 160
    defined, 159, 214
    effects on construction, 160–161
        building design, 160
        local safety amendments, 160–161
    enforcement of, 115, 153, 154, 160
    International Building Code, 159, 214
    International Code Council, 134–135, 159
    model codes, 160
    National Building Code of Canada, 159
    Southern Building Code Congress International, 135
Building construction, 213–232
    balloon-frame, 219–220
    building codes. *See* Building codes
    fire resistance, 214
    firefighter hazards, 225–232
        building collapse, 228–229
        construction, renovation, demolition hazards, 231–232
        fire loading, 225–226
        furnishings and finishes, 226
        large, open spaces, 228
        lightweight and truss hazards, 229–230
        new building technologies, 231
        roof coverings, 226–227
        wooden floors and ceilings, 228
    inspections, 55
    materials, 220–224
        cast iron, 223
        glass/fiberglass, 224
        gypsum, 224
        masonry, 222–223
        reinforced concrete, 224
        steel, 223–224
        wood, 220–222
    openings in a structure, 216
    platform framing, 220
    Type I construction (fire-resistive), 214, 215, 216
    Type II construction (noncombustible or limited-combustible), 214, 215, 217
    Type III construction (ordinary), 214, 215, 217–218
    Type IV construction (heavy timber or mill), 214, 215, 218–219
    Type V construction (wood-frame), 214, 215, 219
Building Construction and Safety Code (NFPA® 5000), 159
Building department, 115
Building Officials and Code Administrators (BOCA), 134–135
Bureau of Alcohol, Tobacco, Firearms, and Explosives (BATFE), 123
Bureau of Explosives (BOE), 139
Bureau of Land Management (BLM), 129
Burn building, 290, 291

C

Cadet program, 35
CAFC (Canadian Association of Fire Chiefs), 131
Camera
    infrared, 273
    thermal, 341
Canada
    Council of Canadian Fire Marshals and Fire Commissioners, 132
    Fire Prevention Canada, 132
    National Building Code of Canada, 159
    National Research Council of Canada, 159
    Public Safety Canada, 127
    Transport Canada, 128
    Underwriters Laboratories of Canada, 139, 245
Canadian Association of Fire Chiefs (CAFC), 131
Canadian Centre for Emergency Preparedness (CCEP), 131
Canadian Fallen Firefighters Foundation (CFFF), 131
Canadian Fire Alarm Association (CFAA), 131

Canadian Fire Safety Association (CFSA), 131
Canadian Medical Association, 126
Canadian Standards Association (CSA), 131
Canadian Volunteer Fire Services Association (CVFSA), 132
Candidate Physical Ability Test (CPAT), 38
Cantilever fire wall, 223
Carbon monoxide (CO)
    detectors, 159, 168, 205, 242-243
    as product of combustion, 205
Career in fire and emergency services, 11-41
    career fire department, 18
    career firefighter, 18
    emergency response organizations, 13
    fire and emergency services culture, 12-13
    fire protection agencies, 13
    firefighter selection, 34-40
        application, 35
        interviews, 39
        physical ability test, 38
        preparation, 35
        probationary period, 39
        recruitment, 35
        volunteers, 39-40
        written examination, 36-38
    firefighters as public figures, 22-27
        conduct, 23-24
        firefighter image, 22
        human relations and customer service, 24-25
        physical fitness, 27
        responsible behaviors, 22-23
        teamwork, 24
        uniform and dress, 25-27
    hours of duty, 40
    pay and salaries, 40
    promotion, 40
    public fire departments, 17-20
        career departments, 18
        combination departments, 19
        paid on-call departments, 19-20
        profile, 17
        public safety departments, 20
        volunteer, 17, 18-19
    retirement, 41
    training and education, 28-34
        certification/credentialing, 33-34
        classroom study, 28
        colleges and universities, 31
        departmental training, 28-29
        drills, 28
        life fire training, 29
        National Fire Academy, 31-32
        non-traditional, 32-33
        regional/county/state/provincial training programs, 29-30
        training vs. education, 28
        value of education, 30
    types of fire departments, 13-21
        airport fire departments, 21
        commercial fire protection services, 21
        county, parish, borough, 15-16
        federal fire departments, 21
        fire districts and fire protection districts, 16-17
        industrial fire departments, 20
        jurisdiction, 14
        municipal, 14-15
        public fire departments, 17-20
Cargo Tank Specialty, 65
Cast iron, 223
CCEP (Canadian Centre for Emergency Preparedness), 131
CDC (Centers for Disease Control and Prevention), 126
Ceiling jet, 198
Ceilings, combustible, 228
Celsius scale, 188, 189
Center for Public Safety Excellence (CPSE), 34
Centers for Disease Control and Prevention (CDC), 126
CERT (Community Emergency Response Team), 118
Certification of firefighters, 33
CFAA (Canadian Fire Alarm Association), 131
CFFF (Canadian Fallen Firefighters Foundation), 131
CFR 1910.156, fire brigade regulations, 20
CFSA (Canadian Fire Safety Association), 131
CFSI (Congressional Fire Services Institute), 122
Chain of command, 321
Channing, William Francis, 103
Chemical engines, 97
Chemical flame inhibition, 209
Chemical heat energy, 148, 189-190
Chemical pellet sprinkler, 246, 247
Chemical reaction, 186
Chemical Safety Hazard Investigation Board (CSB), 123
Chief of Department, 63-64
Chief of Training/Drillmaster, 58
Chimney maintenance, 87
Church bells, ringing as warning, 82
Cider mill, 93, 94
CISM (critical incident stress management), 344-345
Civil Support Team (CST), 125-126
Class I standpipe systems, 252
Class II standpipe systems, 252, 253
Class III standpipe systems, 252, 253
Class A fire extinguishers, 102
Class A fires, 206
Class B fires, 206
Class C fires, 206
Class D fires, 207
Class K fires, 207
Classroom study, 28, 58, 289, 290
Clean Air Act Amendments of 1990, 123
Closed-circuit SCBA, 280, 300
Clothing. *See also* Uniform and dress of firefighters
    hazardous materials, 295-296
    improvements, 99
    personal protective clothing, 280, 294
Coats
    closure system, 26
    coat collars, 26
    loops, 26
Cocoanut Grove Nightclub Fire, Boston, Massachusetts (1942), 106
Code of ethics for firefighters, 22-23
Codes. *See* Building codes
Coffee mill, 93
Collapse of buildings, 228-229
College education of firefighters, 31
Combination detector, 242

Combination fire departments, 19
Combustible materials storage as fire hazard, 149
Combustion
   chemical reaction, 186
   defined, 186
   fire development stages, 196–202
      decay, 201
      factors affecting, 197
      flashover, 199–200
      fully developed, 200–201
      growth, 199
      incipient, 198
      rapid fire development, 201–202
   fire tetrahedron, 187–196
      fire triangle, 187, 188
      fuel, 192–195
      heat and temperature, 188–189
      heat energy sources, 189–190
      oxygen, 195–196
      self-sustained chemical chain reaction, 196, 197
      transmission of heat, 190–192
   potential energy of fuel, 185
   products of, 191, 205, 206
Command, defined, 323
Command staff, 338
Commercial fire protection services, 21
Commercial Fire Rating Schedule, 132
Commercial occupancy fire hazards, 150
Commission form of government, 325
Commission on Professional Credentialing (CPC), 34
Communications, formal, 334–335
Communications systems. *See* Alarm and communications systems
Community Emergency Response Team (CERT), 118
Company, defined, 47. *See also* Fire company
Company officer, 54–55
Composite Fire Department, 19
Computer-assisted preincident planning, 157
Computer-based electronic accountability, 343
Concrete construction, 224
Conduct of firefighters, 23–24, 174–175
Conduction, 191
Conflagration, 83
Congressional Fire Services Institute (CFSI), 122
Construction alteration hazards, 173
Consumer Product Safety Commission (CPSC), 123
Control valves
   outside stem and yoke, 247, 248
   placement, 246–247, 248
   post indicator valve, 247, 248
   post indicator valve assembly, 249
   purpose of, 246
   wall post indicator valve, 248, 249
Convection, 191, 192
Corps of Vigiles, 81–82
Council (Board) Manager form of government, 325
Council of Canadian Fire Marshals and Fire Commissioners, 132
County fire departments, 15–16
County training programs, 29–30
CPAT (Candidate Physical Ability Test), 38
CPC (Commission on Professional Credentialing), 34
CPSC (Consumer Product Safety Commission), 123

CPSE (Center for Public Safety Excellence), 34
Crane, Alanson, 102
"Crawl Low Under Smoke," 165
Credentialing of firefighters, 33
Critical incident stress management (CISM), 344–345
CSA (Canadian Standards Association), 131
CSB (Chemical Safety Hazard Investigation Board), 123
CST (Civil Support Team), 125–126
Ctesibius, 82
Culture of fire and emergency services, 12–13
Customary System, 188
Customer service function of firefighters, 24–25
CVFSA (Canadian Volunteer Fire Services Association), 132

D

Dalmatian dogs, 96
DDP (Degrees at a Distance Program), 32
Deaths
   from fires, 146
      line of duty deaths, 108
Decay fire development stage, 201
Degrees at a Distance Program (DDP), 32
Deluge sprinkler system, 251
Demographics of fire risk, 153
Demolition hazards, 173, 231–232
Density of matter, 184–185
Department of Agriculture, 130
Department of Defense, 20
Department of Homeland Security (DHS)
   divisions, 128–129
   Emergency Management Institute, 123
   Federal Emergency Management Agency, 124. *See also* Federal Emergency Management Agency (FEMA)
   National Fire Academy, 31
   purpose of, 128–129
   USFA, 31–32, 129
Department of Labor (DOL), 127, 129
Department of the Interior (DOI), 129
Department of Transportation (DOT)
   National Highway Traffic Safety Administration, 126
   purpose of, 129
   responsibilities, 120
   subdivisions, 129
Deputy Chief, 62–63
A Description of the Newly Invented and Patented Fire Engines with Flexible Hose, and Their Manner of Extinguishing Fires, 85
Detection systems. *See* Fire detection, alarm, and suppression systems
DHS. *See* Department of Homeland Security (DHS)
Diesel powered equipment, 98–99
Disciplinary procedures, 334
Discipline, 325
Dispatchers, 72–73
Distance learning, 32
District Chief, 55–57
Division, defined, 339
Division of labor, 324
Division Supervisor, 339
DOE (United States Department of Energy), 128
DOI (Department of the Interior), 129
DOL (Department of Labor), 127, 129
DOT. *See* Department of Transportation (DOT)

Double-piston pump, 91
Dress. *See* Uniform and dress of firefighters
Drill tower, 290, 291
Drillmaster, 58
Drills, training, 28
Dry-pipe sprinkler system, 250

E

EDITH (exit drills in the home), 159
Education
    fire and life safety, 165-168
        "Crawl Low Under Smoke," 165
        educator roles, 166
        fire station tours, 168
        presenting, 165-166
        Risk Watch®, 165
        smoke and carbon monoxide alarms, 168
        "Stop, Drop, and Roll," 165, 167
    fire and life safety education, 153
    fire prevention, 146
    of firefighters. *See also* Training and education of firefighters
        colleges and universities, 31
        training vs. education, 28
        value of education, 30
Electrical heat energy, 148, 190
Elevated master stream, 50
Elevating platforms, 268
Emergency Management Institute (EMI), 123
Emergency medical services (EMS)
    Emergency Medical Technician, 70-71
    EMS Chief/officer, 71-72
    first responder, 70
    levels of emergency medical care, 69
    mission, 126
    paramedics, 71
    personnel, 69-72
Emergency Medical Technician (EMT), 70-71
Emergency Planning and Community Right to Know Act (EPCRA), 160
Emergency Services District, 328
EMI (Emergency Management Institute), 123
EMS. *See* Emergency medical services (EMS)
EMT (Emergency Medical Technician), 70-71
Endothermic reaction, 185
Energy
    chemical heat energy, 148, 189-190
    electrical heat energy, 148, 190
    heat energy sources, 189-190
    mass and, 185
    mechanical heat energy, 148, 190
    nuclear heat energy, 148
    potential energy, 185
    solar energy panels, 231
    U.S. Department of Energy, 128
Enforcement for fire prevention, 146
Engine (pumper), 260-262
Engine company duties, 48
Engineer, fire protection, 162-165
Engineering for fire prevention, 146
Enterprise funds, 329
Environmental concerns of fires, 146
Environmental Protection Agency (EPA)
    purpose of, 124
    state agencies, 120
EPA. *See* Environmental Protection Agency (EPA)
EPCRA (Emergency Planning and Community Right to Know Act), 160
Equipment
    carried on rescue apparatus, 270
    fire hazards, 149
    gasoline and diesel powered, 98-99
    maintenance personnel, 74-75
    outdated, 330, 331
    personal protective equipment, 25, 280, 288
    storage of, 287-288
    uses for, 281
Ericsson, John, 95
Everyone Goes Home Program, 137
Evidence protection and preservation, 177-178
Exit drills in the home (EDITH), 159
Exit interview, 157, 158
Exothermic reaction, 185
Explorer Scout program, 35
Extinguishing techniques, history of, 82
Extinguishment, fire, 207-209

F

FAA (Federal Aviation Administration), 124
Facilities
    administrative offices and buildings, 288
    fire stations. *See* Fire stations
    maintenance of, 292
    maintenance personnel, 74
    surveys, 155-159
        conducting the survey, 156-157
        exit interview, 157, 158
        maps and sketch making, 157
        personnel requirements, 155
        residential safety survey, 158-159
        scheduling, 155
    telecommunication centers, 289
    training centers, 289-292
        burn building, 290, 291
        drill tower, 290, 291
        smokehouse, 290, 291
        training pads, 292
Fahrenheit scale, 188, 189
Familia Publicia, 81
Fayol, Henri, 100
FCC (Federal Communications Commission), 104
FDC (Fire Department Connection), 246, 251
Federal Aviation Administration (FAA), 124
Federal Communications Commission (FCC), 104
Federal Emergency Management Agency (FEMA)
    agencies included, 129
    Assistance to Firefighters Grants, 59, 330
    Emergency Management Institute, 123
    programs, 124
    purpose of, 124
    SAFER grant, 330
Federal fire departments, 21
Federal Fire Prevention and Control Act of 1974, 125, 129
Federal Specification KKK-A-1822, 271
FEMA. *See* Federal Emergency Management Agency (FEMA)

FESHE (Fire and Emergency Services Higher Education) Model, 31
Fiberglass used in construction, 224
Finance/Administrative Unit, 337
Fire
    Class A fires, 206
    Class B fires, 206
    Class C fires, 206
    Class D fires, 207
    Class K fires, 207
    color of flames, 172, 173
    development stages, 196–202
        decay, 201
        factors affecting, 197–198
        flashover, 199–200
        fully developed, 200–201
        growth, 199
        incipient, 198
        rapid fire development, 201–202
    extinguishment theory, 207–209
        chemical flame inhibition, 209
        fuel removal, 208
        oxygen exclusion, 209
        temperature reduction, 208
    factors for starting, 170
    observations by emergency responders, 170–173
    patterns, 172
Fire alarm bell, 238
Fire Alarm Interior Levels I and II, 135
Fire Alarm Municipal Levels I and II, 135
Fire alarm speaker, 238
Fire and Emergency Services Higher Education (FESHE) Model, 31
Fire and life safety education, 165–168
    "Crawl Low Under Smoke," 165
    educator roles, 166
    fire station tours, 168
    presenting, 165–166
    Risk Watch®, 165
    smoke and carbon monoxide alarms, 168
    "Stop, Drop, and Roll," 165, 167
Fire apparatus driver/operator duties, 53–54
Fire boats, 276–277
Fire brigade, 85
Fire Chief, 63–64, 121
Fire Code (NFPA® 1), 159
Fire commission, 119
Fire company, 47–51
    defined, 47
    engine company, 48
    ladder (truck) company, 48, 50
    organizational chart, 49
    rescue/squad company, 51
Fire department apparatus. *See* Apparatus
Fire Department Connection (FDC), 246, 251
Fire Department Pumper, 260
Fire departments
    airport fire departments, 21
    commercial fire protection services, 21
    county, parish, borough, 15–16
    federal fire departments, 21
    fire districts and fire protection districts, 16–17, 326
    industrial fire departments, 20
    jurisdiction, 14
    municipal, 14–15
    organizational chart, 49
    public fire departments, 17–20
        career departments, 18
        combination departments, 19
        paid on-call departments, 19–20
        public safety departments, 20
        volunteer, 17, 18–19
Fire detection, alarm, and suppression systems, 237–254
    automatic alarm systems, 243–244
    automatic sprinkler systems, 244–253
        applications, 250–251
        defined, 244
        failure or incomplete operation, 245
        history of, 245
        life safety, 246
        reliability of, 238
        sprinkler position, 246–249
        standpipe systems, 251–253
    auxiliary services, 244
    firefighter observations at fires, 173
    heat detectors, 239–240
        fixed-temperature heat detectors, 239
        rate-of-rise heat detectors, 239–240
    local alarm systems, 238
    manual pull station, 238
    protected premises alarm systems, 238
    reasons for installing systems, 237–238
    smoke detectors and alarms, 240–243
        carbon monoxide detectors, 242–243
        combination detectors, 242
        flame detectors, 242
        ionization, 241
        photoelectric, 240–241
        tamper-proof, 243
    sprinklers. *See* Sprinklers
Fire districts, 16–17, 326
Fire engine development, 92–94
    cider mill, 93, 94
    coffee mill, 93
    hand engine improvements, 94
    history of, 85–86, 92
    Hunneman hand tub, 93, 94
    New York style hand engines, 92–93
    Philadelphia style engine, 93
    piano-type engines, 93, 94
Fire extinguishers
    history of, 102
    modern fire extinguishers, 102
Fire hall, 287
Fire hazards, 147–152
    common hazards, 148–149
    defined, 148
    fire elements, 147
    fuel hazards, 147–148
    heat sources, 148
    multiple hazards, 152
    pyrotechnics, 151
    special considerations, 149–150
    target hazard properties, 151
Fire hook, 83

Fire hose and tools, 282, 286
Fire inspector
    defined, 160
    roles, 161–162
Fire insurance companies, 91–92
Fire investigation. *See* Investigation
Fire investigator, 169–170
Fire loading, 225–226
Fire mark, 85
Fire marshal, 169
Fire marshals association, state, 122
Fire plug, 89
Fire prevention, 145–154
    components, 153–154
        fire and life safety education, 153
        fire investigation, 154
        surveys, inspection, and code enforcement, 153, 154
    defined, 155
    environmental concerns, 146
    fire department roles, 145
    fire hazards, 147–152
        common hazards, 148–149
        defined, 148
        fuel hazards, 147–148
        heat sources, 148
        multiple hazards, 152
        pyrotechnics, 151
        special considerations, 149–150
        target hazard properties, 151
    loss of life and property, 146
    monetary loss from, 146–147
    need for, 146–147
    risk analysis, 152–153
        demographics, 153
        preincident survey, 152
        trends, 152–153
    Three E's, 146
Fire Prevention Canada, 132
Fire Prevention Week, 137
Fire protection districts, 17
Fire protection engineer (FPE), 138, 162–165
Fire Protection Handbook, 137
Fire pump
    defined, 82
    double-piston pump, 91
    history of, 82
    Siphona Pump, 82
Fire retardant chemical, 221
Fire risk analysis, 152–153
    demographics, 153
    trends, 152–153
Fire squirt, 84
Fire stations
    defined, 287
    fire fighter quarters, 288
    garage or apparatus bay, 287
    PPE storage, 288
    purpose of, 287
    SCBA cylinder refilling, 288
    tool and equipment storage, 287–288
    tours, 168
Fire streams, 282, 286

Fire Suppression Rating Schedule, 132
Fire tetrahedron, 187–196
    fire triangle, 187, 188
    fuel, 192–195
    heat and temperature, 188–189
    heat energy sources, 189–190
    oxygen, 195–196
    self-sustained chemical chain reaction, 196, 197
    transmission of heat, 190–192
Fire triangle, 187, 188
Fire wardens, 86–87
Fire wards, 86
Firefighter Fatality Investigation and Prevention Program, 126
Firefighters
    duties of, 52–53
    as public figures, 22–27
        conduct, 23–24
        human relations and customer service, 24–25
        image, 22
        physical fitness, 27
        responsible behaviors, 22–23
        teamwork, 24
        uniform and dress, 25–27
    selection of, 34–40
        application, 35
        interviews, 39
        physical ability test, 38
        preparation, 35
        probationary period, 39
        recruitment, 35
        volunteers, 39–40
        written examination, 36–38
Firefighters association, state, 121
Firehouse, 287
Fire-resistance rating, 214
Fires impacting firefighters and public safety in North America, 105–108
    Beverly Hills Supper Club Fire, Southgate, Kentucky (1977), 107
    Cocoanut Grove Nightclub Fire, Boston, Massachusetts (1942), 106
    Great Fire of 1904, Toronto, Ontario, Canada (1904), 106
    Hartford Hospital Fire, Hartford, Connecticut (1961), 107
    impact on firefighter safety, 108
    Iroquois Theatre Fire, Chicago, Illinois (1903), 106
    MGM Grand Hotel Fire, Las Vegas, Nevada (1980), 107
    Our Lady of the Angels School Fire, Chicago, Illinois (1958), 107
    Ringling Brothers Barnum and Bailey Circus Fire, Hartford, Connecticut (1944), 106
    Station Nightclub Fire, West Warwick, Rhode Island (2003), 107–108, 151
    Triangle Shirtwaist Fire, New York City, New York (1911), 106
Firewall, 222
First responder, 70
Fixed-temperature heat detector, 239
Fixed-temperature/rate-of-rise heat detector, 242
Fixed-wing air tanker, 278
Flame detectors, 242
Flammable liquids as fire hazard, 149
Flammable/explosive range, 195
Flashover, 199–200, 201

Flat roof, 226
Floors, combustible, 228
FM Global Research, 132, 245
Foam application, 68
Foam application training, 290
Forcible entry, 48, 50
Forest fires, 83
Forestry departments, 121
Formal communications, 334–335
FPE (fire protection engineer), 138, 162–165
Frangible bulb sprinkler, 246, 247
Franklin, Benjamin, 90, 91, 92
Franklin, Thomas, 91
Fuel
    defined, 185, 192
    flammable/explosive range, 195
    fuel-controlled fire, 198
    gaseous, 195
    hazards, 147–148
    inorganic, 192
    lower flammable limit, 195
    organic, 192
    potential energy of, 185
    pyrolysis, 193
    removal, for fire extinguishment, 208
    surface-to-mass ratio, 194
    upper flammable limit, 195
    vaporization, 194–195
    volatility, 195
Fully developed stage of fire development, 200–201
Funding of fire departments, 328–330
    bond sales, 329
    budget constraints, 330
    enterprise funds, 329
    fundraisers, 329
    grants/gifts, 329
    municipal fire departments, 15
    subscriptions/fees, 330
    tax revenues, 328
    trust funds, 328
Fundraisers, 329
Furnishings as firefighter hazard, 225, 226
Fusible link sprinkler, 246, 247

G

Gamewell, John Nelson, 103
Gamewell Alarm System, 103
Gaseous fuels, 195
Gasoline powered equipment, 98–99
GED (General Equivalency Diploma), 35
General Equivalency Diploma (GED), 35
Gibbs closed circuit oxygen breathing apparatus, 101
Gifts to fire service organizations, 329
Glass construction, 224
Gloves, 26
Government. *See* Support organizations, federal; Support organizations, local; Support organizations, state/provincial
Grants administrator, 59
Grants to fire service organizations, 329
Great Fire of 1904, Toronto, Ontario, Canada (1904), 106
Ground ladders, 282, 283–285
Group, defined, 340

Group Supervisor, 340
Growth fire development stage, 199
Guide for Fire and Explosion Investigations (NFPA® 921), 169, 185, 201, 186
Guide for Training Fire Service Personnel to Conduct Community Risk Reduction (NFPA® 1452), 158
Gusset plates, 229
Gypsum board, 224

H

Halon, 196
Hand tools, 281
    cutting tools, 303–306
    lifting tools, 307
    mechanic's tools, 310
    prying tools, 302
    specialized tools, 309
    stabilizing tools, 308
    striking tools, 301
Harnesses, 281–282, 317
Hartford Hospital Fire, Hartford, Connecticut (1961), 107
Hayes Aerial, 97–98
Hazardous material protective clothing, 295–296
Hazardous material response unit, 274–275
Hazardous Materials Code (NFPA® 400-2010), 195
Hazardous materials incident response, 13
Hazardous materials technician (haz mat tech), 65
Health and Safety Officer, 60–61
Health department, 118, 121
Health Insurance Portability and Accounting Act of 1996 (HIPAA), 23
Heat
    chemical heat energy, 148, 189–190
    detectors, 239–240
        fixed-temperature heat detectors, 239
        rate-of-rise heat detectors, 239–240
    electrical heat energy, 148, 190
    heating fire hazards, 149
    intensity, 173
    mechanical heat energy, 148, 190
    nuclear heat energy, 148
    sources for fires, 148
    temperature and, 188–189
    transmission, 190–192
        conduction, 191
        convection, 191
        Law of Heat Flow, 190
        radiation, 191–192
Heavy rescue vehicle, 270–271
Helicopter tanker, 278
Helicopters, 278
Helmet, 99
Highway department, 120
HIPAA (Health Insurance Portability and Accountability Act), 23
History of firefighting, 81–108
    alarm and communications systems, 103–105
        mobile data communications systems, 105
        pagers/personal alerting systems, 105
        radio communications, 104
        telegraph alarm systems, 103
        telephone systems, 104
        town crier, 103
        watchmen and watchtowers, 103

American fire engines, 92–94
    cider mill, 93, 94
    coffee mill, 93
    hand engine improvements, 94
    Hunneman hand tub, 93, 94
    New York style hand engines, 92–93
    Philadelphia style engine, 93
    piano-type engines, 93, 94
chemical engine, 97
early fire services, 81–86
    Britain's first fire brigade, 84–85
    causes of fire, 83
    extinguishing techniques, 82
    fire engine development, 85–86
    fire laws and ordinances, 83
    fire pump, 82
    Rome fires, 81–82
fire extinguishers, 102
fire protection in early America, 86–87
gasoline and diesel powered equipment, 98–99
historic fires in North America, 105–108
ladder trucks, 97–98
protective clothing, 99
self-contained breathing apparatus, 99, 100–102
steam engines, 95–96
volunteer fire service, 87–92
    fire insurance companies, 91–92
    first hose company, 89
    Mutual Fire Society of Boston, 90
    Union Volunteer Fire Company of Philadelphia, 90
    Volunteer Fire Department of New York, 90–91
    volunteers, patriots, and competitors, 87–89
Hodge, Paul, 95
Home Fire Sprinkler Coalition, 132
Homeland Security Presidential Directive/HSPD-5, 335
Hook and ladder companies, 83
Hook and ladder truck, 83
Horse-drawn steamers, 98
Hose appliances, 282, 287
Hose company, 89
Hose couplings, 287
Hose wagon, 89
Hoseline, 261
Hours of duty, 40
Housekeeping fire hazards, 149
Human relations function of firefighters, 24–25
Human resources, 59
Hunneman hand tub, 93, 94

I

IAAI (International Association of Arson Investigators), 133, 169
IABPFF (International Association of Black Professional Firefighters), 133
IAFC (International Association of Fire Chiefs), 38, 133
IAFF (International Association of Fire Fighters), 38, 134
IAP. *See* Incident Action Plan (IAP)
IBC (International Building Code), 159, 214
IC (Incident Commander), 336, 338
ICBO (International Congress of Building Officials), 135
ICC (International Code Council), 134–135, 159
ICMA (International City/County Management Association), 134
ICS. *See* Incident Command System (ICS)
IDLH (immediately dangerous to life and health), 100, 101
IFMA (International Fire Marshals Association), 135
IFSAC (International Fire Service Accreditation Congress), 33, 34
IFSTA (International Fire Service Training Association), 135
Ignition of fires, 189
Image of firefighters, 22
Immediately dangerous to life and health (IDLH), 100, 101
IMSA (International Municipal Signal Association), 135
Incendiary devices, 173
Incident Action Plan (IAP)
    described, 339
    fire risk analysis, 153
Incident Command System (ICS), 335–345
    command staff, 338
    critical incident stress management, 344–345
    defined, 333
    emergency operations, 341
    functional areas, 336–337
        command, 336
        Finance/Administrative Unit, 337
        Logistics Section, 337
        Operations Section, 337
        Planning Section, 337
    Incident Commander, 338
    National Incident Management System-Incident Command System
        chain of command, 321
        defined, 335
        terms, 337–340
        training, 341
    NIMS terms, 337–340
        Branch, 339
        Branch Director, 339
        command staff, 338
        Division, 339
        Division Supervisor, 339
        Group, 340
        Group Supervisor, 340
        Incident Action Plan, 339
        Incident Commander, 338
        resources, 340
        Section, 339
        Section Chief, 339
        Strike Team, 48, 340, 341
        Task Force, 340
        Task Force Leader, 340
        Unit, 340
        Unit Leader, 340
    NIMS-ICS, 335
    NIMS-ICS training, 341
    personnel accountability systems, 342–343
    rapid intervention crew (RIC/RIT), 344
Incident Commander (IC), 336, 338
Incipient fire development stage, 198
Industrial accident response, 13
Industrial fire brigades, 20
Industrial fire departments, 20
Industrial firefighter, 68
Information Officer (IO), 61–62
Information technology (IT) professionals, 73
Infrared camera, 273
Inspection

building construction, 55
fire inspector, 160, 161–162
fire risk analysis, 153
purpose of, 159
typical inspection items, 161–162
Instructor, 57–58
Insurance companies, 91–92
Insurance Services Office, Inc. (ISO), 132, 327
Intermodal Tank Specialty, 65
International Association of Arson Investigators (IAAI), 133, 169
International Association of Black Professional Firefighters (IABPFF), 133
International Association of Fire Chiefs (IAFC), 38, 133
International Association of Fire Fighters, 38, 134
International Association of Women in Fire and Emergency Services (iWOMEN), 134
International Building Code (IBC), 159, 214
International City/County Management Association (ICMA), 134
International Code Council (ICC), 134–135, 159
International Congress of Building Officials (ICBO), 135
International Fire Code, 159
International Fire Marshals Association (IFMA), 135
International Fire Service Accreditation Congress (IFSAC), 33, 34
International Fire Service Training Association (IFSTA), 135
International Municipal Signal Association (IMSA), 135
International Society of Fire Service Instructors (ISFSI), 136
International System of Units (SI), 188
Interviews
    exit interview, 157, 158
    firefighter candidates, 39
Investigation
    arson, 169–170
    Chemical Safety Hazard Investigation Board, 123
    conduct at the scene, 174–175
    evidence protection and preservation, 177–178
    explosions, 169, 185, 186, 201
    Firefighter Fatality Investigation and Prevention Program, 126
    firefighter roles, 168
    International Association of Arson Investigators, 133, 169
    investigator roles, 169–170
    legal considerations, 176–177
    media inquiries, 175
    mobile fire investigation unit, 277
    National Association of Fire Investigators, 169
    observations by emergency responders, 170–174
        after the fire, 174
        en route, 170–171
        firefighters, 172–173
        overhaul, 170
        upon arrival, 171
        written account, 174
    purpose of, 154
    search warrants, 177
    securing the scene, 175–176
    statements at the scene, 174–175
IO (Information Officer), 61–62
Ionization detectors, 241
IR detectors, 242
Iroquois Theatre Fire, Chicago, Illinois (1903), 106
ISFSI (International Society of Fire Service Instructors), 136
ISO (Insurance Services Office, Inc.), 132, 327
IT (information technology) professionals, 73

iWOMEN (International Association of Women in Fire and Emergency Services), 134

J
Joint Council of Fire Service Organizations, 33
Joule, 188
Judicial system, 117
Jurisdiction of public fire and emergency services organizations, 14

L
Ladder trucks, 48, 50, 97–98
Ladders
    aerial ladder apparatus, 266
    aerial ladder platform apparatus, 266–267
    ground ladders, 282, 283–285
    hook and ladder companies/trucks, 83
    pompier ladder, 97, 98
    scaling ladder, 98
Lamella roof, 227
Latta steam engine, 95
Law enforcement, 114, 119–120
Law of Heat Flow, 190
Laws and ordinances, history of, 83
LeCour, A., 101
Length of service awards program (LOSAP), 41
LFL (lower flammable limit), 195
Life net, 98
Life Safety Code® (NFPA® 101), 106, 159
Light detectors, 242
Light rescue vehicle, 270
Lighting fire hazards, 149
Lightweight construction hazards, 229–230
Lightweight steel truss, 229, 230
Lightweight wood truss, 229
Line of duty deaths (LODD), 108
Load-bearing wall, 220–221
Local alarm systems, 238
Local government structure, 325–326
    commission, 325
    Council (Board) Manager, 325
    fire districts, 326
    Mayor/Council, 325
LODD (line of duty deaths), 108
Logistics Section, 337
LOSAP (length of service awards program), 41
Lower flammable limit (LFL), 195
Lyon, Pat, 93

M
Maintenance
    apparatus, 53, 74–75
    equipment, 74–75
    facilities for, 292
    facilities maintenance personnel, 74
    fire station area for, 287–288
    outside providers, 293
Manual pull station, 238
Manufacturing fire hazards, 150
Maps, 157
Marine firefighting, 21, 67
Marine Tank Vessel Specialty, 65
Martin, Thomas J., 102

Masonry construction, 222-223
Mass and energy, 185
Mass of an object, 184-185
Math examinations, 36
Matter
    defined, 184
    endothermic reaction, 185
    exothermic reaction, 185
    fuel, 185
    mass and energy, 185
    physical and chemical changes, 185
    physical states of, 184
    potential energy, 185
    specific gravity, 184-185
    vapor density, 184-185
Mayor/Council form of government, 325
MDCS (mobile data communications system), 105
Mechanical aptitude examinations, 37
Mechanical heat energy, 148, 190
Medium rescue vehicle, 270
Merriman's Smoke mask, 101
MGM Grand Hotel Fire, Las Vegas, Nevada (1980), 107
Michigan v. Tyler, 177
Military firefighter, 69
Mine Safety Appliance, 101
Mobile air supply unit, 275
Mobile Command Post, 276
Mobile communications center, 289
Mobile data communications system (MDCS), 105
Mobile fire investigation unit, 277
Mobile water supply apparatus, 264
Model codes, 160
Monetary loss of fires, 146-147
Morse Code, 103
Municipal fire department, 14-15
    fire brigade, 85
    funding, 15
    level and type of services, 15
    separate facilities for separate functions, 15
Mutual aid, 326-328
Mutual Assurance Company, 92
Mutual Fire Society of Boston, 90

N
NAFI (National Association of Fire Investigators), 169
NAFTD (North American Fire Training Directors), 34, 138
NAHF (National Association of Hispanic Firefighters), 136
National Association of Fire Investigators (NAFI), 169
National Association of Hispanic Firefighters (NAHF), 136
National Association of State Directors of Fire Training, 34
National Association of State Fire Marshals, 136
National Board on Fire Service Professional Qualifications (NBFSPQ or Pro-Board), 33-34
National Building Code of Canada, 159
National Bureau of Standards, 126
National Emergency Training Center (NETC), 31-32, 129
National Fallen Firefighters Foundation (NFFF), 136-137
National Fire Academy (NFA), 31-32, 125
National Fire Incident Reporting System (NFIRS), 153
National Fire Protection Association® (NFPA®). *See also specific NFPA®*
    Fire Prevention Week, 137
    Fire Protection Handbook, 137
    International Fire Marshals Association, 135
    member sections, 137
    Professional Qualifications Standards, 33
    purpose of, 137
    Risk Watch®, 165
    SCBA standards, 101
National Fire Service Programs Committee (NFSPC), 31
National Guard/Civil Support Team, 125-126
National Highway Traffic Safety Administration (NHTSA), 126
National Incident Management System-Incident Command System (NIMS-ICS)
    chain of command, 321
    defined, 335
    terms, 337-340
    training, 341
National Institute for Occupational Safety and Health (NIOSH), 126, 129, 281
National Institute of Standards and Technology (NIST), 126-127, 197
National Interagency Fire Center (NIFC), 129, 137
National Oceanic and Atmospheric Administration (NOAA), 127
National Park Service, 14, 129
National Propane Gas Association (NPGA), 138
National Research Council of Canada, 159
National Tank Truck Carriers (NTCC), 139
National Transportation Safety Board (NTSB), 127
National Volunteer Fire Council (NVFC), 138
National Weather Service, 127
National Wildfire Coordinating Group (NWCG), 138
Natural resources department, 121
NBFSPQ (National Board on Fire Service Professional Qualifications), 33-34
Nealy Smoke Mask, 101
Nero, Emperor, 82
NETC (National Emergency Training Center), 31-32, 129
New York style hand engines, 92-93
New York Volunteer Fire Department, 90-91
Newsham pumper, 85-86
NFA (National Fire Academy), 31-32, 125
NFFF (National Fallen Firefighters Foundation), 136-137
NFIRS (National Fire Incident Reporting System), 153
NFPA®. *See* National Fire Protection Association® (NFPA®)
NFPA® 1, *Fire Code*, 159
NFPA® 13D, *Standard for Installation of Sprinkler Systems in One and Two Family Dwellings and Manufactured Homes*, 251
NFPA® 14, *Standard for the Installation of Standpipe and Hose Systems*, 251
NFPA® 96, *Standard for Ventilation Control and Fire Protection of Commercial Cooking Operations*, 207
NFPA® 101, *Life Safety Code*, 106, 159
NFPA® 220, *Standard on Types of Building Construction*, 214
NFPA® 400-2010, *Hazardous Materials Code*, 195
NFPA® 921, *Guide for Fire and Explosion Investigations*, 169, 185, 186, 201
NFPA® 1001, *Standard for Fire Fighter Professional Qualifications*
    airport firefighter, 65
    duties of firefighters, 53
    fire prevention and life safety education, 155
    firefighter certificates, 35
    occupational health and safety, 29

NFPA® 1002, *Standard for Fire Apparatus Driver/Operator Professional Qualifications*, 54
NFPA® 1003, *Standard for Airport Fire Fighter Professional Qualifications*, 65
NFPA® 1005, *Standard for Professional Qualifications for Marine Fire Fighting for Land-Based Fire Fighters*, 67
NFPA® 1006, *Standard for Technical Rescuer Professional Qualifications*, 67
NFPA® 1021, *Standard for Fire Officer Professional Qualifications*
    Assistant/Deputy Chief, 63
    Battalion Chief, 57
    Company Officer, 55
    Fire Chief, 64
NFPA® 1031, *Standard for Professional Qualifications for Fire Inspector and Plan Examiner*, 161, 162
NFPA® 1033, *Standard for Professional Qualifications for Fire Investigator*, 169
NFPA® 1035, *Standard on Fire and Life Safety Educator, Public Information Officer, Youth Firesetter Intervention Specialist and Youth Firesetter Program Manager Professional Qualifications*, 62, 166
NFPA® 1041, *Standard for Fire Service Instructor Professional Qualifications*, 58
NFPA® 1051, *Standard for Wildland Fire Fighter Professional Qualifications*, 67
NFPA® 1061, *Professional Qualifications for Public Safety Telecommunications Personnel*, 73
NFPA® 1071, *Standard for Emergency Vehicle Technician Professional Qualifications*, 75
NFPA® 1072, *Standard for Hazardous Materials/Weapons of Mass Destruction Emergency Response Personnel Professional Qualifications*, 65
NFPA® 1081, *Standard for Industrial Fire Brigade Member Professional Qualifications*, 20
NFPA® 1403, *Standard on Live Fire Training Evolutions*, 290
NFPA® 1452, *Guide for Training Fire Service Personnel to Conduct Community Risk Reduction*, 158
NFPA® 1500, *Standard on Fire Department Occupational Safety and Health Program*, 29, 53, 341
NFPA® 1521, *Standard for Fire Department Safety Officer Professional Qualifications*, 61
NFPA® 1582, *Standard on Comprehensive Occupational Medical Program for Fire Departments*, 27, 53
NFPA® 1583, *Standard on Health-Related Fitness Programs for Fire Department Members*, 53
NFPA® 1901, *Standard for Automotive Fire Apparatus*
    aerial apparatus, 265
    aerial ladder apparatus, 266
    aerial ladder platform apparatus, 266
    aerial ladders, 268
    ambulance design, performance, and testing, 271
    elevating platforms, 268
    hoseline recommendations, 261
    mobile water supply apparatus, 264
    pumper design, 260
    quintuple apparatus, 269
    water towers, 268
NFPA® 1906, *Standard for Wildland Fire Apparatus*, 264–265
NFPA® 1917, *Standard for Automotive Ambulances*, 271
NFPA® 1975, *Standard on Station/Work Uniforms for Emergency Services*, 25, 279

NFPA® 5000, *Building Construction and Safety Code*, 159
NFSPC (National Fire Service Programs Committee), 31
NHTSA (National Highway Traffic Safety Administration), 126
NIFC (National Interagency Fire Center), 129, 137
NIMS-ICS. *See* National Incident Management System-Incident Command System (NIMS-ICS)
9-1-1 calls, 104, 289
NIOSH (National Institute for Occupational Safety and Health Act of 1970), 126, 129, 281
NIST (National Institute of Standards and Technology), 126–127, 197
NOAA (National Oceanic and Atmospheric Administration), 127
Nomex®, 99, 196
Non-load-bearing wall, 220–221
North American Fire Training Directors (NAFTD), 34, 138
Nozzles, 282, 286
NPGA (National Propane Gas Association), 138
NRC (Nuclear Regulatory Commission), 127
NTCC (National Tank Truck Carriers), 139
NTSB (National Transportation Safety Board), 127
Nuclear heat energy, 148
Nuclear Regulatory Commission (NRC), 127
NVFC (National Volunteer Fire Council), 138
NWCG (National Wildfire Coordinating Group), 138

O

Observation/recognition examinations, 37
Observations by emergency responders, 170–174
    after the fire, 174
    en route, 170–171
    firefighters, 172–173
    overhaul, 170
    upon arrival, 171
    written account, 174
Occupational and Safety Act, 47
Occupational Safety and Health Act of 1970, 100, 127, 129
Occupational Safety and Health Administration (OSHA)
    industrial fire brigade standards, 20
    oxygen-enriched atmosphere, defined, 196
    purpose of, 127
    SCBA standards, 101
    state agencies, 120
Office of Emergency Management, 117, 121
Office of EMS, 126
Oklahoma State University fire service training, 31
Online training, 32
On-site training, 58
Open spaces as firefighter hazard, 228
Open-circuit SCBA, 280, 299
Operations Section, 337
Operations Section Chief, 337
Opticom™, 116
Oral board, 39
Oral interview of firefighter candidates, 39
Organization, fire department, 321–346
    chain of command, 321, 323
    challenges of fire protection, 330–333
        funding/budget constraints, 330
        outdated equipment, 330, 331
        personnel recruitment, 331
        personnel retention, 331–332
        water supply, 332–333

discipline, 325
division of labor, 324
fire department funding, 328-330
  bond sales, 329
  enterprise funds, 329
  fundraisers, 329
  grants/gifts, 329
  subscriptions/fees, 330
  tax revenues, 328
  trust funds, 328
Incident Command System, 335-345
  critical incident stress management, 344-345
  defined, 335
  emergency operations, 341
  functional areas, 336-337
  NIMS terms, 337-340
  NIMS-ICS training, 341
  personnel accountability systems, 342-343
  rapid intervention crew, 344
local government structure, 325-326
  commission, 325
  Council (Board) Manager, 325
  fire districts, 326
  Mayor/Council, 325
organization chart sample, 322
policies and procedures, 333-335
  disciplinary procedures, 334
  formal communications, 334-335
  standard operating procedures/guidelines, 333-334
response considerations, 326-328
  automatic aid, 326, 327-328
  mutual aid, 326-328
  outside aid, 326
span of control, 324
unity of command, 323
Organizational chart, fire department, 49
Organizations. *See* Support organizations
OSHA. *See* Occupational Safety and Health Administration (OSHA)
Our Lady of the Angels School Fire, Chicago, Illinois (1958), 107
Outside aid, 326
Outside stem and yoke (OS&Y) valve, 247, 248
Overcurrent, 190
Overhaul, 170
Overload, 190
Oxidizer, 187, 195
Oxygen (oxidizing agent), 187, 195-196
Oxygen exclusion, for fire extinguishment, 209
Oxygen-enriched atmosphere, 196

P

Pagers, 105
Paid on-call fire departments, 19-20
Paid on-call firefighter, 19
Paramedic, 71
Parish fire departments, 15-16
Party walls, 83
PASS (personal alert safety system), 27
Pay of firefighters, 40
PBI Gold, 99
Permits, 162-165
  application, 164

defined, 162
expiration, 165
issuance, 164
process, 164
review, 164
sample, 163
types, 162, 164
Personal alert safety system (PASS), 27
Personal alerting systems, 105
Personal protective clothing (PPC), 280
Personal protective equipment (PPE)
  defined, 280
  storage, 25, 288
Personnel
  recruitment, 331
  retaining, 331-332
  roles. *See* Personnel roles
  services, 59
Personnel accountability systems
  computer-based electronic accountability, 343
  defined, 342
  PASS system, 27
  SCBA tag system, 342-343
  tag system, 342
Personnel roles, 47-75
  administration, 59-64
    assistant/deputy chief, 62-63
    Fire Chief/Chief of Department, 63-64
    grants administrator, 59
    human resources/personnel services, 59
    Public Information Officer, 61-62
    Safety Officer, 60-61
  apparatus/equipment maintenance personnel, 74-75
  battalion/district chief, 55-57
  company officer, 54-55
  emergency medical services, 69-72
    Emergency Medical Technician, 70-71
    EMS Chief/officer, 71-72
    first responder, 70
    levels of emergency medical care, 69
    paramedics, 71
  facilities maintenance personnel, 74
  fire apparatus driver/operator, 53-54
  fire companies, 47-51
    defined, 47
    engine company, 48
    ladder (truck) company, 48, 50
    organizational chart, 49
    rescue/squad company, 51
  firefighter selection, 52-53
  information technology personnel, 73
  special operations personnel, 64-69
    airport firefighter, 64-65
    hazardous materials technician, 65
    industrial firefighters, 68
    marine firefighters, 67
    military firefighter, 69
    technical rescuer, 65-67
    wildland firefighters, 67, 68
  telecommunications/dispatch personnel, 72-73
  training division personnel, 57-58
    instructor, 57-58

Training Officer, 58
Philadelphia, Union Volunteer Fire Company, 90
Philadelphia Contributorship of the Assurance of Houses from Loss by Fire, 91
Philadelphia style engine, 93
Philadelphia water system, 89
Photoelectric smoke detectors, 240-241
Physical ability test, 38
Physical fitness of firefighters, 27
Piano-type engines, 93, 94
Piatt, Jacob Wykoff, 95
Pike pole, 83
Piloted ignition, 189
PIO (Public Information Officer), 61-62, 166
Pitched roof, 226
Planning commission, 115
Planning Section, 337
Plans examiner, 115
Platform framing, 220
Pneumatic rate-of-rise line detector, 239
Pneumatic rate-of-rise spot detector, 239
Policies and procedures, 333-335
    disciplinary procedures, 334
    formal communications, 334-335
    standard operating procedures/guidelines, 325, 333-334
Polybenzimidazole, 99
Pompier ladder, 97, 98
Post indicator valve assembly (PIVA), 249
Post indicator valve (PIV), 247, 248
Potential energy, 185
Power and light unit, 277
Power equipment fire hazards, 149
Power tools, 281
    electric powered, 311
    hydraulic, manually operated, 313
    hydraulic, power driven, 314
    pneumatic powered, 312
PPC (personal protective clothing), 280
PPE. *See* Personal protective equipment (PPE)
Preaction sprinkler system, 251
Preincident planning
    computer-assisted, 157
    defined, 151
Preincident surveys
    conducting the survey, 156-157
    defined, 153
    exit interview, 157, 158
    facility surveys, 155-159
    gathering additional information, 157
    maps and sketch making, 157
    personnel requirements, 155
    purpose of, 154
    residential safety survey, 154, 158-159
    risk analysis, 153
    sample survey plan, 154
    scheduling, 155
Pressure and matter, 184
Pressure regulating devices, 253
Prevention of fires. *See* Fire prevention
Private fire departments, 20
Probationary period of new firefighters, 39
Pro-Board, 33-34

Products of combustion, 191
Professional Qualifications for Public Safety Telecommunications Personnel (NFPA® 1061), 73
Professional Qualifications Standards (Pro-Qual), 33
Promotion of firefighters, 40
Property loss from fires, 146
Pro-Qual (Professional Qualifications Standards), 33
Protected premises alarm systems, 238
Protective clothing, 99
Provincial training programs, 29-30
Proximity protective clothing, 294
PS (Public Safety Canada), 127
Psychological testing, 38
Public assembly occupancy fire hazards, 150
Public fire departments, 17-20
    career departments, 18
    combination departments, 19
    paid on-call departments, 19-20
    profile, 17
    public safety departments, 20
    volunteer, 17, 18-19
Public Information Officer (PIO), 61-62, 166
Public Law 93-498, 125, 129
Public Safety Canada (PS), 127
Public safety departments, 20
Public works department, 116
Pull station box, 103
Pump-and-roll capability, 273
Pumper, 260-262, 263
Pyrolysis, 193, 198
Pyrotechnics, 151

Q
Quintuple apparatus, 269

R
Radiation, 191-192
Radio communications, 104
Rapid intervention crew (RIC/RIT), 344
Rapid intervention truck, 263
Rate-compensation detector, 239-240
Rate-of-rise heat detector, 239-240
Reading comprehension examinations, 36
Recognition/observation examinations, 37
Recruitment of firefighters, 35, 331
Reflective trim on uniforms, 26
Refueling truck, 279
Regional training programs, 29-30
Rehabilitation vehicles, 279
Reinforced concrete, 224
Renovation hazards, 173, 231-232
Rescue apparatus, 269-271
    heavy rescue vehicle, 270-271
    light rescue vehicle, 270
    medium rescue vehicle, 270
    purpose of, 269
    tools and equipment carried on, 270
Rescue Squad, 51
Rescue Truck, 51
Rescue/squad company roles, 51
Residential safety survey, 154, 158-159
Residential sprinkler systems, 159, 251

Resistance heating, 190
Resources, 340
Retention of personnel, 331–332
Retirement of firefighters, 41
Ribbed arched roof, 227
RIC/RIT, 344
Ringling Brothers Barnum and Bailey Circus Fire, Hartford, Connecticut (1944), 106
Risk Watch® program, 165
Roles of personnel. *See* Personnel roles
Rollover, 202
Rome fires, 81–82
Roof coverings as firefighter hazard, 226–227
Ropes, webbing, hardware, harnesses, 281–282, 315–317

S

SAFER grant, 330
Safety amendments, 160–161
Safety Officer, 60–61
Salaries of firefighters, 40
Salvage cover, 175
Salvation Army, 118
SAR (supplied air respirator), 280, 298
SARA Title III, 47
SBCCI (Southern Building Code Congress International), 135
Scaling ladder, 98
SCBA. *See* Self-contained breathing apparatus (SCBA)
Schedules of firefighters, 40
Scientific terminology, 183–209
    backdraft, 203–204
    combustion, 186–202
        defined, 186
        fire development stages, 196–202
        fire tetrahedron, 187–196
        products of, 205
    fire classifications, 205–207
    fire extinguishment theory, 207–209
    matter, 184–185
    rollover, 202
    smoke explosion, 205
    thermal layering of gases, 203
Search and rescue boats, 276–277
Search warrant, 177
Second interview of firefighter candidates, 39
Section, defined, 339
Section Chief, 339
Securing the fire scene, 175–176
Self-contained breathing apparatus (SCBA)
    closed-circuit, 280, 300
    defined, 100
    early attitudes regarding use of, 101
    early design, 101
    improvements, 99, 100–102
    modern technological advances, 102
    open-circuit, 280, 299
    OSHA and NFPA® standards, 101
    PASS device on, 27
    purpose of, 52, 280
    refilling cylinders, 288
    tag system, 342–343
Self-sustained chemical chain reaction, 196, 197
SFPE (Society of Fire Protection Engineers), 138

Shaw, E. M., 100
SI (International System of Units), 188
Siphona Pump, 82
16 Firefighter Life Safety Initiatives, 137
Size-up, 37
Sketch making, 157
Smaller pumper truck, 263
Smartphones, 105
Smoke detectors/alarms, 240–243
    carbon monoxide detectors, 242–243
    combination detectors, 242
    fire and life safety education, 168
    flame detectors, 242
    giveaway programs, 159
    ionization, 241
    legal minimum requirements, 241
    photoelectric, 240–241
    purpose of, 168
    tamper-proof, 243
Smoke explosion, 205
Smokehouse, 290, 291
Society of Fire Protection Engineers (SFPE), 138
Solar energy panels, 231
SOP. *See* Standard operating procedures (SOPs)
Southern Building Code Congress International (SBCCI), 135
Spalling, 224
Span of control, 324
Spanish language program for firefighters, 136
Sparking, 190
Special fire hazard, 149–150
Special operations personnel, 64–69
    airport firefighter, 64–65
    hazardous materials technician, 65
    industrial firefighters, 68
    marine firefighters, 67
    military firefighter, 69
    technical rescuer, 65–67
    wildland firefighters, 67, 68
Special operations vehicles, 278–279
Special rescue technician, 65–67
Special task forces, 121
Specific gravity, 184–185
Spontaneous ignition, 189, 190
Sprinklers
    automatic sprinkler systems, 244–253
        applications, 250–251
        defined, 244
        failure or incomplete operation, 245
        Fire Sprinkler Initiative, 251
        history of, 245
        life safety, 246
        reliability of, 238
        sprinkler position, 246–249
        standpipe systems, 251–253
    chemical pellet, 246, 247
    color coding, 247
    control valves, 246–249
    defined, 245
    deluge, 251
    diagram, 246
    dry-pipe systems, 250
    frangible bulb, 246, 247

fusible link, 246, 247
Home Fire Sprinkler Coalition, 132
preaction, 251
residential sprinklers, 159, 251
temperature ratings and classifications, 247
waterflow alarms, 249
wet-pipe systems, 250
Staffing for Adequate Fire and Emergency Response (SAFER) grant, 330
Standard, defined, 161
Standard for Airport Fire Fighter Professional Qualifications (NFPA® 1003), 65
Standard for Automotive Ambulances (NFPA® 1917), 271
Standard for Automotive Fire Apparatus. See NFPA® 1901, *Standard for Automotive Fire Apparatus*
Standard for Emergency Vehicle Technician Professional Qualifications (NFPA® 1071), 75
Standard for Fire Apparatus Driver/Operator Professional Qualifications (NFPA® 1002), 54
Standard for Fire Department Safety Officer Professional Qualifications (NFPA® 1521), 61
Standard for Fire Fighter Professional Qualifications. See NFPA® 1001, *Standard for Fire Fighter Professional Qualifications*
Standard for Fire Officer Professional Qualifications. See NFPA® 1021, *Standard for Fire Officer Professional Qualifications*
Standard for Fire Service Instructor Professional Qualifications (NFPA® 1041), 58
Standard for Hazardous Materials/Weapons of Mass Destruction Emergency Response Personnel Professional Qualifications (NFPA® 1072), 65
Standard for Industrial Fire Brigade Member Professional Qualifications (NFPA® 1081), 20
Standard for Installation of Sprinkler Systems in One and Two Family Dwellings and Manufactured Homes (NFPA® 13D), 251
Standard for Professional Qualifications for Fire Inspector and Plan Examiner (NFPA® 1031), 161, 162
Standard for Professional Qualifications for Fire Investigator (NFPA® 1033), 169
Standard for Professional Qualifications for Marine Fire Fighting for Land-Based Fire Fighters (NFPA® 1005), 67
Standard for Technical Rescuer Professional Qualifications (NFPA® 1006), 67
Standard for the Installation of Standpipe and Hose Systems (NFPA® 14), 251
Standard for Ventilation Control and Fire Protection of Commercial Cooking Operations (NFPA® 96), 207
Standard for Wildland Fire Apparatus (NFPA® 1906), 264-265
Standard for Wildland Fire Fighter Professional Qualifications (NFPA® 1051), 67
Standard on Comprehensive Occupational Medical Program for Fire Departments (NFPA® 1582), 27, 53
Standard on Fire and Life Safety Educator, Public Information Officer, Youth Firesetter Intervention Specialist and Youth Firesetter Program Manager Professional Qualifications (NFPA® 1035), 62, 166, 169, 201
Standard on Fire Department Occupational Safety and Health Program (NFPA® 1500), 29, 53, 341
Standard on Health-Related Fitness Programs for Fire Department Members (NFPA® 1583), 53
Standard on Live Fire Training Evolutions (NFPA® 1403), 290
Standard on Station/Work Uniforms for Emergency Services (NFPA® 1975), 25, 279

Standard on Types of Building Construction (NFPA® 220), 214
Standard operating procedures (SOPs)
defined, 325
fire department, 333-334
Standpipe systems, 251-253
Class I, 252
Class II, 252, 253
Class III, 252, 253
pressure regulating devices, 253
types of, 253
State fire marshal or commissioner, 119
State forestry departments, 121
State training programs, 29-30
Statements at the scene, 174-175
Station Nightclub Fire, West Warwick, Rhode Island (2003), 107-108, 151
Steam, 184
Steam engines, 95-96
Steel construction material, 223-224
"Stop, Drop, and Roll," 165, 167
Storage
clothing, 25
fire hazards, 149
personal protective equipment, 25, 288
tools and equipment, 287-288
Strike team, 48, 340, 341
Strobe light and fire alarm speaker, 238
Structural alteration hazards, 173
Structural protective clothing, 294
Stuyvesant, Peter, 87
Subscription fees for fire or emergency services, 330
Supplied air respirator (SAR), 280, 298
Support organizations, 113-139
American Association of Railroads Bureau of Explosives, 139
American Trucking Association, 139
federal, 122-130
local, 114-118
National Tank Truck Carriers, 139
North American, 130-139
state/provincial, 118-122
Support organizations, federal, 122-130
Army Corps of Engineers, 123
Bureau of Alcohol, Tobacco, Firearms, and Explosives, 123
Chemical Safety Hazard Investigation Board, 123
Consumer Product Safety Commission, 123
Emergency Management Institute, 123
Environmental Protection Agency, 120, 124
Federal Aviation Administration, 124
Federal Emergency Management Agency, 124. *See also* Federal Emergency Management Agency (FEMA)
National Fire Academy, 31-32, 125
National Guard/Civil Support Team, 125-126
National Highway Traffic Safety Administration, 126
National Institute for Occupational Safety and Health, 126 *****, 129 *****, 281 *****
National Institute of Standards and Technology, 126-127, 197
National Transportation Safety Board, 127
National Weather Service, 127
Nuclear Regulatory Commission, 127
Occupational Safety and Health Administration, 127. *See also* Occupational Safety and Health Administration (OSHA)
Public Safety Canada, 127

Transport Canada, 128
United States Coast Guard, 128
United States Department of Energy, 128
United States Department of Homeland Security, 128-129. *See also* Department of Homeland Security (DHS)
United States Department of Labor, 127, 129
United States Department of the Interior, 129
United States Department of Transportation, 129. *See also* Department of Transportation (DOT)
United States Fire Administration, 31-32, 129
United States Forest Service, 130
Support organizations, local, 114-118
building department (code enforcement), 115
Community Emergency Response Team, 118
health department, 118
judicial system, 117
law enforcement, 114
local government, 114
nongovernment organizations, 118
Office of Emergency Management, 117, 121
public works department, 116
utilities, 116
water department, 115
zoning/planning commission, 115
Support organizations, North American, 130-139
American National Standards Institute, 131
Canadian Association of Fire Chiefs, 131
Canadian Centre for Emergency Preparedness, 131
Canadian Fallen Firefighters Foundation, 131
Canadian Fire Alarm Association, 131
Canadian Fire Safety Association, 131
Canadian Standards Association, 131
Canadian Volunteer Fire Services Association, 132
Council of Canadian Fire Marshals and Fire Commissioners, 132
Fire Prevention Canada, 132
FM Global Research, 132, 245
Home Fire Sprinkler Coalition, 132
Insurance Services Office, Inc., 132, 327
International Association of Arson Investigators, 133, 169
International Association of Black Professional Firefighters, 133
International Association of Fire Chiefs, 38, 133
International Association of Fire Fighters, 38, 134
International Association of Women in Fire and Emergency Services, 134
International City/County Management Association, 134
International Code Council, 134-135, 159
International Fire Marshals Association, 135
International Fire Service Training Association, 135
International Municipal Signal Association, 135
International Society of Fire Service Instructors, 136
National Association of Hispanic Firefighters, 136
National Association of State Fire Marshals, 136
National Fallen Firefighters Foundation, 136-137
National Fire Protection Association®, 137. *See also* National Fire Protection Association® (NFPA®)
National Interagency Fire Center, 129, 137
National Propane Gas Association, 138
National Volunteer Fire Council, 138
National Wildfire Coordinating Group, 138
North American Fire Training Directors, 34, 138
Society of Fire Protection Engineers, 138

Underwriters Laboratories, 138. *See also* Underwriters Laboratories (UL)
Underwriters Laboratories of Canada, 139, 245
Support organizations, state/provincial, 118-122
department of transportation, 120
Environmental Protection Agency, 120
fire chiefs association, 121
fire commission, 119
fire marshals association, 122
fire training, 119
firefighters association, 121
forestry/natural resources, 121
health department, 121
highway department, 120
law enforcement, 119-120
Occupational Safety and Health Administration, 120
Office of Emergency Management, 121
other state agencies, 122
special task forces, 121
state fire marshal or commissioner, 119
turnpike commission, 120
Suppression systems. *See* Fire detection, alarm, and suppression systems
Surface-to-mass ratio, 194
Surveys of facilities, 155-159. *See also* Preincident surveys
conducting the survey, 156-157
exit interview, 157, 158
maps and sketch making, 157
personnel requirements, 155
residential safety survey, 158-159
scheduling, 155
Syringes, 84

T

Tactical Spanish for Firefighters and EMS, 136
Tag system, 342
Tank Car Specialty, 65
Tanker, 264
Target hazards, 151
Task force, 48, 121, 340
Task Force Leader, 340
Tax revenues, 328
Teamwork of firefighters, 24
Technical rescue specialist, 65-67
Technical rescuer, 65-67
Telecommunication centers, 289
Telecommunications personnel, 72-73
Telegraph alarm systems, 103
Telephone systems, 104
Telescoping aerial platform apparatus, 268
Temperature
ambient temperature, 239
heat and, 188-189
reduction, for fire extinguishment, 208
Tender, 264
Testing
ambulance, 271
Candidate Physical Ability Test, 38
physical ability, 38
psychological, 38
written examination, 36-38
Thermal camera, 341

Thermal layering of gases, 203
Thermoelectric detector, 240
Three E's, 146
Tools
    apparatus, 272
    carried on rescue apparatus, 270
    equipment, 281
    fire hose, 282, 286
    hand tools, 281
        cutting tools, 303-306
        lifting tools, 307
        mechanic's tools, 310
        prying tools, 302
        specialized tools, 309
        stabilizing tools, 308
        striking tools, 301
    power tools, 281
        electric powered, 311
        hydraulic, manually operated, 313
        hydraulic, power driven, 314
        pneumatic powered, 312
    storage, 287-288
Tours of fire stations, 168
Town crier, 103
TRADE (Training Resources and Data Exchange), 125
Trailers, 173
Training and education of firefighters, 28-34
    certification/credentialing, 33-34
    classroom study, 28, 58
    colleges and universities, 31
    departmental training, 28-29
    drills, 28
    foam application training, 290
    live fire training, 29
    National Fire Academy, 31-32
    NIMS-ICS, 341
    non-traditional, 32-33
    regional/county/state/provincial training programs, 29-30
    on-site training, 58
    state/provincial fire training, 119
    training centers, 289-292
        burn building, 290, 291
        drill tower, 290, 291
        smokehouse, 290, 291
        training pads, 292
    training vs. education, 28
    value of education, 30
    volunteer fire departments, 332
Training division personnel, 57-58
    instructor, 57-58
    Training Officer, 58
Training Officer, 58
Training Resources and Data Exchange (TRADE), 125
Transport Canada, 128
Trends in fire protection and prevention, 152-153
Triangle Shirtwaist Fire, New York City, New York (1911), 106
Trousers, 26
Truck (ladder) company duties, 48, 50
Truman, Harry S., 146
Truss
    construction hazards, 229-230
    defined, 218

Trust funds, 328
Turn out, 88
Turnpike Commission, 120
Tyndall, John, 100
Type I - AD (Additional Duty) Ambulance, 271, 272
Type I Ambulance, 271, 272
Type I construction (fire-resistive), 214, 215, 216
Type II Ambulance, 271, 272
Type II construction (noncombustible or limited-combustible), 214, 215, 217
Type III - AD (Additional Duty) Ambulance, 271, 272
Type III Ambulance, 271, 272
Type III construction (ordinary), 214, 215, 217-218
Type IV construction (heavy timber or mill), 214, 215, 218-219
Type N Universal gas mask, 100
Type V construction (wood-frame), 214, 215, 219

U

UC (Unified Command), 336
UFL (upper flammable limit), 195
UL. *See* Underwriters Laboratories (UL)
ULC (Underwriters Laboratories of Canada), 139, 245
Underwriters Laboratories of Canada (ULC), 139, 245
Underwriters Laboratories (UL)
    fire behavior studies, 197
    purpose of, 127, 138
    sprinkler listing, 245
Unified command system, 336
Unified Command (UC), 336
Uniform and dress of firefighters, 25-27
    boots, 26
    Class A uniforms, 279
    coat closure system, 26
    coat collars, 26
    department policy, 25
    dress uniform (Class A), 25, 279
    gloves, 26
    loop on the back, 26
    NFPA® standards, 279
    personal protective equipment, 26, 27
    reflective trim, 26
    SCBA with PASS device, 27
    storage, 25
    styles, 25
    trousers, 26
    volunteer firefighters, 25
    work uniform standards, 25, 279
    wristlets, 26
Union Volunteer Fire Company of Philadelphia, 90
Unit, defined, 340
Unit Leader, 340
United States Coast Guard (USCG), 128
United States Department of Energy (DOE), 128
United States Department of Homeland Security. *See* Department of Homeland Security (DHS)
United States Department of Labor (DOL), 127, 129
United States Department of the Interior (DOI), 129
United States Department of Transportation (US DOT). *See* Department of Transportation (DOT)
United States Fire Administration (USFA)
    National Fire Academy, 31-32, 125
    purpose of, 129

United States Forest Service (USFS), 24, 130
Unity of command, 323
Universal emergency number, 104
University education of firefighters, 31
Upper flammable limit (UFL), 195
US DOT. *See* Department of Transportation (DOT)
USAR emergency response, 13
USCG (United States Coast Guard), 128
USFA. *See* United States Fire Administration (USFA)
USFS (United States Forest Service), 24, 130
Utility services, 116
UV detectors, 242

V

Valves. *See* Control valves
Van der Heijden, Jan, 85
Vapor, 193
Vapor density, 184–185
Vaporization, 194–195
Vehicle accident response, 13
Veneer walls, 222
Ventilation-controlled fire, 201
Volatility of fuel, 195
Volunteer fire departments
    Canadian Volunteer Fire Services Association, 132
    defined, 18, 19
    fire insurance companies, 91–92
    firefighter selection, 39–40
    firefighters, 19
    first hose company, 89
    funding, 18–19
    history of, 87–92
    Mutual Fire Society of Boston, 90
    National Volunteer Fire Council, 138
    organization, 18
    personnel, 19
    profile, 17
    recruiting personnel, 331
    retirement benefits, 41
    training, 332
    uniforms, 25
    Union Volunteer Fire Company of Philadelphia, 90
    Volunteer Fire Department of New York, 90–91
    volunteers, patriots, and competitors, 87–89

W

Wall post indicator valve (WPIV), 248, 249
Walls
    cantilever fire wall, 223
    collapse indicators, 228
    firewall, 222
    load-bearing, 220–221
    non-load-bearing, 220–221
    party walls, 83
    veneer walls, 222
Watchmen, 103
Watchtowers, 103
Water department/water authority, 115
Water rescue, 13
Water supply, 332–333
Water towers, 268–269
Waterflow alarms, 249
Webbing, 281–282, 315
Weight of an object, 184–185
Wet-pipe sprinkler system, 250
Wildland fire fighting
    brush trucks, 265
    fire apparatus, 264–265
    fire fighting aircraft, 278
    firefighters, 67–68
    levels of proficiency, 67
    National Wildfire Coordinating Group, 138
    NFPA® 1051, *Standard for Wildland Fire Fighter Professional Qualifications*, 67
    protective clothing, 294
Wood construction, 220–222
Wooden I-beams, 230
Wristlets, 26
Written examination for firefighters, 36–38
    math, 36
    mechanical aptitude, 37
    psychological testing, 38
    reading comprehension, 36
    recognition/observation, 37

Z

Zoning commission, 115

Indexed by Nancy Kopper

NOTE:

NOTE:

NOTE:

NOTE:

NOTE: